Title 40

Protection of Environment

Parts 100 to 135

Revised as of July 1, 2012

Containing a codification of documents
of general applicability and future effect

As of July 1, 2012

Published by the Office of the Federal Register
National Archives and Records Administration
as a Special Edition of the Federal Register

U.S. GOVERNMENT OFFICIAL EDITION NOTICE

Legal Status and Use of Seals and Logos

The seal of the National Archives and Records Administration (NARA) authenticates the Code of Federal Regulations (CFR) as the official codification of Federal regulations established under the Federal Register Act. Under the provisions of 44 U.S.C. 1507, the contents of the CFR, a special edition of the Federal Register, shall be judicially noticed. The CFR is prima facie evidence of the original documents published in the Federal Register (44 U.S.C. 1510).

It is prohibited to use NARA's official seal and the stylized Code of Federal Regulations logo on any republication of this material without the express, written permission of the Archivist of the United States or the Archivist's designee. Any person using NARA's official seals and logos in a manner inconsistent with the provisions of 36 CFR part 1200 is subject to the penalties specified in 18 U.S.C. 506, 701, and 1017.

Use of ISBN Prefix

This is the Official U.S. Government edition of this publication and is herein identified to certify its authenticity. Use of the 0–16 ISBN prefix is for U.S. Government Printing Office Official Editions only. The Superintendent of Documents of the U.S. Government Printing Office requests that any reprinted edition clearly be labeled as a copy of the authentic work with a new ISBN.

 U.S. GOVERNMENT PRINTING OFFICE

U.S. Superintendent of Documents • Washington, DC 20402–0001

http://bookstore.gpo.gov

Phone: toll-free (866) 512-1800; DC area (202) 512-1800

Table of Contents

iv

Explanation

The Code of Federal Regulations is a codification of the general and permanent rules published in the Federal Register by the Executive departments and agencies of the Federal Government. The Code is divided into 50 titles which represent broad areas subject to Federal regulation. Each title is divided into chapters which usually bear the name of the issuing agency. Each chapter is further subdivided into parts covering specific regulatory areas.

Each volume of the Code is revised at least once each calendar year and issued on a quarterly basis approximately as follows:

Title 1 through Title 16...as of January 1
Title 17 through Title 27 ....as of April 1
Title 28 through Title 41 ....as of July 1
Title 42 through Title 50..as of October 1

The appropriate revision date is printed on the cover of each volume.

LEGAL STATUS

The contents of the Federal Register are required to be judicially noticed (44 U.S.C. 1507). The Code of Federal Regulations is prima facie evidence of the text of the original documents (44 U.S.C. 1510).

HOW TO USE THE CODE OF FEDERAL REGULATIONS

The Code of Federal Regulations is kept up to date by the individual issues of the Federal Register. These two publications must be used together to determine the latest version of any given rule.

To determine whether a Code volume has been amended since its revision date (in this case, July 1, 2012), consult the "List of CFR Sections Affected (LSA)," which is issued monthly, and the "Cumulative List of Parts Affected," which appears in the Reader Aids section of the daily Federal Register. These two lists will identify the Federal Register page number of the latest amendment of any given rule.

EFFECTIVE AND EXPIRATION DATES

Each volume of the Code contains amendments published in the Federal Register since the last revision of that volume of the Code. Source citations for the regulations are referred to by volume number and page number of the Federal Register and date of publication. Publication dates and effective dates are usually not the same and care must be exercised by the user in determining the actual effective date. In instances where the effective date is beyond the cut-off date for the Code a note has been inserted to reflect the future effective date. In those instances where a regulation published in the Federal Register states a date certain for expiration, an appropriate note will be inserted following the text.

OMB CONTROL NUMBERS

The Paperwork Reduction Act of 1980 (Pub. L. 96–511) requires Federal agencies to display an OMB control number with their information collection request.

Many agencies have begun publishing numerous OMB control numbers as amendments to existing regulations in the CFR. These OMB numbers are placed as close as possible to the applicable recordkeeping or reporting requirements.

OBSOLETE PROVISIONS

Provisions that become obsolete before the revision date stated on the cover of each volume are not carried. Code users may find the text of provisions in effect on a given date in the past by using the appropriate numerical list of sections affected. For the period before January 1, 2001, consult either the List of CFR Sections Affected, 1949–1963, 1964–1972, 1973–1985, or 1986–2000, published in eleven separate volumes. For the period beginning January 1, 2001, a "List of CFR Sections Affected" is published at the end of each CFR volume.

"[RESERVED]" TERMINOLOGY

The term "[Reserved]" is used as a place holder within the Code of Federal Regulations. An agency may add regulatory information at a "[Reserved]" location at any time. Occasionally "[Reserved]" is used editorially to indicate that a portion of the CFR was left vacant and not accidentally dropped due to a printing or computer error.

INCORPORATION BY REFERENCE

What is incorporation by reference? Incorporation by reference was established by statute and allows Federal agencies to meet the requirement to publish regulations in the Federal Register by referring to materials already published elsewhere. For an incorporation to be valid, the Director of the Federal Register must approve it. The legal effect of incorporation by reference is that the material is treated as if it were published in full in the Federal Register (5 U.S.C. 552(a)). This material, like any other properly issued regulation, has the force of law.

What is a proper incorporation by reference? The Director of the Federal Register will approve an incorporation by reference only when the requirements of 1 CFR part 51 are met. Some of the elements on which approval is based are:

(a) The incorporation will substantially reduce the volume of material published in the Federal Register.

(b) The matter incorporated is in fact available to the extent necessary to afford fairness and uniformity in the administrative process.

(c) The incorporating document is drafted and submitted for publication in accordance with 1 CFR part 51.

What if the material incorporated by reference cannot be found? If you have any problem locating or obtaining a copy of material listed as an approved incorporation by reference, please contact the agency that issued the regulation containing that incorporation. If, after contacting the agency, you find the material is not available, please notify the Director of the Federal Register, National Archives and Records Administration, 8601 Adelphi Road, College Park, MD 20740-6001, or call 202-741-6010.

CFR INDEXES AND TABULAR GUIDES

A subject index to the Code of Federal Regulations is contained in a separate volume, revised annually as of January 1, entitled CFR INDEX AND FINDING AIDS. This volume contains the Parallel Table of Authorities and Rules. A list of CFR titles, chapters, subchapters, and parts and an alphabetical list of agencies publishing in the CFR are also included in this volume.

An index to the text of "Title 3—The President" is carried within that volume.

The Federal Register Index is issued monthly in cumulative form. This index is based on a consolidation of the "Contents" entries in the daily Federal Register.

A List of CFR Sections Affected (LSA) is published monthly, keyed to the revision dates of the 50 CFR titles.

REPUBLICATION OF MATERIAL

There are no restrictions on the republication of material appearing in the Code of Federal Regulations.

INQUIRIES

For a legal interpretation or explanation of any regulation in this volume, contact the issuing agency. The issuing agency's name appears at the top of odd-numbered pages.

For inquiries concerning CFR reference assistance, call 202–741–6000 or write to the Director, Office of the Federal Register, National Archives and Records Administration, 8601 Adelphi Road, College Park, MD 20740-6001 or e-mail *fedreg.info@nara.gov.*

SALES

The Government Printing Office (GPO) processes all sales and distribution of the CFR. For payment by credit card, call toll-free, 866-512-1800, or DC area, 202-512-1800, M-F 8 a.m. to 4 p.m. e.s.t. or fax your order to 202-512-2104, 24 hours a day. For payment by check, write to: US Government Printing Office – New Orders, P.O. Box 979050, St. Louis, MO 63197-9000.

ELECTRONIC SERVICES

The full text of the Code of Federal Regulations, the LSA (List of CFR Sections Affected), The United States Government Manual, the Federal Register, Public Laws, Public Papers of the Presidents of the United States, Compilation of Presidential Documents and the Privacy Act Compilation are available in electronic format via *www.ofr.gov.* For more information, contact the GPO Customer Contact Center, U.S. Government Printing Office. Phone 202-512-1800, or 866-512-1800 (toll-free). E-mail, *gpo@custhelp.com.*

The Office of the Federal Register also offers a free service on the National Archives and Records Administration's (NARA) World Wide Web site for public law numbers, Federal Register finding aids, and related information. Connect to NARA's web site at *www.archives.gov/federal-register.*

CHARLES A. BARTH,
Director,
Office of the Federal Register.
July 1, 2012.

THIS TITLE

Title 40—PROTECTION OF ENVIRONMENT is composed of thirty-four volumes. The parts in these volumes are arranged in the following order: Parts 1–49, parts 50–51, part 52 (52.01–52.1018), part 52 (52.1019–52.2019), part 52 (52.2020–end of part 52), parts 53–59, part 60 (60.1–end of part 60, sections), part 60 (Appendices), parts 61–62, part 63 (63.1–63.599), part 63 (63.600–63.1199), part 63 (63.1200–63.1439), part 63 (63.1440–63.6175), part 63 (63.6580–63.8830), part 63 (63.8980–end of part 63) parts 64–71, parts 72–80, parts 81–84, part 85–§ 86.599–99, part 86 (86.600–1–end of part 86), parts 87–95, parts 96–99, parts 100–135, parts 136–149, parts 150–189, parts 190–259, parts 260–265, parts 266–299, parts 300–399, parts 400–424, parts 425–699, parts 700–789, parts 790–999, and part 1000 to end. The contents of these volumes represent all current regulations codified under this title of the CFR as of July 1, 2012.

Chapter I—Environmental Protection Agency appears in all thirty-four volumes. Regulations issued by the Council on Environmental Quality, including an Index to Parts 1500 through 1508, appear in the volume containing part 1000 to end. The OMB control numbers for title 40 appear in §9.1 of this chapter.

For this volume, Michele Bugenhagen was Chief Editor. The Code of Federal Regulations publication program is under the direction of Michael L. White, assisted by Ann Worley.

Title 40—Protection of Environment

(This book contains parts 100 to 135)

CHAPTER I—ENVIRONMENTAL PROTECTION AGENCY (CONTINUED)

EDITORIAL NOTE: Nomenclature changes to chapter I appear at 65 FR 47324, 47325, Aug. 2, 2000; 66 FR 34375, 34376, June 28, 2001.

SUBCHAPTER D—WATER PROGRAMS

SUBCHAPTER D—WATER PROGRAMS

AUTHORITY: Secs. 501 and 307(a) of the Federal Water Pollution Control Act, as amended (33 U.S.C. 1251 et seq., Pub. L. 92–500, 86 Stat. 816).

SOURCE: 41 FR 17902, Apr. 29, 1976, unless otherwise noted.

§ 104.1 Applicability.

This part shall be applicable to hearings required by statute to be held in connection with the establishment of toxic pollutant effluent standards under section 307(a) of the Act.

§ 104.2 Definitions.

As used in this part, the term:

(a) *Act* means the Federal Water Pollution Control Act, as amended, 33 U.S.C. 1251 et seq., Public Law 92–500, 86 Stat. 816.

(b) *Administrator* means the Administrator of the Environmental Protection Agency, or any employee of the Agency to whom the Administrator may by order delegate his authority to carry out his functions under section 307(a) of the Act, or any person who shall by operation of law be authorized to carry out such functions.

(c) *Agency* means the Environmental Protection Agency.

(d) *Hearing Clerk* means the Hearing Clerk, U.S. Environmental Protection Agency, 401 M Street SW., Washington, DC 20460.

(e) *Party* means the Environmental Protection Agency as the proponent of an effluent standard or standards, and any person who files an objection pursuant to § 104.3 hereof.

(f) *Person* means an individual, corporation, partnership, association, state, municipality or other political subdivision of a state, or any interstate body.

(g) *Effluent standard* means any effluent standard or limitation, which may include a prohibition of any discharge, established or proposed to be established for any toxic pollutant under section 307(a) of the Act.

(h) *Presiding Officer* means the Chief Administrative Law Judge of the Agency or a person designated by the Chief Administrative Law Judge or by the Administrator to preside at a hearing under this part, in accordance with § 104.6 hereof.

§ 104.3 Notice of hearing; objection; public comment.

(a) *Notice of hearing.* Whenever the Administrator publishes any proposed effluent standard, he shall simultaneously publish a notice of a public hearing to be held within thirty days following the date of publication of the proposed standard. Any person who has any objection to a proposed standard may file with the hearing clerk a concise statement of any such objection. No person may participate in the hearing on the proposed toxic pollutant effluent standards unless the hearing clerk has received within 25 days of the publication of the notice of the proposed standards a statement of objection as herein described. In exceptional circumstances and for good cause shown the Presiding Officer may allow an objection to be filed after the filing deadline prescribed in the preceding sentence, which good cause must include at a minimum lack of actual notice on the part of the objector or any representative of such objector of the proposed standards despite his exercise of due diligence, so long as such later filing will not cause undue delay in the proceedings or prejudice to any of the parties.

(b) *Objections.* Any objection to a proposed standard which is filed pursuant to paragraph (a) of this section shall meet the following requirements:

(1) It shall be filed in triplicate with the hearing clerk within the time prescribed in paragraph (a) of this section;

(2) It shall state concisely and with particularity each portion of the proposed standard to which objection is taken; to the greatest extent feasible it shall state the basis for such objection;

(3) To the greatest extent feasible it shall (i) state specifically the objector's proposed modification to any such standard proposed by the Agency to which objection is taken, (ii) set forth the reasons why such modification is sought, and (iii) identify and describe the scientific or other basis for such proposed modification, including reference to any pertinent scientific data or authority in support thereof.

Any objection which fails to comply with the foregoing provisions shall not be accepted for filing. The Presiding Officer shall promptly notify any person whose objection is not accepted for any of the reasons set forth in this section, stating the reasons therefor.

(c) *Data in support of objection or modification.* In the event that the time prescribed for filing objections pursuant to paragraphs (a) and (b) of this section is insufficient to permit an objecting party to fully set forth with such objection the basis therefor together with the information and data specified in paragraph (b)(3) of this section, he may so state at the time of the filing of such objection, and file a more complete statement of such basis, information, and data (hereinafter referred to as "supplemental data") within the time prescribed by this paragraph (c). The supplemental data herein described shall be filed not later than 40 days following publication of the proposed effluent standards.

(d) *Public comment.* The notice required under paragraph (a) of this section shall also provide for the submission to the Agency of written comments on the proposed rulemaking by interested persons not filing objections pursuant to this section as hereinabove described, and hence not participating in the hearing as parties. The notice shall fix a time deadline for the submission of such comments which shall be not later than the date set for commencement of the hearing. Such comments shall be received in evidence at the commencement of the hearing. The Administrator in making any decision based upon the record shall take into account the unavailability of cross-examination in determining the weight to be accorded such comments.

(e) *Promulgation in absence of objection.* If no objection is filed pursuant to this section, then the Administrator shall promulgate the final standards on the basis of the Agency's statement of basis and purpose and any public comments received pursuant to paragraph (d) of this section.

§ 104.4 Statement of basis and purpose.

Whenever the Administrator publishes a proposed effluent standard, the notice thereof published in the FEDERAL REGISTER shall include a statement of the basis and purpose of the standard or a summary thereof. This statement shall include:

(a) The purpose of the proposed standard;

(b) An explanation of how the proposed standard was derived;

(c) Scientific and technical data and studies supporting the proposed standard or references thereto if the materials are published or otherwise readily available; and

(d) Such other information as may be reasonably required to set forth fully the basis of the standard.

Where the notice of the proposed rulemaking summarizes the full statement of basis and purpose, or incorporates documents by reference, the documents thus summarized or incorporated by reference shall thereupon be made available by the Agency for inspection and copying by any interested person.

§ 104.5 Docket and record.

Whenever the Administrator publishes a notice of hearing under this part, the hearing clerk shall promptly establish a docket for the hearing. The docket shall include all written objections filed by any party, any public comments received pursuant to § 104.3(d), a verbatim transcript of the hearing, the statement of basis and purpose required by § 104.4, and any supporting documents referred to therein, and other documents of exhibits that may be received in evidence or marked for identification by or at the direction of the Presiding Officer, or filed by any party in connection with the hearing. Copies of documents in the docket shall be available to any person upon payment to the Agency of such charges as the Agency may prescribe to cover the costs of duplication. The materials contained in the docket shall constitute the record.

§ 104.6 Designation of Presiding Officer.

The Chief Administrative Law Judge of the Agency may preside personally at any hearing under this part, or he may designate another Administrative Law Judge as Presiding Officer for the hearing. In the event of the unavailability of any such Administrative Law Judge, the Administrator may designate a Presiding Officer. No person who has any personal pecuniary interest in the outcome of a proceeding under this part, or who has participated in the development or enforcement of any standard or proposed standard at issue in a proceeding hereunder, shall serve as Presiding Officer in such proceeding.

§ 104.7 Powers of Presiding Officer.

The Presiding Officer shall have the duty to conduct a fair hearing within the time constraints imposed by section 307(a) of the Act. He shall take all necessary action to avoid delay and to maintain order. He shall have all powers necessary to these ends, including but not limited to the power to:

(a) Rule upon motions and requests;

(b) Change the time and place of the hearing, and adjourn the hearing from time to time or from place to place;

(c) Examine and cross-examine witnesses;

(d) Admit or exclude evidence; and

(e) Require any part or all of the evidence to be submitted in writing and by a certain date.

§ 104.8 Prehearing conferences.

Prehearing conferences are encouraged for the purposes of simplification of issues, identification and scheduling of evidence and witnesses, the establishment of an orderly framework for the proceedings, the expediting of the hearing, and such other purposes of a similar nature as may be appropriate.

(a) The Presiding Officer on his own motion may, and at the request of any party made within 20 days of the proposal of standards hereunder shall, direct all parties to appear at a specified time and place for an initial hearing session in the nature of a prehearing conference. Matters taken up at the conference may include, without limitation:

(1) Consideration and simplification of any issues of law or fact;

(2) Identification, advance submission, marking for identification, consideration of any objections to admission, and admission of documentary evidence;

(3) Possible stipulations of fact;

(4) The identification of each witness expected to be called by each party, and the nature and substance of his expected testimony;

(5) Scheduling of witnesses where practicable, and limitation of the number of witnesses where appropriate in order to avoid delay or repetition;

(6) If desirable, the segregation of the hearing into separate segments for different provisions of the proposed effluent standards and the establishment of separate service lists;

(7) Encouragement of objecting parties to agree upon and designate lead counsel for objectors with common interests so as to avoid repetitious questioning of witnesses.

(b) The Presiding Officer may, following a prehearing conference, issue an order setting forth the agreements reached by the parties or representatives, the schedule of witnesses, and a statement of issues for the hearing. In addition such order may direct the parties to file and serve copies of documents or materials, file and serve lists of witnesses which may include a short summary of the expected testimony of each and, in the case of an expert witness, his curriculum vitae, and may contain such other directions as may be appropriate to facilitate the proceedings.

§ 104.9 Admission of evidence.

(a) Where the Presiding Officer has directed identification of witnesses and production of documentation evidence by a certain date, the Presiding Officer may exclude any such evidence, or refuse to allow any witness to testify, when the witness was not identified or the document was not served by the time set by the Presiding Officer. Any such direction with respect to a party's case in chief shall not preclude the use of such evidence or testimony on rebuttal or response, or upon a showing satisfactory to the Presiding Officer that good cause existed for failure to serve testimony or a document or identify a witness by the time required. The Presiding Officer may require direct testimony to be in writing under oath and served by a certain date, and may exclude testimony not so served.

(b) At the first prehearing conference, or at another time before the beginning of the taking of oral testimony to be set by the Presiding Officer, the statement of basis and purpose, together with any publications or reference materials cited therein, except where excluded by stipulation, shall be received in evidence.

(c) The Presiding Officer may exclude evidence which is immaterial, irrelevant, unduly repetitious or cumulative, or would involve undue delay, or which, if hearsay, is not of the sort upon which responsible persons are accustomed to rely.

(d) If relevant and material evidence is contained in a report or document containing immaterial or irrelevant matter, such immaterial or irrelevant matter may be excluded.

(e) Whenever written testimony or a document or object is excluded from evidence by the Presiding Officer, it shall at the request of the proponent be marked for identification. Where oral testimony is permitted by the Presiding Officer, but the Presiding Officer excludes particular oral testimony, the party offering such testimony may make a brief offer of proof.

(f) Any relevant and material documentary evidence, including but not limited to affidavits, published articles, and official documents, regardless of the availability of the affiant or author for cross-examination, may be admitted in evidence, subject to the provisions of paragraphs (a), (c), and (d) of this section. The availability or non-availability of cross-examination shall be considered as affecting the weight to be accorded such evidence in any decision based upon the record.

(g) Official notice may be taken by the Presiding Officer or the Administrator of any matter which could be judicially noticed in the United States District Courts, and of other facts within the specialized knowledge and experience of the Agency. Opposing parties shall be given adequate opportunity to show the contrary.

§ 104.10 Hearing procedures.

(a) Following the admission in evidence of the materials described in § 104.9(b), the Agency shall have the right at the commencement of the hearing to supplement that evidence or to introduce additional relevant evidence. Thereafter the evidence of each objector shall be presented in support of its objection and any proposed modification. The Agency staff shall then be given an opportunity to rebut or respond to the objectors' presentation, including at its option the introduction of evidence which tends to support a standard or standards other than as set forth in the Agency's own initially proposed standards. In the event that evidence which tends to support such other standard or standards is offered and received in evidence, then the objectors may thereafter rebut or respond to any such new evidence.

(b) The burden of proof as to any modification of any standard proposed by the Agency shall be upon the party who advocates such modification to show that the proposed modification is justified based upon a preponderance of the evidence.

(c) Where necessary in order to prevent undue prolongation of the hearing, or to comply with time limitations set forth in the Act, the Presiding Officer may limit the number of witnesses who may testify, and the scope and extent of cross-examination.

8

(d) A verbatim transcript of the hearing shall be maintained and shall constitute a part of the record.

(e) If a party objects to the admission or rejection of any evidence or to any other ruling of the Presiding Officer during the hearing, he shall state briefly the grounds of such objection. With respect to any ruling on evidence, it shall not be necessary for any party to claim an exception in order to preserve any right of subsequent review.

(f) Any party may at any time withdraw his objection to a proposed effluent standard.

§104.11 Briefs and findings of fact.

At the conclusion of the hearing, the Presiding Officer shall set a schedule for the submission by the parties of briefs and proposed findings of fact and conclusions. In establishing the aforesaid time schedule, the Presiding Officer shall consider the time constraints placed upon the parties and the Administrator by the statutory deadlines.

§104.12 Certification of record.

As soon as possible after the hearing, the Presiding Officer shall transmit to the hearing clerk the transcript of the testimony and exhibits introduced in the hearing. The Presiding Officer shall attach to the original transcript his certificate stating that, to the best of his knowledge and belief, the transcript is a true transcript of the testimony given at the hearing except in such particulars as he shall specify, and that the exhibits transmitted are all the exhibits as introduced at the hearing with such exceptions as he shall specify.

§104.13 Interlocutory and post-hearing review of rulings of the Presiding Officer; motions.

(a) The Presiding Officer may certify a ruling for interlocutory review by the Administrator where a party so requests and the Presiding Officer concludes that (1) the ruling from which review is sought involves an important question as to which there is substantial ground for difference of opinion, and (2) either (i) a subsequent reversal of his ruling would be likely to result in substantial delay or expense if left to the conclusion of the proceedings, or

(ii) a ruling on the question by the Administrator would be of material assistance in expediting the hearing. The certificate shall be in writing and shall specify the material relevant to the ruling certified. If the Administrator determines that interlocutory review is not warranted, he may decline to consider the ruling which has been certified.

(b) Where the Presiding Officer declines to certify a ruling the party who had requested certification may apply to the Administrator for interlocutory review, or the Administrator may on his own motion direct that any matter be submitted to him for review, subject to the standards for review set forth in paragraph (a) of this section. An application for review shall be in writing and shall briefly state the grounds relied on. If the Administrator takes no action with respect to such application for interlocutory review within 15 days of its filing, such application shall be deemed to have been denied.

(c) Unless otherwise ordered by the Presiding Officer or the Administrator, the hearing shall continue pending consideration by the Administrator of any ruling or request for interlocutory review.

(d) Unless otherwise ordered by the Presiding Officer or the Administrator, briefs in response to any application for interlocutory review may be filed by any party within five days of the filing of the application for review.

(e) Failure to request or obtain interlocutory review does not waive the rights of any party to complain of a ruling following completion of the hearing. Within five days following the close of a hearing under this part, any party may apply to the Administrator for post-hearing review of any procedural ruling, or any ruling made by the Presiding Officer concerning the admission or exclusion of evidence to which timely objection was made. Within seven days following the filing of any such application any other party may file a brief in response thereto.

(f) If the Administrator on review under paragraph (e) of this section determines that evidence was improperly excluded, he may order its admission without remand for further proceedings, or may remand with such instructions as he deems appropriate concerning cross-examination, or opportunity for any party to submit further evidence, with respect to such evidence as he directs should be admitted. In making his determination whether to remand, the Administrator shall consider whether the statutory time restraints permit a remand, and whether it would be constructive to allow cross-examination or further evidence with respect to the newly admitted evidence. If evidence is admitted without cross-examination, the Administrator shall consider the lack of opportunity for cross-examination in determining the weight to be given such evidence.

(g) Motions shall be brief, in writing, and may be filed at any time following the publication of the proposed effluent standards, unless otherwise ordered by the Presiding Officer or the Administrator. Unless otherwise ordered or provided in these rules, responses to motions may be filed within seven days of the actual filing of the motion with the hearing clerk.

§ 104.14 Tentative and final decision by the Administrator.

(a) As soon as practicable following the certification of the record and the filing by the parties of briefs and proposed findings of fact and conclusions under § 104.11, the Administrator, with such staff assistance as he deems necessary and appropriate, shall review the entire record and prepare and file a tentative decision based thereon. The tentative decision shall include findings of fact and conclusions, and shall be filed with the hearing clerk who shall at once transmit a copy thereof to each party who participated at the hearing, or his attorney or other representative.

(b) Upon filing of the tentative decision, the Administrator may allow a reasonable time for the parties to file with him any exceptions to the tenative decision, a brief in support of such exceptions containing appropriate references to the record, and any proposed changes in the tentative decision. Such materials shall, upon submission, become part of the record. As soon as practicable after the filing

thereof the Administrator shall prepare and file a final decision, copies of which shall be transmitted to the parties or their representatives in the manner prescribed in paragraph (a) of this section.

(c) In the event that the Administrator determines that due and timely execution of his functions, including compliance with time limitations established by law, imperatively and unavoidably so requires, he may omit the preparation and filing of the tentative decision and related procedures set forth in paragraph (b) of this section, and shall instead prepare and file a final decision, copies of which shall be transmitted to the parties or their representatives in the manner prescribed in paragraph (a) of this section.

(d) Any decision rendered by the Administrator pursuant to this section shall include a statement of his findings and conclusions, and the reasons and basis therefor, and shall indicate the toxic pollutant effluent standard or standards which the Administrator is promulgating or intends to promulgate based thereon.

§ 104.15 Promulgation of standards.

Upon consideration of the record, at the time of his final decision the Administrator shall determine whether the proposed effluent standard or standards should be promulgated as proposed, or whether any modification thereof is justified based upon a preponderance of the evidence adduced at the hearing, regardless of whether or not such modification was actually proposed by any objecting party. If he determines that a modification is not justified, he shall promulgate the standard or standards as proposed. If he determines that a modification is justified, he shall promulgate a standard or standards as so modified.

§ 104.16 Filing and time.

(a) All documents or papers required or authorized by the foregoing provisions of this part including, but not limited to, motions, applications for review, and briefs, shall be filed in duplicate with the hearing clerk, except as otherwise expressly provided in these rules. Any document or paper so required or authorized to be filed with the hearing clerk, if it is filed during the course of the hearing, shall be also filed with the Presiding Officer. A copy of each document or paper filed by any party with the Presiding Officer, with the hearing clerk, or with the Administrator shall be served upon all other parties, except to the extent that the list of parties to be so served may be modified by order of the Presiding Officer, and each such document or paper shall be accompanied by a certificate of such service.

(b) A party may be represented in any proceeding under this part by an attorney or other authorized representative. When any document or paper is required under these rules to be served upon a party such service shall be made upon such attorney or other representative.

(c) Except where these rules or an order of the Presiding Officer require receipt of a document by a certain date, any document or paper required or authorized to be filed by this part shall be deemed to be filed when postmarked, or in the case of papers delivered other than by mail, when received by the hearing clerk.

(d) Sundays and legal holidays shall be included in computing the time allowed for the filing of any document or paper, provided, that when such time expires on a Sunday or legal holiday, such period shall be extended to include the next following business day.

PART 105—RECOGNITION AWARDS UNDER THE CLEAN WATER ACT

AUTHORITY: Section 501(a) and (e) of the Clean Water Act (CWA), 33 U.S.C. 1361(a) and (e).

SOURCE: 67 FR 6142, Feb. 8, 2002, unless otherwise noted.

GENERAL

§ 105.1 Background.

The Environmental Protection Agency's (EPA) Clean Water Act (CWA) Recognition Awards Program is authorized by CWA section 501(e). The Administrator may provide official recognition to industrial organizations and political subdivisions of States which during the preceding year demonstrated an outstanding technological achievement or an innovative process, method or device in their waste treatment and pollution abatement programs. The wastewater management programs can generally be characterized as waste treatment and/or pollution abatement programs. Individual EPA Regional Administrators (and Regional officials they may designate) also may conduct Regional CWA Recognition Awards Programs according to and consistent with the provisions of this part.

§ 105.2 Definitions.

Applicant means the person authorized to complete the application on behalf of an industrial organization or political subdivision of States.

Application means a completed questionnaire, nomination form, or other documentation submitted to or by the States, EPA Regions or headquarters for consideration of a national CWA Recognition Award.

I means the applicant for an award.

Industrial organization means any company, corporation, association, partnership, firm, university, not-for-profit organization, or wastewater treatment facility, as well as a Federal, State or Tribal government wastewater treatment facility, or U.S. military command to the extent such government and other organizations operate in an "industrial" capacity in the treatment of wastes or abatement of pollution.

Nominee means a candidate recommended by the State or Tribe or EPA for consideration for a CWA Recognition Award.

Political subdivision of State means a municipality, city, town, borough, county, parish, district, association, or other public body (including an intermunicipal agency of two or more of the foregoing entities) created by or pursuant to State law.

State means any of the 50 States, the District of Columbia, the Commonwealth of Puerto Rico, the U.S. Virgin Islands, Guam, American Samoa, and the Commonwealth of Northern Mariana Islands.

State water pollution control agency means the State agency designated by the Governing Authority having responsibility for enforcing State laws relating to the abatement of water pollution.

You means the applicant for an award.

§ 105.3 Title.

The awards are known as the National Clean Water Act Recognition Awards (hereinafter, the Awards Program).

ELIGIBILITY REQUIREMENTS

§ 105.4 What are the requirements for the Awards Program?

(a) EPA will administer the Awards Program, and should establish annual guidance as necessary to administer the Awards Program. EPA will request from the various offices, and States and Tribes as appropriate, nominations for the Awards Program.

(b) Nominees must be in total compliance with all applicable water quality requirements under the CWA in order to be eligible for an award, and otherwise have a satisfactory record with respect to environmental quality.

(c) Nominees must provide written documentation as evidence to support

their outstanding technological achievement or innovative process, method or device in their waste treatment and/or pollution abatement programs.

(d) EPA may issue annual guidance memoranda to administer each year's awards programs. For information on the availability of additional guidance, contact the U.S. Environmental Protection Agency, Municipal Assistance Branch, 1200 Pennsylvania Avenue, NW., Mail Code 4204–M, Washington, DC 20460. You may also visit EPA's Web site at *www.epa.gov/owm*.

§ 105.5 Who is eligible to win an award?

A municipality, city, town, borough, county, parish, district, association, government agency, or other public body, (including an intermunicipal agency of two or more of the foregoing entities) created by or pursuant to State law; a company, corporation, association, partnership, firm, university, not-for-profit organization, or wastewater treatment facility, as well as a Federal, State or Tribal government wastewater treatment facility, or U.S. military command to the extent such government and other organizations operate in an industrial capacity in the treatment of wastes or abatement of pollution may be considered for a recognition award.

§ 105.6 What are the Awards Program categories for which I may be eligible?

EPA will publish from time to time, a notice in the FEDERAL REGISTER announcing the current Awards Program categories. EPA also may subsequently discontinue, combine, or rename categories by notice published in the FEDERAL REGISTER.

APPLICATION AND NOMINATION PROCESS

§ 105.7 How do I apply for an award?

You may contact your local EPA Regional office for information on the Awards Program guidance each year, or check the Web site at *http://www.epa.gov/owm/intnet.htm*. EPA may use an application or nomination process, as appropriate for the program or Region.

§ 105.8 When can I apply for an award?

You can contact your local EPA Regional office for award submission deadline information which may vary for the award categories, or check the Web site at *http://www.epa.gov/owm/intnet.htm.*

§ 105.9 How can I get nominated for an award?

You may apply to, or ask your State, Tribe or EPA Region to nominate you for an award. Only applications or nominations recommended by EPA Regions are considered for the national award. EPA personnel conduct compliance evaluations prior to presenting a national award.

SELECTION CRITERIA

§ 105.10 What do I need to be considered for an award?

Your facility or pollution abatement program must be in total compliance with all applicable water quality requirements, and otherwise have a satisfactory record with respect to environmental quality. Additionally, your facility or pollution abatement program must provide written documentation as evidence of an outstanding technological achievement or an innovative process, method or device demonstrated in the preceding year, which resulted in environmental benefits, cost savings and/or public acceptance.

§ 105.11 Who selects the award winners?

After EPA receives the completed application, the application is evaluated by a review committee. After the review committee completes its evaluation of the programs that have been nominated, they make recommendations for the national awards. EPA then analyzes the results and selects the award winners.

§ 105.12 How is the awards review committee selected?

EPA review committee members are selected by the EPA and in some cases, State or Tribal water pollution control agencies. The number of participants in a nominations review process is based on staff availability, and may be one person.

§ 105.13 How are the award winners selected?

Nominees and applications are recommended by EPA regions. EPA personnel conduct compliance evaluations prior to presenting a national award. EPA selects national award winners based on demonstrated evidence of outstanding and/or innovative wastewater treatment and pollution abatement programs or projects which result in environmental benefits, cost savings and/or public acceptance. Based upon results of review committee evaluations, the Agency selects first place winners for a national award in the appropriate awards categories. A second place winner may or may not be selected. EPA may or may not select an award winner for every awards program category. Award decisions are not subject to administrative review.

AWARDS RECOGNITION

§ 105.14 How are award winners notified?

EPA notifies national award winners by letter.

§ 105.15 How are award winners recognized?

EPA presents national award winners with a certificate or plaque at an awards presentation ceremony as recognition for an outstanding technological achievement or an innovative process, method or device in wastewater treatment and/or pollution abatement programs. The President of the United States, the Governor of the State, or Tribal leader of the jurisdiction reservation in which the awardee is situated, the Speaker of the House of Representatives and the President pro tempore of the Senate are notified by the Administrator.

§ 105.16 How are award winners publicized?

EPA announces the annual national recognition award winners through notice published in the FEDERAL REGISTER.

PART 108—EMPLOYEE PROTECTION HEARINGS

AUTHORITY: Sec. 507(e), Pub. L. 92-500, 86 Stat. 816 (33 U.S.C. 1251 *et seq.*).

SOURCE: 39 FR 15398, May 3, 1974, unless otherwise noted.

§ 108.1 Applicability.

This part shall be applicable to investigations and hearings required by section 507(e) of the Federal Water Pollution Control Act, as amended, 33 U.S.C. 1251 *et seq.* (Pub. L. 92-500).

§ 108.2 Definitions.

As used in this part, the term:

(a) *Act* means the Federal Water Pollution Control Act, as amended;

(b) *Effluent limitation* means any effluent limitation which is established as a condition of a permit issued or proposed to be issued by a State or by the Environmental Protection Agency pursuant to section 402 of the Act; any toxic or pretreatment effluent standard established under section 307 of the Act; any standard of performance established under section 306 of the Act; and any effluent limitation established under section 302, section 316, or section 318 of the Act.

(c) *Order* means any order issued by the Administrator under section 309 of the Act; any order issued by a State to secure compliance with a permit, or condition thereof, issued under a program approved pursuant to section 402 of the Act; or any order issued by a court in an action brought pursuant to section 309 or section 505 of the Act.

(d) *Party* means an employee filing a request under § 108.3, any employee similarly situated, the employer of any such employee, and the Regional Administrator or his designee.

(e) *Administrator* or *Regional Administrator* means the Administrator or a Regional Administrator of the Environmental Protection Agency.

§ 108.3 Request for investigation.

Any employee who is discharged or laid-off, threatened with discharge or lay-off, or otherwise discriminated against by any person because of the alleged results of any effluent limitation or order issued under the Act, or any representative of such employee, may submit a request for an investigation under this part to the Regional Administrator of the region in which such discrimination is alleged to have occurred.

§ 108.4 Investigation by Regional Administrator.

Upon receipt of any request meeting the requirements of § 108.3, the Regional Administrator shall conduct a full investigation of the matter, in order to determine whether the request may be related to an effluent limitation or order under the Act. Following the investigation, the Regional Administrator shall notify the employee requesting the investigation (or the employee's representative) and the employer of such employee, in writing, of his preliminary findings and conclusions. The employee, the representative of such employee, or the employer may within fifteen days following receipt of the preliminary findings and conclusions of the Regional Administrator request a hearing under this part. Upon receipt of such a request, the Regional Administrator, with the concurrence of the Chief Administrative Law Judge, shall publish notice of a hearing to be held not less than 30 days following the date of such publication where he determines that there are factual issues concerning the existence of the alleged discrimination or its relationship to an effluent limitation or order under the Act. The notice shall specify a date before which any party (or representative of such party) may submit a request to appear.

§ 108.5 Procedure.

Any hearing held pursuant to this part shall be of record and shall be conducted according to the requirements of 5 U.S.C. 554. The Administrative Law Judge shall conduct the hearing in an orderly and expeditious manner. By agreement of the parties, he may dismiss the hearing. The Administrative

Law Judge, on his own motion, or at the request of any party, shall have the power to hold prehearing conferences, to issue subpoenas for the attendance and testimony of witnesses and the production of relevant papers, books, and documents, and he may administer oaths. The Regional Administrator, and any party submitting a request pursuant to §108.3 or §108.4, or counsel or other representative of such party or the Regional Administrator, may appear and offer evidence at the hearing.

§108.6 Recommendations.

At the conclusion of any hearing under this part, the Administrative Law Judge shall, based on the record, issue tentative findings of fact and recommendations concerning the alleged discrimination, and shall submit such tentative findings and recommendations to the Administrator. The Administrator shall adopt or modify the findings and recommendations of the Administrative Law Judge, and shall make copies of such findings and recommendations available to the complaining employee, the employer, and the public.

§108.7 Hearing before Administrator.

At his option, the Administrator may exercise any powers of an Administrative Law Judge with respect to hearings under this part.

PART 109—CRITERIA FOR STATE, LOCAL AND REGIONAL OIL REMOVAL CONTINGENCY PLANS

AUTHORITY: Sec. 11(j)(1)(B), 84 Stat. 96, 33 U.S.C. 1161(j)(1)(B).

SOURCE: 36 FR 22485, Nov. 25, 1971, unless otherwise noted.

§109.1 Applicability.

The criteria in this part are provided to assist State, local and regional agencies in the development of oil removal contingency plans for the inland navigable waters of the United States and all areas other than the high seas, coastal and contiguous zone waters, coastal and Great Lakes ports and harbors and such other areas as may be agreed upon between the Environmental Protection Agency and the Department of Transportation in accordance with section 11(j)(1)(B) of the Federal Act, Executive Order No. 11548 dated July 20, 1970 (35 FR 11677) and §306.2 of the National Oil and Hazardous Materials Pollution Contingency Plan (35 FR 8511).

§109.2 Definitions.

As used in these guidelines, the following terms shall have the meaning indicated below:

(a) *Oil* means oil of any kind or in any form, including, but not limited to, petroleum, fuel oil, sludge, oil refuse, and oil mixed with wastes other than dredged spoil.

(b) *Discharge* includes, but is not limited to, any spilling, leaking, pumping, pouring, emitting, emptying, or dumping.

(c) *Remove* or *removal* refers to the removal of the oil from the water and shorelines or the taking of such other actions as may be necessary to minimize or mitigate damage to the public health or welfare, including, but not limited to, fish, shellfish, wildlife, and public and private property, shorelines, and beaches.

(d) *Major disaster* means any hurricane, tornado, storm, flood, high water, wind-driven water, tidal wave, earthquake, drought, fire, or other catastrophe in any part of the United States which, in the determination of the President, is or threatens to become of sufficient severity and magnitude to warrant disaster assistance by the Federal Government to supplement the efforts and available resources of States and local governments and relief organizations in alleviating the damage, loss, hardship, or suffering caused thereby.

15

(e) *United States* means the States, the District of Columbia, the Commonwealth of Puerto Rico, the Canal Zone, Guam, American Samoa, the Virgin Islands, and the Trust Territory of the Pacific Islands.

(f) *Federal Act* means the Federal Water Pollution Control Act, as amended, 33 U.S.C. 1151 *et seq.*

§ 109.3 Purpose and scope.

The guidelines in this part establish minimum criteria for the development and implementation of State, local, and regional contingency plans by State and local governments in consultation with private interests to insure timely, efficient, coordinated and effective action to minimize damage resulting from oil discharges. Such plans will be directed toward the protection of the public health or welfare of the United States, including, but not limited to, fish, shellfish, wildlife, and public and private property, shorelines, and beaches. The development and implementation of such plans shall be consistent with the National Oil and Hazardous Materials Pollution Contingency Plan. State, local and regional oil removal contingency plans shall provide for the coordination of the total response to an oil discharge so that contingency organizations established thereunder can function independently, in conjunction with each other, or in conjunction with the National and Regional Response Teams established by the National Oil and Hazardous Materials Pollution Contingency Plan.

§ 109.4 Relationship to Federal response actions.

The National Oil and Hazardous Materials Pollution Contingency Plan provides that the Federal on-scene commander shall investigate all reported spills. If such investigation shows that appropriate action is being taken by either the discharger or non-Federal entities, the Federal on-scene commander shall monitor and provide advice or assistance, as required. If appropriate containment or cleanup action is not being taken by the discharger or non-Federal entities, the Federal on-scene commander will take control of the response activity in accordance with section 11(c)(1) of the Federal Act.

§ 109.5 Development and implementation criteria for State, local and regional oil removal contingency plans.

Criteria for the development and implementation of State, local and regional oil removal contingency plans are:

(a) Definition of the authorities, responsibilities and duties of all persons, organizations or agencies which are to be involved or could be involved in planning or directing oil removal operations, with particular care to clearly define the authorities, responsibilities and duties of State and local governmental agencies to avoid unnecessary duplication of contingency planning activities and to minimize the potential for conflict and confusion that could be generated in an emergency situation as a result of such duplications.

(b) Establishment of notification procedures for the purpose of early detection and timely notification of an oil discharge including:

(1) The identification of critical water use areas to facilitate the reporting of and response to oil discharges.

(2) A current list of names, telephone numbers and addresses of the responsible persons and alternates on call to receive notification of an oil discharge as well as the names, telephone numbers and addresses of the organizations and agencies to be notified when an oil discharge is discovered.

(3) Provisions for access to a reliable communications system for timely notification of an oil discharge and incorporation in the communications system of the capability for interconnection with the communications systems established under related oil removal contingency plans, particularly State and National plans.

(4) An established, prearranged procedure for requesting assistance during a major disaster or when the situation exceeds the response capability of the State, local or regional authority.

(c) Provisions to assure that full resource capability is known and can be committed during an oil discharge situation including:

(1) The identification and inventory of applicable equipment, materials and supplies which are available locally and regionally.

(2) An estimate of the equipment, materials and supplies which would be required to remove the maximum oil discharge to be anticipated.

(3) Development of agreements and arrangements in advance of an oil discharge for the acquisition of equipment, materials and supplies to be used in responding to such a discharge.

(d) Provisions for well defined and specific actions to be taken after discovery and notification of an oil discharge including:

(1) Specification of an oil discharge response operating team consisting of trained, prepared and available operating personnel.

(2) Predesignation of a properly qualified oil discharge response coordinator who is charged with the responsibility and delegated commensurate authority for directing and coordinating response operations and who knows how to request assistance from Federal authorities operating under existing national and regional contingency plans.

(3) A preplanned location for an oil discharge response operations center and a reliable communications system for directing the coordinated overall response operations.

(4) Provisions for varying degrees of response effort depending on the severity of the oil discharge.

(5) Specification of the order of priority in which the various water uses are to be protected where more than one water use may be adversely affected as a result of an oil discharge and where response operations may not be adequate to protect all uses.

(e) Specific and well defined procedures to facilitate recovery of damages and enforcement measures as provided for by State and local statutes and ordinances.

§ 109.6 Coordination.

For the purposes of coordination, the contingency plans of State and local governments should be developed and implemented in consultation with private interests. A copy of any oil removal contingency plan developed by

State and local governments should be forwarded to the Council on Environmental Quality upon request to facilitate the coordination of these contingency plans with the National Oil and Hazardous Materials Pollution Contingency Plan.

PART 110—DISCHARGE OF OIL

Sec.
110.1 Definitions.
110.2 Applicability.
110.3 Discharge of oil in such quantities as "may be harmful" pursuant to section 311(b)(4) of the Act.
110.4 Dispersants.
110.5 Discharges of oil not determined "as may be harmful" pursuant to section 311(b)(3) of the Act.
110.6 Notice.

AUTHORITY: 33 U.S.C. 1321(b)(3) and (b)(4) and 1361(a); E.O. 11735, 38 FR 21243, 3 CFR Parts 1971–1975 Comp., p. 793.

SOURCE: 52 FR 10719, Apr. 2, 1987, unless otherwise noted.

§ 110.1 Definitions.

Terms not defined in this section have the same meaning given by the Section 311 of the Act. As used in this part, the following terms shall have the meaning indicated below:

Act means the Federal Water Pollution Control Act, as amended, 33 U.S.C. 1251 *et seq.*, also known as the Clean Water Act;

Administrator means the Administrator of the Environmental Protection Agency (EPA);

Applicable water quality standards means State water quality standards adopted by the State pursuant to section 303 of the Act or promulgated by EPA pursuant to that section;

MARPOL 73/78 means the International Convention for the Prevention of Pollution from Ships, 1973, as modified by the Protocol of 1978 relating thereto, Annex I, which regulates pollution from oil and which entered into force on October 2, 1983;

Navigable waters means the waters of the United States, including the territorial seas. The term includes:

(a) All waters that are currently used, were used in the past, or may be susceptible to use in interstate or foreign commerce, including all waters

17

that are subject to the ebb and flow of the tide;

(b) Interstate waters, including interstate wetlands;

(c) All other waters such as intrastate lakes, rivers, streams (including intermittent streams), mudflats, sandflats, and wetlands, the use, degradation, or destruction of which would affect or could affect interstate or foreign commerce including any such waters:

(1) That are or could be used by interstate or foreign travelers for recreational or other purposes;

(2) From which fish or shellfish are or could be taken and sold in interstate or foreign commerce;

(3) That are used or could be used for industrial purposes by industries in interstate commerce;

(d) All impoundments of waters otherwise defined as navigable waters under this section;

(e) Tributaries of waters identified in paragraphs (a) through (d) of this section, including adjacent wetlands; and

(f) Wetlands adjacent to waters identified in paragraphs (a) through (e) of this section: Provided, That waste treatment systems (other than cooling ponds meeting the criteria of this paragraph) are not waters of the United States;

Navigable waters do not include prior converted cropland. Notwithstanding the determination of an area's status as prior converted cropland by any other federal agency, for the purposes of the Clean Water Act, the final authority regarding Clean Water Act jurisdiction remains with EPA.

NPDES means National Pollutant Discharge Elimination System;

Sheen means an iridescent appearance on the surface of water;

Sludge means an aggregate of oil or oil and other matter of any kind in any form other than dredged spoil having a combined specific gravity equivalent to or greater than water;

United States means the States, the District of Columbia, the Commonwealth of Puerto Rico, Guam, American Samoa, the Virgin Islands, and the Trust Territory of the Pacific Islands;

Wetlands means those areas that are inundated or saturated by surface or ground water at a frequency or duration sufficient to support, and that under normal circumstances do support, a prevalence of vegetation typically adapted for life in saturated soil conditions. Wetlands generally include playa lakes, swamps, marshes, bogs and similar areas such as sloughs, prairie potholes, wet meadows, prairie river overflows, mudflats, and natural ponds.

[52 FR 10719, Apr. 2, 1987, as amended at 58 FR 45039, Aug. 25, 1993; 61 FR 7421, Feb. 28, 1996]

§ 110.2   Applicability.

The regulations of this part apply to the discharge of oil prohibited by section 311(b)(3) of the Act.

[61 FR 7421, Feb. 28, 1996]

§ 110.3   Discharge of oil in such quantities as "may be harmful" pursuant to section 311(b)(4) of the Act.

For purposes of section 311(b)(4) of the Act, discharges of oil in such quantities that the Administrator has determined may be harmful to the public health or welfare or the environment of the United States include discharges of oil that:

(a) Violate applicable water quality standards; or

(b) Cause a film or sheen upon or discoloration of the surface of the water or adjoining shorelines or cause a sludge or emulsion to be deposited beneath the surface of the water or upon adjoining shorelines.

[61 FR 7421, Feb. 28, 1996]

§ 110.4   Dispersants.

Addition of dispersants or emulsifiers to oil to be discharged that would circumvent the provisions of this part is prohibited.

[52 FR 10719, Apr. 2, 1987. Redesignated at 61 FR 7421, Feb. 28, 1996]

§ 110.5   Discharges of oil not determined "as may be harmful" pursuant to Section 311(b)(3) of the Act.

Notwithstanding any other provisions of this part, the Administrator has not determined the following discharges of oil "as may be harmful" for purposes of section 311(b) of the Act:

(a) Discharges of oil from a properly functioning vessel engine (including an engine on a public vessel) and any discharges of such oil accumulated in the bilges of a vessel discharged in compliance with MARPOL 73/78, Annex I, as provided in 33 CFR part 151, subpart A;

(b) Other discharges of oil permitted under MARPOL 73/78, Annex I, as provided in 33 CFR part 151, subpart A; and

(c) Any discharge of oil explicitly permitted by the Administrator in connection with research, demonstration projects, or studies relating to the prevention, control, or abatement of oil pollution.

[61 FR 7421, Feb. 28, 1996]

§ 110.6 Notice.

Any person in charge of a vessel or of an onshore or offshore facility shall, as soon as he or she has knowledge of any discharge of oil from such vessel or facility in violation of section 311(b)(3) of the Act, immediately notify the National Response Center (NRC) (800–424–8802; in the Washington, DC metropolitan area, 202–426–2675). If direct reporting to the NRC is not practicable, reports may be made to the Coast Guard or EPA predesignated On-Scene Coordinator (OSC) for the geographic area where the discharge occurs. All such reports shall be promptly relayed to the NRC. If it is not possible to notify the NRC or the predesignated OCS immediately, reports may be made immediately to the nearest Coast Guard unit, provided that the person in charge of the vessel or onshore or offshore facility notifies the NRC as soon as possible. The reports shall be made in accordance with such procedures as the Secretary of Transportation may prescribe. The procedures for such notice are set forth in U.S. Coast Guard regulations, 33 CFR part 153, subpart B and in the National Oil and Hazardous Substances Pollution Contingency Plan, 40 CFR part 300, subpart E.

(Approved by the Office of Management and Budget under control number 2050–0046)

[52 FR 10719, Apr. 2, 1987. Redesignated and amended at 61 FR 7421, Feb. 28, 1996; 61 FR 14032, Mar. 29, 1996]

PART 112—OIL POLLUTION PREVENTION

Subpart A—Applicability, Definitions, and General Requirements For All Facilities and All Types of Oils

Subpart B—Requirements for Petroleum Oils and Non-Petroleum Oils, Except Animal Fats and Oils and Greases, and Fish and Marine Mammal Oils; and Vegetable Oils (Including Oils from Seeds, Nuts, Fruits, and Kernels)

Subpart C—Requirements for Animal Fats and Oils and Greases, and Fish and Marine Mammal Oils; and for Vegetable Oils, Including Oils from Seeds, Nuts, Fruits and Kernels

Subpart D—Response Requirements

AUTHORITY: 33 U.S.C. 1251 *et seq.*; 33 U.S.C.
2720; E.O. 12777 (October 18, 1991), 3 CFR, 1991
Comp., p. 351.

SOURCE: 38 FR 34165, Dec. 11, 1973, unless
otherwise noted.

EDITORIAL NOTE: Nomenclature changes to
part 112 appear at 65 FR 40798, June 30, 2000.

Subpart A—Applicability, Definitions, and General Requirements for All Facilities and All Types of Oils

SOURCE: 67 FR 47140, July 17, 2002, unless
otherwise noted.

§ 112.1 General applicability.

(a)(1) This part establishes procedures, methods, equipment, and other
requirements to prevent the discharge
of oil from non-transportation-related
onshore and offshore facilities into or
upon the navigable waters of the
United States or adjoining shorelines,
or into or upon the waters of the contiguous zone, or in connection with activities under the Outer Continental
Shelf Lands Act or the Deepwater Port
Act of 1974, or that may affect natural
resources belonging to, appertaining
to, or under the exclusive management
authority of the United States (including resources under the Magnuson
Fishery Conservation and Management
Act).

(2) As used in this part, words in the
singular also include the plural and
words in the masculine gender also include the feminine and vice versa, as
the case may require.

(b) Except as provided in paragraph
(d) of this section, this part applies to
any owner or operator of a non-transportation-related onshore or offshore
facility engaged in drilling, producing,
gathering, storing, processing, refining,
transferring, distributing, using, or
consuming oil and oil products, which
due to its location, could reasonably be
expected to discharge oil in quantities
that may be harmful, as described in
part 110 of this chapter, into or upon
the navigable waters of the United
States or adjoining shorelines, or into
or upon the waters of the contiguous
zone, or in connection with activities
under the Outer Continental Shelf
Lands Act or the Deepwater Port Act
of 1974, or that may affect natural resources belonging to, appertaining to,
or under the exclusive management authority of the United States (including
resources under the Magnuson Fishery
Conservation and Management Act)
that has oil in:

(1) Any aboveground container;

(2) Any completely buried tank as defined in § 112.2;

(3) Any container that is used for
standby storage, for seasonal storage,
or for temporary storage, or not otherwise "permanently closed" as defined
in § 112.2;

(4) Any "bunkered tank" or "partially buried tank" as defined in § 112.2,
or any container in a vault, each of
which is considered an aboveground
storage container for purposes of this
part.

(c) As provided in section 313 of the
Clean Water Act (CWA), departments,
agencies, and instrumentalities of the
Federal government are subject to this
part to the same extent as any person.

(d) Except as provided in paragraph
(f) of this section, this part does not
apply to:

(1) The owner or operator of any fa-
cility, equipment, or operation that is
not subject to the jurisdiction of the
Environmental Protection Agency
(EPA) under section 311(j)(1)(C) of the
CWA, as follows:

(i) Any onshore or offshore facility,
that due to its location, could not reasonably be expected to have a discharge as described in paragraph (b) of

this section. This determination must be based solely upon consideration of the geographical and location aspects of the facility (such as proximity to navigable waters or adjoining shorelines, land contour, drainage, etc.) and must exclude consideration of manmade features such as dikes, equipment or other structures, which may serve to restrain, hinder, contain, or otherwise prevent a discharge as described in paragraph (b) of this section.

(ii) Any equipment, or operation of a vessel or transportation-related onshore or offshore facility which is subject to the authority and control of the U.S. Department of Transportation, as defined in the Memorandum of Understanding between the Secretary of Transportation and the Administrator of EPA, dated November 24, 1971 (appendix A of this part).

(iii) Any equipment, or operation of a vessel or onshore or offshore facility which is subject to the authority and control of the U.S. Department of Transportation or the U.S. Department of the Interior, as defined in the Memorandum of Understanding between the Secretary of Transportation, the Secretary of the Interior, and the Administrator of EPA, dated November 8, 1993 (appendix B of this part).

(2) Any facility which, although otherwise subject to the jurisdiction of EPA, meets both of the following requirements:

(i) The completely buried storage capacity of the facility is 42,000 U.S. gallons or less of oil. For purposes of this exemption, the completely buried storage capacity of a facility excludes the capacity of a completely buried tank, as defined in §112.2, and connected underground piping, underground ancillary equipment, and containment systems, that is currently subject to all of the technical requirements of part 280 of this chapter or all of the technical requirements of a State program approved under part 281 of this chapter, or the capacity of any underground oil storage tanks deferred under 40 CFR part 280 that supply emergency diesel generators at a nuclear power generation facility licensed by the Nuclear Regulatory Commission and subject to any Nuclear Regulatory Commission provision regarding design and quality

criteria, including, but not limited to, 10 CFR part 50. The completely buried storage capacity of a facility also excludes the capacity of a container that is "permanently closed," as defined in §112.2 and the capacity of intra-facility gathering lines subject to the regulatory requirements of 49 CFR part 192 or 195.

(ii) The aggregate aboveground storage capacity of the facility is 1,320 U.S. gallons or less of oil. For the purposes of this exemption, only containers with a capacity of 55 U.S. gallons or greater are counted. The aggregate aboveground storage capacity of a facility excludes:

(A) The capacity of a container that is "permanently closed" as defined in §112.2;

(B) The capacity of a "motive power container" as defined in §112.2;

(C) The capacity of hot-mix asphalt or any hot-mix asphalt container;

(D) The capacity of a container for heating oil used solely at a single-family residence;

(E) The capacity of pesticide application equipment and related mix containers.

(F) The capacity of any milk and milk product container and associated piping and appurtenances.

(3) Any offshore oil drilling, production, or workover facility that is subject to the notices and regulations of the Minerals Management Service, as specified in the Memorandum of Understanding between the Secretary of Transportation, the Secretary of the Interior, and the Administrator of EPA, dated November 8, 1993 (appendix B of this part).

(4) Any completely buried storage tank, as defined in §112.2, and connected underground piping, underground ancillary equipment, and containment systems, at any facility, that is subject to all of the technical requirements of part 280 of this chapter or a State program approved under part 281 of this chapter, or any underground oil storage tanks including below-grade vaulted tanks, deferred under 40 CFR part 280, as originally promulgated, that supply emergency diesel generators at a nuclear power generation facility licensed by the Nuclear Regulatory Commission, provided

that such a tank is subject to any Nuclear Regulatory Commission provision regarding design and quality criteria, including, but not limited to, 10 CFR part 50. Such emergency generator tanks must be marked on the facility diagram as provided in § 112.7(a)(3), if the facility is otherwise subject to this part.

(5) Any container with a storage capacity of less than 55 gallons of oil.

(6) Any facility or part thereof used exclusively for wastewater treatment and not used to satisfy any requirement of this part. The production, recovery, or recycling of oil is not wastewater treatment for purposes of this paragraph.

(7) Any "motive power container," as defined in § 112.2. The transfer of fuel or other oil into a motive power container at an otherwise regulated facility is not eligible for this exemption.

(8) Hot-mix asphalt, or any hot-mix asphalt container.

(9) Any container for heating oil used solely at a single-family residence.

(10) Any pesticide application equipment or related mix containers.

(11) Intra-facility gathering lines subject to the regulatory requirements of 49 CFR part 192 or 195, except that such a line's location must be identified and marked as "exempt" on the facility diagram as provided in § 112.7(a)(3), if the facility is otherwise subject to this part.

(12) Any milk and milk product container and associated piping and appurtenances.

(e) This part establishes requirements for the preparation and implementation of Spill Prevention, Control, and Countermeasure (SPCC) Plans. SPCC Plans are designed to complement existing laws, regulations, rules, standards, policies, and procedures pertaining to safety standards, fire prevention, and pollution prevention rules. The purpose of an SPCC Plan is to form a comprehensive Federal/State spill prevention program that minimizes the potential for discharges. The SPCC Plan must address all relevant spill prevention, control, and countermeasures necessary at the specific facility. Compliance with this part does not in any way relieve the owner or operator of an onshore or an offshore facility from compliance with other Federal, State, or local laws.

(f) Notwithstanding paragraph (d) of this section, the Regional Administrator may require that the owner or operator of any facility subject to the jurisdiction of EPA under section 311(j) of the CWA prepare and implement an SPCC Plan, or any applicable part, to carry out the purposes of the CWA.

(1) Following a preliminary determination, the Regional Administrator must provide a written notice to the owner or operator stating the reasons why he must prepare an SPCC Plan, or applicable part. The Regional Administrator must send such notice to the owner or operator by certified mail or by personal delivery. If the owner or operator is a corporation, the Regional Administrator must also mail a copy of such notice to the registered agent, if any and if known, of the corporation in the State where the facility is located.

(2) Within 30 days of receipt of such written notice, the owner or operator may provide information and data and may consult with the Agency about the need to prepare an SPCC Plan, or applicable part.

(3) Within 30 days following the time under paragraph (b)(2) of this section within which the owner or operator may provide information and data and consult with the Agency about the need to prepare an SPCC Plan, or applicable part, the Regional Administrator must make a final determination regarding whether the owner or operator is required to prepare and implement an SPCC Plan, or applicable part. The Regional Administrator must send the final determination to the owner or operator by certified mail or by personal delivery. If the owner or operator is a corporation, the Regional Administrator must also mail a copy of the final determination to the registered agent, if any and if known, of the corporation in the State where the facility is located.

(4) If the Regional Administrator makes a final determination that an SPCC Plan, or applicable part, is necessary, the owner or operator must prepare the Plan, or applicable part, within six months of that final determination and implement the Plan, or applicable part, as soon as possible, but not

later than one year after the Regional Administrator has made a final determination.

(5) The owner or operator may appeal a final determination made by the Regional Administrator requiring preparation and implementation of an SPCC Plan, or applicable part, under this paragraph. The owner or operator must make the appeal to the Administrator of EPA within 30 days of receipt of the final determination under paragraph (b)(3) of this section from the Regional Administrator requiring preparation and/or implementation of an SPCC Plan, or applicable part. The owner or operator must send a complete copy of the appeal to the Regional Administrator at the time he makes the appeal to the Administrator. The appeal must contain a clear and concise statement of the issues and points of fact in the case. In the appeal, the owner or operator may also provide additional information. The additional information may be from any person. The Administrator may request additional information from the owner or operator. The Administrator must render a decision within 60 days of receiving the appeal or additional information submitted by the owner or operator and must serve the owner or operator with the decision made in the appeal in the manner described in paragraph (f)(1) of this section.

[67 FR 47140, July 17, 2002, as amended at 71 FR 77290, Dec. 26, 2006; 73 FR 74300, Dec. 5, 2008; 74 FR 58809, Nov. 13, 2009; 76 FR 21660, Apr. 18, 2011]

§112.2 Definitions.

For the purposes of this part:

Adverse weather means weather conditions that make it difficult for response equipment and personnel to clean up or remove spilled oil, and that must be considered when identifying response systems and equipment in a response plan for the applicable operating environment. Factors to consider include significant wave height as specified in appendix E to this part (as appropriate), ice conditions, temperatures, weather-related visibility, and currents within the area in which the systems or equipment is intended to function.

Alteration means any work on a container involving cutting, burning, welding, or heating operations that changes the physical dimensions or configuration of the container.

Animal fat means a non-petroleum oil, fat, or grease of animal, fish, or marine mammal origin.

Breakout tank means a container used to relieve surges in an oil pipeline system or to receive and store oil transported by a pipeline for reinjection and continued transportation by pipeline.

Bulk storage container means any container used to store oil. These containers are used for purposes including, but not limited to, the storage of oil prior to use, while being used, or prior to further distribution in commerce. Oil-filled electrical, operating, or manufacturing equipment is not a bulk storage container.

Bunkered tank means a container constructed or placed in the ground by cutting the earth and re-covering the container in a manner that breaks the surrounding natural grade, or that lies above grade, and is covered with earth, sand, gravel, asphalt, or other material. A bunkered tank is considered an aboveground storage container for purposes of this part.

Completely buried tank means any container completely below grade and covered with earth, sand, gravel, asphalt, or other material. Containers in vaults, bunkered tanks, or partially buried tanks are considered aboveground storage containers for purposes of this part.

Complex means a facility possessing a combination of transportation-related and non-transportation-related components that is subject to the jurisdiction of more than one Federal agency under section 311(j) of the CWA.

Contiguous zone means the zone established by the United States under Article 24 of the Convention of the Territorial Sea and Contiguous Zone, that is contiguous to the territorial sea and that extends nine miles seaward from the outer limit of the territorial area.

Contract or other approved means means:

(1) A written contractual agreement with an oil spill removal organization that identifies and ensures the availability of the necessary personnel and

equipment within appropriate response times; and/or

(2) A written certification by the owner or operator that the necessary personnel and equipment resources, owned or operated by the facility owner or operator, are available to respond to a discharge within appropriate response times; and/or

(3) Active membership in a local or regional oil spill removal organization that has identified and ensures adequate access through such membership to necessary personnel and equipment to respond to a discharge within appropriate response times in the specified geographic area; and/or

(4) Any other specific arrangement approved by the Regional Administrator upon request of the owner or operator.

Discharge includes, but is not limited to, any spilling, leaking, pumping, pouring, emitting, emptying, or dumping of oil, but excludes discharges in compliance with a permit under section 402 of the CWA; discharges resulting from circumstances identified, reviewed, and made a part of the public record with respect to a permit issued or modified under section 402 of the CWA, and subject to a condition in such permit; or continuous or anticipated intermittent discharges from a point source, identified in a permit or permit application under section 402 of the CWA, that are caused by events occurring within the scope of relevant operating or treatment systems. For purposes of this part, the term discharge shall not include any discharge of oil that is authorized by a permit issued under section 13 of the River and Harbor Act of 1899 (33 U.S.C. 407).

Facility means any mobile or fixed, onshore or offshore building, property, parcel, lease, structure, installation, equipment, pipe, or pipeline (other than a vessel or a public vessel) used in oil well drilling operations, oil production, oil refining, oil storage, oil gathering, oil processing, oil transfer, oil distribution, and oil waste treatment, or in which oil is used, as described in appendix A to this part. The boundaries of a facility depend on several site-specific factors, including but not limited to, the ownership or operation of buildings, structures, and equipment on the same site and types of activity at the site. Contiguous or non-contiguous buildings, properties, parcels, leases, structures, installations, pipes, or pipelines under the ownership or operation of the same person may be considered separate facilities. Only this definition governs whether a facility is subject to this part.

Farm means a facility on a tract of land devoted to the production of crops or raising of animals, including fish, which produced and sold, or normally would have produced and sold, $1,000 or more of agricultural products during a year.

Fish and wildlife and sensitive environments means areas that may be identified by their legal designation or by evaluations of Area Committees (for planning) or members of the Federal On-Scene Coordinator's spill response structure (during responses). These areas may include wetlands, National and State parks, critical habitats for endangered or threatened species, wilderness and natural resource areas, marine sanctuaries and estuarine reserves, conservation areas, preserves, wildlife areas, wildlife refuges, wild and scenic rivers, recreational areas, national forests, Federal and State lands that are research national areas, heritage program areas, land trust areas, and historical and archaeological sites and parks. These areas may also include unique habitats such as aquaculture sites and agricultural surface water intakes, bird nesting areas, critical biological resource areas, designated migratory routes, and designated seasonal habitats.

Injury means a measurable adverse change, either long- or short-term, in the chemical or physical quality or the viability of a natural resource resulting either directly or indirectly from exposure to a discharge, or exposure to a product of reactions resulting from a discharge.

Loading/unloading rack means a fixed structure (such as a platform, gangway) necessary for loading or unloading a tank truck or tank car, which is located at a facility subject to the requirements of this part. A loading/unloading rack includes a loading or unloading arm, and may include any combination of the following: piping assemblages, valves, pumps, shut-off devices, overfill sensors, or personnel safety devices.

Maximum extent practicable means within the limitations used to determine oil spill planning resources and response times for on-water recovery, shoreline protection, and cleanup for worst case discharges from onshore non-transportation-related facilities in adverse weather. It includes the planned capability to respond to a worst case discharge in adverse weather, as contained in a response plan that meets the requirements in §112.20 or in a specific plan approved by the Regional Administrator.

Mobile refueler means a bulk storage container onboard a vehicle or towed, that is designed or used solely to store and transport fuel for transfer into or from an aircraft, motor vehicle, locomotive, vessel, ground service equipment, or other oil storage container.

Motive power container means any onboard bulk storage container used primarily to power the movement of a motor vehicle, or ancillary onboard oil-filled operational equipment. An onboard bulk storage container which is used to store or transfer oil for further distribution is not a motive power container. The definition of motive power container does not include oil drilling or workover equipment, including rigs.

Navigable waters of the United States means "navigable waters" as defined in section 502(7) of the FWPCA, and includes:

(1) All navigable waters of the United States, as defined in judicial decisions prior to passage of the 1972 Amendments to the FWPCA (Pub. L. 92–500), and tributaries of such waters;

(2) Interstate waters;

(3) Intrastate lakes, rivers, and streams which are utilized by interstate travelers for recreational or other purposes; and

(4) Intrastate lakes, rivers, and streams from which fish or shellfish are taken and sold in interstate commerce.

Non-petroleum oil means oil of any kind that is not petroleum-based, including but not limited to: Fats, oils, and greases of animal, fish, or marine mammal origin; and vegetable oils, including oils from seeds, nuts, fruits, and kernels.

Offshore facility means any facility of any kind (other than a vessel or public vessel) located in, on, or under any of the navigable waters of the United States, and any facility of any kind that is subject to the jurisdiction of the United States and is located in, on, or under any other waters.

Oil means oil of any kind or in any form, including, but not limited to: fats, oils, or greases of animal, fish, or marine mammal origin; vegetable oils, including oils from seeds, nuts, fruits, or kernels; and, other oils and greases, including petroleum, fuel oil, sludge, synthetic oils, mineral oils, oil refuse, or oil mixed with wastes other than dredged spoil.

Oil-filled operational equipment means equipment that includes an oil storage container (or multiple containers) in which the oil is present solely to support the function of the apparatus or the device. Oil-filled operational equipment is not considered a bulk storage container, and does not include oil-filled manufacturing equipment (flow-through process). Examples of oil-filled operational equipment include, but are not limited to, hydraulic systems, lubricating systems (*e.g.*, those for pumps, compressors and other rotating equipment, including pumpjack lubrication systems), gear boxes, machining coolant systems, heat transfer systems, transformers, circuit breakers, electrical switches, and other systems containing oil solely to enable the operation of the device.

Oil Spill Removal Organization means an entity that provides oil spill response resources, and includes any for-profit or not-for-profit contractor, co-operative, or in-house response resources that have been established in a geographic area to provide required response resources.

Onshore facility means any facility of any kind located in, on, or under any land within the United States, other than submerged lands.

Owner or operator means any person owning or operating an onshore facility or an offshore facility, and in the case of any abandoned offshore facility, the person who owned or operated or maintained the facility immediately prior to such abandonment.

Partially buried tank means a storage container that is partially inserted or constructed in the ground, but not entirely below grade, and not completely covered with earth, sand, gravel, asphalt, or other material. A partially buried tank is considered an aboveground storage container for purposes of this part.

Permanently closed means any container or facility for which:

(1) All liquid and sludge has been removed from each container and connecting line; and

(2) All connecting lines and piping have been disconnected from the container and blanked off, all valves (except for ventilation valves) have been closed and locked, and conspicuous signs have been posted on each container stating that it is a permanently closed container and noting the date of closure.

Person includes an individual, firm, corporation, association, or partnership.

Petroleum oil means petroleum in any form, including but not limited to crude oil, fuel oil, mineral oil, sludge, oil refuse, and refined products.

Produced water container means a storage container at an oil production facility used to store the produced water after initial oil/water separation, and prior to reinjection, beneficial reuse, discharge, or transfer for disposal.

Production facility means all structures (including but not limited to wells, platforms, or storage facilities), piping (including but not limited to flowlines or intra-facility gathering lines), or equipment (including but not limited to workover equipment, separation equipment, or auxiliary non-transportation-related equipment) used in the production, extraction, recovery, lifting, stabilization, separation or treating of oil (including condensate), or associated storage or measurement, and is located in an oil or gas field, at a facility. This definition governs whether such structures, piping, or equipment are subject to a specific section of this part.

Regional Administrator means the Regional Administrator of the Environmental Protection Agency, in and for the Region in which the facility is located.

Repair means any work necessary to maintain or restore a container to a condition suitable for safe operation, other than that necessary for ordinary, day-to-day maintenance to maintain the functional integrity of the container and that does not weaken the container.

Spill Prevention, Control, and Countermeasure Plan; SPCC Plan, or Plan means the document required by § 112.3 that details the equipment, workforce, procedures, and steps to prevent, control, and provide adequate countermeasures to a discharge.

Storage capacity of a container means the shell capacity of the container.

Transportation-related and non-transportation-related, as applied to an onshore or offshore facility, are defined in the Memorandum of Understanding between the Secretary of Transportation and the Administrator of the Environmental Protection Agency, dated November 24, 1971, (appendix A of this part).

United States means the States, the District of Columbia, the Commonwealth of Puerto Rico, the Commonwealth of the Northern Mariana Islands, Guam, American Samoa, the U.S. Virgin Islands, and the Pacific Island Governments.

Vegetable oil means a non-petroleum oil or fat of vegetable origin, including but not limited to oils and fats derived from plant seeds, nuts, fruits, and kernels.

Vessel means every description of watercraft or other artificial contrivance used, or capable of being used, as a means of transportation on water, other than a public vessel.

Wetlands means those areas that are inundated or saturated by surface or groundwater at a frequency or duration sufficient to support, and that under normal circumstances do support, a prevalence of vegetation typically adapted for life in saturated soil conditions. Wetlands generally include playa lakes, swamps, marshes, bogs, and similar areas such as sloughs, prairie potholes, wet meadows, prairie river overflows, mudflats, and natural ponds.

Worst case discharge for an onshore non-transportation-related facility means the largest foreseeable discharge in adverse weather conditions as determined using the worksheets in appendix D to this part.

[67 FR 47140, July 17, 2002, as amended at 71 FR 77290, Dec. 26, 2006; 73 FR 71943, Nov. 26, 2008; 73 FR 74300, Dec. 5, 2008]

§ 112.3 Requirement to prepare and implement a Spill Prevention, Control, and Countermeasure Plan.

The owner or operator or an onshore or offshore facility subject to this section must prepare in writing and implement a Spill Prevention Control and Countermeasure Plan (hereafter "SPCC Plan" or "Plan")," in accordance with § 112.7 and any other applicable section of this part.

(a)(1) Except as otherwise provided in this section, if your facility, or mobile or portable facility, was in operation on or before August 16, 2002, you must maintain your Plan, but must amend it, if necessary to ensure compliance with this part, and implement the amended Plan no later than November 10, 2011. If such a facility becomes operational after August 16, 2002, through November 10, 2011, and could reasonably be expected to have a discharge as described in § 112.1(b), you must prepare and implement a Plan on or before November 10, 2011. If such a facility (excluding oil production facilities) becomes operational after November 10, 2011, and could reasonably be expected to have a discharge as described in § 112.1(b), you must prepare and implement a Plan before you begin operations. You are not required to prepare a new Plan each time you move a mobile or portable facility to a new site; the Plan may be general. When you move the mobile or portable facility, you must locate and install it using the discharge prevention practices outlined in the Plan for the facility. The Plan is applicable only while the mobile or portable facility is in a fixed (non-transportation) operating mode.

(2) If your drilling, production or workover facility, including a mobile or portable facility, is offshore or has an offshore component; or your onshore facility is required to have and submit a Facility Response Plan pursuant to 40 CFR 112.20(a), and was in operation on or before August 16, 2002, you must maintain your Plan, but must amend it, if necessary to ensure compliance with this part, and implement the amended Plan no later than November 10, 2010. If such a facility becomes operational after August 16, 2002, through November 10, 2010, and could reasonably be expected to have a discharge as described in § 112.1(b), you must prepare and implement a Plan on or before November 10, 2010. If such a facility (excluding oil production facilities) becomes operational after November 10, 2010, and could reasonably be expected to have a discharge as described in § 112.1(b), you must prepare and implement a Plan before you begin operations. You are not required to prepare a new Plan each time you move a mobile or portable facility to a new site; the Plan may be general. When you move the mobile or portable facility, you must locate and install it using the discharge prevention practices outlined in the Plan for the facility. The Plan is applicable only while the mobile or portable facility is in a fixed (non-transportation) operating mode.

(3) If your farm, as defined in § 112.2, was in operation on or before August 16, 2002, you must maintain your Plan, but must amend it, if necessary to ensure compliance with this part, and implement the amended Plan on or before May 10, 2013. If your farm becomes operational after August 16, 2002, through May 10, 2013, and could reasonably be expected to have a discharge as described in § 112.1(b), you must prepare and implement a Plan on or before May 10, 2013. If your farm becomes operational after May 10, 2013, and could reasonably be expected to have a discharge as described in § 112.1(b), you must prepare and implement a Plan before you begin operations.

(b) If your oil production facility as described in paragraph (a)(1) of this section becomes operational after November 10, 2011, or as described in paragraph (a)(2) of this section becomes operational after November 10, 2010, and could reasonably be expected to have a discharge as described in § 112.1(b), you must prepare and implement a Plan within six months after you begin operations.

(c) [Reserved]

(d) Except as provided in § 112.6, a licensed Professional Engineer must review and certify a Plan for it to be effective to satisfy the requirements of this part.

(1) By means of this certification the Professional Engineer attests:

(i) That he is familiar with the requirements of this part ;

(ii) That he or his agent has visited and examined the facility;

(iii) That the Plan has been prepared in accordance with good engineering practice, including consideration of applicable industry standards, and with the requirements of this part;

(iv) That procedures for required inspections and testing have been established; and

(v) That the Plan is adequate for the facility.

(vi) That, if applicable, for a produced water container subject to § 112.9(c)(6), any procedure to minimize the amount of free-phase oil is designed to reduce the accumulation of free-phase oil and the procedures and frequency for required inspections, maintenance and testing have been established and are described in the Plan.

(2) Such certification shall in no way relieve the owner or operator of a facility of his duty to prepare and fully implement such Plan in accordance with the requirements of this part.

(e) If you are the owner or operator of a facility for which a Plan is required under this section, you must:

(1) Maintain a complete copy of the Plan at the facility if the facility is normally attended at least four hours per day, or at the nearest field office if the facility is not so attended, and

(2) Have the Plan available to the Regional Administrator for on-site review during normal working hours.

(f) *Extension of time.* (1) The Regional Administrator may authorize an extension of time for the preparation and full implementation of a Plan, or any amendment thereto, beyond the time permitted for the preparation, implementation, or amendment of a Plan

under this part, when he finds that the owner or operator of a facility subject to this section, cannot fully comply with the requirements as a result of either nonavailability of qualified personnel, or delays in construction or equipment delivery beyond the control and without the fault of such owner or operator or his agents or employees.

(2) If you are an owner or operator seeking an extension of time under paragraph (f)(1) of this section, you may submit a written extension request to the Regional Administrator. Your request must include:

(i) A full explanation of the cause for any such delay and the specific aspects of the Plan affected by the delay;

(ii) A full discussion of actions being taken or contemplated to minimize or mitigate such delay; and

(iii) A proposed time schedule for the implementation of any corrective actions being taken or contemplated, including interim dates for completion of tests or studies, installation and operation of any necessary equipment, or other preventive measures. In addition you may present additional oral or written statements in support of your extension request.

(3) The submission of a written extension request under paragraph (f)(2) of this section does not relieve you of your obligation to comply with the requirements of this part. The Regional Administrator may request a copy of your Plan to evaluate the extension request. When the Regional Administrator authorizes an extension of time for particular equipment or other specific aspects of the Plan, such extension does not affect your obligation to comply with the requirements related to other equipment or other specific aspects of the Plan for which the Regional Administrator has not expressly authorized an extension.

(g) *Qualified Facilities.* The owner or operator of a qualified facility as defined in this subparagraph may self-certify his facility's Plan, as provided in § 112.6. A qualified facility is one that meets the following Tier I or Tier II qualified facility criteria:

(1) A Tier I qualified facility meets the qualification criteria in paragraph

(g)(2) of this section and has no individual aboveground oil storage container with a capacity greater than 5,000 U.S. gallons.

(2) A Tier II qualified facility is one that has had no single discharge as described in §112.1(b) exceeding 1,000 U.S. gallons or no two discharges as described in §112.1(b) each exceeding 42 U.S. gallons within any twelve month period in the three years prior to the SPCC Plan self-certification date, or since becoming subject to this part if the facility has been in operation for less than three years (other than discharges as described in §112.1(b) that are the result of natural disasters, acts of war, or terrorism), and has an aggregate aboveground oil storage capacity of 10,000 U.S. gallons or less.

[67 FR 47140, July 17, 2002, as amended at 68 FR 1351, Jan. 9, 2003; 68 FR 18894, Apr. 17, 2003; 69 FR 48798, Aug. 11, 2004; 71 FR 8466, Feb. 17, 2006; 71 FR 77290, Dec. 26, 2006; 72 FR 27447, May 16, 2007; 73 FR 74301, Dec. 5, 2008, 74 FR 29141, June 19, 2009; 74 FR 58809, Nov. 13, 2009; 75 FR 63102, Oct. 14, 2010; 76 FR 21660, Apr. 18, 2011; 76 FR 64248, Oct. 18, 2011; 76 FR 72124, Nov. 22, 2011]

§112.4 Amendment of Spill Prevention, Control, and Countermeasure Plan by Regional Administrator.

If you are the owner or operator of a facility subject to this part, you must:

(a) Notwithstanding compliance with §112.3, whenever your facility has discharged more than 1,000 U.S. gallons of oil in a single discharge as described in §112.1(b), or discharged more than 42 U.S. gallons of oil in each of two discharges as described in §112.1(b), occurring within any twelve month period, submit the following information to the Regional Administrator within 60 days from the time the facility becomes subject to this section:

(1) Name of the facility;

(2) Your name;

(3) Location of the facility;

(4) Maximum storage or handling capacity of the facility and normal daily throughput;

(5) Corrective action and countermeasures you have taken, including a description of equipment repairs and replacements;

(6) An adequate description of the facility, including maps, flow diagrams, and topographical maps, as necessary;

(7) The cause of such discharge as described in §112.1(b), including a failure analysis of the system or subsystem in which the failure occurred;

(8) Additional preventive measures you have taken or contemplated to minimize the possibility of recurrence; and

(9) Such other information as the Regional Administrator may reasonably require pertinent to the Plan or discharge.

(b) Take no action under this section until it applies to your facility. This section does not apply until the expiration of the time permitted for the initial preparation and implementation of the Plan under §112.3, but not including any amendments to the Plan.

(c) Send to the appropriate agency or agencies in charge of oil pollution control activities in the State in which the facility is located a complete copy of all information you provided to the Regional Administrator under paragraph (a) of this section. Upon receipt of the information such State agency or agencies may conduct a review and make recommendations to the Regional Administrator as to further procedures, methods, equipment, and other requirements necessary to prevent and to contain discharges from your facility.

(d) Amend your Plan, if after review by the Regional Administrator of the information you submit under paragraph (a) of this section, or submission of information to EPA by the State agency under paragraph (c) of this section, or after on-site review of your Plan, the Regional Administrator requires that you do so. The Regional Administrator may require you to amend your Plan if he finds that it does not meet the requirements of this part or that amendment is necessary to prevent and contain discharges from your facility.

(e) Act in accordance with this paragraph when the Regional Administrator proposes by certified mail or by personal delivery that you amend your SPCC Plan. If the owner or operator is a corporation, he must also notify by mail the registered agent of such corporation, if any and if known, in the State in which the facility is located. The Regional Administrator must specify the terms of such proposed amendment. Within 30 days from receipt of such notice, you may submit written information, views, and arguments on the proposed amendment. After considering all relevant material presented, the Regional Administrator must either notify you of any amendment required or rescind the notice. You must amend your Plan as required within 30 days after such notice, unless the Regional Administrator, for good cause, specifies another effective date. You must implement the amended Plan as soon as possible, but not later than six months after you amend your Plan, unless the Regional Administrator specifies another date.

(f) If you appeal a decision made by the Regional Administrator requiring an amendment to an SPCC Plan, send the appeal to the EPA Administrator in writing within 30 days of receipt of the notice from the Regional Administrator requiring the amendment under paragraph (e) of this section. You must send a complete copy of the appeal to the Regional Administrator at the time you make the appeal. The appeal must contain a clear and concise statement of the issues and points of fact in the case. It may also contain additional information from you, or from any other person. The EPA Administrator may request additional information from you, or from any other person. The EPA Administrator must render a decision within 60 days of receiving the appeal and must notify you of his decision.

§ 112.5 Amendment of Spill Prevention, Control, and Countermeasure Plan by owners or operators.

If you are the owner or operator of a facility subject to this part, you must:

(a) Amend the SPCC Plan for your facility in accordance with the general requirements in § 112.7, and with any specific section of this part applicable to your facility, when there is a change in the facility design, construction, operation, or maintenance that materially affects its potential for a discharge as described in § 112.1(b). Examples of changes that may require amendment of the Plan include, but are not limited to: commissioning or decommissioning containers; replacement, reconstruction, or movement of

containers; reconstruction, replacement, or installation of piping systems; construction or demolition that might alter secondary containment structures; changes of product or service; or revision of standard operation or maintenance procedures at a facility. An amendment made under this section must be prepared within six months, and implemented as soon as possible, but not later than six months following preparation of the amendment.

(b) Notwithstanding compliance with paragraph (a) of this section, complete a review and evaluation of the SPCC Plan at least once every five years from the date your facility becomes subject to this part; or, if your facility was in operation on or before August 16, 2002, five years from the date your last review was required under this part. As a result of this review and evaluation, you must amend your SPCC Plan within six months of the review to include more effective prevention and control technology if the technology has been field-proven at the time of the review and will significantly reduce the likelihood of a discharge as described in § 112.1(b) from the facility. You must implement any amendment as soon as possible, but not later than six months following preparation of any amendment. You must document your completion of the review and evaluation, and must sign a statement as to whether you will amend the Plan, either at the beginning or end of the Plan or in a log or an appendix to the Plan. The following words will suffice, "I have completed review and evaluation of the SPCC Plan for (name of facility) on (date), and will (will not) amend the Plan as a result."

(c) Except as provided in § 112.6, have a Professional Engineer certify any technical amendments to your Plan in accordance with § 112.3(d).

[67 FR 47140, July 17, 2002, as amended at 71 FR 77291, Dec. 26, 2006; 73 FR 74301, Dec. 5, 2008; 74 FR 58809, Nov. 13, 2009]

§ 112.6 Qualified Facilities Plan Requirements.

Qualified facilities meeting the Tier I applicability criteria in § 112.3(g)(1) are subject to the requirements in paragraph (a) of this section. Qualified facilities meeting the Tier II applicability criteria in §112.3(g)(2) are subject to the requirements in paragraph (b) of this section.

(a) *Tier I Qualified Facilities*—(1) *Preparation and Self-Certification of the Plan.* If you are an owner or operator of a facility that meets the Tier I qualified facility criteria in §112.3(g)(1), you must either: comply with the requirements of paragraph (a)(3) of this section; or prepare and implement a Plan meeting requirements of paragraph (b) of this section; or prepare and implement a Plan meeting the general Plan requirements in §112.7 and applicable requirements in subparts B and C, including having the Plan certified by a Professional Engineer as required under §112.3(d). If you do not follow the appendix G template, you must prepare an equivalent Plan that meets all of the applicable requirements listed in this part, and you must supplement it with a section cross-referencing the location of requirements listed in this part and the equivalent requirements in the other prevention plan. To complete the template in appendix G, you must certify that:

(i) You are familiar with the applicable requirements of 40 CFR part 112;

(ii) You have visited and examined the facility;

(iii) You prepared the Plan in accordance with accepted and sound industry practices and standards;

(iv) You have established procedures for required inspections and testing in accordance with industry inspection and testing standards or recommended practices;

(v) You will fully implement the Plan;

(vi) The facility meets the qualification criteria in §112.3(g)(1);

(vii) The Plan does not deviate from any requirement of this part as allowed by §112.7(a)(2) and 112.7(d) or include measures pursuant to §112.9(c)(6) for produced water containers and any associated piping; and

(viii) The Plan and individual(s) responsible for implementing this Plan have the approval of management, and the facility owner or operator has committed the necessary resources to fully implement this Plan.

(2) *Technical Amendments.* You must certify any technical amendments to your Plan in accordance with paragraph (a)(1) of this section when there is a change in the facility design, construction, operation, or maintenance that affects its potential for a discharge as described in §112.1(b). If the facility change results in the facility no longer meeting the Tier I qualifying criteria in §112.3(g)(1) because an individual oil storage container capacity exceeds 5,000 U.S. gallons or the facility capacity exceeds 10,000 U.S. gallons in aggregate aboveground storage capacity, within six months following preparation of the amendment, you must either:

(i) Prepare and implement a Plan in accordance with §112.6(b) if you meet the Tier II qualified facility criteria in §112.3(g)(2); or

(ii) Prepare and implement a Plan in accordance with the general Plan requirements in §112.7, and applicable requirements in subparts B and C, including having the Plan certified by a Professional Engineer as required under §112.3(d).

(3) *Plan Template and Applicable Requirements.* Prepare and implement an SPCC Plan that meets the following requirements under §112.7 and in subparts B and C of this part: introductory paragraph of §§112.7, 112.7(a)(3)(i), 112.7(a)(3)(iv), 112.7(a)(3)(vi), 112.7(a)(4), 112.7(a)(5), 112.7(c), 112.7(e), 112.7(f), 112.7(g), 112.7(k), 112.8(b)(1), 112.8(b)(2), 112.8(c)(1), 112.8(c)(3), 112.8(c)(4), 112.8(c)(5), 112.8(c)(6), 112.8(c)(10), 112.8(d)(4), 112.9(b), 112.9(c)(1), 112.9(c)(2), 112.9(c)(3), 112.9(c)(4), 112.9(c)(5), 112.9(d)(1), 112.9(d)(3), 112.9(d)(4), 112.10(b), 112.10(c), 112.10(d), 112.12(b)(1), 112.12(b)(2), 112.12(c)(1), 112.12(c)(3), 112.12(c)(4), 112.12(c)(5), 112.12(c)(6), 112.12(c)(10), and 112.12(c)(4). The template in appendix G to this part has been developed to meet the requirements of 40 CFR part 112 and, when completed and signed by the owner or operator, may be used as the SPCC Plan. Additionally, you must meet the following requirements:

(i) *Failure analysis, in lieu of the requirements in §112.7(b).* Where experience indicates a reasonable potential for equipment failure (such as loading or unloading equipment, tank overflow, rupture, or leakage, or any other equipment known to be a source of discharge), include in your Plan a prediction of the direction and total quantity of oil which could be discharged from the facility as a result of each type of major equipment failure.

(ii) *Bulk storage container secondary containment, in lieu of the requirements in §§ 112.8(c)(2) and (c)(11) and 112.12(c)(2) and (c)(11).* Construct all bulk storage container installations (except mobile refuelers and other nontransportation-related tank trucks), including mobile or portable oil storage containers, so that you provide a secondary means of containment for the entire capacity of the largest single container plus additional capacity to contain precipitation. Dikes, containment curbs, and pits are commonly employed for this purpose. You may also use an alternative system consisting of a drainage trench enclosure that must be arranged so that any discharge will terminate and be safely confined in a catchment basin or holding pond. Position or locate mobile or portable oil storage containers to prevent a discharge as described in § 112.1(b).

(iii) *Overfill prevention, in lieu of the requirements in §§ 112.8(c)(8) and 112.12(c)(8).* Ensure that each container is provided with a system or documented procedure to prevent overfills of the container, describe the system or procedure in the SPCC Plan and regularly test to ensure proper operation or efficacy.

(b) *Tier II Qualified Facilities—*(1) *Preparation and Self-Certification of Plan.* If you are the owner or operator of a facility that meets the Tier II qualified facility criteria in § 112.3(g)(2), you may choose to self-certify your Plan. You must certify in the Plan that:

(i) You are familiar with the requirements of this part;

(ii) You have visited and examined the facility;

(iii) The Plan has been prepared in accordance with accepted and sound industry practices and standards, and with the requirements of this part;

(iv) Procedures for required inspections and testing have been established;

(v) You will fully implement the Plan;

(vi) The facility meets the qualification criteria set forth under § 112.3(g)(2);

(vii) The Plan does not deviate from any requirement of this part as allowed by § 112.7(a)(2) and 112.7(d) or include measures pursuant to § 112.9(c)(6) for produced water containers and any associated piping, except as provided in paragraph (b)(3) of this section; and

(viii) The Plan and individual(s) responsible for implementing the Plan have the full approval of management and the facility owner or operator has committed the necessary resources to fully implement the Plan.

(2) *Technical Amendments.* If you self-certify your Plan pursuant to paragraph (b)(1) of this section, you must certify any technical amendments to your Plan in accordance with paragraph (b)(1) of this section when there is a change in the facility design, construction, operation, or maintenance that affects its potential for a discharge as described in § 112.1(b), except:

(i) If a Professional Engineer certified a portion of your Plan in accordance with paragraph (b)(4) of this section, and the technical amendment affects this portion of the Plan, you must have the amended provisions of your Plan certified by a Professional Engineer in accordance with paragraph (b)(4)(ii) of this section.

(ii) If the change is such that the facility no longer meets the Tier II qualifying criteria in § 112.3(g)(2) because it exceeds 10,000 U.S. gallons in aggregate aboveground storage capacity you must, within six months following the change, prepare and implement a Plan in accordance with the general Plan requirements in § 112.7 and the applicable requirements in subparts B and C of this part, including having the Plan certified by a Professional Engineer as required under § 112.3(d).

(3) *Applicable Requirements.* Except as provided in this paragraph, your self-certified SPCC Plan must comply with § 112.7 and the applicable requirements in subparts B and C of this part:

(i) *Environmental Equivalence.* Your Plan may not include alternate methods which provide environmental equivalence pursuant to §112.7(a)(2), unless each alternate method has been reviewed and certified in writing by a Professional Engineer, as provided in paragraph (b)(4) of this section.

(ii) *Impracticability.* Your Plan may not include any determinations that secondary containment is impracticable and provisions in lieu of secondary containment pursuant to §112.7(d), unless each such determination and alternate measure has been reviewed and certified in writing by a Professional Engineer, as provided in paragraph (b)(4) of this section.

(iii) *Produced Water Containers.* Your Plan may not include any alternative procedures for skimming produced water containers in lieu of sized secondary containment pursuant to §112.9(c)(6), unless they have been reviewed and certified in writing by a Professional Engineer, as provided in paragraph (b)(4) of this section.

(4) *Professional Engineer Certification of Portions of a Qualified Facility's Self-Certified Plan.*

(i) As described in paragraph (b)(3) of this section, the facility owner or operator may not self-certify alternative measures allowed under §112.7(a)(2) or (d), that are included in the facility's Plan. Such measures must be reviewed and certified, in writing, by a licensed Professional Engineer. For each alternative measure allowed under §112.7(a)(2), the Plan must be accompanied by a written statement by a Professional Engineer that states the reason for nonconformance and describes the alternative method and how it provides equivalent environmental protection in accordance with §112.7(a)(2). For each determination of impracticability of secondary containment pursuant to §112.7(d), the Plan must clearly explain why secondary containment measures are not practicable at this facility and provide the alternative measures required in §112.7(d) in lieu of secondary containment. By certifying each measure allowed under §112.7(a)(2) and (d), the Professional Engineer attests:

(A) That he is familiar with the requirements of this part;

(B) That he or his agent has visited and examined the facility; and

(C) That the alternative method of environmental equivalence in accordance with §112.7(a)(2) or the determination of impracticability and alternative measures in accordance with §112.7(d) is consistent with good engineering practice, including consideration of applicable industry standards, and with the requirements of this part.

(ii) As described in paragraph (b)(3) of this section, the facility owner or operator may not self-certify measures as described in §112.9(c)(6) for produced water containers and any associated piping. Such measures must be reviewed and certified, in writing, by a licensed Professional Engineer, in accordance with §112.3(d)(1)(vi).

(iii) The review and certification by the Professional Engineer under this paragraph is limited to the alternative method which achieves equivalent environmental protection pursuant to §112.7(a)(2); to the impracticability determination and measures in lieu of secondary containment pursuant to §112.7(d); or the measures pursuant to §112.9(c)(6) for produced water containers and any associated piping and appurtenances downstream from the container.

[73 FR 74302, Dec. 5, 2008, as amended at 74 FR 58810, Nov. 13, 2009]

§112.7 **General requirements for Spill Prevention, Control, and Countermeasure Plans.**

If you are the owner or operator of a facility subject to this part you must prepare a Plan in accordance with good engineering practices. The Plan must have the full approval of management at a level of authority to commit the necessary resources to fully implement the Plan. You must prepare the Plan in writing. If you do not follow the sequence specified in this section for the Plan, you must prepare an equivalent Plan acceptable to the Regional Administrator that meets all of the applicable requirements listed in this part, and you must supplement it with a section cross-referencing the location of requirements listed in this part and the equivalent requirements in the other prevention plan. If the Plan calls for additional facilities or procedures, methods, or equipment not yet fully operational, you must discuss these items in separate paragraphs, and must explain separately the details of installation and operational start-up. As detailed elsewhere in this section, you must also:

(a)(1) Include a discussion of your facility's conformance with the requirements listed in this part.

(2) Comply with all applicable requirements listed in this part. Except as provided in § 112.6, your Plan may deviate from the requirements in paragraphs (g), (h)(2) and (3), and (i) of this section and the requirements in subparts B and C of this part, except the secondary containment requirements in paragraphs (c) and (h)(1) of this section, and §§ 112.8(c)(2), 112.8(c)(11), 112.9(c)(2), 112.9(d)(3), 112.10(c), 112.12(c)(2), and 112.12(c)(11), where applicable to a specific facility, if you provide equivalent environmental protection by some other means of spill prevention, control, or countermeasure. Where your Plan does not conform to the applicable requirements in paragraphs (g), (h)(2) and (3), and (i) of this section, or the requirements of subparts B and C of this part, except the secondary containment requirements in paragraph (c) and (h)(1) of this section, and §§ 112.8(c)(2), 112.8(c)(11), 112.9(c)(2), 112.10(c), 112.12(c)(2), and 112.12(c)(11), you must state the reasons for nonconformance in your Plan and describe in detail alternate methods and how you will achieve equivalent environmental protection. If the Regional Administrator determines that the measures described in your Plan do not provide equivalent environmental protection, he may require that you amend your Plan, following the procedures in § 112.4(d) and (e).

(3) Describe in your Plan the physical layout of the facility and include a facility diagram, which must mark the location and contents of each fixed oil storage container and the storage area where mobile or portable containers are located. The facility diagram must identify the location of and mark as "exempt" underground tanks that are otherwise exempted from the requirements of this part under § 112.1(d)(4). The facility diagram must also include

all transfer stations and connecting pipes, including intra-facility gathering lines that are otherwise exempted from the requirements of this part under § 112.1(d)(11). You must also address in your Plan:

(i) The type of oil in each fixed container and its storage capacity. For mobile or portable containers, either provide the type of oil and storage capacity for each container or provide an estimate of the potential number of mobile or portable containers, the types of oil, and anticipated storage capacities;

(ii) Discharge prevention measures including procedures for routine handling of products (loading, unloading, and facility transfers, etc.);

(iii) Discharge or drainage controls such as secondary containment around containers and other structures, equipment, and procedures for the control of a discharge;

(iv) Countermeasures for discharge discovery, response, and cleanup (both the facility's capability and those that might be required of a contractor);

(v) Methods of disposal of recovered materials in accordance with applicable legal requirements; and

(vi) Contact list and phone numbers for the facility response coordinator, National Response Center, cleanup contractors with whom you have an agreement for response, and all appropriate Federal, State, and local agencies who must be contacted in case of a discharge as described in § 112.1(b).

(4) Unless you have submitted a response plan under § 112.20, provide information and procedures in your Plan to enable a person reporting a discharge as described in § 112.1(b) to relate information on the exact address or location and phone number of the facility; the date and time of the discharge; the type of material discharged; estimates of the total quantity discharged; estimates of the quantity discharged as described in § 112.1(b); the source of the discharge; a description of all affected media; the cause of the discharge; any damages or injuries caused by the discharge; actions being used to stop, remove, and mitigate the effects of the discharge; whether an evacuation may be needed;

and, the names of individuals and/or organizations who have also been contacted.

(5) Unless you have submitted a response plan under §112.20, organize portions of the Plan describing procedures you will use when a discharge occurs in a way that will make them readily usable in an emergency, and include appropriate supporting material as appendices.

(b) Where experience indicates a reasonable potential for equipment failure (such as loading or unloading equipment, tank overflow, rupture, or leakage, or any other equipment known to be a source of a discharge), include in your Plan a prediction of the direction, rate of flow, and total quantity of oil which could be discharged from the facility as a result of each type of major equipment failure.

(c) Provide appropriate containment and/or diversionary structures or equipment to prevent a discharge as described in §112.1(b), except as provided in paragraph (k) of this section for qualified oil-filled operational equipment, and except as provided in §112.9(d)(3) for flowlines and intra-facility gathering lines at an oil production facility. The entire containment system, including walls and floor, must be capable of containing oil and must be constructed so that any discharge from a primary containment system, such as a tank, will not escape the containment system before cleanup occurs. In determining the method, design, and capacity for secondary containment, you need only to address the typical failure mode, and the most likely quantity of oil that would be discharged. Secondary containment may be either active or passive in design. At a minimum, you must use one of the following prevention systems or its equivalent:

(1) For onshore facilities:

(i) Dikes, berms, or retaining walls sufficiently impervious to contain oil;

(ii) Curbing or drip pans;

(iii) Sumps and collection systems;

(iv) Culverting, gutters, or other drainage systems;

(v) Weirs, booms, or other barriers;

(vi) Spill diversion ponds;

(vii) Retention ponds; or

(viii) Sorbent materials.

(2) For offshore facilities:

(i) Curbing or drip pans; or

(ii) Sumps and collection systems.

(d) Provided your Plan is certified by a licensed Professional Engineer under §112.3(d), or, in the case of a qualified facility that meets the criteria in §112.3(g), the relevant sections of your Plan are certified by a licensed Professional Engineer under §112.6(d), if you determine that the installation of any of the structures or pieces of equipment listed in paragraphs (c) and (h)(1) of this section, and §§112.8(c)(2), 112.8(c)(11), 112.9(c)(2), 112.10(c), 112.12(c)(2), and 112.12(c)(11) to prevent a discharge as described in §112.1(b) from any onshore or offshore facility is not practicable, you must clearly explain in your Plan why such measures are not practicable; for bulk storage containers, conduct both periodic integrity testing of the containers and periodic integrity and leak testing of the valves and piping; and, unless you have submitted a response plan under §112.20, provide in your Plan the following:

(1) An oil spill contingency plan following the provisions of part 109 of this chapter.

(2) A written commitment of manpower, equipment, and materials required to expeditiously control and remove any quantity of oil discharged that may be harmful.

(e) *Inspections, tests, and records.* Conduct inspections and tests required by this part in accordance with written procedures that you or the certifying engineer develop for the facility. You must keep these written procedures and a record of the inspections and tests, signed by the appropriate supervisor or inspector, with the SPCC Plan for a period of three years. Records of inspections and tests kept under usual and customary business practices will suffice for purposes of this paragraph.

(f) *Personnel, training, and discharge prevention procedures.* (1) At a minimum, train your oil-handling personnel in the operation and maintenance of equipment to prevent discharges; discharge procedure protocols; applicable pollution control laws, rules, and regulations; general facility operations; and, the contents of the facility SPCC Plan.

(2) Designate a person at each applicable facility who is accountable for discharge prevention and who reports to facility management.

(3) Schedule and conduct discharge prevention briefings for your oil-handling personnel at least once a year to assure adequate understanding of the SPCC Plan for that facility. Such briefings must highlight and describe known discharges as described in § 112.1(b) or failures, malfunctioning components, and any recently developed precautionary measures.

(g) *Security (excluding oil production facilities).* Describe in your Plan how you secure and control access to the oil handling, processing and storage areas; secure master flow and drain valves; prevent unauthorized access to starter controls on oil pumps; secure out-of-service and loading/unloading connections of oil pipelines; and address the appropriateness of security lighting to both prevent acts of vandalism and assist in the discovery of oil discharges.

(h) *Facility tank car and tank truck loading/unloading rack (excluding offshore facilities).*

(1) Where loading/unloading rack drainage does not flow into a catchment basin or treatment facility designed to handle discharges, use a quick drainage system for tank car or tank truck loading/unloading racks. You must design any containment system to hold at least the maximum capacity of any single compartment of a tank car or tank truck loaded or unloaded at the facility.

(2) Provide an interlocked warning light or physical barrier system, warning signs, wheel chocks or vehicle brake interlock system in the area adjacent to a loading/unloading rack, to prevent vehicles from departing before complete disconnection of flexible or fixed oil transfer lines.

(3) Prior to filling and departure of any tank car or tank truck, closely inspect for discharges the lowermost drain and all outlets of such vehicles, and if necessary, ensure that they are tightened, adjusted, or replaced to prevent liquid discharge while in transit.

(i) If a field-constructed aboveground container undergoes a repair, alteration, reconstruction, or a change in service that might affect the risk of a discharge or failure due to brittle fracture or other catastrophe, or has discharged oil or failed due to brittle fracture failure or other catastrophe, evaluate the container for risk of discharge or failure due to brittle fracture or other catastrophe, and as necessary, take appropriate action.

(j) In addition to the minimal prevention standards listed under this section, include in your Plan a complete discussion of conformance with the applicable requirements and other effective discharge prevention and containment procedures listed in this part or any applicable more stringent State rules, regulations, and guidelines.

(k) *Qualified Oil-filled Operational Equipment.* The owner or operator of a facility with oil-filled operational equipment that meets the qualification criteria in paragraph (k)(1) of this subsection may choose to implement for this qualified oil-filled operational equipment the alternate requirements as described in paragraph (k)(2) of this sub-section in lieu of general secondary containment required in paragraph (c) of this section.

(1) *Qualification Criteria—Reportable Discharge History:* The owner or operator of a facility that has had no single discharge as described in § 112.1(b) from any oil-filled operational equipment exceeding 1,000 U.S. gallons or no two discharges as described in § 112.1(b) from any oil-filled operational equipment each exceeding 42 U.S. gallons within any twelve month period in the three years prior to the SPCC Plan certification date, or since becoming subject to this part if the facility has been in operation for less than three years (other than oil discharges as described in § 112.1(b) that are the result of natural disasters, acts of war or terrorism); and

(2) *Alternative Requirements to General Secondary Containment.* If secondary containment is not provided for qualified oil-filled operational equipment pursuant to paragraph (c) of this section, the owner or operator of a facility with qualified oil-filled operational equipment must:

(i) Establish and document the facility procedures for inspections or a monitoring program to detect equipment failure and/or a discharge; and

(ii) Unless you have submitted a response plan under §112.20, provide in your Plan the following:

(A) An oil spill contingency plan following the provisions of part 109 of this chapter.

(B) A written commitment of manpower, equipment, and materials required to expeditiously control and remove any quantity of oil discharged that may be harmful.

[67 FR 47140, July 17, 2002, as amended at 71 FR 77292, Dec. 26, 2006; 73 FR 74303, Dec. 5, 2008; 74 FR 58810, Nov. 13, 2009]

Subpart B—Requirements for Petroleum Oils and Non-Petroleum Oils, Except Animal Fats and Oils and Greases, and Fish and Marine Mammal Oils; and Vegetable Oils (Including Oils from Seeds, Nuts, Fruits, and Kernels)

Source: 67 FR 47146, July 17, 2002, unless otherwise noted.

§112.8 Spill Prevention, Control, and Countermeasure Plan requirements for onshore facilities (excluding production facilities).

If you are the owner or operator of an onshore facility (excluding a production facility), you must:

(a) Meet the general requirements for the Plan listed under §112.7, and the specific discharge prevention and containment procedures listed in this section.

(b) *Facility drainage.* (1) Restrain drainage from diked storage areas by valves to prevent a discharge into the drainage system or facility effluent treatment system, except where facility systems are designed to control such discharge. You may empty diked areas by pumps or ejectors; however, you must manually activate these pumps or ejectors and must inspect the condition of the accumulation before starting, to ensure no oil will be discharged.

(2) Use valves of manual, open-and-closed design, for the drainage of diked areas. You may not use flapper-type drain valves to drain diked areas. If your facility drainage drains directly into a watercourse and not into an on-site wastewater treatment plant, you must inspect and may drain uncontaminated retained stormwater, as provided in paragraphs (c)(3)(ii), (iii), and (iv) of this section.

(3) Design facility drainage systems from undiked areas with a potential for a discharge (such as where piping is located outside containment walls or where tank truck discharges may occur outside the loading area) to flow into ponds, lagoons, or catchment basins designed to retain oil or return it to the facility. You must not locate catchment basins in areas subject to periodic flooding.

(4) If facility drainage is not engineered as in paragraph (b)(3) of this section, equip the final discharge of all ditches inside the facility with a diversion system that would, in the event of an uncontrolled discharge, retain oil in the facility.

(5) Where drainage waters are treated in more than one treatment unit and such treatment is continuous, and pump transfer is needed, provide two "lift" pumps and permanently install at least one of the pumps. Whatever techniques you use, you must engineer facility drainage systems to prevent a discharge as described in §112.1(b) in case there is an equipment failure or human error at the facility.

(c) *Bulk storage containers.* (1) Not use a container for the storage of oil unless its material and construction are compatible with the material stored and conditions of storage such as pressure and temperature.

(2) Construct all bulk storage tank installations (except mobile refuelers and other non-transportation-related tank trucks) so that you provide a secondary means of containment for the entire capacity of the largest single container and sufficient freeboard to contain precipitation. You must ensure that diked areas are sufficiently impervious to contain discharged oil. Dikes, containment curbs, and pits are commonly employed for this purpose. You may also use an alternative system consisting of a drainage trench enclosure that must be arranged so that any discharge will terminate and be safely confined in a facility catchment basin or holding pond.

(3) Not allow drainage of uncontaminated rainwater from the diked area into a storm drain or discharge of an effluent into an open watercourse, lake, or pond, bypassing the facility treatment system unless you:

(i) Normally keep the bypass valve sealed closed.

(ii) Inspect the retained rainwater to ensure that its presence will not cause a discharge as described in § 112.1(b).

(iii) Open the bypass valve and reseal it following drainage under responsible supervision; and

(iv) Keep adequate records of such events, for example, any records required under permits issued in accordance with §§ 122.41(j)(2) and 122.41(m)(3) of this chapter.

(4) Protect any completely buried metallic storage tank installed on or after January 10, 1974 from corrosion by coatings or cathodic protection compatible with local soil conditions. You must regularly leak test such completely buried metallic storage tanks.

(5) Not use partially buried or bunkered metallic tanks for the storage of oil, unless you protect the buried section of the tank from corrosion. You must protect partially buried and bunkered tanks from corrosion by coatings or cathodic protection compatible with local soil conditions.

(6) Test or inspect each aboveground container for integrity on a regular schedule and whenever you make material repairs. You must determine, in accordance with industry standards, the appropriate qualifications for personnel performing tests and inspections, the frequency and type of testing and inspections, which take into account container size, configuration, and design (such as containers that are: shop-built, field-erected, skid-mounted, elevated, equipped with a liner, double-walled, or partially buried). Examples of these integrity tests include, but are not limited to: visual inspection, hydrostatic testing, radiographic testing, ultrasonic testing, acoustic emissions testing, or other systems of non-destructive testing. You must keep comparison records and you must also inspect the container's supports and foundations. In addition, you must frequently inspect the outside of the container for signs of deterioration, discharges, or accumulation of oil inside diked areas. Records of inspections and tests kept under usual and customary business practices satisfy the recordkeeping requirements of this paragraph.

(7) Control leakage through defective internal heating coils by monitoring the steam return and exhaust lines for contamination from internal heating coils that discharge into an open watercourse, or pass the steam return or exhaust lines through a settling tank, skimmer, or other separation or retention system.

(8) Engineer or update each container installation in accordance with good engineering practice to avoid discharges. You must provide at least one of the following devices:

(i) High liquid level alarms with an audible or visual signal at a constantly attended operation or surveillance station. In smaller facilities an audible air vent may suffice.

(ii) High liquid level pump cutoff devices set to stop flow at a predetermined container content level.

(iii) Direct audible or code signal communication between the container gauger and the pumping station.

(iv) A fast response system for determining the liquid level of each bulk storage container such as digital computers, telepulse, or direct vision gauges. If you use this alternative, a person must be present to monitor gauges and the overall filling of bulk storage containers.

(v) You must regularly test liquid level sensing devices to ensure proper operation.

(9) Observe effluent treatment facilities frequently enough to detect possible system upsets that could cause a discharge as described in § 112.1(b).

(10) Promptly correct visible discharges which result in a loss of oil from the container, including but not limited to seams, gaskets, piping, pumps, valves, rivets, and bolts. You must promptly remove any accumulations of oil in diked areas.

(11) Position or locate mobile or portable oil storage containers to prevent a discharge as described in § 112.1(b). Except for mobile refuelers and other non-transportation-related   tank

trucks, you must furnish a secondary means of containment, such as a dike or catchment basin, sufficient to contain the capacity of the largest single compartment or container with sufficient freeboard to contain precipitation.

(d) *Facility transfer operations, pumping, and facility process.* (1) Provide buried piping that is installed or replaced on or after August 16, 2002, with a protective wrapping and coating. You must also cathodically protect such buried piping installations or otherwise satisfy the corrosion protection standards for piping in part 280 of this chapter or a State program approved under part 281 of this chapter. If a section of buried line is exposed for any reason, you must carefully inspect it for deterioration. If you find corrosion damage, you must undertake additional examination and corrective action as indicated by the magnitude of the damage.

(2) Cap or blank-flange the terminal connection at the transfer point and mark it as to origin when piping is not in service or is in standby service for an extended time.

(3) Properly design pipe supports to minimize abrasion and corrosion and allow for expansion and contraction.

(4) Regularly inspect all aboveground valves, piping, and appurtenances. During the inspection you must assess the general condition of items, such as flange joints, expansion joints, valve glands and bodies, catch pans, pipeline supports, locking of valves, and metal surfaces. You must also conduct integrity and leak testing of buried piping at the time of installation, modification, construction, relocation, or replacement.

(5) Warn all vehicles entering the facility to be sure that no vehicle will endanger aboveground piping or other oil transfer operations.

[67 FR 47146, July 17, 2002, as amended at 71 FR 77293, Dec. 26, 2006; 73 FR 74304, Dec. 5, 2008]

§112.9 Spill Prevention, Control, and Countermeasure Plan Requirements for onshore oil production facilities (excluding drilling and workover facilities).

If you are the owner or operator of an onshore oil production facility (excluding a drilling or workover facility), you must:

(a) Meet the general requirements for the Plan listed under §112.7, and the specific discharge prevention and containment procedures listed under this section.

(b) *Oil production facility drainage.* (1) At tank batteries and separation and treating areas where there is a reasonable possibility of a discharge as described in §112.1(b), close and seal at all times drains of dikes or drains of equivalent measures required under §112.7(c)(1), except when draining uncontaminated rainwater. Prior to drainage, you must inspect the diked area and take action as provided in §112.8(c)(3)(ii), (iii), and (iv). You must remove accumulated oil on the rainwater and return it to storage or dispose of it in accordance with legally approved methods.

(2) Inspect at regularly scheduled intervals field drainage systems (such as drainage ditches or road ditches), and oil traps, sumps, or skimmers, for an accumulation of oil that may have resulted from any small discharge. You must promptly remove any accumulations of oil.

(c) *Oil production facility bulk storage containers.* (1) Not use a container for the storage of oil unless its material and construction are compatible with the material stored and the conditions of storage.

(2) Except as described in paragraph (c)(5) of this section for flow-through process vessels and paragraph (c)(6) of this section for produced water containers and any associated piping and appurtenances downstream from the container, construct all tank battery, separation, and treating facility installations, so that you provide a secondary means of containment for the entire capacity of the largest single container and sufficient freeboard to contain precipitation. You must safely confine drainage from undiked areas in a catchment basin or holding pond.

(3) Except as described in paragraph (c)(5) of this section for flow-through process vessels and paragraph (c)(6) of this section for produced water containers and any associated piping and appurtenances downstream from the

container, periodically and upon a regular schedule visually inspect each container of oil for deterioration and maintenance needs, including the foundation and support of each container that is on or above the surface of the ground.

(4) Engineer or update new and old tank battery installations in accordance with good engineering practice to prevent discharges. You must provide at least one of the following:

(i) Container capacity adequate to assure that a container will not overfill if a pumper/gauger is delayed in making regularly scheduled rounds.

(ii) Overflow equalizing lines between containers so that a full container can overflow to an adjacent container.

(iii) Vacuum protection adequate to prevent container collapse during a pipeline run or other transfer of oil from the container.

(iv) High level sensors to generate and transmit an alarm signal to the computer where the facility is subject to a computer production control system.

(5) *Flow-through process vessels.* The owner or operator of a facility with flow-through process vessels may choose to implement the alternate requirements as described below in lieu of sized secondary containment required in paragraphs (c)(2) and (c)(3) of this section.

(i) Periodically and on a regular schedule visually inspect and/or test flow-through process vessels and associated components (such as dump valves) for leaks, corrosion, or other conditions that could lead to a discharge as described in § 112.1(b).

(ii) Take corrective action or make repairs to flow-through process vessels and any associated components as indicated by regularly scheduled visual inspections, tests, or evidence of an oil discharge.

(iii) Promptly remove or initiate actions to stabilize and remediate any accumulations of oil discharges associated with flow-through process vessels.

(iv) If your facility discharges more than 1,000 U.S. gallons of oil in a single discharge as described in § 112.1(b), or discharges more than 42 U.S. gallons of oil in each of two discharges as described in § 112.1(b) within any twelve

month period, from flow-through process vessels (excluding discharges that are the result of natural disasters, acts of war, or terrorism) then you must, within six months from the time the facility becomes subject to this paragraph, ensure that all flow-through process vessels subject to this subpart comply with § 112.9(c)(2) and (c)(3).

(6) *Produced water containers.* For each produced water container, comply with § 112.9(c)(1) and (c)(4); and § 112.9(c)(2) and (c)(3), or comply with the provisions of the following paragraphs (c)(6)(i) through (v):

(i) Implement, on a regular schedule, a procedure for each produced water container that is designed to separate the free-phase oil that accumulates on the surface of the produced water. Include in the Plan a description of the procedures, frequency, amount of free-phase oil expected to be maintained inside the container, and a Professional Engineer certification in accordance with § 112.3(d)(1)(vi). Maintain records of such events in accordance with § 112.7(e). Records kept under usual and customary business practices will suffice for purposes of this paragraph. If this procedure is not implemented as described in the Plan or no records are maintained, then you must comply with § 112.9(c)(2) and (c)(3).

(ii) On a regular schedule, visually inspect and/or test the produced water container and associated piping for leaks, corrosion, or other conditions that could lead to a discharge as described in § 112.1(b) in accordance with good engineering practice.

(iii) Take corrective action or make repairs to the produced water container and any associated piping as indicated by regularly scheduled visual inspections, tests, or evidence of an oil discharge.

(iv) Promptly remove or initiate actions to stabilize and remediate any accumulations of oil discharges associated with the produced water container.

(v) If your facility discharges more than 1,000 U.S. gallons of oil in a single discharge as described in § 112.1(b), or discharges more than 42 U.S. gallons of oil in each of two discharges as described in § 112.1(b) within any twelve month period from a produced water container subject to this subpart (excluding discharges that are the result of natural disasters, acts of war, or terrorism) then you must, within six months from the time the facility becomes subject to this paragraph, ensure that all produced water containers subject to this subpart comply with §112.9(c)(2) and (c)(3).

(d) *Facility transfer operations, oil production facility.* (1) Periodically and upon a regular schedule inspect all aboveground valves and piping associated with transfer operations for the general condition of flange joints, valve glands and bodies, drip pans, pipe supports, pumping well polish rod stuffing boxes, bleeder and gauge valves, and other such items.

(2) Inspect saltwater (oil field brine) disposal facilities often, particularly following a sudden change in atmospheric temperature, to detect possible system upsets capable of causing a discharge.

(3) For flowlines and intra-facility gathering lines that are not provided with secondary containment in accordance with §112.7(c), unless you have submitted a response plan under §112.20, provide in your Plan the following:

(i) An oil spill contingency plan following the provisions of part 109 of this chapter.

(ii) A written commitment of manpower, equipment, and materials required to expeditiously control and remove any quantity of oil discharged that might be harmful.

(4) Prepare and implement a written program of flowline/intra-facility gathering line maintenance. The maintenance program must address your procedures to:

(i) Ensure that flowlines and intra-facility gathering lines and associated valves and equipment are compatible with the type of production fluids, their potential corrosivity, volume, and pressure, and other conditions expected in the operational environment.

(ii) Visually inspect and/or test flowlines and intra-facility gathering lines and associated appurtenances on a periodic and regular schedule for leaks, oil discharges, corrosion, or other conditions that could lead to a

discharge as described in §112.1(b). For flowlines and intra-facility gathering lines that are not provided with secondary containment in accordance with §112.7(c), the frequency and type of testing must allow for the implementation of a contingency plan as described under part 109 of this chapter.

(iii) Take corrective action or make repairs to any flowlines and intra-facility gathering lines and associated appurtenances as indicated by regularly scheduled visual inspections, tests, or evidence of a discharge.

(iv) Promptly remove or initiate actions to stabilize and remediate any accumulations of oil discharges associated with flowlines, intra-facility gathering lines, and associated appurtenances.

[73 FR, 74304, Dec. 5, 2008, as amended at 74 FR 58810, Nov. 13, 2009]

§ 112.10 Spill Prevention, Control, and Countermeasure Plan requirements for onshore oil drilling and workover facilities.

If you are the owner or operator of an onshore oil drilling and workover facility, you must:

(a) Meet the general requirements listed under §112.7, and also meet the specific discharge prevention and containment procedures listed under this section.

(b) Position or locate mobile drilling or workover equipment so as to prevent a discharge as described in §112.1(b).

(c) Provide catchment basins or diversion structures to intercept and contain discharges of fuel, crude oil, or oily drilling fluids.

(d) Install a blowout prevention (BOP) assembly and well control system before drilling below any casing string or during workover operations. The BOP assembly and well control system must be capable of controlling any well-head pressure that may be encountered while that BOP assembly and well control system are on the well.

§ 112.11 Spill Prevention, Control, and Countermeasure Plan requirements for offshore oil drilling, production, or workover facilities.

If you are the owner or operator of an offshore oil drilling, production, or workover facility, you must:

(a) Meet the general requirements listed under § 112.7, and also meet the specific discharge prevention and containment procedures listed under this section.

(b) Use oil drainage collection equipment to prevent and control small oil discharges around pumps, glands, valves, flanges, expansion joints, hoses, drain lines, separators, treaters, tanks, and associated equipment. You must control and direct facility drains toward a central collection sump to prevent the facility from having a discharge as described in § 112.1(b). Where drains and sumps are not practicable, you must remove oil contained in collection equipment as often as necessary to prevent overflow.

(c) For facilities employing a sump system, provide adequately sized sump and drains and make available a spare pump to remove liquid from the sump and assure that oil does not escape. You must employ a regularly scheduled preventive maintenance inspection and testing program to assure reliable operation of the liquid removal system and pump start-up device. Redundant automatic sump pumps and control devices may be required on some installations.

(d) At facilities with areas where separators and treaters are equipped with dump valves which predominantly fail in the closed position and where pollution risk is high, specially equip the facility to prevent the discharge of oil. You must prevent the discharge of oil by:

(1) Extending the flare line to a diked area if the separator is near shore;

(2) Equipping the separator with a high liquid level sensor that will automatically shut in wells producing to the separator; or

(3) Installing parallel redundant dump valves.

(e) Equip atmospheric storage or surge containers with high liquid level sensing devices that activate an alarm or control the flow, or otherwise prevent discharges.

(f) Equip pressure containers with high and low pressure sensing devices that activate an alarm or control the flow.

(g) Equip containers with suitable corrosion protection.

(h) Prepare and maintain at the facility a written procedure within the Plan for inspecting and testing pollution prevention equipment and systems.

(i) Conduct testing and inspection of the pollution prevention equipment and systems at the facility on a scheduled periodic basis, commensurate with the complexity, conditions, and circumstances of the facility and any other appropriate regulations. You must use simulated discharges for testing and inspecting human and equipment pollution control and countermeasure systems.

(j) Describe in detailed records surface and subsurface well shut-in valves and devices in use at the facility for each well sufficiently to determine their method of activation or control, such as pressure differential, change in fluid or flow conditions, combination of pressure and flow, manual or remote control mechanisms.

(k) Install a BOP assembly and well control system during workover operations and before drilling below any casing string. The BOP assembly and well control system must be capable of controlling any well-head pressure that may be encountered while the BOP assembly and well control system are on the well.

(l) Equip all manifolds (headers) with check valves on individual flowlines.

(m) Equip the flowline with a high pressure sensing device and shut-in valve at the wellhead if the shut-in well pressure is greater than the working pressure of the flowline and manifold valves up to and including the header valves. Alternatively you may provide a pressure relief system for flowlines.

(n) Protect all piping appurtenant to the facility from corrosion, such as with protective coatings or cathodic protection.

(o) Adequately protect sub-marine piping appurtenant to the facility against environmental stresses and

42

other activities such as fishing operations.

(p) Maintain sub-marine piping appurtenant to the facility in good operating condition at all times. You must periodically and according to a schedule inspect or test such piping for failures. You must document and keep a record of such inspections or tests at the facility.

Subpart C—Requirements for Animal Fats and Oils and Greases, and Fish and Marine Mammal Oils; and for Vegetable Oils, including Oils from Seeds, Nuts, Fruits, and Kernels

Source: 67 FR 57149, July 17, 2002, unless otherwise noted.

§112.12 Spill Prevention, Control, and Countermeasure Plan requirements.

If you are the owner or operator of an onshore facility, you must:

(a) Meet the general requirements for the Plan listed under §112.7, and the specific discharge prevention and containment procedures listed in this section.

(b) *Facility drainage.* (1) Restrain drainage from diked storage areas by valves to prevent a discharge into the drainage system or facility effluent treatment system, except where facility systems are designed to control such discharge. You may empty diked areas by pumps or ejectors; however, you must manually activate these pumps or ejectors and must inspect the condition of the accumulation before starting, to ensure no oil will be discharged.

(2) Use valves of manual, open-and-closed design, for the drainage of diked areas. You may not use flapper-type drain valves to drain diked areas. If your facility drainage drains directly into a watercourse and not into an on-site wastewater treatment plant, you must inspect and may drain uncontaminated retained stormwater, subject to the requirements of paragraphs (c)(3)(ii), (iii), and (iv) of this section.

(3) Design facility drainage systems from undiked areas with a potential for a discharge (such as where piping is located outside containment walls or where tank truck discharges may occur outside the loading area) to flow into ponds, lagoons, or catchment basins designed to retain oil or return it to the facility. You must not locate catchment basins in areas subject to periodic flooding.

(4) If facility drainage is not engineered as in paragraph (b)(3) of this section, equip the final discharge of all ditches inside the facility with a diversion system that would, in the event of an uncontrolled discharge, retain oil in the facility.

(5) Where drainage waters are treated in more than one treatment unit and such treatment is continuous, and pump transfer is needed, provide two "lift" pumps and permanently install at least one of the pumps. Whatever techniques you use, you must engineer facility drainage systems to prevent a discharge as described in §112.1(b) in case there is an equipment failure or human error at the facility.

(c) *Bulk storage containers.* (1) Not use a container for the storage of oil unless its material and construction are compatible with the material stored and conditions of storage such as pressure and temperature.

(2) Construct all bulk storage tank installations (except mobile refuelers and other non-transportation-related tank trucks) so that you provide a secondary means of containment for the entire capacity of the largest single container and sufficient freeboard to contain precipitation. You must ensure that diked areas are sufficiently impervious to contain discharged oil. Dikes, containment curbs, and pits are commonly employed for this purpose. You may also use an alternative system consisting of a drainage trench enclosure that must be arranged so that any discharge will terminate and be safely confined in a facility catchment basin or holding pond.

(3) Not allow drainage of uncontaminated rainwater from the diked area into a storm drain or discharge of an effluent into an open watercourse, lake, or pond, bypassing the facility treatment system unless you:

(i) Normally keep the bypass valve sealed closed.

(ii) Inspect the retained rainwater to ensure that its presence will not cause a discharge as described in § 112.1(b).

(iii) Open the bypass valve and reseal it following drainage under responsible supervision; and

(iv) Keep adequate records of such events, for example, any records required under permits issued in accordance with §§ 122.41(j)(2) and 122.41(m)(3) of this chapter.

(4) Protect any completely buried metallic storage tank installed on or after January 10, 1974 from corrosion by coatings or cathodic protection compatible with local soil conditions. You must regularly leak test such completely buried metallic storage tanks.

(5) Not use partially buried or bunkered metallic tanks for the storage of oil, unless you protect the buried section of the tank from corrosion. You must protect partially buried and bunkered tanks from corrosion by coatings or cathodic protection compatible with local soil conditions.

(6) *Bulk storage container inspections.*

(i) Except for containers that meet the criteria provided in paragraph (c)(6)(ii) of this section, test or inspect each aboveground container for integrity on a regular schedule and whenever you make material repairs. You must determine, in accordance with industry standards, the appropriate qualifications for personnel performing tests and inspections, the frequency and type of testing and inspections, which take into account container size, configuration, and design (such as containers that are: shop-built, field-erected, skid-mounted, elevated, equipped with a liner, double-walled, or partially buried). Examples of these integrity tests include, but are not limited to: Visual inspection, hydrostatic testing, radiographic testing, ultrasonic testing, acoustic emissions testing, or other systems of non-destructive testing. You must keep comparison records and you must also inspect the container's supports and foundations. In addition, you must frequently inspect the outside of the container for signs of deterioration, discharges, or accumulation of oil inside diked areas. Records

of inspections and tests kept under usual and customary business practices satisfy the recordkeeping requirements of this paragraph.

(ii) For bulk storage containers that are subject to 21 CFR part 110, are elevated, constructed of austenitic stainless steel, have no external insulation, and are shop-fabricated, conduct formal visual inspection on a regular schedule. In addition, you must frequently inspect the outside of the container for signs of deterioration, discharges, or accumulation of oil inside diked areas. You must determine and document in the Plan the appropriate qualifications for personnel performing tests and inspections. Records of inspections and tests kept under usual and customary business practices satisfy the recordkeeping requirements of this paragraph (c)(6).

(7) Control leakage through defective internal heating coils by monitoring the steam return and exhaust lines for contamination from internal heating coils that discharge into an open watercourse, or pass the steam return or exhaust lines through a settling tank, skimmer, or other separation or retention system.

(8) Engineer or update each container installation in accordance with good engineering practice to avoid discharges. You must provide at least one of the following devices:

(i) High liquid level alarms with an audible or visual signal at a constantly attended operation or surveillance station. In smaller facilities an audible air vent may suffice.

(ii) High liquid level pump cutoff devices set to stop flow at a predetermined container content level.

(iii) Direct audible or code signal communication between the container gauger and the pumping station.

(iv) A fast response system for determining the liquid level of each bulk storage container such as digital computers, telepulse, or direct vision gauges. If you use this alternative, a person must be present to monitor gauges and the overall filling of bulk storage containers.

(v) You must regularly test liquid level sensing devices to ensure proper operation.

(9) Observe effluent treatment facilities frequently enough to detect possible system upsets that could cause a discharge as described in §112.1(b).

(10) Promptly correct visible discharges which result in a loss of oil from the container, including but not limited to seams, gaskets, piping, pumps, valves, rivets, and bolts. You must promptly remove any accumulations of oil in diked areas.

(11) Position or locate mobile or portable oil storage containers to prevent a discharge as described in §112.1(b). Except for mobile refuelers and other non-transportation-related tank trucks, you must furnish a secondary means of containment, such as a dike or catchment basin, sufficient to contain the capacity of the largest single compartment or container with sufficient freeboard to contain precipitation.

(d) *Facility transfer operations, pumping, and facility process.* (1) Provide buried piping that is installed or replaced on or after August 16, 2002, with a protective wrapping and coating. You must also cathodically protect such buried piping installations or otherwise satisfy the corrosion protection standards for piping in part 280 of this chapter or a State program approved under part 281 of this chapter. If a section of buried line is exposed for any reason, you must carefully inspect it for deterioration. If you find corrosion damage, you must undertake additional examination and corrective action as indicated by the magnitude of the damage.

(2) Cap or blank-flange the terminal connection at the transfer point and mark it as to origin when piping is not in service or is in standby service for an extended time.

(3) Properly design pipe supports to minimize abrasion and corrosion and allow for expansion and contraction.

(4) Regularly inspect all aboveground valves, piping, and appurtenances. During the inspection you must assess the general condition of items, such as flange joints, expansion joints, valve glands and bodies, catch pans, pipeline supports, locking of valves, and metal surfaces. You must also conduct integrity and leak testing of buried piping at the time of installation, modification, construction, relocation, or replacement.

(5) Warn all vehicles entering the facility to be sure that no vehicle will endanger aboveground piping or other oil transfer operations.

[67 FR 57149, July 17, 2002, as amended at 71 FR 77293, Dec. 26, 2006; 73 FR 74305, Dec. 5, 2008]

§§112.13–112.15 [Reserved]

Subpart D—Response Requirements

§112.20 Facility response plans.

(a) The owner or operator of any non-transportation-related onshore facility that, because of its location, could reasonably be expected to cause substantial harm to the environment by discharging oil into or on the navigable waters or adjoining shorelines shall prepare and submit a facility response plan to the Regional Administrator, according to the following provisions:

(1) For the owner or operator of a facility in operation on or before February 18, 1993 who is required to prepare and submit a response plan under 33 U.S.C. 1321(j)(5), the Oil Pollution Act of 1990 (Pub. L. 101–380, 33 U.S.C. 2701 *et seq.*) requires the submission of a response plan that satisfies the requirements of 33 U.S.C. 1321(j)(5) no later than February 18, 1993.

(i) The owner or operator of an existing facility that was in operation on or before February 18, 1993 who submitted a response plan by February 18, 1993 shall revise the response plan to satisfy the requirements of this section and resubmit the response plan or updated portions of the response plan to the Regional Administrator by February 18, 1995.

(ii) The owner or operator of an existing facility in operation on or before February 18, 1993 who failed to submit a response plan by February 18, 1993 shall prepare and submit a response plan that satisfies the requirements of this section to the Regional Administrator before August 30, 1994.

(2) The owner or operator of a facility in operation on or after August 30, 1994 that satisfies the criteria in paragraph (f)(1) of this section or that is notified

by the Regional Administrator pursuant to paragraph (b) of this section shall prepare and submit a facility response plan that satisfies the requirements of this section to the Regional Administrator.

(i) For a facility that commenced operations after February 18, 1993 but prior to August 30, 1994, and is required to prepare and submit a response plan based on the criteria in paragraph (f)(1) of this section, the owner or operator shall submit the response plan or updated portions of the response plan, along with a completed version of the response plan cover sheet contained in appendix F to this part, to the Regional Administrator prior to August 30, 1994.

(ii) For a newly constructed facility that commences operation after August 30, 1994, and is required to prepare and submit a response plan based on the criteria in paragraph (f)(1) of this section, the owner or operator shall submit the response plan, along with a completed version of the response plan cover sheet contained in appendix F to this part, to the Regional Administrator prior to the start of operations (adjustments to the response plan to reflect changes that occur at the facility during the start-up phase of operations must be submitted to the Regional Administrator after an operational trial period of 60 days).

(iii) For a facility required to prepare and submit a response plan after August 30, 1994, as a result of a planned change in design, construction, operation, or maintenance that renders the facility subject to the criteria in paragraph (f)(1) of this section, the owner or operator shall submit the response plan, along with a completed version of the response plan cover sheet contained in appendix F to this part, to the Regional Administrator before the portion of the facility undergoing change commences operations (adjustments to the response plan to reflect changes that occur at the facility during the start-up phase of operations must be submitted to the Regional Administrator after an operational trial period of 60 days).

(iv) For a facility required to prepare and submit a response plan after August 30, 1994, as a result of an un-

planned event or change in facility characteristics that renders the facility subject to the criteria in paragraph (f)(1) of this section, the owner or operator shall submit the response plan, along with a completed version of the response plan cover sheet contained in appendix F to this part, to the Regional Administrator within six months of the unplanned event or change.

(3) In the event the owner or operator of a facility that is required to prepare and submit a response plan uses an alternative formula that is comparable to one contained in appendix C to this part to evaluate the criterion in paragraph (f)(1)(ii)(B) or (f)(1)(ii)(C) of this section, the owner or operator shall attach documentation to the response plan cover sheet contained in appendix F to this part that demonstrates the reliability and analytical soundness of the alternative formula.

(4) *Preparation and submission of response plans—Animal fat and vegetable oil facilities.* The owner or operator of any non-transportation-related facility that handles, stores, or transports animal fats and vegetable oils must prepare and submit a facility response plan as follows:

(i) *Facilities with approved plans.* The owner or operator of a facility with a facility response plan that has been approved under paragraph (c) of this section by July 31, 2000 need not prepare or submit a revised plan except as otherwise required by paragraphs (b), (c), or (d) of this section.

(ii) *Facilities with plans that have been submitted to the Regional Administrator.* Except for facilities with approved plans as provided in paragraph (a)(4)(i) of this section, the owner or operator of a facility that has submitted a response plan to the Regional Administrator prior to July 31, 2000 must review the plan to determine if it meets or exceeds the applicable provisions of this part. An owner or operator need not prepare or submit a new plan if the existing plan meets or exceeds the applicable provisions of this part. If the plan does not meet or exceed the applicable provisions of this part, the owner or operator must prepare and submit a new plan by September 28, 2000.

(iii) *Newly regulated facilities.* The owner or operator of a newly constructed facility that commences operation after July 31, 2000 must prepare and submit a plan to the Regional Administrator in accordance with paragraph (a)(2)(ii) of this section. The plan must meet or exceed the applicable provisions of this part. The owner or operator of an existing facility that must prepare and submit a plan after July 31, 2000 as a result of a planned or unplanned change in facility characteristics that causes the facility to become regulated under paragraph (f)(1) of this section, must prepare and submit a plan to the Regional Administrator in accordance with paragraph (a)(2)(iii) or (iv) of this section, as appropriate. The plan must meet or exceed the applicable provisions of this part.

(iv) *Facilities amending existing plans.* The owner or operator of a facility submitting an amended plan in accordance with paragraph (d) of this section after July 31, 2000, including plans that had been previously approved, must also review the plan to determine if it meets or exceeds the applicable provisions of this part. If the plan does not meet or exceed the applicable provisions of this part, the owner or operator must revise and resubmit revised portions of an amended plan to the Regional Administrator in accordance with paragraph (d) of this section, as appropriate. The plan must meet or exceed the applicable provisions of this part.

(b)(1) The Regional Administrator may at any time require the owner or operator of any non-transportation-related onshore facility to prepare and submit a facility response plan under this section after considering the factors in paragraph (f)(2) of this section. If such a determination is made, the Regional Administrator shall notify the facility owner or operator in writing and shall provide a basis for the determination. If the Regional Administrator notifies the owner or operator in writing of the requirement to prepare and submit a response plan under this section, the owner or operator of the facility shall submit the response plan to the Regional Administrator within six months of receipt of such written notification.

(2) The Regional Administrator shall review plans submitted by such facilities to determine whether the facility could, because of its location, reasonably be expected to cause significant and substantial harm to the environment by discharging oil into or on the navigable waters or adjoining shorelines.

(c) The Regional Administrator shall determine whether a facility could, because of its location, reasonably be expected to cause significant and substantial harm to the environment by discharging oil into or on the navigable waters or adjoining shorelines, based on the factors in paragraph (f)(3) of this section. If such a determination is made, the Regional Administrator shall notify the owner or operator of the facility in writing and:

(1) Promptly review the facility response plan;

(2) Require amendments to any response plan that does not meet the requirements of this section;

(3) Approve any response plan that meets the requirements of this section; and

(4) Review each response plan periodically thereafter on a schedule established by the Regional Administrator provided that the period between plan reviews does not exceed five years.

(d)(1) The owner or operator of a facility for which a response plan is required under this part shall revise and resubmit revised portions of the response plan within 60 days of each facility change that materially may affect the response to a worst case discharge, including:

(i) A change in the facility's configuration that materially alters the information included in the response plan;

(ii) A change in the type of oil handled, stored, or transferred that materially alters the required response resources;

(iii) A material change in capabilities of the oil spill removal organization(s) that provide equipment and personnel to respond to discharges of oil described in paragraph (h)(5) of this section;

(iv) A material change in the facility's spill prevention and response equipment or emergency response procedures; and

(v) Any other changes that materially affect the implementation of the response plan.

(2) Except as provided in paragraph (d)(1) of this section, amendments to personnel and telephone number lists included in the response plan and a change in the oil spill removal organization(s) that does not result in a material change in support capabilities do not require approval by the Regional Administrator. Facility owners or operators shall provide a copy of such changes to the Regional Administrator as the revisions occur.

(3) The owner or operator of a facility that submits changes to a response plan as provided in paragraph (d)(1) or (d)(2) of this section shall provide the EPA-issued facility identification number (where one has been assigned) with the changes.

(4) The Regional Administrator shall review for approval changes to a response plan submitted pursuant to paragraph (d)(1) of this section for a facility determined pursuant to paragraph (f)(3) of this section to have the potential to cause significant and substantial harm to the environment.

(e) If the owner or operator of a facility determines pursuant to paragraph (a)(2) of this section that the facility could not, because of its location, reasonably be expected to cause substantial harm to the environment by discharging oil into or on the navigable waters or adjoining shorelines, the owner or operator shall complete and maintain at the facility the certification form contained in appendix C to this part and, in the event an alternative formula that is comparable to one contained in appendix C to this part is used to evaluate the criterion in paragraph (f)(1)(ii)(B) or (f)(1)(ii)(C) of this section, the owner or operator shall attach documentation to the certification form that demonstrates the reliability and analytical soundness of the comparable formula and shall notify the Regional Administrator in writing that an alternative formula was used.

(f)(1) A facility could, because of its location, reasonably be expected to cause substantial harm to the environment by discharging oil into or on the navigable waters or adjoining shore-

lines pursuant to paragraph (a)(2) of this section, if it meets any of the following criteria applied in accordance with the flowchart contained in attachment C–I to appendix C to this part:

(i) The facility transfers oil over water to or from vessels and has a total oil storage capacity greater than or equal to 42,000 gallons; or

(ii) The facility's total oil storage capacity is greater than or equal to 1 million gallons, and one of the following is true:

(A) The facility does not have secondary containment for each aboveground storage area sufficiently large to contain the capacity of the largest aboveground oil storage tank within each storage area plus sufficient freeboard to allow for precipitation;

(B) The facility is located at a distance (as calculated using the appropriate formula in appendix C to this part or a comparable formula) such that a discharge from the facility could cause injury to fish and wildlife and sensitive environments. For further description of fish and wildlife and sensitive environments, see Appendices I, II, and III of the "Guidance for Facility and Vessel Response Plans: Fish and Wildlife and Sensitive Environments" (see appendix E to this part, section 13, for availability) and the applicable Area Contingency Plan prepared pursuant to section 311(j)(4) of the Clean Water Act;

(C) The facility is located at a distance (as calculated using the appropriate formula in appendix C to this part or a comparable formula) such that a discharge from the facility would shut down a public drinking water intake; or

(D) The facility has had a reportable oil discharge in an amount greater than or equal to 10,000 gallons within the last 5 years.

(2)(i) To determine whether a facility could, because of its location, reasonably be expected to cause substantial harm to the environment by discharging oil into or on the navigable waters or adjoining shorelines pursuant to paragraph (b) of this section, the Regional Administrator shall consider the following:

(A) Type of transfer operation;

(B) Oil storage capacity;

(C) Lack of secondary containment;

(D) Proximity to fish and wildlife and sensitive environments and other areas determined by the Regional Administrator to possess ecological value;

(E) Proximity to drinking water intakes;

(F) Spill history; and

(G) Other site-specific characteristics and environmental factors that the Regional Administrator determines to be relevant to protecting the environment from harm by discharges of oil into or on navigable waters or adjoining shorelines.

(ii) Any person, including a member of the public or any representative from a Federal, State, or local agency who believes that a facility subject to this section could, because of its location, reasonably be expected to cause substantial harm to the environment by discharging oil into or on the navigable waters or adjoining shorelines may petition the Regional Administrator to determine whether the facility meets the criteria in paragraph (f)(2)(i) of this section. Such petition shall include a discussion of how the factors in paragraph (f)(2)(i) of this section apply to the facility in question. The RA shall consider such petitions and respond in an appropriate amount of time.

(3) To determine whether a facility could, because of its location, reasonably be expected to cause significant and substantial harm to the environment by discharging oil into or on the navigable waters or adjoining shorelines, the Regional Administrator may consider the factors in paragraph (f)(2) of this section as well as the following:

(i) Frequency of past discharges;

(ii) Proximity to navigable waters;

(iii) Age of oil storage tanks; and

(iv) Other facility-specific and Region-specific information, including local impacts on public health.

(g)(1) All facility response plans shall be consistent with the requirements of the National Oil and Hazardous Substance Pollution Contingency Plan (40 CFR part 300) and applicable Area Contingency Plans prepared pursuant to section 311(j)(4) of the Clean Water Act. The facility response plan should be coordinated with the local emergency response plan developed by the local emergency planning committee under section 303 of Title III of the Superfund Amendments and Reauthorization Act of 1986 (42 U.S.C. 11001 et seq.). Upon request, the owner or operator should provide a copy of the facility response plan to the local emergency planning committee or State emergency response commission.

(2) The owner or operator shall review relevant portions of the National Oil and Hazardous Substances Pollution Contingency Plan and applicable Area Contingency Plan annually and, if necessary, revise the facility response plan to ensure consistency with these plans.

(3) The owner or operator shall review and update the facility response plan periodically to reflect changes at the facility.

(h) A response plan shall follow the format of the model facility-specific response plan included in appendix F to this part, unless you have prepared an equivalent response plan acceptable to the Regional Administrator to meet State or other Federal requirements. A response plan that does not follow the specified format in appendix F to this part shall have an emergency response action plan as specified in paragraphs (h)(1) of this section and be supplemented with a cross-reference section to identify the location of the elements listed in paragraphs (h)(2) through (h)(10) of this section. To meet the requirements of this part, a response plan shall address the following elements, as further described in appendix F to this part:

(1) *Emergency response action plan.* The response plan shall include an emergency response action plan in the format specified in paragraphs (h)(1)(i) through (viii) of this section that is maintained in the front of the response plan, or as a separate document accompanying the response plan, and that includes the following information:

(i) The identity and telephone number of a qualified individual having full authority, including contracting authority, to implement removal actions;

(ii) The identity of individuals or organizations to be contacted in the event of a discharge so that immediate communications between the qualified individual identified in paragraph (h)(1) of this section and the appropriate Federal officials and the persons providing response personnel and equipment can be ensured;

(iii) A description of information to pass to response personnel in the event of a reportable discharge;

(iv) A description of the facility's response equipment and its location;

(v) A description of response personnel capabilities, including the duties of persons at the facility during a response action and their response times and qualifications;

(vi) Plans for evacuation of the facility and a reference to community evacuation plans, as appropriate;

(vii) A description of immediate measures to secure the source of the discharge, and to provide adequate containment and drainage of discharged oil; and

(viii) A diagram of the facility.

(2) *Facility information.* The response plan shall identify and discuss the location and type of the facility, the identity and tenure of the present owner and operator, and the identity of the qualified individual identified in paragraph (h)(1) of this section.

(3) *Information about emergency response.* The response plan shall include:

(i) The identity of private personnel and equipment necessary to remove to the maximum extent practicable a worst case discharge and other discharges of oil described in paragraph (h)(5) of this section, and to mitigate or prevent a substantial threat of a worst case discharge (To identify response resources to meet the facility response plan requirements of this section, owners or operators shall follow appendix E to this part or, where not appropriate, shall clearly demonstrate in the response plan why use of appendix E of this part is not appropriate at the facility and make comparable arrangements for response resources);

(ii) Evidence of contracts or other approved means for ensuring the availability of such personnel and equipment;

(iii) The identity and the telephone number of individuals or organizations to be contacted in the event of a discharge so that immediate communications between the qualified individual identified in paragraph (h)(1) of this

section and the appropriate Federal official and the persons providing response personnel and equipment can be ensured;

(iv) A description of information to pass to response personnel in the event of a reportable discharge;

(v) A description of response personnel capabilities, including the duties of persons at the facility during a response action and their response times and qualifications;

(vi) A description of the facility's response equipment, the location of the equipment, and equipment testing;

(vii) Plans for evacuation of the facility and a reference to community evacuation plans, as appropriate;

(viii) A diagram of evacuation routes; and

(ix) A description of the duties of the qualified individual identified in paragraph (h)(1) of this section, that include:

(A) Activate internal alarms and hazard communication systems to notify all facility personnel;

(B) Notify all response personnel, as needed;

(C) Identify the character, exact source, amount, and extent of the release, as well as the other items needed for notification;

(D) Notify and provide necessary information to the appropriate Federal, State, and local authorities with designated response roles, including the National Response Center, State Emergency Response Commission, and Local Emergency Planning Committee;

(E) Assess the interaction of the discharged substance with water and/or other substances stored at the facility and notify response personnel at the scene of that assessment;

(F) Assess the possible hazards to human health and the environment due to the release. This assessment must consider both the direct and indirect effects of the release (i.e., the effects of any toxic, irritating, or asphyxiating gases that may be generated, or the effects of any hazardous surface water runoffs from water or chemical agents used to control fire and heat-induced explosion);

(G) Assess and implement prompt removal actions to contain and remove the substance released;

(H) Coordinate rescue and response actions as previously arranged with all response personnel;

(I) Use authority to immediately access company funding to initiate cleanup activities; and

(J) Direct cleanup activities until properly relieved of this responsibility.

(4) *Hazard evaluation.* The response plan shall discuss the facility's known or reasonably identifiable history of discharges reportable under 40 CFR part 110 for the entire life of the facility and shall identify areas within the facility where discharges could occur and what the potential effects of the discharges would be on the affected environment. To assess the range of areas potentially affected, owners or operators shall, where appropriate, consider the distance calculated in paragraph (f)(1)(ii) of this section to determine whether a facility could, because of its location, reasonably be expected to cause substantial harm to the environment by discharging oil into or on the navigable waters or adjoining shorelines.

(5) *Response planning levels.* The response plan shall include discussion of specific planning scenarios for:

(i) A worst case discharge, as calculated using the appropriate worksheet in appendix D to this part. In cases where the Regional Administrator determines that the worst case discharge volume calculated by the facility is not appropriate, the Regional Administrator may specify the worst case discharge amount to be used for response planning at the facility. For complexes, the worst case planning quantity shall be the larger of the amounts calculated for each component of the facility;

(ii) A discharge of 2,100 gallons or less, provided that this amount is less than the worst case discharge amount. For complexes, this planning quantity shall be the larger of the amounts calculated for each component of the facility; and

(iii) A discharge greater than 2,100 gallons and less than or equal to 36,000 gallons or 10 percent of the capacity of the largest tank at the facility, whichever is less, provided that this amount is less than the worst case discharge amount. For complexes, this planning quantity shall be the larger of the amounts calculated for each component of the facility.

(6) *Discharge detection systems.* The response plan shall describe the procedures and equipment used to detect discharges.

(7) *Plan implementation.* The response plan shall describe:

(i) Response actions to be carried out by facility personnel or contracted personnel under the response plan to ensure the safety of the facility and to mitigate or prevent discharges described in paragraph (h)(5) of this section or the substantial threat of such discharges;

(ii) A description of the equipment to be used for each scenario;

(iii) Plans to dispose of contaminated cleanup materials; and

(iv) Measures to provide adequate containment and drainage of discharged oil.

(8) *Self-inspection, drills/exercises, and response training.* The response plan shall include:

(i) A checklist and record of inspections for tanks, secondary containment, and response equipment;

(ii) A description of the drill/exercise program to be carried out under the response plan as described in §112.21;

(iii) A description of the training program to be carried out under the response plan as described in §112.21; and

(iv) Logs of discharge prevention meetings, training sessions, and drills/exercises. These logs may be maintained as an annex to the response plan.

(9) *Diagrams.* The response plan shall include site plan and drainage plan diagrams.

(10) *Security systems.* The response plan shall include a description of facility security systems.

(11) *Response plan cover sheet.* The response plan shall include a completed response plan cover sheet provided in section 2.0 of appendix F to this part.

(i)(1) In the event the owner or operator of a facility does not agree with the Regional Administrator's determination that the facility could, because of its location, reasonably be expected to cause substantial harm or significant and substantial harm to the environment by discharging oil into or on the navigable waters or adjoining shorelines, or that amendments to the facility response plan are necessary prior to approval, such as changes to the worst case discharge planning volume, the owner or operator may submit a request for reconsideration to the Regional Administrator and provide additional information and data in writing to support the request. The request and accompanying information must be submitted to the Regional Administrator within 60 days of receipt of notice of the Regional Administrator's original decision. The Regional Administrator shall consider the request and render a decision as rapidly as practicable.

(2) In the event the owner or operator of a facility believes a change in the facility's classification status is warranted because of an unplanned event or change in the facility's characteristics (i.e., substantial harm or significant and substantial harm), the owner or operator may submit a request for reconsideration to the Regional Administrator and provide additional information and data in writing to support the request. The Regional Administrator shall consider the request and render a decision as rapidly as practicable.

(3) After a request for reconsideration under paragraph (i)(1) or (i)(2) of this section has been denied by the Regional Administrator, an owner or operator may appeal a determination made by the Regional Administrator. The appeal shall be made to the EPA Administrator and shall be made in writing within 60 days of receipt of the decision from the Regional Administrator that the request for reconsideration was denied. A complete copy of the appeal must be sent to the Regional Administrator at the time the appeal is made. The appeal shall contain a clear and concise statement of the issues and points of fact in the case. It also may contain additional information from the owner or operator, or from any other person. The EPA Administrator may request additional information from the owner or operator, or from any other person. The EPA Administrator shall render a decision as

rapidly as practicable and shall notify the owner or operator of the decision.

[59 FR 34098, July 1, 1994, as amended at 65 FR 40798, June 30, 2000; 66 FR 34560, June 29, 2001; 67 FR 47151, July 17, 2002]

§112.21 Facility response training and drills/exercises.

(a) The owner or operator of any facility required to prepare a facility response plan under §112.20 shall develop and implement a facility response training program and a drill/exercise program that satisfy the requirements of this section. The owner or operator shall describe the programs in the response plan as provided in §112.20(h)(8).

(b) The facility owner or operator shall develop a facility response training program to train those personnel involved in oil spill response activities. It is recommended that the training program be based on the USCG's Training Elements for Oil Spill Response, as applicable to facility operations. An alternative program can also be acceptable subject to approval by the Regional Administrator.

(1) The owner or operator shall be responsible for the proper instruction of facility personnel in the procedures to respond to discharges of oil and in applicable oil spill response laws, rules, and regulations.

(2) Training shall be functional in nature according to job tasks for both supervisory and non-supervisory operational personnel.

(3) Trainers shall develop specific lesson plans on subject areas relevant to facility personnel involved in oil spill response and cleanup.

(c) The facility owner or operator shall develop a program of facility response drills/exercises, including evaluation procedures. A program that follows the National Preparedness for Response Exercise Program (PREP) (see appendix E to this part, section 13, for availability) will be deemed satisfactory for purposes of this section. An alternative program can also be acceptable subject to approval by the Regional Administrator.

[59 FR 34101, July 1, 1994, as amended at 65 FR 40798, June 30, 2000]

APPENDIX A TO PART 112—MEMORANDUM OF UNDERSTANDING BETWEEN THE SECRETARY OF TRANSPORTATION AND THE ADMINISTRATOR OF THE ENVIRONMENTAL PROTECTION AGENCY

SECTION II—DEFINITIONS

The Environmental Protection Agency and the Department of Transportation agree that for the purposes of Executive Order 11548, the term:

(1) *Non-transportation-related onshore and offshore facilities* means:

(A) Fixed onshore and offshore oil well drilling facilities including all equipment and appurtenances related thereto used in drilling operations for exploratory or development wells, but excluding any terminal facility, unit or process integrally associated with the handling or transferring of oil in bulk to or from a vessel.

(B) Mobile onshore and offshore oil well drilling platforms, barges, trucks, or other mobile facilities including all equipment and appurtenances related thereto when such mobile facilities are fixed in position for the purpose of drilling operations for exploratory or development wells, but excluding any terminal facility, unit or process integrally associated with the handling or transferring of oil in bulk to or from a vessel.

(C) Fixed onshore and offshore oil production structures, platforms, derricks, and rigs including all equipment and appurtenances related thereto, as well as completed wells and the wellhead separators, oil separators, and storage facilities used in the production of oil, but excluding any terminal facility, unit or process integrally associated with the handling or transferring of oil in bulk to or from a vessel.

(D) Mobile onshore and offshore oil production facilities including all equipment and appurtenances related thereto as well as completed wells and wellhead equipment, piping from wellheads to oil separators, oil separators, and storage facilities used in the production of oil when such mobile facilities are fixed in position for the purpose of oil production operations, but excluding any terminal facility, unit or process integrally associated with the handling or transferring of oil in bulk to or from a vessel.

(E) Oil refining facilities including all equipment and appurtenances related thereto as well as in-plant processing units, storage units, piping, drainage systems and waste treatment units used in the refining of oil, but excluding any terminal facility, unit or process integrally associated with the handling or transferring of oil in bulk to or from a vessel.

(F) Oil storage facilities including all equipment and appurtenances related thereto as well as fixed bulk plant storage, terminal oil storage facilities, consumer storage, pumps and drainage systems used in the storage of oil, but excluding inline or break-out storage tanks needed for the continuous operation of a pipeline system and any terminal facility, unit or process integrally associated with the handling or transferring of oil in bulk to or from a vessel.

(G) Industrial, commercial, agricultural or public facilities which use and store oil, but excluding any terminal facility, unit or process integrally associated with the handling or transferring of oil in bulk to or from a vessel.

(H) Waste treatment facilities including in-plant pipelines, effluent discharge lines, and storage tanks, but excluding waste treatment facilities located on vessels and terminal storage tanks and appurtenances for the reception of oily ballast water or tank washings from vessels and associated systems used for off-loading vessels.

(I) Loading racks, transfer hoses, loading arms and other equipment which are appurtenant to a nontransportation-related facility or terminal facility and which are used to transfer oil in bulk to or from highway vehicles or railroad cars.

(J) Highway vehicles and railroad cars which are used for the transport of oil exclusively within the confines of a nontransportation-related facility and which are not intended to transport oil in interstate or intrastate commerce.

(K) Pipeline systems which are used for the transport of oil exclusively within the confines of a nontransportation-related facility or terminal facility and which are not intended to transport oil in interstate or intrastate commerce, but excluding pipeline systems used to transfer oil in bulk to or from a vessel.

(2) *Transportation-related onshore and offshore facilities* means:

(A) Onshore and offshore terminal facilities including transfer hoses, loading arms and other equipment and appurtenances used for the purpose of handling or transferring oil in bulk to or from a vessel as well as storage tanks and appurtenances for the reception of oily ballast water or tank washings from vessels, but excluding terminal waste treatment facilities and terminal oil storage facilities.

(B) Transfer hoses, loading arms and other equipment appurtenant to a non-transportation-related facility which is used to transfer oil in bulk to or from a vessel.

(C) Interstate and intrastate onshore and offshore pipeline systems including pumps and appurtenances related thereto as well as in-line or breakout storage tanks needed for the continuous operation of a pipeline system, and pipelines from onshore and offshore oil production facilities, but excluding onshore and offshore piping from wellheads to oil separators and pipelines which are used for the transport of oil exclusively within the confines of a nontransportation-related facility or terminal facility and which are not intended to transport oil in interstate or intrastate commerce or to transfer oil in bulk to or from a vessel.

(D) Highway vehicles and railroad cars which are used for the transport of oil in interstate or intrastate commerce and the equipment and appurtenances related thereto, and equipment used for the fueling of locomotive units, as well as the rights-of-way on which they operate. Excluded are highway vehicles and railroad cars and motive power used exclusively within the confines of a nontransportation-related facility or terminal facility and which are not intended for use in interstate or intrastate commerce.

APPENDIX B TO PART 112—MEMORANDUM OF UNDERSTANDING AMONG THE SECRETARY OF THE INTERIOR, SECRETARY OF TRANSPORTATION, AND ADMINISTRATOR OF THE ENVIRONMENTAL PROTECTION AGENCY

PURPOSE

This Memorandum of Understanding (MOU) establishes the jurisdictional responsibilities for offshore facilities, including pipelines, pursuant to section 311 (j)(1)(c), (j)(5), and (j)(6)(A) of the Clean Water Act (CWA), as amended by the Oil Pollution Act of 1990 (Public Law 101–380). The Secretary of the Department of the Interior (DOI), Secretary of the Department of Transportation (DOT), and Administrator of the Environmental Protection Agency (EPA) agree to the division of responsibilities set forth below for spill prevention and control, response planning, and equipment inspection activities pursuant to those provisions.

BACKGROUND

Executive Order (E.O.) 12777 (56 FR 54757) delegates to DOI, DOT, and EPA various responsibilities identified in section 311(j) of the CWA. Sections 2(b)(3), 2(d)(3), and 2(e)(3) of E.O. 12777 assigned to DOI spill prevention and control, contingency planning, and equipment inspection activities associated with offshore facilities. Section 311(a)(11) defines the term "offshore facility" to include facilities of any kind located in, on, or under navigable waters of the United States. By using this definition, the traditional DOI role of regulating facilities on the Outer Continental Shelf is expanded by E.O. 12777 to include inland lakes, rivers, streams, and any other inland waters.

RESPONSIBILITIES

Pursuant to section 2(i) of E.O. 12777, DOI redelegates, and EPA and DOT agree to assume, the functions vested in DOI by sections 2(b)(3), 2(d)(3), and 2(e)(3) of E.O. 12777

as set forth below. For purposes of this MOU, the term "coast line" shall be defined as in the Submerged Lands Act (43 U.S.C. 1301(c)) to mean "the line of ordinary low water along that portion of the coast which is in direct contact with the open sea and the line marking the seaward limit of inland waters."

1. To EPA, DOI redelegates responsibility for non-transportation-related offshore facilities located landward of the coast line.

2. To DOT, DOI redelegates responsibility for transportation-related facilities, including pipelines, located landward of the coast line. The DOT retains jurisdiction for deepwater ports and their associated seaward pipelines, as delegated by E.O. 12777.

3. The DOI retains jurisdiction over facilities, including pipelines, located seaward of the coast line, except for deepwater ports and associated seaward pipelines delegated by E.O. 12777 to DOT.

EFFECTIVE DATE

This MOU is effective on the date of the final execution by the indicated signatories.

LIMITATIONS

1. The DOI, DOT, and EPA may agree in writing to exceptions to this MOU on a facility-specific basis. Affected parties will receive notification of the exceptions.

2. Nothing in this MOU is intended to replace, supersede, or modify any existing agreements between or among DOI, DOT, or EPA.

MODIFICATION AND TERMINATION

Any party to this agreement may propose modifications by submitting them in writing to the heads of the other agency/department. No modification may be adopted except with the consent of all parties. All parties shall indicate their consent to or disagreement with any proposed modification within 60 days of receipt. Upon the request of any party, representatives of all parties shall meet for the purpose of considering exceptions or modifications to this agreement. This MOU may be terminated only with the mutual consent of all parties.

Dated: November 8, 1993.
Bruce Babbitt,
Secretary of the Interior.
Dated: December 14, 1993.
Federico Peña,
Secretary of Transportation.
Dated: February 3, 1994.
Carol M. Browner,
Administrator, Environmental Protection Agency.

[59 FR 34102, July 1, 1994]

APPENDIX C TO PART 112—SUBSTANTIAL
HARM CRITERIA

1.0 INTRODUCTION

The flowchart provided in Attachment C–I
to this appendix shows the decision tree with
the criteria to identify whether a facility
"could reasonably be expected to cause substantial harm to the environment by discharging into or on the navigable waters or
adjoining shorelines." In addition, the Regional Administrator has the discretion to
identify facilities that must prepare and submit facility-specific response plans to EPA.

1.1 Definitions

1.1.1 *Great Lakes* means Lakes Superior,
Michigan, Huron, Erie, and Ontario, their
connecting and tributary waters, the Saint
Lawrence River as far as Saint Regis, and
adjacent port areas.

1.1.2 Higher Volume Port Areas include

(1) Boston, MA;

(2) New York, NY;

(3) Delaware Bay and River to Philadelphia, PA;

(4) St. Croix, VI;

(5) Pascagoula, MS;

(6) Mississippi River from Southwest Pass,
LA to Baton Rouge, LA;

(7) Louisiana Offshore Oil Port (LOOP),
LA;

(8) Lake Charles, LA;

(9) Sabine-Neches River, TX;

(10) Galveston Bay and Houston Ship Channel, TX;

(11) Corpus Christi, TX;

(12) Los Angeles/Long Beach Harbor, CA;

(13) San Francisco Bay, San Pablo Bay,
Carquinez Strait, and Suisun Bay to Anti-
och, CA;

(14) Straits of Juan de Fuca from Port An-
geles, WA to and including Puget Sound,
WA;

(15) Prince William Sound, AK; and

(16) Others as specified by the Regional Administrator for any EPA Region.

1.1.3 *Inland Area* means the area shoreward of the boundary lines defined in 46 CFR
part 7, except in the Gulf of Mexico. In the
Gulf of Mexico, it means the area shoreward
of the lines of demarcation (COLREG lines as
defined in 33 CFR 80.740–80.850). The inland
area does not include the Great Lakes.

1.1.4 *Rivers and Canals* means a body of
water confined within the inland area, including the Intracoastal Waterways and
other waterways artificially created for
navigating that have project depths of 12 feet
or less.

2.0 DESCRIPTION OF SCREENING CRITERIA FOR THE SUBSTANTIAL HARM FLOWCHART

A facility that has the potential to cause
substantial harm to the environment in the
event of a discharge must prepare and sub-

mit a facility-specific response plan to EPA
in accordance with appendix F to this part.
A description of the screening criteria for
the substantial harm flowchart is provided
below:

2.1 *Non-Transportation-Related Facilities
With a Total Oil Storage Capacity Greater Than
or Equal to 42,000 Gallons Where Operations Include Over-Water Transfers of Oil.* A non-
transportation-related facility with a total
oil storage capacity greater than or equal to
42,000 gallons that transfers oil over water to
or from vessels must submit a response plan
to EPA. Daily oil transfer operations at
these types of facilities occur between barges
and vessels and onshore bulk storage tanks
over open water. These facilities are located
adjacent to navigable water.

2.2 *Lack of Adequate Secondary Containment at Facilities With a Total Oil Storage Capacity Greater Than or Equal to 1 Million Gallons.* Any facility with a total oil storage capacity greater than or equal to 1 million gallons without secondary containment sufficiently large to contain the capacity of the
largest aboveground oil storage tank within
each area plus sufficient freeboard to allow
for precipitation must submit a response
plan to EPA. Secondary containment structures that meet the standard of good engineering practice for the purposes of this part
include berms, dikes, retaining walls, curbing, culverts, gutters, or other drainage systems.

2.3 *Proximity to Fish and Wildlife and Sensitive Environments at Facilities With a Total
Oil Storage Capacity Greater Than or Equal to
1 Million Gallons.* A facility with a total oil
storage capacity greater than or equal to 1
million gallons must submit its response
plan if it is located at a distance such that
a discharge from the facility could cause injury (as defined at 40 CFR 112.2) to fish and
wildlife and sensitive environments. For further description of fish and wildlife and sensitive environments, see Appendices I, II, and
III to DOC/NOAA's "Guidance for Facility
and Vessel Response Plans: Fish and Wildlife
and Sensitive Environments" (see appendix
E to this part, section 13, for availability)
and the applicable Area Contingency Plan.
Facility owners or operators must determine
the distance at which an oil discharge could
cause injury to fish and wildlife and sensitive environments using the appropriate
formula presented in Attachment C–III to
this appendix or a comparable formula.

2.4 *Proximity to Public Drinking Water Intakes at Facilities with a Total Oil Storage Capacity Greater than or Equal to 1 Million Gallons* A facility with a total oil storage capacity greater than or equal to 1 million gallons
must submit its response plan if it is located
at a distance such that a discharge from the
facility would shut down a public drinking
water intake, which is analogous to a public
water system as described at 40 CFR 143.2(c).

55

The distance at which an oil discharge from an SPCC-regulated facility would shut down a public drinking water intake shall be calculated using the appropriate formula presented in Attachment C–III to this appendix or a comparable formula.

2.5 *Facilities That Have Experienced Reportable Oil Discharges in an Amount Greater Than or Equal to 10,000 Gallons Within the Past 5 Years and That Have a Total Oil Storage Capacity Greater Than or Equal to 1 Million Gallons.* A facility's oil spill history within the past 5 years shall be considered in the evaluation for substantial harm. Any facility with a total oil storage capacity greater than or equal to 1 million gallons that has experienced a reportable oil discharge in an amount greater than or equal to 10,000 gallons within the past 5 years must submit a response plan to EPA.

3.0 CERTIFICATION FOR FACILITIES THAT DO NOT POSE SUBSTANTIAL HARM

If the facility does not meet the substantial harm criteria listed in Attachment C–I

to this appendix, the owner or operator shall complete and maintain at the facility the certification form contained in Attachment C–II to this appendix. In the event an alternative formula that is comparable to the one in this appendix is used to evaluate the substantial harm criteria, the owner or operator shall attach documentation to the certification form that demonstrates the reliability and analytical soundness of the comparable formula and shall notify the Regional Administrator in writing that an alternative formula was used.

4.0 REFERENCES

Chow, V.T. 1959. Open Channel Hydraulics. McGraw Hill.

USCG IFR (58 FR 7353, February 5, 1993). This document is available through EPA's rulemaking docket as noted in appendix E to this part, section 13.

ATTACHMENTS TO APPENDIX C

Attachment C-I

Flowchart of Criteria for Substantial Harm

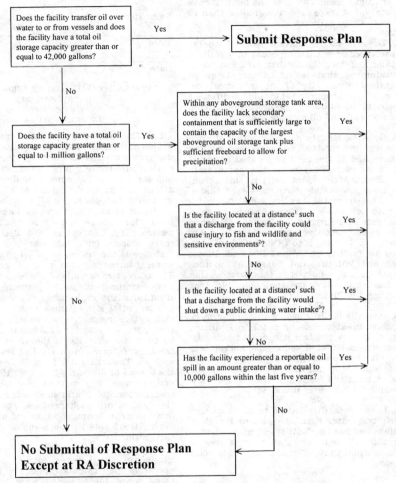

[1] Calculated using the appropriate formula in Attachment C-III to this appendix or a comparable formula.

[2] For further description of fish and wildlife and sensitive environments, see Appendices I,II, and III to DOC/NOAA's "Guidance for Facility and vessel response Plans: Fish and Wildlife and Sensitive Environments" (59 FR 14713, March 29, 1994) and the applicable Area Contingency Plan.

[3] Public drinking water intakes are analogous to public water systems as described at CFR 143.2(c).

ATTACHMENT C-II—CERTIFICATION OF THE AP-
PLICABILITY OF THE SUBSTANTIAL HARM CRI-
TERIA

Facility Name: _____

Facility Address: _____

1. Does the facility transfer oil over water
to or from vessels and does the facility have
a total oil storage capacity greater than or
equal to 42,000 gallons?

Yes _____   No _____

2. Does the facility have a total oil storage
capacity greater than or equal to 1 million
gallons and does the facility lack secondary
containment that is sufficiently large to
contain the capacity of the largest aboveground oil storage tank plus sufficient
freeboard to allow for precipitation within
any aboveground oil storage tank area?

Yes _____   No _____

3. Does the facility have a total oil storage
capacity greater than or equal to 1 million
gallons and is the facility located at a distance (as calculated using the appropriate
formula in Attachment C–III to this appendix or a comparable formula [1]) such that a
discharge from the facility could cause injury to fish and wildlife and sensitive environments? For further description of fish and
wildlife and sensitive environments, see Appendices I, II, and III to DOC/NOAA's "Guidance for Facility and Vessel Response Plans:
Fish and Wildlife and Sensitive Environments" (see appendix E to this part, section
13, for availability) and the applicable Area
Contingency Plan.

Yes _____   No _____

4. Does the facility have a total oil storage
capacity greater than or equal to 1 million
gallons and is the facility located at a distance (as calculated using the appropriate
formula in Attachment C–III to this appendix
or a comparable formula [1]) such that a discharge from the facility would shut down a
public drinking water intake [2]?

Yes _____   No _____

5. Does the facility have a total oil storage
capacity greater than or equal to 1 million
gallons and has the facility experienced a reportable oil discharge in an amount greater
than or equal to 10,000 gallons within the last
5 years?

Yes _____   No _____

Certification

I certify under penalty of law that I have
personally examined and am familiar with
the information submitted in this document,

and that based on my inquiry of those individuals responsible for obtaining this information, I believe that the submitted information is true, accurate, and complete.

Signature _____

Name (please type or print) _____

Title _____

Date _____

ATTACHMENT C-III—CALCULATION OF THE
PLANNING DISTANCE

1.0 Introduction

1.1 The facility owner or operator must
evaluate whether the facility is located at a
distance such that a discharge from the facility could cause injury to fish and wildlife
and sensitive environments or disrupt operations at a public drinking water intake. To
quantify that distance, EPA considered oil
transport mechanisms over land and on still,
tidal influence, and moving navigable
waters. EPA has determined that the primary concern for calculation of a planning
distance is the transport of oil in navigable
waters during adverse weather conditions.
Therefore, two formulas have been developed
to determine distances for planning purposes
from the point of discharge at the facility to
the potential site of impact on moving and
still waters, respectively. The formula for oil
transport on moving navigable water is
based on the velocity of the water body and
the time interval for arrival of response resources. The still water formula accounts for
the spread of discharged oil over the surface
of the water. The method to determine oil
transport on tidal influence areas is based on
the type of oil discharged and the distance
down current during ebb tide and up current
during flood tide to the point of maximum
tidal influence.

1.2 EPA's formulas were designed to be
simple to use. However, facility owners or
operators may calculate planning distances
using more sophisticated formulas, which
take into account broader scientific or engineering principles, or local conditions. Such
comparable formulas may result in different
planning distances than EPA's formulas. In
the event that an alternative formula that is
comparable to one contained in this appendix is used to evaluate the criterion in 40
CFR 112.20(f)(1)(ii)(B) or (f)(1)(ii)(C), the
owner or operator shall attach documentation to the response plan cover sheet contained in appendix F to this part that demonstrates the reliability and analytical
soundness of the alternative formula and
shall notify the Regional Administrator in

---

[1] If a comparable formula is used, documentation of the reliability and analytical
soundness of the comparable formula must
be attached to this form.

[2] For the purposes of 40 CFR part 112, public drinking water intakes are analogous to
public water systems as described at 40 CFR
143.2(c).

writing that an alternative formula was used.[1]

1.3 A regulated facility may meet the criteria for the potential to cause substantial harm to the environment without having to perform a planning distance calculation. For facilities that meet the substantial harm criteria because of inadequate secondary containment or oil spill history, as listed in the flowchart in Attachment C–I to this appendix, calculation of the planning distance is unnecessary. For facilities that do not meet the substantial harm criteria for secondary containment or oil spill history as listed in the flowchart, calculation of a planning distance for proximity to fish and wildlife and sensitive environments and public drinking water intakes is required, unless it is clear without performing the calculation (e.g., the facility is located in a wetland) that these areas would be impacted.

1.4 A facility owner or operator who must perform a planning distance calculation on navigable water is only required to do so for the type of navigable water conditions (i.e., moving water, still water, or tidal- influenced water) applicable to the facility. If a facility owner or operator determines that more than one type of navigable water condition applies, then the facility owner or operator is required to perform a planning distance calculation for each navigable water type to determine the greatest single distance that oil may be transported. As a result, the final planning distance for oil transport on water shall be the greatest individual distance rather than a summation of each calculated planning distance.

1.5 The planning distance formula for transport on moving waterways contains three variables: the velocity of the navigable water (v), the response time interval (t), and a conversion factor (c). The velocity, v, is determined by using the Chezy-Manning equation, which, in this case, models the flood flow rate of water in open channels. The Chezy-Manning equation contains three variables which must be determined by facility owners or operators. Manning's Roughness

Coefficient (for flood flow rates), n, can be determined from Table 1 of this attachment. The hydraulic radius, r, can be estimated using the average mid-channel depth from charts provided by the sources listed in Table 2 of this attachment. The average slope of the river, s, can be determined using topographic maps that can be ordered from the U.S. Geological Survey, as listed in Table 2 of this attachment.

1.6 Table 3 of this attachment contains specified time intervals for estimating the arrival of response resources at the scene of a discharge. Assuming no prior planning, response resources should be able to arrive at the discharge site within 12 hours of the discovery of any oil discharge in Higher Volume Port Areas and within 24 hours in Great Lakes and all other river, canal, inland, and nearshore areas. The specified time intervals in Table 3 of appendix C are to be used only to aid in the identification of whether a facility could cause substantial harm to the environment. Once it is determined that a plan must be developed for the facility, the owner or operator shall reference appendix E to this part to determine appropriate resource levels and response times. The specified time intervals of this appendix include a 3-hour time period for deployment of boom and other response equipment. The Regional Administrator may identify additional areas as appropriate.

2.0 Oil Transport on Moving Navigable Waters

2.1 The facility owner or operator must use the following formula or a comparable formula as described in §112.20(a)(3) to calculate the planning distance for oil transport on moving navigable water:

d=v×t×c; where

d: the distance downstream from a facility within which fish and wildlife and sensitive environments or a public drinking water intake would be shut down in the event of an oil discharge (in miles);

v: the velocity of the river/navigable water of concern (in ft/sec) as determined by Chezy-Manning's equation (see below and Tables 1 and 2 of this attachment);

t: the time interval specified in Table 3 based upon the type of water body and location (in hours); and

c: constant conversion factor 0.68 secω mile/hrω ft (3600 sec/hr ÷ 5280 ft/mile).

2.2 Chezy-Manning's equation is used to determine velocity:

v=1.5/n×r$^{2/3}$×s$^{1/2}$; where

v=the velocity of the river of concern (in ft/sec);

n=Manning's Roughness Coefficient from Table 1 of this attachment;

r=the hydraulic radius; the hydraulic radius can be approximated for parabolic channels by multiplying the average mid-channel depth of the river (in feet) by 0.667

(sources for obtaining the mid-channel depth are listed in Table 2 of this attachment); and

s=the average slope of the river (unitless) obtained from U.S. Geological Survey topographic maps at the address listed in Table 2 of this attachment.

TABLE 1—MANNING'S ROUGHNESS COEFFICIENT FOR NATURAL STREAMS

[NOTE: Coefficients are presented for high flow rates at or near flood stage.]

Stream description	Roughness coefficient (n)
Minor Streams (Top Width <100 ft.)	
Clean:	
Straight	0.03
Winding	0.04
Sluggish (Weedy, deep pools):	
No trees or brush	0.06
Trees and/or brush	0.10
Major Streams (Top Width >100 ft.)	
Regular section:	
(No boulders/brush)	0.035
Irregular section:	
(Brush)	0.05

TABLE 2—SOURCES OF R AND S FOR THE CHEZY-MANNING EQUATION

All of the charts and related publications for navigational waters may be ordered from:

Distribution Branch
(N/CG33)
National Ocean Service
Riverdale, Maryland 20737-1199
Phone: (301) 436-6990

There will be a charge for materials ordered and a VISA or Mastercard will be accepted. The mid-channel depth to be used in the calculation of the hydraulic radius (r) can be obtained directly from the following sources:

Charts of Canadian Coastal and Great Lakes Waters:

Canadian Hydrographic Service
Department of Fisheries and Oceans Institute
P.O. Box 8080
1675 Russell Road
Ottawa, Ontario KIG 3H6
Canada
Phone: (613) 998-4931

Charts and Maps of Lower Mississippi River (Gulf of Mexico to Ohio River and St. Francis, White, Big Sunflower, Atchafalaya, and other rivers):

U.S. Army Corps of Engineers
Vicksburg District
P.O. Box 60
Vicksburg, Mississippi 39180
Phone: (601) 634-5000

Charts of Upper Mississippi River and Illinois Waterway to Lake Michigan:

U.S. Army Corps of Engineers
Rock Island District
P.O. Box 2004

Rock Island, Illinois 61204
Phone: (309) 794-5552

Charts of Missouri River:

U.S. Army Corps of Engineers
Omaha District
6014 U.S. Post Office and Courthouse
Omaha, Nebraska 68102
Phone: (402) 221-3900

Charts of Ohio River:

U.S. Army Corps of Engineers
Ohio River Division
P.O. Box 1159
Cincinnati, Ohio 45201
Phone: (513) 684-3002

Charts of Tennessee Valley Authority Reservoirs, Tennessee River and Tributaries:

Tennessee Valley Authority
Maps and Engineering Section
416 Union Avenue
Knoxville, Tennessee 37902
Phone: (615) 632-2921

Charts of Black Warrior River, Alabama River, Tombigbee River, Apalachicola River and Pearl River:

U.S. Army Corps of Engineers
Mobile District
P.O. Box 2288
Mobile, Alabama 36628-0001
Phone: (205) 690-2511

The average slope of the river (s) may be obtained from topographic maps:

U.S. Geological Survey
Map Distribution
Federal Center
Bldg. 41
Box 25286
Denver, Colorado 80225

Additional information can be obtained from the following sources:

1. The State's Department of Natural Resources (DNR) or the State's Aids to Navigation office;

2. A knowledgeable local marina operator; or

3. A knowledgeable local water authority (e.g., State water commission)

2.3 The average slope of the river (s) can be determined from the topographic maps using the following steps:

(1) Locate the facility on the map.

(2) Find the Normal Pool Elevation at the point of discharge from the facility into the water (A).

(3) Find the Normal Pool Elevation of the public drinking water intake or fish and wildlife and sensitive environment located downstream (B) (Note: The owner or operator should use a minimum of 20 miles downstream as a cutoff to obtain the average slope if the location of a specific public drinking water intake or fish and wildlife and sensitive environment is unknown).

(4) If the Normal Pool Elevation is not available, the elevation contours can be used to find the slope. Determine elevation of the water at the point of discharge from the facility (A). Determine the elevation of the

water at the appropriate distance downstream (B). The formula presented below can be used to calculate the slope.

(5) Determine the distance (in miles) between the facility and the public drinking water intake or fish and wildlife and sensitive environments (C).

(6) Use the following formula to find the slope, which will be a unitless value: Average Slope=[(A−B) (ft)/C (miles)] × [1 mile/5280 feet]

2.4 If it is not feasible to determine the slope and mid-channel depth by the Chezy-Manning equation, then the river velocity can be approximated on- site. A specific length, such as 100 feet, can be marked off along the shoreline. A float can be dropped into the stream above the mark, and the time required for the float to travel the distance can be used to determine the velocity in feet per second. However, this method will not yield an average velocity for the length of the stream, but a velocity only for the specific location of measurement. In addition, the flow rate will vary depending on weather conditions such as wind and rainfall. It is recommended that facility owners or operators repeat the measurement under a variety of conditions to obtain the most accurate estimate of the surface water velocity under adverse weather conditions.

2.5 The planning distance calculations for moving and still navigable waters are based on worst case discharges of persistent oils. Persistent oils are of concern because they can remain in the water for significant periods of time and can potentially exist in large quantities downstream. Owners or operators of facilities that store persistent as well as non-persistent oils may use a comparable formula. The volume of oil discharged is not included as part of the planning distance calculation for moving navigable waters. Facilities that will meet this substantial harm criterion are those with facility capacities greater than or equal to 1 million gallons. It is assumed that these facilities are capable of having an oil discharge of sufficient quantity to cause injury to fish and wildlife and sensitive environments or shut down a public drinking water intake. While owners or operators of transfer facilities that store greater than or equal to 42,000 gallons are not required to use a planning distance formula for purposes of the substantial harm criteria, they should use a planning distance calculation in the development of facility-specific response plans.

TABLE 3—SPECIFIED TIME INTERVALS

Operating areas	Substantial harm planning time (hrs)
Higher volume port area.	12 hour arrival+3 hour deployment=15 hours.
Great Lakes ...	24 hour arrival+3 hour deployment=27 hours.

TABLE 3—SPECIFIED TIME INTERVALS— Continued

Operating areas	Substantial harm planning time (hrs)
All other rivers and canals, inland, and nearshore areas.	24 hour arrival+3 hour deployment=27 hours.

2.6 *Example of the Planning Distance Calculation for Oil Transport on Moving Navigable Waters.* The following example provides a sample calculation using the planning distance formula for a facility discharging oil into the Monongahela River:

(1) Solve for v by evaluating n, r, and s for the Chezy-Manning equation:

Find the roughness coefficient, n, on Table 1 of this attachment for a regular section of a major stream with a top width greater than 100 feet. The top width of the river can be found from the topographic map.

n=0.035.

Find slope, s, where A=727 feet, B=710 feet, and C=25 miles.

Solving:
s=[(727 ft−1710 ft)/25 · miles]×[1 mile/5280 feet]=1.3×10^{-4}

The average mid-channel depth is found by averaging the mid-channel depth for each mile along the length of the river between the facility and the public drinking water intake or the fish or wildlife or sensitive environment (or 20 miles downstream if applicable). This value is multiplied by 0.667 to obtain the hydraulic radius. The mid-channel depth is found by obtaining values for r and s from the sources shown in Table 2 for the Monongahela River.

Solving:
r=0.667×20 feet=13.33 feet
Solve for v using:
v=1.5/n×r$^{2/3}$×s$^{1/2}$:
v=[1.5/0.035]×(13.33)$^{2/3}$×(1.3×10^{-4})$^{1/2}$
v=2.73 feet/second

(2) Find t from Table 3 of this attachment. The Monongahela River's resource response time is 27 hours.

(3) Solve for planning distance, d:

d=v×t×c
d=(2.73 ft/sec)×(27 hours)×(0.68 secω mile/hrω ft)
d=50 miles

Therefore, 50 miles downstream is the appropriate planning distance for this facility.

3.0 Oil Transport on Still Water

3.1 For bodies of water including lakes or ponds that do not have a measurable velocity, the spreading of the oil over the surface must be considered. Owners or operators of facilities located next to still water bodies may use a comparable means of calculating the planning distance. If a comparable formula is used, documentation of the reliability and analytical soundness of the comparable calculation must be attached to the response plan cover sheet.

3.2 *Example of the Planning Distance Calculation for Oil Transport on Still Water.* To assist those facilities which could potentially discharge into a still body of water, the following analysis was performed to provide an example of the type of formula that may be used to calculate the planning distance. For this example, a worst case discharge of 2,000,000 gallons is used.

(1) The surface area in square feet covered by an oil discharge on still water, A1, can be determined by the following formula,[2] where V is the volume of the discharge in gallons and C is a constant conversion factor:

$A_1 = 10^5 \times V^{3/4} \times C$

$C = 0.1643$

$A_1 = 10^5 \times (2,000,000 \text{ gallons})^{3/4} \times (0.1643)$

$A_1 = 8.74 \times 10^8 \text{ ft}^2$

(2) The spreading formula is based on the theoretical condition that the oil will spread uniformly in all directions forming a circle. In reality, the outfall of the discharge will direct the oil out to the surface of the water where it intersects the shoreline. Although the oil will not spread uniformly in all directions, it is assumed that the discharge will spread from the shoreline into a semi-circle (this assumption does not account for winds or wave action).

(3) The area of a circle = $\dagger \ r^2$

(4) To account for the assumption that oil will spread in a semi-circular shape, the area of a circle is divided by 2 and is designated as A_2.

$A_2 = (\dagger \ r^2)/2$

Solving for the radius, r, using the relationship $A_1 = A_2$: $8.74 \times 10^8 \text{ ft}^2 = (\dagger^2)/2$

Therefore, r = 23,586 ft

r = 23,586 ft ÷ 5,280 ft/mile = 4.5 miles

Assuming a 20 knot wind under storm conditions:

1 knot = 1.15 miles/hour

20 knots × 1.15 miles/hour/knot = 23 miles/hr

Assuming that the oil slick moves at 3 percent of the wind's speed:[3]

23 miles/hour × 0.03 = 0.69 miles/hour

(5) To estimate the distance that the oil will travel, use the times required for response resources to arrive at different geographic locations as shown in Table 3 of this attachment.

For example:

---

[2] Huang, J.C. and Monastero, F.C., 1982. *Review of the State-of-the-Art of Oil Pollution Models.* Final report submitted to the American Petroleum Institute by Raytheon Ocean Systems, Co., East Providence, Rhode Island.

[3] *Oil Spill Prevention & Control.* National Spill Control School, Corpus Christi State University, Thirteenth Edition, May 1990.

For Higher Volume Port Areas: 15 hrs × 0.69 miles/hr = 10.4 miles

For Great Lakes and all other areas: 27 hrs × 0.69 miles/hr = 18.6 miles

(6) The total distance that the oil will travel from the point of discharge, including the distance due to spreading, is calculated as follows:

Higher Volume Port Areas: d = 10.4 + 4.5 miles or approximately 15 miles

Great Lakes and all other areas: d = 18.6 + 4.5 miles or approximately 23 miles

4.0 Oil Transport on Tidal-Influence Areas

4.1 The planning distance method for tidal influence navigable water is based on worst case discharges of persistent and non-persistent oils. Persistent oils are of primary concern because they can potentially cause harm over a greater distance. For persistent oils discharged into tidal waters, the planning distance is 15 miles from the facility down current during ebb tide and to the point of maximum tidal influence or 15 miles, whichever is less, during flood tide.

4.2 For non-persistent oils discharged into tidal waters, the planning distance is 5 miles from the facility down current during ebb tide and to the point of maximum tidal influence or 5 miles, whichever is less, during flood tide.

4.3 *Example of Determining the Planning Distance for Two Types of Navigable Water Conditions.* Below is an example of how to determine the proper planning distance when a facility could impact two types of navigable water conditions: moving water and tidal water.

(1) Facility X stores persistent oil and is located downstream from locks along a slow moving river which is affected by tides. The river velocity, v, is determined to be 0.5 feet/second from the Chezy-Manning equation used to calculate oil transport on moving navigable waters. The specified time interval, t, obtained from Table 3 of this attachment for river areas is 27 hours. Therefore, solving for the planning distance, d:

d = v × t × c

d = (0.5 ft/sec) × (27 hours) × (0.68 secmile/hrft)

d = 9.18 miles.

(2) However, the planning distance for maximum tidal influence down current during ebb tide is 15 miles, which is greater than the calculated 9.18 miles. Therefore, 15 miles downstream is the appropriate planning distance for this facility.

5.0 Oil Transport Over Land

5.1 Facility owners or operators must evaluate the potential for oil to be transported over land to navigable waters of the United States. The owner or operator must evaluate the likelihood that portions of a worst case discharge would reach navigable waters via open channel flow or from sheet flow across the land, or be prevented from reaching navigable waters when trapped in natural or man-made depressions excluding secondary containment structures.

5.2 As discharged oil travels over land, it may enter a storm drain or open concrete channel intended for drainage. It is assumed that once oil reaches such an inlet, it will flow into the receiving navigable water. During a storm event, it is highly probable that the oil will either flow into the drainage structures or follow the natural contours of the land and flow into the navigable water. Expected minimum and maximum velocities are provided as examples of open concrete channel and pipe flow. The ranges listed below reflect minimum and maximum velocities used as design criteria.[4] The calculation below demonstrates that the time required for oil to travel through a storm drain or open concrete channel to navigable water is negligible and can be considered instantaneous. The velocities are:

For open concrete channels:
maximum velocity=25 feet per second
minimum velocity=3 feet per second
For storm drains:
maximum velocity=25 feet per second
minimum velocity=2 feet per second

5.3 Assuming a length of 0.5 mile from the point of discharge through an open concrete channel or concrete storm drain to a navigable water, the travel times (distance/velocity) are:

1.8 minutes at a velocity of 25 feet per second
14.7 minutes at a velocity of 3 feet per second
22.0 minutes for at a velocity of 2 feet per second

5.4 The distances that shall be considered to determine the planning distance are illustrated in Figure C–I of this attachment. The relevant distances can be described as follows:

D1=Distance from the nearest opportunity for discharge, X_1, to a storm drain or an open concrete channel leading to navigable water.

D2=Distance through the storm drain or open concrete channel to navigable water.

D3=Distance downstream from the outfall within which fish and wildlife and sensitive

---

[4] The design velocities were obtained from Howard County, Maryland Department of Public Works' Storm Drainage Design Manual.

environments could be injured or a public drinking water intake would be shut down as determined by the planning distance formula.

D4=Distance from the nearest opportunity for discharge, X_2, to fish and wildlife and sensitive environments not bordering navigable water.

5.5 A facility owner or operator whose nearest opportunity for discharge is located within 0.5 mile of a navigable water must complete the planning distance calculation (D3) for the type of navigable water near the facility or use a comparable formula.

5.6 A facility that is located at a distance greater than 0.5 mile from a navigable water must also calculate a planning distance (D3) if it is in close proximity (i.e., D1 is less than 0.5 mile and other factors are conducive to oil travel over land) to storm drains that flow to navigable waters. Factors to be considered in assessing oil transport over land to storm drains shall include the topography of the surrounding area, drainage patterns, man-made barriers (excluding secondary containment structures), and soil distribution and porosity. Storm drains or concrete drainage channels that are located in close proximity to the facility can provide a direct pathway to navigable waters, regardless of the length of the drainage pipe. If D1 is less than or equal to 0.5 mile, a discharge from the facility could pose substantial harm because the time to travel the distance from the storm drain to the navigable water (D2) is virtually instantaneous.

5.7 A facility's proximity to fish and wildlife and sensitive environments not bordering a navigable water, as depicted as D4 in Figure C–I of this attachment, must also be considered, regardless of the distance from the facility to navigable waters. Factors to be considered in assessing oil transport over land to fish and wildlife and sensitive environments should include the topography of the surrounding area, drainage patterns, man-made barriers (excluding secondary containment structures), and soil distribution and porosity.

5.8 If a facility is not found to pose substantial harm to fish and wildlife and sensitive environments not bordering navigable waters via oil transport on land, then supporting documentation should be maintained at the facility. However, such documentation should be submitted with the response plan if a facility is found to pose substantial harm.

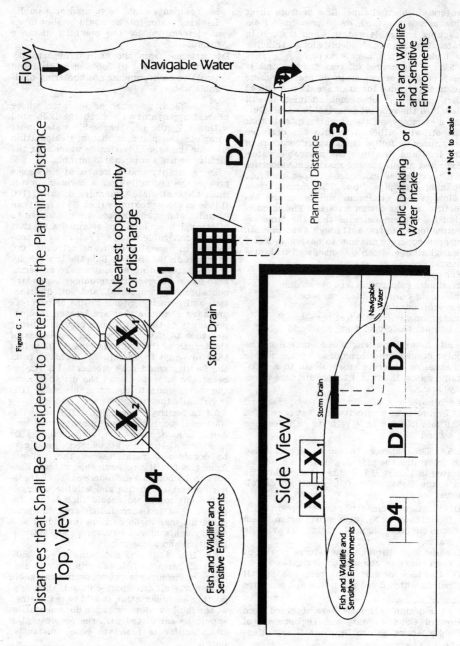

Figure C - 1

Distances that Shall Be Considered to Determine the Planning Distance

[59 FR 34102, July 1, 1994, as amended at 65 FR 40798, June 30, 2000; 67 FR 47152, July 17, 2002]

APPENDIX D TO PART 112—DETERMINA-
TION OF A WORST CASE DISCHARGE
PLANNING VOLUME

1.0 Instructions

1.1 An owner or operator is required to
complete this worksheet if the facility meets
the criteria, as presented in appendix C to
this part, or it is determined by the RA that
the facility could cause substantial harm to
the environment. The calculation of a worst
case discharge planning volume is used for
emergency planning purposes, and is required in 40 CFR 112.20 for facility owners or
operators who must prepare a response plan.
When planning for the amount of resources
and equipment necessary to respond to the
worst case discharge planning volume, adverse weather conditions must be taken into
consideration. An owner or operator is required to determine the facility's worst case
discharge planning volume from either part
A of this appendix for an onshore storage facility, or part B of this appendix for an onshore production facility. The worksheet
considers the provision of adequate secondary containment at a facility.

1.2 For onshore storage facilities and production facilities, permanently manifolded
oil storage tanks are defined as tanks that
are designed, installed, and/or operated in
such a manner that the multiple tanks function as one storage unit (i.e., multiple tank
volumes are equalized). In a worst case discharge scenario, a single failure could cause
the discharge of the contents of more than
one tank. The owner or operator must provide evidence in the response plan that tanks
with common piping or piping systems are
not operated as one unit. If such evidence is
provided and is acceptable to the RA, the
worst case discharge planning volume would
be based on the capacity of the largest oil
storage tank within a common secondary
containment area or the largest oil storage
tank within a single secondary containment
area, whichever is greater. For permanently
manifolded tanks that function as one oil
storage unit, the worst case discharge planning volume would be based on the combined
oil storage capacity of all manifolded tanks
or the capacity of the largest single oil storage tank within a secondary containment
area, whichever is greater. For purposes of
this rule, permanently manifolded tanks
that are separated by internal divisions for
each tank are considered to be single tanks
and individual manifolded tank volumes are
not combined.

1.3 For production facilities, the presence
of exploratory wells, production wells, and
oil storage tanks must be considered in the
calculation. Part B of this appendix takes
these additional factors into consideration
and provides steps for their inclusion in the
total worst case discharge planning volume.

Onshore oil production facilities may include
all wells, flowlines, separation equipment,
storage facilities, gathering lines, and auxiliary non-transportation-related equipment
and facilities in a single geographical oil or
gas field operated by a single operator. Although a potential worst case discharge
planning volume is calculated within each
section of the worksheet, the final worst
case amount depends on the risk parameter
that results in the greatest volume.

1.4 Marine transportation-related transfer
facilities that contain fixed aboveground onshore structures used for bulk oil storage are
jointly regulated by EPA and the U.S. Coast
Guard (USCG), and are termed "complexes."
Because the USCG also requires response
plans from transportation-related facilities
to address a worst case discharge of oil, a
separate calculation for the worst case discharge planning volume for USCG-related facilities is included in the USCG IFR (see appendix E to this part, section 13, for availability). All complexes that are jointly regulated by EPA and the USCG must compare
both calculations for worst case discharge
planning volume derived by using the EPA
and USCG methodologies and plan for whichever volume is greater.

PART A: WORST CASE DISCHARGE PLAN-
NING VOLUME CALCULATION FOR ON-
SHORE STORAGE FACILITIES [1]

Part A of this worksheet is to be completed by the owner or operator of an SPCC-
regulated facility (excluding oil production
facilities) if the facility meets the criteria as
presented in appendix C to this part, or if it
is determined by the RA that the facility
could cause substantial harm to the environment. If you are the owner or operator of a
production facility, please proceed to part B
of this worksheet.

A.1 SINGLE-TANK FACILITIES

For facilities containing only one aboveground oil storage tank, the worst case discharge planning volume equals the capacity
of the oil storage tank. If adequate secondary containment (sufficiently large to
contain the capacity of the aboveground oil
storage tank plus sufficient freeboard to
allow for precipitation) exists for the oil
storage tank, multiply the capacity of the
tank by 0.8.

(1) FINAL WORST CASE VOLUME:
_____ GAL

(2) Do not proceed further.

_____

[1] "Storage facilities" represent all facilities subject to this part, excluding oil production facilities.

*A.2 SECONDARY CONTAINMENT—
MULTIPLE-TANK FACILITIES*

Are *all* aboveground oil storage tanks or groups of aboveground oil storage tanks at the facility *without* adequate secondary containment?[2]

_____ (Y/N)

A.2.1 If the answer is yes, the final worst case discharge planning volume equals the *total aboveground oil storage capacity at the facility.*

(1) FINAL WORST CASE VOLUME: _____ GAL

(2) Do not proceed further.

A.2.2 If the answer is no, calculate the total aboveground oil storage capacity of tanks without adequate secondary containment. If *all* aboveground oil storage tanks or groups of aboveground oil storage tanks at the facility have adequate secondary containment, ENTER "0" (zero).

_____ GAL

A.2.3 Calculate the capacity of the largest single aboveground oil storage tank within an adequate secondary containment area or the combined capacity of a group of aboveground oil storage tanks permanently manifolded together, whichever is greater, PLUS THE VOLUME FROM QUESTION A.2.2.

FINAL WORST CASE VOLUME:[3] _____ GAL

PART B: WORST CASE DISCHARGE PLANNING VOLUME CALCULATION FOR ONSHORE PRODUCTION FACILITIES

Part B of this worksheet is to be completed by the owner or operator of an SPCC-regulated oil production facility if the facility meets the criteria presented in appendix C to this part, or if it is determined by the RA that the facility could cause substantial harm. A production facility consists of all wells (producing and exploratory) and related equipment in a single geographical oil or gas field operated by a single operator.

B.1 SINGLE-TANK FACILITIES

B.1.1 For facilities containing only one aboveground oil storage tank, the worst case discharge planning volume equals the capacity of the aboveground oil storage tank plus the production volume of the well with the highest output at the facility. If adequate

_____
[2] Secondary containment is described in 40 CFR part 112, subparts A through C. Acceptable methods and structures for containment are also given in 40 CFR 112.7(c)(1).

[3] All complexes that are jointly regulated by EPA and the USCG must also calculate the worst case discharge planning volume for the transportation-related portions of the facility and plan for whichever volume is greater.

secondary containment (sufficiently large to contain the capacity of the aboveground oil storage tank plus sufficient freeboard to allow for precipitation) exists for the storage tank, multiply the capacity of the tank by 0.8.

B.1.2 For facilities with production wells producing by pumping, if the rate of the well with the highest output is known and the number of days the facility is unattended can be predicted, then the production volume is equal to the pumping rate of the well multiplied by the greatest number of days the facility is unattended.

B.1.3 If the pumping rate of the well with the highest output is estimated or the maximum number of days the facility is unattended is estimated, then the production volume is determined from the pumping rate of the well multiplied by 1.5 times the greatest number of days that the facility has been or is expected to be unattended.

B.1.4 Attachment D–1 to this appendix provides methods for calculating the production volume for exploratory wells and production wells producing under pressure.

(1) FINAL WORST CASE VOLUME: _____ GAL

(2) Do not proceed further.

*B.2 SECONDARY CONTAINMENT—
MULTIPLE-TANK FACILITIES*

Are *all* aboveground oil storage tanks or groups of aboveground oil storage tanks at the facility *without* adequate secondary containment?

_____ (Y/N)

B.2.1 If the answer is yes, the final worst case volume equals the total aboveground oil storage capacity without adequate secondary containment plus the production volume of the well with the highest output at the facility.

(1) For facilities with production wells producing by pumping, if the rate of the well with the highest output is known and the number of days the facility is unattended can be predicted, then the production volume is equal to the pumping rate of the well multiplied by the greatest number of days the facility is unattended.

(2) If the pumping rate of the well with the highest output is estimated or the maximum number of days the facility is unattended is estimated, then the production volume is determined from the pumping rate of the well multiplied by 1.5 times the greatest number of days that the facility has been or is expected to be unattended.

(3) Attachment D–1 to this appendix provides methods for calculating the production volumes for exploratory wells and production wells producing under pressure.

(A) FINAL WORST CASE VOLUME: _____ GAL

(B) Do not proceed further.

B.2.2 If the answer is no, calculate the total aboveground oil storage capacity of tanks without adequate secondary containment. If *all* aboveground oil storage tanks or groups of aboveground oil storage tanks at the facility have adequate secondary containment, ENTER "0" (zero).

_____ GAL

B.2.3 Calculate the capacity of the largest single aboveground oil storage tank within an adequate secondary containment area or the combined capacity of a group of aboveground oil storage tanks permanently manifolded together, whichever is greater, plus the production volume of the well with the highest output, PLUS THE VOLUME FROM QUESTION B.2.2. Attachment D–1 provides methods for calculating the production volumes for exploratory wells and production wells producing under pressure.

(1) FINAL WORST CASE VOLUME:[4]
_____ GAL

(2) Do not proceed further.

ATTACHMENTS TO APPENDIX D

ATTACHMENT D–I—METHODS TO CALCULATE PRODUCTION VOLUMES FOR PRODUCTION FACILITIES WITH EXPLORATORY WELLS OR PRODUCTION WELLS PRODUCING UNDER PRESSURE

1.0 Introduction

The owner or operator of a production facility with exploratory wells or production wells producing under pressure shall compare the well rate of the highest output well (rate of well), in barrels per day, to the ability of response equipment and personnel to recover the volume of oil that could be discharged (rate of recovery), in barrels per day. The result of this comparison will determine the method used to calculate the production volume for the production facility. This production volume is to be used to calculate the worst case discharge planning volume in part B of this appendix.

2.0 Description of Methods

2.1 Method A

If the well rate would overwhelm the response efforts (i.e., rate of well/rate of recovery ≥1), then the production volume would be the 30-day forecasted well rate for a well 10,000 feet deep or less, or the 45-day forecasted well rate for a well deeper than 10,000 feet.

(1) For wells 10,000 feet deep or less:
Production volume=30 days × rate of well.

[4] All complexes that are jointly regulated by EPA and the USCG must also calculate the worst case discharge planning volume for the transportation-related portions of the facility and plan for whichever volume is greater.

(2) For wells deeper than 10,000 feet:
Production volume=45 days × rate of well.

2.2 Method B

2.2.1 If the rate of recovery would be greater than the well rate (i.e., rate of well/rate of recovery <1), then the production volume would equal the sum of two terms:

Production volume=discharge volume$_1$ + discharge volume$_2$

2.2.2 The first term represents the volume of the oil discharged from the well between the time of the blowout and the time the response resources are on scene and recovering oil (discharge volume$_1$).

Discharge volume$_1$=(days unattended+days to respond) × (rate of well)

2.2.3 The second term represents the volume of oil discharged from the well after the response resources begin operating until the discharge is stopped, adjusted for the recovery rate of the response resources (discharge volume$_2$).

(1) For wells 10,000 feet deep or less:
Discharge volume$_2$=[30 days−(days unattended + days to respond)] × (rate of well) × (rate of well/rate of recovery)

(2) For wells deeper than 10,000 feet:
Discharge volume$_2$=[45 days−(days unattended + days to respond)] × (rate of well) × (rate of well/rate of recovery)

3.0 Example

3.1 A facility consists of two production wells producing under pressure, which are both less than 10,000 feet deep. The well rate of well A is 5 barrels per day, and the well rate of well B is 10 barrels per day. The facility is unattended for a maximum of 7 days. The facility operator estimates that it will take 2 days to have response equipment and personnel on scene and responding to a blowout, and that the projected rate of recovery will be 20 barrels per day.

(1) First, the facility operator determines that the highest output well is well B. The facility operator calculates the ratio of the rate of well to the rate of recovery:

10 barrels per day/20 barrels per day=0.5 Because the ratio is less than one, the facility operator will use Method B to calculate the production volume.

(2) The first term of the equation is:

Discharge volume$_1$=(7 days + 2 days) × (10 barrels per day)=90 barrels

(3) The second term of the equation is:

Discharge volume$_2$=[30 days—(7 days + 2 days)] × (10 barrels per day) × (0.5)=105 barrels

(4) Therefore, the production volume is:

Production volume=90 barrels + 105 barrels=195 barrels

3.2 If the recovery rate was 5 barrels per day, the ratio of rate of well to rate of recovery would be 2, so the facility operator would use Method A. The production volume would have been:

30 days × 10 barrels per day=300 barrels

[59 FR 34110, July 1, 1994; 59 FR 49006, Sept. 26, 1994, as amended at 65 FR 40800, June 30, 2000; 67 FR 47152, July 17, 2002]

APPENDIX E TO PART 112—DETERMINA-TION AND EVALUATION OF REQUIRED RESPONSE RESOURCES FOR FACILITY RESPONSE PLANS

1.0 Purpose and Definitions

1.1 The purpose of this appendix is to describe the procedures to identify response resources to meet the requirements of §112.20. To identify response resources to meet the facility response plan requirements of 40 CFR 112.20(h), owners or operators shall follow this appendix or, where not appropriate, shall clearly demonstrate in the response plan why use of this appendix is not appropriate at the facility and make comparable arrangements for response resources.

1.2 Definitions.

1.2.1 *Animal fat* means a non-petroleum oil, fat, or grease of animal, fish, or marine mammal origin. Animal fats are further classified based on specific gravity as follows:

(1) Group A—specific gravity less than 0.8.

(2) Group B—specific gravity equal to or greater than 0.8 and less than 1.0.

(3) Group C—specific gravity equal to or greater than 1.0.

1.2.2 *Nearshore* is an operating area defined as extending seaward 12 miles from the boundary lines defined in 46 CFR part 7, except in the Gulf of Mexico. In the Gulf of Mexico, it means the area extending 12 miles from the line of demarcation (COLREG lines) defined in 49 CFR 80.740 and 80.850.

1.2.3 *Non-persistent oils* or *Group 1 oils* include:

(1) A petroleum-based oil that, at the time of shipment, consists of hydrocarbon fractions:

(A) At least 50 percent of which by volume, distill at a temperature of 340 degrees C (645 degrees F); and

(B) At least 95 percent of which by volume, distill at a temperature of 370 degrees C (700 degrees F); and

(2) A non-petroleum oil, other than an animal fat or vegetable oil, with a specific gravity less than 0.8.

1.2.4 *Non-petroleum oil* means oil of any kind that is not petroleum-based, including but not limited to: fats, oils, and greases of animal, fish, or marine mammal origin; and vegetable oils, including oils from seeds, nuts, fruits, and kernels.

1.2.5 *Ocean* means the nearshore area.

1.2.6 *Operating area* means Rivers and Canals, Inland, Nearshore, and Great Lakes geographic location(s) in which a facility is handling, storing, or transporting oil.

1.2.7 *Operating environment* means Rivers and Canals, Inland, Great Lakes, or Ocean. These terms are used to define the conditions in which response equipment is designed to function.

1.2.8 *Persistent oils* include:

(1) A petroleum-based oil that does not meet the distillation criteria for a non-persistent oil. Persistent oils are further classified based on specific gravity as follows:

(A) Group 2—specific gravity less than 0.85;

(B) Group 3—specific gravity equal to or greater than 0.85 and less than 0.95;

(C) Group 4—specific gravity equal to or greater than 0.95 and less than 1.0; or

(D) Group 5—specific gravity equal to or greater than 1.0.

(2) A non-petroleum oil, other than an animal fat or vegetable oil, with a specific gravity of 0.8 or greater. These oils are further classified based on specific gravity as follows:

(A) Group 2—specific gravity equal to or greater than 0.8 and less than 0.85;

(B) Group 3—specific gravity equal to or greater than 0.85 and less than 0.95;

(C) Group 4—specific gravity equal to or greater than 0.95 and less than 1.0; or

(D) Group 5—specific gravity equal to or greater than 1.0.

1.2.9 *Vegetable oil* means a non-petroleum oil or fat of vegetable origin, including but not limited to oils and fats derived from plant seeds, nuts, fruits, and kernels. Vegetable oils are further classified based on specific gravity as follows:

(1) Group A—specific gravity less than 0.8.

(2) Group B—specific gravity equal to or greater than 0.8 and less than 1.0.

(3) Group C—specific gravity equal to or greater than 1.0.

1.2.10 Other definitions are included in §112.2, section 1.1 of appendix C, and section 3.0 of appendix F.

2.0 Equipment Operability and Readiness

2.1 All equipment identified in a response plan must be designed to operate in the conditions expected in the facility's geographic area (i.e., operating environment). These conditions vary widely based on location and season. Therefore, it is difficult to identify a single stockpile of response equipment that will function effectively in each geographic location (i.e., operating area).

2.2 Facilities handling, storing, or transporting oil in more than one operating environment as indicated in Table 1 of this appendix must identify equipment capable of successfully functioning in each operating environment.

2.3 When identifying equipment for the response plan (based on the use of this appendix), a facility owner or operator must consider the inherent limitations of the operability of equipment components and response systems. The criteria in Table 1 of this appendix shall be used to evaluate the operability in a given environment. These criteria reflect the general conditions in certain operating environments.

2.3.1 The Regional Administrator may require documentation that the boom identified in a facility response plan meets the criteria in Table 1 of this appendix. Absent acceptable documentation, the Regional Administrator may require that the boom be tested to demonstrate that it meets the criteria in Table 1 of this appendix. Testing must be in accordance with ASTM F 715, ASTM F 989, or other tests approved by EPA as deemed appropriate (see appendix E to this part, section 13, for general availability of documents).

2.4 Table 1 of this appendix lists criteria for oil recovery devices and boom. All other equipment necessary to sustain or support response operations in an operating environment must be designed to function in the same conditions. For example, boats that deploy or support skimmers or boom must be capable of being safely operated in the significant wave heights listed for the applicable operating environment.

2.5 A facility owner or operator shall refer to the applicable Area Contingency Plan (ACP), where available, to determine if ice, debris, and weather-related visibility are significant factors to evaluate the operability of equipment. The ACP may also identify the average temperature ranges expected in the facility's operating area. All equipment identified in a response plan must be designed to operate within those conditions or ranges.

2.6 This appendix provides information on response resource mobilization and response times. The distance of the facility from the storage location of the response resources must be used to determine whether the resources can arrive on-scene within the stated time. A facility owner or operator shall include the time for notification, mobilization, and travel of resources identified to meet the medium and Tier 1 worst case discharge requirements identified in sections 4.3 and 9.3 of this appendix (for medium discharges) and section 5.3 of this appendix (for worst case discharges). The facility owner or operator must plan for notification and mobilization of Tier 2 and 3 response resources as necessary to meet the requirements for arrival on-scene in accordance with section 5.3 of this appendix. An on-water speed of 5 knots and a land speed of 35 miles per hour is assumed, unless the facility owner or operator can demonstrate otherwise.

2.7 In identifying equipment, the facility owner or operator shall list the storage location, quantity, and manufacturer's make and model. For oil recovery devices, the effective daily recovery capacity, as determined using section 6 of this appendix, must be included. For boom, the overall boom height (draft and freeboard) shall be included. A facility owner or operator is responsible for ensuring that the identified boom has compatible connectors.

3.0 Determining Response Resources Required for Small Discharges—Petroleum Oils and Non-Petroleum Oils Other Than Animal Fats and Vegetable Oils

3.1 A facility owner or operator shall identify sufficient response resources available, by contract or other approved means as described in § 112.2, to respond to a small discharge. A small discharge is defined as any discharge volume less than or equal to 2,100 gallons, but not to exceed the calculated worst case discharge. The equipment must be designed to function in the operating environment at the point of expected use.

3.2 Complexes that are regulated by EPA and the United States Coast Guard (USCG) must also consider planning quantities for the transportation-related transfer portion of the facility.

3.2.1 *Petroleum oils.* The USCG planning level that corresponds to EPA's "small discharge" is termed "the average most probable discharge." A USCG rule found at 33 CFR 154.1020 defines "the average most probable discharge" as the lesser of 50 barrels (2,100 gallons) or 1 percent of the volume of the worst case discharge. Owners or operators of complexes that handle, store, or transport petroleum oils must compare oil discharge volumes for a small discharge and an average most probable discharge, and plan for whichever quantity is greater.

3.2.2 *Non-petroleum oils other than animal fats and vegetable oils.* Owners or operators of complexes that handle, store, or transport non-petroleum oils other than animal fats and vegetable oils must plan for oil discharge volumes for a small discharge. There is no USCG planning level that directly corresponds to EPA's "small discharge." However, the USCG (at 33 CFR 154.545) has requirements to identify equipment to contain oil resulting from an operational discharge.

3.3 The response resources shall, as appropriate, include:

3.3.1 One thousand feet of containment boom (or, for complexes with marine transfer components, 1,000 feet of containment boom or two times the length of the largest vessel that regularly conducts oil transfers to or from the facility, whichever is greater), and a means of deploying it within 1 hour of the discovery of a discharge;

3.3.2 Oil recovery devices with an effective daily recovery capacity equal to the amount of oil discharged in a small discharge or greater which is available at the facility within 2 hours of the detection of an oil discharge; and

3.3.3 Oil storage capacity for recovered oily material indicated in section 12.2 of this appendix.

4.0 Determining Response Resources Required for Medium Discharges—Petroleum Oils and Non-Petroleum Oils Other Than Animal Fats and Vegetable Oils

4.1 A facility owner or operator shall identify sufficient response resources available, by contract or other approved means as described in §112.2, to respond to a medium discharge of oil for that facility. This will require response resources capable of containing and collecting up to 36,000 gallons of oil or 10 percent of the worst case discharge, whichever is less. All equipment identified must be designed to operate in the applicable operating environment specified in Table 1 of this appendix.

4.2 Complexes that are regulated by EPA and the USCG must also consider planning quantities for the transportation-related transfer portion of the facility.

4.2.1 *Petroleum oils.* The USCG planning level that corresponds to EPA's "medium discharge" is termed "the maximum most probable discharge." The USCG rule found at 33 CFR part 154 defines "the maximum most probable discharge" as a discharge of 1,200 barrels (50,400 gallons) or 10 percent of the worst case discharge, whichever is less. Owners or operators of complexes that handle, store, or transport petroleum oils must compare calculated discharge volumes for a medium discharge and a maximum most probable discharge, and plan for whichever quantity is greater.

4.2.2 *Non-petroleum oils other than animal fats and vegetable oils.* Owners or operators of complexes that handle, store, or transport non-petroleum oils other than animal fats and vegetable oils must plan for oil discharge volumes for a medium discharge. For non-petroleum oils, there is no USCG planning level that directly corresponds to EPA's "medium discharge."

4.3 Oil recovery devices identified to meet the applicable medium discharge volume planning criteria must be located such that they are capable of arriving on-scene within 6 hours in higher volume port areas and the Great Lakes and within 12 hours in all other areas. Higher volume port areas and Great Lakes areas are defined in section 1.1 of appendix C to this part.

4.4 Because rapid control, containment, and removal of oil are critical to reduce discharge impact, the owner or operator must determine response resources using an effective daily recovery capacity for oil recovery devices equal to 50 percent of the planning volume applicable for the facility as determined in section 4.1 of this appendix. The effective daily recovery capacity for oil recovery devices identified in the plan must be determined using the criteria in section 6 of this appendix.

4.5 In addition to oil recovery capacity, the plan shall, as appropriate, identify sufficient quantity of containment boom available, by contract or other approved means as described in §112.2, to arrive within the required response times for oil collection and containment and for protection of fish and wildlife and sensitive environments. For further description of fish and wildlife and sensitive environments, see Appendices I, II, and III to DOC/NOAA's "Guidance for Facility and Vessel Response Plans: Fish and Wildlife and Sensitive Environments" (see appendix E to this part, section 13, for availability) and the applicable ACP. Although 40 CFR part 112 does not set required quantities of boom for oil collection and containment, the response plan shall identify and ensure, by contract or other approved means as described in §112.2, the availability of the quantity of boom identified in the plan for this purpose.

4.6 The plan must indicate the availability of temporary storage capacity to meet section 12.2 of this appendix. If available storage capacity is insufficient to meet this level, then the effective daily recovery capacity must be derated (downgraded) to the limits of the available storage capacity.

4.7 The following is an example of a medium discharge volume planning calculation for equipment identification in a higher volume port area: The facility's largest aboveground storage tank volume is 840,000 gallons. Ten percent of this capacity is 84,000 gallons. Because 10 percent of the facility's largest tank, or 84,000 gallons, is greater than 36,000 gallons, 36,000 gallons is used as the planning volume. The effective daily recovery capacity is 50 percent of the planning volume, or 18,000 gallons per day. The ability of oil recovery devices to meet this capacity must be calculated using the procedures in section 6 of this appendix. Temporary storage capacity available on-scene must equal twice the daily recovery capacity as indicated in section 12.2 of this appendix, or 36,000 gallons per day. This is the information the facility owner or operator must use to identify and ensure the availability of the required response resources, by contract or other approved means as described in §112.2. The facility owner shall also identify how much boom is available for use.

5.0 Determining Response Resources Required for the Worst Case Discharge to the Maximum Extent Practicable

5.1 A facility owner or operator shall identify and ensure the availability of, by contract or other approved means as described in §112.2, sufficient response resources to respond to the worst case discharge of oil to the maximum extent practicable. Sections 7 and 10 of this appendix describe the method to determine the necessary response resources. Worksheets are provided as Attachments E–1 and E–2 at the end of this appendix to simplify the procedures involved in calculating the planning volume for response resources for the worst case discharge.

5.2 Complexes that are regulated by EPA and the USCG must also consider planning for the worst case discharge at the transportation-related portion of the facility. The USCG requires that transportation-related facility owners or operators use a different calculation for the worst case discharge in the revisions to 33 CFR part 154. Owners or operators of complex facilities that are regulated by EPA and the USCG must compare both calculations of worst case discharge derived by EPA and the USCG and plan for whichever volume is greater.

5.3 Oil discharge response resources identified in the response plan and available, by contract or other approved means as described in §112.2, to meet the applicable worst case discharge planning volume must be located such that they are capable of arriving at the scene of a discharge within the times specified for the applicable response tier listed as follows

	Tier 1 (in hours)	Tier 2 (in hours)	Tier 3 (in hours)
Higher volume port areas	6	30	54
Great Lakes	12	36	60
All other river and canal, inland, and nearshore areas	12	36	60

The three levels of response tiers apply to the amount of time in which facility owners or operators must plan for response resources to arrive at the scene of a discharge to respond to the worst case discharge planning volume. For example, at a worst case discharge in an inland area, the first tier of response resources (*i.e.*, that amount of on-water and shoreline cleanup capacity necessary to respond to the fraction of the worst case discharge as indicated through the series of steps described in sections 7.2 and 7.3 or sections 10.2 and 10.3 of this appendix) would arrive at the scene of the discharge within 12 hours; the second tier of response resources would arrive within 36 hours; and the third tier of response resources would arrive within 60 hours.

5.4 The effective daily recovery capacity for oil recovery devices identified in the response plan must be determined using the criteria in section 6 of this appendix. A facility owner or operator shall identify the storage locations of all response resources used for each tier. The owner or operator of a facility whose required daily recovery capacity exceeds the applicable contracting caps in Table 5 of this appendix, as appropriate, identify sources of additional equipment, their location, and the arrangements made to obtain this equipment during a response. The owner or operator of a facility whose calculated planning volume exceeds the applicable contracting caps in Table 5 of this appendix shall, as appropriate, identify sources of additional equipment equal to twice the cap listed in Tier 3 or the amount necessary to reach the calculated planning volume, whichever is lower. The resources identified above the cap shall be capable of arriving on-scene not later than the Tier 3 response times in section 5.3 of this appendix. No contract is required. While general listings of available response equipment may be used to identify additional sources (i.e., "public" resources vs. "private" resources), the response plan shall identify the specific sources, locations, and quantities of equipment that a facility owner or operator has considered in his or her planning. When listing USCG-classified oil spill removal organization(s) that have sufficient removal capacity to recover the volume above the response capacity cap for the specific facility, as specified in Table 5 of this appendix, it is not necessary to list specific quantities of equipment.

5.5 A facility owner or operator shall identify the availability of temporary storage capacity to meet section 12.2 of this appendix. If available storage capacity is insufficient, then the effective daily recovery capacity must be derated (downgraded) to the limits of the available storage capacity.

5.6 When selecting response resources necessary to meet the response plan requirements, the facility owner or operator shall, as appropriate, ensure that a portion of those resources is capable of being used in close-to-shore response activities in shallow water. For any EPA-regulated facility that is required to plan for response in shallow water, at least 20 percent of the on-water response equipment identified for the applicable operating area shall, as appropriate, be capable of operating in water of 6 feet or less depth.

5.7 In addition to oil spill recovery devices, a facility owner or operator shall identify sufficient quantities of boom that are available, by contract or other approved means as described in §112.2, to arrive on-

71

scene within the specified response times for oil containment and collection. The specific quantity of boom required for collection and containment will depend on the facility-specific information and response strategies employed. A facility owner or operator shall, as appropriate, also identify sufficient quantities of oil containment boom to protect fish and wildlife and sensitive environments. For further description of fish and wildlife and sensitive environments, see Appendices I, II, and III to DOC/NOAA's "Guidance for Facility and Vessel Response Plans: Fish and Wildlife and Sensitive Environments" (see appendix E to this part, section 13, for availability), and the applicable ACP. Refer to this guidance document for the number of days and geographic areas (*i.e.*, operating environments) specified in Table 2 and Table 6 of this appendix.

5.8 A facility owner or operator shall also identify, by contract or other approved means as described in §112.2, the availability of an oil spill removal organization(s) (as described in §112.2) capable of responding to a shoreline cleanup operation involving the calculated volume of oil and emulsified oil that might impact the affected shoreline. The volume of oil that shall, as appropriate, be planned for is calculated through the application of factors contained in Tables 2, 3, 6, and 7 of this appendix. The volume calculated from these tables is intended to assist the facility owner or operator to identify an oil spill removal organization with sufficient resources and expertise.

6.0 Determining Effective Daily Recovery Capacity for Oil Recovery Devices

6.1 Oil recovery devices identified by a facility owner or operator must be identified by the manufacturer, model, and effective daily recovery capacity. These capacities must be used to determine whether there is sufficient capacity to meet the applicable planning criteria for a small discharge, a medium discharge, and a worst case discharge to the maximum extent practicable.

6.2 To determine the effective daily recovery capacity of oil recovery devices, the formula listed in section 6.2.1 of this appendix shall be used. This formula considers potential limitations due to available daylight, weather, sea state, and percentage of emulsified oil in the recovered material. The RA may assign a lower efficiency factor to equipment listed in a response plan if it is determined that such a reduction is warranted.

6.2.1 The following formula shall be used to calculate the effective daily recovery capacity:

$$R = T \times 24 \text{ hours} \times E$$

where:

R—Effective daily recovery capacity;

T—Throughput rate in barrels per hour (nameplate capacity); and

E—20 percent efficiency factor (or lower factor as determined by the Regional Administrator).

6.2.2 For those devices in which the pump limits the throughput of liquid, throughput rate shall be calculated using the pump capacity.

6.2.3 For belt or moptype devices, the throughput rate shall be calculated using the speed of the belt or mop through the device, assumed thickness of oil adhering to or collected by the device, and surface area of the belt or mop. For purposes of this calculation, the assumed thickness of oil will be ¼ inch.

6.2.4 Facility owners or operators that include oil recovery devices whose throughput is not measurable using a pump capacity or belt/mop speed may provide information to support an alternative method of calculation. This information must be submitted following the procedures in section 6.3.2 of this appendix.

6.3 As an alternative to section 6.2 of this appendix, a facility owner or operator may submit adequate evidence that a different effective daily recovery capacity should be applied for a specific oil recovery device. Adequate evidence is actual verified performance data in discharge conditions or tests using American Society of Testing and Materials (ASTM) Standard F 631-99, F 808-83 (1999), or an equivalent test approved by EPA as deemed appropriate (see Appendix E to this part, section 13, for general availability of documents).

6.3.1 The following formula must be used to calculate the effective daily recovery capacity under this alternative:

$$R = D \times U$$

where:

R—Effective daily recovery capacity;

D—Average Oil Recovery Rate in barrels per hour (Item 26 in F 808-83; Item 13.2.16 in F 631-99; or actual performance data); and

U—Hours per day that equipment can operate under discharge conditions. Ten hours per day must be used unless a facility owner or operator can demonstrate that the recovery operation can be sustained for longer periods.

6.3.2 A facility owner or operator submitting a response plan shall provide data that supports the effective daily recovery capacities for the oil recovery devices listed. The following is an example of these calculations:

(1) A weir skimmer identified in a response plan has a manufacturer's rated throughput at the pump of 267 gallons per minute (gpm).

267 gpm=381 barrels per hour (bph)

R=381 bph×24 hr/day×0.2=1,829 barrels per day

(2) After testing using ASTM procedures, the skimmer's oil recovery rate is determined to be 220 gpm. The facility owner or operator identifies sufficient resources available to support operations for 12 hours per day.

220 gpm=314 bph
R=314 bph×12 hr/day=3,768 barrels per day

(3) The facility owner or operator will be able to use the higher capacity if sufficient temporary oil storage capacity is available. Determination of alternative efficiency factors under section 6.2 of this appendix or the acceptability of an alternative effective daily recovery capacity under section 6.3 of this appendix will be made by the Regional Administrator as deemed appropriate.

7.0 Calculating Planning Volumes for a Worst Case Discharge—Petroleum Oils and Non-Petroleum Oils Other Than Animal Fats and Vegetable Oils

7.1 A facility owner or operator shall plan for a response to the facility's worst case discharge. The planning for on-water oil recovery must take into account a loss of some oil to the environment due to evaporative and natural dissipation, potential increases in volume due to emulsification, and the potential for deposition of oil on the shoreline. The procedures for non-petroleum oils other than animal fats and vegetable oils are discussed in section 7.7 of this appendix.

7.2 The following procedures must be used by a facility owner or operator in determining the required on-water oil recovery capacity:

7.2.1 The following must be determined: the worst case discharge volume of oil in the facility; the appropriate group(s) for the types of oil handled, stored, or transported at the facility [persistent (Groups 2, 3, 4, 5) or non-persistent (Group 1)]; and the facility's specific operating area. See sections 1.2.3 and 1.2.8 of this appendix for the definitions of non-persistent and persistent oils, respectively. Facilities that handle, store, or transport oil from different oil groups must calculate each group separately, unless the oil group constitutes 10 percent or less by volume of the facility's total oil storage capacity. This information is to be used with Table 2 of this appendix to determine the percentages of the total volume to be used for removal capacity planning. Table 2 of this appendix divides the volume into three categories: oil lost to the environment; oil deposited on the shoreline; and oil available for on-water recovery.

7.2.2 The on-water oil recovery volume shall, as appropriate, be adjusted using the appropriate emulsification factor found in Table 3 of this appendix. Facilities that handle, store, or transport oil from different petroleum groups must compare the on-water recovery volume for each oil group (unless

the oil group constitutes 10 percent or less by volume of the facility's total storage capacity) and use the calculation that results in the largest on-water oil recovery volume to plan for the amount of response resources for a worst case discharge.

7.2.3 The adjusted volume is multiplied by the on-water oil recovery resource mobilization factor found in Table 4 of this appendix from the appropriate operating area and response tier to determine the total on-water oil recovery capacity in barrels per day that must be identified or contracted to arrive on-scene within the applicable time for each response tier. Three tiers are specified. For higher volume port areas, the contracted tiers of resources must be located such that they are capable of arriving on-scene within 6 hours for Tier 1, 30 hours for Tier 2, and 54 hours for Tier 3 of the discovery of an oil discharge. For all other rivers and canals, inland, nearshore areas, and the Great Lakes, these tiers are 12, 36, and 60 hours.

7.2.4 The resulting on-water oil recovery capacity in barrels per day for each tier is used to identify response resources necessary to sustain operations in the applicable operating area. The equipment shall be capable of sustaining operations for the time period specified in Table 2 of this appendix. The facility owner or operator shall identify and ensure the availability, by contract or other approved means as described in §112.2, of sufficient oil spill recovery devices to provide the effective daily oil recovery capacity required. If the required capacity exceeds the applicable cap specified in Table 5 of this appendix, then a facility owner or operator shall ensure, by contract or other approved means as described in §112.2, only for the quantity of resources required to meet the cap, but shall identify sources of additional resources as indicated in section 5.4 of this appendix. The owner or operator of a facility whose planning volume exceeded the cap in 1993 must make arrangements to identify and ensure the availability, by contract or other approved means as described in §112.2, for additional capacity to be under contract by 1998 or 2003, as appropriate. For a facility that handles multiple groups of oil, the required effective daily recovery capacity for each oil group is calculated before applying the cap. The oil group calculation resulting in the largest on-water recovery volume must be used to plan for the amount of response resources for a worst case discharge, unless the oil group comprises 10 percent or less by volume of the facility's total oil storage capacity.

7.3 The procedures discussed in sections 7.3.1–7.3.3 of this appendix must be used to calculate the planning volume for identifying shoreline cleanup capacity (for Group 1 through Group 4 oils).

7.3.1 The following must be determined: the worst case discharge volume of oil for

73

the facility; the appropriate group(s) for the types of oil handled, stored, or transported at the facility [persistent (Groups 2, 3, or 4) or non-persistent (Group 1)]; and the geographic area(s) in which the facility operates (*i.e.*, operating areas). For a facility handling, storing, or transporting oil from different groups, each group must be calculated separately. Using this information, Table 2 of this appendix must be used to determine the percentages of the total volume to be used for shoreline cleanup resource planning.

7.3.2 The shoreline cleanup planning volume must be adjusted to reflect an emulsification factor using the same procedure as described in section 7.2.2 of this appendix.

7.3.3 The resulting volume shall be used to identify an oil spill removal organization with the appropriate shoreline cleanup capability.

7.4 A response plan must identify response resources with fire fighting capability. The owner or operator of a facility that handles, stores, or transports Group 1 through Group 4 oils that does not have adequate fire fighting resources located at the facility or that cannot rely on sufficient local fire fighting resources must identify adequate fire fighting resources. The facility owner or operator shall ensure, by contract or other approved means as described in §112.2, the availability of these resources. The response plan must also identify an individual located at the facility to work with the fire department for Group 1 through Group 4 oil fires. This individual shall also verify that sufficient well-trained fire fighting resources are available within a reasonable response time to a worst case scenario. The individual may be the qualified individual identified in the response plan or another appropriate individual located at the facility.

7.5 The following is an example of the procedure described above in sections 7.2 and 7.3 of this appendix: A facility with a 270,000 barrel (11.3 million gallons) capacity for #6 oil (specific gravity 0.96) is located in a higher volume port area. The facility is on a peninsula and has docks on both the ocean and bay sides. The facility has four aboveground oil storage tanks with a combined total capacity of 80,000 barrels (3.36 million gallons) and no secondary containment. The remaining facility tanks are inside secondary containment structures. The largest aboveground oil storage tank (90,000 barrels or 3.78 million gallons) has its own secondary containment. Two 50,000 barrel (2.1 million gallon) tanks (that are not connected by a manifold) are within a common secondary containment tank area, which is capable of holding 100,000 barrels (4.2 million gallons) plus sufficient freeboard.

7.5.1 The worst case discharge for the facility is calculated by adding the capacity of all aboveground oil storage tanks without secondary containment (80,000 barrels) plus the capacity of the largest aboveground oil storage tank inside secondary containment. The resulting worst case discharge volume is 170,000 barrels or 7.14 million gallons.

7.5.2 Because the requirements for Tiers 1, 2, and 3 for inland and nearshore exceed the caps identified in Table 5 of this appendix, the facility owner will contract for a response to 10,000 barrels per day (bpd) for Tier 1, 20,000 bpd for Tier 2, and 40,000 bpd for Tier 3. Resources for the remaining 7,850 bpd for Tier 1, 9,750 bpd for Tier 2, and 7,600 bpd for Tier 3 shall be identified but need not be contracted for in advance. The facility owner or operator shall, as appropriate, also identify or contract for quantities of boom identified in their response plan for the protection of fish and wildlife and sensitive environments within the area potentially impacted by a worst case discharge from the facility. For further description of fish and wildlife and sensitive environments, see Appendices I, II, and III to DOC/NOAA's "Guidance for Facility and Vessel Response Plans: Fish and Wildlife and Sensitive Environments," (see appendix E to this part, section 13, for availability) and the applicable ACP. Attachment C-III to Appendix C provides a method for calculating a planning distance to fish and wildlife and sensitive environments and public drinking water intakes that may be impacted in the event of a worst case discharge.

7.6 The procedures discussed in sections 7.6.1–7.6.3 of this appendix must be used to determine appropriate response resources for facilities with Group 5 oils.

7.6.1 The owner or operator of a facility that handles, stores, or transports Group 5 oils shall, as appropriate, identify the response resources available by contract or other approved means, as described in §112.2. The equipment identified in a response plan shall, as appropriate, include:

(1) Sonar, sampling equipment, or other methods for locating the oil on the bottom or suspended in the water column;

(2) Containment boom, sorbent boom, silt curtains, or other methods for containing the oil that may remain floating on the surface or to reduce spreading on the bottom;

(3) Dredges, pumps, or other equipment necessary to recover oil from the bottom and shoreline;

(4) Equipment necessary to assess the impact of such discharges; and

(5) Other appropriate equipment necessary to respond to a discharge involving the type of oil handled, stored,, or transported.

7.6.2 Response resources identified in a response plan for a facility that handles, stores, or transports Group 5 oils under section 7.6.1 of this appendix shall be capable of being deployed (on site) within 24 hours of discovery of a discharge to the area where the facility is operating.

7.6.3 A response plan must identify response resources with fire fighting capability. The owner or operator of a facility that handles, stores, or transports Group 5 oils that does not have adequate fire fighting resources located at the facility or that cannot rely on sufficient local fire fighting resources must identify adequate fire fighting resources. The facility owner or operator shall ensure, by contract or other approved means as described in §112.2, the availability of these resources. The response plan shall also identify an individual located at the facility to work with the fire department for Group 5 oil fires. This individual shall also verify that sufficient well-trained fire fighting resources are available within a reasonable response time to respond to a worst case discharge. The individual may be the qualified individual identified in the response plan or another appropriate individual located at the facility.

7.7 *Non-petroleum oils other than animal fats and vegetable oils.* The procedures described in sections 7.7.1 through 7.7.5 of this appendix must be used to determine appropriate response plan development and evaluation criteria for facilities that handle, store, or transport non-petroleum oils other than animal fats and vegetable oils. Refer to section 11 of this appendix for information on the limitations on the use of chemical agents for inland and nearshore areas.

7.7.1 An owner or operator of a facility that handles, stores, or transports non-petroleum oils other than animal fats and vegetable oils must provide information in his or her plan that identifies:

(1) Procedures and strategies for responding to a worst case discharge to the maximum extent practicable; and

(2) Sources of the equipment and supplies necessary to locate, recover, and mitigate such a discharge.

7.7.2 An owner or operator of a facility that handles, stores, or transports non-petroleum oils other than animal fats and vegetable oils must ensure that any equipment identified in a response plan is capable of operating in the conditions expected in the geographic area(s) (*i.e.,* operating environments) in which the facility operates using the criteria in Table 1 of this appendix. When evaluating the operability of equipment, the facility owner or operator must consider limitations that are identified in the appropriate ACPs, including:

(1) Ice conditions;

(2) Debris;

(3) Temperature ranges; and

(4) Weather-related visibility.

7.7.3 The owner or operator of a facility that handles, stores, or transports non-petroleum oils other than animal fats and vegetable oils must identify the response resources that are available by contract or other approved means, as described in §112.2.

The equipment described in the response plan shall, as appropriate, include:

(1) Containment boom, sorbent boom, or other methods for containing oil floating on the surface or to protect shorelines from impact;

(2) Oil recovery devices appropriate for the type of non-petroleum oil carried; and

(3) Other appropriate equipment necessary to respond to a discharge involving the type of oil carried.

7.7.4 Response resources identified in a response plan according to section 7.7.3 of this appendix must be capable of commencing an effective on-scene response within the applicable tier response times in section 5.3 of this appendix.

7.7.5 A response plan must identify response resources with fire fighting capability. The owner or operator of a facility that handles, stores, or transports non-petroleum oils other than animal fats and vegetable oils that does not have adequate fire fighting resources located at the facility or that cannot rely on sufficient local fire fighting resources must identify adequate fire fighting resources. The owner or operator shall ensure, by contract or other approved means as described in §112.2, the availability of these resources. The response plan must also identify an individual located at the facility to work with the fire department for fires of these oils. This individual shall also verify that sufficient well-trained fire fighting resources are available within a reasonable response time to a worst case scenario. The individual may be the qualified individual identified in the response plan or another appropriate individual located at the facility.

8.0 Determining Response Resources Required for Small Discharges—Animal Fats and Vegetable Oils

8.1 A facility owner or operator shall identify sufficient response resources available, by contract or other approved means as described in §112.2, to respond to a small discharge of animal fats or vegetable oils. A small discharge is defined as any discharge volume less than or equal to 2,100 gallons, but not to exceed the calculated worst case discharge. The equipment must be designed to function in the operating environment at the point of expected use.

8.2 Complexes that are regulated by EPA and the USCG must also consider planning quantities for the marine transportation-related portion of the facility.

8.2.1 The USCG planning level that corresponds to EPA's "small discharge" is termed "the average most probable discharge." A USCG rule found at 33 CFR 154.1020 defines "the average most probable discharge" as the lesser of 50 barrels (2,100 gallons) or 1 percent of the volume of the worst case discharge. Owners or operators of complexes that handle, store, or transport animal fats and vegetable oils must compare oil discharge volumes for a small discharge and an average most probable discharge, and plan for whichever quantity is greater.

8.3 The response resources shall, as appropriate, include:

8.3.1 One thousand feet of containment boom (or, for complexes with marine transfer components, 1,000 feet of containment boom or two times the length of the largest vessel that regularly conducts oil transfers to or from the facility, whichever is greater), and a means of deploying it within 1 hour of the discovery of a discharge;

8.3.2 Oil recovery devices with an effective daily recovery capacity equal to the amount of oil discharged in a small discharge or greater which is available at the facility within 2 hours of the detection of a discharge; and

8.3.3 Oil storage capacity for recovered oily material indicated in section 12.2 of this appendix.

9.0 Determining Response Resources Required for Medium Discharges—Animal Fats and Vegetable Oils

9.1 A facility owner or operator shall identify sufficient response resources available, by contract or other approved means as described in §112.2, to respond to a medium discharge of animal fats or vegetable oils for that facility. This will require response resources capable of containing and collecting up to 36,000 gallons of oil or 10 percent of the worst case discharge, whichever is less. All equipment identified must be designed to operate in the applicable operating environment specified in Table 1 of this appendix.

9.2 Complexes that are regulated by EPA and the USCG must also consider planning quantities for the transportation-related transfer portion of the facility. Owners or operators of complexes that handle, store, or transport animal fats or vegetable oils must plan for oil discharge volumes for a medium discharge. For non-petroleum oils, there is no USCG planning level that directly corresponds to EPA's "medium discharge." Although the USCG does not have planning requirements for medium discharges, they do have requirements (at 33 CFR 154.545) to identify equipment to contain oil resulting from an operational discharge.

9.3 Oil recovery devices identified to meet the applicable medium discharge volume planning criteria must be located such that they are capable of arriving on-scene within 6 hours in higher volume port areas and the Great Lakes and within 12 hours in all other areas. Higher volume port areas and Great Lakes areas are defined in section 1.1 of appendix C to this part.

9.4 Because rapid control, containment, and removal of oil are critical to reduce discharge impact, the owner or operator must determine response resources using an effective daily recovery capacity for oil recovery devices equal to 50 percent of the planning volume applicable for the facility as determined in section 9.1 of this appendix. The effective daily recovery capacity for oil recovery devices identified in the plan must be determined using the criteria in section 6 of this appendix.

9.5 In addition to oil recovery capacity, the plan shall, as appropriate, identify sufficient quantity of containment boom available, by contract or other approved means as described in §112.2, to arrive within the required response times for oil collection and containment and for protection of fish and wildlife and sensitive environments. For further description of fish and wildlife and sensitive environments, see Appendices I, II, and III to DOC/NOAA's "Guidance for Facility and Vessel Response Plans: Fish and Wildlife and Sensitive Environments" (59 FR 14713–22, March 29, 1994) and the applicable ACP. Although 40 CFR part 112 does not set required quantities of boom for oil collection and containment, the response plan shall identify and ensure, by contract or other approved means as described in §112.2, the availability of the quantity of boom identified in the plan for this purpose.

9.6 The plan must indicate the availability of temporary storage capacity to meet section 12.2 of this appendix. If available storage capacity is insufficient to meet this level, then the effective daily recovery capacity must be derated (downgraded) to the limits of the available storage capacity.

9.7 The following is an example of a medium discharge volume planning calculation for equipment identification in a higher volume port area:

The facility's largest aboveground storage tank volume is 840,000 gallons. Ten percent of this capacity is 84,000 gallons. Because 10 percent of the facility's largest tank, or 84,000 gallons, is greater than 36,000 gallons, 36,000 gallons is used as the planning volume. The effective daily recovery capacity is 50 percent of the planning volume, or 18,000 gallons per day. The ability of oil recovery devices to meet this capacity must be calculated using the procedures in section 6 of this appendix. Temporary storage capacity available on-scene must equal twice the daily recovery capacity as indicated in section 12.2 of this appendix, or 36,000 gallons per day. This is the information the facility owner or operator must use to identify and ensure the availability of the required response resources, by contract or other approved means as described in §112.2. The facility owner shall also identify how much boom is available for use.

10.0 Calculating Planning Volumes for a Worst Case Discharge—Animal Fats and Vegetable Oils.

10.1 A facility owner or operator shall plan for a response to the facility's worst case discharge. The planning for on-water oil recovery must take into account a loss of some oil to the environment due to physical, chemical, and biological processes, potential increases in volume due to emulsification, and the potential for deposition of oil on the shoreline or on sediments. The response planning procedures for animal fats and vegetable oils are discussed in section 10.7 of this appendix. You may use alternate response planning procedures for animal fats and vegetable oils if those procedures result in environmental protection equivalent to that provided by the procedures in section 10.7 of this appendix.

10.2 The following procedures must be used by a facility owner or operator in determining the required on-water oil recovery capacity:

10.2.1 The following must be determined: the worst case discharge volume of oil in the facility; the appropriate group(s) for the types of oil handled, stored, or transported at the facility (Groups A, B, C); and the facility's specific operating area. See sections 1.2.1 and 1.2.9 of this appendix for the definitions of animal fats and vegetable oils and groups thereof. Facilities that handle, store, or transport oil from different oil groups must calculate each group separately, unless the oil group constitutes 10 percent or less by volume of the facility's total oil storage capacity. This information is to be used with Table 6 of this appendix to determine the percentages of the total volume to be used for removal capacity planning. Table 6 of this appendix divides the volume into three categories: oil lost to the environment; oil deposited on the shoreline; and oil available for on-water recovery.

10.2.2 The on-water oil recovery volume shall, as appropriate, be adjusted using the appropriate emulsification factor found in Table 7 of this appendix. Facilities that handle, store, or transport oil from different groups must compare the on-water recovery volume for each oil group (unless the oil group constitutes 10 percent or less by volume of the facility's total storage capacity) and use the calculation that results in the largest on-water oil recovery volume to plan for the amount of response resources for a worst case discharge.

10.2.3 The adjusted volume is multiplied by the on-water oil recovery resource mobilization factor found in Table 4 of this appendix from the appropriate operating area and response tier to determine the total on-water oil recovery capacity in barrels per day that must be identified or contracted to arrive on-scene within the applicable time for each response tier. Three tiers are specified. For higher volume port areas, the contracted tiers of resources must be located such that they are capable of arriving on-scene within 6 hours for Tier 1, 30 hours for Tier 2, and 54 hours for Tier 3 of the discovery of a discharge. For all other rivers and canals, inland, nearshore areas, and the Great Lakes, these tiers are 12, 36, and 60 hours.

10.2.4 The resulting on-water oil recovery capacity in barrels per day for each tier is used to identify response resources necessary to sustain operations in the applicable operating area. The equipment shall be capable of sustaining operations for the time period specified in Table 6 of this appendix. The facility owner or operator shall identify and ensure, by contract or other approved means as described in §112.2, the availability of sufficient oil spill recovery devices to provide the effective daily oil recovery capacity required. If the required capacity exceeds the applicable cap specified in Table 5 of this appendix, then a facility owner or operator shall ensure, by contract or other approved means as described in §112.2, only for the quantity of resources required to meet the cap, but shall identify sources of additional resources as indicated in section 5.4 of this appendix. The owner or operator of a facility whose planning volume exceeded the cap in 1998 must make arrangements to identify and ensure, by contract or other approved means as described in §112.2, the availability of additional capacity to be under contract by 2003, as appropriate. For a facility that handles multiple groups of oil, the required effective daily recovery capacity for each oil group is calculated before applying the cap. The oil group calculation resulting in the largest on-water recovery volume must be used to plan for the amount of response resources for a worst case discharge, unless the oil group comprises 10 percent or less by volume of the facility's oil storage capacity.

10.3 The procedures discussed in sections 10.3.1 through 10.3.3 of this appendix must be used to calculate the planning volume for identifying shoreline cleanup capacity (for Groups A and B oils).

10.3.1 The following must be determined: the worst case discharge volume of oil for the facility; the appropriate group(s) for the types of oil handled, stored, or transported at the facility (Groups A or B); and the geographic area(s) in which the facility operates (i.e., operating areas). For a facility handling, storing, or transporting oil from different groups, each group must be calculated separately. Using this information, Table 6 of this appendix must be used to determine the percentages of the total volume to be used for shoreline cleanup resource planning.

10.3.2 The shoreline cleanup planning volume must be adjusted to reflect an emulsification factor using the same procedure as described in section 10.2.2 of this appendix.

10.3.3 The resulting volume shall be used to identify an oil spill removal organization with the appropriate shoreline cleanup capability.

10.4 A response plan must identify response resources with fire fighting capability appropriate for the risk of fire and explosion at the facility from the discharge or threat of discharge of oil. The owner or operator of a facility that handles, stores, or transports Group A or B oils that does not have adequate fire fighting resources located at the facility or that cannot rely on sufficient local fire fighting resources must identify adequate fire fighting resources. The facility owner or operator shall ensure, by contract or other approved means as described in §112.2, the availability of these resources. The response plan must also identify an individual to work with the fire department for Group A or B oil fires. This individual shall also verify that sufficient well-trained fire fighting resources are available within a reasonable response time to a worst case scenario. The individual may be the qualified individual identified in the response plan or another appropriate individual located at the facility.

10.5 The following is an example of the procedure described in sections 10.2 and 10.3 of this appendix. A facility with a 37.04 million gallon (881,904 barrel) capacity of several types of vegetable oils is located in the Inland Operating Area. The vegetable oil with the highest specific gravity stored at the facility is soybean oil (specific gravity 0.922, Group B vegetable oil). The facility has ten aboveground oil storage tanks with a combined total capacity of 18 million gallons (428,571 barrels) and without secondary containment. The remaining facility tanks are inside secondary containment structures. The largest aboveground oil storage tank (3 million gallons or 71,428 barrels) has its own secondary containment. Two 2.1 million gallon (50,000 barrel) tanks (that are not connected by a manifold) are within a common secondary containment tank area, which is capable of holding 4.2 million gallons (100,000 barrels) plus sufficient freeboard.

10.5.1 The worst case discharge for the facility is calculated by adding the capacity of all aboveground vegetable oil storage tanks without secondary containment (18.0 million gallons) plus the capacity of the largest aboveground storage tank inside secondary containment (3.0 million gallons). The resulting worst case discharge is 21 million gallons or 500,000 barrels.

10.5.2 With a specific worst case discharge identified, the planning volume for on-water recovery can be identified as follows:

Worst case discharge: 21 million gallons (500,000 barrels) of Group B vegetable oil
Operating Area: Inland
Planned percent recovered floating vegetable oil (from Table 6, column Nearshore/Inland/Great Lakes): Inland, Group B is 20%
Emulsion factor (from Table 7): 2.0
Planning volumes for on-water recovery: 21,000,000 gallons × 0.2 × 2.0 = 8,400,000 gallons or 200,000 barrels.
Determine required resources for on-water recovery for each of the three tiers using mobilization factors (from Table 4, column Inland/Nearshore/Great Lakes)

Inland Operating Area	Tier 1	Tier 2	Tier 3
Mobilization factor by which you multiply planning volume	.15	.25	.40
Estimated Daily Recovery Capacity (bbls)	30,000	50,000	80,000

10.5.3 Because the requirements for On-Water Recovery Resources for Tiers 1, 2, and 3 for Inland Operating Area exceed the caps identified in Table 5 of this appendix, the facility owner will contract for a response of 12,500 barrels per day (bpd) for Tier 1, 25,000 bpd for Tier 2, and 50,000 bpd for Tier 3. Resources for the remaining 17,500 bpd for Tier 1, 25,000 bpd for Tier 2, and 30,000 bpd for Tier 3 shall be identified but need not be contracted for in advance.

10.5.4 With the specific worst case discharge identified, the planning volume of onshore recovery can be identified as follows:

Worst case discharge: 21 million gallons (500,000 barrels) of Group B vegetable oil
Operating Area: Inland
Planned percent recovered floating vegetable oil from onshore (from Table 6, column Nearshore/Inland/Great Lakes): Inland, Group B is 65%
Emulsion factor (from Table 7): 2.0
Planning volumes for shoreline recovery: 21,000,000 gallons × 0.65 × 2.0 = 27,300,000 gallons or 650,000 barrels

10.5.5 The facility owner or operator shall, as appropriate, also identify or contract for quantities of boom identified in the response plan for the protection of fish and wildlife and sensitive environments within the area potentially impacted by a worst case discharge from the facility. For further description of fish and wildlife and sensitive environments, see Appendices I, II, and III to DOC/NOAA's "Guidance for Facility and Vessel Response Plans: Fish and Wildlife and Sensitive Environments," (see Appendix E to this part, section 13, for availability) and the applicable ACP. Attachment C–III to Appendix C provides a method for calculating a planning distance to fish and wildlife and sensitive environments and public drinking water intakes that may be adversely affected in the event of a worst case discharge.

10.6 The procedures discussed in sections 10.6.1 through 10.6.3 of this appendix must be used to determine appropriate response resources for facilities with Group C oils.

10.6.1 The owner or operator of a facility that handles, stores, or transports Group C oils shall, as appropriate, identify the response resources available by contract or other approved means, as described in §112.2. The equipment identified in a response plan shall, as appropriate, include:

(1) Sonar, sampling equipment, or other methods for locating the oil on the bottom or suspended in the water column;

(2) Containment boom, sorbent boom, silt curtains, or other methods for containing the oil that may remain floating on the surface or to reduce spreading on the bottom;

(3) Dredges, pumps, or other equipment necessary to recover oil from the bottom and shoreline;

(4) Equipment necessary to assess the impact of such discharges; and

(5) Other appropriate equipment necessary to respond to a discharge involving the type of oil handled, stored, or transported.

10.6.2 Response resources identified in a response plan for a facility that handles, stores, or transports Group C oils under section 10.6.1 of this appendix shall be capable of being deployed on scene within 24 hours of discovery of a discharge.

10.6.3 A response plan must identify response resources with fire fighting capability. The owner or operator of a facility that handles, stores, or transports Group C oils that does not have adequate fire fighting resources located at the facility or that cannot rely on sufficient local fire fighting resources must identify adequate fire fighting resources. The owner or operator shall ensure, by contract or other approved means as described in §112.2, the availability of these resources. The response plan shall also identify an individual located at the facility to work with the fire department for Group C oil fires. This individual shall also verify that sufficient well-trained fire fighting resources are available within a reasonable response time to respond to a worst case discharge. The individual may be the qualified individual identified in the response plan or another appropriate individual located at the facility.

10.7 The procedures described in sections 10.7.1 through 10.7.5 of this appendix must be used to determine appropriate response plan development and evaluation criteria for facilities that handle, store, or transport animal fats and vegetable oils. Refer to section 11 of this appendix for information on the limitations on the use of chemical agents for inland and nearshore areas.

10.7.1 An owner or operator of a facility that handles, stores, or transports animal fats and vegetable oils must provide information in the response plan that identifies:

(1) Procedures and strategies for responding to a worst case discharge of animal fats and vegetable oils to the maximum extent practicable; and

(2) Sources of the equipment and supplies necessary to locate, recover, and mitigate such a discharge.

10.7.2 An owner or operator of a facility that handles, stores, or transports animal fats and vegetable oils must ensure that any equipment identified in a response plan is capable of operating in the geographic area(s) (i.e., operating environments) in which the facility operates using the criteria in Table 1 of this appendix. When evaluating the operability of equipment, the facility owner or operator must consider limitations that are identified in the appropriate ACPs, including:

(1) Ice conditions;

(2) Debris;

(3) Temperature ranges; and

(4) Weather-related visibility.

10.7.3. The owner or operator of a facility that handles, stores, or transports animal fats and vegetable oils must identify the response resources that are available by contract or other approved means, as described in §112.2. The equipment described in the response plan shall, as appropriate, include:

(1) Containment boom, sorbent boom, or other methods for containing oil floating on the surface or to protect shorelines from impact;

(2) Oil recovery devices appropriate for the type of animal fat or vegetable oil carried; and

(3) Other appropriate equipment necessary to respond to a discharge involving the type of oil carried.

10.7.4 Response resources identified in a response plan according to section 10.7.3 of this appendix must be capable of commencing an effective on-scene response within the applicable tier response times in section 5.3 of this appendix.

10.7.5 A response plan must identify response resources with fire fighting capability. The owner or operator of a facility that handles, stores, or transports animal fats and vegetable oils that does not have adequate fire fighting resources located at the facility or that cannot rely on sufficient local fire fighting resources must identify adequate fire fighting resources. The owner or operator shall ensure, by contract or other approved means as described in §112.2, the availability of these resources. The response plan shall also identify an individual located at the facility to work with the fire department for animal fat and vegetable oil fires. This individual shall also verify that sufficient well-trained fire fighting resources are available within a reasonable response time to respond to a worst case discharge.

The individual may be the qualified individual identified in the response plan or another appropriate individual located at the facility.

11.0 Determining the Availability of Alternative Response Methods

11.1 For chemical agents to be identified in a response plan, they must be on the NCP Product Schedule that is maintained by EPA. (Some States have a list of approved dispersants for use within State waters. Not all of these State-approved dispersants are listed on the NCP Product Schedule.)

11.2 Identification of chemical agents in the plan does not imply that their use will be authorized. Actual authorization will be governed by the provisions of the NCP and the applicable ACP.

12.0 Additional Equipment Necessary to Sustain Response Operations

12.1 A facility owner or operator shall identify sufficient response resources available, by contract or other approved means as described in §112.2, to respond to a medium discharge of animal fats or vegetables oils for that facility. This will require response resources capable of containing and collecting up to 36,000 gallons of oil or 10 percent of the worst case discharge, whichever is less. All equipment identified must be designed to operate in the applicable operating environment specified in Table 1 of this appendix.

12.2 A facility owner or operator shall evaluate the availability of adequate temporary storage capacity to sustain the effective daily recovery capacities from equipment identified in the plan. Because of the inefficiencies of oil spill recovery devices, response plans must identify daily storage capacity equivalent to twice the effective daily recovery capacity required on-scene. This temporary storage capacity may be reduced if a facility owner or operator can demonstrate by waste stream analysis that the efficiencies of the oil recovery devices, ability to decant waste, or the availability of alternative temporary storage or disposal locations will reduce the overall volume of oily material storage.

12.3 A facility owner or operator shall ensure that response planning includes the capability to arrange for disposal of recovered oil products. Specific disposal procedures will be addressed in the applicable ACP.

13.0 References and Availability

13.1 All materials listed in this section are part of EPA's rulemaking docket and are located in the Superfund Docket, 1235 Jefferson Davis Highway, Crystal Gateway 1, Arlington, Virginia 22202, Suite 105 (Docket Numbers SPCC–2P, SPCC–3P, and SPCC–9P). The docket is available for inspection between 9 a.m. and 4 p.m., Monday through Friday, excluding Federal holidays.

Appointments to review the docket can be made by calling 703–603–9232. Docket hours are subject to change. As provided in 40 CFR part 2, a reasonable fee may be charged for copying services.

13.2 The docket will mail copies of materials to requestors who are outside the Washington, DC metropolitan area. Materials may be available from other sources, as noted in this section. As provided in 40 CFR part 2, a reasonable fee may be charged for copying services. The RCRA/Superfund Hotline at 800–424–9346 may also provide additional information on where to obtain documents. To contact the RCRA/Superfund Hotline in the Washington, DC metropolitan area, dial 703–412–9810. The Telecommunications Device for the Deaf (TDD) Hotline number is 800–553–7672, or, in the Washington, DC metropolitan area, 703–412–3323.

13.3 Documents

(1) National Preparedness for Response Exercise Program (PREP). The PREP draft guidelines are available from United States Coast Guard Headquarters (G-MEP-4), 2100 Second Street, SW., Washington, DC 20593. (See 58 FR 53990–91, October 19, 1993, Notice of Availability of PREP Guidelines).

(2) "Guidance for Facility and Vessel Response Plans: Fish and Wildlife and Sensitive Environments (published in the FEDERAL REGISTER by DOC/NOAA at 59 FR 14713–22, March 29, 1994.). The guidance is available in the Superfund Docket (see sections 13.1 and 13.2 of this appendix).

(3) ASTM Standards. ASTM F 715, ASTM F 989, ASTM F 631–99, ASTM F 808–83 (1999). The ASTM standards are available from the American Society for Testing and Materials, 100 Barr Harbor Drive, West Conshohocken, PA 19428–2959.

(4) Response Plans for Marine Transportation-Related Facilities, Interim Final Rule. Published by USCG, DOT at 58 FR 7330–76, February 5, 1993.

TABLE 1 TO APPENDIX E—RESPONSE RESOURCE OPERATING CRITERIA

Oil Recovery Devices		
Operating environment	Significant wave height [1]	Sea state
Rivers and Canals	≤ 1 foot	1
Inland	≤ 3 feet	2

80

TABLE 1 TO APPENDIX E—RESPONSE RESOURCE OPERATING CRITERIA—Continued

Oil Recovery Devices

Operating environment	Significant wave height [1]	Sea state
Great Lakes	≤ 4 feet	2–3
Ocean	≤ 6 feet	3–4

Boom

Boom property	Use			
	Rivers and canals	Inland	Great Lakes	Ocean
Significant Wave Height [1]	≤ 1	≤ 3	≤ 4	≤ 6
Sea State	1	2	2–3	3–4
Boom height—inches (draft plus freeboard)	6–18	18–42	18–42	≥42
Reserve Buoyancy to Weight Ratio	2:1	2:1	2:1	3:1 to 4:1
Total Tensile Strength—pounds	4,500	15,000–20,000.	15,000–20,000.	≥20,000
Skirt Fabric Tensile Strength—pounds	200	300	300	500
Skirt Fabric Tear Strength—pounds	100	100	100	125

[1] Oil recovery devices and boom *shall* be at least capable of operating in wave heights up to and including the values listed in Table 1 for each operating environment.

TABLE 2 TO APPENDIX E—REMOVAL CAPACITY PLANNING TABLE FOR PETROLEUM OILS

Spill location	Rivers and canals			Nearshore/Inland/Great Lakes		
Sustainability of on-water oil recovery	3 days			4 days		
Oil group [1]	Percent natural dissipation	Percent recovered floating oil	Percent oil onshore	Percent natural dissipation	Percent recovered floating oil	Percent oil onshore
1—Non-persistent oils	80	10	10	80	20	10
2—Light crudes	40	15	45	50	50	30
3—Medium crudes and fuels	20	15	65	30	50	50
4—Heavy crudes and fuels	5	20	75	10	50	70

[1] The response resource considerations for non-petroleum oils other than animal fats and vegetable oils are outlined in section 7.7 of this appendix.

NOTE: Group 5 oils are defined in section 1.2.8 of this appendix; the response resource considerations are outlined in section 7.6 of this appendix.

TABLE 3 TO APPENDIX E—EMULSIFICATION FACTORS FOR PETROLEUM OIL GROUPS [1]

Non-Persistent Oil:
  Group 1 ... 1.0
Persistent Oil:
  Group 2 ... 1.8
  Group 3 ... 2.0
  Group 4 ... 1.4

Group 5 oils are defined in section 1.2.7 of this appendix; the response resource considerations are outlined in section 7.6 of this appendix.

[1] See sections 1.2.2 and 1.2.7 of this appendix for group designations for non-persistent and persistent oils, respectively.

TABLE 4 TO APPENDIX E—ON-WATER OIL RECOVERY RESOURCE MOBILIZATION FACTORS

Operating area	Tier 1	Tier 2	Tier 3
Rivers and Canals	0.30	0.40	0.60
Inland/Nearshore Great Lakes	0.15	0.25	0.40

Note: These mobilization factors are for total resources mobilized, not incremental response resources.

TABLE 5 TO APPENDIX E—RESPONSE CAPABILITY CAPS BY OPERATING AREA

	Tier 1	Tier 2	Tier 3
February 18, 1993:			
All except Rivers & Canals, Great Lakes	10K bbls/day	20K bbls/day	40K bbls/day.
Great Lakes	5K bbls/day	10K bbls/day	20K bbls/day.
Rivers & Canals	1.5K bbls/day	3.0K bbls/day	6.0K bbls/day.

TABLE 5 TO APPENDIX E—RESPONSE CAPABILITY CAPS BY OPERATING AREA—Continued

	Tier 1	Tier 2	Tier 3
February 18, 1998:			
All except Rivers & Canals, Great Lakes	12.5K bbls/day	25K bbls/day	50K bbls/day.
Great Lakes	6.35K bbls/day	12.3K bbls/day	25K bbls/day.
Rivers & Canals	1.875K bbls/day	3.75K bbls/day	7.5K bbls/day.
February 18, 2003:			
All except Rivers & Canals, Great Lakes	TBD	TBD	TBD.
Great Lakes	TBD	TBD	TBD.
Rivers & Canals	TBD	TBD	TBD.

Note: The caps show cumulative overall effective daily recovery capacity, not incremental increases.
TBD=To Be Determined.

TABLE 6 TO APPENDIX E—REMOVAL CAPACITY PLANNING TABLE FOR ANIMAL FATS AND VEGETABLE OILS

Spill location	Rivers and canals			Nearshore/Inland/Great Lakes		
Sustainability of on-water oil recovery	3 days			4 days		
Oil group [1]	Percent natural loss	Percent recovered floating oil	Percent recovered oil from onshore	Percent natural loss	Percent recovered floating oil	Percent recovered oil from onshore
Group A	40	15	45	50	20	30
Group B	20	15	65	30	20	50

[1] Substances with a specific gravity greater than 1.0 generally sink below the surface of the water. Response resource considerations are outlined in section 10.6 of this appendix. The owner or operator of the facility is responsible for determining appropriate response resources for Group C oils including locating oil on the bottom or suspended in the water column; containment boom or other appropriate methods for containing oil that may remain floating on the surface; and dredges, pumps, or other equipment to recover animal fats or vegetable oils from the bottom and shoreline.

NOTE: Group C oils are defined in sections 1.2.1 and 1.2.9 of this appendix; the response resource procedures are discussed in section 10.6 of this appendix.

TABLE 7 TO APPENDIX E—EMULSIFICATION FACTORS FOR ANIMAL FATS AND VEGETABLE OILS

Oil Group[1]:	
Group A	1.0
Group B	2.0

[1] Substances with a specific gravity greater than 1.0 generally sink below the surface of the water. Response resource considerations are outlined in section 10.6 of this appendix. The owner or operator of the facility is responsible for determining appropriate response resources for Group C oils including locating oil on the bottom or suspended in the water column; containment boom or other appropriate methods for containing oil that may remain floating on the surface; and dredges, pumps, or other equipment to recover animal fats or vegetable oils from the bottom and shoreline.

NOTE: Group C oils are defined in sections 1.2.1 and 1.2.9 of this appendix; the response resource procedures are discussed in section 10.6 of this appendix.

ATTACHMENTS TO APPENDIX E

Attachment E-1 --
Worksheet to Plan Volume of Response Resources
for Worst Case Discharge - Petroleum Oils

Part I Background Information

Step (A) Calculate Worst Case Discharge in barrels (Appendix D)

```
                                                              (A)
```

Step (B) Oil Group[1] (Table 3 and section 1.2 of this appendix) .

```
                                                              (B)
```

Step (C) Operating Area (choose one) . . . .   ☐ Near   ☐ or Rivers
                                                   shore/Inla   and
                                                   nd Great   Canals
                                                   Lakes

Step (D) Percentages of Oil (Table 2 of this appendix)

Percent Lost to Natural Dissipation	Percent Recovered Floating Oil	Percent Oil Onshore
(D1)	(D2)	(D3)

Step (E1) On-Water Oil Recovery $\dfrac{\text{Step (D2) x Step(A)}}{100}$

```
                                                              (E1)
```

Step (E2) Shoreline Recovery $\dfrac{\text{Step (D3) x Step (A)}}{100}$ . . . .

```
                                                              (E2)
```

Step (F) Emulsification Factor
  (Table 3 of this appendix) . . . .

```
                                                              (F)
```

Step (G) On-Water Oil Recovery Resource Mobilization Factor
  (Table 4 of this appendix)

Tier 1	Tier 2	Tier 3
(G1)	(G2)	(G3)

[1] A facility that handles, stores, or transports multiple groups of oil must do separate calculations for each oil group on site except for those oil groups that constitute 10 percent or less by volume of the total oil storage capacity at the facility. For purposes of this calculation, the volumes of all products in an oil group must be summed to determine the percentage of the facility's total oil storage capacity.

83

Attachment E-1 (continued) --
Worksheet to Plan Volume of Response Resources
for Worst Case Discharge - Petroleum Oils

Part II <u>On-Water Oil Recovery Capacity</u> (barrels/day)

Tier 1	Tier 2	Tier 3

Step (E1) x Step (F) x   Step (E1) x Step (F) x   Step (E1) x Step (F) x
        Step (G1)   Step (G2)   Step (G3)

Part III <u>Shoreline Cleanup Volume</u> (barrels) . . . .

Step (E2) x Step (F)

Part IV <u>On-Water Response Capacity By Operating Area</u>
(Table 5 of this appendix)
(Amount needed to be contracted for in barrels/day)

Tier 1	Tier 2	Tier 3

(J1)   (J2)   (J3)

Part V <u>On-Water Amount Needed to be Identified, but not Contracted for in</u>
<u>Advance</u> (barrels/day)

Tier 1	Tier 2	Tier 3

Part II Tier 1 - Step (J1)   Part II Tier 2 - Step (J2)   Part II Tier 3 - Step (J3)

NOTE: To convert from barrels/day to gallons/day, multiply the quantities in
Parts II through V by 42 gallons/barrel.

Attachment E-1 Example --
Worksheet to Plan Volume of Response Resources
for Worst Case Discharge - Petroleum Oils

Part I Background Information

Step (A) Calculate Worst Case Discharge in barrels (Appendix D)

170,000
(A)

Step (B) Oil Group[1] (Table 3 and section 1.2 of this appendix) .

4

Step (C) Operating Area (choose one) . .

X	Near shore/Inla nd Great Lakes		or Rivers and Canals

Step (D) Percentages of Oil (Table 2 of this appendix)

Percent Lost to Natural Dissipation	Percent Recovered Floating Oil	Percent Oil Onshore
10	50	70
(D1)	(D2)	(D3)

Step (E1) On-Water Oil Recovery $\dfrac{\text{Step (D2) x Step (A)}}{100}$

85,000
(E1)

Step (E2) Shoreline Recovery $\dfrac{\text{Step (D3) x Step (A)}}{100}$

119,000
(E2)

Step (F) Emulsification Factor
(Table 3 of this appendix) . . . .

1.4
(F)

Step (G) On-Water Oil Recovery Resource Mobilization Factor
(Table 4 of this appendix)

Tier 1	Tier 2	Tier 3
0.15	0.25	0.40
(G1)	(G2)	(G3)

[1] A facility that handles, stores, or transports multiple groups of oil must do separate calculations for each oil group on site except for those oil groups that constitute 10 percent or less by volume of the total oil storage capacity at the facility. For purposes of this calculation, the volumes of all products in an oil group must be summed to determine the percentage of the facility's total oil storage capacity.

Attachment E-1 Example (continued) --
Worksheet to Plan Volume of Response Resources
for Worst Case Discharge - Petroleum Oils

Part II <u>On-Water Oil Recovery Capacity</u> (barrels/day)

Tier 1	Tier 2	Tier 3
17,850	29,750	47,600
Step (E1) x Step (F) x Step (G1)	Step (E1) x Step (F) x Step (G2)	Step (E1) x Step (F) x Step (G3)

Part III <u>Shoreline Cleanup Volume</u> (barrels) . . . . . .

166,600
Step (E2) x Step (F)

Part IV <u>On-Water Response Capacity By Operating Area</u>
(Table 5 of this appendix)
(Amount needed to be contracted for in barrels/day)

Tier 1	Tier 2	Tier 3
10,000	20,000	40,000
(J1)	(J2)	(J3)

Part V <u>On-Water Amount Needed to be Identified, but not Contracted for in Advance</u> (barrels/day)

Tier 1	Tier 2	Tier 3
7,850	9,750	7,600
Part II Tier 1 - Step (J1)	Part II Tier 2 - Step (J2)	Part II Tier 3 - Step (J3)

NOTE: To convert from barrels/day to gallons/day, multiply the quantities in Parts II through V by 42 gallons/barrel.

Attachment E-2 --
Worksheet to Plan Volume of Response Resources
for Worst Case Discharge - Animal Fats and Vegetable Oils

Part I Background Information

Step (A) Calculate Worst Case Discharge in barrels (Appendix D)

```
                                                        ┌──────────────┐
                                                        │              │
                                                        └──────────────┘
                                                              (A)
```

Step (B) Oil Group[1] (Table 7 and section 1.2 of this appendix) .

```
                                                        ┌──────────────┐
                                                        │              │
                                                        └──────────────┘
```

Step (C) Operating Area (choose one) . . . .

| Near shore/Inland Great Lakes | or Rivers and Canals |

Step (D) Percentages of Oil (Table 6 of this appendix)

Percent Lost to Natural Dissipation	Percent Recovered Floating Oil	Percent Oil Onshore
(D1)	(D2)	(D3)

Step (E1) On-Water Oil Recovery $\dfrac{\text{Step (D2) x Step (A)}}{100}$

```
                                                        ┌──────────────┐
                                                        │              │
                                                        └──────────────┘
                                                              (E1)
```

Step (E2) Shoreline Recovery $\dfrac{\text{Step (D3) x Step (A)}}{100}$ . . .

```
                                                        ┌──────────────┐
                                                        │              │
                                                        └──────────────┘
                                                              (E2)
```

Step (F) Emulsification Factor
(Table 7 of this appendix) . . . .

```
                                                        ┌──────────────┐
                                                        │              │
                                                        └──────────────┘
                                                              (F)
```

Step (G) On-Water Oil Recovery Resource Mobilization Factor
(Table 4 of this appendix)

Tier 1	Tier 2	Tier 3
(G1)	(G2)	(G3)

[1] A facility that handles, stores, or transports multiple groups of oil must do separate calculations for each oil group on site except for those oil groups that constitute 10 percent or less by volume of the total oil storage capacity at the facility. For purposes of this calculation, the volumes of all products in an oil group must be summed to determine the percentage of the facility's total oil storage capacity.

87

Attachment E-2 (continued) --
Worksheet to Plan Volume of Response Resources
for Worst Case Discharge - Animal Fats and Vegetable Oils

Part II <u>On-Water Oil Recovery Capacity</u> (barrels/day)

Tier 1	Tier 2	Tier 3
Step (E1) x Step (F) x Step (G1)	Step (E1) x Step (F) x Step (G2)	Step (E1) x Step (F) x Step (G3)

Part III <u>Shoreline Cleanup Volume</u> (barrels) . . . .

Step (E2) x Step (F)

Part IV <u>On-Water Response Capacity By Operating Area</u>
(Table 5 of this appendix)
(Amount needed to be contracted for in barrels/day)

Tier 1	Tier 2	Tier 3
(J1)	(J2)	(J3)

Part V <u>On-Water Amount Needed to be Identified, but not Contracted for
in Advance</u> (barrels/day)

Tier 1	Tier 2	Tier 3
Part II Tier 1 - Step (J1)	Part II Tier 2 - Step (J2)	Part II Tier 3 - Step (J3)

NOTE: To convert from barrels/day to gallons/day, multiply the
quantities in Parts II through V by 42 gallons/barrel.

Attachment E-2 Example --
Worksheet to Plan Volume of Response Resources
for Worst Case Discharge - Animal Fats and Vegetable Oils

Part I <u>Background Information</u>

Step (A) Calculate Worst Case Discharge in barrels
(Appendix D) . . . .

500,000
(A)

Step (B) Oil Group[1] (Table 7 and section 1.2 of this
appendix) . . . .

B

Step (C) Operating Area (choose
one)

X	Near shore/Inl and Great Lakes		or Rivers and Canals

Step (D) Percentages of Oil (Table 6 of this appendix)

Percent Lost to Natural Dissipation	Percent Recovered Floating Oil	Percent Oil Onshore
30	20	50
(D1)	(D2)	(D3)

Step (E1) On-Water Oil Recovery $\dfrac{\text{Step (D2) x Step (A)}}{100}$

100,000
(E1)

Step (E2) Shoreline Recovery $\dfrac{\text{Step (D3) x Step (A)}}{100}$

250,000
(E2)

Step (F) Emulsification Factor
(Table 7 of this appendix) . . . .

2.0
(F)

Step (G) On-Water Oil Recovery Resource Mobilization Factor
(Table 4 of this appendix)

Tier 1	Tier 2	Tier 3
0.15	0.25	0.40
(G1)	(G2)	(G3)

[1] A facility that handles, stores, or transports multiple groups of oil must do separate calculations for each oil group on site except for those oil groups that constitute 10 percent or less by volume of the total oil storage capacity at the facility. For purposes of this calculation, the volumes of all products in an oil group must be summed to determine the percentage of the facility's total oil storage capacity.

Attachment E-2 Example (continued) --
Worksheet to Plan Volume of Response Resources
for Worst Case Discharge - Animal Fats and Vegetable Oils (continued)

Part II <u>On-Water Oil Recovery Capacity</u> (barrels/day)

Tier 1	Tier 2	Tier 3
30,000	50,000	80,000
Step (E1) x Step (F) x Step (G1)	Step (E1) x Step (F) x Step (G2)	Step (E1) x Step (F) x Step (G3)

Part III <u>Shoreline Cleanup Volume</u> (barrels) . . . .

500,000
Step (E2) x Step (F)

Part IV <u>On-Water Response Capacity By Operating Area</u>
(Table 5 of this appendix)
(Amount needed to be contracted for in barrels/day)

Tier 1	Tier 2	Tier 3
12,500	25,000	50,000
(J1)	(J2)	(J3)

Part V <u>On-Water Amount Needed to be Identified, but not Contracted for in Advance</u> (barrels/day)

Tier 1	Tier 2	Tier 3
17,500	25,000	30,000
Part II Tier 1 - Step (J1)	Part II Tier 2 - Step (J2)	Part II Tier 3 - Step (J3)

NOTE: To convert from barrels/day to gallons/day, multiply the quantities in Parts II through V by 42 gallons/barrel.

[59 FR 34111, July 1, 1994; 59 FR 49006, Sept. 26, 1994, as amended at 65 FR 40806, 40807, June 30, 2000; 65 FR 47325, Aug. 2, 2000; 66 FR 34560, June 29, 2001]

APPENDIX F TO PART 112—FACILITY-
SPECIFIC RESPONSE PLAN

Table of Contents

1.0 Model Facility-Specific Response Plan

(A) Owners or operators of facilities regulated under this part which pose a threat of substantial harm to the environment by discharging oil into or on navigable waters or adjoining shorelines are required to prepare and submit facility-specific response plans to EPA in accordance with the provisions in this appendix. This appendix further describes the required elements in §112.20(h).

(B) Response plans must be sent to the appropriate EPA Regional office. Figure F–1 of this Appendix lists each EPA Regional office and the address where owners or operators must submit their response plans. Those facilities deemed by the Regional Administrator (RA) to pose a threat of significant and substantial harm to the environment will have their plans reviewed and approved by EPA. In certain cases, information required in the model response plan is similar to information currently maintained in the facility's Spill Prevention, Control, and Countermeasures (SPCC) Plan as required by 40 CFR 112.3. In these cases, owners or operators may reproduce the information and include a photocopy in the response plan.

(C) A complex may develop a single response plan with a set of core elements for all regulating agencies and separate sections for the non-transportation-related and transportation-related components, as described in §112.20(h). Owners or operators of large facilities that handle, store, or transport oil at more than one geographically distinct location (e.g., oil storage areas at opposite ends of a single, continuous parcel of property) shall, as appropriate, develop separate sections of the response plan for each storage area.

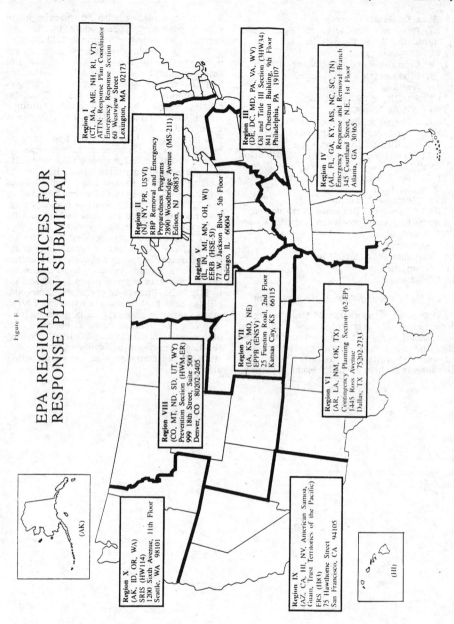

Figure F-1

EPA REGIONAL OFFICES FOR RESPONSE PLAN SUBMITTAL

Region I
(CT, MA, ME, NH, RI, VT)
ATTN: Response Plan Coordinator
Emergency Response Section
60 Westview Street
Lexington, MA 02173

Region II
(NJ, NY, PR, USVI)
RBP Removal and Emergency
Preparedness Programs
2890 Woodbridge Avenue (MS-211)
Edison, NJ 08837

Region III
(DE, DC, MD, PA, VA, WV)
Oil and Title III Section (3HW34)
841 Chestnut Building, 9th Floor
Philadelphia, PA 19107

Region IV
(AL, FL, GA, KY, MS, NC, SC, TN)
Emergency Response and Removal Branch
345 Courtland Street, N.E., 1st Floor
Atlanta, GA 30365

Region V
(IL, IN, MI, MN, OH, WI)
EERB (HSE-5J)
77 W. Jackson Blvd., 5th Floor
Chicago, IL 60604

Region VI
(AR, LA, NM, OK, TX)
Contingency Planning Section (6Z-EP)
1445 Ross Avenue
Dallas, TX 75202-2733

Region VII
(IA, KS, MO, NE)
EPPB (ENSV)
25 Funston Road, 2nd Floor
Kansas City, KS 66115

Region VIII
(CO, MT, ND, SD, UT, WY)
Prevention Section (HWM-ER)
999 18th Street, Suite 500
Denver, CO 80202-2405

Region IX
(AZ, CA, HI, NV, American Samoa,
Guam, Trust Territories of the Pacific)
ERS (HR3)
75 Hawthorne Street
San Francisco, CA 94105

Region X
(AK, ID, OR, WA)
SRIS (HW114)
1200 Sixth Avenue, 11th Floor
Seattle, WA 98101

(AK)

(HI)

1.1 Emergency Response Action Plan

Several sections of the response plan shall be co-located for easy access by response personnel during an actual emergency or oil discharge. This collection of sections shall be called the Emergency Response Action Plan. The Agency intends that the Action Plan contain only as much information as is necessary to combat the discharge and be arranged so response actions are not delayed. The Action Plan may be arranged in a number of ways. For example, the sections of the Emergency Response Action Plan may be photocopies or condensed versions of the forms included in the associated sections of the response plan. Each Emergency Response Action Plan section may be tabbed for quick reference. The Action Plan shall be maintained in the front of the same binder that contains the complete response plan or it shall be contained in a separate binder. In the latter case, both binders shall be kept together so that the entire plan can be accessed by the qualified individual and appropriate spill response personnel. The Emergency Response Action Plan shall be made up of the following sections:

1. Qualified Individual Information (Section 1.2) partial
2. Emergency Notification Phone List (Section 1.3.1) partial
3. Spill Response Notification Form (Section 1.3.1) partial
4. Response Equipment List and Location (Section 1.3.2) complete
5. Response Equipment Testing and Deployment (Section 1.3.3) complete
6. Facility Response Team (Section 1.3.4) partial
7. Evacuation Plan (Section 1.3.5) condensed
8. Immediate Actions (Section 1.7.1) complete
9. Facility Diagram (Section 1.9) complete

1.2 Facility Information

The facility information form is designed to provide an overview of the site and a description of past activities at the facility. Much of the information required by this section may be obtained from the facility's existing SPCC Plan.

1.2.1 *Facility name and location:* Enter facility name and street address. Enter the address of corporate headquarters only if corporate headquarters are physically located at the facility. Include city, county, state, zip code, and phone number.

1.2.2 *Latitude and Longitude:* Enter the latitude and longitude of the facility. Include degrees, minutes, and seconds of the main entrance of the facility.

1.2.3 *Wellhead Protection Area:* Indicate if the facility is located in or drains into a wellhead protection area as defined by the Safe Drinking Water Act of 1986 (SDWA).[1] The response plan requirements in the Wellhead Protection Program are outlined by the

---

[1] A wellhead protection area is defined as the surface and subsurface area surrounding a water well or wellfield, supplying a public water system, through which contaminants are reasonably likely to move toward and reach such water well or wellfield. For further information regarding State and territory protection programs, facility owners or operators may contact the SDWA Hotline at 1–800–426–4791.

State or Territory in which the facility resides.

1.2.4 *Owner/operator:* Write the name of the company or person operating the facility and the name of the person or company that owns the facility, if the two are different. List the address of the owner, if the two are different.

1.2.5 *Qualified Individual:* Write the name of the qualified individual for the entire facility. If more than one person is listed, each individual indicated in this section shall have full authority to implement the facility response plan. For each individual, list: name, position, home and work addresses (street addresses, not P.O. boxes), emergency phone number, and specific response training experience.

1.2.6 *Date of Oil Storage Start-up:* Enter the year which the present facility first started storing oil.

1.2.7 *Current Operation:* Briefly describe the facility's operations and include the North American Industrial Classification System (NAICS) code.

1.2.8 *Dates and Type of Substantial Expansion:* Include information on expansions that have occurred at the facility. Examples of such expansions include, but are not limited to: Throughput expansion, addition of a product line, change of a product line, and installation of additional oil storage capacity. The data provided shall include all facility historical information and detail the expansion of the facility. An example of substantial expansion is any material alteration of the facility which causes the owner or operator of the facility to re-evaluate and increase the response equipment necessary to adequately respond to a worst case discharge from the facility.

Date of Last Update: _____

FACILITY INFORMATION FORM

Facility Name: _____
    Location (Street Address): _____
    City: _____ State: _____ Zip: _____
    County: _____ Phone Number: (   ) _____
    Latitude: _____ Degrees _____ Minutes
    _____ Seconds
    Longitude: _____ Degrees _____ Minutes
    _____ Seconds
Wellhead Protection Area: _____
Owner: _____
    Owner Location (Street Address): _____

    (if different from Facility Address)
    City: _____ State: _____ Zip: _____
    County: _____ Phone Number: (   ) _____
Operator (if not Owner): _____
Qualified Individual(s): (attach additional sheets if more than one)
    Name: _____
    Position: _____
    Work Address: _____
    Home Address: _____
    Emergency Phone Number: (   ) _____

Date of Oil Storage Start-up: _____
Current Operations: _____

_____

Date(s) and Type(s) of Substantial Expansion(s): _____

(Attach additional sheets if necessary)

1.3 Emergency Response Information

(A) The information provided in this section shall describe what will be needed in an actual emergency involving the discharge of oil or a combination of hazardous substances and oil discharge. The Emergency Response Information section of the plan must include the following components:

(1) The information provided in the Emergency Notification Phone List in section 1.3.1 identifies and prioritizes the names and phone numbers of the organizations and personnel that need to be notified immediately in the event of an emergency. This section shall include all the appropriate phone numbers for the facility. These numbers must be verified each time the plan is updated. The contact list must be accessible to all facility employees to ensure that, in case of a discharge, any employee on site could immediately notify the appropriate parties.

(2) The Spill Response Notification Form in section 1.3.1 creates a checklist of information that shall be provided to the National Response Center (NRC) and other response personnel. All information on this checklist must be known at the time of notification, or be in the process of being collected. This notification form is based on a similar form used by the NRC. Note: Do not delay spill notification to collect the information on the list.

(3) Section 1.3.2 provides a description of the facility's list of emergency response equipment and location of the response equipment. When appropriate, the amount of oil that emergency response equipment can handle and any limitations (e.g., launching sites) must be described.

(4) Section 1.3.3 provides information regarding response equipment tests and deployment drills. Response equipment deployment exercises shall be conducted to ensure that response equipment is operational and the personnel who would operate the equipment in a spill response are capable of deploying and operating it. Only a representative sample of each type of response equipment needs to be deployed and operated, as long as the remainder is properly maintained. If appropriate, testing of response equipment may be conducted while it is being deployed. Facilities without facility-owned response equipment must ensure that the oil spill removal organization that is identified in the response plan to provide this response equipment certifies that the deployment exercises have been met. Refer

to the National Preparedness for Response Exercise Program (PREP) Guidelines (see appendix E to this part, section 13, for availability), which satisfy Oil Pollution Act (OPA) response exercise requirements.

(5) Section 1.3.4 lists the facility response personnel, including those employed by the facility and those under contract to the facility for response activities, the amount of time needed for personnel to respond, their responsibility in the case of an emergency, and their level of response training. Three different forms are included in this section. The Emergency Response Personnel List shall be composed of all personnel employed by the facility whose duties involve responding to emergencies, including oil discharges, even when they are not physically present at the site. An example of this type of person would be the Building Engineer-in-Charge or Plant Fire Chief. The second form is a list of the Emergency Response Contractors (both primary and secondary) retained by the facility. Any changes in contractor status must be reflected in updates to the response plan. Evidence of contracts with response contractors shall be included in this section so that the availability of resources can be verified. The last form is the Facility Response Team List, which shall be composed of both emergency response personnel (referenced by job title/position) and emergency response contractors, included in one of the two lists described above, that will respond immediately upon discovery of an oil discharge or other emergency (i.e., the first people to respond). These are to be persons normally on the facility premises or primary response contractors. Examples of these personnel would be the Facility Hazardous Materials (HAZMAT) Spill Team 1, Facility Fire Engine Company 1, Production Supervisor, or Transfer Supervisor. Company personnel must be able to respond immediately and adequately if contractor support is not available.

(6) Section 1.3.5 lists factors that must, as appropriate, be considered when preparing an evacuation plan.

(7) Section 1.3.6 references the responsibilities of the qualified individual for the facility in the event of an emergency.

(B) The information provided in the emergency response section will aid in the assessment of the facility's ability to respond to a worst case discharge and will identify additional assistance that may be needed. In addition, the facility owner or operator may want to produce a wallet-size card containing a checklist of the immediate response and notification steps to be taken in the event of an oil discharge.

1.3.1 Notification

Date of Last Update: _____

EMERGENCY NOTIFICATION PHONE LIST WHOM
TO NOTIFY

Reporter's Name: _____
Date: _____
Facility Name: _____
Owner Name: _____
Facility Identification Number: _____
Date and Time of Each NRC Notification: ___

Organization	Phone No.
1. National Response Center (NRC):	1–800–424–8802
2. Qualified Individual:	
Evening Phone:	
3. Company Response Team:	
Evening Phone:	
4. Federal On-Scene Coordinator (OSC) and/or Regional Response Center (RRC):	
Evening Phone(s):	
Pager Number(s):	
5. Local Response Team (Fire Dept./Co-operatives):	
6. Fire Marshall:	
Evening Phone:	
7. State Emergency Response Commission (SERC):	
Evening Phone:	
8. State Police:	
9. Local Emergency Planning Committee (LEPC):	
10. Local Water Supply System:	
Evening Phone:	
11. Weather Report:	
12. Local Television/Radio Station for Evacuation Notification:	
13. Hospitals:	

SPILL RESPONSE NOTIFICATION FORM

Reporter's Last Name: _____
First: _____
M.I.: _____
Position: _____
Phone Numbers:
  Day (　) 　–
  Evening (　) 　–
Company: _____
Organization Type: _____
Address: _____
_____
City: _____
State: _____
Zip: _____
Were Materials Discharged? _____ (Y/N) Confidential? _____ (Y/N)
Meeting Federal Obligations to Report? _____ (Y/N) Date Called: _____
Calling for Responsible Party? _____ (Y/N) Time Called: _____

Incident Description

Source and/or Cause of Incident: _____
_____
_____
_____
_____

Date of Incident: _____
Time of Incident: _____ AM/PM
Incident Address/Location: _____
_____
Nearest City: _____ State: _____
  County: _____ Zip: _____
Distance from City: _____ Units of Measure: _____ Direction from City: _____
Section: _____ Township: _____ Range: _____ Borough: _____
Container Type: _____ Tank Oil Storage Capacity: _____ Units of Measure: _____
Facility Oil Storage Capacity: _____ Units of Measure: _____
Facility Latitude: _____ Degrees _____ Minutes _____ Seconds
Facility Longitude: _____ Degrees _____ Minutes _____ Seconds

Material

CHRIS Code	Discharged quantity	Unit of measure	Material Discharged in water	Quantity	Unit of measure

Response Action

Actions Taken to Correct, Control or Mitigate Incident:

_____

_____

_____

_____

Impact

Number of Injuries: _____ Number of Deaths:

_____

Were there Evacuations? _____ (Y/N) Number Evacuated: _____

Was there any Damage? _____ (Y/N)

Damage in Dollars (approximate): _____

Medium Affected: _____

Description: _____

More Information about Medium: _____

_____

_____

Additional Information

Any information about the incident not recorded elsewhere in the report:

_____

_____

Caller Notifications

EPA? _____ (Y/N) USCG? _____ (Y/N) State? _____ (Y/N)

Other? _____ (Y/N) Describe: _____

1.3.2 Response Equipment List

Date of Last Update:_____

FACILITY RESPONSE EQUIPMENT LIST

1. Skimmers/Pumps—Operational Status: _

Type, Model, and Year: _____

Type   Model   Year

Number: _____

Capacity: _____ gal./min.

Daily Effective Recovery Rate: _____

Storage Location(s): _____

Date Fuel Last Changed: _____

2. Boom—Operational Status: _____

Type, Model, and Year: _____

Type   Model   Year

Number: _____

Size (length): _____ ft.

Containment Area: _____ sq. ft.

Storage Location: _____

3. Chemicals Stored (Dispersants listed on EPA's NCP Product Schedule)

Type	Amount	Date purchased	Treatment capacity	Storage location

Were appropriate procedures used to receive approval for use of dispersants in accordance with the NCP (40 CFR 300.910) and the Area Contingency Plan (ACP), where applicable?_____ (Y/N).

Name and State of On-Scene Coordinator (OSC) authorizing use: _____ .

Date Authorized: _____ .

4. Dispersant Dispensing Equipment—Operational Status: _____ .

Type and year	Capacity	Storage location	Response time (minutes)

5. Sorbents—Operational Status: _____

Type and Year Purchased: _____

Amount: _____

Absorption Capacity (gal.): _____

Storage Location(s): _____

6. Hand Tools—Operational Status: _____

Type and year	Quantity	Storage location

Type and year	Quantity	Storage location

7. Communication Equipment (include operating frequency and channel and/or cellular phone numbers)—Operational Status: ___

Type and year	Quantity	Storage location/ number

8. Fire Fighting and Personnel Protective Equipment—Operational Status: ___

Type and year	Quantity	Storage location

9. Other (e.g., Heavy Equipment, Boats and Motors)—Operational Status: ___

Type and year	Quantity	Storage location

1.3.3 Response Equipment Testing/Deployment

Date of Last Update: ___

Response Equipment Testing and Deployment Drill Log

Last Inspection or Response Equipment Test Date: ___
Inspection Frequency: ___
Last Deployment Drill Date: ___
Deployment Frequency: ___
Oil Spill Removal Organization Certification (if applicable): ___

1.3.4 Personnel

Date of Last Update: ___

EMERGENCY RESPONSE PERSONNEL

Company Personnel

	Name	Phone [1]	Response time	Responsibility during response action	Response training type/date
1.					
2.					
3.					
4.					
5.					
6.					
7.					
8.					
9.					
10.					
11.					
12.					

[1] Phone number to be used when person is not on-site.

EMERGENCY RESPONSE CONTRACTORS

Date of Last Update: ___

	Contractor	Phone	Response time	Contract responsibility [1]
1.				

EMERGENCY RESPONSE CONTRACTORS—Continued
Date of Last Update: _____

Contractor	Phone	Response time	Contract responsibility [1]
2.			
3.			
4.			

[1] Include evidence of contracts/agreements with response contractors to ensure the availability of personnel and response equipment.

FACILITY RESPONSE TEAM
Date of Last Update: _____

Team member	Response time (minutes)	Phone or pager number (day/evening)
Qualified Individual:		/
		/
		/
		/
		/
		/
		/
		/
		/
		/
		/
		/
		/
		/
		/
		/
		/
		/

NOTE: If the facility uses contracted help in an emergency response situation, the owner or operator must provide the contractors' names and review the contractors' capacities to provide adequate personnel and response equipment.

1.3.5 Evacuation Plans

1.3.5.1 Based on the analysis of the facility, as discussed elsewhere in the plan, a facility-wide evacuation plan shall be developed. In addition, plans to evacuate parts of the facility that are at a high risk of exposure in the event of a discharge or other release must be developed. Evacuation routes must be shown on a diagram of the facility (see section 1.9 of this appendix). When developing evacuation plans, consideration must be given to the following factors, as appropriate:

(1) Location of stored materials;

(2) Hazard imposed by discharged material;

(3) Discharge flow direction;

(4) Prevailing wind direction and speed;

(5) Water currents, tides, or wave conditions (if applicable);

(6) Arrival route of emergency response personnel and response equipment;

(7) Evacuation routes;

(8) Alternative routes of evacuation;

(9) Transportation of injured personnel to nearest emergency medical facility;

(10) Location of alarm/notification systems;

(11) The need for a centralized check-in area for evacuation validation (roll call);

(12) Selection of a mitigation command center; and

(13) Location of shelter at the facility as an alternative to evacuation.

1.3.5.2 One resource that may be helpful to owners or operators in preparing this section of the response plan is The *Handbook of Chemical Hazard Analysis Procedures* by the Federal Emergency Management Agency (FEMA), Department of Transportation (DOT), and EPA. *The Handbook of Chemical Hazard Analysis Procedures* is available from: FEMA , Publication Office, 500 C. Street, S.W., Washington, DC 20472, (202) 646–3484.

1.3.5.3 As specified in §112.20(h)(1)(vi), the facility owner or operator must reference existing community evacuation plans, as appropriate.

1.3.6 Qualified Individual's Duties

The duties of the designated qualified individual are specified in §112.20(h)(3)(ix). The qualified individual's duties must be described and be consistent with the minimum requirements in §112.20(h)(3)(ix). In addition, the qualified individual must be identified with the Facility Information in section 1.2 of the response plan.

1.4 Hazard Evaluation

This section requires the facility owner or operator to examine the facility's operations closely and to predict where discharges could occur. Hazard evaluation is a widely used industry practice that allows facility owners or operators to develop a complete understanding of potential hazards and the response actions necessary to address these hazards. *The Handbook of Chemical Hazard Analysis Procedures*, prepared by the EPA, DOT, and the FEMA and the *Hazardous Materials Emergency Planning Guide* (NRT–1), prepared by the National Response Team are good references for conducting a hazard analysis. Hazard identification and evaluation will assist facility owners or operators in planning for potential discharges, thereby reducing the severity of discharge impacts that may occur in the future. The evaluation also may help the operator identify and correct potential sources of discharges. In addition, special hazards to workers and emergency response personnel's health and safety shall be evaluated, as well as the facility's oil spill history.

1.4.1 Hazard Identification

The Tank and Surface Impoundment (SI) forms, or their equivalent, that are part of this section must be completed according to the directions below. ("Surface Impoundment" means a facility or part of a facility which is a natural topographic depression, man-made excavation, or diked area formed primarily of earthen materials (although it may be lined with man-made materials), which is designed to hold an accumulation of liquid wastes or wastes containing free liquids, and which is not an injection well or a seepage facility.) Similar worksheets, or their equivalent, must be developed for any other type of storage containers.

(1) List each tank at the facility with a separate and distinct identifier. Begin aboveground tank identifiers with an "A" and belowground tank identifiers with a "B", or submit multiple sheets with the aboveground tanks and belowground tanks on separate sheets.

(2) Use gallons for the maximum capacity of a tank; and use square feet for the area.

(3) Using the appropriate identifiers and the following instructions, fill in the appropriate forms:

(a) Tank or SI number—Using the aforementioned identifiers (A or B) or multiple reporting sheets, identify each tank or SI at the facility that stores oil or hazardous materials.

(b) Substance Stored—For each tank or SI identified, record the material that is stored therein. If the tank or SI is used to store more than one material, list all of the stored materials.

(c) Quantity Stored—For each material stored in each tank or SI, report the average volume of material stored on any given day.

(d) Tank Type or Surface Area/Year—For each tank, report the type of tank (e.g., floating top), and the year the tank was originally installed. If the tank has been refabricated, the year that the latest refabrication was completed must be recorded in parentheses next to the year installed. For each SI, record the surface area of the impoundment and the year it went into service.

(e) Maximum Capacity—Record the operational maximum capacity for each tank and SI. If the maximum capacity varies with the season, record the upper and lower limits.

(f) Failure/Cause—Record the cause and date of any tank or SI failure which has resulted in a loss of tank or SI contents.

(4) Using the numbers from the tank and SI forms, label a schematic drawing of the facility. This drawing shall be identical to any schematic drawings included in the SPCC Plan.

(5) Using knowledge of the facility and its operations, describe the following in writing:

(a) The loading and unloading of transportation vehicles that risk the discharge of oil or release of hazardous substances during transport processes. These operations may include loading and unloading of trucks, railroad cars, or vessels. Estimate the volume of material involved in transfer oper-

ations, if the exact volume cannot be determined.

(b) Day-to-day operations that may present a risk of discharging oil or releasing a hazardous substance. These activities include scheduled venting, piping repair or replacement, valve maintenance, transfer of tank contents from one tank to another, etc. (not including transportation-related activities). Estimate the volume of material involved in these operations, if the exact volume cannot be determined.

(c) The secondary containment volume associated with each tank and/or transfer point at the facility. The numbering scheme developed on the tables, or an equivalent system, must be used to identify each containment area. Capacities must be listed for each individual unit (tanks, slumps, drainage traps, and ponds), as well as the facility total.

(d) Normal daily throughput for the facility and any effect on potential discharge volumes that a negative or positive change in that throughput may cause.

HAZARD IDENTIFICATION TANKS [1]

Date of Last Update: _____

Tank No.	Substance Stored (Oil and Hazardous Substance)	Quantity Stored (gallons)	Tank Type/Year	Maximum Capacity (gallons)	Failure/Cause

[1] Tank = any container that stores oil.
Attach as many sheets as necessary.

HAZARD IDENTIFICATION SURFACE IMPOUNDMENTS (SIs)

Date of Last Update: _____

SI No.	Substance Stored	Quantity Stored (gallons)	Surface Area/Year	Maximum Capacity (gallons)	Failure/Cause

HAZARD IDENTIFICATION SURFACE IMPOUNDMENTS (SIs)—Continued

Date of Last Update: _____

SI No.	Substance Stored	Quantity Stored (gallons)	Surface Area/Year	Maximum Capacity (gallons)	Failure/Cause

Attach as many sheets as necessary.

1.4.2 Vulnerability Analysis

The vulnerability analysis shall address the potential effects (i.e., to human health, property, or the environment) of an oil discharge. Attachment C–III to Appendix C to this part provides a method that owners or operators shall use to determine appropriate distances from the facility to fish and wildlife and sensitive environments. Owners or operators can use a comparable formula that is considered acceptable by the RA. If a comparable formula is used, documentation of the reliability and analytical soundness of the formula must be attached to the response plan cover sheet. This analysis must be prepared for each facility and, as appropriate, must discuss the vulnerability of:

(1) Water intakes (drinking, cooling, or other);

(2) Schools;

(3) Medical facilities;

(4) Residential areas;

(5) Businesses;

(6) Wetlands or other sensitive environments;[2]

(7) Fish and wildlife;

(8) Lakes and streams;

(9) Endangered flora and fauna;

(10) Recreational areas;

(11) Transportation routes (air, land, and water);

(12) Utilities; and

(13) Other areas of economic importance (e.g., beaches, marinas) including terrestrially sensitive environments, aquatic environments, and unique habitats.

1.4.3 Analysis of the Potential for an Oil Discharge

Each owner or operator shall analyze the probability of a discharge occurring at the facility. This analysis shall incorporate factors such as oil discharge history, horizontal range of a potential discharge, and vulnerability to natural disaster, and shall, as appropriate, incorporate other factors such as tank age. This analysis will provide information for developing discharge scenarios for a worst case discharge and small and medium discharges and aid in the development of techniques to reduce the size and frequency of discharges. The owner or operator may need to research the age of the tanks the oil discharge history at the facility.

1.4.4 Facility Reportable Oil Spill History

Briefly describe the facility's reportable oil spill[3] history for the entire life of the facility to the extent that such information is reasonably identifiable, including:

(1) Date of discharge(s);

(2) List of discharge causes;

(3) Material(s) discharged;

(4) Amount discharged in gallons;

(5) Amount of discharge that reached navigable waters, if applicable;

(6) Effectiveness and capacity of secondary containment;

(7) Clean-up actions taken;

(8) Steps taken to reduce possibility of recurrence;

(9) Total oil storage capacity of the tank(s) or impoundment(s) from which the material discharged;

(10) Enforcement actions;

(11) Effectiveness of monitoring equipment; and

(12) Description(s) of how each oil discharge was detected.

---

[2] Refer to the DOC/NOAA "Guidance for Facility and Vessel Response Plans: Fish and Wildlife and Sensitive Environments" (See appendix E to this part, section 13, for availability).

[3] As described in 40 CFR part 110, reportable oil spills are those that: (a) violate applicable water quality standards, or (b) cause a film or sheen upon or discoloration of the surface of the water or adjoining shorelines or cause a sludge or emulsion to be deposited beneath the surface of the water or upon adjoining shorelines.

The information solicited in this section may be similar to requirements in 40 CFR 112.4(a). Any duplicate information required by § 112.4(a) may be photocopied and inserted.

1.5 Discharge Scenarios

In this section, the owner or operator is required to provide a description of the facility's worst case discharge, as well as a small and medium discharge, as appropriate. A multi-level planning approach has been chosen because the response actions to a discharge (i.e., necessary response equipment, products, and personnel) are dependent on the magnitude of the discharge. Planning for lesser discharges is necessary because the nature of the response may be qualitatively different depending on the quantity of the discharge. The facility owner or operator shall discuss the potential direction of the discharge pathway.

1.5.1 Small and Medium Discharges

1.5.1.1 To address multi-level planning requirements, the owner or operator must consider types of facility-specific discharge scenarios that may contribute to a small or medium discharge. The scenarios shall account for all the operations that take place at the facility, including but not limited to:

(1) Loading and unloading of surface transportation;

(2) Facility maintenance;

(3) Facility piping;

(4) Pumping stations and sumps;

(5) Oil storage tanks;

(6) Vehicle refueling; and

(7) Age and condition of facility and components.

1.5.1.2 The scenarios shall also consider factors that affect the response efforts required by the facility. These include but are not limited to:

(1) Size of the discharge;

(2) Proximity to downgradient wells, waterways, and drinking water intakes;

(3) Proximity to fish and wildlife and sensitive environments;

(4) Likelihood that the discharge will travel offsite (i.e., topography, drainage);

(5) Location of the material discharged (i.e., on a concrete pad or directly on the soil);

(6) Material discharged;

(7) Weather or aquatic conditions (i.e., river flow);

(8) Available remediation equipment;

(9) Probability of a chain reaction of failures; and

(10) Direction of discharge pathway.

1.5.2 Worst Case Discharge

1.5.2.1 In this section, the owner or operator must identify the worst case discharge volume at the facility. Worksheets for production and non-production facility owners

or operators to use when calculating worst case discharge are presented in appendix D to this part. When planning for the worst case discharge response, all of the aforementioned factors listed in the small and medium discharge section of the response plan shall be addressed.

1.5.2.2 For onshore storage facilities and production facilities, permanently manifolded oil storage tanks are defined as tanks that are designed, installed, and/or operated in such a manner that the multiple tanks function as one storage unit (i.e., multiple tank volumes are equalized). In this section of the response plan, owners or operators must provide evidence that oil storage tanks with common piping or piping systems are not operated as one unit. If such evidence is provided and is acceptable to the RA, the worst case discharge volume shall be based on the combined oil storage capacity of all manifold tanks or the oil storage capacity of the largest single oil storage tank within the secondary containment area, whichever is greater. For permanently manifolded oil storage tanks that function as one storage unit, the worst case discharge shall be based on the combined oil storage capacity of all manifolded tanks or the oil storage capacity of the largest single tank within a secondary containment area, whichever is greater. For purposes of the worst case discharge calculation, permanently manifolded oil storage tanks that are separated by internal divisions for each tank are considered to be single tanks and individual manifolded tank volumes are not combined.

1.6 Discharge Detection Systems

In this section, the facility owner or operator shall provide a detailed description of the procedures and equipment used to detect discharges. A section on discharge detection by personnel and a discussion of automated discharge detection, if applicable, shall be included for both regular operations and after hours operations. In addition, the facility owner or operator shall discuss how the reliability of any automated system will be checked and how frequently the system will be inspected.

1.6.1 Discharge Detection by Personnel

In this section, facility owners or operators shall describe the procedures and personnel that will detect any discharge of oil or release of a hazardous substance. A thorough discussion of facility inspections must be included. In addition, a description of initial response actions shall be addressed. This section shall reference section 1.3.1 of the response plan for emergency response information.

1.6.2 Automated Discharge Detection

In this section, facility owners or operators must describe any automated discharge detection equipment that the facility has in place. This section shall include a discussion of overfill alarms, secondary containment sensors, etc. A discussion of the plans to verify an automated alarm and the actions to be taken once verified must also be included.

1.7 Plan Implementation

In this section, facility owners or operators must explain in detail how to implement the facility's emergency response plan by describing response actions to be carried out under the plan to ensure the safety of the facility and to mitigate or prevent discharges described in section 1.5 of the response plan. This section shall include the identification of response resources for small, medium, and worst case discharges; disposal plans; and containment and drainage planning. A list of those personnel who would be involved in the cleanup shall be identified. Procedures that the facility will use, where appropriate or necessary, to update their plan after an oil discharge event and the time frame to update the plan must be described.

1.7.1 Response Resources for Small, Medium, and Worst Case Discharages

1.7.1.1 Once the discharge scenarios have been identified in section 1.5 of the response plan, the facility owner or operator shall identify and describe implementation of the response actions. The facility owner or operator shall demonstrate accessibility to the proper response personnel and equipment to effectively respond to all of the identified discharge scenarios. The determination and demonstration of adequate response capability are presented in appendix E to this part. In addition, steps to expedite the cleanup of oil discharges must be discussed. At a minimum, the following items must be addressed:

(1) Emergency plans for spill response;

(2) Additional response training;

(3) Additional contracted help;

(4) Access to additional response equipment/experts; and

(5) Ability to implement the plan including response training and practice drills.

1.7.1.2A recommended form detailing immediate actions follows.

OIL SPILL RESPONSE—IMMEDIATE ACTIONS

1. Stop the product flow	Act quickly to secure pumps, close valves, etc.

OIL SPILL RESPONSE—IMMEDIATE ACTIONS— Continued

2. Warn personnel	Enforce safety and security measures.
3. Shut off ignition sources.	Motors, electrical circuits, open flames, etc.
4. Initiate containment	Around the tank and/or in the water with oil boom.
5. Notify NRC	1–800–424–8802
6. Notify OSC	
7. Notify, as appropriate	

Source: FOSS, Oil Spill Response—Emergency Procedures, Revised December 3, 1992.

1.7.2 Disposal Plans

1.7.2.1 Facility owners or operators must describe how and where the facility intends to recover, reuse, decontaminate, or dispose of materials after a discharge has taken place. The appropriate permits required to transport or dispose of recovered materials according to local, State, and Federal requirements must be addressed. Materials that must be accounted for in the disposal plan, as appropriate, include:

(1) Recovered product;

(2) Contaminated soil;

(3) Contaminated equipment and materials, including drums, tank parts, valves, and shovels;

(4) Personnel protective equipment;

(5) Decontamination solutions;

(6) Adsorbents; and

(7) Spent chemicals.

1.7.2.2 These plans must be prepared in accordance with Federal (e.g., the Resource Conservation and Recovery Act [RCRA]), State, and local regulations, where applicable. A copy of the disposal plans from the facility's SPCC Plan may be inserted with this section, including any diagrams in those plans.

Material	Disposal facility	Location	RCRA permit/manifest
1.			
2.			
3.			
4.			

1.7.3 Containment and Drainage Planning

A proper plan to contain and control a discharge through drainage may limit the threat of harm to human health and the environment. This section shall describe how to contain and control a discharge through drainage, including:

103

(1) The available volume of containment (use the information presented in section 1.4.1 of the response plan);

(2) The route of drainage from oil storage and transfer areas;

(3) The construction materials used in drainage troughs;

(4) The type and number of valves and separators used in the drainage system;

(5) Sump pump capacities;

(6) The containment capacity of weirs and booms that might be used and their location (see section 1.3.2 of this appendix); and

(7) Other cleanup materials.

In addition, a facility owner or operator must meet the inspection and monitoring requirements for drainage contained in 40 CFR part 112, subparts A through C. A copy of the containment and drainage plans that are required in 40 CFR part 112, subparts A through C may be inserted in this section, including any diagrams in those plans.

NOTE: The general permit for stormwater drainage may contain additional requirements.

1.8 Self-Inspection, Drills/Exercises, and Response Training

The owner or operator must develop programs for facility response training and for drills/exercises according to the requirements of 40 CFR 112.21. Logs must be kept for facility drills/exercises, personnel response training, and spill prevention meetings. Much of the recordkeeping information required by this section is also contained in the SPCC Plan required by 40 CFR 112.3. These logs may be included in the facility response plan or kept as an annex to the facility response plan.

1.8.1 Facility Self-Inspection

Under 40 CFR 112.7(e), you must include the written procedures and records of inspections for each facility in the SPCC Plan. You must include the inspection records for each container, secondary containment, and item of response equipment at the facility. You must cross-reference the records of inspec-tions of each container and secondary containment required by 40 CFR 112.7(e) in the facility response plan. The inspection record of response equipment is a new requirement in this plan. Facility self-inspection requires two-steps: (1) a checklist of things to inspect; and (2) a method of recording the actual inspection and its findings. You must note the date of each inspection. You must keep facility response plan records for five years. You must keep SPCC records for three years.

1.8.1.1. Tank Inspection

The tank inspection checklist presented below has been included as guidance during inspections and monitoring. Similar requirements exist in 40 CFR part 112, subparts A through C. Duplicate information from the SPCC Plan may be photocopied and inserted in this section. The inspection checklist consists of the following items:

TANK INSPECTION CHECKLIST

1. Check tanks for leaks, specifically looking for:
   A. drip marks;
   B. discoloration of tanks;
   C. puddles containing spilled or leaked material;
   D. corrosion;
   E. cracks; and
   F. localized dead vegetation.
2. Check foundation for:
   A. cracks;
   B. discoloration;
   C. puddles containing spilled or leaked material;
   D. settling;
   E. gaps between tank and foundation; and
   F. damage caused by vegetation roots.
3. Check piping for:
   A. droplets of stored material;
   B. discoloration;
   C. corrosion;
   D. bowing of pipe between supports;
   E. evidence of stored material seepage from valves or seals; and
   F. localized dead vegetation.

TANK/SURFACE IMPOUNDMENT INSPECTION LOG

Inspector	Tank or SI#	Date	Comments

TANK/SURFACE IMPOUNDMENT INSPECTION LOG—Continued

Inspector	Tank or SI#	Date	Comments

1.8.1.2 Response Equipment Inspection

Using the Emergency Response Equipment List provided in section 1.3.2 of the response plan, describe each type of response equipment, checking for the following:

Response Equipment Checklist

1. Inventory (item and quantity);
2. Storage location;
3. Accessibility (time to access and respond);
4. Operational status/condition;
5. Actual use/testing (last test date and frequency of testing); and
6. Shelf life (present age, expected replacement date).

Please note any discrepancies between this list and the available response equipment.

RESPONSE EQUIPMENT INSPECTION LOG
[Use section 1.3.2 of the response plan as a checklist]

Inspector	Date	Comments

105

RESPONSE EQUIPMENT INSPECTION LOG—Continued

[Use section 1.3.2 of the response plan as a checklist]

Inspector	Date	Comments

1.8.1.3 Secondary Containment Inspection

Inspect the secondary containment (as described in sections 1.4.1 and 1.7.2 of the response plan), checking the following:

Secondary Containment Checklist

1. Dike or berm system.
   A. Level of precipitation in dike/available capacity;
   B. Operational status of drainage valves;
   C. Dike or berm permeability;
   D. Debris;
   E. Erosion;
   F. Permeability of the earthen floor of diked area; and
   G. Location/status of pipes, inlets, drainage beneath tanks, etc.
2. Secondary containment
   A. Cracks;
   B. Discoloration;
   C. Presence of spilled or leaked material (standing liquid);
   D. Corrosion; and
   E. Valve conditions.
3. Retention and drainage ponds
   A. Erosion;
   B. Available capacity;
   C. Presence of spilled or leaked material;
   D. Debris; and
   E. Stressed vegetation.

The tank inspection checklist presented below has been included as guidance during inspections and monitoring. Similar requirements exist in 40 CFR part 112, subparts A through C. Similar requirements exist in 40 CFR 112.7(e). Duplicate information from the SPCC Plan may be photocopied and inserted in this section.

1.8.2 Facility Drills/Exercises

(A) CWA section 311(j)(5), as amended by OPA, requires the response plan to contain a description of facility drills/exercises. According to 40 CFR 112.21(c), the facility owner or operator shall develop a program of facility response drills/exercises, including evaluation procedures. Following the PREP guidelines (see appendix E to this part, section 13, for availability) would satisfy a facility's requirements for drills/exercises under this part. Alternately, under §112.21(c), a facility owner or operator may develop a program that is not based on the PREP guidelines. Such a program is subject to approval by the Regional Administrator based on the description of the program provided in the response plan.

(B) The PREP Guidelines specify that the facility conduct internal and external drills/exercises. The internal exercises include: qualified individual notification drills, spill management team tabletop exercises, equipment deployment exercises, and unannounced exercises. External exercises include Area Exercises. Credit for an Area or Facility-specific Exercise will be given to the facility for an actual response to a discharge in the area if the plan was utilized for response to the discharge and the objectives of the Exercise were met and were properly evaluated, documented, and self-certified.

(C) Section 112.20(h)(8)(ii) requires the facility owner or operator to provide a description of the drill/exercise program to be carried out under the response plan. Qualified Individual Notification Drill and Spill Management Team Tabletop Drill logs shall be provided in sections 1.8.2.1 and 1.8.2.2, respectively. These logs may be included in the facility response plan or kept as an annex to the facility response plan. See section 1.3.3 of this appendix for Equipment Deployment Drill Logs.

106

1.8.2.1   Qualified Individual Notification Drill Logs

Qualified Individual Notification Drill Log

Date: _____

Company: _____

Qualified Individual(s): _____

Emergency Scenario: _____

_____

Evaluation: _____

_____

_____

Changes to be Implemented: _____

_____

Time Table for Implementation: _____

_____

1.8.2.2   Spill Management Team Tabletop Exercise Logs

Spill Management Team Tabletop Exercise Log

Date: _____

Company: _____

Qualified Individual(s): _____

Emergency Scenario: _____

_____

Evaluation: _____

_____

_____

Changes to be Implemented: _____

_____

Time Table for Implementation: _____

1.8.3   Response Training

Section 112.21(a) requires facility owners or operators to develop programs for facility response training. Facility owners or operators are required by §112.20(h)(8)(iii) to provide a description of the response training program to be carried out under the response plan. A facility's training program can be based on the USCG's Training Elements for Oil Spill Response, to the extent applicable to facility operations, or another response training program acceptable to the RA. The training elements are available from the USCG Office of Response (G-MOR) at (202) 267–0518 or fax (202) 267–4085. Personnel response training logs and discharge prevention meeting logs shall be included in sections 1.8.3.1 and 1.8.3.2 of the response plan respectively. These logs may be included in the facility response plan or kept as an annex to the facility response plan.

1.8.3.1   Personnel Response Training Logs

PERSONNEL RESPONSE TRAINING LOG

Name	Response training/date and number of hours	Prevention training/date and number of hours

1.8.3.2   Discharge Prevention Meetings Logs

DISCHARGE PREVENTION MEETING LOG

Date: _____

Attendees: _____

_____

_____

_____

_____

Subject/issue identified	Required action	Implementation date

1.9 Diagrams

The facility-specific response plan shall include the following diagrams. Additional diagrams that would aid in the development of response plan sections may also be included.

(1) The Site Plan Diagram shall, as appropriate, include and identify:

(A) the entire facility to scale;

(B) above and below ground bulk oil storage tanks;

(C) the contents and capacities of bulk oil storage tanks;

(D) the contents and capacity of drum oil storage areas;

(E) the contents and capacities of surface impoundments;

(F) process buildings;

(G) transfer areas;

(H) secondary containment systems (location and capacity);

(I) structures where hazardous materials are stored or handled, including materials stored and capacity of storage;

(J) location of communication and emergency response equipment;

(K) location of electrical equipment which contains oil; and

(L) for complexes only, the interface(s) (i.e., valve or component) between the portion of the facility regulated by EPA and the portion(s) regulated by other Agencies. In most cases, this interface is defined as the last valve inside secondary containment before piping leaves the secondary containment area to connect to the transportation-related portion of the facility (i.e., the structure used or intended to be used to transfer oil to or from a vessel or pipeline). In the absence of secondary containment, this interface is the valve manifold adjacent to the tank nearest the transfer structure as described above. The interface may be defined differently at a specific facility if agreed to by the RA and the appropriate Federal official.

(2) The Site Drainage Plan Diagram shall, as appropriate, include:

(A) major sanitary and storm sewers, manholes, and drains;

(B) weirs and shut-off valves;

(C) surface water receiving streams;

(D) fire fighting water sources;

(E) other utilities;

(F) response personnel ingress and egress;

(G) response equipment transportation routes; and

(H) direction of discharge flow from discharge points.

(3) The Site Evacuation Plan Diagram shall, as appropriate, include:

(A) site plan diagram with evacuation route(s); and

(B) location of evacuation regrouping areas.

1.10 Security

According to 40 CFR 112.7(g) facilities are required to maintain a certain level of security, as appropriate. In this section, a description of the facility security shall be provided and include, as appropriate:

(1) emergency cut-off locations (automatic or manual valves);

(2) enclosures (e.g., fencing, etc.);

(3) guards and their duties, day and night;

(4) lighting;

(5) valve and pump locks; and

(6) pipeline connection caps.

The SPCC Plan contains similar information. Duplicate information may be photocopied and inserted in this section.

2.0 Response Plan Cover Sheet

A three-page form has been developed to be completed and submitted to the RA by owners or operators who are required to prepare and submit a facility-specific response plan. The cover sheet (Attachment F–1) must accompany the response plan to provide the Agency with basic information concerning the facility. This section will describe the Response Plan Cover Sheet and provide instructions for its completion.

2.1 General Information

Owner/Operator of Facility: Enter the name of the owner of the facility (if the owner is the operator). Enter the operator of the facility if otherwise. If the owner/operator of the facility is a corporation, enter the name of the facility's principal corporate executive. Enter as much of the name as will fit in each section.

(1) *Facility Name:* Enter the proper name of the facility.

(2) *Facility Address:* Enter the street address, city, State, and zip code.

(3) *Facility Phone Number:* Enter the phone number of the facility.

(4) *Latitude and Longitude:* Enter the facility latitude and longitude in degrees, minutes, and seconds.

(5) *Dun and Bradstreet Number:* Enter the facility's Dun and Bradstreet number if available (this information may be obtained from public library resources).

(6) North American Industrial Classification System (NAICS) Code: Enter the facility's NAICS code as determined by the Office of Management and Budget (this information may be obtained from public library resources.)

(7) *Largest Oil Storage Tank Capacity:* Enter the capacity in GALLONS of the largest aboveground oil storage tank at the facility.

(8) *Maximum Oil Storage Capacity:* Enter the total maximum capacity in GALLONS of all aboveground oil storage tanks at the facility.

(9) *Number of Oil Storage Tanks:* Enter the number of all aboveground oil storage tanks at the facility.

(10) *Worst Case Discharge Amount:* Using information from the worksheets in appendix D, enter the amount of the worst case discharge in GALLONS.

(11) *Facility Distance to Navigable Waters:* Mark the appropriate line for the nearest distance between an opportunity for discharge (i.e., oil storage tank, piping, or flowline) and a navigable water.

2.2 Applicability of Substantial Harm Criteria

Using the flowchart provided in Attachment C–I to appendix C to this part, mark the appropriate answer to each question. Explanations of referenced terms can be found in Appendix C to this part. If a comparable formula to the ones described in Attachment C–III to appendix C to this part is used to calculate the planning distance, documentation of the reliability and analytical soundness of the formula must be attached to the response plan cover sheet.

2.3 Certification

Complete this block after all other questions have been answered.

3.0 Acronyms

ACP: Area Contingency Plan
ASTM: American Society of Testing Materials
bbls: Barrels
bpd: Barrels per Day

bph: Barrels per Hour
CHRIS: Chemical Hazards Response Information System
CWA: Clean Water Act
DOI: Department of Interior
DOC: Department of Commerce
DOT: Department of Transportation
EPA: Environmental Protection Agency
FEMA: Federal Emergency Management Agency
FR: Federal Register
gal: Gallons
gpm: Gallons per Minute
HAZMAT: Hazardous Materials
LEPC: Local Emergency Planning Committee
MMS: Minerals Management Service (part of DOI)
NAICS: North American Industrial Classification System
NCP: National Oil and Hazardous Substances Pollution Contingency Plan
NOAA: National Oceanic and Atmospheric Administration (part of DOC)
NRC: National Response Center
NRT: National Response Team
OPA: Oil Pollution Act of 1990
OSC: On-Scene Coordinator
PREP: National Preparedness for Response Exercise Program
RA: Regional Administrator
RCRA: Resource Conservation and Recovery Act
RRC: Regional Response Centers
RRT: Regional Response Team
RSPA: Research and Special Programs Administration
SARA: Superfund Amendments and Reauthorization Act
SERC: State Emergency Response Commission
SDWA: Safe Drinking Water Act of 1986
SI: Surface Impoundment
SPCC: Spill Prevention, Control, and Countermeasures
USCG: United States Coast Guard

4.0 References

CONCAWE. 1982. Methodologies for Hazard Analysis and Risk Assessment in the Petroleum Refining and Storage Industry. Prepared by CONCAWE's Risk Assessment Ad-hoc Group.

U.S. Department of Housing and Urban Development. 1987. Siting of HUD-Assisted Projects Near Hazardous Facilities: Acceptable Separation Distances from Explosive and Flammable Hazards. Prepared by the Office of Environment and Energy, Environmental Planning Division, Department of Housing and Urban Development. Washington, DC.

U.S. DOT, FEMA and U.S. EPA. Handbook of Chemical Hazard Analysis Procedures.

U.S. DOT, FEMA and U.S. EPA. Technical Guidance for Hazards Analysis: Emergency

Planning for Extremely Hazardous Substances.

The National Response Team. 1987. Hazardous Materials Emergency Planning Guide. Washington, DC.

The National Response Team. 1990. Oil Spill Contingency Planning, National Status: A Report to the President. Washington, DC. U.S. Government Printing Office.

Offshore Inspection and Enforcement Division. 1988. Minerals Management Service, Offshore Inspection Program: National Potential Incident of Noncompliance (PINC) List. Reston, VA.

ATTACHMENTS TO APPENDIX F

Attachment F-1—Response Plan Cover Sheet

This cover sheet will provide EPA with basic information concerning the facility. It must accompany a submitted facility response plan. Explanations and detailed instructions can be found in appendix F. Please type or write legibly in blue or black ink. Public reporting burden for the collection of this information is estimated to vary from 1 hour to 270 hours per response in the first year, with an average of 5 hours per response. This estimate includes time for reviewing instructions, searching existing data sources, gathering the data needed, and completing and reviewing the collection of information. Send comments regarding the burden estimate of this information, including suggestions for reducing this burden to: Chief, Information Policy Branch, Mail Code: PM-2822, U.S. Environmental Protection Agency, Ariel Rios Building, 1200 Pennsylvania Avenue, NW., Washington, DC 20460; and to the Office of Information and Regulatory Affairs, Office of Management and Budget, Washington D.C. 20503.

GENERAL INFORMATION

Owner/Operator of Facility:

_____

Facility Name: _____
Facility Address (street address or route):

_____
_____

City, State, and U.S. Zip Code:

_____

Facility Phone No.: _____
Latitude (Degrees: North):

_____

degrees, minutes, seconds _____
Dun & Bradstreet Number:[1]

_____

Largest Aboveground Oil Storage Tank Capacity (Gallons):

_____
_____

[1] These numbers may be obtained from public library resources.

Number of Aboveground Oil Storage Tanks:

_____

Longitude (Degrees: West):

_____

degrees, minutes, seconds _____
North American Industrial Classification System (NAICS) Code:[1] _____

_____

Maximum Oil Storage Capacity (Gallons): __
Worst Case Oil Discharge Amount (Gallons):
Facility Distance to Navigable Water. Mark the appropriate line. _____
0-¼ mile ___ ¼-½ mile ___ ½-1 mile ___ >1 mile ___

APPLICABILITY OF SUBSTANTIAL HARM CRITERIA

Does the facility transfer oil over-water[2] to or from vessels and does the facility have a total oil storage capacity greater than or equal to 42,000 gallons?
Yes _____
No _____

Does the facility have a total oil storage capacity greater than or equal to 1 million gallons and, within any storage area, does the facility lack secondary containment[2] that is sufficiently large to contain the capacity of the largest aboveground oil storage tank plus sufficient freeboard to allow for precipitation?
Yes _____
No _____

Does the facility have a total oil storage capacity greater than or equal to 1 million gallons and is the facility located at a distance[2] (as calculated using the appropriate formula in appendix C or a comparable formula) such that a discharge from the facility could cause injury to fish and wildlife and sensitive environments?[3]
Yes _____
No _____

Does the facility have a total oil storage capacity greater than or equal to 1 million

[2] Explanations of the above-referenced terms can be found in appendix C to this part. If a comparable formula to the ones contained in Attachment C-III is used to establish the appropriate distance to fish and wildlife and sensitive environments or public drinking water intakes, documentation of the reliability and analytical soundness of the formula must be attached to this form.

[3] For further description of fish and wildlife and sensitive environments, see Appendices I, II, and III to DOC/NOAA's "Guidance for Facility and Vessel Response Plans: Fish and Wildlife and Sensitive Environments" (see appendix E to this part, section 13, for availability) and the applicable ACP.

gallons and is the facility located at a distance[2] (as calculated using the appropriate formula in appendix C or a comparable formula) such that a discharge from the facility would shut down a public drinking water intake?[2]

Yes _____

No _____

Does the facility have a total oil storage capacity greater than or equal to 1 million gallons and has the facility experienced a reportable oil spill[2] in an amount greater than or equal to 10,000 gallons within the last 5 years?

Yes _____

No _____

CERTIFICATION

I certify under penalty of law that I have personally examined and am familiar with the information submitted in this document, and that based on my inquiry of those individuals responsible for obtaining information, I believe that the submitted information is true, accurate, and complete.

Signature: _____

Name (Please type or print): _____

Title: _____

Date: _____

[59 FR 34122, July 1, 1994; 59 FR 49006, Sept. 26, 1994, as amended at 65 FR 40816, June 30, 2000; 65 FR 43840, July 14, 2000; 66 FR 34561, June 29, 2001; 67 FR 47152, July 17, 2002]

APPENDIX G TO PART 112—TIER I QUALIFIED FACILITY SPCC PLAN

Tier I Qualified Facility SPCC Plan

This template constitutes the SPCC Plan for the facility, when completed and signed by the owner or operator of a facility that meets the applicability criteria in §112.3(g)(1). This template addresses the requirements of 40 CFR part 112. Maintain a complete copy of the Plan at the facility if the facility is normally attended at least four hours per day, or for a facility attended fewer than four hours per day, at the nearest field office. When making operational changes at a facility that are necessary to comply with the rule requirements, the owner/operator should follow state and local requirements (such as for permitting, design and construction) and obtain professional assistance, as appropriate.

Facility Description

Facility Name	_____				
Facility Address	_____				
City	_____	State	_____	ZIP	_____
County	_____	Tel. Number	( ) -		

Owner or operator Name	_____				
Owner or operator Address	_____				
City	_____	State	_____	ZIP	_____
County	_____	Tel. Number	( ) -		

I. Self-Certification Statement (§112.6(a)(1))

The owner or operator of a facility certifies that each of the following is true in order to utilize this template to comply with the SPCC requirements:

I _____, certify that the following is accurate:

1. I am familiar with the applicable requirements of 40 CFR part 112;
2. I have visited and examined the facility;
3. This Plan was prepared in accordance with accepted and sound industry practices and standards;
4. Procedures for required inspections and testing have been established in accordance with industry inspection and testing standards or recommended practices;
5. I will fully implement the Plan;
6. This facility meets the following qualification criteria (under §112.3(g)(1)):
   a. The aggregate aboveground oil storage capacity of the facility is 10,000 U.S. gallons or less; and
   b. The facility has had no single discharge as described in §112.1(b) exceeding 1,000 U.S. gallons and no two discharges as described in §112.1(b) each exceeding 42 U.S. gallons within any twelve month period in the three years prior to the SPCC Plan self-certification date, or since becoming subject to 40 CFR part 112 if the facility has been in operation for less than three years (not including oil discharges as described in §112.1(b) that are the result of natural disasters, acts of war, or terrorism); and

    c.   There is no individual oil storage container at the facility with an aboveground capacity greater than 5,000 U.S. gallons.

7.   This Plan does not deviate from any requirement of 40 CFR part 112 as allowed by §112.7(a)(2) (environmental equivalence) and §112.7(d) (impracticability of secondary containment) or include an measures pursuant to §112.9(c)(6) for produced water containers and any associated piping;

8.   This Plan and individual(s) responsible for implementing this Plan have the full approval of management and I have committed the necessary resources to fully implement this Plan.

I also understand my other obligations relating to the storage of oil at this facility, including, among others:

1.   To report any oil discharge to navigable waters or adjoining shorelines to the appropriate authorities. Notification information is included in this Plan.

2.   To review and amend this Plan whenever there is a material change at the facility that affects the potential for an oil discharge, and at least once every five years. Reviews and amendments are recorded in an attached log [See Five Year Review Log and Technical Amendment Log in Attachments 1.1 and 1.2.]

3.   Optional use of a contingency plan. A contingency plan:

    a.   May be used in lieu of secondary containment for qualified oil-filled operational equipment, in accordance with the requirements under §112.7(k), and;

    b.   Must be prepared for flowlines and/or intra-facility gathering lines which do not have secondary containment at an oil production facility, and;

    c.   Must include an established and documented inspection or monitoring program; must follow the provisions of 40 CFR part 109; and must include a written commitment of manpower, equipment and materials to expeditiously remove any quantity of oil discharged that may be harmful. If applicable, a copy of the contingency plan and any additional documentation will be attached to this Plan as Attachment 2.

I certify that I have satisfied the requirement to prepare and implement a Plan under §112.3 and all of the requirements under §112.6(a). I certify that the information contained in this Plan is true.

Signature _____   Title: _____

Name _____   Date: ___/____/20___

II. Record of Plan Review and Amendments

Five Year Review (§112.5(b)):

Complete a review and evaluation of this SPCC Plan at least once every five years. As a result of the review, amend this Plan within six months to include more effective prevention and control measures for the facility, if applicable. Implement any SPCC Plan amendment as soon as possible, but no later than six months following Plan amendment. Document completion of the review and evaluation, and complete the Five Year Review Log in Attachment 1.1. If the facility no longer meets Tier I qualified facility eligibility, the owner or operator must revise the Plan to meet Tier II qualified facility requirements, or complete a full PE certified Plan.

Table G-1 Technical Amendments (§§112.5(a), (c) and 112.6(a)(2))	
This SPCC Plan will be amended when there is a change in the facility design, construction, operation, or maintenance that materially affects the potential for a discharge to navigable waters or adjoining shorelines. Examples include adding or removing containers, reconstruction, replacement, or installation of piping systems, changes to secondary containment systems, changes in product stored at this facility, or revisions to standard operating procedures.	☐
Any technical amendments to this Plan will be re-certified in accordance with Section I of this Plan template. *[§112.6(a)(2)]* [See Technical Amendment Log in Attachment 1.2]	☐

III. Plan Requirements

1. Oil Storage Containers (§112.7(a)(3)(i)):

Table G-2 Oil Storage Containers and Capacities		
This table includes a complete list of all oil storage containers (aboveground containers[a] and completely buried tanks[b]) with capacity of 55 U.S. gallons or more, unless otherwise exempt from the rule. For mobile/portable containers, an estimate number of containers, types of oil, and anticipated capacities are provided.		☐
Oil Storage Container *(indicate whether aboveground (A) or completely buried (B))*	**Type of Oil**	**Shell Capacity (gallons)**
	Total Aboveground Storage Capacity [c]	_____ gallons
	Total Completely Buried Storage Capacity	_____ gallons
	Facility Total Oil Storage Capacity	_____ gallons

[a] Aboveground storage containers that must be included when calculating total facility oil storage capacity include: tanks and mobile or portable containers; oil-filled operational equipment (e.g. transformers); other oil-filled equipment, such as flow-through process equipment. Exempt containers that are not included in the capacity calculation include: any container with a storage capacity of less than 55 gallons of oil; containers used exclusively for wastewater treatment; permanently closed containers; motive power containers; hot-mix asphalt containers; heating oil containers used solely at a single-family residence; and pesticide application equipment or related mix containers.

[b] Although the criteria to determine eligibility for qualified facilities focuses on the aboveground oil storage containers at the facility, the completely buried tanks at a qualified facility are still subject to the rule requirements and must be addressed in the template; however, they are not counted toward the qualified facility applicability threshold.

[c] Counts toward qualified facility applicability threshold.

2. Secondary Containment and Oil Spill Control (§§112.6(a)(3)(i) and (ii), 112.7(c) and 112.9(c)(2)):

Table G-3 Secondary Containment and Oil Spill Control	
Appropriate secondary containment and/or diversionary structures or equipment[a] is provided for all oil handling containers, equipment, and transfer areas to prevent a discharge to navigable waters or adjoining shorelines. The entire secondary containment system, including walls and floor, is capable of containing oil and is constructed so that any discharge from a primary containment system, such as a tank or pipe, will not escape the containment system before cleanup occurs.	☐

[a] Use one of the following methods of secondary containment or its equivalent: (1) Dikes, berms, or retaining walls sufficiently impervious to contain oil; (2) Curbing; (3) Culverting, gutters, or other drainage systems; (4) Weirs, booms, or other barriers; (5) Spill diversion ponds; (6) Retention ponds; or (7) Sorbent materials.

Table G-4 below identifies the tanks and containers at the facility with the potential for an oil discharge; the mode of failure; the flow direction and potential quantity of the discharge; and the secondary containment method and containment capacity that is provided.

Table G-4 Containers with Potential for an Oil Discharge

Area	Type of failure (discharge scenario)	Potential discharge volume (gallons)	Direction of flow for uncontained discharge	Secondary containment method[a]	Secondary containment capacity (gallons)
Bulk Storage Containers and Mobile/Portable Containers[b]					
Oil-filled Operational Equipment (e.g., hydraulic equipment, transformers)[c]					
Piping, Valves, etc.					
Product Transfer Areas (location where oil is loaded to or from a container, pipe or other piece of equipment.)					
Other Oil-Handling Areas or Oil-Filled Equipment (e.g. flow-through process vessels at an oil production facility)					

[a] Use one of the following methods of secondary containment or its equivalent: (1) Dikes, berms, or retaining walls sufficiently impervious to contain oil; (2) Curbing; (3) Culverting, gutters, or other drainage systems; (4) Weirs, booms, or other barriers; (5) Spill diversion ponds; (6) Retention ponds; or (7) Sorbent materials.
[b] For storage tanks and bulk storage containers, the secondary containment capacity must be at least the capacity of the largest container plus additional capacity to contain rainfall or other precipitation.
[c] For oil-filled operational equipment: Document in the table above if alternative measures to secondary containment (as described in §112.7(k)) are implemented at the facility.

3. **Inspections, Testing, Recordkeeping and Personnel Training (§§112.7(e) and (f), 112.8(c)(6) and (d)(4), 112.9(c)(3), 112.12(c)(6) and (d)(4)):**

Table G-5 Inspections, Testing, Recordkeeping and Personnel Training	
An inspection and/or testing program is implemented for all aboveground bulk storage containers and piping at this facility. *[§§112.8(c)(6) and (d)(4), 112.9(c)(3), 112.12(c)(6) and (d)(4)]*	☐
The following is a description of the inspection and/or testing program (e.g. reference to industry standard utilized, scope, frequency, method of inspection or test, and person conducting the inspection) for all aboveground bulk storage containers and piping at this facility:	
Inspections, tests, and records are conducted in accordance with written procedures developed for the facility. Records of inspections and tests kept under usual and customary business practices will suffice for purposes of this paragraph. *[§112.7(e)]*	☐
A record of the inspections and tests are kept at the facility or with the SPCC Plan for a period of three years. *[§112.7(e)]* [See Inspection Log and Schedule in Attachment 3.1]	☐
Inspections and tests are signed by the appropriate supervisor or inspector. *[§112.7(e)]*	☐
Personnel, training, and discharge prevention procedures [§112.7(f)]	
Oil-handling personnel are trained in the operation and maintenance of equipment to prevent discharges; discharge procedure protocols; applicable pollution control laws, rules, and regulations; general facility operations; and, the contents of the facility SPCC Plan. *[§112.7(f)]*	☐
A person who reports to facility management is designated and accountable for discharge prevention. *[§112.7(f)]* Name/Title:_____	☐
Discharge prevention briefings are conducted for oil-handling personnel annually to assure adequate understanding of the SPCC Plan for that facility. Such briefings highlight and describe past reportable discharges or failures, malfunctioning components, and any recently developed precautionary measures. *[§112.7(f)]* [See Oil-handling Personnel Training and Briefing Log in Attachment 3.4]	☐

4. Security (excluding oil production facilities) §112.7(g):

Table G-6 Implementation and Description of Security Measures	
Security measures are implemented at this facility to prevent unauthorized access to oil handling, processing, and storage area.	☐
The following is a description of how you secure and control access to the oil handling, processing and storage areas; secure master flow and drain valves; prevent unauthorized access to starter controls on oil pumps; secure out-of-service and loading/unloading connections of oil pipelines; address the appropriateness of security lighting to both prevent acts of vandalism and assist in the discovery of oil discharges:	

5. Emergency Procedures and Notifications (§112.7(a)(3)(iv) and 112.7(a)(5)):

Table G-7 Description of Emergency Procedures and Notifications
The following is a description of the immediate actions to be taken by facility personnel in the event of a discharge to navigable waters or adjoining shorelines *[§112.7(a)(3)(iv) and 112.7(a)(5)]*:

6. Contact List (§112.7(a)(3)(vi)):

Table G-8 Contact List	
Contact Organization / Person	**Telephone Number**
National Response Center (NRC)	1-800-424-8802
Cleanup Contractor(s)	
Key Facility Personnel	
Designated Person Accountable for Discharge Prevention:	Office:
	Emergency:
	Office:
	Emergency:
	Office:
	Emergency:
	Office:
	Emergency:
State Oil Pollution Control Agencies	
Other State, Federal, and Local Agencies	
Local Fire Department	
Local Police Department	
Hospital	
Other Contact References (e.g., downstream water intakes or neighboring facilities)	

7. NRC Notification Procedure (§112.7(a)(4) and (a)(5)):

Table G-9 NRC Notification Procedure	
In the event of a discharge of oil to navigable waters or adjoining shorelines, the following information identified in Attachment 4 will be provided to the National Response Center immediately following identification of a discharge to navigable waters or adjoining shorelines [See Discharge Notification Form in Attachment 4]: *[§112.7(a)(4)]*	☐

- The exact address or location and phone number of the facility;
- Date and time of the discharge;
- Type of material discharged;
- Estimate of the total quantity discharged;
- Estimate of the quantity discharged to navigable waters;
- Source of the discharge;
- Description of all affected media;
- Cause of the discharge;
- Any damages or injuries caused by the discharge;
- Actions being used to stop, remove, and mitigate the effects of the discharge;
- Whether an evacuation may be needed; and
- Names of individuals and/or organizations who have also been contacted.

8. SPCC Spill Reporting Requirements (Report within 60 days) (§112.4):

Submit information to the EPA Regional Administrator (RA) and the appropriate agency or agencies in charge of oil pollution control activities in the State in which the facility is located within 60 days from one of the following discharge events:

- A single discharge of more than 1,000 U.S. gallons of oil to navigable waters or adjoining shorelines or
- Two discharges to navigable waters or adjoining shorelines each more than 42 U.S. gallons of oil occurring within any twelve month period

You must submit the following information to the RA:
(1) Name of the facility;
(2) Your name;
(3) Location of the facility;
(4) Maximum storage or handling capacity of the facility and normal daily throughput;
(5) Corrective action and countermeasures you have taken, including a description of equipment repairs and replacements;
(6) An adequate description of the facility, including maps, flow diagrams, and topographical maps, as necessary;
(7) The cause of the reportable discharge, including a failure analysis of the system or subsystem in which the failure occurred; and
(8) Additional preventive measures you have taken or contemplated to minimize the possibility of recurrence
(9) Such other information as the Regional Administrator may reasonably require pertinent to the Plan or discharge

* * * * *

NOTE: Complete one of the following sections (A, B or C)

as appropriate for the facility type.

A. Onshore Facilities (excluding production) (§§112.8(b) through (d), 112.12(b) through (d)):

The owner or operator must meet the general rule requirements as well as requirements under this section. Note that not all provisions may be applicable to all owners/operators. For example, a facility may not maintain completely buried metallic storage tanks installed after January 10, 1974, and thus would not have to abide by requirements in §§112.8(c)(4) and 112.12(c)(4), listed below. In cases where a provision is not applicable, write "N/A".

Table G-10 General Rule Requirements for Onshore Facilities	
Drainage from diked storage areas is restrained by valves to prevent a discharge into the drainage system or facility effluent treatment system, except where facility systems are designed to control such discharge. Diked areas may be emptied by pumps or ejectors that must be manually activated after inspecting the condition of the accumulation to ensure no oil will be discharged. *[§§112.8(b)(1) and 112.12(b)(1)]*	☐
Valves of manual, open-and-closed design are used for the drainage of diked areas. *[§§112.8(b)(2) and 112.12(b)(2)]*	☐
The containers at the facility are compatible with materials stored and conditions of storage such as pressure and temperature. *[§§112.8(c)(1) and 112.12(c)(1)]*	☐
Secondary containment for the bulk storage containers (including mobile/portable oil storage containers) holds the capacity of the largest container plus additional capacity to contain precipitation. Mobile or portable oil storage containers are positioned to prevent a discharge as described in §112.1(b). *[§112.6(a)(3)(ii)]*	☐
If uncontaminated rainwater from diked areas drains into a storm drain or open watercourse the following procedures will be implemented at the facility: *[§§112.8(c)(3) and 112.12(c)(3)]*	
• Bypass valve is normally sealed closed	☐
• Retained rainwater is inspected to ensure that its presence will not cause a discharge to navigable waters or adjoining shorelines	☐
• Bypass valve is opened and resealed under responsible supervision	☐
• Adequate records of drainage are kept [See Dike Drainage Log in Attachment 3.3]	☐
For completely buried metallic tanks installed on or after January 10, 1974 at this facility *[§§112.8(c)(4) and 112.12(c)(4)]*: • Tanks have corrosion protection with coatings or cathodic protection compatible with local soil conditions.	☐
• Regular leak testing is conducted.	☐
For partially buried or bunkered metallic tanks *[§112.8(c)(5) and §112.12(c)(5)]*: • Tanks have corrosion protection with coatings or cathodic protection compatible with local soil conditions.	☐
Each aboveground bulk container is tested or inspected for integrity on a regular schedule and whenever material repairs are made. Scope and frequency of the inspections and inspector qualifications are in accordance with industry standards. Container supports and foundations are regularly inspected. [See Inspection Log and Schedule and Bulk Storage Container Inspection Schedule in Attachments 3.1 and 3.2] *[§112.8(c)(6) and §112.12(c)(6)(i)]*	☐
Outsides of bulk storage containers are frequently inspected for signs of deterioration, discharges, or accumulation of oil inside diked areas. [See Inspection Log and Schedule in Attachment 3.1] *[§§112.8(c)(6) and 112.12(c)(6)]*	☐
For bulk storage containers that are subject to 21 CFR part 110 which are shop-fabricated, constructed of austenitic stainless steel, elevated and have no external insulation, formal visual inspection is conducted on a regular schedule. Appropriate qualifications for personnel performing tests and inspections are documented. [See Inspection Log and Schedule and Bulk	☐

Table G-10 General Rule Requirements for Onshore Facilities	
Storage Container Inspection Schedule in Attachments 3.1 and 3.2] *[§112.12(c)(6)(ii)]*	
Each container is provided with a system or documented procedure to prevent overfills for the container. Describe:	☐
Liquid level sensing devices are regularly tested to ensure proper operation [See Inspection Log and Schedule in Attachment 3.1]. *[§112.6(a)(3)(iii)]*	☐
Visible discharges which result in a loss of oil from the container, including but not limited to seams, gaskets, piping, pumps, valves, rivets, and bolts are promptly corrected and oil in diked areas is promptly removed. *[§§112.8(c)(10) and 112.12(c)(10)]*	☐
Aboveground valves, piping, and appurtenances such as flange joints, expansion joints, valve glands and bodies, catch pans, pipeline supports, locking of valves, and metal surfaces are inspected regularly. [See Inspection Log and Schedule in Attachment 3.1] *[§§112.8(d)(4) and 112.12(d)(4)]*	☐
Integrity and leak testing are conducted on buried piping at the time of installation, modification, construction, relocation, or replacement. [See Inspection Log and Schedule in Attachment 3.1] *[§§112.8(d)(4) and 112.12(d)(4)]*	☐

B. Onshore Oil Production Facilities (excluding drilling and workover facilities) (§112.9(b), (c), and (d)):

The owner or operator must meet the general rule requirements as well as the requirements under this section. Note that not all provisions may be applicable to all owners/operators. In cases where a provision is not applicable, write "N/A".

Table G-11 General Rule Requirements for Onshore Oil Production Facilities	
At tank batteries, separation and treating areas, drainage is closed and sealed except when draining uncontaminated rainwater. Accumulated oil on the rainwater is returned to storage or disposed of in accordance with legally approved methods. *[§112.9(b)(1)]*	☐
Prior to drainage, diked areas are inspected and *[§112.9(b)(1)]*: • Retained rainwater is inspected to ensure that its presence will not cause a discharge to navigable waters	☐
• Bypass valve is opened and resealed under responsible supervision	☐
• Adequate records of drainage are kept [See Dike Drainage Log in Attachment 3.3]	☐
Field drainage systems and oil traps, sumps, or skimmers are inspected at regularly scheduled intervals for oil, and accumulations of oil are promptly removed [See Inspection Log and Schedule in Attachment 3.1] *[§112.9(b)(2)]*	☐
The containers used at this facility are compatible with materials stored and conditions of storage. *[§112.9(c)(1)]*	☐
All tank battery, separation, and treating facility installations (except for flow-through process vessels) are constructed with a capacity to hold the largest single container plus additional capacity to contain rainfall. Drainage from undiked areas is safely confined in a catchment basin or holding pond. *[§112.9(c)(2)]*	☐
Except for flow-through process vessels, containers that are on or above the surface of the ground, including foundations and supports, are visually inspected for deterioration and maintenance needs on a regular schedule. [See Inspection Log and Schedule in Attachment 3.1] *[§112.9(c)(3)]*	☐
New and old tank batteries at this facility are engineered/updated in accordance with good engineering practices to prevent discharges including at least one of the following: (i) adequate container capacity to prevent overfill if regular pumping/gauging is delayed; (ii) overflow equalizing lines between containers so that a full container can overflow to an adjacent container; (iii) vacuum protection to prevent container collapse; or (iv) high level sensors to generate and transmit an alarm to the computer where the facility is subject to a computer production control system. *[§112.9(c)(4)]*	☐
Flow-through process vessels and associated components are: • Are constructed with a capacity to hold the largest single container plus additional capacity to contain rainfall. Drainage from undiked areas is safely confined in a catchment basin or holding pond; *[§112.9(c)(2)]* and	☐
• That are on or above the surface of the ground, including foundations and supports, are visually inspected for deterioration and maintenance needs on a regular schedule. [See Inspection Log and Schedule in Attachment 3.1] *[§112.9(c)(3)]*	☐
Or • Visually inspected and/or tested periodically and on a regular schedule for leaks, corrosion, or other conditions that could lead to a discharge to navigable waters; and	☐
• Corrective action or repairs are applied to flow-through process vessels and any associated components as indicated by regularly scheduled visual inspections, tests, or evidence of an oil discharge; and	☐
• Any accumulations of oil discharges associated with flow-through process vessels are promptly removed; and	☐

Table G-11 General Rule Requirements for Onshore Oil Production Facilities	
• Flow-through process vessels are provided with a secondary means of containment for the entire capacity of the largest single container and sufficient freeboard to contain precipitation within six months of a discharge from flow-through process vessels of more than 1,000 U.S. gallons of oil in a single discharge as described in §112.1(b), or a discharge more than 42 U.S. gallons of oil in each of two discharges as described in §112.1(b) within any twelve month period. *[§112.9(c)(5)]* *(Leave blank until such time that this provision is applicable.)*	☐
All aboveground valves and piping associated with transfer operations are inspected periodically and upon a regular schedule. The general condition of flange joints, valve glands and bodies, drip pans, pipe supports, pumping well polish rod stuffing boxes, bleeder and gauge valves, and other such items are included in the inspection. [See Inspection Log and Schedule in Attachment 3.1] *[§112.9(d)(1)]*	☐
An oil spill contingency plan and written commitment of resources are provided for flowlines and intra-facility gathering lines [See Oil Spill Contingency Plan and Checklist in Attachment 2 and Inspection Log and Schedule in Attachment 3.1] *[§112.9(d)(3)]* or	☐
Appropriate secondary containment and/or diversionary structures or equipment is provided for flowlines and intra-facility gathering lines to prevent a discharge to navigable waters or adjoining shorelines. The entire secondary containment system, including walls and floor, is capable of containing oil and is constructed so that any discharge from the pipe, will not escape the containment system before cleanup occurs.	☐
A flowline/intra-facility gathering line maintenance program to prevent discharges from each flowline has been established at this facility. The maintenance program addresses each of the following:	☐
• Flowlines and intra-facility gathering lines and associated valves and equipment are compatible with the type of production fluids, their potential corrosivity, volume, and pressure, and other conditions expected in the operational environment;	☐
• Flowlines, intra-facility gathering lines and associated appurtenances are visually inspected and/or tested on a periodic and regular schedule for leaks, oil discharges, corrosion, or other conditions that could lead to a discharge as described in §112.1(b). The frequency and type of testing allows for the implementation of a contingency plan as described under part 109 of this chapter.	☐
• Corrective action and repairs to any flowlines and intra-facility gathering lines and associated appurtenances as indicated by regularly scheduled visual inspections, tests, or evidence of a discharge.	☐
• Accumulations of oil discharges associated with flowlines, intra-facility gathering lines, and associated appurtenances are promptly removed. *[§112.9(d)(4)]*	☐
The following is a description of the flowline/intra-facility gathering line maintenance program implemented at this facility:	

124

C. Onshore Oil Drilling and Workover Facilities (§112.10(b), (c) and (d)):

The owner or operator must meet the general rule requirements as well as the requirements under this section.

Table G-12 General Rule Requirements for Onshore Oil Drilling and Workover Facilities	
Mobile drilling or worker equipment is positioned or located to prevent discharge as described in §112.1(b). [§112.10(b)]	☐
Catchment basins or diversion structures are provided to intercept and contain discharges of fuel, crude oil, or oily drilling fluids. [§112.10(c)]	☐
A blowout prevention (BOP) assembly and well control system was installed before drilling below any casing string or during Workover operations. [§112.10(d)]	☐
The BOP assembly and well control system is capable of controlling any well-head pressure that may be encountered while the BOP assembly and well control system are on the well. [§112.10(d)]	☐

ATTACHMENT 1 – Five Year Review and Technical Amendment Logs

ATTACHMENT 1.1 – Five Year Review Log

I have completed a review and evaluation of the SPCC Plan for this facility, and will/will not amend this Plan as a result.

Table G-13 Review and Evaluation of SPCC Plan for Facility			
Review Date	Plan Amendment		Name and signature of person authorized to review this Plan
	Will Amend	Will Not Amend	
	☐	☐	
	☐	☐	
	☐	☐	
	☐	☐	
	☐	☐	
	☐	☐	
	☐	☐	
	☐	☐	

ATTACHMENT 1.2 – Technical Amendment Log
Any technical amendments to this Plan will be re-certified in accordance with Section I of this Plan template.

Table G-14 Description and Certification of Technical Amendments		
Review Date	Description of Technical Amendment	Name and signature of person certifying this technical amendment

ATTACHMENT 2 – Oil Spill Contingency Plan and Checklist

An oil spill contingency plan and written commitment of resources is required for:

- Flowlines and intra-facility gathering lines at oil production facilities and

- Qualified oil-filled operational equipment which has no secondary containment.

An oil spill contingency plan meeting the provisions of 40 CFR part 109, as described below, and a written commitment of manpower, equipment and materials required to expeditiously control and remove any quantity of oil discharged that may be harmful is attached to this Plan.	☐

Complete the checklist below to verify that the necessary operations outlined in 40 CFR part 109 - Criteria for State, Local and Regional Oil Removal Contingency Plans - have been included.

Table G-15 Checklist of Development and Implementation Criteria for State, Local and Regional Oil Removal Contingency Plans (§109.5)[a]	
(a) Definition of the authorities, responsibilities and duties of all persons, organizations or agencies which are to be involved in planning or directing oil removal operations.	☐
(b) Establishment of notification procedures for the purpose of early detection and timely notification of an oil discharge including:	
(1) The identification of critical water use areas to facilitate the reporting of and response to oil discharges.	☐
(2) A current list of names, telephone numbers and addresses of the responsible persons (with alternates) and organizations to be notified when an oil discharge is discovered.	☐
(3) Provisions for access to a reliable communications system for timely notification of an oil discharge, and the capability of interconnection with the communications systems established under related oil removal contingency plans, particularly State and National plans (e.g., NCP).	☐
(4) An established, prearranged procedure for requesting assistance during a major disaster or when the situation exceeds the response capability of the State, local or regional authority.	☐
(c) Provisions to assure that full resource capability is known and can be committed during an oil discharge situation including:	
(1) The identification and inventory of applicable equipment, materials and supplies which are available locally and regionally.	☐
(2) An estimate of the equipment, materials and supplies which would be required to remove the maximum oil discharge to be anticipated.	☐
(3) Development of agreements and arrangements in advance of an oil discharge for the acquisition of equipment, materials and supplies to be used in responding to such a discharge.	☐
(d) Provisions for well defined and specific actions to be taken after discovery and notification of an oil discharge including:	
(1) Specification of an oil discharge response operating team consisting of trained, prepared and available operating personnel.	☐
(2) Predesignation of a properly qualified oil discharge response coordinator who is charged with the responsibility and delegated commensurate authority for directing and coordinating response operations and who knows how to request assistance from Federal authorities operating under existing national and regional contingency plans.	☐

Table G-15 Checklist of Development and Implementation Criteria for State, Local and Regional Oil Removal Contingency Plans (§109.5)[a]	
(3) A preplanned location for an oil discharge response operations center and a reliable communications system for directing the coordinated overall response operations.	☐
(4) Provisions for varying degrees of response effort depending on the severity of the oil discharge.	☐
(5) Specification of the order of priority in which the various water uses are to be protected where more than one water use may be adversely affected as a result of an oil discharge and where response operations may not be adequate to protect all uses.	☐
(6) Specific and well defined procedures to facilitate recovery of damages and enforcement measures as provided for by State and local statutes and ordinances.	☐

[a] The contingency plan must be consistent with all applicable state and local plans, Area Contingency Plans, and the National Contingency Plan (NCP).

ATTACHMENT 3 – Inspections, Dike Drainage and Personnel Training Logs

ATTACHMENT 3.1 – Inspection Log and Schedule

Table G-16 Inspection Log and Schedule

This log is intended to document compliance with §§112.6(a)(3)(iii), 112.8(c)(6), 112.8(d)(4), 112.9(b)(2), 112.9(c)(3), 112.9(d)(1), 112.9(d)(4), 112.12.(c)(6), and 112.12(d)(4), as applicable.

Date of Inspection	Container / Piping / Equipment	Describe Scope (or cite Industry Standard)	Observations	Name/ Signature of Inspector	Records maintained separately [a]
					☐
					☐
					☐
					☐
					☐

[a] Indicate in the table above if records of facility inspections are maintained separately at this facility.

129

ATTACHMENT 3.2 – Bulk Storage Container Inspection Schedule – onshore facilities (excluding production):

To comply with integrity inspection requirement for bulk storage containers, inspect/test each shop-built aboveground bulk storage container on a regular schedule in accordance with a recognized container inspection standard based on the minimum requirements in the following table.

Table G-17 Bulk Storage Container Inspection Schedule	
Container Size and Design Specification	**Inspection requirement**
Portable containers (including drums, totes, and intermodal bulk containers (IBC))	Visually inspect monthly for signs of deterioration, discharges or accumulation of oil inside diked areas
55 to 1,100 gallons with sized secondary containment 1,101 to 5,000 gallons with sized secondary containment and a means of leak detection[a]	Visually inspect monthly for signs of deterioration, discharges or accumulation of oil inside diked areas plus any annual inspection elements per industry inspection standards
1,101 to 5,000 gallons with sized secondary containment and no method of leak detection[a]	Visually inspect monthly for signs of deterioration, discharges or accumulation of oil inside diked areas, plus any annual inspection elements and other specific integrity tests that may be required per industry inspection standards

[a] Examples of leak detection include, but are not limited to, double-walled tanks and elevated containers where a leak can be visually identified.

ATTACHMENT 3.3 – Dike Drainage Log

Table G-18 Dike Drainage Log

Date	Bypass valve sealed closed	Rainwater inspected to be sure no oil (or sheen) is visible	Open bypass valve and reseal it following drainage	Drainage activity supervised	Observations	Signature of Inspector
	☐	☐	☐	☐		
	☐	☐	☐	☐		
	☐	☐	☐	☐		
	☐	☐	☐	☐		
	☐	☐	☐	☐		
	☐	☐	☐	☐		
	☐	☐	☐	☐		
	☐	☐	☐	☐		

ATTACHMENT 3.4 – Oil-handling Personnel Training and Briefing Log

Table G-19 Oil-Handling Personnel Training and Briefing Log		
Date	Description / Scope	Attendees

ATTACHMENT 4 – Discharge Notification Form

In the event of a discharge of oil to navigable waters or adjoining shorelines, the following information will be provided to the National Response Center [also see the notification information provided in Section 7 of the Plan]:

Table G-20 Information provided to the National Response Center in the Event of a Discharge			
Discharge/Discovery Date		Time	
Facility Name			
Facility Location (Address/Lat-Long/Section Township Range)			
Name of reporting individual		Telephone #	
Type of material discharged		Estimated total quantity discharged	Gallons/Barrels
Source of the discharge		Media affected	☐ Soil
			☐ Water (specify)
			☐ Other (specify)
Actions taken			
Damage or injuries	☐ No ☐ Yes (specify)	Evacuation needed?	☐ No ☐ Yes (specify)
Organizations and individuals contacted	☐ National Response Center 800-424-8802 Time _____		
	☐ Cleanup contractor (Specify) Time _____		
	☐ Facility personnel (Specify) Time _____		
	☐ State Agency (Specify) Time _____		
	☐ Other (Specify) Time _____		

[74 FR 58811, Nov. 13, 2009]

PART 113—LIABILITY LIMITS FOR SMALL ONSHORE STORAGE FACILITIES

Subpart A—Oil Storage Facilities

Sec.
113.1 Purpose.
113.2 Applicability.
113.3 Definitions.
113.4 Size classes and associated liability limits for fixed onshore oil storage facilities, 1,000 barrels or less capacity.
113.5 Exclusions.
113.6 Effect on other laws.

AUTHORITY: Sec. 311(f)(2), 86 Stat. 867 (33 U.S.C. 1251 (1972)).

SOURCE: 38 FR 25440, Sept. 13, 1973, unless otherwise noted.

Subpart A—Oil Storage Facilities

§ 113.1 Purpose.

This subpart establishes size classifications and associated liability limits for small onshore oil storage facilities with fixed capacity of 1,000 barrels or less.

§ 113.2 Applicability.

This subpart applies to all onshore oil storage facilities with fixed capacity of 1,000 barrels or less. When a discharge to the waters of the United States occurs from such facilities and when removal of said discharge is performed by the United States Government pursuant to the provisions of subsection 311(c)(1) of the Act, the liability of the owner or operator and the facility will be limited to the amounts specified in § 113.4.

§ 113.3 Definitions.

As used in this subpart, the following terms shall have the meanings indicated below:

(a) *Aboveground* storage facility means a tank or other container, the bottom of which is on a plane not more than 6 inches below the surrounding surface.

(b) *Act* means the Federal Water Pollution Control Act, as amended, 33 U.S.C. 1151, *et seq.*

(c) *Barrel* means 42 United States gallons at 60 degrees Fahrenheit.

(d) *Belowground* storage facility means a tank or other container located other than as defined as "Aboveground".

(e) *Discharge* includes, but is not limited to any spilling, leaking, pumping, pouring, emitting, emptying or dumping.

(f) *Onshore Oil Storage Facility* means any facility (excluding motor vehicles and rolling stock) of any kind located in, on, or under, any land within the United States, other than submerged land.

(g) *On-Scene Coordinator* is the single Federal representative designated pursuant to the National Oil and Hazardous Substances Pollution Contingency Plan and identified in approved Regional Oil and Hazardous Substances Pollution Contingency Plans.

(h) *Oil* means oil of any kind or in any form, including but not limited to, petroleum, fuel oil, sludge, oil refuse, and oil mixed with wastes other than dredged spoil.

(i) *Remove* or *removal* means the removal of the oil from the water and shorelines or the taking of such other actions as the Federal On-Scene Coordinator may determine to be necessary to minimize or mitigate damage to the public health or welfare, including but not limited to, fish, shellfish, wildlife, and public and private property, shorelines, and beaches.

Additionally, the terms not otherwise defined herein shall have the meanings assigned them by section 311(a) of the Act.

§ 113.4 Size classes and associated liability limits for fixed onshore oil storage facilities, 1,000 barrels or less capacity.

Unless the United States can show that oil was discharged as a result of willful negligence or willful misconduct within the privity and knowledge of the owner or operator, the following limits of liability are established for fixed onshore facilities in the classes specified:

(a) Aboveground storage.

Size class	Capacity (barrels)	Limit (dollars)
I	Up to 10	4,000
II	11 to 170	60,000
III	171 to 500	150,000
IV	501 to 1,000	200,000

(b) Belowground storage.

Size class	Capacity (barrels)	Limit (dollars)
I	Up to 10	5,200
II	11 to 170	78,000
III	171 to 500	195,000
IV	501 to 1,000	260,000

§ 113.5 Exclusions.

This subpart does not apply to:

(a) Those facilities whose average daily oil throughout is more than their fixed oil storage capacity.

(b) Vehicles and rolling stock.

§ 113.6 Effect on other laws.

Nothing herein shall be construed to limit the liability of any facility under State or local law or under any Federal law other than section 311 of the Act, nor shall the liability of any facility for any charges or damages under State or local law reduce its liability to the Federal Government under section 311 of the Act, as limited by this subpart.

PART 116—DESIGNATION OF HAZARDOUS SUBSTANCES

AUTHORITY: Secs. 311(b)(2)(A) and 501(a), Federal Water Pollution Control Act (33 U.S.C. 1251 et seq.).

§116.1 Applicability.

This regulation designates hazardous substances under section 311(b)(2)(A) of the Federal Water Pollution Control Act (the Act). The regulation applies to discharges of substances designated in Table 116.4.

[43 FR 10474, Mar. 13, 1978]

§116.2 Abbreviations.

ppm=parts per million
mg=milligram(s)
kg=kilogram(s)
mg/l=milligrams(s) per liter= (approx.) ppm
mg/kg=milligram(s) per kilogram= (approx.) ppm

[43 FR 10474, Mar. 13, 1978]

§116.3 Definitions.

As used in this part, all terms shall have the meaning defined in the Act and as given below:

The Act means the Federal Water Pollution Control Act, as amended by the Federal Water Pollution Control Act Amendments of 1972 (Pub. L. 92–500), and as further amended by the Clean Water Act of 1977 (Pub. L. 95–217), 33 U.S.C. 1251 et seq.; and as further amended by the Clean Water Act Amendments of 1978 (Pub. L. 95–676);

Animals means appropriately sensitive animals which carry out respiration by means of a lung structure permitting gaseous exchange between air and the circulatory system;

Aquatic animals means appropriately sensitive wholly aquatic animals which carry out respiration by means of a gill structure permitting gaseous exchange between the water and the circulatory system;

Aquatic flora means plant life associated with the aquatic eco-system including, but not limited to, algae and higher plants;

Contiguous zone means the entire zone established or to be established by the United States under article 24 of the Convention of the Territorial Sea and the Contiguous Zone;

Discharge includes, but is not limited to, any spilling, leaking, pumping, pouring, emitting, emptying or dumping, but excludes (A) discharges in compliance with a permit under section 402 of this Act, (B) discharges resulting from circumstances identified and reviewed and made a part of the public record with respect to a permit issued or modified under section 402 of this Act, and subject to a condition in such permit, and (C) continuous or anticipated intermittent discharges from a point source, identified in a permit or permit application under section 402 of this Act, which are caused by events occurring within the scope of relevant operating or treatment systems;

LC50 means that concentration of material which is lethal to one-half of the test population of aquatic animals upon continuous exposure for 96 hours or less.

Mixture means any combination of two or more elements and/or compounds in solid, liquid, or gaseous form except where such substances have undergone a chemical reaction so as to become inseparable by physical means.

Navigable waters is defined in section 502(7) of the Act to mean "waters of the United States, including the territorial seas," and includes, but is not limited to:

(1) All waters which are presently used, or were used in the past, or may be susceptible to use as a means to transport interstate or foreign commerce, including all waters which are subject to the ebb and flow of the tide, and including adjacent wetlands; the term *wetlands* as used in this regulation shall include those areas that are inundated or saturated by surface or ground water at a frequency and duration sufficient to support, and that under normal circumstances do support, a prevelance of vegetation typically adapted for life in saturated soil conditions. Wetlands generally include swamps, marshes, bogs and similar areas; the term *adjacent* means bordering, contiguous or neighboring;

(2) Tributaries of navigable waters of the United States, including adjacent wetlands;

(3) Interstate waters, including wetlands; and

(4) All other waters of the United States such as intrastate lakes, rivers, streams, mudflats, sandflats and wetlands, the use, degradation or destruction of which affect interstate commerce including, but not limited to:

(i) Intrastate lakes, rivers, streams, and wetlands which are utilized by

135

interstate travelers for recreational or other purposes; and

(ii) Intrastate lakes, rivers, streams, and wetlands from which fish or shellfish are or could be taken and sold in interstate commerce; and

(iii) Intrastate lakes, rivers, streams, and wetlands which are utilized for industrial purposes by industries in interstate commerce.

Navigable waters do not include prior converted cropland. Notwithstanding the determination of an area's status as prior converted cropland by any other federal agency, for the purposes of the Clean Water Act, the final authority regarding Clean Water Act jurisdiction remains with EPA.

Offshore facility means any facility of any kind located in, on, or under, any of the navigable waters of the United States, and any facility of any kind which is subject to the jurisdiction of the United States and is located in, on, or under any other waters, other than a vessel or a public vessel;

Onshore facility means any facility (including, but not limited to, motor vehicles and rolling stock) of any kind located in, on, or under, any land within the United States other than submerged land;

Otherwise subject to the jurisdiction of the United States means subject to the jurisdiction of the United States by virtue of United States citizenship, United States vessel documentation or numbering, or as provided for by international agreement to which the United States is a party.

A discharge *in connection with activities under the Outer Continental Shelf Lands Act or the Deepwater Port Act of 1974, or which may affect natural resources belonging to, appertaining to, or under the exclusive management authority of the United States (including resources under the Fishery Conservation and Management Act of 1976)*, means: (1) A discharge into any waters beyond the contiguous zone from any vessel or on-shore or offshore facility, which vessel or facility is subject to or is engaged in activities under the Outer Continental Shelf Lands Act or the Deepwater Port Act of 1974, and (2) any discharge into any waters beyond the contiguous zone which contain, cover, or support any natural resource belonging to, appertaining to, or under the exclusive management authority of the United States (including resources under the Fishery Conservation and Management Act of 1976).

Public vessel means a vessel owned or bareboat-chartered and operated by the United States, or a State or political subdivision thereof, or by a foreign nation, except when such vessel is engaged in commerce.

Territorial seas means the belt of the seas measured from the line of ordinary low water along that portion of the coast which is in direct contact with the open sea and the line marking the seaward limit of inland waters, and extending seaward a distance of 3 miles.

Vessel means every description of watercraft or other artificial contrivance used, or capable of being used, as a means of transportation on water other than a public vessel;

[43 FR 10474, Mar. 13, 1978; 43 FR 27533, June 26, 1978, as amended at 44 FR 10266, Feb. 16, 1979; 58 FR 45039, Aug. 25, 1993]

§ 116.4 Designation of hazardous substances.

The elements and compounds appearing in Tables 116.4 A and B are designated as hazardous substances in accordance with section 311(b)(2)(A) of the Act. This designation includes any isomers and hydrates, as well as any solutions and mixtures containing these substances. Synonyms and Chemical Abstract System (CAS) numbers have been added for convenience of the user only. In case of any disparity the common names shall be considered the designated substance.

TABLE 116.4A—LIST OF HAZARDOUS SUBSTANCES

Common name	CAS No.	Synonyms	Isomers	CAS No.
Acetaldehyde	75070	Ethanal, ethyl aldehyde, acetic aldehyde.		
Acetic acid	64197	Glacial acetic acid, vinegar acid.		
Acetic anhydride	108247	Acetic oxide, acetyl oxide.		
Acetone cyanohydrin	75865	2-methyllactonitrile, alpha-hydroxyisobutyronitrile.		

TABLE 116.4A—LIST OF HAZARDOUS SUBSTANCES—Continued

Common name	CAS No.	Synonyms	Isomers	CAS No.
Acetyl bromide	506967			
Acetyl chloride	79367			
Acrolein	107028	2-propenal, acrylic aldehyde, acrylaldehyde, acraldehyde.		
Acrylonitrile	107131	Cyanoethylene, Fumigrain, Ventox, propeneitrile, vinyl cyanide.		
Adipic acid	124049	Hexanedioic acid.		
Aldrin	309002	Octalene, HHDN.		
Allyl alcohol	107186	2-propen-1-ol, 1-propenol-3, vinyl carbinol.		
Allyl chloride	107051	3-chloropropene, 3-chloropropylene, Chlorallylene.		
Aluminum sulfate	10043013	Alum.		
Ammonia	7664417			
Ammonium acetate	631618	Acetic acid ammonium, salt.		
Ammonium benzoate	1863634			
Ammonium bicarbonate	1066337	Acid ammonium carbonate, ammonium hydrogen carbonate.		
Ammonium bichromate	7789095			
Ammonium bifluoride	1341497	Acid ammonium fluoride, ammonium hydrogen fluoride.		
Ammonium bisulfite	10192300			
Ammonium carbamate	1111780	Ammonium aminoformate.		
Ammonium carbonate	506876			
Ammonium chloride	12125029	Ammonium muriate, sal ammoniac, salmiac, Amchlor.		
Ammonium chromate	7788989			
Ammonium citrate dibasic	3012655	Diammonium citrate, citric acid diammonium salt.		
Ammonium fluoborate	13826830	Ammonium fluoroborate, ammonium borofluoride.		
Ammonium fluoride	12125018	Neutral ammonium fluoride.		
Ammonium hydroxide	1336216			
Ammonium oxalate	6009707			
	5972736			
	14258492			
Ammonium silicofluoride	16919190	Ammonium fluosilicate.		
Ammonium sulfamate	7773060	Ammate, AMS, ammonium amidosulfate.		
Ammonium sulfide	12135761			
Ammonium sulfite	10196040			
	10192300			
Ammonium tartrate	3164292	Tartaric acid ammonium salt.		
	14307438			
Ammonium thiocyanate	1762954	Ammonium rhodanide, ammonium sulfocyanate, ammonium sulfocyanide.		
Amly acetate	628637	Amylacetic ester	iso-	123922
		Pear oil	sec-	626380
		Banana oil	tert-	625161
Aniline	62533	Aniline oil, phenylamine, aminobenzene, aminophen, kyanol.		
Antimony pentachloride	7647189			
Antimony potassium tartrate	28300745	Tartar emetic, tartrated antimony, tartarized antimony, potassium antimonyltartrate.		
Antimony tribromide	7789619			
Antimony trichloride	10025919	Butter of antimony.		
Antimony trifluoride	7783564	Antimony fluoride.		
Antimony trioxide	1309644	Diantimony trioxide, flowers of antimony.		
Arsenic disulfide	1303328	Red arsenic sulfide.		
Arsenic pentoxide	1303282	Arsenic acid anhydride, arsenic oxide.		
Arsenic trichloride	7784341	Arsenic chloride, arsenious chloride, arsenous chloride, butter of arsenic.		
Arsenic trioxide	1327533	Arsenious acid, arsenious oxide, white arsenic.		
Arsenic trisulfide	1303339	Arsenious sulfide, yellow arsenic sulfide.		
Barium cyanide	542621			
Benzene	71432	Cyclohexatriene, benzol.		
Benzoic acid	65850	Benzenecarboxylic acid, phenylformic acid, dracylic acid.		
Benzonitrile	100470	Phenyl cyanide, cyanobenzene.		
Benzoyl chloride	98884	Benzenecarbonyl chloride.		
Benzyl chloride	100447			
Beryllium chloride	7787475			
Beryllium fluoride	7787497			
Beryllium nitrate	7787555			

137

TABLE 116.4A—LIST OF HAZARDOUS SUBSTANCES—Continued

Common name	CAS No.	Synonyms	Isomers	CAS No.
Butyl acetate	13597994 123864	Acetic acid butyl ester	iso- sec- tert-	110190 105464 540885
Butylamine	109739	1-aminobutane	iso- sec- sec- tert-	78819 513495 13952846 75649
n/butyl phthalate	84742	1.2-benzenedicarboxylic acid, dibutyl ester, dibutyl phthalate.		
Butyric acid	107926	Butanoic acid, ethylacetic acid	iso-	79312
Cadmium acetate	543908			
Cadmium bromide	7789426			
Cadmium chloride	10108642			
Calcium arsenate	7778441	Tricalcium orthoarsenate.		
Calcium arsenite	52740166			
Calcium carbide	75207	Carbide, acetylenogen.		
Calcium chromate	13765190	Calcium chrome yellow, geblin, yellow ultra-marine.		
Calcium cyanide	592018			
Calcium dodecylbenzenesulfonate	26264062			
Calcium hypochlorite	7778543			
Captan	133062	Orthocide-406, SR–406, Vancide-89.		
Carbaryl	63252	Sevin.		
Carbofuran	1563662	Furadan.		
Carbon disulfide	75150	Carbon bisulfide, dithiocarbonic anhydride.		
Carbon tetrachloride	56235	Tetrachloromethane Perchloromethane.		
Chlordane	57749	Toxichlor, chlordan.		
Chlorine	75003			
Chlorobenzene	108907	Monochlorobenzene, benzene chloride.		
Chloroform	67663	Trichloromethane.		
Chlorpyrifos	2921882	Dursban.		
Chlorosulfonic acid	7790945	Sulfuric chlorohydrin.		
Chromic acetate	1066304			
Chromic acid	11115745	Chromic anhydride, chromium trioxide.		
Chromic sulfate	10101538			
Chromous chloride	10049055			
Cobaltous bromide	7789437	Cobalt bromide.		
Coabaltous formate	544183	Cobalt formate.		
Cobaltous sulfamate	14017415	Cobalt sulfamate.		
Coumaphos	56724	Co-Ral.		
Cresol	1319773	Cresylic acid Hydroxytoluene	m- o- p-	108394 95487 106445
Crotonaldehyde	4170303	2-butenal propylene aldelhyde.		
Cupric acetate	142712	Copper acetate, crystalized verdigris.		
Cupric acetoarsenite	12002038	Copper acetoarsenite, copper acetate arsenite, Paris green.		
Cupric chloride	7447394	Copper chloride.		
Cupric nitrate	3251238	Copper nitrate.		
Cupric oxalate	5893663	Copper oxalate.		
Cupric sulfate	7758987	Copper sulfate.		
Cupric sulfate, ammoniated	10380297	Ammoniated copper sulfate.		
Cupric tartrate	815827	Copper tartrate.		
Cyanogen chloride	506774			
Cyclohexane	110827	Hexahydrobenzene, hexamethylene, hexanaphthene.		
2,4-D acid	94757	2,4-dichlorophenoxyacetic acid.		
2,4-D ester	94111 94791 94804 1320189 1928387 1928616 1929733 2971382 25168267 53467111	2,4-dichlorophenoxyacetic acid ester.		
DDT	50293	p,p′-DDT.		
Diazinon	333415	Dipofene, Diazitol, Basudin, Spectracide.		
Dicamba	1918009	2-methoxy-3,6-dichlorobenzoic acid.		
Dichlobenil	1194656	2,6-dichlorobenzonitrile, 2,6-DBN.		

TABLE 116.4A—LIST OF HAZARDOUS SUBSTANCES—Continued

Common name	CAS No.	Synonyms	Isomers	CAS No.
Dichlone	117806	Phygon, dichloronaphthoquinone.		
Dichlorobenzene	25321226	Di-chloricide	Ortho	95501
		Paramoth (Para)	Para	106467
Dichloropropane	26638197	Propylene dichloride	1,1	78999
			1,2	78875
			1,3	142289
Dichloropropene	26952238	....	1,3	542756
			2,3	78886
Dichloropropene-dichloropropane (mixture).	8003198	D-D mixture Vidden D.		
2,2-Dichloropropionic acid	75990	Dalapon.		
Dichlorvos	62737	2,2-dichlorovinyl dimethyl phosphate, Vapona.		
Dicofol	115322	Di(p-chlorophenyl)-trichloromethylcarbinol, DTMC, dicofol.		
Dieldrin	60571	Alvit.		
Diethylamine	109897			
Dimethylamine	124403			
Dinitrobenzene (mixed)	25154545	Dinitrobenzol	m-	99650
			o-	528290
			p-	100254
Dinitrophenol	51285	Aldifen	(2,5-)	329715
			(2,4-).	
			(2,6-)	573568
Dinitrotoluene	25321146	DNT	2,4	121142
			2,6	606202
			3,4	610399
Diquat	85007	Aquacide.		
	2764729	Dextrone, Reglone, Diquat dibromide.		
Disulfoton	298044	Di-syston.		
Diuron	330541	DCMU, DMU.		
Dodecylbenzenesulfonic acid	27176870			
Endosulfan	115297	Thiodan.		
Endrin	72208	Mendrin, Compound 269.		
Epichlorohydrin	106898	-chloropropylene oxide.		
Ethion	563122	Nialate, ethyl methylene, phosphorodithioate.		
Ethylbenzene	100414	Phenylethane.		
Ethylenediamine	107153	1,2-diaminoethane.		
Ethylenediamine-tetraacetic acid (EDTA).	60004	Edetic acid, Havidote, (ethylenedinitrilo)-tetraacetic acid.		
Ethylene dibromide	106934	1,2-dibromoethane acetylene dibromide sym-dibromoethylene.		
Ethylene dichloride	107062	1,2-dichloroethane sym-bichloroethane.		
Ferric ammonium citrate	1185575	Ammonium ferric citrate.		
Ferric ammonium oxalate	2944674	Ammonium ferric oxalate.		
	55488874			
Ferric chloride	7705080	Flores martis, iron trichloride.		
Ferric fluoride	7783508			
Ferric nitrate	10421484	Iron nitrate.		
Ferric sulfate	10028225	Ferric persulfate, ferric sesquisulfate, ferric tersulfate.		
Ferrous ammonium sulfate	10045893	Mohr's salt, iron ammonium sulfate.		
Ferrous chloride	7758943	Iron chloride, iron dichloride, iron protochloride.		
Ferrous sulfate	7720787	Green vitriol.		
	7782630	Iron vitriol, iron sulfate, iron protosulfate.		
Formaldehyde	50000	Methyl aldehyde, methanal, formalin.		
Formic acid	64186	Methanoic acid.		
Fumaric acid	110178	Trans-butenedioic acid, trans-1,2-ethylenedicarboxylic acid, boletic acid, allomaleic acid.		
Furfural	98011	2-furaldehyde, pyromucic aldehyde.		
Guthion	86500	Gusathion, azinphos-methyl.		
Heptachlor	76448	Velsicol-104, Drinox, Heptagran.		
Hexachlorocyclopentadiene	77474	Perchlorocyclopentadiene.		
Hydrochloric acid	7647010	Hydrogen chloride, muriatic acid.		
Hydrofluoric acid	7664393	Fluohydric acid.		
Hydrogen cyanide	74908	Hydrocyanic acid.		
Hydrogen sulfide	7783064	Hydrosulfuric acid sulfur hydride.		
Isoprene	78795	2-methyl-1,3-butadiene.		
Isopropanolamine dodecylbenzenesulfonate.	42504461			
Kepone	143500	Chlordecone 1,1a,3,3a,4,5,5,5a,5b,6-decachlorooctahydro-1,3,4-metheno-2H-cyclobuta(cd)pentalen-2-one.		

TABLE 116.4A—LIST OF HAZARDOUS SUBSTANCES—Continued

Common name	CAS No.	Synonyms	Isomers	CAS No.
Lead acetate	301042	Sugar of lead.		
Lead arsenate	7784409			
	7645252			
	10102484			
Lead chloride	7758954			
Lead fluoborate	13814965	Lead fluoroborate.		
Lead fluoride	7783462	Lead difluoride, plumbous fluoride.		
Lead iodide	10101630			
Lead nitrate	10099748			
Lead stearate	7428480	Stearic acid lead salt.		
	1072351			
	52652592			
Lead sulfate	7446142			
Lead sulfide	1314870	Galena.		
Lead thiocyanate	592870	Lead sulfocyanate.		
Lindane	58899	Gamma-BHC, gamma-benzene hexachloride.		
Lithium chromate	14307358			
Malathion	121755	Phospothion.		
Maleic acid	110167	Cis-butenedioic acid, cis-1,2-ethylenedicarboxylic acid, toxilic acid.		
Maleic anhydride	108316	2,5-furandione, cis-butenedioic anhydride, toxilic anhydride.		
Mercaptodimethur	203657	Mesurol.		
Mercuric cyanide	592041	Mercury cyanide.		
Mercuric nitrate	10045940	Mercury nitrate, mercury pernitrate.		
Mercuric sulfate	7783359	Mercury sulfate, mercury persulfate.		
Mercuric thiocyanate	592858	Mercury thiocyanate, mercuric sulfocyanate, mercuric sulfocyanide.		
Mercurous nitrate	7782867			
	10415755	Mercury protonitrate.		
Methoxychlor	72435	DMDT, methoxy-DDT.		
Methyl mercaptan	74931	Methanethiol, mercaptomethane, methyl sulfhydrate, thiomethyl alcohol.		
Methyl methacrylate	80626	Methacrylic acid methyl ester, methyl-2-methyl-2-propenoate.		
Methyl parathion	298000	Nitrox-80.		
Mevinphos	7786347	Phosdrin.		
Mexacarbate	315184	Zectran.		
Monoethylamine	75047	Ethylamine, aminoethane.		
Monomethylamine	74895	Methylamine, aminomethane.		
Naled	300765	Dibrom.		
Naphthalene	91203	White tar, tar camphor, naphthalin.		
Naphthenic acid	1338245	Cyclohexanecarboxylic acid, hexahydrobenzoic acid.		
Nickel ammonium sulfate	15699180	Ammonium nickel sulfate.		
Nickel chloride	37211055	Nickelous chloride.		
	7718549			
Nickel hydroxide	12054487	Nickelous hydroxide.		
Nickel nitrate	14216752			
Nickel sulfate	7786814	Nickelous sulfate.		
Nitric acid	7697372	Aqua fortis.		
Nitrobenzene	98953	Nitrobenzol, oil of mirbane.		
Nitrogen dioxide	10102440	Nitrogen tetraoxide.		
Nitrophenol (mixed)	25154556	Mononitrophenol	m-	554847
			o-	88755
			p-	100027
Nitrotoluene	1321126		Ortho	88722
			Meta	99081
			Para	99990
Paraformaldehyde	30525894	Paraform, Formagene, Triformol, polymerized formaldehyde, polyoxymethylene.		
Parathion	56382	DNTP, Niran.		
Pentachlorophenol	87865	PCP, Penta.		
Phenol	108952	Carbolic acid, phenyl hydroxide, hydroxybenzene, oxybenzene.		
Phosgene	75445	Diphosgene, carbonyl chloride, chloroformyl chloride.		
Phosphoric acid	7664382	Orthophosphoric acid.		
Phosphorus	7723140	Black phosphorus, red phosphorus, white phosphorus, yellow phosphorus.		
Phosphorus oxychloride	10025873	Phosphoryl chloride, phosphorus chloride.		
Phosphorus pentasulfide	1314803	Phosphoric sulfide, thiophosphoric anhydride, phosphorus persulfide.		

TABLE 116.4A—LIST OF HAZARDOUS SUBSTANCES—Continued

Common name	CAS No.	Synonyms	Isomers	CAS No.
Phosphorus trichloride	7719122	Phosphorous chloride.		
Polychorinated biphenyls	1336363	PCB, Aroclor, polychlorinated diphenyls.		
Potassium arsenate	7784410			
Potassium arsenite	10124502	Potassium metaarsenite.		
Potassium bichromate	7778509	Potassium dichromate.		
Potassium chromate	7789006			
Potassium cyanide	151508			
Potassium hydroxide	1310583	Potassium hydrate, caustic potash, potassa.		
Potassium permanganate	7722647	Chameleon mineral.		
Propargite	2312358	Omite.		
Propionic acid	79094	Propanoic acid, methylacetic acid, ethylformic acid.		
Propionic anhydride	123626	Propanoic anhydride, methylacetic anhydride.		
Propylene oxide	75569	Propene oxide.		
Pyrethrins	121299	Pyrethrin I.		
	121211	Pyrethrin II.		
Quinoline	91225	1-benzazine, benzo(b)pyridine, leuocoline, chinoleine, leucol.		
Resorcinol	108463	Resorcin, 1,3-benzenediol, meta-dihydroxybenzene.		
Selenium oxide	7446084	Selenium dioxide.		
Silver nitrate	7761888	Nitric acid silver (1+) salt lunar caustic.		
Sodium	7440235	Natrium.		
Sodium arsenate	7631892	Disodium arsenate.		
Sodium arsenite	7784465	Sodium metaarsenite.		
Sodium bichromate	10588019	Sodium dichromate.		
Sodium bifluoride	1333831			
Sodium bisulfite	7631905	Sodium acid sulfite, sodium hydrogen sulfite.		
Sodium chromate	7775113			
Sodium cyanide	143339			
Sodium dodecylbenzene-sulfonate	25155300			
Sodium fluoride	7681494	Villiaumite.		
Sodium hydrosulfide	16721805	Sodium hydrogen sulfide.		
Sodium hydroxide	1310732	Caustic soda, soda lye, sodium hydrate.		
Sodium hypochlorite	7681529	Bleach.		
	10022705			
Sodium methylate	124414	Sodium methoxide.		
Sodium nitrite	7632000			
Sodium phosphate, dibasic	7558794			
	10039324			
	10140655			
Sodium phosphate, tribasic	7601549			
	10101890			
	10361894			
Sodium selenite	10102188			
	7782823			
Strontium chromate	7789062			
Strychnine	57249			
Styrene	100425	Vinylbenzene, phenylethylene, styrol, styrolene, cinnamene, cinnamol.		
Sulfuric acid	7664939	Oil of vitriol, oleum.		
Sulfur monochloride	12771083	Sulfur chloride.		
2,4,5-T acid	93765	2,4,5-trichlorophenoxyacetic acid.		
2,4,5-T amines	6369966	Acetic acid (2,4,5-trichlorophenoxy)-compound with N,N-dimethylmethanamine (1:1).		
	6369977	Acetic acid (2,4,5-trichlorophenoxy)-compound with N-methylmethanamine (1:1).		
	1319728	Acetic acid (2,4,5-trichlorophenoxy)-compound with 1-amino-2-propanol (1:1).		
	3813147	Acetic acid (2,4,5-trichlorophenoxy)-compound with 2,2'2''-nitrilotris [ethanol] (1:1).		
2,4,5-T esters	2545597	2,4,5-trichlorophenoxyacetic esters.		
	93798			
	61792072			
	1928478			
	25168154			
2,4,5-T salts	13560991	Acetic acid (2,4,5-trichlorophenoxy)-sodium salt.		
TDE	72548	DDD.		
2,4,5-TP acid	93721	Propanoic acid 2-(2,4,5-trichlorophenoxy).		
2,4,5-TP esters	32534955	Propanoic acid, 2-(2,4,5-trichlorophenoxy)-, isooctyl ester.		

TABLE 116.4A—LIST OF HAZARDOUS SUBSTANCES—Continued

Common name	CAS No.	Synonyms	Isomers	CAS No.
Tetraethyl lead	78002	Lead tetraethyl, TEL.		
Tetraethyl pyrophosphate	107493	TEPP.		
Thallium sulfate	10031591			
	7446186			
Toluene	108883	Toluol, methylbenzene, phenylmethane, Methacide.		
Toxaphene	8001352	Camphechlor.		
Trichlorfon	52686	Dipterex		
		Dylox.		
Trichlorethylene	79016	Ethylene trichloride.		
Trichlorophenol	25167822	Collunosol, Dowicide 2 or 2S, Omal, Phenachlor.	(2,3,4-)	15950660
			(2,3,5-)	933788
			(2,3,6-)	933755
			(2,4,5-)	95954
			(2,4,6-)	88062
			(3,4,5-)	609198
Triethanolamine dodecylbenzenesulfonate.	27323417			
Triethylamine	121448			
Trimethylamine	75503	TMA.		
Uranyl acetate	541093			
Uranyl nitrate	10102064			
	36478769			
Vanadium pentoxide	1314621	Vanadic anhydride, vanadic acid anhydride.		
Vanadyl sulfate	27774136	Vanadic sulfate, vanadium sulfate.		
Vinyl acetate	108054	Acetic acid ethylene ether.		
Vinylidene chloride	75354	1,1–dichlorethylene.		
		1,1–dichloroethene.		
Xylene (mixed)	1330207	Dimethylbenzene	m-	108383
		Xylol	o-	95476
			p-	106423
Xylenol	1300716	Dimethylphenol, hydroxydimethylbenzene.		
Zinc acetate	557346			
Zinc ammonium chloride	14639975			
	14639986			
	52628258			
Zinc borate	1332076			
Zinc bromide	7699458			
Zinc carbonate	3486359			
Zinc chloride	7646857	Butter of zinc.		
Zinc cyanide	557211			
Zinc fluoride	7783495			
Zinc formate	557415			
Zinc hydrosulfite	7779864			
Zinc nitrate	7779886			
Zinc phenolsulfonate	127822	Zinc sulfocarbolate.		
Zinc phosphide	1314847			
Zinc silicofluoride	16871719	Zinc fluosilicate.		
Zinc sulfate	7733020	White vitriol, zinc vitriol, white copperas.		
Zirconium nitrate	13746899			
Zirconium potassium fluoride	16923958			
Zirconium sulfate	14644612	Disulfatozirconic acid.		
Zirconium tetrachloride	10026116			

TABLE 116.4B—LIST OF HAZARDOUS SUBSTANCES BY CAS NUMBER

TABLE 116.4B—LIST OF HAZARDOUS SUBSTANCES BY CAS NUMBER—Continued

CAS No.	Common name
50000	Formaldehyde
50293	DDT
51285	2,4-Dinitrophenol
52686	Trichlorfon
56382	Parathion
56724	Coumaphos
57249	Strychnine
57749	Chlordane
58899	Lindane
60004	Ethylenediaminetetraacetic acid (EDTA)
60571	Dieldrin

CAS No.	Common name
62533	Aniline
62737	Dichlorvos
63252	Carbaryl
64186	Formic acid
64197	Acetic acid
65850	Benzoic acid
67663	Chloroform
71432	Benzene
72208	Endrin
72435	Methoxychlor
72548	TDE
74895	Monomethylamine

TABLE 116.4B—LIST OF HAZARDOUS SUBSTANCES BY CAS NUMBER—Continued

CAS No.	Common name
74908	Hydrogen cyanide
74931	Methyl mercaptan
75047	Monoethylamine
75070	Acetaldehyde
75150	Carbon disulfide
75207	Calcium carbide
75445	Phosgene
75503	Trimethylamine
75649	tert-Butylamine
75865	Acetone cyanohydrin
75990	2,2-Dichloropropionic acid
76448	Heptachlor
78002	Tetraethyl lead
78795	Isoprene
78819	iso-Butylamine
79094	Propionic acid
79312	iso-Butyric acid
79367	Acetyl chloride
80626	Methyl methacrylate
85007	Diquat
86500	Guthion
87865	Pentachlorophenol
88755	o-Nitrophenol
91203	Naphthalene
91225	Quinoline
93765	2,4,5-T acid
93798	2,4,5-T ester
94111	2,4-D ester
94757	2,4-D acid
94791	2,4-D ester
94804	2,4-D Butyl ester
95476	o-Xylene
95487	o-Cresol
98011	Furfural
98884	Benzoyl chloride
98953	Nitrobenzene
99650	m-Dinitrobenzene
100027	p-Nitrophenol
100254	p-Dinitrobenzene
100414	Ethylbenzene
100425	Styrene
100447	Benzyl chloride
100470	Benzonitrile
105464	sec-Butyl acetate
106423	p-Xylene
106445	p-Cresol
107028	Acrolein
107051	Allyl chloride
107131	Acrylonitrile
107153	Ethylenediamine
107186	Allyl alcohol
107493	Tetraethyl pyrophosphate
107926	n-Butyric acid
108054	Vinyl acetate
108247	Acetic anhydride
108316	Maleic anhydride
108383	m-Xylene
108394	m-Cresol
108463	Resorcinol
108883	Toluene
108907	Chlorobenzene
108952	Phenol
109739	n-Butylamine
109897	Diethylamine
110167	Maleic acid
110178	Fumaric acid
110190	iso-Butyl acetate
110827	Cyclohexane
115297	Endosulfan
115322	Dicofol
117806	Dichlone

TABLE 116.4B—LIST OF HAZARDOUS SUBSTANCES BY CAS NUMBER—Continued

CAS No.	Common name
121211	Pyrethrin
121299	Pyrethrin
121448	Triethylamine
121755	Malathion
123626	Propionic anhydride
123864	n-Butyl acetate
123922	iso-Amyl acetate
124403	Dimethylamine
124414	Sodium methylate
127822	Zinc phenolsulfonate
133062	Captan
142712	Cupric acetate
143339	Sodium cyanide
151508	Potassium cyanide
298000	Methyl parathion
298044	Disulfoton
300765	Naled
301042	Lead acetate
309002	Aldrin
315184	Mexacarbate
329715	2,5-Dinitrophenol
330541	Diuron
333415	Diazinon
506774	Cyanogen chloride
506876	Ammonium carbonate
506967	Acetyl bromide
513495	sec-Butylamine
528290	o-Dinitrobenzene
540885	tert-Butyl acetate
541093	Uranyl acetate
542621	Barium cyanide
543908	Cadmium acetate
544183	Cobaltous formate
554847	m-Nitrophenol
557211	Zinc cyanide
557346	Zinc acetate
557415	Zinc formate
563122	Ethion
573568	2,6-Dinitrophenol
592018	Calcium cyanide
592041	Mercuric cyanide
592858	Mercuric thiocyanate
592870	Lead thiocyanate
625161	tert-Amyl acetate
626380	sec-Amyl acetate
628637	n-Amyl acetate
631618	Ammonium acetate
815827	Cupric tartrate
1066304	Chromic acetate
1066337	Ammonium bicarbonate
1072351	Lead stearate
1111780	Ammonium carbamate
1185575	Ferric ammonium citrate
1194656	Dichlobenil
1300716	Xylenol
1303282	Arsenic pentoxide
1303328	Arsenic disulfide
1303339	Arsenic trisulfide
1309644	Antimony trioxide
1310583	Potassium hydroxide
1310732	Sodium hydroxide
1314621	Vanadium pentoxide
1314803	Phosphorus pentasulfide
1314847	Zinc phosphide
1314870	Lead sulfide
1319773	Cresol (mixed)
1320189	2,4-D ester
1327533	Arsenic trioxide
1330207	Xylene
1332076	Zinc borate
1333831	Sodium bifluoride

143

TABLE 116.4B—LIST OF HAZARDOUS
SUBSTANCES BY CAS NUMBER—Continued

TABLE 116.4B—LIST OF HAZARDOUS
SUBSTANCES BY CAS NUMBER—Continued

CAS No.	Common name	CAS No.	Common name
1336216	Ammonium hydroxide	7783564	Antimony trifluoride
1336363	Polychlorinated biphenyls	7784341	Arsenic trichloride
1338245	Naphthenic acid	7784409	Lead arsenate
1341497	Ammonium bifluoride	7784410	Potassium arsenate
1762954	Ammonium thiocyanate	7784465	Sodium arsenite
1863634	Ammonium benzoate	7786347	Mevinphos
1918009	Dicamba	7786814	Nickel sulfate
1928387	2,4-D esters	7787475	Beryllium chloride
1928478	2,4,5-T esters	7787497	Beryllium fluoride
1928616	2,4-D ester	7787555	Beryllium nitrate
1929733	2,4-D ester	7788989	Ammonium chromate
2545597	2,4,5-T ester	7789006	Potassium chromate
2764729	Diquat	7789062	Strontium chromate
2921882	Chlorpyrifos	7789095	Ammonium bichromate
2944674	Ferric ammonium oxalate	7789426	Cadmium bromide
2971382	2,4-D ester	7789437	Cobaltous bromide
3012655	Ammonium citrate, dibasic	7789619	Antimony tribromide
3164292	Ammonium tartrate	7790945	Chlorosulfonic acid
3251238	Cupric nitrate	8001352	Toxaphene
3486359	Zinc carbonate	10022705	Sodium hypochlorite
5893663	Cupric oxalate	10025873	Phosphorus oxychloride
5972736	Ammonium oxalate	10025919	Antimony trichloride
6009707	Ammonium oxalate	10026116	Zirconium tetrachloride
6369966	2,4,5-T ester	10028225	Ferric sulfate
7428480	Lead stearate	10028247	Sodium phosphate, dibasic
7440235	Sodium	10039324	Sodium phosphate, dibasic
7446084	Selenium oxide	10043013	Aluminum sulfate
7446142	Lead sulfate	10045893	Ferrous ammonium sulfate
7447394	Cupric chloride	10045940	Mercuric nitrate
7558794	Sodium phosphate, dibasic	10049055	Chromous chloride
7601549	Sodium phosphate, tribasic	10099748	Lead nitrate
7631892	Sodium arsenate	10101538	Chromic sulfate
7631905	Sodium bisulfite	10101630	Lead iodide
7632000	Sodium nitrite	10101890	Sodium phosphate, tribasic
7645252	Lead arsenate	10102064	Uranyl nitrate
7646857	Zinc chloride	10102188	Sodium selenite
7647010	Hydrochloric acid	10102440	Nitrogen dioxide
7647189	Antimony pentachloride	10102484	Lead arsenate
7664382	Phosphoric acid	10108642	Cadmium chloride
7664393	Hydrofluoric acid	10124502	Potassium arsenite
7664417	Ammonia	10140655	Sodium phosphate, dibasic
7664939	Sulfuric acid	10192300	Ammonium bisulfite
7681494	Sodium fluoride	10196040	Ammonium sulfite
7681529	Sodium hypochlorite	10361894	Sodium phosphate, tribasic
7697372	Nitric acid	10380297	Cupric sulfate, ammoniated
7699458	Zinc bromide	10415755	Mercurous nitrate
7705080	Ferric chloride	10421484	Ferric nitrate
7718549	Nickel chloride	10588019	Sodium bichromate
7719122	Phosphorus trichloride	11115745	Chromic acid
7720787	Ferrous sulfate	12002038	Cupric acetoarsenite
7722647	Potassium permanganate	12054487	Nickel hydroxide
7723140	Phosphorus	12125018	Ammonium fluoride
7733020	Zinc sulfate	12125029	Ammonium chloride
7758943	Ferrous chloride	12135761	Ammonium sulfide
7758954	Lead chloride	12771083	Sulfur chloride
7758987	Cupric sulfate	13597994	Beryllium nitrate
7773060	Ammonium sulfamate	13746899	Zirconium nitrate
7775113	Sodium chromate	13765190	Calcium chromate
7778441	Calcium arsenate	13814965	Lead fluoborate
7778509	Potassium bichromate	13826830	Ammonium fluoborate
7778543	Calcium hypochlorite	13952846	sec-Butylamine
7779864	Zinc hydrosulfite	14017415	Cobaltous sulfamate
7779886	Zinc nitrate	14216752	Nickel nitrate
7782505	Chlorine	14258492	Ammonium oxalate
7782630	Ferrous sulfate	14307358	Lithium chromate
7782823	Sodium selenite	14307438	Ammonium tartrate
7782867	Mercurous nitrate	14639975	Zinc ammonium chloride
7783359	Mercuric sulfate	14639986	Zinc ammonium chloride
7783462	Lead fluoride	14644612	Zirconium sulfate
7783495	Zinc fluoride	15699180	Nickel ammonium sulfate
7783508	Ferric fluoride	16721805	Sodium hydrosulfide

TABLE 116.4B—LIST OF HAZARDOUS
SUBSTANCES BY CAS NUMBER—Continued

CAS No.	Common name
16871719	Zinc silicofluoride
16919190	Ammonium silicofluoride
16923958	Zirconium potassium fluoride
25154545	Dinitrobenzene
25154556	Nitrophenol
25155300	Sodium dodecylbenzenesulfonate
25167822	Trichlorophenol
25168154	2,4,5-T ester
25168267	2,4-D ester
26264062	Calcium dodecylbenzenesulfonate
27176870	Dodecylbenzenesulfonic acid
27323417	Triethanolamine dodecylbenzenesulfonate
27774136	Vanadyl sulfate
28300745	Antimony potassium tartrate
30525894	Paraformaldehyde
36478769	Uranyl nitrate
37211055	Nickel chloride
42504461	Dodecylbenzenesulfonate isopropanolamine
52628258	Zinc ammonium chloride
52740166	Calcium arsenite
53467111	2,4-D ester
55488874	Ferric ammonium oxalate
61792072	2,4,5-T ester

[43 FR 10474, Mar. 13, 1978; 43 FR 27533, June 26, 1978, as amended at 44 FR 10268, Feb. 16, 1979; 44 FR 65400, Nov. 13, 1979; 44 FR 66602, Nov. 20, 1979; 54 FR 33482, Aug. 14, 1989; 76 FR 55584, Sept. 8, 2011]

PART 117—DETERMINATION OF RE-PORTABLE QUANTITIES FOR HAZARDOUS SUBSTANCES

Subpart A—General Provisions

Sec.
117.1 Definitions.
117.2 Abbreviations.
117.3 Determination of reportable quantities.

Subpart B—Applicability

117.11 General applicability.
117.12 Applicability to discharges from facilities with NPDES permits.
117.13 Applicability to discharges from publicly owned treatment works and their users.
117.14 Demonstration projects.

Subpart C—Notice of Discharge of a Reportable Quantity

117.21 Notice.
117.23 Liabilities for removal.

AUTHORITY: Secs. 311 and 501(a), Federal Water Pollution Control Act (33 U.S.C. 1251 et seq.), ("the Act") and Executive Order 11735, superseded by Executive Order 12777, 56 FR 54757.

SOURCE: 44 FR 50776, Aug. 29, 1979, unless otherwise noted.

Subpart A—General Provisions

§117.1 Definitions.

As used in this part, all terms shall have the meanings stated in 40 CFR part 116.

(a) *Reportable quantities* means quantities that may be harmful as set forth in §117.3, the discharge of which is a violation of section 311(b)(3) and requires notice as set forth in §117.21.

(b) *Administrator* means the Administrator of the Environmental Protection Agency ("EPA").

(c) *Mobile source* means any vehicle, rolling stock, or other means of transportation which contains or carries a reportable quantity of a hazardous substance.

(d) *Public record* means the NPDES permit application or the NPDES permit itself and the materials comprising the administrative record for the permit decision specified in §124.18 of this chapter.

(e) *National Pretreatment Standard* or *Pretreatment Standard* means any regulation containing pollutant discharge limits promulgated by the EPA in accordance with section 307 (b) and (c) of the Act, which applies to industrial users of a publicly owned treatment works. It further means any State or local pretreatment requirement applicable to a discharge and which is incorporated into a permit issued to a publicly owned treatment works under section 402 of the Act.

(f) *Publicly Owned Treatment Works* or *POTW* means a treatment works as defined by section 212 of the Act, which is owned by a State or municipality (as defined by section 502(4) of the Act). This definition includes any sewers that convey wastewater to such a treatment works, but does not include pipes, sewers or other conveyances not connected to a facility providing treatment. The term also means the municipality as defined in section 502(4) of the Act, which has jurisdiction over the indirect discharges to and the discharges from such a treatment works.

(g) *Remove* or *removal* refers to removal of the oil or hazardous substances from the water and shoreline or the taking of such other actions as may be necessary to minimize or mitigate damage to the public health or welfare, including, but not limited to, fish, shellfish, wildlife, and public and private property, shorelines, and beaches.

(h) *Contiguous zone* means the entire zone established by the United States under Article 24 of the Convention on the Territorial Sea and Contiguous Zone.

(i) *Navigable waters* means "waters of the United States, including the territorial seas." This term includes:

(1) All waters which are currently used, were used in the past, or may be susceptible to use in interstate or foreign commerce, including all waters which are subject to the ebb and flow of the tide;

(2) Interstate waters, including interstate wetlands;

(3) All other waters such as intrastate lakes, rivers, streams, (including intermittent streams), mudflats, sandflats, and wetlands, the use, degradation or destruction of which would affect or could affect interstate or foreign commerce including any such waters:

(i) Which are or could be used by interstate or foreign travelers for recreational or other purposes;

(ii) From which fish or shellfish are or could be taken and sold in interstate or foreign commerce;

(iii) Which are used or could be used for industrial purposes by industries in interstate commerce;

(4) All impoundments of waters otherwise defined as navigable waters under this paragraph;

(5) Tributaries of waters identified in paragraphs (i) (1) through (4) of this section, including adjacent wetlands; and

(6) Wetlands adjacent to waters identified in paragraphs (i) (1) through (5) of this section ("Wetlands" means those areas that are inundated or saturated by surface or ground water at a frequency and duration sufficient to support, and that under normal circumstances do support, a prevalence of vegetation typically adapted for life in saturated soil conditions. Wetlands generally included playa lakes, swamps, marshes, bogs, and similar areas such as sloughs, prairie potholes, wet meadows, prairie river overflows, mudflats, and natural ponds): *Provided,* That waste treatment systems (other than cooling ponds meeting the criteria of this paragraph) are not waters of the United States.

Navigable waters do not include prior converted cropland. Notwithstanding the determination of an area's status as prior converted cropland by any other federal agency, for the purposes of the Clean Water Act, the final authority regarding Clean Water Act jurisdiction remains with EPA.

(j) *Process waste water* means any water which, during manufacturing or processing, comes into direct contact with or results from the production or use of any raw material, intermediate product, finished product, byproduct, or waste product.

[44 FR 50776, Aug. 29, 1979, as amended at 58 FR 45039, Aug. 25, 1993; 65 FR 30904, May 15, 2000]

§ 117.2 Abbreviations.

NPDES equals National Pollutant Discharge Elimination System. RQ equals reportable quantity.

§ 117.3 Determination of reportable quantities.

Each substance in Table 117.3 that is listed in Table 302.4, 40 CFR part 302, is assigned the reportable quantity listed in Table 302.4 for that substance.

TABLE 117.3—REPORTABLE QUANTITIES OF HAZARDOUS SUBSTANCES DESIGNATED PURSUANT TO SECTION 311 OF THE CLEAN WATER ACT

NOTE: The first number under the column headed "RQ" is the reportable quantity in pounds. The number in parentheses is the metric equivalent in kilograms. For convenience, the table contains a column headed "Category" which lists the code letters "X", "A", "B", "C", and "D" associated with reportable quantities of 1, 10, 100, 1000, and 5000 pounds, respectively.

TABLE 117.3—REPORTABLE QUANTITIES OF HAZARDOUS SUBSTANCES DESIGNATED PURSUANT TO SECTION 311 OF THE CLEAN WATER ACT

Material	Category	RQ in pounds (kilograms)
Acetaldehyde	C	1,000 (454)
Acetic acid	D	5,000 (2,270)
Acetic anhydride	D	5,000 (2,270)
Acetone cyanohydrin	A	10 (4.54)
Acetyl bromide	D	5,000 (2,270)
Acetyl chloride	D	5,000 (2,270)
Acrolein	X	1 (0.454)
Acrylonitrile	B	100 (45.4)
Adipic acid	D	5,000 (2,270)
Aldrin	X	1 (0.454)
Allyl alcohol	B	100 (45.4)
Allyl chloride	C	1,000 (454)
Aluminum sulfate	D	5,000 (2,270)
Ammonia	B	100 (45.4)
Ammonium acetate	D	5,000 (2,270)
Ammonium benzoate	D	5,000 (2,270)
Ammonium bicarbonate	D	5,000 (2,270)
Ammonium bichromate	A	10 (4.54)
Ammonium bifluoride	B	100 (45.4)
Ammonium bisulfite	D	5,000 (2,270)
Ammonium carbamate	D	5,000 (2,270)
Ammonium carbonate	D	5,000 (2,270)
Ammonium chloride	D	5,000 (2,270)
Ammonium chromate	A	10 (4.54)
Ammonium citrate dibasic	D	5,000 (2,270)
Ammonium fluoborate	D	5,000 (2,270)
Ammonium fluoride	B	100 (45.4)
Ammonium hydroxide	C	1,000 (454)
Ammonium oxalate	D	5,000 (2,270)
Ammonium silicofluoride	C	1,000 (454)
Ammonium sulfamate	D	5,000 (2,270)
Ammonium sulfide	B	100 (45.4)
Ammonium sulfite	D	5,000 (2,270)
Ammonium tartrate	D	5,000 (2,270)
Ammonium thiocyanate	D	5,000 (2,270)
Amyl acetate	D	5,000 (2,270)
Aniline	D	5,000 (2,270)
Antimony pentachloride	C	1,000 (454)
Antimony potassium tartrate	B	100 (45.4)
Antimony tribromide	C	1,000 (454)
Antimony trichloride	C	1,000 (454)
Antimony trifluoride	C	1,000 (454)
Antimony trioxide	C	1,000 (454)
Arsenic disulfide	X	1 (0.454)
Arsenic pentoxide	X	1 (0.454)
Arsenic trichloride	X	1 (0.454)
Arsenic trioxide	X	1 (0.454)
Arsenic trisulfide	X	1 (0.454)
Barium cyanide	A	10 (4.54)
Benzene	A	10 (4.54)
Benzoic acid	D	5,000 (2,270)
Benzonitrile	D	5,000 (2,270)
Benzoyl chloride	C	1,000 (454)
Benzyl chloride	B	100 (45.4)
Beryllium chloride	X	1 (0.454)
Beryllium fluoride	X	1 (0.454)
Beryllium nitrate	X	1 (0.454)
Butyl acetate	D	5,000 (2,270)
Butylamine	C	1,000 (454)
n-Butyl phthalate	A	10 (4.54)
Butyric acid	D	5,000 (2,270)
Cadmium acetate	A	10 (4.54)
Cadmium bromide	A	10 (4.54)
Cadmium chloride	A	10 (4.54)
Calcium arsenate	X	1 (0.454)
Calcium arsenite	X	1 (0.454)
Calcium carbide	A	10 (4.54)
Calcium chromate	A	10 (4.54)
Calcium cyanide	A	10 (4.54)
Calcium dodecylbenzenesulfonate	C	1,000 (454)
Calcium hypochlorite	A	10 (4.54)

TABLE 117.3—REPORTABLE QUANTITIES OF HAZARDOUS SUBSTANCES DESIGNATED PURSUANT TO SECTION 311 OF THE CLEAN WATER ACT—Continued

Material	Category	RQ in pounds (kilograms)
Captan	A	10 (4.54)
Carbaryl	B	100 (45.4)
Carbofuran	A	10 (4.54)
Carbon disulfide	B	100 (45.4)
Carbon tetrachloride	A	10 (4.54)
Chlordane	X	1 (0.454)
Chlorine	A	10 (4.54)
Chlorobenzene	B	100 (45.4)
Chloroform	A	10 (4.54)
Chlorosulfonic acid	C	1,000 (454)
Chlorpyrifos	X	1 (0.454)
Chromic acetate	C	1,000 (454)
Chromic acid	A	10 (4.54)
Chromic sulfate	C	1,000 (454)
Chromous chloride	C	1,000 (454)
Cobaltous bromide	C	1,000 (454)
Cobaltous formate	C	1,000 (454)
Cobaltous sulfamate	C	1,000 (454)
Coumaphos	A	10 (4.54)
Cresol	B	100 (45.4)
Crotonaldehyde	B	100 (45.4)
Cupric acetate	B	100 (45.4)
Cupric acetoarsenite	X	1 (0.454)
Cupric chloride	A	10 (4.54)
Cupric nitrate	B	100 (45.4)
Cupric oxalate	B	100 (45.4)
Cupric sulfate	A	10 (4.54)
Cupric sulfate, ammoniated	B	100 (45.4)
Cupric tartrate	B	100 (45.4)
Cyanogen chloride	A	10 (4.54)
Cyclohexane	C	1,000 (454)
2,4-D Acid	B	100 (45.4)
2,4-D Esters	B	100 (45.4)
DDT	X	1 (0.454)
Diazinon	X	1 (0.454)
Dicamba	C	1,000 (454)
Dichlobenil	B	100 (45.4)
Dichlone	X	1 (0.454)
Dichlorobenzene	B	100 (45.4)
Dichloropropane	C	1,000 (454)
Dichloropropene	B	100 (45.4)
Dichloropropene-Dichloropropane (mixture)	B	100 (45.4)
2,2-Dichloropropionic acid	D	5,000 (2,270)
Dichlorvos	A	10 (4.54)
Dicofol	A	10 (4.54)
Dieldrin	X	1 (0.454)
Diethylamine	B	100 (45.4)
Dimethylamine	C	1,000 (454)
Dinitrobenzene (mixed)	B	100 (45.4)
Dinitrophenol	A	10 (45.4)
Dinitrotoluene	A	10 (4.54)
Diquat	C	1,000 (454)
Disulfoton	X	1 (0.454)
Diuron	B	100 (45.4)
Dodecylbenzenesulfonic acid	C	1,000 (454)
Endosulfan	X	1 (0.454)
Endrin	X	1 (0.454)
Epichlorohydrin	B	100 (45.4)
Ethion	A	10 (4.54)
Ethylbenzene	C	1,000 (454)
Ethylenediamine	D	5,000 (2,270)
Ethylenediamine-tetraacetic acid (EDTA)	D	5,000 (2,270)
Ethylene dibromide	X	1 (0.454)
Ethylene dichloride	B	100 (45.4)
Ferric ammonium citrate	C	1,000 (454)
Ferric ammonium oxalate	C	1,000 (454)
Ferric chloride	C	1,000 (454)
Ferric fluoride	B	100 (45.4)
Ferric nitrate	C	1,000 (454)
Ferric sulfate	C	1,000 (454)
Ferrous ammonium sulfate	C	1,000 (454)

TABLE 117.3—REPORTABLE QUANTITIES OF HAZARDOUS SUBSTANCES DESIGNATED PURSUANT TO SECTION 311 OF THE CLEAN WATER ACT—Continued

Material	Category	RQ in pounds (kilograms)
Ferrous chloride	B	100 (45.4)
Ferrous sulfate	C	1,000 (454)
Formaldehyde	B	100 (45.4)
Formic acid	D	5,000 (2,270)
Fumaric acid	D	5,000 (2,270)
Furfural	D	5,000 (2,270)
Guthion	X	1 (0.454)
Heptachlor	X	1 (0.454)
Hexachlorocyclopentadiene	A	10 (4.54)
Hydrochloric acid	D	5,000 (2,270)
Hydrofluoric acid	B	100 (45.4)
Hydrogen cyanide	A	10 (4.54)
Hydrogen sulfide	B	100 (45.4)
Isoprene	B	100 (45.4)
Isopropanolamine dodecylbenzenesulfonate	C	1,000 (454)
Kepone	X	1 (0.454)
Lead acetate	A	10 (4.54)
Lead arsenate	X	1 (0.454)
Lead chloride	A	10 (4.54)
Lead fluoborate	A	10 (4.54)
Lead fluoride	A	10 (4.54)
Lead iodide	A	10 (4.54)
Lead nitrate	A	10 (4.54)
Lead stearate	A	10 (4.54)
Lead sulfate	A	10 (4.54)
Lead sulfide	A	10 (4.54)
Lead thiocyanate	A	10 (4.54)
Lindane	X	1 (0.454)
Lithium chromate	A	10 (4.54)
Malathion	B	100 (45.4)
Maleic acid	D	5,000 (2,270)
Maleic anhydride	D	5,000 (2,270)
Mercaptodimethur	A	10 (4.54)
Mercuric cyanide	X	1 (0.454)
Mercuric nitrate	A	10 (4.54)
Mercuric sulfate	A	10 (4.54)
Mercuric thiocyanate	A	10 (4.54)
Mercurous nitrate	A	10 (4.54)
Methoxychlor	X	1 (0.454)
Methyl mercaptan	B	100 (45.4)
Methyl methacrylate	C	1,000 (454)
Methyl parathion	B	100 (45.4)
Mevinphos	A	10 (4.54)
Mexacarbate	C	1,000 (454)
Monoethylamine	B	100 (45.4)
Monomethylamine	B	100 (45.4)
Naled	A	10 (4.54)
Naphthalene	B	100 (45.4)
Naphthenic acid	B	100 (45.4)
Nickel ammonium sulfate	B	100 (45.4)
Nickel chloride	B	100 (45.4)
Nickel hydroxide	A	10 (4.54)
Nickel nitrate	B	100 (45.4)
Nickel sulfate	B	100 (45.4)
Nitric acid	C	1,000 (454)
Nitrobenzene	C	1,000 (454)
Nitrogen dioxide	A	10 (4.54)
Nitrophenol (mixed)	B	100 (45.4)
Nitrotoluene	C	1,000 (454)
Paraformaldehyde	C	1,000 (454)
Parathion	A	10 (4.54)
Pentachlorophenol	A	10 (4.54)
Phenol	C	1,000 (454)
Phosgene	A	10 (4.54)
Phosphoric acid	D	5,000 (2,270)
Phosphorus	X	1 (0.454)
Phosphorus oxychloride	C	1,000 (454)
Phosphorus pentasulfide	B	100 (45.4)
Phosphorus trichloride	C	1,000 (454)
Polychlorinated biphenyls	X	1 (0.454)
Potassium arsenate	X	1 (0.454)

TABLE 117.3—REPORTABLE QUANTITIES OF HAZARDOUS SUBSTANCES DESIGNATED PURSUANT TO SECTION 311 OF THE CLEAN WATER ACT—Continued

Material	Category	RQ in pounds (kilograms)
Potassium arsenite	X	1 (0.454)
Potassium bichromate	A	10 (4.54)
Potassium chromate	A	10 (4.54)
Potassium cyanide	A	10 (4.54)
Potassium hydroxide	C	1,000 (454)
Potassium permanganate	B	100 (45.4)
Propargite	A	10 (4.54)
Propionic acid	D	5,000 (2,270)
Propionic anhydride	D	5,000 (2,270)
Propylene oxide	B	100 (45.4)
Pyrethrins	X	1 (0.454)
Quinoline	D	5,000 (2,270)
Resorcinol	D	5,000 (2,270)
Selenium oxide	A	10 (4.54)
Silver nitrate	X	1 (0.454)
Sodium	A	10 (4.54)
Sodium arsenate	X	1 (0.454)
Sodium arsenite	X	1 (0.454)
Sodium bichromate	A	10 (4.54)
Sodium bifluoride	B	100 (45.4)
Sodium bisulfite	D	5,000 (2,270)
Sodium chromate	A	10 (4.54)
Sodium cyanide	A	10 (4.54)
Sodium dodecylbenzenesulfonate	C	1,000 (454)
Sodium fluoride	C	1,000 (454)
Sodium hydrosulfide	D	5,000 (2,270)
Sodium hydroxide	C	1,000 (454)
Sodium hypochlorite	B	100 (45.4)
Sodium methylate	C	1,000 (454)
Sodium nitrite	B	100 (45.4)
Sodium phosphate, dibasic	D	5,000 (2,270)
Sodium phosphate, tribasic	D	5,000 (2,270)
Sodium selenite	B	100 (45.4)
Strontium chromate	A	10 (4.54)
Strychnine	A	10 (4.54)
Styrene	C	1,000 (454)
Sulfuric acid	C	1,000 (454)
Sulfur monochloride	C	1,000 (454)
2,4,5-T acid	C	1,000 (454)
2,4,5-T amines	D	5,000 (2,270)
2,4,5-T esters	C	1,000 (454)
2,4,5-T salts	C	1,000 (454)
TDE	X	1 (0.454)
2,4,5-TP acid	B	100 (45.4)
2,4,5-TP acid esters	B	100 (45.4)
Tetraethyl lead	A	10 (4.54)
Tetraethyl pyrophosphate	A	10 (4.54)
Thallium sulfate	B	100 (45.4)
Toluene	C	1,000 (454)
Toxaphene	X	1 (0.454)
Trichlorfon	B	100 (45.4)
Trichloroethylene	B	100 (45.4)
Trichlorophenol	A	10 (4.54)
Triethanolamine dodecylbenzenesulfonate	C	1,000 (454)
Triethylamine	D	5,000 (2,270)
Trimethylamine	B	100 (45.4)
Uranyl acetate	B	100 (45.4)
Uranyl nitrate	B	100 (45.4)
Vanadium pentoxide	C	1,000 (454)
Vanadyl sulfate	C	1,000 (454)
Vinyl acetate	D	5,000 (2,270)
Vinylidene chloride	B	100 (45.4)
Xylene (mixed)	B	100 (45.4)
Xylenol	C	1,000 (454)
Zinc acetate	C	1,000 (454)
Zinc ammonium chloride	C	1,000 (454)
Zinc borate	C	1,000 (454)
Zinc bromide	C	1,000 (454)
Zinc carbonate	C	1,000 (454)
Zinc chloride	C	1,000 (454)
Zinc cyanide	A	10 (4.54)

TABLE 117.3—REPORTABLE QUANTITIES OF HAZARDOUS SUBSTANCES DESIGNATED PURSUANT TO
SECTION 311 OF THE CLEAN WATER ACT—Continued

Material	Category	RQ in pounds (kilograms)
Zinc fluoride	C	1,000 (454)
Zinc formate	C	1,000 (454)
Zinc hydrosulfite	C	1,000 (454)
Zinc nitrate	C	1,000 (454)
Zinc phenolsulfonate	D	5,000 (2,270)
Zinc phosphide	B	100 (45.4)
Zinc silicofluoride	D	5,000 (2,270)
Zinc sulfate	C	1,000 (454)
Zirconium nitrate	D	5,000 (2,270)
Zirconium potassium fluoride	C	1,000 (454)
Zirconium sulfate	D	5,000 (2,270)
Zirconium tetrachloride	D	5,000 (2,270)

[50 FR 13513, Apr. 4, 1985, as amended at 51 FR 34547, Sept. 29, 1986; 54 FR 33482, Aug. 14, 1989; 58 FR 35327, June 30, 1993; 60 FR 30937, June 12, 1995]

Subpart B—Applicability

§117.11 General applicability.

This regulation sets forth a determination of the reportable quantity for each substance designated as hazardous in 40 CFR part 116. The regulation applies to quantities of designated substances equal to or greater than the reportable quantities, when discharged into or upon the navigable waters of the United States, adjoining shorelines, into or upon the contiguous zone, or beyond the contiguous zone as provided in section 311(b)(3) of the Act, except to the extent that the owner or operator can show such that discharges are made:

(a) In compliance with a permit issued under the Marine Protection, Research and Sanctuaries Act of 1972 (33 U.S.C. 1401 *et seq.*);

(b) In compliance with approved water treatment plant operations as specified by local or State regulations pertaining to safe drinking water;

(c) Pursuant to the label directions for application of a pesticide product registered under section 3 or section 24 of the Federal Insecticide, Fungicide, and Rodenticide Act (FIFRA), as amended (7 U.S.C. 136 *et seq.*), or pursuant to the terms and conditions of an experimental use permit issued under section 5 of FIFRA, or pursuant to an exemption granted under section 18 of FIFRA;

(d) In compliance with the regulations issued under section 3004 or with permit conditions issued pursuant to section 3005 of the Resource Conservation and Recovery Act (90 Stat. 2795; 42 U.S.C. 6901);

(e) In compliance with instructions of the On-Scene Coordinator pursuant to 40 CFR part 1510 (the National Oil and Hazardous Substances Pollution Plan) or 33 CFR 153.10(e) (Pollution by Oil and Hazardous Substances) or in accordance with applicable removal regulations as required by section 311(j)(1)(A);

(f) In compliance with a permit issued under §165.7 of Title 14 of the State of California Administrative Code;

(g) From a properly functioning inert gas system when used to provide inert gas to the cargo tanks of a vessel;

(h) From a permitted source and are excluded by §117.12 of this regulation;

(i) To a POTW and are specifically excluded or reserved in §117.13; or

(j) In compliance with a permit issued under section 404(a) of the Clean Water Act or when the discharges are exempt from such requirements by section 404(f) or 404(r) of the Act (33 U.S.C. 1344(a), (f), (r)).

§117.12 Applicability to discharges from facilities with NPDES permits.

(a) This regulation does not apply to:

(1) Discharges in compliance with a permit under section 402 of this Act;

(2) Discharges resulting from circumstances identified, reviewed and made a part of the public record with respect to a permit issued or modified

under section 402 of this Act, and subject to a condition in such permit;

(3) Continuous or anticipated intermittent discharges from a point source, identified in a permit or permit application under section 402 of this Act, which are caused by events occurring within the scope of the relevant operating or treatment systems; or

(b) A discharge is "in compliance with a permit issued under section 402 of this Act" if the permit contains an effluent limitation specifically applicable to the substance discharged or an effluent limitation applicable to another waste parameter which has been specifically identified in the permit as intended to limit such substance, and the discharge is in compliance with the effluent limitation.

(c) A discharge results "from circumstances identified, reviewed and made a part of the public record with respect to a permit issued or modified under section 402 of the Act, and subject to a condition in such permit," whether or not the discharge is in compliance with the permit, where:

(1) The permit application, the permit, or another portion of the public record contains documents that specifically identify:

(i) The substance and the amount of the substance; and

(ii) The origin and source of the substance; and

(iii) The treatment which is to be provided for the discharge either by:

(A) An on-site treatment system separate from any treatment system treating the permittee's normal discharge; or

(B) A treatment system designed to treat the permittee's normal discharge and which is additionally capable of treating the identified amount of the identified substance; or

(C) Any combination of the above; and

(2) The permit contains a requirement that the substance and amounts of the substance, as identified in §117.12(c)(1)(i) and §117.12(c)(1)(ii) be treated pursuant to §117.12(c)(1)(iii) in the event of an on-site release; and

(3) The treatment to be provided is in place.

(d) A discharge is a "continuous or anticipated intermittent discharge from a point source, identified in a permit or permit application under section 402 of this Act, and caused by events occurring within the scope of the relevant operating or treatment systems," whether or not the discharge is in compliance with the permit, if:

(1) The hazardous substance is discharged from a point source for which a valid permit exists or for which a permit application has been submitted; and

(2) The discharge of the hazardous substance results from:

(i) The contamination of noncontact cooling water or storm water, provided that such cooling water or storm water is not contaminated by an on-site spill of a hazardous substance; or

(ii) A continuous or anticipated intermittent discharge of process waste water, and the discharge originates within the manufacturing or treatment systems; or

(iii) An upset or failure of a treatment system or of a process producing a continuous or anticipated intermittent discharge where the upset or failure results from a control problem, an operator error, a system failure or malfunction, an equipment or system startup or shutdown, an equipment wash, or a production schedule change, provided that such upset or failure is not caused by an on-site spill of a hazardous substance.

[44 FR 50776, Aug. 29, 1979, as amended at 44 FR 58910, Oct. 12, 1979]

§ 117.13 Applicability to discharges from publicly owned treatment works and their users.

(a) [Reserved]

(b) These regulations apply to all discharges of reportable quantities to a POTW, where the discharge originates from a mobile source, except where such source has contracted with, or otherwise received written permission from the owners or operators of the POTW to discharge that quantity, and the mobile source can show that prior to accepting the substance from an industrial discharger, the substance had been treated to comply with any effluent limitation under sections 301, 302 or 306 or pretreatment standard under section 307 applicable to that facility.

§117.14 Demonstration projects.

Notwithstanding any other provision of this part, the Administrator of the Environmental Protection Agency may, on a case-by-case basis, allow the discharge of designated hazardous substances in connection with research or demonstration projects relating to the prevention, control, or abatement of hazardous substance pollution. The Administrator will allow such a discharge only where he determines that the expected environmental benefit from such a discharge will outweigh the potential hazard associated with the discharge.

Subpart C—Notice of Discharge of a Reportable Quantity

§117.21 Notice.

Any person in charge of a vessel or an onshore or an offshore facility shall, as soon as he has knowledge of any discharge of a designated hazardous substance from such vessel or facility in quantities equal to or exceeding in any 24-hour period the reportable quantity determined by this part, immediately notify the appropriate agency of the United States Government of such discharge. Notice shall be given in accordance with such procedures as the Secretary of Transportation has set forth in 33 CFR 153.203. This provision applies to all discharges not specifically excluded or reserved by another section of these regulations.

§117.23 Liabilities for removal.

In any case where a substance designated as hazardous in 40 CFR part 116 is discharged from any vessel or onshore or offshore facility in a quantity equal to or exceeding the reportable quantity determined by this part, the owner, operator or person in charge will be liable, pursuant to section 311 (f) and (g) of the Act, to the United States Government for the actual costs incurred in the removal of such substance, subject only to the defenses and monetary limitations enumerated in section 311 (f) and (g) of the Act. The Administrator may act to mitigate the damage to the public health or welfare caused by a discharge and the cost of such mitigation shall be considered a cost incurred under section 311(c) for the removal of that substance by the United States Government.

PART 121—STATE CERTIFICATION OF ACTIVITIES REQUIRING A FEDERAL LICENSE OR PERMIT

Subpart A—General

AUTHORITY: Sec. 21 (b) and (c), 84 Stat. 91 (33 U.S.C. 1171(b) (1970)); Reorganization Plan No. 3 of 1970.

SOURCE: 36 FR 22487, Nov. 25, 1971, unless otherwise noted. Redesignated at 37 FR 21441, Oct. 11, 1972, and further redesignated at 44 FR 32899, June 7, 1979.

Subpart A—General

§121.1 Definitions.

As used in this part, the following terms shall have the meanings indicated below:

(a) *License or permit* means any license or permit granted by an agency of the Federal Government to conduct any activity which may result in any discharge into the navigable waters of the United States.

(b) *Licensing or permitting agency* means any agency of the Federal Government to which application is made for a license or permit.

(c) *Administrator* means the Administrator, Environmental Protection Agency.

(d) *Regional Administrator* means the Regional designee appointed by the Administrator, Environmental Protection Agency.

(e) *Certifying agency* means the person or agency designated by the Governor of a State, by statute, or by other governmental act, to certify compliance with applicable water quality standards. If an interstate agency has sole authority to so certify for the area within its jurisdiction, such interstate agency shall be the certifying agency. Where a State agency and an interstate agency have concurrent authority to certify, the State agency shall be the certifying agency. Where water quality standards have been promulgated by the Administrator pursuant to section 10(c)(2) of the Act, or where no State or interstate agency has authority to certify, the Administrator shall be the certifying agency.

(f) *Act* means the Federal Water Pollution Control Act, 33 U.S.C. 1151 *et seq.*

(g) *Water quality standards* means standards established pursuant to section 10(c) of the Act, and State-adopted water quality standards for navigable waters which are not interstate waters.

§ 121.2 Contents of certification.

(a) A certification made by a certifying agency shall include the following:

(1) The name and address of the applicant;

(2) A statement that the certifying agency has either (i) examined the application made by the applicant to the licensing or permitting agency (specifically identifying the number or code affixed to such application) and bases its certification upon an evaluation of the information contained in such application which is relevant to water quality considerations, or (ii) examined other information furnished by the applicant sufficient to permit the certifying agency to make the statement described in paragraph (a)(3) of this section;

(3) A statement that there is a reasonable assurance that the activity will be conducted in a manner which will not violate applicable water quality standards;

(4) A statement of any conditions which the certifying agency deems necessary or desirable with respect to the discharge of the activity; and

(5) Such other information as the certifying agency may determine to be appropriate.

(b) The certifying agency may modify the certification in such manner as may be agreed upon by the certifying agency, the licensing or permitting agency, and the Regional Administrator.

§ 121.3 Contents of application.

A licensing or permitting agency shall require an applicant for a license or permit to include in the form of application such information relating to water quality considerations as may be agreed upon by the licensing or permitting agency and the Administrator.

Subpart B—Determination of Effect on Other States

§ 121.11 Copies of documents.

(a) Upon receipt from an applicant of an application for a license or permit without an accompanying certification, the licensing or permitting agency shall either: (1) Forward one copy of the application to the appropriate certifying agency and two copies to the Regional Administrator, or (2) forward three copies of the application to the Regional Administrator, pursuant to an agreement between the licensing or permitting agency and the Administrator that the Regional Administrator will transmit a copy of the application to the appropriate certifying agency. Upon subsequent receipt from an applicant of a certification, the licensing or permitting agency shall forward a copy of such certification to the Regional Administrator, unless such certification shall have been made by the Regional Administrator pursuant to §121.24.

(b) Upon receipt from an applicant of an application for a license or permit with an accompanying certification, the licensing or permitting agency shall forward two copies of the application and certification to the Regional Administrator.

(c) Only those portions of the application which relate to water quality considerations shall be forwarded to the Regional Administrator.

§ 121.12 Supplemental information.

If the documents forwarded to the Regional Administrator by the licensing or permitting agency pursuant to § 121.11 do not contain sufficient information for the Regional Administrator to make the determination provided for in § 121.13, the Regional Administrator may request, and the licensing or permitting agency shall obtain from the applicant and forward to the Regional Administrator, any supplemental information as may be required to make such determination.

§ 121.13 Review by Regional Administrator and notification.

The Regional Administrator shall review the application, certification, and any supplemental information provided in accordance with §§ 121.11 and 121.12 and if the Regional Administrator determines there is reason to believe that a discharge may affect the quality of the waters of any State or States other than the State in which the discharge originates, the Regional Administrator shall, no later than 30 days of the date of receipt of the application and certification from the licensing or permitting agency as provided in § 121.11, so notify each affected State, the licensing or permitting agency, and the applicant.

§ 121.14 Forwarding to affected State.

The Regional Administrator shall forward to each affected State a copy of the material provided in accordance with § 121.11.

§ 121.15 Hearings on objection of affected State.

When a licensing or permitting agency holds a public hearing on the objection of an affected State, notice of such objection, including the grounds for such objection, shall be forwarded to the Regional Administrator by the licensing or permitting agency no later than 30 days prior to such hearing. The Regional Administrator shall at such hearing submit his evaluation with respect to such objection and his recommendations as to whether and under what conditions the license or permit should be issued.

§ 121.16 Waiver.

The certification requirement with respect to an application for a license or permit shall be waived upon:

(a) Written notification from the State or interstate agency concerned that it expressly waives its authority to act on a request for certification; or

(b) Written notification from the licensing or permitting agency to the Regional Administrator of the failure of the State or interstate agency concerned to act on such request for certification within a reasonable period of time after receipt of such request, as determined by the licensing or permitting agency (which period shall generally be considered to be 6 months, but in any event shall not exceed 1 year).

In the event of a waiver hereunder, the Regional Administrator shall consider such waiver as a substitute for a certification, and as appropriate, shall conduct the review, provide the notices, and perform the other functions identified in §§ 121.13, 121.14, and 121.15. The notices required by § 121.13 shall be provided not later than 30 days after the date of receipt by the Regional Administrator of either notification referred to herein.

Subpart C—Certification by the Administrator

§ 121.21 When Administrator certifies.

Certification by the Administrator that the discharge resulting from an activity requiring a license or permit will not violate applicable water quality standards will be required where:

(a) Standards have been promulgated, in whole or in part, by the Administrator pursuant to section 10(c)(2) of the Act: *Provided, however,* That the Administrator will certify compliance only with respect to those water quality standards promulgated by him; or

155

(b) Water quality standards have been established, but no State or interstate agency has authority to give such a certification.

§ 121.22 Applications.

An applicant for certification from the Administrator shall submit to the Regional Administrator a complete description of the discharge involved in the activity for which certification is sought, with a request for certification signed by the applicant. Such description shall include the following:

(a) The name and address of the applicant;

(b) A description of the facility or activity, and of any discharge into navigable waters which may result from the conduct of any activity including, but not limited to, the construction or operation of the facility, including the biological, chemical, thermal, and other characteristics of the discharge, and the location or locations at which such discharge may enter navigable waters;

(c) A description of the function and operation of equipment or facilities to treat wastes or other effluents which may be discharged, including specification of the degree of treatment expected to be attained;

(d) The date or dates on which the activity will begin and end, if known, and the date or dates on which the discharge will take place;

(e) A description of the methods and means being used or proposed to monitor the quality and characteristics of the discharge and the operation of equipment or facilities employed in the treatment or control of wastes or other effluents.

§ 121.23 Notice and hearing.

The Regional Administrator will provide public notice of each request for certification by mailing to State, County, and municipal authorities, heads of State agencies responsible for water quality improvement, and other parties known to be interested in the matter, including adjacent property owners and conservation organizations, or may provide such notice in a newspaper of general circulation in the area in which the activity is proposed to be conducted if the Regional Adminis-

trator deems mailed notice to be impracticable. Interested parties shall be provided an opportunity to comment on such request in such manner as the Regional Administrator deems appropriate. All interested and affected parties will be given reasonable opportunity to present evidence and testimony at a public hearing on the question whether to grant or deny certification if the Regional Administrator determines that such a hearing is necessary or appropriate.

§ 121.24 Certification.

If, after considering the complete description, the record of a hearing, if any, held pursuant to § 121.23, and such other information and data as the Regional Administrator deems relevant, the Regional Administrator determines that there is reasonable assurance that the proposed activity will not result in a violation of applicable water quality standards, he shall so certify. If the Regional Administrator determines that no water quality standards are applicable to the waters which might be affected by the proposed activity, he shall so notify the applicant and the licensing or permitting agency in writing and shall provide the licensing or permitting agency with advice, suggestions, and recommendations with respect to conditions to be incorporated in any license or permit to achieve compliance with the purpose of this Act. In such case, no certification shall be required.

§ 121.25 Adoption of new water quality standards.

(a) In any case where:

(1) A license or permit was issued without certification due to the absence of applicable water quality standards; and

(2) Water quality standards applicable to the waters into which the licensed or permitted activity may discharge are subsequently established; and

(3) The Administrator is the certifying agency because:

(i) No State or interstate agency has authority to certify; or

(ii) Such new standards were promulgated by the Administrator pursuant to section 10(c)(2) of the Act; and

(4) The Regional Administrator determines that such uncertified activity is violating water quality standards;

Then the Regional Administrator shall notify the licensee or permittee of such violation, including his recommendations as to actions necessary for compliance. If the licensee or permittee fails within 6 months of the date of such notice to take action which in the opinion of the Regional Administrator will result in compliance with applicable water quality standards, the Regional Administrator shall notify the licensing or permitting agency that the licensee or permittee has failed, after reasonable notice, to comply with such standards and that suspension of the applicable license or permit is required by section 21(b)(9)(B) of the Act.

(b) Where a license or permit is suspended pursuant to paragraph (a) of this section, and where the licensee or permittee subsequently takes action which in the Regional Administrator's opinion will result in compliance with applicable water quality standards, the Regional Administrator shall then notify the licensing or permitting agency that there is reasonable assurance that the licensed or permitted activity will comply with applicable water quality standards.

§ 121.26 **Inspection of facility or activity before operation.**

Where any facility or activity has received certification pursuant to § 121.24 in connection with the issuance of a license or permit for construction, and where such facility or activity is not required to obtain an operating license or permit, the Regional Administrator or his representative, prior to the initial operation of such facility or activity, shall be afforded the opportunity to inspect such facility or activity for the purpose of determining if the manner in which such facility or activity will be operated or conducted will violate applicable water quality standards.

§ 121.27 **Notification to licensing or permitting agency.**

If the Regional Administrator, after an inspection pursuant to § 121.26, determines that operation of the proposed facility or activity will violate applicable water quality standards, he shall so notify the applicant and the licensing or permitting agency, including his recommendations as to remedial measures necessary to bring the operation of the proposed facility into compliance with such standards.

§ 121.28 **Termination of suspension.**

Where a licensing or permitting agency, following a public hearing, suspends a license or permit after receiving the Regional Administrator's notice and recommendation pursuant to § 121.27, the applicant may submit evidence to the Regional Administrator that the facility or activity or the operation or conduct thereof has been modified so as not to violate water quality standards. If the Regional Administrator determines that water quality standards will not be violated, he shall so notify the licensing or permitting agency.

Subpart D—Consultations

§ 121.30 **Review and advice.**

The Regional Administrator may, and upon request shall, provide licensing and permitting agencies with determinations, definitions and interpretations with respect to the meaning and content of water quality standards where they have been federally approved under section 10 of the Act, and findings with respect to the application of all applicable water quality standards in particular cases and in specific circumstances relative to an activity for which a license or permit is sought. The Regional Administrator may, and upon request shall, also advise licensing and permitting agencies as to the status of compliance by dischargers with the conditions and requirements of applicable water quality standards. In cases where an activity for which a license or permit is sought will affect water quality, but for which there are no applicable water quality standards, the Regional Administrator may advise licensing or permitting agencies with respect to conditions of such license or permit to achieve compliance with the purpose of the Act.

PART 122—EPA ADMINISTERED PERMIT PROGRAMS: THE NATIONAL POLLUTANT DISCHARGE ELIMINATION SYSTEM

APPENDIX I TO PART 122—COUNTIES WITH UN-INCORPORATED URBANIZED AREAS GREATER THAN 100,000, BUT LESS THAN 250,000 ACCORDING TO THE 1990 DECENNIAL CENSUS BY THE BUREAU OF THE CENSUS

APPENDIX J TO PART 122—NPDES PERMIT TESTING REQUIREMENTS FOR PUBLICLY OWNED TREATMENT WORKS (§122.21(j))

AUTHORITY: The Clean Water Act, 33 U.S.C. 1251 *et seq.*

SOURCE: 48 FR 14153, Apr. 1, 1983, unless otherwise noted.

Subpart A—Definitions and General Program Requirements

§122.1 Purpose and scope.

(a) *Coverage.* (1) The regulatory provisions contained in this part and parts 123, and 124 of this chapter implement the National Pollutant Discharge Elimination System (NPDES) Program under sections 318, 402, and 405 of the Clean Water Act (CWA) (Public Law 92–500, as amended, 33 U.S.C. 1251 *et seq.*)

(2) These provisions cover basic EPA permitting requirements (this part 122), what a State must do to obtain approval to operate its program in lieu of a Federal program and minimum requirements for administering the approved State program (part 123 of this chapter), and procedures for EPA processing of permit applications and appeals (part 124 of this chapter).

(3) These provisions also establish the requirements for public participation in EPA and State permit issuance and enforcement and related variance proceedings, and in the approval of State NPDES programs. These provisions carry out the purposes of the public participation requirements of part 25 of this chapter, and supersede the requirements of that part as they apply to actions covered under this part and parts 123, and 124 of this chapter.

(4) Regulatory provisions in Parts 125, 129, 133, 136 of this chapter and 40 CFR subchapter N and subchapter O of this chapter also implement the NPDES permit program.

(5) Certain requirements set forth in parts 122 and 124 of this chapter are made applicable to approved State programs by reference in part 123 of this chapter. These references are set forth in §123.25 of this chapter. If a section or paragraph of part 122 or 124 of this chapter is applicable to States, through reference in §123.25 of this chapter, that fact is signaled by the following words at the end of the section or paragraph heading: (Applicable to State programs, see §123.25 of this chapter). If these words are absent, the section (or paragraph) applies only to EPA administered permits. Nothing in this part and parts 123, or 124 of this chapter precludes more stringent State regulation of any activity covered by the regulations in 40 CFR parts 122, 123, and 124, whether or not under an approved State program.

(b) *Scope of the NPDES permit requirement.* (1) The NPDES program requires permits for the discharge of "pollutants" from any "point source" into "waters of the United States." The terms "pollutant", "point source" and "waters of the United States" are defined at §122.2.

(2) The permit program established under this part also applies to owners or operators of any treatment works treating domestic sewage, whether or not the treatment works is otherwise required to obtain an NPDES permit, unless all requirements implementing section 405(d) of the CWA applicable to the treatment works treating domestic sewage are included in a permit issued under the appropriate provisions of subtitle C of the Solid Waste Disposal Act, Part C of the Safe Drinking Water Act, the Marine Protection, Research, and Sanctuaries Act of 1972, or the Clean Air Act, or under State permit programs approved by the Administrator as adequate to assure compliance with section 405 of the CWA.

(3) The Regional Administrator may designate any person subject to the standards for sewage sludge use and disposal as a "treatment works treating domestic sewage" as defined in §122.2, where the Regional Administrator finds that a permit is necessary to protect public health and the environment from the adverse effects of sewage sludge or to ensure compliance with the technical standards for sludge use and disposal developed under CWA section 405(d). Any person designated as a "treatment works treating domestic sewage" shall submit an application for a permit under §122.21 within 180 days of being notified by the Regional

Administrator that a permit is required. The Regional Administrator's decision to designate a person as a "treatment works treating domestic sewage" under this paragraph shall be stated in the fact sheet or statement of basis for the permit.

[NOTE TO § 122.1: Information concerning the NPDES program and its regulations can be obtained by contacting the Water Permits Division(4203), Office of Wastewater Management, U.S.E.P.A., Ariel Rios Building, 1200 Pennsylvania Avenue, NW., Washington, DC 20460 at (202) 260-9545 and by visiting the homepage at *http://www.epa.gov/owm/*]

[65 FR 30904, May 15, 2000, as amended at 72 FR 11211, Mar. 12, 2007]

§ 122.2 Definitions.

The following definitions apply to parts 122, 123, and 124. Terms not defined in this section have the meaning given by CWA. When a defined term appears in a definition, the defined term is sometimes placed in quotation marks as an aid to readers.

Administrator means the Administrator of the United States Environmental Protection Agency, or an authorized representative.

Animal feeding operation is defined at § 122.23.

Applicable standards and limitations means all State, interstate, and federal standards and limitations to which a "discharge," a "sewage sludge use or disposal practice," or a related activity is subject under the CWA, including "effluent limitations," water quality standards, standards of performance, toxic effluent standards or prohibitions, "best management practices," pretreatment standards, and "standards for sewage sludge use or disposal" under sections 301, 302, 303, 304, 306, 307, 308, 403 and 405 of CWA.

Application means the EPA standard national forms for applying for a permit, including any additions, revisions or modifications to the forms; or forms approved by EPA for use in "approved States," including any approved modifications or revisions.

Approved program or *approved State* means a State or interstate program which has been approved or authorized by EPA under part 123.

Aquaculture project is defined at § 122.25.

Average monthly discharge limitation means the highest allowable average of "daily discharges" over a calendar month, calculated as the sum of all "daily discharges" measured during a calendar month divided by the number of "daily discharges" measured during that month.

Average weekly discharge limitation means the highest allowable average of "daily discharges" over a calendar week, calculated as the sum of all "daily discharges" measured during a calendar week divided by the number of "daily discharges" measured during that week.

Best management practices ("BMPs") means schedules of activities, prohibitions of practices, maintenance procedures, and other management practices to prevent or reduce the pollution of "waters of the United States." BMPs also include treatment requirements, operating procedures, and practices to control plant site runoff, spillage or leaks, sludge or waste disposal, or drainage from raw material storage.

BMPs means "best management practices."

Class I sludge management facility means any POTW identified under 40 CFR 403.8(a) as being required to have an approved pretreatment program (including such POTWs located in a State that has elected to assume local program responsibilities pursuant to 40 CFR 403.10(e)) and any other treatment works treating domestic sewage classified as a Class I sludge management facility by the Regional Administrator, or, in the case of approved State programs, the Regional Administrator in conjunction with the State Director, because of the potential for its sludge use or disposal practices to adversely affect public health and the environment.

Bypass is defined at § 122.41(m).

Concentrated animal feeding operation is defined at § 122.23.

Concentrated aquatic animal feeding operation is defined at § 122.24.

Contiguous zone means the entire zone established by the United States under Article 24 of the Convention on the Territorial Sea and the Contiguous Zone.

Continuous discharge means a "discharge" which occurs without interruption throughout the operating hours of the facility, except for infrequent shutdowns for maintenance, process changes, or other similar activities.

CWA means the Clean Water Act (formerly referred to as the Federal Water Pollution Control Act or Federal Water Pollution Control Act Amendments of 1972) Public Law 92–500, as amended by Public Law 95–217, Public Law 95–576, Public Law 96–483 and Public Law 97–117, 33 U.S.C. 1251 *et seq.*

CWA and regulations means the Clean Water Act (CWA) and applicable regulations promulgated thereunder. In the case of an approved State program, it includes State program requirements.

Daily discharge means the "discharge of a pollutant" measured during a calendar day or any 24-hour period that reasonably represents the calendar day for purposes of sampling. For pollutants with limitations expressed in units of mass, the "daily discharge" is calculated as the total mass of the pollutant discharged over the day. For pollutants with limitations expressed in other units of measurement, the "daily discharge" is calculated as the average measurement of the pollutant over the day.

Direct discharge means the "discharge of a pollutant."

Director means the Regional Administrator or the State Director, as the context requires, or an authorized representative. When there is no "approved State program," and there is an EPA administered program, "Director" means the Regional Administrator. When there is an approved State program, "Director" normally means the State Director. In some circumstances, however, EPA retains the authority to take certain actions even when there is an approved State program. (For example, when EPA has issued an NPDES permit prior to the approval of a State program, EPA may retain jurisdiction over that permit after program approval, see §123.1.) In such cases, the term "Director" means the Regional Administrator and not the State Director.

Discharge when used without qualification means the "discharge of a pollutant."

Discharge of a pollutant means:

(a) Any addition of any "pollutant" or combination of pollutants to "waters of the United States" from any "point source," or

(b) Any addition of any pollutant or combination of pollutants to the waters of the "contiguous zone" or the ocean from any point source other than a vessel or other floating craft which is being used as a means of transportation.

This definition includes additions of pollutants into waters of the United States from: surface runoff which is collected or channelled by man; discharges through pipes, sewers, or other conveyances owned by a State, municipality, or other person which do not lead to a treatment works; and discharges through pipes, sewers, or other conveyances, leading into privately owned treatment works. This term does not include an addition of pollutants by any "indirect discharger."

Discharge Monitoring Report ("DMR") means the EPA uniform national form, including any subsequent additions, revisions, or modifications for the reporting of self-monitoring results by permittees. DMRs must be used by "approved States" as well as by EPA. EPA will supply DMRs to any approved State upon request. The EPA national forms may be modified to substitute the State Agency name, address, logo, and other similar information, as appropriate, in place of EPA's.

DMR means "Discharge Monitoring Report."

Draft permit means a document prepared under §124.6 indicating the Director's tentative decision to issue or deny, modify, revoke and reissue, terminate, or reissue a "permit." A notice of intent to terminate a permit, and a notice of intent to deny a permit, as discussed in §124.5, are types of "draft permits." A denial of a request for modification, revocation and reissuance, or termination, as discussed in §124.5, is not a "draft permit." A "proposed permit" is not a "draft permit."

Effluent limitation means any restriction imposed by the Director on quantities, discharge rates, and concentrations of "pollutants" which are "discharged" from "point sources" into "waters of the United States," the waters of the "contiguous zone," or the ocean.

Effluent limitations guidelines means a regulation published by the Administrator under section 304(b) of CWA to adopt or revise "effluent limitations."

Environmental Protection Agency ("EPA") means the United States Environmental Protection Agency.

EPA means the United States "Environmental Protection Agency."

Facility or activity means any NPDES "point source" or any other facility or activity (including land or appurtenances thereto) that is subject to regulation under the NPDES program.

Federal Indian reservation means all land within the limits of any Indian reservation under the jurisdiction of the United States Government, notwithstanding the issuance of any patent, and including rights-of-way running through the reservation.

General permit means an NPDES "permit" issued under § 122.28 authorizing a category of discharges under the CWA within a geographical area.

Hazardous substance means any substance designated under 40 CFR part 116 pursuant to section 311 of CWA.

Indian country means:

(1) All land within the limits of any Indian reservation under the jurisdiction of the United States Government, notwithstanding the issuance of any patent, and, including rights-of-way running through the reservation;

(2) All dependent Indian communities with the borders of the United States whether within the originally or subsequently acquired territory thereof, and whether within or without the limits of a state; and

(3) All Indian allotments, the Indian titles to which have not been extinguished, including rights-of-way running through the same.

Indian Tribe means any Indian Tribe, band, group, or community recognized by the Secretary of the Interior and exercising governmental authority over a Federal Indian reservation.

Indirect discharger means a nondomestic discharger introducing "pollutants" to a "publicly owned treatment works."

Individual control strategy is defined at 40 CFR 123.46(c).

Interstate agency means an agency of two or more States established by or under an agreement or compact approved by the Congress, or any other agency of two or more States having substantial powers or duties pertaining to the control of pollution as determined and approved by the Administrator under the CWA and regulations.

Major facility means any NPDES "facility or activity" classified as such by the Regional Administrator, or, in the case of "approved State programs," the Regional Administrator in conjunction with the State Director.

Maximum daily discharge limitation means the highest allowable "daily discharge."

Municipality means a city, town, borough, county, parish, district, association, or other public body created by or under State law and having jurisdiction over disposal of sewage, industrial wastes, or other wastes, or an Indian tribe or an authorized Indian tribal organization, or a designated and approved management agency under section 208 of CWA.

Municipal separate storm sewer system is defined at § 122.26 (b)(4) and (b)(7).

National Pollutant Discharge Elimination System (NPDES) means the national program for issuing, modifying, revoking and reissuing, terminating, monitoring and enforcing permits, and imposing and enforcing pretreatment requirements, under sections 307, 402, 318, and 405 of CWA. The term includes an "approved program."

New discharger means any building, structure, facility, or installation:

(a) From which there is or may be a "discharge of pollutants;"

(b) That did not commence the "discharge of pollutants" at a particular "site" prior to August 13, 1979;

(c) Which is not a "new source;" and

(d) Which has never received a finally effective NDPES permit for discharges at that "site."

This definition includes an "indirect discharger" which commences discharging into "waters of the United

States" after August 13, 1979. It also includes any existing mobile point source (other than an offshore or coastal oil and gas exploratory drilling rig or a coastal oil and gas developmental drilling rig) such as a seafood processing rig, seafood processing vessel, or aggregate plant, that begins discharging at a "site" for which it does not have a permit; and any offshore or coastal mobile oil and gas exploratory drilling rig or coastal mobile oil and gas developmental drilling rig that commences the discharge of pollutants after August 13, 1979, at a "site" under EPA's permitting jurisdiction for which it is not covered by an individual or general permit and which is located in an area determined by the Regional Administrator in the issuance of a final permit to be an area or biological concern. In determining whether an area is an area of biological concern, the Regional Administrator shall consider the factors specified in 40 CFR 125.122(a) (1) through (10).

An offshore or coastal mobile exploratory drilling rig or coastal mobile developmental drilling rig will be considered a "new discharger" only for the duration of its discharge in an area of biological concern.

New source means any building, structure, facility, or installation from which there is or may be a "discharge of pollutants," the construction of which commenced:

(a) After promulgation of standards of performance under section 306 of CWA which are applicable to such source, or

(b) After proposal of standards of performance in accordance with section 306 of CWA which are applicable to such source, but only if the standards are promulgated in accordance with section 306 within 120 days of their proposal.

NPDES means "National Pollutant Discharge Elimination System."

Owner or operator means the owner or operator of any "facility or activity" subject to regulation under the NPDES program.

Permit means an authorization, license, or equivalent control document issued by EPA or an "approved State" to implement the requirements of this part and parts 123 and 124. "Permit" includes an NPDES "general permit" (§ 122.28). Permit does not include any permit which has not yet been the subject of final agency action, such as a "draft permit" or a "proposed permit."

Person means an individual, association, partnership, corporation, municipality, State or Federal agency, or an agent or employee thereof.

Point source means any discernible, confined, and discrete conveyance, including but not limited to, any pipe, ditch, channel, tunnel, conduit, well, discrete fissure, container, rolling stock, concentrated animal feeding operation, landfill leachate collection system, vessel or other floating craft from which pollutants are or may be discharged. This term does not include return flows from irrigated agriculture or agricultural storm water runoff. (See § 122.3).

Pollutant means dredged spoil, solid waste, incinerator residue, filter backwash, sewage, garbage, sewage sludge, munitions, chemical wastes, biological materials, radioactive materials (except those regulated under the Atomic Energy Act of 1954, as amended (42 U.S.C. 2011 *et seq.*)), heat, wrecked or discarded equipment, rock, sand, cellar dirt and industrial, municipal, and agricultural waste discharged into water. It does not mean:

(a) Sewage from vessels; or

(b) Water, gas, or other material which is injected into a well to facilitate production of oil or gas, or water derived in association with oil and gas production and disposed of in a well, if the well used either to facilitate production or for disposal purposes is approved by authority of the State in which the well is located, and if the State determines that the injection or disposal will not result in the degradation of ground or surface water resources.

NOTE: Radioactive materials covered by the Atomic Energy Act are those encompassed in its definition of source, byproduct, or special nuclear materials. Examples of materials not covered include radium and accelerator-produced isotopes. See *Train v. Colorado Public Interest Research Group, Inc.,* 426 U.S. 1 (1976).

POTW is defined at § 403.3 of this chapter.

Primary industry category means any industry category listed in the NRDC settlement agreement (*Natural Resources Defense Council et al.* v. *Train,* 8 E.R.C. 2120 (D.D.C. 1976), *modified* 12 E.R.C. 1833 (D.D.C. 1979)); also listed in appendix A of part 122.

Privately owned treatment works means any device or system which is (a) used to treat wastes from any facility whose operator is not the operator of the treatment works and (b) not a "POTW."

Process wastewater means any water which, during manufacturing or processing, comes into direct contact with or results from the production or use of any raw material, intermediate product, finished product, byproduct, or waste product.

Proposed permit means a State NPDES "permit" prepared after the close of the public comment period (and, when applicable, any public hearing and administrative appeals) which is sent to EPA for review before final issuance by the State. A "proposed permit" is not a "draft permit."

Publicly owned treatment works is defined at 40 CFR 403.3.

Recommencing discharger means a source which recommences discharge after terminating operations.

Regional Administrator means the Regional Administrator of the appropriate Regional Office of the Environmental Protection Agency or the authorized representative of the Regional Administrator.

Schedule of compliance means a schedule of remedial measures included in a "permit", including an enforceable sequence of interim requirements (for example, actions, operations, or milestone events) leading to compliance with the CWA and regulations.

Secondary industry category means any industry category which is not a "primary industry category."

Secretary means the Secretary of the Army, acting through the Chief of Engineers.

Septage means the liquid and solid material pumped from a septic tank, cesspool, or similar domestic sewage treatment system, or a holding tank when the system is cleaned or maintained.

Sewage from vessels means human body wastes and the wastes from toilets and other receptacles intended to receive or retain body wastes that are discharged from vessels and regulated under section 312 of CWA, except that with respect to commercial vessels on the Great Lakes this term includes graywater. For the purposes of this definition, "graywater" means galley, bath, and shower water.

Sewage Sludge means any solid, semisolid, or liquid residue removed during the treatment of municipal waste water or domestic sewage. Sewage sludge includes, but is not limited to, solids removed during primary, secondary, or advanced waste water treatment, scum, septage, portable toilet pumpings, type III marine sanitation device pumpings (33 CFR part 159), and sewage sludge products. Sewage sludge does not include grit or screenings, or ash generated during the incineration of sewage sludge.

Sewage sludge use or disposal practice means the collection, storage, treatment, transportation, processing, monitoring, use, or disposal of sewage sludge.

Silvicultural point source is defined at § 122.27.

Site means the land or water area where any "facility or activity" is physically located or conducted, including adjacent land used in connection with the facility or activity.

Sludge-only facility means any "treatment works treating domestic sewage" whose methods of sewage sludge use or disposal are subject to regulations promulgated pursuant to section 405(d) of the CWA and is required to obtain a permit under § 122.1(b)(2).

Standards for sewage sludge use or disposal means the regulations promulgated pursuant to section 405(d) of the CWA which govern minimum requirements for sludge quality, management practices, and monitoring and reporting applicable to sewage sludge or the use or disposal of sewage sludge by any person.

State means any of the 50 States, the District of Columbia, Guam, the Commonwealth of Puerto Rico, the Virgin Islands, American Samoa, the Commonwealth of the Northern Mariana Islands, the Trust Territory of the Pacific Islands, or an Indian Tribe as defined in these regulations which meets the requirements of §123.31 of this chapter.

State Director means the chief administrative officer of any State or interstate agency operating an "approved program," or the delegated representative of the State Director. If responsibility is divided among two or more State or interstate agencies, "State Director" means the chief administrative officer of the State or interstate agency authorized to perform the particular procedure or function to which reference is made.

State/EPA Agreement means an agreement between the Regional Administrator and the State which coordinates EPA and State activities, responsibilities and programs including those under the CWA programs.

Storm water is defined at §122.26(b)(13).

Storm water discharge associated with industrial activity is defined at §122.26(b)(14).

Total dissolved solids means the total dissolved (filterable) solids as determined by use of the method specified in 40 CFR part 136.

Toxic pollutant means any pollutant listed as toxic under section 307(a)(1) or, in the case of "sludge use or disposal practices," any pollutant identified in regulations implementing section 405(d) of the CWA.

Treatment works treating domestic sewage means a POTW or any other sewage sludge or waste water treatment devices or systems, regardless of ownership (including federal facilities), used in the storage, treatment, recycling, and reclamation of municipal or domestic sewage, including land dedicated for the disposal of sewage sludge. This definition does not include septic tanks or similar devices. For purposes of this definition, "domestic sewage" includes waste and waste water from humans or household operations that are discharged to or otherwise enter a treatment works. In States where there is no approved State sludge management program under section 405(f) of the CWA, the Regional Administrator may designate any person subject to the standards for sewage sludge use and disposal in 40 CFR part 503 as a "treatment works treating domestic sewage," where he or she finds that there is a potential for adverse effects on public health and the environment from poor sludge quality or poor sludge handling, use or disposal practices, or where he or she finds that such designation is necessary to ensure that such person is in compliance with 40 CFR part 503.

TWTDS means "treatment works treating domestic sewage."

Upset is defined at §122.41(n).

Variance means any mechanism or provision under section 301 or 316 of CWA or under 40 CFR part 125, or in the applicable "effluent limitations guidelines" which allows modification to or waiver of the generally applicable effluent limitation requirements or time deadlines of CWA. This includes provisions which allow the establishment of alternative limitations based on fundamentally different factors or on sections 301(c), 301(g), 301(h), 301(i), or 316(a) of CWA.

Waters of the United States or *waters of the U.S.* means:

(a) All waters which are currently used, were used in the past, or may be susceptible to use in interstate or foreign commerce, including all waters which are subject to the ebb and flow of the tide;

(b) All interstate waters, including interstate "wetlands;"

(c) All other waters such as intrastate lakes, rivers, streams (including intermittent streams), mudflats, sandflats, "wetlands," sloughs, prairie potholes, wet meadows, playa lakes, or natural ponds the use, degradation, or destruction of which would affect or could affect interstate or foreign commerce including any such waters:

(1) Which are or could be used by interstate or foreign travelers for recreational or other purposes;

(2) From which fish or shellfish are or could be taken and sold in interstate or foreign commerce; or

(3) Which are used or could be used for industrial purposes by industries in interstate commerce;

(d) All impoundments of waters otherwise defined as waters of the United States under this definition;

(e) Tributaries of waters identified in paragraphs (a) through (d) of this definition;

(f) The territorial sea; and

(g) "Wetlands" adjacent to waters (other than waters that are themselves wetlands) identified in paragraphs (a) through (f) of this definition.

Waste treatment systems, including treatment ponds or lagoons designed to meet the requirements of CWA (other than cooling ponds as defined in 40 CFR 423.11(m) which also meet the criteria of this definition) are not waters of the United States. This exclusion applies only to manmade bodies of water which neither were originally created in waters of the United States (such as disposal area in wetlands) nor resulted from the impoundment of waters of the United States. [See Note 1 of this section.] Waters of the United States do not include prior converted cropland. Notwithstanding the determination of an area's status as prior converted cropland by any other federal agency, for the purposes of the Clean Water Act, the final authority regarding Clean Water Act jurisdiction remains with EPA.

Wetlands means those areas that are inundated or saturated by surface or groundwater at a frequency and duration sufficient to support, and that under normal circumstances do support, a prevalence of vegetation typically adapted for life in saturated soil conditions. Wetlands generally include swamps, marshes, bogs, and similar areas.

Whole effluent toxicity means the aggregate toxic effect of an effluent measured directly by a toxicity test.

NOTE: At 45 FR 48620, July 21, 1980, the Environmental Protection Agency suspended until further notice in § 122.2, the last sentence, beginning "This exclusion applies . . ." in the definition of "Waters of the

United States." This revision continues that suspension.[1]

(Clean Water Act (33 U.S.C. 1251 *et seq.*), Safe Drinking Water Act (42 U.S.C. 300f *et seq.*), Clean Air Act (42 U.S.C. 7401 *et seq.*), Resource Conservation and Recovery Act (42 U.S.C. 6901 *et seq.*))

[48 FR 14153, Apr. 1, 1983, as amended at 48 FR 39619, Sept. 1, 1983; 50 FR 6940, 6941, Feb. 19, 1985; 54 FR 254, Jan. 4, 1989; 54 FR 18781, May 2, 1989; 54 FR 23895, June 2, 1989; 58 FR 45039, Aug. 25, 1993; 58 FR 67980, Dec. 22, 1993; 64 FR 42462, Aug. 4, 1999; 65 FR 30905, May 15, 2000]

§ 122.3 Exclusions.

The following discharges do not require NPDES permits:

(a) Any discharge of sewage from vessels, effluent from properly functioning marine engines, laundry, shower, and galley sink wastes, or any other discharge incidental to the normal operation of a vessel. This exclusion does not apply to rubbish, trash, garbage, or other such materials discharged overboard; nor to other discharges when the vessel is operating in a capacity other than as a means of transportation such as when used as an energy or mining facility, a storage facility or a seafood processing facility, or when secured to a storage facility or a seafood processing facility, or when secured to the bed of the ocean, contiguous zone or waters of the United States for the purpose of mineral or oil exploration or development.

(b) Discharges of dredged or fill material into waters of the United States which are regulated under section 404 of CWA.

(c) The introduction of sewage, industrial wastes or other pollutants into publicly owned treatment works by indirect dischargers. Plans or agreements to switch to this method of disposal in the future do not relieve dischargers of the obligation to have and comply with permits until all discharges of pollutants to waters of the United States are eliminated. (See also § 122.47(b)). This exclusion does not apply to the introduction of pollutants to privately owned treatment works or to other discharges through pipes, sewers, or other

_____
[1] EDITORIAL NOTE: The words "This revision" refer to the document published at 48 FR 14153, Apr. 1, 1983.

conveyances owned by a State, municipality, or other party not leading to treatment works.

(d) Any discharge in compliance with the instructions of an On-Scene Coordinator pursuant to 40 CFR part 300 (The National Oil and Hazardous Substances Pollution Contingency Plan) or 33 CFR 153.10(e) (Pollution by Oil and Hazardous Substances).

(e) Any introduction of pollutants from non point-source agricultural and silvicultural activities, including storm water runoff from orchards, cultivated crops, pastures, range lands, and forest lands, but not discharges from concentrated animal feeding operations as defined in §122.23, discharges from concentrated aquatic animal production facilities as defined in §122.24, discharges to aquaculture projects as defined in §122.25, and discharges from silvicultural point sources as defined in §122.27.

(f) Return flows from irrigated agriculture.

(g) Discharges into a privately owned treatment works, except as the Director may otherwise require under §122.44(m).

(h) The application of pesticides consistent with all relevant requirements under FIFRA (i.e., those relevant to protecting water quality), in the following two circumstances:

(1) The application of pesticides directly to waters of the United States in order to control pests. Examples of such applications include applications to control mosquito larvae, aquatic weeds, or other pests that are present in waters of the United States.

(2) The application of pesticides to control pests that are present over waters of the United States, including near such waters, where a portion of the pesticides will unavoidably be deposited to waters of the United States in order to target the pests effectively; for example, when insecticides are aerially applied to a forest canopy where waters of the United States may be present below the canopy or when pesticides are applied over or near water for control of adult mosquitoes or other pests.

(i) Discharges from a water transfer. Water transfer means an activity that conveys or connects waters of the United States without subjecting the transferred water to intervening industrial, municipal, or commercial use. This exclusion does not apply to pollutants introduced by the water transfer activity itself to the water being transferred.

[48 FR 14153, Apr. 1, 1983, as amended at 54 FR 254, 258, Jan. 4, 1989; 71 FR 68492, Nov. 27, 2006; 73 FR 33708, June 13, 2008]

§ 122.4 Prohibitions (applicable to State NPDES programs, see § 123.25).

No permit may be issued:

(a) When the conditions of the permit do not provide for compliance with the applicable requirements of CWA, or regulations promulgated under CWA;

(b) When the applicant is required to obtain a State or other appropriate certification under section 401 of CWA and §124.53 and that certification has not been obtained or waived;

(c) By the State Director where the Regional Administrator has objected to issuance of the permit under §123.44;

(d) When the imposition of conditions cannot ensure compliance with the applicable water quality requirements of all affected States;

(e) When, in the judgment of the Secretary, anchorage and navigation in or on any of the waters of the United States would be substantially impaired by the discharge;

(f) For the discharge of any radiological, chemical, or biological warfare agent or high-level radioactive waste;

(g) For any discharge inconsistent with a plan or plan amendment approved under section 208(b) of CWA;

(h) For any discharge to the territorial sea, the waters of the contiguous zone, or the oceans in the following circumstances:

(1) Before the promulgation of guidelines under section 403(c) of CWA (for determining degradation of the waters of the territorial seas, the contiguous zone, and the oceans) unless the Director determines permit issuance to be in the public interest; or

(2) After promulgation of guidelines under section 403(c) of CWA, when insufficient information exists to make a reasonable judgment whether the discharge complies with them.

(i) To a new source or a new discharger, if the discharge from its construction or operation will cause or contribute to the violation of water quality standards. The owner or operator of a new source or new discharger proposing to discharge into a water segment which does not meet applicable water quality standards or is not expected to meet those standards even after the application of the effluent limitations required by sections 301(b)(1)(A) and 301(b)(1)(B) of CWA, and for which the State or interstate agency has performed a pollutants load allocation for the pollutant to be discharged, must demonstrate, before the close of the public comment period, that:

(1) There are sufficient remaining pollutant load allocations to allow for the discharge; and

(2) The existing dischargers into that segment are subject to compliance schedules designed to bring the segment into compliance with applicable water quality standards. The Director may waive the submission of information by the new source or new discharger required by paragraph (i) of this section if the Director determines that the Director already has adequate information to evaluate the request. An explanation of the development of limitations to meet the criteria of this paragraph (i)(2) is to be included in the fact sheet to the permit under § 124.56(b)(1) of this chapter.

[48 FR 14153, Apr. 1, 1983, as amended at 50 FR 6940, Feb. 19, 1985; 65 FR 30905, May 15, 2000]

§ 122.5　Effect of a permit.

(a) *Applicable to State programs, see § 123.25.* (1) Except for any toxic effluent standards and prohibitions imposed under section 307 of the CWA and "standards for sewage sludge use or disposal" under 405(d) of the CWA, compliance with a permit during its term constitutes compliance, for purposes of enforcement, with sections 301, 302, 306, 307, 318, 403, and 405 (a)–(b) of CWA. However, a permit may be modified, revoked and reissued, or terminated during its term for cause as set forth in §§ 122.62 and 122.64.

(2) Compliance with a permit condition which implements a particular "standard for sewage sludge use or disposal" shall be an affirmative defense in any enforcement action brought for a violation of that "standard for sewage sludge use or disposal" pursuant to sections 405(e) and 309 of the CWA.

(b) *Applicable to State programs, See § 123.25.* The issuance of a permit does not convey any property rights of any sort, or any exclusive privilege.

(c) The issuance of a permit does not authorize any injury to persons or property or invasion of other private rights, or any infringement of State or local law or regulations.

[48 FR 14153, Apr. 1, 1983, as amended at 54 FR 18782, May 2, 1989]

§ 122.6　Continuation of expiring permits.

(a) *EPA permits.* When EPA is the permit-issuing authority, the conditions of an expired permit continue in force under 5 U.S.C. 558(c) until the effective date of a new permit (see § 124.15) if:

(1) The permittee has submitted a timely application under § 122.21 which is a complete (under § 122.21(e)) application for a new permit; and

(2) The Regional Administrator, through no fault of the permittee does not issue a new permit with an effective date under § 124.15 on or before the expiration date of the previous permit (for example, when issuance is impracticable due to time or resource constraints).

(b) *Effect.* Permits continued under this section remain fully effective and enforceable.

(c) *Enforcement.* When the permittee is not in compliance with the conditions of the expiring or expired permit the Regional Administrator may choose to do any or all of the following:

(1) Initiate enforcement action based upon the permit which has been continued;

(2) Issue a notice of intent to deny the new permit under § 124.6. If the permit is denied, the owner or operator would then be required to cease the activities authorized by the continued permit or be subject to enforcement action for operating without a permit;

(3) Issue a new permit under part 124 with appropriate conditions; or

(4) Take other actions authorized by these regulations.

(d) *State continuation.* (1) An EPA-issued permit does not continue in force beyond its expiration date under Federal law if at that time a State is the permitting authority. States authorized to administer the NPDES program may continue either EPA or State-issued permits until the effective date of the new permits, if State law allows. Otherwise, the facility or activity is operating without a permit from the time of expiration of the old permit to the effective date of the State-issued new permit.

[48 FR 14153, Apr. 1, 1983, as amended at 50 FR 6940, Feb. 19, 1985]

§122.7 Confidentiality of information.

(a) In accordance with 40 CFR part 2, any information submitted to EPA pursuant to these regulations may be claimed as confidential by the submitter. Any such claim must be asserted at the time of submission in the manner prescribed on the application form or instructions or, in the case of other submissions, by stamping the words "confidential business information" on each page containing such information. If no claim is made at the time of submission, EPA may make the information available to the public without further notice. If a claim is asserted, the information will be treated in accordance with the procedures in 40 CFR part 2 (Public Information).

(b) *Applicable to State programs, see §123.25.* Claims of confidentiality for the following information will be denied:

(1) The name and address of any permit applicant or permittee;

(2) Permit applications, permits, and effluent data.

(c) *Applicable to State programs, see §123.25.* Information required by NPDES application forms provided by the Director under §122.21 may not be claimed confidential. This includes information submitted on the forms themselves and any attachments used to supply information required by the forms.

Subpart B—Permit Application and Special NPDES Program Requirements

§122.21 Application for a permit (applicable to State programs, see §123.25).

(a) *Duty to apply.* (1) Any person who discharges or proposes to discharge pollutants or who owns or operates a "sludge-only facility" whose sewage sludge use or disposal practice is regulated by part 503 of this chapter, and who does not have an effective permit, except persons covered by general permits under §122.28, excluded under §122.3, or a user of a privately owned treatment works unless the Director requires otherwise under §122.44(m), must submit a complete application to the Director in accordance with this section and part 124 of this chapter. The requirements for concentrated animal feeding operations are described in §122.23(d).

(2) *Application Forms:* (i) All applicants for EPA-issued permits must submit applications on EPA permit application forms. More than one application form may be required from a facility depending on the number and types of discharges or outfalls found there. Application forms may be obtained by contacting the EPA water resource center at (202) 260–7786 or Water Resource Center, U.S. EPA, Mail Code 4100, 1200 Pennsylvania Ave., NW., Washington, DC 20460 or at the EPA Internet site *www.epa.gov/owm/npdes.htm.* Applications for EPA-issued permits must be submitted as follows:

(A) All applicants, other than POTWs and TWTDS, must submit Form 1.

(B) Applicants for new and existing POTWs must submit the information contained in paragraph (j) of this section using Form 2A or other form provided by the director.

(C) Applicants for concentrated animal feeding operations or aquatic animal production facilities must submit Form 2B.

(D) Applicants for existing industrial facilities (including manufacturing facilities, commercial facilities, mining activities, and silvicultural activities), must submit Form 2C.

(E) Applicants for new industrial facilities that discharge process wastewater must submit Form 2D.

(F) Applicants for new and existing industrial facilities that discharge only nonprocess wastewater must submit Form 2E.

(G) Applicants for new and existing facilities whose discharge is composed entirely of storm water associated with industrial activity must submit Form 2F, unless exempted by § 122.26(c)(1)(ii). If the discharge is composed of storm water and non-storm water, the applicant must also submit, Forms 2C, 2D, and/or 2E, as appropriate (in addition to Form 2F).

(H) Applicants for new and existing TWTDS, subject to paragraph (c)(2)(i) of this section must submit the application information required by paragraph (q) of this section, using Form 2S or other form provided by the director.

(ii) The application information required by paragraph (a)(2)(i) of this section may be electronically submitted if such method of submittal is approved by EPA or the Director.

(iii) Applicants can obtain copies of these forms by contacting the Water Management Divisions (or equivalent division which contains the NPDES permitting function) of the EPA Regional Offices. The Regional Offices' addresses can be found at § 1.7 of this chapter.

(iv) Applicants for State-issued permits must use State forms which must require at a minimum the information listed in the appropriate paragraphs of this section.

(b) *Who applies?* When a facility or activity is owned by one person but is operated by another person, it is the operator's duty to obtain a permit.

(c) *Time to apply.* (1) Any person proposing a new discharge, shall submit an application at least 180 days before the date on which the discharge is to commence, unless permission for a later date has been granted by the Director. Facilities proposing a new discharge of storm water associated with industrial activity shall submit an application 180 days before that facility commences industrial activity which may result in a discharge of storm water associated with that industrial activity. Facilities described under § 122.26(b)(14)(x) or

(b)(15)(i) shall submit applications at least 90 days before the date on which construction is to commence. Different submittal dates may be required under the terms of applicable general permits. Persons proposing a new discharge are encouraged to submit their applications well in advance of the 90 or 180 day requirements to avoid delay. See also paragraph (k) of this section and § 122.26(c)(1)(i)(G) and (c)(1)(ii).

(2) *Permits under section 405(f) of CWA.* All TWTDS whose sewage sludge use or disposal practices are regulated by part 503 of this chapter must submit permit applications according to the applicable schedule in paragraphs (c)(2)(i) or (ii) of this section.

(i) A TWTDS with a currently effective NPDES permit must submit a permit application at the time of its next NPDES permit renewal application. Such information must be submitted in accordance with paragraph (d) of this section.

(ii) Any other TWTDS not addressed under paragraph (c)(2)(i) of this section must submit the information listed in paragraphs (c)(2)(ii)(A) through (E) of this section to the Director within 1 year after publication of a standard applicable to its sewage sludge use or disposal practice(s), using Form 2S or another form provided by the Director. The Director will determine when such TWTDS must submit a full permit application.

(A) The TWTDS's name, mailing address, location, and status as federal, State, private, public or other entity;

(B) The applicant's name, address, telephone number, and ownership status;

(C) A description of the sewage sludge use or disposal practices. Unless the sewage sludge meets the requirements of paragraph (q)(8)(iv) of this section, the description must include the name and address of any facility where sewage sludge is sent for treatment or disposal, and the location of any land application sites;

(D) Annual amount of sewage sludge generated, treated, used or disposed (estimated dry weight basis); and

(E) The most recent data the TWTDS may have on the quality of the sewage sludge.

(iii) Notwithstanding paragraphs (c)(2)(i) or (ii) of this section, the Director may require permit applications from any TWTDS at any time if the Director determines that a permit is necessary to protect public health and the environment from any potential adverse effects that may occur from toxic pollutants in sewage sludge.

(iv) Any TWTDS that commences operations after promulgation of an applicable "standard for sewage sludge use or disposal" must submit an application to the Director at least 180 days prior to the date proposed for commencing operations.

(d) *Duty to reapply.* (1) Any POTW with a currently effective permit shall submit a new application at least 180 days before the expiration date of the existing permit, unless permission for a later date has been granted by the Director. (The Director shall not grant permission for applications to be submitted later than the expiration date of the existing permit.)

(2) All other permittees with currently effective permits shall submit a new application 180 days before the existing permit expires, except that:

(i) The Regional Administrator may grant permission to submit an application later than the deadline for submission otherwise applicable, but no later than the permit expiration date; and

(3) [Reserved]

(e) *Completeness.* (1) The Director shall not issue a permit before receiving a complete application for a permit except for NPDES general permits. An application for a permit is complete when the Director receives an application form and any supplemental information which are completed to his or her satisfaction. The completeness of any application for a permit shall be judged independently of the status of any other permit application or permit for the same facility or activity. For EPA administered NPDES programs, an application which is reviewed under §124.3 of this chapter is complete when the Director receives either a complete application or the information listed in a notice of deficiency.

(2) A permit application shall not be considered complete if a permitting authority has waived application requirements under paragraphs (j) or (q) of this section and EPA has disapproved the waiver application. If a waiver request has been submitted to EPA more than 210 days prior to permit expiration and EPA has not disapproved the waiver application 181 days prior to permit expiration, the permit application lacking the information subject to the waiver application shall be considered complete.

(f) *Information requirements.* All applicants for NPDES permits, other than POTWs and other TWTDS, must provide the following information to the Director, using the application form provided by the Director. Additional information required of applicants is set forth in paragraphs (g) through (k) of this section.

(1) The activities conducted by the applicant which require it to obtain an NPDES permit.

(2) Name, mailing address, and location of the facility for which the application is submitted.

(3) Up to four SIC codes which best reflect the principal products or services provided by the facility.

(4) The operator's name, address, telephone number, ownership status, and status as Federal, State, private, public, or other entity.

(5) Whether the facility is located on Indian lands.

(6) A listing of all permits or construction approvals received or applied for under any of the following programs:

(i) Hazardous Waste Management program under RCRA.

(ii) UIC program under SDWA.

(iii) NPDES program under CWA.

(iv) Prevention of Significant Deterioration (PSD) program under the Clean Air Act.

(v) Nonattainment program under the Clean Air Act.

(vi) National Emission Standards for Hazardous Pollutants (NESHAPS) preconstruction approval under the Clean Air Act.

(vii) Ocean dumping permits under the Marine Protection Research and Sanctuaries Act.

(viii) Dredge or fill permits under section 404 of CWA.

(ix) Other relevant environmental permits, including State permits.

(7) A topographic map (or other map if a topographic map is unavailable) extending one mile beyond the property boundaries of the source, depicting the facility and each of its intake and discharge structures; each of its hazardous waste treatment, storage, or disposal facilities; each well where fluids from the facility are injected underground; and those wells, springs, other surface water bodies, and drinking water wells listed in public records or otherwise known to the applicant in the map area.

(8) A brief description of the nature of the business.

(g) *Application requirements for existing manufacturing, commercial, mining, and silvicultural dischargers.* Existing manufacturing, commercial mining, and silvicultural dischargers applying for NPDES permits, except for those facilities subject to the requirements of §122.21(h), shall provide the following information to the Director, using application forms provided by the Director.

(1) *Outfall location.* The latitude and longitude to the nearest 15 seconds and the name of the receiving water.

(2) *Line drawing.* A line drawing of the water flow through the facility with a water balance, showing operations contributing wastewater to the effluent and treatment units. Similar processes, operations, or production areas may be indicated as a single unit, labeled to correspond to the more detailed identification under paragraph (g)(3) of this section. The water balance must show approximate average flows at intake and discharge points and between units, including treatment units. If a water balance cannot be determined (for example, for certain mining activities), the applicant may provide instead a pictorial description of the nature and amount of any sources of water and any collection and treatment measures.

(3) *Average flows and treatment.* A narrative identification of each type of process, operation, or production area which contributes wastewater to the effluent for each outfall, including process wastewater, cooling water, and stormwater runoff; the average flow which each process contributes; and a description of the treatment the waste-water receives, including the ultimate disposal of any solid or fluid wastes other than by discharge. Processes, operations, or production areas may be described in general terms (for example, "dye-making reactor", "distillation tower"). For a privately owned treatment works, this information shall include the identity of each user of the treatment works. The average flow of point sources composed of storm water may be estimated. The basis for the rainfall event and the method of estimation must be indicated.

(4) *Intermittent flows.* If any of the discharges described in paragraph (g)(3) of this section are intermittent or seasonal, a description of the frequency, duration and flow rate of each discharge occurrence (except for stormwater runoff, spillage or leaks).

(5) *Maximum production.* If an effluent guideline promulgated under section 304 of CWA applies to the applicant and is expressed in terms of production (or other measure of operation), a reasonable measure of the applicant's actual production reported in the units used in the applicable effluent guideline. The reported measure must reflect the actual production of the facility as required by §122.45(b)(2).

(6) *Improvements.* If the applicant is subject to any present requirements or compliance schedules for construction, upgrading or operation of waste treatment equipment, an identification of the abatement requirement, a description of the abatement project, and a listing of the required and projected final compliance dates.

(7) *Effluent characteristics.* (i) Information on the discharge of pollutants specified in this paragraph (g)(7) (except information on storm water discharges which is to be provided as specified in §122.26). When "quantitative data" for a pollutant are required, the applicant must collect a sample of effluent and analyze it for the pollutant in accordance with analytical methods approved under Part 136 of this chapter unless use of another method is required for the pollutant under 40 CFR subchapters N or O. When no analytical method is approved under Part 136 or required under subchapters N or O, the applicant may use any suitable method

but must provide a description of the method. When an applicant has two or more outfalls with substantially identical effluents, the Director may allow the applicant to test only one outfall and report that quantitative data as applying to the substantially identical outfall. The requirements in paragraphs (g)(7)(vi) and (vii) of this section state that an applicant must provide quantitative data for certain pollutants known or believed to be present do not apply to pollutants present in a discharge solely as the result of their presence in intake water; however, an applicant must report such pollutants as present. When paragraph (g)(7) of this section requires analysis of pH, temperature, cyanide, total phenols, residual chlorine, oil and grease, fecal coliform (including *E. coli*), and Enterococci (previously known as fecal streptococcus at §122.26 (d)(2)(iii)(A)(3)), or volatile organics, grab samples must be collected for those pollutants. For all other pollutants, a 24-hour composite sample, using a minimum of four (4) grab samples, must be used unless specified otherwise at 40 CFR Part 136. However, a minimum of one grab sample may be taken for effluents from holding ponds or other impoundments with a retention period greater than 24 hours. In addition, for discharges other than storm water discharges, the Director may waive composite sampling for any outfall for which the applicant demonstrates that the use of an automatic sampler is infeasible and that the minimum of four (4) grab samples will be a representative sample of the effluent being discharged. Results of analyses of individual grab samples for any parameter may be averaged to obtain the daily average. Grab samples that are not required to be analyzed immediately (see Table II at 40 CFR 136.3 (e)) may be composited in the laboratory, provided that container, preservation, and holding time requirements are met (see Table II at 40 CFR 136.3 (e)) and that sample integrity is not compromised by compositing.

(ii) *Storm water discharges.* For storm water discharges, all samples shall be collected from the discharge resulting from a storm event that is greater than 0.1 inch and at least 72 hours from the previously measurable (greater than 0.1 inch rainfall) storm event. Where feasible, the variance in the duration of the event and the total rainfall of the event should not exceed 50 percent from the average or median rainfall event in that area. For all applicants, a flow-weighted composite shall be taken for either the entire discharge or for the first three hours of the discharge. The flow-weighted composite sample for a storm water discharge may be taken with a continuous sampler or as a combination of a minimum of three sample aliquots taken in each hour of discharge for the entire discharge or for the first three hours of the discharge, with each aliquot being separated by a minimum period of fifteen minutes (applicants submitting permit applications for storm water discharges under §122.26(d) may collect flow-weighted composite samples using different protocols with respect to the time duration between the collection of sample aliquots, subject to the approval of the Director). However, a minimum of one grab sample may be taken for storm water discharges from holding ponds or other impoundments with a retention period greater than 24 hours. For a flow-weighted composite sample, only one analysis of the composite of aliquots is required. For storm water discharge samples taken from discharges associated with industrial activities, quantitative data must be reported for the grab sample taken during the first thirty minutes (or as soon thereafter as practicable) of the discharge for all pollutants specified in §122.26(c)(1). For all storm water permit applicants taking flow-weighted composites, quantitative data must be reported for all pollutants specified in §122.26 except pH, temperature, cyanide, total phenols, residual chlorine, oil and grease, fecal coliform, and fecal streptococcus. The Director may allow or establish appropriate site-specific sampling procedures or requirements, including sampling locations, the season in which the sampling takes place, the minimum duration between the previous measurable storm event and the storm event sampled, the minimum or maximum level of precipitation required for an appropriate storm event, the form of precipitation sampled (snow melt or rain fall), protocols for collecting samples under part 136 of this chapter, and additional time for submitting data on a case-by-case basis. An applicant is expected to "know or have reason to believe" that a pollutant is present in an effluent based on an evaluation of the expected use, production, or storage of the pollutant, or on any previous analyses for the pollutant. (For example, any pesticide manufactured by a facility may be expected to be present in contaminated storm water runoff from the facility.)

(iii) *Reporting requirements.* Every applicant must report quantitative data for every outfall for the following pollutants:

> Biochemical Oxygen Demand (BOD5)
> Chemical Oxygen Demand
> Total Organic Carbon
> Total Suspended Solids
> Ammonia (as N)
> Temperature (both winter and summer)
> pH

(iv) The Director may waive the reporting requirements for individual point sources or for a particular industry category for one or more of the pollutants listed in paragraph (g)(7)(iii) of this section if the applicant has demonstrated that such a waiver is appropriate because information adequate to support issuance of a permit can be obtained with less stringent requirements.

(v) Each applicant with processes in one or more primary industry category (see appendix A of this part) contributing to a discharge must report quantitative data for the following pollutants in each outfall containing process wastewater:

(A) The organic toxic pollutants in the fractions designated in table I of appendix D of this part for the applicant's industrial category or categories unless the applicant qualifies as a small business under paragraph (g)(8) of this section. Table II of appendix D of this part lists the organic toxic pollutants in each fraction. The fractions result from the sample preparation required by the analytical procedure which uses gas chromatography/mass spectrometry. A determination that an applicant falls within a particular industrial category for the purposes of

selecting fractions for testing is not conclusive as to the applicant's inclusion in that category for any other purposes. See Notes 2, 3, and 4 of this section.

(B) The pollutants listed in table III of appendix D of this part (the toxic metals, cyanide, and total phenols).

(vi)(A) Each applicant must indicate whether it knows or has reason to believe that any of the pollutants in table IV of appendix D of this part (certain conventional and nonconventional pollutants) is discharged from each outfall. If an applicable effluent limitations guideline either directly limits the pollutant or, by its express terms, indirectly limits the pollutant through limitations on an indicator, the applicant must report quantitative data. For every pollutant discharged which is not so limited in an effluent limitations guideline, the applicant must either report quantitative data or briefly describe the reasons the pollutant is expected to be discharged.

(B) Each applicant must indicate whether it knows or has reason to believe that any of the pollutants listed in table II or table III of appendix D of this part (the toxic pollutants and total phenols) for which quantitative data are not otherwise required under paragraph (g)(7)(v) of this section are discharged from each outfall. For every pollutant expected to be discharged in concentrations of 10 ppb or greater the applicant must report quantitative data. For acrolein, acrylonitrile, 2,4 dinitrophenol, and 2-methyl-4, 6 dinitrophenol, where any of these four pollutants are expected to be discharged in concentrations of 100 ppb or greater the applicant must report quantitative data. For every pollutant expected to be discharged in concentrations less than 10 ppb, or in the case of acrolein, acrylonitrile, 2,4 dinitrophenol, and 2-methyl-4, 6 dinitrophenol, in concentrations less than 100 ppb, the applicant must either submit quantitative data or briefly describe the reasons the pollutant is expected to be discharged. An applicant qualifying as a small business under paragraph (g)(8) of this section is not required to analyze for pollutants listed in table II of appendix D of this part (the organic toxic pollutants).

(vii) Each applicant must indicate whether it knows or has reason to believe that any of the pollutants in table V of appendix D of this part (certain hazardous substances and asbestos) are discharged from each outfall. For every pollutant expected to be discharged, the applicant must briefly describe the reasons the pollutant is expected to be discharged, and report any quantitative data it has for any pollutant.

(viii) Each applicant must report qualitative data, generated using a screening procedure not calibrated with analytical standards, for 2,3,7,8-tetrachlorodibenzo-p-dioxin (TCDD) if it:

(A) Uses or manufactures 2,4,5-trichlorophenoxy acetic acid (2,4,5,-T); 2-(2,4,5-trichlorophenoxy) propanoic acid (Silvex, 2,4,5,-TP); 2-(2,4,5-trichlorophenoxy) ethyl, 2,2-dichloropropionate (Erbon); O,O-dimethyl O-(2,4,5-trichlorophenyl) phosphorothioate (Ronnel); 2,4,5-trichlorophenol (TCP); or hexachlorophene (HCP); or

(B) Knows or has reason to believe that TCDD is or may be present in an effluent.

(8) *Small business exemption.* An application which qualifies as a small business under one of the following criteria is exempt from the requirements in paragraph (g)(7)(v)(A) or (g)(7)(vi)(A) of this section to submit quantitative data for the pollutants listed in table II of appendix D of this part (the organic toxic pollutants):

(i) For coal mines, a probable total annual production of less than 100,000 tons per year.

(ii) For all other applicants, gross total annual sales averaging less than $100,000 per year (in second quarter 1980 dollars).

(9) *Used or manufactured toxics.* A listing of any toxic pollutant which the applicant currently uses or manufactures as an intermediate or final product or byproduct. The Director may waive or modify this requirement for any applicant if the applicant demonstrates that it would be unduly burdensome to identify each toxic pollutant and the Director has adequate information to issue the permit.

(10) [Reserved]

(11) *Biological toxicity tests.* An identification of any biological toxicity tests which the applicant knows or has reason to believe have been made within the last 3 years on any of the applicant's discharges or on a receiving water in relation to a discharge.

(12) *Contract analyses.* If a contract laboratory or consulting firm performed any of the analyses required by paragraph (g)(7) of this section, the identity of each laboratory or firm and the analyses performed.

(13) *Additional information.* In addition to the information reported on the application form, applicants shall provide to the Director, at his or her request, such other information as the Director may reasonably require to assess the discharges of the facility and to determine whether to issue an NPDES permit. The additional information may include additional quantitative data and bioassays to assess the relative toxicity of discharges to aquatic life and requirements to determine the cause of the toxicity.

(h) *Application requirements for manufacturing, commercial, mining and silvicultural facilities which discharge only non-process wastewater.* Except for stormwater discharges, all manufacturing, commercial, mining and silvicultural dischargers applying for NPDES permits which discharge only non-process wastewater not regulated by an effluent limitations guideline or new source performance standard shall provide the following information to the Director, using application forms provided by the Director:

(1) *Outfall location.* Outfall number, latitude and longitude to the nearest 15 seconds, and the name of the receiving water.

(2) *Discharge date* (for new dischargers). Date of expected commencement of discharge.

(3) *Type of waste.* An identification of the general type of waste discharged, or expected to be discharged upon commencement of operations, including sanitary wastes, restaurant or cafeteria wastes, or noncontact cooling water. An identification of cooling water additives (if any) that are used or expected to be used upon commencement of operations, along with their

composition if existing composition is available.

(4) *Effluent characteristics.* (i) Quantitative data for the pollutants or parameters listed below, unless testing is waived by the Director. The quantitative data may be data collected over the past 365 days, if they remain representative of current operations, and must include maximum daily value, average daily value, and number of measurements taken. The applicant must collect and analyze samples in accordance with 40 CFR Part 136. When analysis of pH, temperature, residual chlorine, oil and grease, or fecal coliform (including *E. coli*), and Enterococci (previously known as fecal streptococcus) and volatile organics is required in paragraphs (h)(4)(i)(A) through (K) of this section, grab samples must be collected for those pollutants. For all other pollutants, a 24-hour composite sample, using a minimum of four (4) grab samples, must be used unless specified otherwise at 40 CFR Part 136. For a composite sample, only one analysis of the composite of aliquots is required. New dischargers must include estimates for the pollutants or parameters listed below instead of actual sampling data, along with the source of each estimate. All levels must be reported or estimated as concentration and as total mass, except for flow, pH, and temperature.

(A) Biochemical Oxygen Demand (BOD_5).

(B) Total Suspended Solids (TSS).

(C) Fecal Coliform (if believed present or if sanitary waste is or will be discharged).

(D) Total Residual Chlorine (if chlorine is used).

(E) Oil and Grease.

(F) Chemical Oxygen Demand (COD) (if non-contact cooling water is or will be discharged).

(G) Total Organic Carbon (TOC) (if non-contact cooling water is or will be discharged).

(H) Ammonia (as N).

(I) Discharge Flow.

(J) pH.

(K) Temperature (Winter and Summer).

(ii) The Director may waive the testing and reporting requirements for any of the pollutants or flow listed in paragraph (h)(4)(i) of this section if the applicant submits a request for such a waiver before or with his application which demonstrates that information adequate to support issuance of a permit can be obtained through less stringent requirements.

(iii) If the applicant is a new discharger, he must complete and submit Item IV of Form 2e (see § 122.21(h)(4)) by providing quantitative data in accordance with that section no later than two years after commencement of discharge. However, the applicant need not complete those portions of Item IV requiring tests which he has already performed and reported under the discharge monitoring requirements of his NPDES permit.

(iv) The requirements of parts i and iii of this section that an applicant must provide quantitative data or estimates of certain pollutants do not apply to pollutants present in a discharge solely as a result of their presence in intake water. However, an applicant must report such pollutants as present. Net credit may be provided for the presence of pollutants in intake water if the requirements of § 122.45(g) are met.

(5) *Flow.* A description of the frequency of flow and duration of any seasonal or intermittent discharge (except for stormwater runoff, leaks, or spills).

(6) *Treatment system.* A brief description of any system used or to be used.

(7) *Optional information.* Any additional information the applicant wishes to be considered, such as influent data for the purpose of obtaining "net" credits pursuant to § 122.45(g).

(8) *Certification.* Signature of certifying official under § 122.22.

(i) *Application requirements for new and existing concentrated animal feeding operations and aquatic animal production facilities.* New and existing concentrated animal feeding operations (defined in § 122.23) and concentrated aquatic animal production facilities (defined in § 122.24) shall provide the following information to the Director, using the application form provided by the Director:

(1) For concentrated animal feeding operations:

(i) The name of the owner or operator;

(ii) The facility location and mailing addresses;

(iii) Latitude and longitude of the production area (entrance to production area);

(iv) A topographic map of the geographic area in which the CAFO is located showing the specific location of the production area, in lieu of the requirements of paragraph (f)(7) of this section;

(v) Specific information about the number and type of animals, whether in open confinement or housed under roof (beef cattle, broilers, layers, swine weighing 55 pounds or more, swine weighing less than 55 pounds, mature dairy cows, dairy heifers, veal calves, sheep and lambs, horses, ducks, turkeys, other);

(vi) The type of containment and storage (anaerobic lagoon, roofed storage shed, storage ponds, underfloor pits, above ground storage tanks, below ground storage tanks, concrete pad, impervious soil pad, other) and total capacity for manure, litter, and process wastewater storage(tons/gallons);

(vii) The total number of acres under control of the applicant available for land application of manure, litter, or process wastewater;

(viii) Estimated amounts of manure, litter, and process wastewater generated per year (tons/gallons);

(ix) Estimated amounts of manure, litter and process wastewater transferred to other persons per year (tons/gallons); and

(x) A nutrient management plan that at a minimum satisfies the requirements specified in §122.42(e), including, for all CAFOs subject to 40 CFR part 412, subpart C or subpart D, the requirements of 40 CFR 412.4(c), as applicable.

(2) For concentrated aquatic animal production facilities:

(i) The maximum daily and average monthly flow from each outfall.

(ii) The number of ponds, raceways, and similar structures.

(iii) The name of the receiving water and the source of intake water.

(iv) For each species of aquatic animals, the total yearly and maximum harvestable weight.

(v) The calendar month of maximum feeding and the total mass of food fed during that month.

(j) *Application requirements for new and existing POTWs.* Unless otherwise indicated, all POTWs and other dischargers designated by the Director must provide, at a minimum, the information in this paragraph to the Director, using Form 2A or another application form provided by the Director. Permit applicants must submit all information available at the time of permit application. The information may be provided by referencing information previously submitted to the Director. The Director may waive any requirement of this paragraph if he or she has access to substantially identical information. The Director may also waive any requirement of this paragraph that is not of material concern for a specific permit, if approved by the Regional Administrator. The waiver request to the Regional Administrator must include the State's justification for the waiver. A Regional Administrator's disapproval of a State's proposed waiver does not constitute final Agency action, but does provide notice to the State and permit applicant(s) that EPA may object to any State-issued permit issued in the absence of the required information.

(1) *Basic application information.* All applicants must provide the following information:

(i) *Facility information.* Name, mailing address, and location of the facility for which the application is submitted;

(ii) *Applicant information.* Name, mailing address, and telephone number of the applicant, and indication as to whether the applicant is the facility's owner, operator, or both;

(iii) *Existing environmental permits.* Identification of all environmental permits or construction approvals received or applied for (including dates) under any of the following programs:

(A) Hazardous Waste Management program under the Resource Conservation and Recovery Act (RCRA), Subpart C;

(B) Underground Injection Control program under the Safe Drinking Water Act (SDWA);

(C) NPDES program under Clean Water Act (CWA);

(D) Prevention of Significant Deterioration (PSD) program under the Clean Air Act;

(E) Nonattainment program under the Clean Air Act;

(F) National Emission Standards for Hazardous Air Pollutants (NESHAPS) preconstruction approval under the Clean Air Act;

(G) Ocean dumping permits under the Marine Protection Research and Sanctuaries Act;

(H) Dredge or fill permits under section 404 of the CWA; and

(I) Other relevant environmental permits, including State permits;

(iv) *Population.* The name and population of each municipal entity served by the facility, including unincorporated connector districts. Indicate whether each municipal entity owns or maintains the collection system and whether the collection system is separate sanitary or combined storm and sanitary, if known;

(v) *Indian country.* Information concerning whether the facility is located in Indian country and whether the facility discharges to a receiving stream that flows through Indian country;

(vi) *Flow rate.* The facility's design flow rate (the wastewater flow rate the plant was built to handle), annual average daily flow rate, and maximum daily flow rate for each of the previous 3 years;

(vii) *Collection system.* Identification of type(s) of collection system(s) used by the treatment works (i.e., separate sanitary sewers or combined storm and sanitary sewers) and an estimate of the percent of sewer line that each type comprises; and

(viii) *Outfalls and other discharge or disposal methods.* The following information for outfalls to waters of the United States and other discharge or disposal methods:

(A) For effluent discharges to waters of the United States, the total number and types of outfalls (e.g, treated effluent, combined sewer overflows, bypasses, constructed emergency overflows);

(B) For wastewater discharged to surface impoundments:

(1) The location of each surface impoundment;

(2) The average daily volume discharged to each surface impoundment; and

(3) Whether the discharge is continuous or intermittent;

(C) For wastewater applied to the land:

(1) The location of each land application site;

(2) The size of each land application site, in acres;

(3) The average daily volume applied to each land application site, in gallons per day; and

(4) Whether land application is continuous or intermittent;

(D) For effluent sent to another facility for treatment prior to discharge:

(1) The means by which the effluent is transported;

(2) The name, mailing address, contact person, and phone number of the organization transporting the discharge, if the transport is provided by a party other than the applicant;

(3) The name, mailing address, contact person, phone number, and NPDES permit number (if any) of the receiving facility; and

(4) The average daily flow rate from this facility into the receiving facility, in millions of gallons per day; and

(E) For wastewater disposed of in a manner not included in paragraphs (j)(1)(viii)(A) through (D) of this section (e.g., underground percolation, underground injection):

(1) A description of the disposal method, including the location and size of each disposal site, if applicable;

(2) The annual average daily volume disposed of by this method, in gallons per day; and

(3) Whether disposal through this method is continuous or intermittent;

(2) *Additional Information.* All applicants with a design flow greater than or equal to 0.1 mgd must provide the following information:

(i) *Inflow and infiltration.* The current average daily volume of inflow and infiltration, in gallons per day, and steps the facility is taking to minimize inflow and infiltration;

(ii) *Topographic map.* A topographic map (or other map if a topographic map is unavailable) extending at least one mile beyond property boundaries of the treatment plant, including all unit processes, and showing:

(A) Treatment plant area and unit processes;

(B) The major pipes or other structures through which wastewater enters the treatment plant and the pipes or other structures through which treated wastewater is discharged from the treatment plant. Include outfalls from bypass piping, if applicable;

(C) Each well where fluids from the treatment plant are injected underground;

(D) Wells, springs, and other surface water bodies listed in public records or otherwise known to the applicant within ¼ mile of the treatment works' property boundaries;

(E) Sewage sludge management facilities (including on-site treatment, storage, and disposal sites); and

(F) Location at which waste classified as hazardous under RCRA enters the treatment plant by truck, rail, or dedicated pipe;

(iii) *Process flow diagram or schematic.* (A) A diagram showing the processes of the treatment plant, including all bypass piping and all backup power sources or redundancy in the system. This includes a water balance showing all treatment units, including disinfection, and showing daily average flow rates at influent and discharge points, and approximate daily flow rates between treatment units; and

(B) A narrative description of the diagram; and

(iv) *Scheduled improvements, schedules of implementation.* The following information regarding scheduled improvements:

(A) The outfall number of each outfall affected;

(B) A narrative description of each required improvement;

(C) Scheduled or actual dates of completion for the following:

(1) Commencement of construction;

(2) Completion of construction;

(3) Commencement of discharge; and

(4) Attainment of operational level;

(D) A description of permits and clearances concerning other Federal and/or State requirements;

(3) *Information on effluent discharges.* Each applicant must provide the following information for each outfall, including bypass points, through which effluent is discharged, as applicable:

(i) *Description of outfall.* The following information about each outfall:

(A) Outfall number;

(B) State, county, and city or town in which outfall is located;

(C) Latitude and longitude, to the nearest second;

(D) Distance from shore and depth below surface;

(E) Average daily flow rate, in million gallons per day;

(F) The following information for each outfall with a seasonal or periodic discharge:

(1) Number of times per year the discharge occurs;

(2) Duration of each discharge;

(3) Flow of each discharge; and

(4) Months in which discharge occurs; and

(G) Whether the outfall is equipped with a diffuser and the type (e.g., high-rate) of diffuser used;

(ii) *Description of receiving waters.* The following information (if known) for each outfall through which effluent is discharged to waters of the United States:

(A) Name of receiving water;

(B) Name of watershed/river/stream system and United States Soil Conservation Service 14-digit watershed code;

(C) Name of State Management/River Basin and United States Geological Survey 8-digit hydrologic cataloging unit code; and

(D) Critical flow of receiving stream and total hardness of receiving stream at critical low flow (if applicable);

(iii) *Description of treatment.* The following information describing the treatment provided for discharges from each outfall to waters of the United States:

(A) The highest level of treatment (e.g., primary, equivalent to secondary, secondary, advanced, other) that is provided for the discharge for each outfall and:

(1) Design biochemical oxygen demand (BOD_5 or $CBOD_5$) removal (percent);

(2) Design suspended solids (SS) removal (percent); and, where applicable,

(3) Design phosphorus (P) removal (percent);

(*4*) Design nitrogen (N) removal (percent); and

(*5*) Any other removals that an advanced treatment system is designed to achieve.

(B) A description of the type of disinfection used, and whether the treatment plant dechlorinates (if disinfection is accomplished through chlorination);

(4) *Effluent monitoring for specific parameters.* (i) As provided in paragraphs (j)(4)(ii) through (x) of this section, all applicants must submit to the Director effluent monitoring information for samples taken from each outfall through which effluent is discharged to waters of the United States, except for CSOs. The Director may allow applicants to submit sampling data for only one outfall on a case-by-case basis, where the applicant has two or more outfalls with substantially identical effluent. The Director may also allow applicants to composite samples from one or more outfalls that discharge into the same mixing zone;

(ii) All applicants must sample and analyze for the pollutants listed in appendix J, Table 1A of this part;

(iii) All applicants with a design flow greater than or equal to 0.1 mgd must sample and analyze for the pollutants listed in appendix J, Table 1 of this part. Facilities that do not use chlorine for disinfection, do not use chlorine elsewhere in the treatment process, and have no reasonable potential to discharge chlorine in their effluent may delete chlorine from Table 1;

(iv) The following applicants must sample and analyze for the pollutants listed in appendix J, Table 2 of this part, and for any other pollutants for which the State or EPA have established water quality standards applicable to the receiving waters:

(A) All POTWs with a design flow rate equal to or greater than one million gallons per day;

(B) All POTWs with approved pretreatment programs or POTWs required to develop a pretreatment program;

(C) Other POTWs, as required by the Director;

(v) The Director should require sampling for additional pollutants, as appropriate, on a case-by-case basis;

(vi) Applicants must provide data from a minimum of three samples taken within four and one-half years prior to the date of the permit application. Samples must be representative of the seasonal variation in the discharge from each outfall. Existing data may be used, if available, in lieu of sampling done solely for the purpose of this application. The Director should require additional samples, as appropriate, on a case-by-case basis.

(vii) All existing data for pollutants specified in paragraphs (j)(4)(ii) through (v) of this section that is collected within four and one-half years of the application must be included in the pollutant data summary submitted by the applicant. If, however, the applicant samples for a specific pollutant on a monthly or more frequent basis, it is only necessary, for such pollutant, to summarize all data collected within one year of the application.

(viii) Applicants must collect samples of effluent and analyze such samples for pollutants in accordance with analytical methods approved under 40 CFR Part 136 unless an alternative is specified in the existing NPDES permit. When analysis of pH, temperature, cyanide, total phenols, residual chlorine, oil and grease, fecal coliform (including *E. coli*), or volatile organics is required in paragraphs (j)(4)(ii) through (iv) of this section, grab samples must be collected for those pollutants. For all other pollutants, 24-hour composite samples must be used. For a composite sample, only one analysis of the composite of aliquots is required.

(ix) The effluent monitoring data provided must include at least the following information for each parameter:

(A) Maximum daily discharge, expressed as concentration or mass, based upon actual sample values.

(B) Average daily discharge for all samples, expressed as concentration or mass, and the number of samples used to obtain this value;

(C) The analytical method used; and

(D) The threshold level (i.e., method detection limit, minimum level, or other designated method endpoints) for the analytical method used.

(x) Unless otherwise required by the Director, metals must be reported as total recoverable.

(5) *Effluent monitoring for whole effluent toxicity.* (i) All applicants must provide an identification of any whole effluent toxicity tests conducted during the four and one-half years prior to the date of the application on any of the applicant's discharges or on any receiving water near the discharge.

(ii) As provided in paragraphs (j)(5)(iii)–(ix) of this section, the following applicants must submit to the Director the results of valid whole effluent toxicity tests for acute or chronic toxicity for samples taken from each outfall through which effluent is discharged to surface waters, except for combined sewer overflows:

(A) All POTWs with design flow rates greater than or equal to one million gallons per day;

(B) All POTWs with approved pretreatment programs or POTWs required to develop a pretreatment program;

(C) Other POTWs, as required by the Director, based on consideration of the following factors:

(*1*) The variability of the pollutants or pollutant parameters in the POTW effluent (based on chemical-specific information, the type of treatment plant, and types of industrial contributors);

(*2*) The ratio of effluent flow to receiving stream flow;

(*3*) Existing controls on point or nonpoint sources, including total maximum daily load calculations for the receiving stream segment and the relative contribution of the POTW;

(*4*) Receiving stream characteristics, including possible or known water quality impairment, and whether the POTW discharges to a coastal water, one of the Great Lakes, or a water designated as an outstanding natural resource water; or .

(*5*) Other considerations (including, but not limited to, the history of toxic impacts and compliance problems at the POTW) that the Director determines could cause or contribute to adverse water quality impacts.

(iii) Where the POTW has two or more outfalls with substantially identical effluent discharging to the same receiving stream segment, the Director may allow applicants to submit whole effluent toxicity data for only one outfall on a case-by-case basis. The Director may also allow applicants to composite samples from one or more outfalls that discharge into the same mixing zone.

(iv) Each applicant required to perform whole effluent toxicity testing pursuant to paragraph (j)(5)(ii) of this section must provide:

(A) Results of a minimum of four quarterly tests for a year, from the year preceding the permit application; or

(B) Results from four tests performed at least annually in the four and one half year period prior to the application, provided the results show no appreciable toxicity using a safety factor determined by the permitting authority.

(v) Applicants must conduct tests with multiple species (no less than two species; e.g., fish, invertebrate, plant), and test for acute or chronic toxicity, depending on the range of receiving water dilution. EPA recommends that applicants conduct acute or chronic testing based on the following dilutions:

(A) Acute toxicity testing if the dilution of the effluent is greater than 1000:1 at the edge of the mixing zone;

(B) Acute or chronic toxicity testing if the dilution of the effluent is between 100:1 and 1000:1 at the edge of the mixing zone. Acute testing may be more appropriate at the higher end of this range (1000:1), and chronic testing may be more appropriate at the lower end of this range (100:1); and

(C) Chronic testing if the dilution of the effluent is less than 100:1 at the edge of the mixing zone.

(vi) Each applicant required to perform whole effluent toxicity testing pursuant to paragraph (j)(5)(ii) of this section must provide the number of chronic or acute whole effluent toxicity tests that have been conducted since the last permit reissuance.

(vii) Applicants must provide the results using the form provided by the Director, or test summaries if available and comprehensive, for each whole effluent toxicity test conducted pursuant to paragraph (j)(5)(ii) of this section for which such information has not been reported previously to the Director.

(viii) Whole effluent toxicity testing conducted pursuant to paragraph (j)(5)(ii) of this section must be conducted using methods approved under 40 CFR part 136. West coast facilities in Washington, Oregon, California, Alaska, Hawaii, and the Pacific Territories are exempted from 40 CFR part 136 chronic methods and must use alternative guidance as directed by the permitting authority.

(ix) For whole effluent toxicity data submitted to the Director within four and one-half years prior to the date of the application, applicants must provide the dates on which the data were submitted and a summary of the results.

(x) Each POTW required to perform whole effluent toxicity testing pursuant to paragraph (j)(5)(ii) of this section must provide any information on the cause of toxicity and written details of any toxicity reduction evaluation conducted, if any whole effluent toxicity test conducted within the past four and one-half years revealed toxicity.

(6) *Industrial discharges.* Applicants must submit the following information about industrial discharges to the POTW:

(i) Number of significant industrial users (SIUs) and categorical industrial users (CIUs) discharging to the POTW; and

(ii) POTWs with one or more SIUs shall provide the following information for each SIU, as defined at 40 CFR 403.3(v), that discharges to the POTW:

(A) Name and mailing address;

(B) Description of all industrial processes that affect or contribute to the SIU's discharge;

(C) Principal products and raw materials of the SIU that affect or contribute to the SIU's discharge;

(D) Average daily volume of wastewater discharged, indicating the amount attributable to process flow and non-process flow;

(E) Whether the SIU is subject to local limits;

(F) Whether the SIU is subject to categorical standards, and if so, under which category(ies) and subcategory(ies); and

(G) Whether any problems at the POTW (e.g., upsets, pass through, in-

terference) have been attributed to the SIU in the past four and one-half years.

(iii) The information required in paragraphs (j)(6)(i) and (ii) of this section may be waived by the Director for POTWs with pretreatment programs if the applicant has submitted either of the following that contain information substantially identical to that required in paragraphs (j)(6)(i) and (ii) of this section.

(A) An annual report submitted within one year of the application; or

(B) A pretreatment program;

(7) *Discharges from hazardous waste generators and from waste cleanup or remediation sites.* POTWs receiving Resource Conservation and Recovery Act (RCRA), Comprehensive Environmental Response, Compensation, and Liability Act (CERCLA), or RCRA Corrective Action wastes or wastes generated at another type of cleanup or remediation site must provide the following information:

(i) If the POTW receives, or has been notified that it will receive, by truck, rail, or dedicated pipe any wastes that are regulated as RCRA hazardous wastes pursuant to 40 CFR part 261, the applicant must report the following:

(A) The method by which the waste is received (i.e., whether by truck, rail, or dedicated pipe); and

(B) The hazardous waste number and amount received annually of each hazardous waste;

(ii) If the POTW receives, or has been notified that it will receive, wastewaters that originate from remedial activities, including those undertaken pursuant to CERCLA and sections 3004(u) or 3008(h) of RCRA, the applicant must report the following:

(A) The identity and description of the site(s) or facility(ies) at which the wastewater originates;

(B) The identities of the wastewater's hazardous constituents, as listed in appendix VIII of part 261 of this chapter; if known; and

(C) The extent of treatment, if any, the wastewater receives or will receive before entering the POTW;

(iii) Applicants are exempt from the requirements of paragraph (j)(7)(ii) of this section if they receive no more than fifteen kilograms per month of hazardous wastes, unless the wastes are acute hazardous wastes as specified in 40 CFR 261.30(d) and 261.33(e).

(8) *Combined sewer overflows.* Each applicant with combined sewer systems must provide the following information:

(i) *Combined sewer system information.* The following information regarding the combined sewer system:

(A) *System map.* A map indicating the location of the following:

(1) All CSO discharge points;

(2) Sensitive use areas potentially affected by CSOs (e.g., beaches, drinking water supplies, shellfish beds, sensitive aquatic ecosystems, and outstanding national resource waters); and

(3) Waters supporting threatened and endangered species potentially affected by CSOs; and

(B) *System diagram.* A diagram of the combined sewer collection system that includes the following information:

(1) The location of major sewer trunk lines, both combined and separate sanitary;

(2) The locations of points where separate sanitary sewers feed into the combined sewer system;

(3) In-line and off-line storage structures;

(4) The locations of flow-regulating devices; and

(5) The locations of pump stations;

(ii) *Information on CSO outfalls.* The following information for each CSO discharge point covered by the permit application:

(A) *Description of outfall.* The following information on each outfall:

(1) Outfall number;

(2) State, county, and city or town in which outfall is located;

(3) Latitude and longitude, to the nearest second; and

(4) Distance from shore and depth below surface;

(5) Whether the applicant monitored any of the following in the past year for this CSO:

(i) Rainfall;

(ii) CSO flow volume;

(iii) CSO pollutant concentrations;

(iv) Receiving water quality;

(v) CSO frequency; and

(6) The number of storm events monitored in the past year;

(B) *CSO events.* The following information about CSO overflows from each outfall:

(1) The number of events in the past year;

(2) The average duration per event, if available;

(3) The average volume per CSO event, if available; and

(4) The minimum rainfall that caused a CSO event, if available, in the last year;

(C) *Description of receiving waters.* The following information about receiving waters:

(1) Name of receiving water;

(2) Name of watershed/stream system and the United States Soil Conservation Service watershed (14-digit) code (if known); and

(3) Name of State Management/River Basin and the United States Geological Survey hydrologic cataloging unit (8-digit) code (if known); and

(D) *CSO operations.* A description of any known water quality impacts on the receiving water caused by the CSO (e.g., permanent or intermittent beach closings, permanent or intermittent shellfish bed closings, fish kills, fish advisories, other recreational loss, or exceedance of any applicable State water quality standard);

(9) *Contractors.* All applicants must provide the name, mailing address, telephone number, and responsibilities of all contractors responsible for any operational or maintenance aspects of the facility; and

(10) *Signature.* All applications must be signed by a certifying official in compliance with §122.22.

(k) *Application requirements for new sources and new discharges.* New manufacturing, commercial, mining and silvicultural dischargers applying for NPDES permits (except for new discharges of facilities subject to the requirements of paragraph (h) of this section or new discharges of storm water associated with industrial activity which are subject to the requirements of §122.26(c)(1) and this section (except as provided by §122.26(c)(1)(ii)) shall provide the following information to the Director, using the application forms provided by the Director:

(1) *Expected outfall location.* The latitude and longitude to the nearest 15

seconds and the name of the receiving water.

(2) *Discharge dates.* The expected date of commencement of discharge.

(3) *Flows, sources of pollution, and treatment technologies*—(i) *Expected treatment of wastewater.* Description of the treatment that the wastewater will receive, along with all operations contributing wastewater to the effluent, average flow contributed by each operation, and the ultimate disposal of any solid or liquid wastes not discharged.

(ii) *Line drawing.* A line drawing of the water flow through the facility with a water balance as described in § 122.21(g)(2).

(iii) *Intermittent flows.* If any of the expected discharges will be intermittent or seasonal, a description of the frequency, duration and maximum daily flow rate of each discharge occurrence (except for stormwater runoff, spillage, or leaks).

(4) *Production.* If a new source performance standard promulgated under section 306 of CWA or an effluent limitation guideline applies to the applicant and is expressed in terms of production (or other measure of operation), a reasonable measure of the applicant's expected actual production reported in the units used in the applicable effluent guideline or new source performance standard as required by § 122.45(b)(2) for each of the first three years. Alternative estimates may also be submitted if production is likely to vary.

(5) *Effluent characteristics.* The requirements in paragraphs (h)(4)(i), (ii), and (iii) of this section that an applicant must provide estimates of certain pollutants expected to be present do not apply to pollutants present in a discharge solely as a result of their presence in intake water; however, an applicant must report such pollutants as present. Net credits may be provided for the presence of pollutants in intake water if the requirements of § 122.45(g) are met. All levels (except for discharge flow, temperature, and pH) must be estimated as concentration and as total mass.

(i) Each applicant must report estimated daily maximum, daily average, and source of information for each outfall for the following pollutants or pa-

rameters. The Director may waive the reporting requirements for any of these pollutants and parameters if the applicant submits a request for such a waiver before or with his application which demonstrates that information adequate to support issuance of the permit can be obtained through less stringent reporting requirements.

(A) Biochemical Oxygen Demand (BOD).

(B) Chemical Oxygen Demand (COD).

(C) Total Organic Carbon (TOC).

(D) Total Suspended Solids (TSS).

(E) Flow.

(F) Ammonia (as N).

(G) Temperature (winter and summer).

(H) pH.

(ii) Each applicant must report estimated daily maximum, daily average, and source of information for each outfall for the following pollutants, if the applicant knows or has reason to believe they will be present or if they are limited by an effluent limitation guideline or new source performance standard either directly or indirectly through limitations on an indicator pollutant: all pollutants in table IV of appendix D of part 122 (certain conventional and nonconventional pollutants).

(iii) Each applicant must report estimated daily maximum, daily average and source of information for the following pollutants if he knows or has reason to believe that they will be present in the discharges from any outfall:

(A) The pollutants listed in table III of appendix D (the toxic metals, in the discharge from any outfall: Total cyanide, and total phenols);

(B) The organic toxic pollutants in table II of appendix D (except bis (chloromethyl) ether, dichlorofluoromethane and trichlorofluoromethane). This requirement is waived for applicants with expected gross sales of less than $100,000 per year for the next three years, and for coal mines with expected average production of less than 100,000 tons of coal per year.

(iv) The applicant is required to report that 2,3,7,8 Tetrachlorodibenzo-P-Dioxin (TCDD) may be discharged if he uses or manufactures one of the following compounds, or if he knows or has reason to believe that TCDD will or may be present in an effluent:

(A) 2,4,5-trichlorophenoxy acetic acid (2,4,5-T) (CAS #93–76–5);

(B) 2-(2,4,5-trichlorophenoxy) propanoic acid (Silvex, 2,4,5-TP) (CAS #93–72–1);

(C) 2-(2,4,5-trichlorophenoxy) ethyl 2,2-dichloropropionate (Erbon) (CAS #136–25–4);

(D) 0,0-dimethyl 0-(2,4,5-trichlorophenyl) phosphorothioate (Ronnel) (CAS #299–84–3);

(E) 2,4,5-trichlorophenol (TCP) (CAS #95–95–4); or

(F) Hexachlorophene (HCP) (CAS #70–30–4);

(v) Each applicant must report any pollutants listed in table V of appendix D (certain hazardous substances) if he believes they will be present in any outfall (no quantitative estimates are required unless they are already available).

(vi) No later than two years after the commencement of discharge from the proposed facility, the applicant is required to complete and submit Items V and VI of NPDES application Form 2c (see § 122.21(g)). However, the applicant need not complete those portions of Item V requiring tests which he has already performed and reported under the discharge monitoring requirements of his NPDES permit.

(6) *Engineering Report.* Each applicant must report the existence of any technical evaluation concerning his wastewater treatment, along with the name and location of similar plants of which he has knowledge.

(7) *Other information.* Any optional information the permittee wishes to have considered.

(8) *Certification.* Signature of certifying official under § 122.22.

(1) *Special provisions for applications from new sources.* (1) The owner or operator of any facility which may be a new source (as defined in § 122.2) and which is located in a State without an approved NPDES program must comply with the provisions of this paragraph (1)(1).

(2)(i) Before beginning any on-site construction as defined in § 122.29, the owner or operator of any facility which

may be a new source must submit information to the Regional Administrator so that he or she can determine if the facility is a new source. The Regional Administrator may request any additional information needed to determine whether the facility is a new source.

(ii) The Regional Administrator shall make an initial determination whether the facility is a new source within 30 days of receiving all necessary information under paragraph (1)(2)(i) of this section.

(3) The Regional Administrator shall issue a public notice in accordance with § 124.10 of this chapter of the new source determination under paragraph (1)(2) of this section. If the Regional Administrator has determined that the facility is a new source, the notice shall state that the applicant must comply with the environmental review requirements of 40 CFR 6.600 through 6.607.

(4) Any interested party may challenge the Regional Administrator's initial new source determination by requesting review of the determination under § 124.19 of this chapter within 30 days of the public notice of the initial determination. If all interested parties agree, the Environmental Appeals Board may defer review until after a final permit decision is made, and consolidate review of the determination with any review of the permit decision.

(m) *Variance requests by non-POTWs.* A discharger which is not a publicly owned treatment works (POTW) may request a variance from otherwise applicable effluent limitations under any of the following statutory or regulatory provisions within the times specified in this paragraph:

(1) *Fundamentally different factors.* (i) A request for a variance based on the presence of "fundamentally different factors" from those on which the effluent limitations guideline was based shall be filed as follows:

(A) For a request from best practicable control technology currently available (BPT), by the close of the public comment period under § 124.10.

(B) For a request from best available technology economically achievable

(BAT) and/or best conventional pollutant control technology (BCT), by no later than:

(1) July 3, 1989, for a request based on an effluent limitation guideline promulgated before February 4, 1987, to the extent July 3, 1989 is not later than that provided under previously promulgated regulations; or

(2) 180 days after the date on which an effluent limitation guideline is published in the FEDERAL REGISTER for a request based on an effluent limitation guideline promulgated on or after February 4, 1987.

(ii) The request shall explain how the requirements of the applicable regulatory and/or statutory criteria have been met.

(2) *Non-conventional pollutants.* A request for a variance from the BAT requirements for CWA section 301(b)(2)(F) pollutants (commonly called "non-conventional" pollutants) pursuant to section 301(c) of CWA because of the economic capability of the owner or operator, or pursuant to section 301(g) of the CWA (provided however that a § 301(g) variance may only be requested for ammonia; chlorine; color; iron; total phenols (4AAP) (when determined by the Administrator to be a pollutant covered by section 301(b)(2)(F)) and any other pollutant which the Administrator lists under section 301(g)(4) of the CWA) must be made as follows:

(i) For those requests for a variance from an effluent limitation based upon an effluent limitation guideline by:

(A) Submitting an initial request to the Regional Administrator, as well as to the State Director if applicable, stating the name of the discharger, the permit number, the outfall number(s), the applicable effluent guideline, and whether the discharger is requesting a section 301(c) or section 301(g) modification or both. This request must have been filed not later than:

(1) September 25, 1978, for a pollutant which is controlled by a BAT effluent limitation guideline promulgated before December 27, 1977; or

(2) 270 days after promulgation of an applicable effluent limitation guideline for guidelines promulgated after December 27, 1977; and

(B) Submitting a completed request no later than the close of the public comment period under § 124.10 demonstrating that the requirements of § 124.13 and the applicable requirements of part 125 have been met. Notwithstanding this provision, the complete application for a request under section 301(g) shall be filed 180 days before EPA must make a decision (unless the Regional Division Director establishes a shorter or longer period).

(ii) For those requests for a variance from effluent limitations not based on effluent limitation guidelines, the request need only comply with paragraph (m)(2)(i)(B) of this section and need not be preceded by an initial request under paragraph (m)(2)(i)(A) of this section.

(3)–(4) [Reserved]

(5) *Water quality related effluent limitations.* A modification under section 302(b)(2) of requirements under section 302(a) for achieving water quality related effluent limitations may be requested no later than the close of the public comment period under § 124.10 on the permit from which the modification is sought.

(6) *Thermal discharges.* A variance under CWA section 316(a) for the thermal component of any discharge must be filed with a timely application for a permit under this section, except that if thermal effluent limitations are established under CWA section 402(a)(1) or are based on water quality standards the request for a variance may be filed by the close of the public comment period under § 124.10. A copy of the request as required under 40 CFR part 125, subpart H, shall be sent simultaneously to the appropriate State or interstate certifying agency as required under 40 CFR part 125. (See § 124.65 for special procedures for section 316(a) thermal variances.)

(n) *Variance requests by POTWs.* A discharger which is a publicly owned treatment works (POTW) may request a variance from otherwise applicable effluent limitations under any of the following statutory provisions as specified in this paragraph:

(1) *Discharges into marine waters.* A request for a modification under CWA section 301(h) of requirements of CWA section 301(b)(1)(B) for discharges into marine waters must be filed in accordance with the requirements of 40 CFR part 125, subpart G.

(2) [Reserved]

(3) *Water quality based effluent limitation.* A modification under CWA section 302(b)(2) of the requirements under section 302(a) for achieving water quality based effluent limitations shall be requested no later than the close of the public comment period under §124.10 on the permit from which the modification is sought.

(o) *Expedited variance procedures and time extensions.* (1) Notwithstanding the time requirements in paragraphs (m) and (n) of this section, the Director may notify a permit applicant before a draft permit is issued under §124.6 that the draft permit will likely contain limitations which are eligible for variances. In the notice the Director may require the applicant as a condition of consideration of any potential variance request to submit a request explaining how the requirements of part 125 applicable to the variance have been met and may require its submission within a specified reasonable time after receipt of the notice. The notice may be sent before the permit application has been submitted. The draft or final permit may contain the alternative limitations which may become effective upon final grant of the variance.

(2) A discharger who cannot file a timely complete request required under paragraph (m)(2)(i)(B) or (m)(2)(ii) of this section may request an extension. The extension may be granted or denied at the discretion of the Director. Extensions shall be no more than 6 months in duration.

(p) *Recordkeeping.* Except for information required by paragraph (d)(3)(ii) of this section, which shall be retained for a period of at least five years from the date the application is signed (or longer as required by 40 CFR part 503), applicants shall keep records of all data used to complete permit applications and any supplemental information submitted under this section for a period of at least 3 years from the date the application is signed.

(q) *Sewage sludge management.* All TWTDS subject to paragraph (c)(2)(i) of this section must provide the information in this paragraph to the Director, using Form 2S or another application form approved by the Director. New ap-plicants must submit all information available at the time of permit application. The information may be provided by referencing information previously submitted to the Director. The Director may waive any requirement of this paragraph if he or she has access to substantially identical information. The Director may also waive any requirement of this paragraph that is not of material concern for a specific permit, if approved by the Regional Administrator. The waiver request to the Regional Administrator must include the State's justification for the waiver. A Regional Administrator's disapproval of a State's proposed waiver does not constitute final Agency action, but does provide notice to the State and permit applicant(s) that EPA may object to any State-issued permit issued in the absence of the required information.

(1) *Facility information.* All applicants must submit the following information:

(i) The name, mailing address, and location of the TWTDS for which the application is submitted;

(ii) Whether the facility is a Class I Sludge Management Facility;

(iii) The design flow rate (in million gallons per day);

(iv) The total population served; and

(v) The TWTDS's status as Federal, State, private, public, or other entity;

(2) *Applicant information.* All applicants must submit the following information:

(i) The name, mailing address, and telephone number of the applicant; and

(ii) Indication whether the applicant is the owner, operator, or both;

(3) *Permit information.* All applicants must submit the facility's NPDES permit number, if applicable, and a listing of all other Federal, State, and local permits or construction approvals received or applied for under any of the following programs:

(i) Hazardous Waste Management program under the Resource Conservation and Recovery Act (RCRA);

(ii) UIC program under the Safe Drinking Water Act (SDWA);

(iii) NPDES program under the Clean Water Act (CWA);

(iv) Prevention of Significant Deterioration (PSD) program under the Clean Air Act;

(v) Nonattainment program under the Clean Air Act;

(vi) National Emission Standards for Hazardous Air Pollutants (NESHAPS) preconstruction approval under the Clean Air Act;

(vii) Dredge or fill permits under section 404 of CWA;

(viii) Other relevant environmental permits, including State or local permits;

(4) *Indian country.* All applicants must identify any generation, treatment, storage, land application, or disposal of sewage sludge that occurs in Indian country;

(5) *Topographic map.* All applicants must submit a topographic map (or other map if a topographic map is unavailable) extending one mile beyond property boundaries of the facility and showing the following information:

(i) All sewage sludge management facilities, including on-site treatment, storage, and disposal sites; and

(ii) Wells, springs, and other surface water bodies that are within ¼ mile of the property boundaries and listed in public records or otherwise known to the applicant;

(6) *Sewage sludge handling.* All applicants must submit a line drawing and/or a narrative description that identifies all sewage sludge management practices employed during the term of the permit, including all units used for collecting, dewatering, storing, or treating sewage sludge, the destination(s) of all liquids and solids leaving each such unit, and all processes used for pathogen reduction and vector attraction reduction;

(7) *Sewage sludge quality.* The applicant must submit sewage sludge monitoring data for the pollutants for which limits in sewage sludge have been established in 40 CFR part 503 for the applicant's use or disposal practices on the date of permit application.

(i) The Director may require sampling for additional pollutants, as appropriate, on a case-by-case basis;

(ii) Applicants must provide data from a minimum of three samples taken within four and one-half years prior to the date of the permit application. Samples must be representative of the sewage sludge and should be taken at least one month apart. Existing data may be used in lieu of sampling done solely for the purpose of this application;

(iii) Applicants must collect and analyze samples in accordance with analytical methods approved under SW-846 unless an alternative has been specified in an existing sewage sludge permit;

(iv) The monitoring data provided must include at least the following information for each parameter:

(A) Average monthly concentration for all samples (mg/kg dry weight), based upon actual sample values;

(B) The analytical method used; and

(C) The method detection level.

(8) *Preparation of sewage sludge.* If the applicant is a "person who prepares" sewage sludge, as defined at 40 CFR 503.9(r), the applicant must provide the following information:

(i) If the applicant's facility generates sewage sludge, the total dry metric tons per 365-day period generated at the facility;

(ii) If the applicant's facility receives sewage sludge from another facility, the following information for each facility from which sewage sludge is received:

(A) The name, mailing address, and location of the other facility;

(B) The total dry metric tons per 365-day period received from the other facility; and

(C) A description of any treatment processes occurring at the other facility, including blending activities and treatment to reduce pathogens or vector attraction characteristics;

(iii) If the applicant's facility changes the quality of sewage sludge through blending, treatment, or other activities, the following information:

(A) Whether the Class A pathogen reduction requirements in 40 CFR 503.32(a) or the Class B pathogen reduction requirements in 40 CFR 503.32(b) are met, and a description of any treatment processes used to reduce pathogens in sewage sludge;

(B) Whether any of the vector attraction reduction options of 40 CFR 503.33(b)(1) through (b)(8) are met, and a description of any treatment processes used to reduce vector attraction properties in sewage sludge; and

(C) A description of any other blending, treatment, or other activities that change the quality of sewage sludge;

(iv) If sewage sludge from the applicant's facility meets the ceiling concentrations in 40 CFR 503.13(b)(1), the pollutant concentrations in §503.13(b)(3), the Class A pathogen requirements in §503.32(a), and one of the vector attraction reduction requirements in §503.33(b)(1) through (b)(8), and if the sewage sludge is applied to the land, the applicant must provide the total dry metric tons per 365-day period of sewage sludge subject to this paragraph that is applied to the land;

(v) If sewage sludge from the applicant's facility is sold or given away in a bag or other container for application to the land, and the sewage sludge is not subject to paragraph (q)(8)(iv) of this section, the applicant must provide the following information:

(A) The total dry metric tons per 365-day period of sewage sludge subject to this paragraph that is sold or given away in a bag or other container for application to the land; and

(B) A copy of all labels or notices that accompany the sewage sludge being sold or given away;

(vi) If sewage sludge from the applicant's facility is provided to another "person who prepares," as defined at 40 CFR 503.9(r), and the sewage sludge is not subject to paragraph (q)(8)(iv) of this section, the applicant must provide the following information for each facility receiving the sewage sludge:

(A) The name and mailing address of the receiving facility;

(B) The total dry metric tons per 365-day period of sewage sludge subject to this paragraph that the applicant provides to the receiving facility;

(C) A description of any treatment processes occurring at the receiving facility, including blending activities and treatment to reduce pathogens or vector attraction characteristic;

(D) A copy of the notice and necessary information that the applicant is required to provide the receiving facility under 40 CFR 503.12(g); and

(E) If the receiving facility places sewage sludge in bags or containers for sale or give-away to application to the land, a copy of any labels or notices that accompany the sewage sludge;

(9) *Land application of bulk sewage sludge.* If sewage sludge from the applicant's facility is applied to the land in bulk form, and is not subject to paragraphs (q)(8)(iv), (v), or (vi) of this section, the applicant must provide the following information:

(i) The total dry metric tons per 365-day period of sewage sludge subject to this paragraph that is applied to the land;

(ii) If any land application sites are located in States other than the State where the sewage sludge is prepared, a description of how the applicant will notify the permitting authority for the State(s) where the land application sites are located;

(iii) The following information for each land application site that has been identified at the time of permit application:

(A) The name (if any), and location for the land application site;

(B) The site's latitude and longitude to the nearest second, and method of determination;

(C) A topographic map (or other map if a topographic map is unavailable) that shows the site's location;

(D) The name, mailing address, and telephone number of the site owner, if different from the applicant;

(E) The name, mailing address, and telephone number of the person who applies sewage sludge to the site, if different from the applicant;

(F) Whether the site is agricultural land, forest, a public contact site, or a reclamation site, as such site types are defined under 40 CFR 503.11;

(G) The type of vegetation grown on the site, if known, and the nitrogen requirement for this vegetation;

(H) Whether either of the vector attraction reduction options of 40 CFR 503.33(b)(9) or (b)(10) is met at the site, and a description of any procedures employed at the time of use to reduce vector attraction properties in sewage sludge; and

(I) Other information that describes how the site will be managed, as specified by the permitting authority.

(iv) The following information for each land application site that has been identified at the time of permit application, if the applicant intends to apply bulk sewage sludge subject to the cumulative pollutant loading rates in 40 CFR 503.13(b)(2) to the site:

(A) Whether the applicant has contacted the permitting authority in the State where the bulk sewage sludge subject to §503.13(b)(2) will be applied, to ascertain whether bulk sewage sludge subject to §503.13(b)(2) has been applied to the site on or since July 20, 1993, and if so, the name of the permitting authority and the name and phone number of a contact person at the permitting authority;

(B) Identification of facilities other than the applicant's facility that have sent, or are sending, sewage sludge subject to the cumulative pollutant loading rates in §503.13(b)(2) to the site since July 20, 1993, if, based on the inquiry in paragraph (q)(iv)(A), bulk sewage sludge subject to cumulative pollutant loading rates in §503.13(b)(2) has been applied to the site since July 20, 1993;

(v) If not all land application sites have been identified at the time of permit application, the applicant must submit a land application plan that, at a minimum:

(A) Describes the geographical area covered by the plan;

(B) Identifies the site selection criteria;

(C) Describes how the site(s) will be managed;

(D) Provides for advance notice to the permit authority of specific land application sites and reasonable time for the permit authority to object prior to land application of the sewage sludge; and

(E) Provides for advance public notice of land application sites in the manner prescribed by State and local law. When State or local law does not require advance public notice, it must be provided in a manner reasonably calculated to apprize the general public of the planned land application.

(10) *Surface disposal.* If sewage sludge from the applicant's facility is placed on a surface disposal site, the applicant must provide the following information:

(i) The total dry metric tons of sewage sludge from the applicant's facility that is placed on surface disposal sites per 365-day period;

(ii) The following information for each surface disposal site receiving sewage sludge from the applicant's facility that the applicant does *not* own or operate:

(A) The site name or number, contact person, mailing address, and telephone number for the surface disposal site; and

(B) The total dry metric tons from the applicant's facility per 365-day period placed on the surface disposal site;

(iii) The following information for each active sewage sludge unit at each surface disposal site that the applicant owns or operates:

(A) The name or number and the location of the active sewage sludge unit;

(B) The unit's latitude and longitude to the nearest second, and method of determination;

(C) If not already provided, a topographic map (or other map if a topographic map is unavailable) that shows the unit's location;

(D) The total dry metric tons placed on the active sewage sludge unit per 365-day period;

(E) The total dry metric tons placed on the active sewage sludge unit over the life of the unit;

(F) A description of any liner for the active sewage sludge unit, including whether it has a maximum permeability of 1×10^{-7} cm/sec;

(G) A description of any leachate collection system for the active sewage sludge unit, including the method used for leachate disposal, and any Federal, State, and local permit number(s) for leachate disposal;

(H) If the active sewage sludge unit is less than 150 meters from the property line of the surface disposal site, the actual distance from the unit boundary to the site property line;

(I) The remaining capacity (dry metric tons) for the active sewage sludge unit;

(J) The date on which the active sewage sludge unit is expected to close, if such a date has been identified;

(K) The following information for any other facility that sends sewage sludge to the active sewage sludge unit:

(*1*) The name, contact person, and mailing address of the facility; and

(*2*) Available information regarding the quality of the sewage sludge received from the facility, including any treatment at the facility to reduce pathogens or vector attraction characteristics;

(L) Whether any of the vector attraction reduction options of 40 CFR 503.33(b)(9) through (b)(11) is met at the active sewage sludge unit, and a description of any procedures employed at the time of disposal to reduce vector attraction properties in sewage sludge;

(M) The following information, as applicable to any ground-water monitoring occurring at the active sewage sludge unit:

(*1*) A description of any ground-water monitoring occurring at the active sewage sludge unit;

(*2*) Any available ground-water monitoring data, with a description of the well locations and approximate depth to ground water;

(*3*) A copy of any ground-water monitoring plan that has been prepared for the active sewage sludge unit;

(*4*) A copy of any certification that has been obtained from a qualified ground-water scientist that the aquifer has not been contaminated; and

(N) If site-specific pollutant limits are being sought for the sewage sludge placed on this active sewage sludge unit, information to support such a request;

(11) *Incineration*. If sewage sludge from the applicant's facility is fired in a sewage sludge incinerator, the applicant must provide the following information:

(i) The total dry metric tons of sewage sludge from the applicant's facility that is fired in sewage sludge incinerators per 365-day period;

(ii) The following information for each sewage sludge incinerator firing the applicant's sewage sludge that the applicant does *not* own or operate:

(A) The name and/or number, contact person, mailing address, and telephone number of the sewage sludge incinerator; and

(B) The total dry metric tons from the applicant's facility per 365-day period fired in the sewage sludge incinerator;

(iii) The following information for each sewage sludge incinerator that the applicant owns or operates:

(A) The name and/or number and the location of the sewage sludge incinerator;

(B) The incinerator's latitude and longitude to the nearest second, and method of determination;

(C) The total dry metric tons per 365-day period fired in the sewage sludge incinerator;

(D) Information, test data, and documentation of ongoing operating parameters indicating that compliance with the National Emission Standard for Beryllium in 40 CFR part 61 will be achieved;

(E) Information, test data, and documentation of ongoing operating parameters indicating that compliance with the National Emission Standard for Mercury in 40 CFR part 61 will be achieved;

(F) The dispersion factor for the sewage sludge incinerator, as well as modeling results and supporting documentation;

(G) The control efficiency for parameters regulated in 40 CFR 503.43, as well as performance test results and supporting documentation;

(H) Information used to calculate the risk specific concentration (RSC) for chromium, including the results of incinerator stack tests for hexavalent and total chromium concentrations, if the applicant is requesting a chromium limit based on a site-specific RSC value;

(I) Whether the applicant monitors total hydrocarbons (THC) or Carbon Monoxide (CO) in the exit gas for the sewage sludge incinerator;

(J) The type of sewage sludge incinerator;

(K) The maximum performance test combustion temperature, as obtained during the performance test of the sewage sludge incinerator to determine pollutant control efficiencies;

(L) The following information on the sewage sludge feed rate used during the performance test:

(*1*) Sewage sludge feed rate in dry metric tons per day;

(*2*) Identification of whether the feed rate submitted is average use or maximum design; and

(3) A description of how the feed rate was calculated;

(M) The incinerator stack height in meters for each stack, including identification of whether actual or creditable stack height was used;

(N) The operating parameters for the sewage sludge incinerator air pollution control device(s), as obtained during the performance test of the sewage sludge incinerator to determine pollutant control efficiencies;

(O) Identification of the monitoring equipment in place, including (but not limited to) equipment to monitor the following:

(1) Total hydrocarbons or Carbon Monoxide;

(2) Percent oxygen;

(3) Percent moisture; and

(4) Combustion temperature; and

(P) A list of all air pollution control equipment used with this sewage sludge incinerator;

(12) *Disposal in a municipal solid waste landfill.* If sewage sludge from the applicant's facility is sent to a municipal solid waste landfill (MSWLF), the applicant must provide the following information for each MSWLF to which sewage sludge is sent:

(i) The name, contact person, mailing address, location, and all applicable permit numbers of the MSWLF;

(ii) The total dry metric tons per 365-day period sent from this facility to the MSWLF;

(iii) A determination of whether the sewage sludge meets applicable requirements for disposal of sewage sludge in a MSWLF, including the results of the paint filter liquids test and any additional requirements that apply on a site-specific basis; and

(iv) Information, if known, indicating whether the MSWLF complies with criteria set forth in 40 CFR part 258;

(13) *Contractors.* All applicants must provide the name, mailing address, telephone number, and responsibilities of all contractors responsible for any operational or maintenance aspects of the facility related to sewage sludge generation, treatment, use, or disposal;

(14) *Other information.* At the request of the permitting authority, the applicant must provide any other information necessary to determine the appropriate standards for permitting under 40 CFR part 503, and must provide any other information necessary to assess the sewage sludge use and disposal practices, determine whether to issue a permit, or identify appropriate permit requirements; and

(15) *Signature.* All applications must be signed by a certifying official in compliance with § 122.22.

[Note 1: At 46 FR 2046, Jan. 8, 1981, the Environmental Protection Agency suspended until further notice § 122.21(g)(7)(v)(A) and the corresponding portions of Item V-C of the NPDES application Form 2C as they apply to coal mines. This suspension continues in effect.]

[Note 2: At 46 FR 22585, Apr. 20, 1981, the Environmental Protection Agency suspended until further notice § 122.21(g)(7)(v)(A) and the corresponding portions of Item V-C of the NPDES application Form 2C as they apply to:

a. Testing and reporting for all four organic fractions in the Greige Mills Subcategory of the Textile Mills industry (subpart C—Low water use processing of 40 CFR part 410), and testing and reporting for the pesticide fraction in all other subcategories of this industrial category.

b. Testing and reporting for the volatile, base/neutral and pesticide fractions in the Base and Precious Metals Subcategory of the Ore Mining and Dressing industry (subpart B of 40 CFR part 440), and testing and reporting for all four fractions in all other subcategories of this industrial category.

c. Testing and reporting for all four GC/MS fractions in the Porcelain Enameling industry.

This revision continues that suspension.][1]

[Note 3: At 46 FR 35090, July 1, 1981, the Environmental Protection Agency suspended until further notice § 122.21(g)(7)(v)(A) and the corresponding portions of Item V-C of the NPDES application Form 2C as they apply to:

a. Testing and reporting for the pesticide fraction in the Tall Oil Rosin Subcategory (subpart D) and Rosin-Based Derivatives Subcategory (subpart F) of the Gum and Wood Chemicals industry (40 CFR part 454), and testing and reporting for the pesticide and base-neutral fractions in all other subcategories of this industrial category.

b. Testing and reporting for the pesticide fraction in the Leather Tanning and Finishing, Paint and Ink Formulation, and Photographic Supplies industrial categories.

c. Testing and reporting for the acid, base/neutral and pesticide fractions in the Petroleum Refining industrial category.

d. Testing and reporting for the pesticide fraction in the Papergrade Sulfite subcategories (subparts J and U) of the Pulp and Paper industry (40 CFR part 430); testing and reporting for the base/neutral and pesticide fractions in the following subcategories: Deink (subpart Q), Dissolving Kraft (subpart F), and Paperboard from Waste Paper (subpart E); testing and reporting for the volatile, base/neutral and pesticide fractions in the following subcategories: BCT Bleached Kraft (subpart H), Semi-Chemical (subparts B and C), and Nonintegrated-Fine Papers (subpart R); and testing and reporting for the acid, base/neutral, and pesticide fractions in the following subcategories: Fine Bleached Kraft (subpart I), Dissolving Sulfite Pulp (subpart K), Groundwood-Fine Papers (subpart O), Market Bleached Kraft (subpart G), Tissue from Wastepaper (subpart T), and Nonintegrated-Tissue Papers (subpart S).

e. Testing and reporting for the base/neutral fraction in the Once-Through Cooling Water, Fly Ash and Bottom Ash Transport Water process wastestreams of the Steam Electric Power Plant industrial category. This revision continues that suspension.][1]

(r) *Application requirements for facilities with cooling water intake structures—* (1)(i) *New facilities with new or modified cooling water intake structures.* New facilities (other than offshore oil and gas extraction facilities) with cooling water intake structures as defined in part 125, subpart I, of this chapter must submit to the Director for review the information required under paragraphs (r)(2) (except (r)(2)(iv)), (3), and (4) of this section and §125.86 of this chapter as part of their application. New offshore oil and gas extraction facilities with cooling water intake structures as defined in part 125, subpart N, of this chapter that are fixed facilities must submit to the Director for review the information required under paragraphs (r)(2) (except (r)(2)(iv)), (3), and (4) of this section and §125.136 of this chapter as part of their application. New offshore oil and gas extraction facilities that are *not* fixed facilities must submit to the Director for review only the information required under paragraphs (r)(2)(iv), (r)(3) (except (r)(3)(ii)), and §125.136 of this chapter as part of their application. Requests for alternative requirements under §125.85 or §125.135 of this chapter must be submitted with your permit application.

(ii) *Phase II existing facilities.* Phase II existing facilities as defined in part 125, subpart J, of this chapter must

submit to the Director for review the information required under paragraphs (r)(2), (3), and (5) of this section and all applicable provisions of §125.95 of this chapter as part of their application except for the Proposal for Information Collection which must be provided in accordance with §125.95(b)(1).

(2) *Source water physical data.* These include:

(i) A narrative description and scaled drawings showing the physical configuration of all source water bodies used by your facility, including areal dimensions, depths, salinity and temperature regimes, and other documentation that supports your determination of the water body type where each cooling water intake structure is located;

(ii) Identification and characterization of the source waterbody's hydrological and geomorphological features, as well as the methods you used to conduct any physical studies to determine your intake's area of influence within the waterbody and the results of such studies;

(iii) Locational maps; and

(iv) For new offshore oil and gas facilities that are not fixed facilities, a narrative description and/or locational maps providing information on predicted locations within the waterbody during the permit term in sufficient detail for the Director to determine the appropriateness of additional impingement requirements under §125.134(b)(4).

(3) *Cooling water intake structure data.* These include:

(i) A narrative description of the configuration of each of your cooling water intake structures and where it is located in the water body and in the water column;

(ii) Latitude and longitude in degrees, minutes, and seconds for each of your cooling water intake structures;

(iii) A narrative description of the operation of each of your cooling water intake structures, including design intake flows, daily hours of operation, number of days of the year in operation and seasonal changes, if applicable;

(iv) A flow distribution and water balance diagram that includes all sources of water to the facility, recirculating flows, and discharges; and

(v) Engineering drawings of the cooling water intake structure.

[1] EDITORIAL NOTE: The words "This revision" refer to the document published at 48 FR 14153, Apr. 1, 1983.

(4) *Source water baseline biological characterization data.* This information is required to characterize the biological community in the vicinity of the cooling water intake structure and to characterize the operation of the cooling water intake structures. The Director may also use this information in subsequent permit renewal proceedings to determine if your Design and Construction Technology Plan as required in § 125.86(b)(4) or § 125.136(b)(3) of this chapter should be revised. This supporting information must include existing data (if they are available). However, you may supplement the data using newly conducted field studies if you choose to do so. The information you submit must include:

(i) A list of the data in paragraphs (r)(4)(ii) through (vi) of this section that are not available and efforts made to identify sources of the data;

(ii) A list of species (or relevant taxa) for all life stages and their relative abundance in the vicinity of the cooling water intake structure;

(iii) Identification of the species and life stages that would be most susceptible to impingement and entrainment. Species evaluated should include the forage base as well as those most important in terms of significance to commercial and recreational fisheries;

(iv) Identification and evaluation of the primary period of reproduction, larval recruitment, and period of peak abundance for relevant taxa;

(v) Data representative of the seasonal and daily activities (e.g., feeding and water column migration) of biological organisms in the vicinity of the cooling water intake structure;

(vi) Identification of all threatened, endangered, and other protected species that might be susceptible to impingement and entrainment at your cooling water intake structures;

(vii) Documentation of any public participation or consultation with Federal or State agencies undertaken in development of the plan; and

(viii) If you supplement the information requested in paragraph (r)(4)(i) of this section with data collected using field studies, supporting documentation for the Source Water Baseline Biological Characterization must include a description of all methods and qual-ity assurance procedures for sampling, and data analysis including a description of the study area; taxonomic identification of sampled and evaluated biological assemblages (including all life stages of fish and shellfish); and sampling and data analysis methods. The sampling and/or data analysis methods you use must be appropriate for a quantitative survey and based on consideration of methods used in other biological studies performed within the same source water body. The study area should include, at a minimum, the area of influence of the cooling water intake structure.

(5) *Cooling water system data.* Phase II existing facilities as defined in part 125, subpart J of this chapter must provide the following information for each cooling water intake structure they use:

(i) A narrative description of the operation of the cooling water system, its relationship to cooling water intake structures, the proportion of the design intake flow that is used in the system, the number of days of the year the cooling water system is in operation and seasonal changes in the operation of the system, if applicable; and

(ii) Design and engineering calculations prepared by a qualified professional and supporting data to support the description required by paragraph (r)(5)(i) of this section.

[48 FR 14153, Apr. 1, 1983]

EDITORIAL NOTE: For FEDERAL REGISTER citations affecting § 122.21, see the List of CFR Sections Affected, which appears in the Finding Aids section of the printed volume and at *www.fdsys.gov.*

EFFECTIVE DATE NOTE: At 72 FR 37109, July 9, 2007, § 122.21(r)(1)(ii) and (r)(5) were suspended.

§ 122.22   Signatories to permit applications and reports (applicable to State programs, see § 123.25).

(a) *Applications.* All permit applications shall be signed as follows:

(1) *For a corporation.* By a responsible corporate officer. For the purpose of this section, a responsible corporate officer means: (i) A president, secretary, treasurer, or vice-president of the corporation in charge of a principal business function, or any other person who performs similar policy- or decision-

making functions for the corporation, or (ii) the manager of one or more manufacturing, production, or operating facilities, provided, the manager is authorized to make management decisions which govern the operation of the regulated facility including having the explicit or implicit duty of making major capital investment recommendations, and initiating and directing other comprehensive measures to assure long term environmental compliance with environmental laws and regulations; the manager can ensure that the necessary systems are established or actions taken to gather complete and accurate information for permit application requirements; and where authority to sign documents has been assigned or delegated to the manager in accordance with corporate procedures.

NOTE: EPA does not require specific assignments or delegations of authority to responsible corporate officers identified in § 122.22(a)(1)(i). The Agency will presume that these responsible corporate officers have the requisite authority to sign permit applications unless the corporation has notified the Director to the contrary. Corporate procedures governing authority to sign permit applications may provide for assignment or delegation to applicable corporate positions under § 122.22(a)(1)(ii) rather than to specific individuals.

(2) *For a partnership or sole proprietorship.* By a general partner or the proprietor, respectively; or

(3) *For a municipality, State, Federal, or other public agency.* By either a principal executive officer or ranking elected official. For purposes of this section, a principal executive officer of a Federal agency includes: (i) The chief executive officer of the agency, or (ii) a senior executive officer having responsibility for the overall operations of a principal geographic unit of the agency (e.g., Regional Administrators of EPA).

(b) All reports required by permits, and other information requested by the Director shall be signed by a person described in paragraph (a) of this section, or by a duly authorized representative of that person. A person is a duly authorized representative only if:

(1) The authorization is made in writing by a person described in paragraph (a) of this section;

(2) The authorization specifies either an individual or a position having re-

sponsibility for the overall operation of the regulated facility or activity such as the position of plant manager, operator of a well or a well field, superintendent, position of equivalent responsibility, or an individual or position having overall responsibility for environmental matters for the company, (A duly authorized representative may thus be either a named individual or any individual occupying a named position.) and,

(3) The written authorization is submitted to the Director.

(c) *Changes to authorization.* If an authorization under paragraph (b) of this section is no longer accurate because a different individual or position has responsibility for the overall operation of the facility, a new authorization satisfying the requirements of paragraph (b) of this section must be submitted to the Director prior to or together with any reports, information, or applications to be signed by an authorized representative.

(d) *Certification.* Any person signing a document under paragraph (a) or (b) of this section shall make the following certification:

I certify under penalty of law that this document and all attachments were prepared under my direction or supervision in accordance with a system designed to assure that qualified personnel properly gather and evaluate the information submitted. Based on my inquiry of the person or persons who manage the system, or those persons directly responsible for gathering the information, the information submitted is, to the best of my knowledge and belief, true, accurate, and complete. I am aware that there are significant penalties for submitting false information, including the possibility of fine and imprisonment for knowing violations.

(Clean Water Act (33 U.S.C. 1251 *et seq.*), Safe Drinking Water Act (42 U.S.C. 300f *et seq.*), Clean Air Act (42 U.S.C. 7401 *et seq.*), Resource Conservation and Recovery Act (42 U.S.C. 6901 *et seq.*))

[48 FR 14153, Apr. 1, 1983, as amended at 48 FR 39619, Sept. 1, 1983; 49 FR 38047, Sept. 29, 1984; 50 FR 6941, Feb. 19, 1985; 55 FR 48063, Nov. 16, 1990; 65 FR 30907, May 15, 2000]

§ 122.23 Concentrated animal feeding operations (applicable to State NPDES programs, see § 123.25).

(a) *Scope.* Concentrated animal feeding operations (CAFOs), as defined in paragraph (b) of this section or designated in accordance with paragraph (c) of this section, are point sources, subject to NPDES permitting requirements as provided in this section. Once an animal feeding operation is defined as a CAFO for at least one type of animal, the NPDES requirements for CAFOs apply with respect to all animals in confinement at the operation and all manure, litter, and process wastewater generated by those animals or the production of those animals, regardless of the type of animal.

(b) Definitions applicable to this section:

(1) *Animal feeding operation* ("AFO") means a lot or facility (other than an aquatic animal production facility) where the following conditions are met:

(i) Animals (other than aquatic animals) have been, are, or will be stabled or confined and fed or maintained for a total of 45 days or more in any 12-month period, and

(ii) Crops, vegetation, forage growth, or post-harvest residues are not sustained in the normal growing season over any portion of the lot or facility.

(2) *Concentrated animal feeding operation* ("CAFO") means an AFO that is defined as a Large CAFO or as a Medium CAFO by the terms of this paragraph, or that is designated as a CAFO in accordance with paragraph (c) of this section. Two or more AFOs under common ownership are considered to be a single AFO for the purposes of determining the number of animals at an operation, if they adjoin each other or if they use a common area or system for the disposal of wastes.

(3) The term *land application area* means land under the control of an AFO owner or operator, whether it is owned, rented, or leased, to which manure, litter or process wastewater from the production area is or may be applied.

(4) *Large concentrated animal feeding operation* ("Large CAFO"). An AFO is defined as a Large CAFO if it stables or confines as many as or more than the numbers of animals specified in any of the following categories:

(i) 700 mature dairy cows, whether milked or dry;

(ii) 1,000 veal calves;

(iii) 1,000 cattle other than mature dairy cows or veal calves. Cattle includes but is not limited to heifers, steers, bulls and cow/calf pairs;

(iv) 2,500 swine each weighing 55 pounds or more;

(v) 10,000 swine each weighing less than 55 pounds;

(vi) 500 horses;

(vii) 10,000 sheep or lambs;

(viii) 55,000 turkeys;

(ix) 30,000 laying hens or broilers, if the AFO uses a liquid manure handling system;

(x) 125,000 chickens (other than laying hens), if the AFO uses other than a liquid manure handling system;

(xi) 82,000 laying hens, if the AFO uses other than a liquid manure handling system;

(xii) 30,000 ducks (if the AFO uses other than a liquid manure handling system); or

(xiii) 5,000 ducks (if the AFO uses a liquid manure handling system).

(5) The term *manure* is defined to include manure, bedding, compost and raw materials or other materials commingled with manure or set aside for disposal.

(6) *Medium concentrated animal feeding operation* ("Medium CAFO"). The term Medium CAFO includes any AFO with the type and number of animals that fall within any of the ranges listed in paragraph (b)(6)(i) of this section and which has been defined or designated as a CAFO. An AFO is defined as a Medium CAFO if:

(i) The type and number of animals that it stables or confines falls within any of the following ranges:

(A) 200 to 699 mature dairy cows, whether milked or dry;

(B) 300 to 999 veal calves;

(C) 300 to 999 cattle other than mature dairy cows or veal calves. Cattle includes but is not limited to heifers, steers, bulls and cow/calf pairs;

(D) 750 to 2,499 swine each weighing 55 pounds or more;

(E) 3,000 to 9,999 swine each weighing less than 55 pounds;

(F) 150 to 499 horses;

(G) 3,000 to 9,999 sheep or lambs;

(H) 16,500 to 54,999 turkeys;

(I) 9,000 to 29,999 laying hens or broilers, if the AFO uses a liquid manure handling system;

(J) 37,500 to 124,999 chickens (other than laying hens), if the AFO uses other than a liquid manure handling system;

(K) 25,000 to 81,999 laying hens, if the AFO uses other than a liquid manure handling system;

(L) 10,000 to 29,999 ducks (if the AFO uses other than a liquid manure handling system); or

(M) 1,500 to 4,999 ducks (if the AFO uses a liquid manure handling system); and

(ii) Either one of the following conditions are met:

(A) Pollutants are discharged into waters of the United States through a man-made ditch, flushing system, or other similar man-made device; or

(B) Pollutants are discharged directly into waters of the United States which originate outside of and pass over, across, or through the facility or otherwise come into direct contact with the animals confined in the operation.

(7) *Process wastewater* means water directly or indirectly used in the operation of the AFO for any or all of the following: spillage or overflow from animal or poultry watering systems; washing, cleaning, or flushing pens, barns, manure pits, or other AFO facilities; direct contact swimming, washing, or spray cooling of animals; or dust control. Process wastewater also includes any water which comes into contact with any raw materials, products, or byproducts including manure, litter, feed, milk, eggs or bedding.

(8) *Production area* means that part of an AFO that includes the animal confinement area, the manure storage area, the raw materials storage area, and the waste containment areas. The animal confinement area includes but is not limited to open lots, housed lots, feedlots, confinement houses, stall barns, free stall barns, milkrooms, milking centers, cowyards, barnyards, medication pens, walkers, animal walkways, and stables. The manure storage area includes but is not limited to lagoons, runoff ponds, storage sheds, stockpiles, under house or pit storages, liquid impoundments, static piles, and composting piles. The raw materials storage area includes but is not limited

to feed silos, silage bunkers, and bedding materials. The waste containment area includes but is not limited to settling basins, and areas within berms and diversions which separate uncontaminated storm water. Also included in the definition of production area is any egg washing or egg processing facility, and any area used in the storage, handling, treatment, or disposal of mortalities.

(9) *Small concentrated animal feeding operation* ("Small CAFO"). An AFO that is designated as a CAFO and is not a Medium CAFO.

(c) *How may an AFO be designated as a CAFO?* The appropriate authority (*i.e.,* State Director or Regional Administrator, or both, as specified in paragraph (c)(1) of this section) may designate any AFO as a CAFO upon determining that it is a significant contributor of pollutants to waters of the United States.

(1) *Who may designate?*—(i) *Approved States.* In States that are approved or authorized by EPA under Part 123, CAFO designations may be made by the State Director. The Regional Administrator may also designate CAFOs in approved States, but only where the Regional Administrator has determined that one or more pollutants in the AFO's discharge contributes to an impairment in a downstream or adjacent State or Indian country water that is impaired for that pollutant.

(ii) *States with no approved program.* The Regional Administrator may designate CAFOs in States that do not have an approved program and in Indian country where no entity has expressly demonstrated authority and has been expressly authorized by EPA to implement the NPDES program.

(2) In making this designation, the State Director or the Regional Administrator shall consider the following factors:

(i) The size of the AFO and the amount of wastes reaching waters of the United States;

(ii) The location of the AFO relative to waters of the United States;

(iii) The means of conveyance of animal wastes and process waste waters into waters of the United States;

(iv) The slope, vegetation, rainfall, and other factors affecting the likelihood or frequency of discharge of animal wastes manure and process waste waters into waters of the United States; and

(v) Other relevant factors.

(3) No AFO shall be designated under this paragraph unless the State Director or the Regional Administrator has conducted an on-site inspection of the operation and determined that the operation should and could be regulated under the permit program. In addition, no AFO with numbers of animals below those established in paragraph (b)(6) of this section may be designated as a CAFO unless:

(i) Pollutants are discharged into waters of the United States through a manmade ditch, flushing system, or other similar manmade device; or

(ii) Pollutants are discharged directly into waters of the United States which originate outside of the facility and pass over, across, or through the facility or otherwise come into direct contact with the animals confined in the operation.

(d) *Who must seek coverage under an NPDES permit?*—(1) *Permit Requirement.* The owner or operator of a CAFO must seek coverage under an NPDES permit if the CAFO discharges or proposes to discharge. A CAFO proposes to discharge if it is designed, constructed, operated, or maintained such that a discharge will occur. Specifically, the CAFO owner or operator must either apply for an individual NPDES permit or submit a notice of intent for coverage under an NPDES general permit. If the Director has not made a general permit available to the CAFO, the CAFO owner or operator must submit an application for an individual permit to the Director.

(2) *Information to submit with permit application or notice of intent.* An application for an individual permit must include the information specified in § 122.21. A notice of intent for a general permit must include the information specified in §§ 122.21 and 122.28.

(3) *Information to submit with permit application.* A permit application for an individual permit must include the information specified in § 122.21. A notice of intent for a general permit must include the information specified in §§ 122.21 and 122.28.

(e) *Land application discharges from a CAFO are subject to NPDES requirements.* The discharge of manure, litter or process wastewater to waters of the United States from a CAFO as a result of the application of that manure, litter or process wastewater by the CAFO to land areas under its control is a discharge from that CAFO subject to NPDES permit requirements, except where it is an agricultural storm water discharge as provided in 33 U.S.C. 1362(14). For purposes of this paragraph, where the manure, litter or process wastewater has been applied in accordance with site specific nutrient management practices that ensure appropriate agricultural utilization of the nutrients in the manure, litter or process wastewater, as specified in § 122.42(e)(1)(vi)–(ix), a precipitation-related discharge of manure, litter or process wastewater from land areas under the control of a CAFO is an agricultural stormwater discharge.

(1) For unpermitted Large CAFOs, a precipitation-related discharge of manure, litter, or process wastewater from land areas under the control of a CAFO shall be considered an agricultural stormwater discharge only where the manure, litter, or process wastewater has been land applied in accordance with site-specific nutrient management practices that ensure appropriate agricultural utilization of the nutrients in the manure, litter, or process wastewater, as specified in § 122.42(e)(1)(vi) through (ix).

(2) Unpermitted Large CAFOs must maintain documentation specified in § 122.42(e)(1)(ix) either on site or at a nearby office, or otherwise make such documentation readily available to the Director or Regional Administrator upon request.

(f) *When must the owner or operator of a CAFO seek coverage under an NPDES permit?* Any CAFO that is required to seek permit coverage under paragraph (d)(1) of this section must seek coverage when the CAFO proposes to discharge, unless a later deadline is specified below.

(1) *Operations defined as CAFOs prior to April 14, 2003.* For operations defined as CAFOs under regulations that were

in effect prior to April 14, 2003, the owner or operator must have or seek to obtain coverage under an NPDES permit as of April 14, 2003, and comply with all applicable NPDES requirements, including the duty to maintain permit coverage in accordance with paragraph (g) of this section.

(2) *Operations defined as CAFOs as of April 14, 2003, that were not defined as CAFOs prior to that date.* For all operations defined as CAFOs as of April 14, 2003, that were not defined as CAFOs prior to that date, the owner or operator of the CAFO must seek to obtain coverage under an NPDES permit by February 27, 2009.

(3) *Operations that become defined as CAFOs after April 14, 2003, but which are not new sources.* For a newly constructed CAFO and for an AFO that makes changes to its operations that result in its becoming defined as a CAFO for the first time after April 14, 2003, but is not a new source, the owner or operator must seek to obtain coverage under an NPDES permit, as follows:

(i) For newly constructed operations not subject to effluent limitations guidelines, 180 days prior to the time CAFO commences operation;

(ii) For other operations (e.g., resulting from an increase in the number of animals), as soon as possible, but no later than 90 days after becoming defined as a CAFO; or

(iii) If an operational change that makes the operation a CAFO would not have made it a CAFO prior to April 14, 2003, the operation has until February 27, 2009, or 90 days after becoming defined as a CAFO, whichever is later.

(4) *New sources.* The owner or operator of a new source must seek to obtain coverage under a permit at least 180 days prior to the time that the CAFO commences operation.

(5) *Operations that are designated as CAFOs.* For operations designated as a CAFO in accordance with paragraph (c) of this section, the owner or operator must seek to obtain coverage under a permit no later than 90 days after receiving notice of the designation.

(g) *Duty to maintain permit coverage.* No later than 180 days before the expiration of the permit, or as provided by the Director, any permitted CAFO must submit an application to renew its permit, in accordance with § 122.21(d), unless the CAFO will not discharge or propose to discharge upon expiration of the permit.

(h) *Procedures for CAFOs seeking coverage under a general permit.* (1) CAFO owners or operators must submit a notice of intent when seeking authorization to discharge under a general permit in accordance with § 122.28(b). The Director must review notices of intent submitted by CAFO owners or operators to ensure that the notice of intent includes the information required by § 122.21(i)(1), including a nutrient management plan that meets the requirements of § 122.42(e) and applicable effluent limitations and standards, including those specified in 40 CFR part 412. When additional information is necessary to complete the notice of intent or clarify, modify, or supplement previously submitted material, the Director may request such information from the owner or operator. If the Director makes a preliminary determination that the notice of intent meets the requirements of §§ 122.21(i)(1) and 122.42(e), the Director must notify the public of the Director's proposal to grant coverage under the permit to the CAFO and make available for public review and comment the notice of intent submitted by the CAFO, including the CAFO's nutrient management plan, and the draft terms of the nutrient management plan to be incorporated into the permit. The process for submitting public comments and hearing requests, and the hearing process if a request for a hearing is granted, must follow the procedures applicable to draft permits set forth in 40 CFR 124.11 through 124.13. The Director may establish, either by regulation or in the general permit, an appropriate period of time for the public to comment and request a hearing that differs from the time period specified in 40 CFR 124.10. The Director must respond to significant comments received during the comment period, as provided in 40 CFR 124.17, and, if necessary, require the CAFO owner or operator to revise the nutrient management plan in order to be granted permit coverage. When the Director authorizes coverage for the

CAFO owner or operator under the general permit, the terms of the nutrient management plan shall become incorporated as terms and conditions of the permit for the CAFO. The Director shall notify the CAFO owner or operator and inform the public that coverage has been authorized and of the terms of the nutrient management plan incorporated as terms and conditions of the permit applicable to the CAFO.

(2) *For EPA-issued permits only.* The Regional Administrator shall notify each person who has submitted written comments on the proposal to grant coverage and the draft terms of the nutrient management plan or requested notice of the final permit decision. Such notification shall include notice that coverage has been authorized and of the terms of the nutrient management plan incorporated as terms and conditions of the permit applicable to the CAFO.

(3) Nothing in this paragraph (h) shall affect the authority of the Director to require an individual permit under § 122.28(b)(3).

(i) *No discharge certification option.* (1) The owner or operator of a CAFO that meets the eligibility criteria in paragraph (i)(2) of this section may certify to the Director that the CAFO does not discharge or propose to discharge. A CAFO owner or operator who certifies that the CAFO does not discharge or propose to discharge is not required to seek coverage under an NPDES permit pursuant to paragraph (d)(1) of this section, provided that the CAFO is designed, constructed, operated, and maintained in accordance with the requirements of paragraphs (i)(2) and (3) of this section, and subject to the limitations in paragraph (i)(4) of this section.

(2) *Eligibility criteria.* In order to certify that a CAFO does not discharge or propose to discharge, the owner or operator of a CAFO must document, based on an objective assessment of the conditions at the CAFO, that the CAFO is designed, constructed, operated, and maintained in a manner such that the CAFO will not discharge, as follows:

(i) The CAFO's production area is designed, constructed, operated, and maintained so as not to discharge. The CAFO must maintain documentation that demonstrates that:

(A) Any open manure storage structures are designed, constructed, operated, and maintained to achieve no discharge based on a technical evaluation in accordance with the elements of the technical evaluation set forth in 40 CFR 412.46(a)(1)(i) through (viii);

(B) Any part of the CAFO's production area that is not addressed by paragraph (i)(2)(i)(A) of this section is designed, constructed, operated, and maintained such that there will be no discharge of manure, litter, or process wastewater; and

(C) The CAFO implements the additional measures set forth in 40 CFR 412.37(a) and (b);

(ii) The CAFO has developed and is implementing an up-to-date nutrient management plan to ensure no discharge from the CAFO, including from all land application areas under the control of the CAFO, that addresses, at a minimum, the following:

(A) The elements of § 122.42(e)(1)(i) through (ix) and 40 CFR 412.37(c); and

(B) All site-specific operation and maintenance practices necessary to ensure no discharge, including any practices or conditions established by a technical evaluation pursuant to paragraph (i)(2)(i)(A) of this section; and

(iii) The CAFO must maintain documentation required by this paragraph either on site or at a nearby office, or otherwise make such documentation readily available to the Director or Regional Administrator upon request.

(3) *Submission to the Director.* In order to certify that a CAFO does not discharge or propose to discharge, the CAFO owner or operator must complete and submit to the Director, by certified mail or equivalent method of documentation, a certification that includes, at a minimum, the following information:

(i) The legal name, address and phone number of the CAFO owner or operator (see § 122.21(b));

(ii) The CAFO name and address, the county name and the latitude and longitude where the CAFO is located;

(iii) A statement that describes the basis for the CAFO's certification that it satisfies the eligibility requirements identified in paragraph (i)(2) of this section; and

(iv) The following certification statement: "I certify under penalty of law that I am the owner or operator of a concentrated animal feeding operation (CAFO), identified as [Name of CAFO], and that said CAFO meets the requirements of 40 CFR 122.23(i). I have read and understand the eligibility requirements of 40 CFR 122.23(i)(2) for certifying that a CAFO does not discharge or propose to discharge and further certify that this CAFO satisfies the eligibility requirements. As part of this certification, I am including the information required by 40 CFR 122.23(i)(3). I also understand the conditions set forth in 40 CFR 122.23(i)(4), (5) and (6) regarding loss and withdrawal of certification. I certify under penalty of law that this document and all other documents required for this certification were prepared under my direction or supervision and that qualified personnel properly gathered and evaluated the information submitted. Based upon my inquiry of the person or persons directly involved in gathering and evaluating the information, the information submitted is to the best of my knowledge and belief true, accurate and complete. I am aware there are significant penalties for submitting false information, including the possibility of fine and imprisonment for knowing violations."; and

(v) The certification must be signed in accordance with the signatory requirements of 40 CFR 122.22.

(4) *Term of certification.* A certification that meets the requirements of paragraphs (i)(2) and (i)(3) of this section shall become effective on the date it is submitted, unless the Director establishes an effective date of up to 30 days after the date of submission. Certification will remain in effect for five years or until the certification is no longer valid or is withdrawn, whichever occurs first. A certification is no longer valid when a discharge has occurred or when the CAFO ceases to meet the eligibility criteria in paragraph (i)(2) of this section.

(5) *Withdrawal of certification.*(i) At any time, a CAFO may withdraw its certification by notifying the Director by certified mail or equivalent method

of documentation. A certification is withdrawn on the date the notification is submitted to the Director. The CAFO does not need to specify any reason for the withdrawal in its notification to the Director.

(ii) If a certification becomes invalid in accordance with paragraph (i)(4) of this section, the CAFO must withdraw its certification within three days of the date on which the CAFO becomes aware that the certification is invalid. Once a CAFO's certification is no longer valid, the CAFO is subject to the requirement in paragraph (d)(1) of this section to seek permit coverage if it discharges or proposes to discharge.

(6) *Recertification.*A previously certified CAFO that does not discharge or propose to discharge may recertify in accordance with paragraph (i) of this section, except that where the CAFO has discharged, the CAFO may only recertify if the following additional conditions are met:

(i) The CAFO had a valid certification at the time of the discharge;

(ii) The owner or operator satisfies the eligibility criteria of paragraph (i)(2) of this section, including any necessary modifications to the CAFO's design, construction, operation, and/or maintenance to permanently address the cause of the discharge and ensure that no discharge from this cause occurs in the future;

(iii) The CAFO has not previously recertified after a discharge from the same cause;

(iv) The owner or operator submits to the Director for review the following documentation: a description of the discharge, including the date, time, cause, duration, and approximate volume of the discharge, and a detailed explanation of the steps taken by the CAFO to permanently address the cause of the discharge in addition to submitting a certification in accordance with paragraph (i)(3) of this section; and

(v) Notwithstanding paragraph (i)(4) of this section, a recertification that meets the requirements of paragraphs (i)(6)(iii) and (i)(6)(iv) of this section shall only become effective 30 days from the date of submission of the recertification documentation.

(j) *Effect of certification.* (1) An unpermitted CAFO certified in accordance with paragraph (i) of this section is presumed not to propose to discharge. If such a CAFO does discharge, it is not in violation of the requirement that CAFOs that propose to discharge seek permit coverage pursuant to paragraphs (d)(1) and (f) of this section, with respect to that discharge. In all instances, the discharge of a pollutant without a permit is a violation of the Clean Water Act section 301(a) prohibition against unauthorized discharges from point sources.

(2) In any enforcement proceeding for failure to seek permit coverage under paragraphs (d)(1) or (f) of this section that is related to a discharge from an unpermitted CAFO, the burden is on the CAFO to establish that it did not propose to discharge prior to the discharge when the CAFO either did not submit certification documentation as provided in paragraph (i)(3) or (i)(6)(iv) of this section within at least five years prior to the discharge, or withdrew its certification in accordance with paragraph (i)(5) of this section. Design, construction, operation, and maintenance in accordance with the criteria of paragraph (i)(2) of this section satisfies this burden.

[68 FR 7265, Feb. 12, 2003, as amended at 71 FR 6984, Feb. 10, 2006; 72 FR 40250, July 24, 2007; 73 FR 70480, Nov. 20, 2008]

§ 122.24 Concentrated aquatic animal production facilities (applicable to State NPDES programs, see § 123.25).

(a) *Permit requirement.* Concentrated aquatic animal production facilities, as defined in this section, are point sources subject to the NPDES permit program.

(b) *Definition. Concentrated aquatic animal production facility* means a hatchery, fish farm, or other facility which meets the criteria in appendix C of this part, or which the Director designates under paragraph (c) of this section.

(c) *Case-by-case designation of concentrated aquatic animal production facilities.* (1) The Director may designate any warm or cold water aquatic animal production facility as a concentrated aquatic animal production facility

upon determining that it is a significant contributor of pollution to waters of the United States. In making this designation the Director shall consider the following factors:

(i) The location and quality of the receiving waters of the United States;

(ii) The holding, feeding, and production capacities of the facility;

(iii) The quantity and nature of the pollutants reaching waters of the United States; and

(iv) Other relevant factors.

(2) A permit application shall not be required from a concentrated aquatic animal production facility designated under this paragraph until the Director has conducted on-site inspection of the facility and has determined that the facility should and could be regulated under the permit program.

[48 FR 14153, Apr. 1, 1983, as amended at 65 FR 30907, May 15, 2000]

§ 122.25 Aquaculture projects (applicable to State NPDES programs, see § 123.25).

(a) *Permit requirement.* Discharges into aquaculture projects, as defined in this section, are subject to the NPDES permit program through section 318 of CWA, and in accordance with 40 CFR part 125, subpart B.

(b) *Definitions.* (1) *Aquaculture project* means a defined managed water area which uses discharges of pollutants into that designated area for the maintenance or production of harvestable freshwater, estuarine, or marine plants or animals.

(2) *Designated project area* means the portions of the waters of the United States within which the permittee or permit applicant plans to confine the cultivated species, using a method or plan or operation (including, but not limited to, physical confinement) which, on the basis of reliable scientific evidence, is expected to ensure that specific individual organisms comprising an aquaculture crop will enjoy increased growth attributable to the discharge of pollutants, and be harvested within a defined geographic area.

§122.26 Storm water discharges (applicable to State NPDES programs, see §123.25).

(a) *Permit requirement.* (1) Prior to October 1, 1994, discharges composed entirely of storm water shall not be required to obtain a NPDES permit except:

(i) A discharge with respect to which a permit has been issued prior to February 4, 1987;

(ii) A discharge associated with industrial activity (see §122.26(a)(4));

(iii) A discharge from a large municipal separate storm sewer system;

(iv) A discharge from a medium municipal separate storm sewer system;

(v) A discharge which the Director, or in States with approved NPDES programs, either the Director or the EPA Regional Administrator, determines to contribute to a violation of a water quality standard or is a significant contributor of pollutants to waters of the United States. This designation may include a discharge from any conveyance or system of conveyances used for collecting and conveying storm water runoff or a system of discharges from municipal separate storm sewers, except for those discharges from conveyances which do not require a permit under paragraph (a)(2) of this section or agricultural storm water runoff which is exempted from the definition of point source at §122.2.

The Director may designate discharges from municipal separate storm sewers on a system-wide or jurisdiction-wide basis. In making this determination the Director may consider the following factors:

(A) The location of the discharge with respect to waters of the United States as defined at 40 CFR 122.2.

(B) The size of the discharge;

(C) The quantity and nature of the pollutants discharged to waters of the United States; and

(D) Other relevant factors.

(2) The Director may not require a permit for discharges of storm water runoff from the following:

(i) Mining operations composed entirely of flows which are from conveyances or systems of conveyances (including but not limited to pipes, conduits, ditches, and channels) used for collecting and conveying precipitation runoff and which are not contaminated by contact with or that have not come into contact with, any overburden, raw material, intermediate products, finished product, byproduct, or waste products located on the site of such operations, except in accordance with paragraph (c)(1)(iv) of this section.

(ii) All field activities or operations associated with oil and gas exploration, production, processing, or treatment operations or transmission facilities, including activities necessary to prepare a site for drilling and for the movement and placement of drilling equipment, whether or not such field activities or operations may be considered to be construction activities, except in accordance with paragraph (c)(1)(iii) of this section. Discharges of sediment from construction activities associated with oil and gas exploration, production, processing, or treatment operations or transmission facilities are not subject to the provisions of paragraph (c)(1)(iii)(C) of this section.

NOTE TO PARAGRAPH (a)(2)(ii): EPA encourages operators of oil and gas field activities or operations to implement and maintain Best Management Practices (BMPs) to minimize discharges of pollutants, including sediment, in storm water both during and after construction activities to help ensure protection of surface water quality during storm events. Appropriate controls would be those suitable to the site conditions and consistent with generally accepted engineering design criteria and manufacturer specifications. Selection of BMPs could also be affected by seasonal or climate conditions.

(3) *Large and medium municipal separate storm sewer systems.* (i) Permits must be obtained for all discharges from large and medium municipal separate storm sewer systems.

(ii) The Director may either issue one system-wide permit covering all discharges from municipal separate storm sewers within a large or medium municipal storm sewer system or issue distinct permits for appropriate categories of discharges within a large or medium municipal separate storm sewer system including, but not limited to: all discharges owned or operated by the same municipality; located within the same jurisdiction; all discharges within a system that discharge to the same watershed; discharges within a system that are similar in nature; or for individual discharges from municipal separate storm sewers within the system.

(iii) The operator of a discharge from a municipal separate storm sewer which is part of a large or medium municipal separate storm sewer system must either:

(A) Participate in a permit application (to be a permittee or a co-permittee) with one or more other operators of discharges from the large or medium municipal storm sewer system which covers all, or a portion of all, discharges from the municipal separate storm sewer system;

(B) Submit a distinct permit application which only covers discharges from the municipal separate storm sewers for which the operator is responsible; or

(C) A regional authority may be responsible for submitting a permit application under the following guidelines:

(1) The regional authority together with co-applicants shall have authority over a storm water management program that is in existence, or shall be in existence at the time part 1 of the application is due;

(2) The permit applicant or co-applicants shall establish their ability to make a timely submission of part 1 and part 2 of the municipal application;

(3) Each of the operators of municipal separate storm sewers within the systems described in paragraphs (b)(4) (i), (ii), and (iii) or (b)(7) (i), (ii), and (iii) of this section, that are under the purview of the designated regional authority, shall comply with the application requirements of paragraph (d) of this section.

(iv) One permit application may be submitted for all or a portion of all municipal separate storm sewers within adjacent or interconnected large or medium municipal separate storm sewer systems. The Director may issue one system-wide permit covering all, or a portion of all municipal separate storm sewers in adjacent or interconnected large or medium municipal separate storm sewer systems.

(v) Permits for all or a portion of all discharges from large or medium municipal separate storm sewer systems that are issued on a system-wide, jurisdiction-wide, watershed or other basis may specify different conditions relating to different discharges covered by the permit, including different management programs for different drainage areas which contribute storm water to the system.

(vi) Co-permittees need only comply with permit conditions relating to discharges from the municipal separate storm sewers for which they are operators.

(4) *Discharges through large and medium municipal separate storm sewer systems.* In addition to meeting the requirements of paragraph (c) of this section, an operator of a storm water discharge associated with industrial activity which discharges through a large or medium municipal separate storm sewer system shall submit, to the operator of the municipal separate storm sewer system receiving the discharge no later than May 15, 1991, or 180 days prior to commencing such discharge: the name of the facility; a contact person and phone number; the location of the discharge; a description, including Standard Industrial Classification, which best reflects the principal products or services provided by each facility; and any existing NPDES permit number.

(5) *Other municipal separate storm sewers.* The Director may issue permits for municipal separate storm sewers that are designated under paragraph (a)(1)(v) of this section on a system-wide basis, jurisdiction-wide basis, watershed basis or other appropriate basis, or may issue permits for individual discharges.

(6) *Non-municipal separate storm sewers.* For storm water discharges associated with industrial activity from point sources which discharge through a non-municipal or non-publicly owned separate storm sewer system, the Director, in his discretion, may issue: a single NPDES permit, with each discharger a co-permittee to a permit issued to the operator of the portion of the system that discharges into waters of the United States; or, individual permits to each discharger of storm water associated with industrial activity through the non-municipal conveyance system.

(i) All storm water discharges associated with industrial activity that discharge through a storm water discharge system that is not a municipal separate storm sewer must be covered by an individual permit, or a permit issued to the operator of the portion of the system that discharges to waters of the United States, with each discharger to the non-municipal conveyance a co-permittee to that permit.

(ii) Where there is more than one operator of a single system of such conveyances, all operators of storm water discharges associated with industrial activity must submit applications.

(iii) Any permit covering more than one operator shall identify the effluent limitations, or other permit conditions, if any, that apply to each operator.

(7) *Combined sewer systems.* Conveyances that discharge storm water runoff combined with municipal sewage are point sources that must obtain NPDES permits in accordance with the procedures of §122.21 and are not subject to the provisions of this section.

(8) Whether a discharge from a municipal separate storm sewer is or is not subject to regulation under this section shall have no bearing on whether the owner or operator of the discharge is eligible for funding under title II, title III or title VI of the Clean Water Act. *See* 40 CFR part 35, subpart I, appendix A(b)H.2.j.

(9)(i) On and after October 1, 1994, for discharges composed entirely of storm water, that are not required by paragraph (a)(1) of this section to obtain a permit, operators shall be required to obtain a NPDES permit only if:

(A) The discharge is from a small MS4 required to be regulated pursuant to §122.32;

(B) The discharge is a storm water discharge associated with small construction activity pursuant to paragraph (b)(15) of this section;

(C) The Director, or in States with approved NPDES programs either the Director or the EPA Regional Administrator, determines that storm water controls are needed for the discharge based on wasteload allocations that are part of "total maximum daily loads" (TMDLs) that address the pollutant(s) of concern; or

(D) The Director, or in States with approved NPDES programs either the Director or the EPA Regional Administrator, determines that the discharge, or category of discharges within a geographic area, contributes to a violation of a water quality standard or is a significant contributor of pollutants to waters of the United States.

(ii) Operators of small MS4s designated pursuant to paragraphs (a)(9)(i)(A), (a)(9)(i)(C), and (a)(9)(i)(D) of this section shall seek coverage under an NPDES permit in accordance with §§122.33 through 122.35. Operators of non-municipal sources designated pursuant to paragraphs (a)(9)(i)(B), (a)(9)(i)(C), and (a)(9)(i)(D) of this section shall seek coverage under an NPDES permit in accordance with paragraph (c)(1) of this section.

(iii) Operators of storm water discharges designated pursuant to paragraphs (a)(9)(i)(C) and (a)(9)(i)(D) of this section shall apply to the Director for a permit within 180 days of receipt of notice, unless permission for a later date is granted by the Director (see §124.52(c) of this chapter).

(b) *Definitions.* (1) *Co-permittee* means a permittee to a NPDES permit that is only responsible for permit conditions relating to the discharge for which it is operator.

(2) *Illicit discharge* means any discharge to a municipal separate storm sewer that is not composed entirely of storm water except discharges pursuant to a NPDES permit (other than the NPDES permit for discharges from the municipal separate storm sewer) and discharges resulting from fire fighting activities.

(3) *Incorporated place* means the District of Columbia, or a city, town, township, or village that is incorporated under the laws of the State in which it is located.

(4) *Large municipal separate storm sewer system* means all municipal separate storm sewers that are either:

(i) Located in an incorporated place with a population of 250,000 or more as determined by the 1990 Decennial Census by the Bureau of the Census (Appendix F of this part); or

(ii) Located in the counties listed in appendix H, except municipal separate storm sewers that are located in the incorporated places, townships or towns within such counties; or

(iii) Owned or operated by a municipality other than those described in paragraph (b)(4) (i) or (ii) of this section and that are designated by the Director as part of the large or medium municipal separate storm sewer system due to the interrelationship between the discharges of the designated storm sewer and the discharges from municipal separate storm sewers described under paragraph (b)(4) (i) or (ii) of this section. In making this determination the Director may consider the following factors:

(A) Physical interconnections between the municipal separate storm sewers;

(B) The location of discharges from the designated municipal separate storm sewer relative to discharges from municipal separate storm sewers described in paragraph (b)(4)(i) of this section;

(C) The quantity and nature of pollutants discharged to waters of the United States;

(D) The nature of the receiving waters; and

(E) Other relevant factors; or

(iv) The Director may, upon petition, designate as a large municipal separate storm sewer system, municipal separate storm sewers located within the boundaries of a region defined by a storm water management regional authority based on a jurisdictional, watershed, or other appropriate basis that includes one or more of the systems described in paragraph (b)(4) (i), (ii), (iii) of this section.

(5) *Major municipal separate storm sewer outfall* (or "major outfall") means a municipal separate storm sewer outfall that discharges from a single pipe with an inside diameter of 36 inches or more or its equivalent (discharge from a single conveyance other than circular pipe which is associated with a drainage area of more than 50 acres); or for municipal separate storm sewers that receive storm water from lands zoned for industrial activity (based on comprehensive zoning plans or the equivalent), an outfall that discharges from a single pipe with an inside diameter of 12 inches or more or from its equiva-

lent (discharge from other than a circular pipe associated with a drainage area of 2 acres or more).

(6) *Major outfall* means a major municipal separate storm sewer outfall.

(7) *Medium municipal separate storm sewer system* means all municipal separate storm sewers that are either:

(i) Located in an incorporated place with a population of 100,000 or more but less than 250,000, as determined by the 1990 Decennial Census by the Bureau of the Census (appendix G of this part); or

(ii) Located in the counties listed in appendix I, except municipal separate storm sewers that are located in the incorporated places, townships or towns within such counties; or

(iii) Owned or operated by a municipality other than those described in paragraph (b)(7) (i) or (ii) of this section and that are designated by the Director as part of the large or medium municipal separate storm sewer system due to the interrelationship between the discharges of the designated storm sewer and the discharges from municipal separate storm sewers described under paragraph (b)(7) (i) or (ii) of this section. In making this determination the Director may consider the following factors:

(A) Physical interconnections between the municipal separate storm sewers;

(B) The location of discharges from the designated municipal separate storm sewer relative to discharges from municipal separate storm sewers described in paragraph (b)(7)(i) of this section;

(C) The quantity and nature of pollutants discharged to waters of the United States;

(D) The nature of the receiving waters; or

(E) Other relevant factors; or

(iv) The Director may, upon petition, designate as a medium municipal separate storm sewer system, municipal separate storm sewers located within the boundaries of a region defined by a storm water management regional authority based on a jurisdictional, watershed, or other appropriate basis that includes one or more of the systems described in paragraphs (b)(7) (i), (ii), (iii) of this section.

(8) *Municipal separate storm sewer* means a conveyance or system of conveyances (including roads with drainage systems, municipal streets, catch basins, curbs, gutters, ditches, manmade channels, or storm drains):

(i) Owned or operated by a State, city, town, borough, county, parish, district, association, or other public body (created by or pursuant to State law) having jurisdiction over disposal of sewage, industrial wastes, storm water, or other wastes, including special districts under State law such as a sewer district, flood control district or drainage district, or similar entity, or an Indian tribe or an authorized Indian tribal organization, or a designated and approved management agency under section 208 of the CWA that discharges to waters of the United States;

(ii) Designed or used for collecting or conveying storm water;

(iii) Which is not a combined sewer; and

(iv) Which is not part of a Publicly Owned Treatment Works (POTW) as defined at 40 CFR 122.2.

(9) *Outfall* means a *point source* as defined by 40 CFR 122.2 at the point where a municipal separate storm sewer discharges to waters of the United States and does not include open conveyances connecting two municipal separate storm sewers, or pipes, tunnels or other conveyances which connect segments of the same stream or other waters of the United States and are used to convey waters of the United States.

(10) *Overburden* means any material of any nature, consolidated or unconsolidated, that overlies a mineral deposit, excluding topsoil or similar naturally-occurring surface materials that are not disturbed by mining operations.

(11) *Runoff coefficient* means the fraction of total rainfall that will appear at a conveyance as runoff.

(12) *Significant materials* includes, but is not limited to: raw materials; fuels; materials such as solvents, detergents, and plastic pellets; finished materials such as metallic products; raw materials used in food processing or production; hazardous substances designated under section 101(14) of CERCLA; any chemical the facility is required to report pursuant to section 313 of title III of SARA; fertilizers; pesticides; and waste products such as ashes, slag and sludge that have the potential to be released with storm water discharges.

(13) *Storm water* means storm water runoff, snow melt runoff, and surface runoff and drainage.

(14) *Storm water discharge associated with industrial activity* means the discharge from any conveyance that is used for collecting and conveying storm water and that is directly related to manufacturing, processing or raw materials storage areas at an industrial plant. The term does not include discharges from facilities or activities excluded from the NPDES program under this part 122. For the categories of industries identified in this section, the term includes, but is not limited to, storm water discharges from industrial plant yards; immediate access roads and rail lines used or traveled by carriers of raw materials, manufactured products, waste material, or by-products used or created by the facility; material handling sites; refuse sites; sites used for the application or disposal of process waste waters (as defined at part 401 of this chapter); sites used for the storage and maintenance of material handling equipment; sites used for residual treatment, storage, or disposal; shipping and receiving areas; manufacturing buildings; storage areas (including tank farms) for raw materials, and intermediate and final products; and areas where industrial activity has taken place in the past and significant materials remain and are exposed to storm water. For the purposes of this paragraph, material handling activities include storage, loading and unloading, transportation, or conveyance of any raw material, intermediate product, final product, by-product or waste product. The term excludes areas located on plant lands separate from the plant's industrial activities, such as office buildings and accompanying parking lots as long as the drainage from the excluded areas is not mixed with storm water drained from the above described areas. Industrial facilities (including industrial facilities that are federally, State, or municipally owned or operated that meet the description of the facilities listed in paragraphs (b)(14)(i) through (xi) of this

section) include those facilities designated under the provisions of paragraph (a)(1)(v) of this section. The following categories of facilities are considered to be engaging in "industrial activity" for purposes of paragraph (b)(14):

(i) Facilities subject to storm water effluent limitations guidelines, new source performance standards, or toxic pollutant effluent standards under 40 CFR subchapter N (except facilities with toxic pollutant effluent standards which are exempted under category (xi) in paragraph (b)(14) of this section);

(ii) Facilities classified as Standard Industrial Classifications 24 (except 2434), 26 (except 265 and 267), 28 (except 283), 29, 311, 32 (except 323), 33, 3441, 373;

(iii) Facilities classified as Standard Industrial Classifications 10 through 14 (mineral industry) including active or inactive mining operations (except for areas of coal mining operations no longer meeting the definition of a reclamation area under 40 CFR 434.11(1) because the performance bond issued to the facility by the appropriate SMCRA authority has been released, or except for areas of non-coal mining operations which have been released from applicable State or Federal reclamation requirements after December 17, 1990) and oil and gas exploration, production, processing, or treatment operations, or transmission facilities that discharge storm water contaminated by contact with or that has come into contact with, any overburden, raw material, intermediate products, finished products, byproducts or waste products located on the site of such operations; (inactive mining operations are mining sites that are not being actively mined, but which have an identifiable owner/operator; inactive mining sites do not include sites where mining claims are being maintained prior to disturbances associated with the extraction, beneficiation, or processing of mined materials, nor sites where minimal activities are undertaken for the sole purpose of maintaining a mining claim);

(iv) Hazardous waste treatment, storage, or disposal facilities, including those that are operating under interim status or a permit under subtitle C of RCRA;

(v) Landfills, land application sites, and open dumps that receive or have received any industrial wastes (waste that is received from any of the facilities described under this subsection) including those that are subject to regulation under subtitle D of RCRA;

(vi) Facilities involved in the recycling of materials, including metal scrapyards, battery reclaimers, salvage yards, and automobile junkyards, including but limited to those classified as Standard Industrial Classification 5015 and 5093;

(vii) Steam electric power generating facilities, including coal handling sites;

(viii) Transportation facilities classified as Standard Industrial Classifications 40, 41, 42 (except 4221-25), 43, 44, 45, and 5171 which have vehicle maintenance shops, equipment cleaning operations, or airport deicing operations. Only those portions of the facility that are either involved in vehicle maintenance (including vehicle rehabilitation, mechanical repairs, painting, fueling, and lubrication), equipment cleaning operations, airport deicing operations, or which are otherwise identified under paragraphs (b)(14) (i)–(vii) or (ix)–(xi) of this section are associated with industrial activity;

(ix) Treatment works treating domestic sewage or any other sewage sludge or wastewater treatment device or system, used in the storage treatment, recycling, and reclamation of municipal or domestic sewage, including land dedicated to the disposal of sewage sludge that are located within the confines of the facility, with a design flow of 1.0 mgd or more, or required to have an approved pretreatment program under 40 CFR part 403. Not included are farm lands, domestic gardens or lands used for sludge management where sludge is beneficially reused and which are not physically located in the confines of the facility, or areas that are in compliance with section 405 of the CWA;

(x) Construction activity including clearing, grading and excavation, except operations that result in the disturbance of less than five acres of total land area. Construction activity also includes the disturbance of less than five acres of total land area that is a part of a larger common plan of development or sale if the larger common plan will ultimately disturb five acres or more;

(xi) Facilities under Standard Industrial Classifications 20, 21, 22, 23, 2434, 25, 265, 267, 27, 283, 285, 30, 31 (except 311), 323, 34 (except 3441), 35, 36, 37 (except 373), 38, 39, and 4221–25;

(15) *Storm water discharge associated with small construction activity* means the discharge of storm water from:

(i) Construction activities including clearing, grading, and excavating that result in land disturbance of equal to or greater than one acre and less than five acres. Small construction activity also includes the disturbance of less than one acre of total land area that is part of a larger common plan of development or sale if the larger common plan will ultimately disturb equal to or greater than one and less than five acres. Small construction activity does not include routine maintenance that is performed to maintain the original line and grade, hydraulic capacity, or original purpose of the facility. The Director may waive the otherwise applicable requirements in a general permit for a storm water discharge from construction activities that disturb less than five acres where:

(A) The value of the rainfall erosivity factor ("R" in the Revised Universal Soil Loss Equation) is less than five during the period of construction activity. The rainfall erosivity factor is determined in accordance with Chapter 2 of *Agriculture Handbook Number 703, Predicting Soil Erosion by Water: A Guide to Conservation Planning With the Revised Universal Soil Loss Equation (RUSLE)*, pages 21–64, dated January 1997. The Director of the Federal Register approves this incorporation by reference in accordance with 5 U.S.C 552(a) and 1 CFR part 51. Copies may be obtained from EPA's Water Resource Center, Mail Code RC4100, 401 M St. SW, Washington, DC 20460. A copy is also available for inspection at the U.S.

EPA Water Docket , 401 M Street SW, Washington, DC 20460, or at the National Archives and Records Administration (NARA). For information on the availability of this material at NARA, call 202–741–6030, or go to: *http://www.archives.gov/federal_register/code_of_federal_regulations/ibr_locations.html*. An operator must certify to the Director that the construction activity will take place during a period when the value of the rainfall erosivity factor is less than five; or

(B) Storm water controls are not needed based on a "total maximum daily load" (TMDL) approved or established by EPA that addresses the pollutant(s) of concern or, for non-impaired waters that do not require TMDLs, an equivalent analysis that determines allocations for small construction sites for the pollutant(s) of concern or that determines that such allocations are not needed to protect water quality based on consideration of existing in-stream concentrations, expected growth in pollutant contributions from all sources, and a margin of safety. For the purpose of this paragraph, the pollutant(s) of concern include sediment or a parameter that addresses sediment (such as total suspended solids, turbidity or siltation) and any other pollutant that has been identified as a cause of impairment of any water body that will receive a discharge from the construction activity. The operator must certify to the Director that the construction activity will take place, and storm water discharges will occur, within the drainage area addressed by the TMDL or equivalent analysis.

(ii) Any other construction activity designated by the Director, or in States with approved NPDES programs either the Director or the EPA Regional Administrator, based on the potential for contribution to a violation of a water quality standard or for significant contribution of pollutants to waters of the United States.

EXHIBIT 1 TO § 122.26(B)(15)—SUMMARY OF COVERAGE OF "STORM WATER DISCHARGES ASSOCIATED WITH SMALL CONSTRUCTION ACTIVITY" UNDER THE NPDES STORM WATER PROGRAM

Automatic Designation: Required Nationwide Coverage.	• Construction activities that result in a land disturbance of equal to or greater than one acre and less than five acres.

EXHIBIT 1 TO § 122.26(B)(15)—SUMMARY OF COVERAGE OF "STORM WATER DISCHARGES ASSOCI-
ATED WITH SMALL CONSTRUCTION ACTIVITY" UNDER THE NPDES STORM WATER PROGRAM—
Continued

Potential Designation: Optional Evaluation and Designation by the NPDES Permitting Authority or EPA Regional Administrator. Potential Waiver: Waiver from Requirements as Determined by the NPDES Permitting Authority..	• Construction activities disturbing less than one acre if part of a larger common plan of development or sale with a planned disturbance of equal to or greater than one acre and less than five acres. (see § 122.26(b)(15)(i).) • Construction activities that result in a land disturbance of less than one acre based on the potential for contribution to a violation of a water quality standard or for significant contribution of pollutants. (see § 122.26(b)(15)(ii).) Any automatically designated construction activity where the operator certifies: (1) A rainfall erosivity factor of less than five, or (2) That the activity will occur within an area where controls are not needed based on a TMDL or, for non-impaired waters that do not require a TMDL, an equivalent analysis for the pollutant(s) of concern. (see § 122.26(b)(15)(i).)

(16) *Small municipal separate storm sewer system* means all separate storm sewers that are:

(i) Owned or operated by the United States, a State, city, town, borough, county, parish, district, association, or other public body (created by or pursuant to State law) having jurisdiction over disposal of sewage, industrial wastes, storm water, or other wastes, including special districts under State law such as a sewer district, flood control district or drainage district, or similar entity, or an Indian tribe or an authorized Indian tribal organization, or a designated and approved management agency under section 208 of the CWA that discharges to waters of the United States.

(ii) Not defined as "large" or "medium" municipal separate storm sewer systems pursuant to paragraphs (b)(4) and (b)(7) of this section, or designated under paragraph (a)(1)(v) of this section.

(iii) This term includes systems similar to separate storm sewer systems in municipalities, such as systems at military bases, large hospital or prison complexes, and highways and other thoroughfares. The term does not include separate storm sewers in very discrete areas, such as individual buildings.

(17) *Small MS4* means a small municipal separate storm sewer system.

(18) *Municipal separate storm sewer system* means all separate storm sewers that are defined as "large" or "medium" or "small" municipal separate

storm sewer systems pursuant to paragraphs (b)(4), (b)(7), and (b)(16) of this section, or designated under paragraph (a)(1)(v) of this section.

(19) *MS4* means a municipal separate storm sewer system.

(20) *Uncontrolled sanitary landfill* means a landill or open dump, whether in operation or closed, that does not meet the requirements for runon or runoff controls established pursuant to subtitle D of the Solid Waste Disposal Act.

(c) *Application requirements for storm water discharges associated with industrial activity and storm water discharges associated with small construction activity*—(1) *Individual application.* Dischargers of storm water associated with industrial activity and with small construction activity are required to apply for an individual permit or seek coverage under a promulgated storm water general permit. Facilities that are required to obtain an individual permit or any dischage of storm water which the Director is evaluating for designation (see § 124.52(c) of this chapter) under paragraph (a)(1)(v) of this section and is not a municipal storm sewer, shall submit an NPDES application in accordance with the requirements of § 122.21 as modified and supplemented by the provisions of this paragraph.

(i) Except as provided in § 122.26(c)(1) (ii)–(iv), the operator of a storm water discharge associated with industrial activity subject to this section shall provide:

(A) A site map showing topography (or indicating the outline of drainage areas served by the outfall(s) covered in the application if a topographic map is unavailable) of the facility including: each of its drainage and discharge structures; the drainage area of each storm water outfall; paved areas and buildings within the drainage area of each storm water outfall, each past or present area used for outdoor storage or disposal of significant materials, each existing structural control measure to reduce pollutants in storm water runoff, materials loading and access areas, areas where pesticides, herbicides, soil conditioners and fertilizers are applied, each of its hazardous waste treatment, storage or disposal facilities (including each area not required to have a RCRA permit which is issued for accumulating hazardous waste under 40 CFR 262.34); each well where fluids from the facility are injected underground; springs, and other surface water bodies which receive storm water discharges from the facility;

(B) An estimate of the area of impervious surfaces (including paved areas and building roofs) and the total area drained by each outfall (within a mile radius of the facility) and a narrative description of the following: Significant materials that in the three years prior to the submittal of this application have been treated, stored or disposed in a manner to allow exposure to storm water; method of treatment, storage or disposal of such materials; materials management practices employed, in the three years prior to the submittal of this application, to minimize contact by these materials with storm water runoff; materials loading and access areas; the location, manner and frequency in which pesticides, herbicides, soil conditioners and fertilizers are applied; the location and a description of existing structural and nonstructural control measures to reduce pollutants in storm water runoff; and a description of the treatment the storm water receives, including the ultimate disposal of any solid or fluid wastes other than by discharge;

(C) A certification that all outfalls that should contain storm water discharges associated with industrial activity have been tested or evaluated for the presence of non-storm water discharges which are not covered by a NPDES permit; tests for such non-storm water discharges may include smoke tests, fluorometric dye tests, analysis of accurate schematics, as well as other appropriate tests. The certification shall include a description of the method used, the date of any testing, and the on-site drainage points that were directly observed during a test;

(D) Existing information regarding significant leaks or spills of toxic or hazardous pollutants at the facility that have taken place within the three years prior to the submittal of this application;

(E) Quantitative data based on samples collected during storm events and collected in accordance with §122.21 of this part from all outfalls containing a storm water discharge associated with industrial activity for the following parameters:

(1) Any pollutant limited in an effluent guideline to which the facility is subject;

(2) Any pollutant listed in the facility's NPDES permit for its process wastewater (if the facility is operating under an existing NPDES permit);

(3) Oil and grease, pH, BOD5, COD, TSS, total phosphorus, total Kjeldahl nitrogen, and nitrate plus nitrite nitrogen;

(4) Any information on the discharge required under §122.21(g)(7)(vi) and (vii);

(5) Flow measurements or estimates of the flow rate, and the total amount of discharge for the storm event(s) sampled, and the method of flow measurement or estimation; and

(6) The date and duration (in hours) of the storm event(s) sampled, rainfall measurements or estimates of the storm event (in inches) which generated the sampled runoff and the duration between the storm event sampled and the end of the previous measurable (greater than 0.1 inch rainfall) storm event (in hours);

(F) Operators of a discharge which is composed entirely of storm water are exempt from the requirements of §122.21 (g)(2), (g)(3), (g)(4), (g)(5), (g)(7)(iii), (g)(7)(iv), (g)(7)(v), and (g)(7)(viii); and

(G) Operators of new sources or new discharges (as defined in § 122.2 of this part) which are composed in part or entirely of storm water must include estimates for the pollutants or parameters listed in paragraph (c)(1)(i)(E) of this section instead of actual sampling data, along with the source of each estimate. Operators of new sources or new discharges composed in part or entirely of storm water must provide quantitative data for the parameters listed in paragraph (c)(1)(i)(E) of this section within two years after commencement of discharge, unless such data has already been reported under the monitoring requirements of the NPDES permit for the discharge. Operators of a new source or new discharge which is composed entirely of storm water are exempt from the requirements of § 122.21 (k)(3)(ii), (k)(3)(iii), and (k)(5).

(ii) An operator of an existing or new storm water discharge that is associated with industrial activity solely under paragraph (b)(14)(x) of this section or is associated with small construction activity solely under paragraph (b)(15) of this section, is exempt from the requirements of § 122.21(g) and paragraph (c)(1)(i) of this section. Such operator shall provide a narrative description of:

(A) The location (including a map) and the nature of the construction activity;

(B) The total area of the site and the area of the site that is expected to undergo excavation during the life of the permit;

(C) Proposed measures, including best management practices, to control pollutants in storm water discharges during construction, including a brief description of applicable State and local erosion and sediment control requirements;

(D) Proposed measures to control pollutants in storm water discharges that will occur after construction operations have been completed, including a brief description of applicable State or local erosion and sediment control requirements;

(E) An estimate of the runoff coefficient of the site and the increase in impervious area after the construction addressed in the permit application is completed, the nature of fill material and existing data describing the soil or the quality of the discharge; and

(F) The name of the receiving water.

(iii) The operator of an existing or new discharge composed entirely of storm water from an oil or gas exploration, production, processing, or treatment operation, or transmission facility is not required to submit a permit application in accordance with paragraph (c)(1)(i) of this section, unless the facility:

(A) Has had a discharge of storm water resulting in the discharge of a reportable quantity for which notification is or was required pursuant to 40 CFR 117.21 or 40 CFR 302.6 at anytime since November 16, 1987; or

(B) Has had a discharge of storm water resulting in the discharge of a reportable quantity for which notification is or was required pursuant to 40 CFR 110.6 at any time since November 16, 1987; or

(C) Contributes to a violation of a water quality standard.

(iv) The operator of an existing or new discharge composed entirely of storm water from a mining operation is not required to submit a permit application unless the discharge has come into contact with, any overburden, raw material, intermediate products, finished product, byproduct or waste products located on the site of such operations.

(v) Applicants shall provide such other information the Director may reasonably require under § 122.21(g)(13) of this part to determine whether to issue a permit and may require any facility subject to paragraph (c)(1)(ii) of this section to comply with paragraph (c)(1)(i) of this section.

(2) [Reserved]

(d) *Application requirements for large and medium municipal separate storm sewer discharges.* The operator of a discharge from a large or medium municipal separate storm sewer or a municipal separate storm sewer that is designated by the Director under paragraph (a)(1)(v) of this section, may submit a jurisdiction-wide or system-wide permit application. Where more than one public entity owns or operates a municipal separate storm sewer within a geographic area (including adjacent or interconnected municipal separate storm sewer systems), such operators may be a coapplicant to the same application. Permit applications for discharges from large and medium municipal storm sewers or municipal storm sewers designated under paragraph (a)(1)(v) of this section shall include;

(1) *Part 1.* Part 1 of the application shall consist of;

(i) *General information.* The applicants' name, address, telephone number of contact person, ownership status and status as a State or local government entity.

(ii) *Legal authority.* A description of existing legal authority to control discharges to the municipal separate storm sewer system. When existing legal authority is not sufficient to meet the criteria provided in paragraph (d)(2)(i) of this section, the description shall list additional authorities as will be necessary to meet the criteria and shall include a schedule and commitment to seek such additional authority that will be needed to meet the criteria.

(iii) *Source identification.* (A) A description of the historic use of ordinances, guidance or other controls which limited the discharge of non-storm water discharges to any Publicly Owned Treatment Works serving the same area as the municipal separate storm sewer system.

(B) A USGS 7.5 minute topographic map (or equivalent topographic map with a scale between 1:10,000 and 1:24,000 if cost effective) extending one mile beyond the service boundaries of the municipal storm sewer system covered by the permit application. The following information shall be provided:

(1) The location of known municipal storm sewer system outfalls discharging to waters of the United States;

(2) A description of the land use activities (e.g. divisions indicating undeveloped, residential, commercial, agricultural and industrial uses) accompanied with estimates of population densities and projected growth for a ten year period within the drainage area served by the separate storm sewer. For each land use type, an estimate of an average runoff coefficient shall be provided;

(3) The location and a description of the activities of the facility of each currently operating or closed municipal landfill or other treatment, storage or disposal facility for municipal waste;

(4) The location and the permit number of any known discharge to the municipal storm sewer that has been issued a NPDES permit;

(5) The location of major structural controls for storm water discharge (retention basins, detention basins, major infiltration devices, etc.); and

(6) The identification of publicly owned parks, recreational areas, and other open lands.

(iv) *Discharge characterization.* (A) Monthly mean rain and snow fall estimates (or summary of weather bureau data) and the monthly average number of storm events.

(B) Existing quantitative data describing the volume and quality of discharges from the municipal storm sewer, including a description of the outfalls sampled, sampling procedures and analytical methods used.

(C) A list of water bodies that receive discharges from the municipal separate storm sewer system, including downstream segments, lakes and estuaries, where pollutants from the system discharges may accumulate and cause water degradation and a brief description of known water quality impacts. At a minimum, the description of impacts shall include a description of whether the water bodies receiving such discharges have been:

(1) Assessed and reported in section 305(b) reports submitted by the State, the basis for the assessment (evaluated or monitored), a summary of designated use support and attainment of Clean Water Act (CWA) goals (fishable and swimmable waters), and causes of nonsupport of designated uses;

(2) Listed under section 304(l)(1)(A)(i), section 304(l)(1)(A)(ii), or section 304(l)(1)(B) of the CWA that is not expected to meet water quality standards or water quality goals;

(3) Listed in State Nonpoint Source Assessments required by section 319(a) of the CWA that, without additional action to control nonpoint sources of pollution, cannot reasonably be expected to attain or maintain water

quality standards due to storm sewers, construction, highway maintenance and runoff from municipal landfills and municipal sludge adding significant pollution (or contributing to a violation of water quality standards);

(4) Identified and classified according to eutrophic condition of publicly owned lakes listed in State reports required under section 314(a) of the CWA (include the following: A description of those publicly owned lakes for which uses are known to be impaired; a description of procedures, processes and methods to control the discharge of pollutants from municipal separate storm sewers into such lakes; and a description of methods and procedures to restore the quality of such lakes);

(5) Areas of concern of the Great Lakes identified by the International Joint Commission;

(6) Designated estuaries under the National Estuary Program under section 320 of the CWA;

(7) Recognized by the applicant as highly valued or sensitive waters;

(8) Defined by the State or U.S. Fish and Wildlife Services's National Wetlands Inventory as wetlands; and

(9) Found to have pollutants in bottom sediments, fish tissue or biosurvey data.

(D) *Field screening.* Results of a field screening analysis for illicit connections and illegal dumping for either selected field screening points or major outfalls covered in the permit application. At a minimum, a screening analysis shall include a narrative description, for either each field screening point or major outfall, of visual observations made during dry weather periods. If any flow is observed, two grab samples shall be collected during a 24 hour period with a minimum period of four hours between samples. For all such samples, a narrative description of the color, odor, turbidity, the presence of an oil sheen or surface scum as well as any other relevant observations regarding the potential presence of non-storm water discharges or illegal dumping shall be provided. In addition, a narrative description of the results of a field analysis using suitable methods to estimate pH, total chlorine, total copper, total phenol, and detergents (or surfactants) shall be provided along with a description of the flow rate. Where the field analysis does not involve analytical methods approved under 40 CFR part 136, the applicant shall provide a description of the method used including the name of the manufacturer of the test method along with the range and accuracy of the test. Field screening points shall be either major outfalls or other outfall points (or any other point of access such as manholes) randomly located throughout the storm sewer system by placing a grid over a drainage system map and identifying those cells of the grid which contain a segment of the storm sewer system or major outfall. The field screening points shall be established using the following guidelines and criteria:

(1) A grid system consisting of perpendicular north-south and east-west lines spaced 1/4 mile apart shall be overlayed on a map of the municipal storm sewer system, creating a series of cells;

(2) All cells that contain a segment of the storm sewer system shall be identified; one field screening point shall be selected in each cell; major outfalls may be used as field screening points;

(3) Field screening points should be located downstream of any sources of suspected illegal or illicit activity;

(4) Field screening points shall be located to the degree practicable at the farthest manhole or other accessible location downstream in the system, within each cell; however, safety of personnel and accessibility of the location should be considered in making this determination;

(5) Hydrological conditions; total drainage area of the site; population density of the site; traffic density; age of the structures or buildings in the area; history of the area; and land use types;

(6) For medium municipal separate storm sewer systems, no more than 250 cells need to have identified field screening points; in large municipal separate storm sewer systems, no more than 500 cells need to have identified field screening points; cells established by the grid that contain no storm sewer segments will be eliminated from consideration; if fewer than 250 cells in medium municipal sewers are created,

and fewer than 500 in large systems are created by the overlay on the municipal sewer map, then all those cells which contain a segment of the sewer system shall be subject to field screening (unless access to the separate storm sewer system is impossible); and

(7) Large or medium municipal separate storm sewer systems which are unable to utilize the procedures described in paragraphs (d)(1)(iv)(D) (*1*) through (*6*) of this section, because a sufficiently detailed map of the separate storm sewer systems is unavailable, shall field screen no more than 500 or 250 major outfalls respectively (or all major outfalls in the system, if less); in such circumstances, the applicant shall establish a grid system consisting of north-south and east-west lines spaced ¼ mile apart as an overlay to the boundaries of the municipal storm sewer system, thereby creating a series of cells; the applicant will then select major outfalls in as many cells as possible until at least 500 major outfalls (large municipalities) or 250 major outfalls (medium municipalities) are selected; a field screening analysis shall be undertaken at these major outfalls.

(E) *Characterization plan.* Information and a proposed program to meet the requirements of paragraph (d)(2)(iii) of this section. Such description shall include: the location of outfalls or field screening points appropriate for representative data collection under paragraph (d)(2)(iii)(A) of this section, a description of why the outfall or field screening point is representative, the seasons during which sampling is intended, a description of the sampling equipment. The proposed location of outfalls or field screening points for such sampling should reflect water quality concerns (*see* paragraph (d)(1)(iv)(C) of this section) to the extent practicable.

(v) *Management programs.* (A) A description of the existing management programs to control pollutants from the municipal separate storm sewer system. The description shall provide information on existing structural and source controls, including operation and maintenance measures for structural controls, that are currently being implemented. Such controls may in-

clude, but are not limited to: Procedures to control pollution resulting from construction activities; floodplain management controls; wetland protection measures; best management practices for new subdivisions; and emergency spill response programs. The description may address controls established under State law as well as local requirements.

(B) A description of the existing program to identify illicit connections to the municipal storm sewer system. The description should include inspection procedures and methods for detecting and preventing illicit discharges, and describe areas where this program has been implemented.

(vi) *Fiscal resources.* (A) A description of the financial resources currently available to the municipality to complete part 2 of the permit application. A description of the municipality's budget for existing storm water programs, including an overview of the municipality's financial resources and budget, including overall indebtedness and assets, and sources of funds for storm water programs.

(2) *Part 2.* Part 2 of the application shall consist of:

(i) *Adequate legal authority.* A demonstration that the applicant can operate pursuant to legal authority established by statute, ordinance or series of contracts which authorizes or enables the applicant at a minimum to:

(A) Control through ordinance, permit, contract, order or similar means, the contribution of pollutants to the municipal storm sewer by storm water discharges associated with industrial activity and the quality of storm water discharged from sites of industrial activity;

(B) Prohibit through ordinance, order or similar means, illicit discharges to the municipal separate storm sewer;

(C) Control through ordinance, order or similar means the discharge to a municipal separate storm sewer of spills, dumping or disposal of materials other than storm water;

(D) Control through interagency agreements among coapplicants the contribution of pollutants from one portion of the municipal system to another portion of the municipal system;

(E) Require compliance with conditions in ordinances, permits, contracts or orders; and

(F) Carry out all inspection, surveillance and monitoring procedures necessary to determine compliance and noncompliance with permit conditions including the prohibition on illicit discharges to the municipal separate storm sewer.

(ii) *Source identification.* The location of any major outfall that discharges to waters of the United States that was not reported under paragraph (d)(1)(iii)(B)(*1*) of this section. Provide an inventory, organized by watershed of the name and address, and a description (such as SIC codes) which best reflects the principal products or services provided by each facility which may discharge, to the municipal separate storm sewer, storm water associated with industrial activity;

(iii) *Characterization data.* When "quantitative data" for a pollutant are required under paragraph (d)(2)(iii)(A)(*3*) of this section, the applicant must collect a sample of effluent in accordance with 40 CFR 122.21(g)(7) and analyze it for the pollutant in accordance with analytical methods approved under part 136 of this chapter. When no analytical method is approved the applicant may use any suitable method but must provide a description of the method. The applicant must provide information characterizing the quality and quantity of discharges covered in the permit application, including:

(A) Quantitative data from representative outfalls designated by the Director (based on information received in part 1 of the application, the Director shall designate between five and ten outfalls or field screening points as representative of the commercial, residential and industrial land use activities of the drainage area contributing to the system or, where there are less than five outfalls covered in the application, the Director shall designate all outfalls) developed as follows:

(*1*) For each outfall or field screening point designated under this subparagraph, samples shall be collected of storm water discharges from three storm events occurring at least one month apart in accordance with the requirements at §122.21(g)(7) (the Director may allow exemptions to sampling three storm events when climatic conditions create good cause for such exemptions);

(*2*) A narrative description shall be provided of the date and duration of the storm event(s) sampled, rainfall estimates of the storm event which generated the sampled discharge and the duration between the storm event sampled and the end of the previous measurable (greater than 0.1 inch rainfall) storm event;

(*3*) For samples collected and described under paragraphs (d)(2)(iii)(A)(*1*) and (A)(*2*) of this section, quantitative data shall be provided for: the organic pollutants listed in Table II; the pollutants listed in Table III (toxic metals, cyanide, and total phenols) of appendix D of 40 CFR part 122, and for the following pollutants:

Total suspended solids (TSS)
Total dissolved solids (TDS)
COD
BOD_5
Oil and grease
Fecal coliform
Fecal streptococcus
pH
Total Kjeldahl nitrogen
Nitrate plus nitrite
Dissolved phosphorus
Total ammonia plus organic nitrogen
Total phosphorus

(*4*) Additional limited quantitative data required by the Director for determining permit conditions (the Director may require that quantitative data shall be provided for additional parameters, and may establish sampling conditions such as the location, season of sample collection, form of precipitation (snow melt, rainfall) and other parameters necessary to insure representativeness);

(B) Estimates of the annual pollutant load of the cumulative discharges to waters of the United States from all identified municipal outfalls and the event mean concentration of the cumulative discharges to waters of the United States from all identified municipal outfalls during a storm event (as described under §122.21(c)(7)) for BOD_5, COD, TSS, dissolved solids, total nitrogen, total ammonia plus organic nitrogen, total phosphorus, dissolved phosphorus, cadmium, copper, lead, and zinc. Estimates shall be accompanied by a description of the procedures for estimating constituent loads and concentrations, including any modelling, data analysis, and calculation methods;

(C) A proposed schedule to provide estimates for each major outfall identified in either paragraph (d)(2)(ii) or (d)(1)(iii)(B)(*1*) of this section of the seasonal pollutant load and of the event mean concentration of a representative storm for any constituent detected in any sample required under paragraph (d)(2)(iii)(A) of this section; and

(D) A proposed monitoring program for representative data collection for the term of the permit that describes the location of outfalls or field screening points to be sampled (or the location of instream stations), why the location is representative, the frequency of sampling, parameters to be sampled, and a description of sampling equipment.

(iv) *Proposed management program.* A proposed management program covers the duration of the permit. It shall include a comprehensive planning process which involves public participation and where necessary intergovernmental coordination, to reduce the discharge of pollutants to the maximum extent practicable using management practices, control techniques and system, design and engineering methods, and such other provisions which are appropriate. The program shall also include a description of staff and equipment available to implement the program. Separate proposed programs may be submitted by each coapplicant. Proposed programs may impose controls on a systemwide basis, a watershed basis, a jurisdiction basis, or on individual outfalls. Proposed programs will be considered by the Director when developing permit conditions to reduce pollutants in discharges to the maximum extent practicable. Proposed management programs shall describe priorities for implementing controls. Such programs shall be based on:

(A) A description of structural and source control measures to reduce pollutants from runoff from commercial and residential areas that are discharged from the municipal storm sewer system that are to be implemented during the life of the permit, accompanied with an estimate of the expected reduction of pollutant loads and a proposed schedule for implementing such controls. At a minimum, the description shall include:

(*1*) A description of maintenance activities and a maintenance schedule for structural controls to reduce pollutants (including floatables) in discharges from municipal separate storm sewers;

(*2*) A description of planning procedures including a comprehensive master plan to develop, implement and enforce controls to reduce the discharge of pollutants from municipal separate storm sewers which receive discharges from areas of new development and significant redevelopment. Such plan shall address controls to reduce pollutants in discharges from municipal separate storm sewers after construction is completed. (Controls to reduce pollutants in discharges from municipal separate storm sewers containing construction site runoff are addressed in paragraph (d)(2)(iv)(D) of this section;

(*3*) A description of practices for operating and maintaining public streets, roads and highways and procedures for reducing the impact on receiving waters of discharges from municipal storm sewer systems, including pollutants discharged as a result of deicing activities;

(*4*) A description of procedures to assure that flood management projects assess the impacts on the water quality of receiving water bodies and that existing structural flood control devices have been evaluated to determine if retrofitting the device to provide additional pollutant removal from storm water is feasible;

(*5*) A description of a program to monitor pollutants in runoff from operating or closed municipal landfills or other treatment, storage or disposal facilities for municipal waste, which shall identify priorities and procedures for inspections and establishing and implementing control measures for such discharges (this program can be coordinated with the program developed under paragraph (d)(2)(iv)(C) of this section); and

(6) A description of a program to reduce to the maximum extent practicable, pollutants in discharges from municipal separate storm sewers associated with the application of pesticides, herbicides and fertilizer which will include, as appropriate, controls such as educational activities, permits, certifications and other measures for commercial applicators and distributors, and controls for application in public right-of-ways and at municipal facilities.

(B) A description of a program, including a schedule, to detect and remove (or require the discharger to the municipal separate storm sewer to obtain a separate NPDES permit for) illicit discharges and improper disposal into the storm sewer. The proposed program shall include:

(1) A description of a program, including inspections, to implement and enforce an ordinance, orders or similar means to prevent illicit discharges to the municipal separate storm sewer system; this program description shall address all types of illicit discharges, however the following category of non-storm water discharges or flows shall be addressed where such discharges are identified by the municipality as sources of pollutants to waters of the United States: water line flushing, landscape irrigation, diverted stream flows, rising ground waters, uncontaminated ground water infiltration (as defined at 40 CFR 35.2005(20)) to separate storm sewers, uncontaminated pumped ground water, discharges from potable water sources, foundation drains, air conditioning condensation, irrigation water, springs, water from crawl space pumps, footing drains, lawn watering, individual residential car washing, flows from riparian habitats and wetlands, dechlorinated swimming pool discharges, and street wash water (program descriptions shall address discharges or flows from fire fighting only where such discharges or flows are identified as significant sources of pollutants to waters of the United States);

(2) A description of procedures to conduct on-going field screening activities during the life of the permit, including areas or locations that will be evaluated by such field screens;

(3) A description of procedures to be followed to investigate portions of the separate storm sewer system that, based on the results of the field screen, or other appropriate information, indicate a reasonable potential of containing illicit discharges or other sources of non-storm water (such procedures may include: sampling procedures for constituents such as fecal coliform, fecal streptococcus, surfactants (MBAS), residual chlorine, fluorides and potassium; testing with fluorometric dyes; or conducting in storm sewer inspections where safety and other considerations allow. Such description shall include the location of storm sewers that have been identified for such evaluation);

(4) A description of procedures to prevent, contain, and respond to spills that may discharge into the municipal separate storm sewer;

(5) A description of a program to promote, publicize, and facilitate public reporting of the presence of illicit discharges or water quality impacts associated with discharges from municipal separate storm sewers;

(6) A description of educational activities, public information activities, and other appropriate activities to facilitate the proper management and disposal of used oil and toxic materials; and

(7) A description of controls to limit infiltration of seepage from municipal sanitary sewers to municipal separate storm sewer systems where necessary;

(C) A description of a program to monitor and control pollutants in storm water discharges to municipal systems from municipal landfills, hazardous waste treatment, disposal and recovery facilities, industrial facilities that are subject to section 313 of title III of the Superfund Amendments and Reauthorization Act of 1986 (SARA), and industrial facilities that the municipal permit applicant determines are contributing a substantial pollutant loading to the municipal storm sewer system. The program shall:

(1) Identify priorities and procedures for inspections and establishing and implementing control measures for such discharges;

(2) Describe a monitoring program for storm water discharges associated

with the industrial facilities identified in paragraph (d)(2)(iv)(C) of this section, to be implemented during the term of the permit, including the submission of quantitative data on the following constituents: any pollutants limited in effluent guidelines subcategories, where applicable; any pollutant listed in an existing NPDES permit for a facility; oil and grease, COD, pH, BOD₅, TSS, total phosphorus, total Kjeldahl nitrogen, nitrate plus nitrite nitrogen, and any information on discharges required under §122.21(g)(7) (vi) and (vii).

(D) A description of a program to implement and maintain structural and non-structural best management practices to reduce pollutants in storm water runoff from construction sites to the municipal storm sewer system, which shall include:

(*1*) A description of procedures for site planning which incorporate consideration of potential water quality impacts;

(*2*) A description of requirements for nonstructural and structural best management practices;

(*3*) A description of procedures for identifying priorities for inspecting sites and enforcing control measures which consider the nature of the construction activity, topography, and the characteristics of soils and receiving water quality; and

(*4*) A description of appropriate educational and training measures for construction site operators.

(v) *Assessment of controls.* Estimated reductions in loadings of pollutants from discharges of municipal storm sewer constituents from municipal storm sewer systems expected as the result of the municipal storm water quality management program. The assessment shall also identify known impacts of storm water controls on ground water.

(vi) *Fiscal analysis.* For each fiscal year to be covered by the permit, a fiscal analysis of the necessary capital and operation and maintenance expenditures necessary to accomplish the activities of the programs under paragraphs (d)(2) (iii) and (iv) of this section. Such analysis shall include a description of the source of funds that are proposed to meet the necessary ex-

penditures, including legal restrictions on the use of such funds.

(vii) Where more than one legal entity submits an application, the application shall contain a description of the roles and responsibilities of each legal entity and procedures to ensure effective coordination.

(viii) Where requirements under paragraph (d)(1)(iv)(E), (d)(2)(ii), (d)(2)(iii)(B) and (d)(2)(iv) of this section are not applicable or are not applicable, the Director may exclude any operator of a discharge from a municipal separate storm sewer which is designated under paragraph (a)(1)(v), (b)(4)(ii) or (b)(7)(ii) of this section from such requirements. The Director shall not exclude the operator of a discharge from a municipal separate storm sewer identified in appendix F, G, H or I of part 122, from any of the permit application requirements under this paragraph except where authorized under this section.

(e) *Application deadlines.* Any operator of a point source required to obtain a permit under this section that does not have an effective NPDES permit authorizing discharges from its storm water outfalls shall submit an application in accordance with the following deadlines:

(1) *Storm water discharges associated with industrial activity.* (i) Except as provided in paragraph (e)(1)(ii) of this section, for any storm water discharge associated with industrial activity identified in paragraphs (b)(14)(i) through (xi) of this section, that is not part of a group application as described in paragraph (c)(2) of this section or that is not authorized by a storm water general permit, a permit application made pursuant to paragraph (c) of this section must be submitted to the Director by October 1, 1992;

(ii) For any storm water discharge associated with industrial activity from a facility that is owned or operated by a municipality with a population of less than 100,000 that is not authorized by a general or individual permit, other than an airport, powerplant, or uncontrolled sanitary landfill, the permit application must be submitted to the Director by March 10, 2003.

(2) For any group application submitted in accordance with paragraph (c)(2) of this section:

(i) *Part 1.* (A) Except as provided in paragraph (e)(2)(i)(B) of this section, part 1 of the application shall be submitted to the Director, Office of Wastewater Enforcement and Compliance by September 30, 1991;

(B) Any municipality with a population of less than 250,000 shall not be required to submit a part 1 application before May 18, 1992.

(C) For any storm water discharge associated with industrial activity from a facility that is owned or operated by a municipality with a population of less than 100,000 other than an airport, powerplant, or uncontrolled sanitary landfill, permit applications requirements are reserved.

(ii) Based on information in the part 1 application, the Director will approve or deny the members in the group application within 60 days after receiving part 1 of the group application.

(iii) *Part 2.* (A) Except as provided in paragraph (e)(2)(iii)(B) of this section, part 2 of the application shall be submittted to the Director, Office of Wastewater Enforcement and Compliance by October 1, 1992;

(B) Any municipality with a population of less than 250,000 shall not be required to submit a part 1 application before May 17, 1993.

(C) For any storm water discharge associated with industrial activity from a facility that is owned or operated by a municipality with a population of less than 100,000 other than an airport, powerplant, or uncontrolled sanitary landfill, permit applications requirements are reserved.

(iv) *Rejected facilities.* (A) Except as provided in paragraph (e)(2)(iv)(B) of this section, facilities that are rejected as members of the group shall submit an individual application (or obtain coverage under an applicable general permit) no later than 12 months after the date of receipt of the notice of rejection or October 1, 1992, whichever comes first.

(B) Facilities that are owned or operated by a municipality and that are rejected as members of part 1 group application shall submit an individual application no later than 180 days after the date of receipt of the notice of rejection or October 1, 1992, whichever is later.

(v) A facility listed under paragraph (b)(14) (i)–(xi) of this section may add on to a group application submitted in accordance with paragraph (e)(2)(i) of this section at the discretion of the Office of Water Enforcement and Permits, and only upon a showing of good cause by the facility and the group applicant; the request for the addition of the facility shall be made no later than February 18, 1992; the addition of the facility shall not cause the percentage of the facilities that are required to submit quantitative data to be less than 10%, unless there are over 100 facilities in the group that are submitting quantitative data; approval to become part of group application must be obtained from the group or the trade association representing the individual facilities.

(3) For any discharge from a large municipal separate storm sewer system;

(i) Part 1 of the application shall be submitted to the Director by November 18, 1991;

(ii) Based on information received in the part 1 application the Director will approve or deny a sampling plan under paragraph (d)(1)(iv)(E) of this section within 90 days after receiving the part 1 application;

(iii) Part 2 of the application shall be submitted to the Director by November 16, 1992.

(4) For any discharge from a medium municipal separate storm sewer system;

(i) Part 1 of the application shall be submitted to the Director by May 18, 1992.

(ii) Based on information received in the part 1 application the Director will approve or deny a sampling plan under paragraph (d)(1)(iv)(E) of this section within 90 days after receiving the part 1 application.

(iii) Part 2 of the application shall be submitted to the Director by May 17, 1993.

(5) A permit application shall be submitted to the Director within 180 days of notice, unless permission for a later date is granted by the Director (see § 124.52(c) of this chapter), for:

(i) A storm water discharge that the Director, or in States with approved NPDES programs, either the Director or the EPA Regional Administrator, determines that the discharge contributes to a violation of a water quality standard or is a significant contributor of pollutants to waters of the United States (see paragraphs (a)(1)(v) and (b)(15)(ii) of this section);

(ii) A storm water discharge subject to paragraph (c)(1)(v) of this section.

(6) Facilities with existing NPDES permits for storm water discharges associated with industrial activity shall maintain existing permits. Facilities with permits for storm water discharges associated with industrial activity which expire on or after May 18, 1992 shall submit a new application in accordance with the requirements of 40 CFR 122.21 and 40 CFR 122.26(c) (Form 1, Form 2F, and other applicable Forms) 180 days before the expiration of such permits.

(7) The Director shall issue or deny permits for discharges composed entirely of storm water under this section in accordance with the following schedule:

(i)(A) Except as provided in paragraph (e)(7)(i)(B) of this section, the Director shall issue or deny permits for storm water discharges associated with industrial activity no later than October 1, 1993, or, for new sources or existing sources which fail to submit a complete permit application by October 1, 1992, one year after receipt of a complete permit application;

(B) For any municipality with a population of less than 250,000 which submits a timely Part I group application under paragraph (e)(2)(i)(B) of this section, the Director shall issue or deny permits for storm water discharges associated with industrial activity no later than May 17, 1994, or, for any such municipality which fails to submit a complete Part II group permit application by May 17, 1993, one year after receipt of a complete permit application;

(ii) The Director shall issue or deny permits for large municipal separate storm sewer systems no later than November 16, 1993, or, for new sources or existing sources which fail to submit a complete permit application by No-vember 16, 1992, one year after receipt of a complete permit application;

(iii) The Director shall issue or deny permits for medium municipal separate storm sewer systems no later than May 17, 1994, or, for new sources or existing sources which fail to submit a complete permit application by May 17, 1993, one year after receipt of a complete permit application.

(8) For any storm water discharge associated with small construction activities identified in paragraph (b)(15)(i) of this section, see §122.21(c)(1). Discharges from these sources require permit authorization by March 10, 2003, unless designated for coverage before then.

(9) For any discharge from a regulated small MS4, the permit application made under §122.33 must be submitted to the Director by:

(i) March 10, 2003 if designated under §122.32(a)(1) unless your MS4 serves a jurisdiction with a population under 10,000 and the NPDES permitting authority has established a phasing schedule under §123.35(d)(3) (see §122.33(c)(1)); or

(ii) Within 180 days of notice, unless the NPDES permitting authority grants a later date, if designated under §122.32(a)(2) (see §122.33(c)(2)).

(f) *Petitions.* (1) Any operator of a municipal separate storm sewer system may petition the Director to require a separate NPDES permit (or a permit issued under an approved NPDES State program) for any discharge into the municipal separate storm sewer system.

(2) Any person may petition the Director to require a NPDES permit for a discharge which is composed entirely of storm water which contributes to a violation of a water quality standard or is a significant contributor of pollutants to waters of the United States.

(3) The owner or operator of a municipal separate storm sewer system may petition the Director to reduce the Census estimates of the population served by such separate system to account for storm water discharged to combined sewers as defined by 40 CFR 35.2005(b)(11) that is treated in a publicly owned treatment works. In municipalities in which combined sewers are operated, the Census estimates of

population may be reduced proportional to the fraction, based on estimated lengths, of the length of combined sewers over the sum of the length of combined sewers and municipal separate storm sewers where an applicant has submitted the NPDES permit number associated with each discharge point and a map indicating areas served by combined sewers and the location of any combined sewer overflow discharge point.

(4) Any person may petition the Director for the designation of a large, medium, or small municipal separate storm sewer system as defined by paragraph (b)(4)(iv), (b)(7)(iv), or (b)(16) of this section.

(5) The Director shall make a final determination on any petition received under this section within 90 days after receiving the petition with the exception of petitions to designate a small MS4 in which case the Director shall make a final determination on the petition within 180 days after its receipt.

(g) *Conditional exclusion for "no exposure" of industrial activities and materials to storm water.* Discharges composed entirely of storm water are not storm water discharges associated with industrial activity if there is "no exposure" of industrial materials and activities to rain, snow, snowmelt and/or runoff, and the discharger satisfies the conditions in paragraphs (g)(1) through (g)(4) of this section. "No exposure" means that all industrial materials and activities are protected by a storm resistant shelter to prevent exposure to rain, snow, snowmelt, and/or runoff. Industrial materials or activities include, but are not limited to, material handling equipment or activities, industrial machinery, raw materials, intermediate products, by-products, final products, or waste products. Material handling activities include the storage, loading and unloading, transportation, or conveyance of any raw material, intermediate product, final product or waste product.

(1) *Qualification.* To qualify for this exclusion, the operator of the discharge must:

(i) Provide a storm resistant shelter to protect industrial materials and activities from exposure to rain, snow, snow melt, and runoff;

(ii) Complete and sign (according to § 122.22) a certification that there are no discharges of storm water contaminated by exposure to industrial materials and activities from the entire facility, except as provided in paragraph (g)(2) of this section;

(iii) Submit the signed certification to the NPDES permitting authority once every five years;

(iv) Allow the Director to inspect the facility to determine compliance with the "no exposure" conditions;

(v) Allow the Director to make any "no exposure" inspection reports available to the public upon request; and

(vi) For facilities that discharge through an MS4, upon request, submit a copy of the certification of "no exposure" to the MS4 operator, as well as allow inspection and public reporting by the MS4 operator.

(2) *Industrial materials and activities not requiring storm resistant shelter.* To qualify for this exclusion, storm resistant shelter is not required for:

(i) Drums, barrels, tanks, and similar containers that are tightly sealed, provided those containers are not deteriorated and do not leak ("Sealed" means banded or otherwise secured and without operational taps or valves);

(ii) Adequately maintained vehicles used in material handling; and

(iii) Final products, other than products that would be mobilized in storm water discharge (e.g., rock salt).

(3) *Limitations.* (i) Storm water discharges from construction activities identified in paragraphs (b)(14)(x) and (b)(15) are not eligible for this conditional exclusion.

(ii) This conditional exclusion from the requirement for an NPDES permit is available on a facility-wide basis only, not for individual outfalls. If a facility has some discharges of storm water that would otherwise be "no exposure" discharges, individual permit requirements should be adjusted accordingly.

(iii) If circumstances change and industrial materials or activities become exposed to rain, snow, snow melt, and/or runoff, the conditions for this exclusion no longer apply. In such cases, the discharge becomes subject to enforcement for un-permitted discharge. Any conditionally exempt discharger who anticipates changes in circumstances should apply for and obtain permit authorization prior to the change of circumstances.

(iv) Notwithstanding the provisions of this paragraph, the NPDES permitting authority retains the authority to require permit authorization (and deny this exclusion) upon making a determination that the discharge causes, has a reasonable potential to cause, or contributes to an instream excursion above an applicable water quality standard, including designated uses.

(4) *Certification.* The no exposure certification must require the submission of the following information, at a minimum, to aid the NPDES permitting authority in determining if the facility qualifies for the no exposure exclusion:

(i) The legal name, address and phone number of the discharger (see §122.21(b));

(ii) The facility name and address, the county name and the latitude and longitude where the facility is located;

(iii) The certification must indicate that none of the following materials or activities are, or will be in the foreseeable future, exposed to precipitation:

(A) Using, storing or cleaning industrial machinery or equipment, and areas where residuals from using, storing or cleaning industrial machinery or equipment remain and are exposed to storm water;

(B) Materials or residuals on the ground or in storm water inlets from spills/leaks;

(C) Materials or products from past industrial activity;

(D) Material handling equipment (except adequately maintained vehicles);

(E) Materials or products during loading/unloading or transporting activities;

(F) Materials or products stored outdoors (except final products intended for outside use, e.g., new cars, where exposure to storm water does not result in the discharge of pollutants);

(G) Materials contained in open, deteriorated or leaking storage drums, barrels, tanks, and similar containers;

(H) Materials or products handled/stored on roads or railways owned or maintained by the discharger;

(I) Waste material (except waste in covered, non-leaking containers, e.g., dumpsters);

(J) Application or disposal of process wastewater (unless otherwise permitted); and

(K) Particulate matter or visible deposits of residuals from roof stacks/vents not otherwise regulated, i.e., under an air quality control permit, and evident in the storm water outflow;

(iv) All "no exposure" certifications must include the following certification statement, and be signed in accordance with the signatory requirements of §122.22: "I certify under penalty of law that I have read and understand the eligibility requirements for claiming a condition of "no exposure" and obtaining an exclusion from NPDES storm water permitting; and that there are no discharges of storm water contaminated by exposure to industrial activities or materials from the industrial facility identified in this document (except as allowed under paragraph (g)(2)) of this section. I understand that I am obligated to submit a no exposure certification form once every five years to the NPDES permitting authority and, if requested, to the operator of the local MS4 into which this facility discharges (where applicable). I understand that I must allow the NPDES permitting authority, or MS4 operator where the discharge is into the local MS4, to perform inspections to confirm the condition of no exposure and to make such inspection reports publicly available upon request. I understand that I must obtain coverage under an NPDES permit prior to any point source discharge of storm water from the facility. I certify under penalty of law that this document and all attachments were prepared under my direction or supervision in accordance with a system designed to assure that qualified personnel properly gathered and evaluated the information submitted. Based upon my inquiry of the person or persons who manage the system, or those persons directly involved in gathering the information, the information submitted is to the best of my knowledge and belief true, accurate and complete. I am aware there are significant penalties for submitting false information, including the possibility of fine and imprisonment for knowing violations."

[55 FR 48063, Nov. 16, 1990]

EDITORIAL NOTE: For FEDERAL REGISTER citations affecting § 122.26, see the List of CFR Sections Affected, which appears in the Finding Aids section of the printed volume and at *www.fdsys.gov.*

§ 122.27 Silvicultural activities (applicable to State NPDES programs, see § 123.25).

(a) *Permit requirement.* Silvicultural point sources, as defined in this section, as point sources subject to the NPDES permit program.

(b) *Definitions.* (1) *Silvicultural point source* means any discernible, confined and discrete conveyance related to rock crushing, gravel washing, log sorting, or log storage facilities which are operated in connection with silvicultural activities and from which pollutants are discharged into waters of the United States. The term does not include non-point source silvicultural activities such as nursery operations, site preparation, reforestation and subsequent cultural treatment, thinning, prescribed burning, pest and fire control, harvesting operations, surface drainage, or road construction and maintenance from which there is natural runoff. However, some of these activities (such as stream crossing for roads) may involve point source discharges of dredged or fill material which may require a CWA section 404 permit (See 33 CFR 209.120 and part 233).

(2) *Rock crushing and gravel washing facilities* means facilities which process crushed and broken stone, gravel, and riprap (See 40 CFR part 436, subpart B, including the effluent limitations guidelines).

(3) *Log sorting and log storage facilities* means facilities whose discharges result from the holding of unprocessed wood, for example, logs or roundwood with bark or after removal of bark held in self-contained bodies of water (mill ponds or log ponds) or stored on land where water is applied intentionally on the logs (wet decking). (See 40 CFR part 429, subpart I, including the effluent limitations guidelines).

§ 122.28 General permits (applicable to State NPDES programs, see § 123.25).

(a) *Coverage.* The Director may issue a general permit in accordance with the following:

(1) *Area.* The general permit shall be written to cover one or more categories or subcategories of discharges or sludge use or disposal practices or facilities described in the permit under paragraph (a)(2)(ii) of this section, except those covered by individual permits, within a geographic area. The area should correspond to existing geographic or political boundaries such as:

(i) Designated planning areas under sections 208 and 303 of CWA;

(ii) Sewer districts or sewer authorities;

(iii) City, county, or State political boundaries;

(iv) State highway systems;

(v) Standard metropolitan statistical areas as defined by the Office of Management and Budget;

(vi) Urbanized areas as designated by the Bureau of the Census according to criteria in 30 FR 15202 (May 1, 1974); or

(vii) Any other appropriate division or combination of boundaries.

(2) *Sources.* The general permit may be written to regulate one or more categories or subcategories of discharges or sludge use or disposal practices or facilities, within the area described in paragraph (a)(1) of this section, where the sources within a covered subcategory of discharges are either:

(i) Storm water point sources; or (ii) One or more categories or subcategories of point sources other than storm water point sources, or one or more categories or subcategories of "treatment works treating domestic sewage", if the sources or "treatment works treating domestic sewage" within each category or subcategory all:

(A) Involve the same or substantially similar types of operations;

(B) Discharge the same types of wastes or engage in the same types of sludge use or disposal practices;

(C) Require the same effluent limitations, operating conditions, or standards for sewage sludge use or disposal;

(D) Require the same or similar monitoring; and (E) In the opinion of the

Director, are more appropriately controlled under a general permit than under individual permits.

(3) *Water quality-based limits.* Where sources within a specific category or subcategory of dischargers are subject to water quality-based limits imposed pursuant to §122.44, the sources in that specific category or subcategory shall be subject to the same water quality-based effluent limitations.

(4) *Other requirements.* (i) The general permit must clearly identify the applicable conditions for each category or subcategory of dischargers or treatment works treating domestic sewage covered by the permit.

(ii) The general permit may exclude specified sources or areas from coverage.

(b) *Administration*—(1) *In general.* General permits may be issued, modified, revoked and reissued, or terminated in accordance with applicable requirements of part 124 of this chapter or corresponding State regulations. Special procedures for issuance are found at §123.44 of this chapter for States.

(2) *Authorization to discharge, or authorization to engage in sludge use and disposal practices.* (i) Except as provided in paragraphs (b)(2)(v) and (b)(2)(vi) of this section, dischargers (or treatment works treating domestic sewage) seeking coverage under a general permit shall submit to the Director a written notice of intent to be covered by the general permit. A discharger (or treatment works treating domestic sewage) who fails to submit a notice of intent in accordance with the terms of the permit is not authorized to discharge, (or in the case of sludge disposal permit, to engage in a sludge use or disposal practice), under the terms of the general permit unless the general permit, in accordance with paragraph (b)(2)(v) of this section, contains a provision that a notice of intent is not required or the Director notifies a discharger (or treatment works treating domestic sewage) that it is covered by a general permit in accordance with paragraph (b)(2)(vi) of this section. A complete and timely, notice of intent (NOI), to be covered in accordance with general permit requirements, fulfills the requirements for permit applica-

tions for purposes of §§122.6, 122.21 and 122.26.

(ii) The contents of the notice of intent shall be specified in the general permit and shall require the submission of information necessary for adequate program implementation, including at a minimum, the legal name and address of the owner or operator, the facility name and address, type of facility or discharges, and the receiving stream(s). General permits for storm water discharges associated with industrial activity from inactive mining, inactive oil and gas operations, or inactive landfills occurring on Federal lands where an operator cannot be identified may contain alternative notice of intent requirements. All notices of intent shall be signed in accordance with §122.22. Notices of intent for coverage under a general permit for concentrated animal feeding operations must include the information specified in §122.21(i)(1), including a topographic map.

(iii) General permits shall specify the deadlines for submitting notices of intent to be covered and the date(s) when a discharger is authorized to discharge under the permit;

(iv) General permits shall specify whether a discharger (or treatment works treating domestic sewage) that has submitted a complete and timely notice of intent to be covered in accordance with the general permit and that is eligible for coverage under the permit, is authorized to discharge, (or in the case of a sludge disposal permit, to engage in a sludge use or disposal practice), in accordance with the permit either upon receipt of the notice of intent by the Director, after a waiting period specified in the general permit, on a date specified in the general permit, or upon receipt of notification of inclusion by the Director. Coverage may be terminated or revoked in accordance with paragraph (b)(3) of this section.

(v) Discharges other than discharges from publicly owned treatment works, combined sewer overflows, municipal separate storm sewer systems, primary industrial facilities, and storm water discharges associated with industrial activity, may, at the discretion of the Director, be authorized to discharge under a general permit without submitting a notice of intent where the Director finds that a notice of intent requirement would be inappropriate. In making such a finding, the Director shall consider: the type of discharge; the expected nature of the discharge; the potential for toxic and conventional pollutants in the discharges; the expected volume of the discharges; other means of identifying discharges covered by the permit; and the estimated number of discharges to be covered by the permit. The Director shall provide in the public notice of the general permit the reasons for not requiring a notice of intent.

(vi) The Director may notify a discharger (or treatment works treating domestic sewage) that it is covered by a general permit, even if the discharger (or treatment works treating domestic sewage) has not submitted a notice of intent to be covered. A discharger (or treatment works treating domestic sewage) so notified may request an individual permit under paragraph (b)(3)(iii) of this section.

(vii) A CAFO owner or operator may be authorized to discharge under a general permit only in accordance with the process described in § 122.23(h).

(3) *Requiring an individual permit.* (i) The Director may require any discharger authorized by a general permit to apply for and obtain an individual NPDES permit. Any interested person may petition the Director to take action under this paragraph. Cases where an individual NPDES permit may be required include the following:

(A) The discharger or "treatment works treating domestic sewage" is not in compliance with the conditions of the general NPDES permit;

(B) A change has occurred in the availability of demonstrated technology or practices for the control or abatement of pollutants applicable to the point source or treatment works treating domestic sewage;

(C) Effluent limitation guidelines are promulgated for point sources covered by the general NPDES permit;

(D) A Water Quality Management plan containing requirements applicable to such point sources is approved;

(E) Circumstances have changed since the time of the request to be covered so that the discharger is no longer appropriately controlled under the general permit, or either a temporary or permanent reduction or elimination of the authorized discharge is necessary;

(F) Standards for sewage sludge use or disposal have been promulgated for the sludge use and disposal practice covered by the general NPDES permit; or

(G) The discharge(s) is a significant contributor of pollutants. In making this determination, the Director may consider the following factors:

(1) The location of the discharge with respect to waters of the United States;

(2) The size of the discharge;

(3) The quantity and nature of the pollutants discharged to waters of the United States; and

(4) Other relevant factors;

(ii) *For EPA issued general permits only,* the Regional Administrator may require any owner or operator authorized by a general permit to apply for an individual NPDES permit as provided in paragraph (b)(3)(i) of this section, only if the owner or operator has been notified in writing that a permit application is required. This notice shall include a brief statement of the reasons for this decision, an application form, a statement setting a time for the owner or operator to file the application, and a statement that on the effective date of the individual NPDES permit the general permit as it applies to the individual permittee shall automatically terminate. The Director may grant additional time upon request of the applicant.

(iii) Any owner or operator authorized by a general permit may request to be excluded from the coverage of the general permit by applying for an individual permit. The owner or operator shall submit an application under § 122.21, with reasons supporting the request, to the Director no later than 90 days after the publication by EPA of the general permit in the FEDERAL REGISTER or the publication by a State in accordance with applicable State law. The request shall be processed under part 124 or applicable State procedures. The request shall be granted by issuing of any individual permit if the reasons cited by the owner or operator are adequate to support the request.

(iv) When an individual NPDES permit is issued to an owner or operator otherwise subject to a general NPDES permit, the applicability of the general permit to the individual NPDES permittee is automatically terminated on the effective date of the individual permit.

(v) A source excluded from a general permit solely because it already has an individual permit may request that the individual permit be revoked, and that it be covered by the general permit. Upon revocation of the individual permit, the general permit shall apply to the source.

(c) *Offshore oil and gas facilities* (Not applicable to State programs). (1) The Regional Administrator shall, except as provided below, issue general permits covering discharges from offshore oil and gas exploration and production facilities within the Region's jurisdiction. Where the offshore area includes areas, such as areas of biological concern, for which separate permit conditions are required, the Regional Administrator may issue separate general permits, individual permits, or both. The reason for separate general permits or individual permits shall be set forth in the appropriate fact sheets or statements of basis. Any statement of basis or fact sheet for a draft permit shall include the Regional Administrator's tentative determination as to whether the permit applies to "new sources," "new dischargers," or existing sources and the reasons for this determination, and the Regional Administrator's proposals as to areas of biological concern subject either to separate individual or general permits. For Federally leased lands, the general permit area should generally be no less extensive than the lease sale area defined by the Department of the Interior.

(2) Any interested person, including any prospective permittee, may petition the Regional Administrator to issue a general permit. Unless the Regional Administrator determines under paragraph (c)(1) of this section that no general permit is appropriate, he shall promptly provide a project decision schedule covering the issuance of the general permit or permits for any lease sale area for which the Department of the Interior has published a draft environmental impact statement. The project decision schedule shall meet the requirements of §124.3(g), and shall include a schedule providing for the issuance of the final general permit or permits not later than the date of the final notice of sale projected by the Department of the Interior or six months after the date of the request, whichever is later. The Regional Administrator may, at his discretion, issue a project decision schedule for offshore oil and gas facilities in the territorial seas.

(3) Nothing in this paragraph (c) shall affect the authority of the Regional Administrator to require an individual permit under §122.28(b)(3)(i) (A) through (G).

(Clean Water Act (33 U.S.C. 1251 *et seq.*), Safe Drinking Water Act (42 U.S.C. 300f *et seq.*), Clean Air Act (42 U.S.C. 7401 *et seq.*), Resource Conservation and Recovery Act (42 U.S.C. 6901 *et seq.*))

[48 FR 14153, Apr. 1, 1983, as amended at 48 FR 39619, Sept. 1, 1983; 49 FR 38048, Sept. 26, 1984; 50 FR 6940, Feb. 19, 1985; 54 FR 18782, May 2, 1989; 55 FR 48072, Nov. 16, 1990; 57 FR 11412 and 11413, Apr. 2, 1992; 64 FR 68841, Dec. 8, 1999; 65 FR 30908, May 15, 2000; 68 FR 7268, Feb. 12, 2003; 73 FR 70483, Nov. 20, 2008]

§122.29 New sources and new dischargers.

(a) *Definitions.* (1) *New source* and *new discharger* are defined in §122.2. [See Note 2.]

(2) *Source* means any building, structure, facility, or installation from which there is or may be a discharge of pollutants.

(3) *Existing source* means any source which is not a new source or a new discharger.

(4) *Site* is defined in §122.2;

(5) *Facilities or equipment* means buildings, structures, process or production equipment or machinery which form a permanent part of the new source and which will be used in its operation, if these facilities or equipment are of such value as to represent a substantial commitment to construct. It excludes facilities or equipment used in connection with feasibility, engineering, and design studies regarding the source or water pollution treatment for the source.

(b) *Criteria for new source determination.* (1) Except as otherwise provided in an applicable new source performance standard, a source is a "new source" if it meets the definition of "new source" in § 122.2, and

(i) It is constructed at a site at which no other source is located; or

(ii) It totally replaces the process or production equipment that causes the discharge of pollutants at an existing source; or

(iii) Its processes are substantially independent of an existing source at the same site. In determining whether these processes are substantially independent, the Director shall consider such factors as the extent to which the new facility is integrated with the existing plant; and the extent to which the new facility is engaged in the same general type of activity as the existing source.

(2) A source meeting the requirements of paragraphs (b)(1) (i), (ii), or (iii) of this section is a new source only if a new source performance standard is independently applicable to it. If there is no such independently applicable standard, the source is a new discharger. See § 122.2.

(3) Construction on a site at which an existing source is located results in a modification subject to § 122.62 rather than a new source (or a new discharger) if the construction does not create a new building, structure, facility, or installation meeting the criteria of paragraph (b)(1) (ii) or (iii) of this section but otherwise alters, replaces, or adds to existing process or production equipment.

(4) Construction of a new source as defined under § 122.2 has commenced if the owner or operator has:

(i) Begun, or caused to begin as part of a continuous on-site construction program:

(A) Any placement, assembly, or installation of facilities or equipment; or

(B) Significant site preparation work including clearing, excavation or removal of existing buildings, structures, or facilities which is necessary for the placement, assembly, or installation of new source facilities or equipment; or

(ii) Entered into a binding contractual obligation for the purchase of facilities or equipment which are intended to be used in its operation with a reasonable time. Options to purchase or contracts which can be terminated or modified without substantial loss, and contracts for feasibility engineering, and design studies do not constitute a contractual obligation under the paragraph.

(c) *Requirement for an environmental impact statement.* (1) The issuance of an NPDES permit to new source:

(i) By EPA may be a major Federal action significantly affecting the quality of the human environment within the meaning of the National Environmental Policy Act of 1969 (NEPA), 33 U.S.C. 4321 *et seq.* and is subject to the environmental review provisions of NEPA as set out in 40 CFR part 6, subpart F. EPA will determine whether an Environmental Impact Statement (EIS) is required under § 122.21(l) (special provisions for applications from new sources) and 40 CFR part 6, subpart F;

(ii) By an NPDES approved State is not a Federal action and therefore does not require EPA to conduct an environmental review.

(2) An EIS prepared under this paragraph shall include a recommendation either to issue or deny the permit.

(i) If the recommendation is to deny the permit, the final EIS shall contain the reasons for the recommendation and list those measures, if any, which the applicant could take to cause the recommendation to be changed;

(ii) If the recommendation is to issue the permit, the final EIS shall recommend the actions, if any, which the permittee should take to prevent or minimize any adverse environmental impacts;

(3) The Regional Administrator, to the extent allowed by law, shall issue, condition (other than imposing effluent limitations), or deny the new source NPDES permit following a complete evaluation of any significant beneficial and adverse impacts of the proposed action and a review of the recommendations contained in the EIS or finding of no significant impact.

(d) *Effect of compliance with new source performance standards.* (The provisions of this paragraph do not apply to existing sources which modify their pollution control facilities or construct new pollution control facilities and achieve performance standards, but which are neither new sources or new dischargers or otherwise do not meet the requirements of this paragraph.)

(1) Except as provided in paragraph (d)(2) of this section, any new discharger, the construction of which commenced after October 18, 1972, or new source which meets the applicable promulgated new source performance standards before the commencement of discharge, may not be subject to any more stringent new source performance standards or to any more stringent technology-based standards under section 301(b)(2) of CWA for the soonest ending of the following periods:

(i) Ten years from the date that construction is completed;

(ii) Ten years from the date the source begins to discharge process or other nonconstruction related wastewater; or

(iii) The period of depreciation or amortization of the facility for the purposes of section 167 or 169 (or both) of the Internal Revenue Code of 1954.

(2) The protection from more stringent standards of performance afforded by paragraph (d)(1) of this section does not apply to:

(i) Additional or more stringent permit conditions which are not technology based; for example, conditions based on water quality standards, or toxic effluent standards or prohibitions under section 307(a) of CWA; or

(ii) Additional permit conditions in accordance with §125.3 controlling toxic pollutants or hazardous substances which are not controlled by new source performance standards. This includes permit conditions controlling pollutants other than those identified as toxic pollutants or hazardous substances when control of these pollutants has been specifically identified as the method to control the toxic pollutants or hazardous substances.

(3) When an NPDES permit issued to a source with a "protection period" under paragraph (d)(1) of this section will expire on or after the expiration of the protection period, that permit shall require the owner or operator of the source to comply with the requirements of section 301 and any other then applicable requirements of CWA immediately upon the expiration of the protection period. No additional period for achieving compliance with these requirements may be allowed except when necessary to achieve compliance with requirements promulgated less than 3 years before the expiration of the protection period.

(4) The owner or operator of a new source, a new discharger which commenced discharge after August 13, 1979, or a recommencing discharger shall install and have in operating condition, and shall "start-up" all pollution control equipment required to meet the conditions of its permits before beginning to discharge. Within the shortest feasible time (not to exceed 90 days), the owner or operator must meet all permit conditions. The requirements of this paragraph do not apply if the owner or operator is issued a permit containing a compliance schedule under §122.47(a)(2).

(5) After the effective date of new source performance standards, it shall be unlawful for any owner or operator of any new source to operate the source in violation of those standards applicable to the source.

[48 FR 14153, Apr. 1, 1983, as amended at 49 FR 38048, Sept. 26, 1984; 50 FR 4514, Jan. 31, 1985; 50 FR 6941, Feb. 19, 1985; 65 FR 30908, May 15, 2000]

§122.30 What are the objectives of the storm water regulations for small MS4s?

(a) Sections 122.30 through 122.37 are written in a "readable regulation" format that includes both rule requirements and EPA guidance that is not legally binding. EPA has clearly distinguished its recommended guidance from the rule requirements by putting the guidance in a separate paragraph headed by the word "guidance".

(b) Under the statutory mandate in section 402(p)(6) of the Clean Water Act, the purpose of this portion of the storm water program is to designate additional sources that need to be regulated to protect water quality and to establish a comprehensive storm water program to regulate these sources. (Because the storm water program is part of the National Pollutant Discharge

Elimination System (NPDES) Program, you should also refer to § 122.1 which addresses the broader purpose of the NPDES program.)

(c) Storm water runoff continues to harm the nation's waters. Runoff from lands modified by human activities can harm surface water resources in several ways including by changing natural hydrologic patterns and by elevating pollutant concentrations and loadings. Storm water runoff may contain or mobilize high levels of contaminants, such as sediment, suspended solids, nutrients, heavy metals, pathogens, toxins, oxygen-demanding substances, and floatables.

(d) EPA strongly encourages partnerships and the watershed approach as the management framework for efficiently, effectively, and consistently protecting and restoring aquatic ecosystems and protecting public health.

[64 FR 68842, Dec. 8, 1999]

§ 122.31 As a Tribe, what is my role under the NPDES storm water program?

As a Tribe you may:

(a) Be authorized to operate the NPDES program including the storm water program, after EPA determines that you are eligible for treatment in the same manner as a State under §§ 123.31 through 123.34 of this chapter. (If you do not have an authorized NPDES program, EPA implements the program for discharges on your reservation as well as other Indian country, generally.);

(b) Be classified as an owner of a regulated small MS4, as defined in § 122.32. (Designation of your Tribe as an owner of a small MS4 for purposes of this part is an approach that is consistent with EPA's 1984 Indian Policy of operating on a government-to-government basis with EPA looking to Tribes as the lead governmental authorities to address environmental issues on their reservations as appropriate. If you operate a separate storm sewer system that meets the definition of a regulated small MS4, you are subject to the requirements under §§ 122.33 through 122.35. If you are not designated as a regulated small MS4, you may ask EPA to designate you as such for the purposes of this part.); or

(c) Be a discharger of storm water associated with industrial activity or small construction activity under §§ 122.26(b)(14) or (b)(15), in which case you must meet the applicable requirements. Within Indian country, the NPDES permitting authority is generally EPA, unless you are authorized to administer the NPDES program.

[64 FR 68842, Dec. 8, 1999]

§ 122.32 As an operator of a small MS4, am I regulated under the NPDES storm water program?

(a) Unless you qualify for a waiver under paragraph (c) of this section, you are regulated if you operate a small MS4, including but not limited to systems operated by federal, State, Tribal, and local governments, including State departments of transportation; and:

(1) Your small MS4 is located in an urbanized area as determined by the latest Decennial Census by the Bureau of the Census. (If your small MS4 is not located entirely within an urbanized area, only the portion that is within the urbanized area is regulated); or

(2) You are designated by the NPDES permitting authority, including where the designation is pursuant to §§ 123.35(b)(3) and (b)(4) of this chapter, or is based upon a petition under § 122.26(f).

(b) You may be the subject of a petition to the NPDES permitting authority to require an NPDES permit for your discharge of storm water. If the NPDES permitting authority determines that you need a permit, you are required to comply with §§ 122.33 through 122.35.

(c) The NPDES permitting authority may waive the requirements otherwise applicable to you if you meet the criteria of paragraph (d) or (e) of this section. If you receive a waiver under this section, you may subsequently be required to seek coverage under an NPDES permit in accordance with § 122.33(a) if circumstances change. (See also § 123.35(b) of this chapter.)

(d) The NPDES permitting authority may waive permit coverage if your MS4 serves a population of less than 1,000 within the urbanized area and you meet the following criteria:

(1) Your system is not contributing substantially to the pollutant loadings

of a physically interconnected MS4 that is regulated by the NPDES storm water program (see § 123.35(b)(4) of this chapter); and

(2) If you discharge any pollutant(s) that have been identified as a cause of impairment of any water body to which you discharge, storm water controls are not needed based on wasteload allocations that are part of an EPA approved or established "total maximum daily load" (TMDL) that addresses the pollutant(s) of concern.

(e) The NPDES permitting authority may waive permit coverage if your MS4 serves a population under 10,000 and you meet the following criteria:

(1) The permitting authority has evaluated all waters of the U.S., including small streams, tributaries, lakes, and ponds, that receive a discharge from your MS4;

(2) For all such waters, the permitting authority has determined that storm water controls are not needed based on wasteload allocations that are part of an EPA approved or established TMDL that addresses the pollutant(s) of concern or, if a TMDL has not been developed or approved, an equivalent analysis that determines sources and allocations for the pollutant(s) of concern;

(3) For the purpose of this paragraph (e), the pollutant(s) of concern include biochemical oxygen demand (BOD), sediment or a parameter that addresses sediment (such as total suspended solids, turbidity or siltation), pathogens, oil and grease, and any pollutant that has been identified as a cause of impairment of any water body that will receive a discharge from your MS4; and

(4) The permitting authority has determined that future discharges from your MS4 do not have the potential to result in exceedances of water quality standards, including impairment of designated uses, or other significant water quality impacts, including habitat and biological impacts.

[64 FR 68842, Dec. 8, 1999]

§ 122.33 If I am an operator of a regulated small MS4, how do I apply for an NPDES permit and when do I have to apply?

(a) If you operate a regulated small MS4 under § 122.32, you must seek coverage under a NPDES permit issued by your NPDES permitting authority. If you are located in an NPDES authorized State, Tribe, or Territory, then that State, Tribe, or Territory is your NPDES permitting authority. Otherwise, your NPDES permitting authority is the EPA Regional Office.

(b) You must seek authorization to discharge under a general or individual NPDES permit, as follows:

(1) If your NPDES permitting authority has issued a general permit applicable to your discharge and you are seeking coverage under the general permit, you must submit a Notice of Intent (NOI) that includes the information on your best management practices and measurable goals required by § 122.34(d). You may file your own NOI, or you and other municipalities or governmental entities may jointly submit an NOI. If you want to share responsibilities for meeting the minimum measures with other municipalities or governmental entities, you must submit an NOI that describes which minimum measures you will implement and identify the entities that will implement the other minimum measures within the area served by your MS4. The general permit will explain any other steps necessary to obtain permit authorization.

(2)(i) If you are seeking authorization to discharge under an individual permit and wish to implement a program under § 122.34, you must submit an application to your NPDES permitting authority that includes the information required under §§ 122.21(f) and 122.34(d), an estimate of square mileage served by your small MS4, and any additional information that your NPDES permitting authority requests. A storm sewer map that satisfies the requirement of § 122.34(b)(3)(i) will satisfy the map requirement in § 122.21(f)(7).

(ii) If you are seeking authorization to discharge under an individual permit and wish to implement a program that is different from the program under § 122.34, you will need to comply with the permit application requirements of § 122.26(d). You must submit both Parts of the application requirements in §§ 122.26(d)(1) and (2) by March 10, 2003. You do not need to submit the information required by §§ 122.26(d)(1)(ii) and (d)(2) regarding your legal authority, unless you intend for the permit writer to take such information into account when developing your other permit conditions.

(iii) If allowed by your NPDES permitting authority, you and another regulated entity may jointly apply under either paragraph (b)(2)(i) or (b)(2)(ii) of this section to be co-permittees under an individual permit.

(3) If your small MS4 is in the same urbanized area as a medium or large MS4 with an NPDES storm water permit and that other MS4 is willing to have you participate in its storm water program, you and the other MS4 may jointly seek a modification of the other MS4 permit to include you as a limited co-permittee. As a limited co-permittee, you will be responsible for compliance with the permit's conditions applicable to your jurisdiction. If you choose this option you will need to comply with the permit application requirements of § 122.26, rather than the requirements of § 122.34. You do not need to comply with the specific application requirements of § 122.26(d)(1)(iii) and (iv) and (d)(2)(iii) (discharge characterization). You may satisfy the requirements in § 122.26 (d)(1)(v) and (d)(2)(iv) (identification of a management program) by referring to the other MS4's storm water management program.

(4) Guidance: In referencing an MS4's storm water management program, you should briefly describe how the existing plan will address discharges from your small MS4 or would need to be supplemented in order to adequately address your discharges. You should also explain your role in coordinating storm water pollutant control activities in your MS4, and detail the resources available to you to accomplish the plan.

(c) If you operate a regulated small MS4:

(1) Designated under § 122.32(a)(1), you must apply for coverage under an NPDES permit, or apply for a modification of an existing NPDES permit under paragraph (b)(3) of this section by March 10, 2003, unless your MS4 serves a jurisdiction with a population under 10,000 and the NPDES permitting authority has established a phasing

schedule under § 123.35(d)(3) of this chapter.

(2) Designated under § 122.32(a)(2), you must apply for coverage under an NPDES permit, or apply for a modification of an existing NPDES permit under paragraph (b)(3) of this section, within 180 days of notice, unless the NPDES permitting authority grants a later date.

[64 FR 68843, Dec. 8, 1999]

§ 122.34 As an operator of a regulated small MS4, what will my NPDES MS4 storm water permit require?

(a) Your NPDES MS4 permit will require at a minimum that you develop, implement, and enforce a storm water management program designed to reduce the discharge of pollutants from your MS4 to the maximum extent practicable (MEP), to protect water quality, and to satisfy the appropriate water quality requirements of the Clean Water Act. Your storm water management program must include the minimum control measures described in paragraph (b) of this section unless you apply for a permit under § 122.26(d). For purposes of this section, narrative effluent limitations requiring implementation of best management practices (BMPs) are generally the most appropriate form of effluent limitations when designed to satisfy technology requirements (including reductions of pollutants to the maximum extent practicable) and to protect water quality. Implementation of best management practices consistent with the provisions of the storm water management program required pursuant to this section and the provisions of the permit required pursuant to § 122.33 constitutes compliance with the standard of reducing pollutants to the "maximum extent practicable." Your NPDES permitting authority will specify a time period of up to 5 years from the date of permit issuance for you to develop and implement your program.

(b) *Minimum control measures*—(1) *Public education and outreach on storm water impacts.* (i) You must implement a public education program to distribute educational materials to the community or conduct equivalent outreach activities about the impacts of storm water discharges on water bodies and the steps that the public can take to reduce pollutants in storm water run-off.

(ii) *Guidance:* You may use storm water educational materials provided by your State, Tribe, EPA, environmental, public interest or trade organizations, or other MS4s. The public education program should inform individuals and households about the steps they can take to reduce storm water pollution, such as ensuring proper septic system maintenance, ensuring the proper use and disposal of landscape and garden chemicals including fertilizers and pesticides, protecting and restoring riparian vegetation, and properly disposing of used motor oil or household hazardous wastes. EPA recommends that the program inform individuals and groups how to become involved in local stream and beach restoration activities as well as activities that are coordinated by youth service and conservation corps or other citizen groups. EPA recommends that the public education program be tailored, using a mix of locally appropriate strategies, to target specific audiences and communities. Examples of strategies include distributing brochures or fact sheets, sponsoring speaking engagements before community groups, providing public service announcements, implementing educational programs targeted at school age children, and conducting community-based projects such as storm drain stenciling, and watershed and beach cleanups. In addition, EPA recommends that some of the materials or outreach programs be directed toward targeted groups of commercial, industrial, and institutional entities likely to have significant storm water impacts. For example, providing information to restaurants on the impact of grease clogging storm drains and to garages on the impact of oil discharges. You are encouraged to tailor your outreach program to address the viewpoints and concerns of all communities, particularly minority and disadvantaged communities, as well as any special concerns relating to children.

(2) *Public involvement/participation.* (i) You must, at a minimum, comply with State, Tribal and local public notice requirements when implementing a public involvement/ participation program.

(ii) *Guidance:* EPA recommends that the public be included in developing, implementing, and reviewing your storm water management program and that the public participation process should make efforts to reach out and engage all economic and ethnic groups. Opportunities for members of the public to participate in program development and implementation include serving as citizen representatives on a local storm water management panel, attending public hearings, working as citizen volunteers to educate other individuals about the program, assisting in program coordination with other pre-existing programs, or participating in volunteer monitoring efforts. (Citizens should obtain approval where necessary for lawful access to monitoring sites.)

(3) *Illicit discharge detection and elimination.* (i) You must develop, implement and enforce a program to detect and eliminate illicit discharges (as defined at §122.26(b)(2)) into your small MS4.

(ii) You must:

(A) Develop, if not already completed, a storm sewer system map, showing the location of all outfalls and the names and location of all waters of the United States that receive discharges from those outfalls;

(B) To the extent allowable under State, Tribal or local law, effectively prohibit, through ordinance, or other regulatory mechanism, non-storm water discharges into your storm sewer system and implement appropriate enforcement procedures and actions;

(C) Develop and implement a plan to detect and address non-storm water discharges, including illegal dumping, to your system; and

(D) Inform public employees, businesses, and the general public of hazards associated with illegal discharges and improper disposal of waste.

(iii) You need address the following categories of non-storm water discharges or flows (i.e., illicit discharges) only if you identify them as significant contributors of pollutants to your small MS4: water line flushing, landscape irrigation, diverted stream flows, rising ground waters, uncontaminated ground water infiltration (as defined at 40 CFR 35.2005(20)), uncontaminated pumped ground water, discharges from potable water sources, foundation drains, air conditioning condensation, irrigation water, springs, water from crawl space pumps, footing drains, lawn watering, individual residential car washing, flows from riparian habitats and wetlands, dechlorinated swimming pool discharges, and street wash water (discharges or flows from fire fighting activities are excluded from the effective prohibition against nonstorm water and need only be addressed where they are identified as significant sources of pollutants to waters of the United States).

(iv) Guidance: EPA recommends that the plan to detect and address illicit discharges include the following four components: procedures for locating priority areas likely to have illicit discharges; procedures for tracing the source of an illicit discharge; procedures for removing the source of the discharge; and procedures for program evaluation and assessment. EPA recommends visually screening outfalls during dry weather and conducting field tests of selected pollutants as part of the procedures for locating priority areas. Illicit discharge education actions may include storm drain stenciling, a program to promote, publicize, and facilitate public reporting of illicit connections or discharges, and distribution of outreach materials.

(4) *Construction site storm water runoff control.* (i) You must develop, implement, and enforce a program to reduce pollutants in any storm water runoff to your small MS4 from construction activities that result in a land disturbance of greater than or equal to one acre. Reduction of storm water discharges from construction activity disturbing less than one acre must be included in your program if that construction activity is part of a larger common plan of development or sale that would disturb one acre or more. If the NPDES permitting authority waives requirements for storm water discharges associated with small construction activity in accordance with § 122.26(b)(15)(i), you are not required to develop, implement, and/or enforce a program to reduce pollutant discharges from such sites.

(ii) Your program must include the development and implementation of, at a minimum:

(A) An ordinance or other regulatory mechanism to require erosion and sediment controls, as well as sanctions to ensure compliance, to the extent allowable under State, Tribal, or local law;

(B) Requirements for construction site operators to implement appropriate erosion and sediment control best management practices;

(C) Requirements for construction site operators to control waste such as discarded building materials, concrete truck washout, chemicals, litter, and sanitary waste at the construction site that may cause adverse impacts to water quality;

(D) Procedures for site plan review which incorporate consideration of potential water quality impacts;

(E) Procedures for receipt and consideration of information submitted by the public, and

(F) Procedures for site inspection and enforcement of control measures.

(iii) Guidance: Examples of sanctions to ensure compliance include non-monetary penalties, fines, bonding requirements and/or permit denials for noncompliance. EPA recommends that procedures for site plan review include the review of individual pre-construction site plans to ensure consistency with local sediment and erosion control requirements. Procedures for site inspections and enforcement of control measures could include steps to identify priority sites for inspection and enforcement based on the nature of the construction activity, topography, and the characteristics of soils and receiving water quality. You are encouraged to provide appropriate educational and training measures for construction site operators. You may wish to require a storm water pollution prevention plan for construction sites within your jurisdiction that discharge into your system. See § 122.44(s) (NPDES permitting authorities' option to incorporate qualifying State, Tribal and local erosion and sediment control programs into NPDES permits for storm water discharges from construction sites).

Also see §122.35(b) (The NPDES permitting authority may recognize that another government entity, including the permitting authority, may be responsible for implementing one or more of the minimum measures on your behalf.)

(5) *Post-construction storm water management in new development and redevelopment.* (i) You must develop, implement, and enforce a program to address storm water runoff from new development and redevelopment projects that disturb greater than or equal to one acre, including projects less than one acre that are part of a larger common plan of development or sale, that discharge into your small MS4. Your program must ensure that controls are in place that would prevent or minimize water quality impacts.

(ii) You must:

(A) Develop and implement strategies which include a combination of structural and/or non-structural best management practices (BMPs) appropriate for your community;

(B) Use an ordinance or other regulatory mechanism to address post-construction runoff from new development and redevelopment projects to the extent allowable under State, Tribal or local law; and

(C) Ensure adequate long-term operation and maintenance of BMPs.

(iii) Guidance: If water quality impacts are considered from the beginning stages of a project, new development and potentially redevelopment provide more opportunities for water quality protection. EPA recommends that the BMPs chosen: be appropriate for the local community; minimize water quality impacts; and attempt to maintain pre-development runoff conditions. In choosing appropriate BMPs, EPA encourages you to participate in locally-based watershed planning efforts which attempt to involve a diverse group of stakeholders including interested citizens. When developing a program that is consistent with this measure's intent, EPA recommends that you adopt a planning process that identifies the municipality's program goals (e.g., minimize water quality impacts resulting from post-construction runoff from new development and redevelopment), implementation strategies (e.g., adopt a combination of structural and/or non-structural BMPs), operation and maintenance policies and procedures, and enforcement procedures. In developing your program, you should consider assessing existing ordinances, policies, programs and studies that address storm water runoff quality. In addition to assessing these existing documents and programs, you should provide opportunities to the public to participate in the development of the program. Non-structural BMPs are preventative actions that involve management and source controls such as: policies and ordinances that provide requirements and standards to direct growth to identified areas, protect sensitive areas such as wetlands and riparian areas, maintain and/or increase open space (including a dedicated funding source for open space acquisition), provide buffers along sensitive water bodies, minimize impervious surfaces, and minimize disturbance of soils and vegetation; policies or ordinances that encourage infill development in higher density urban areas, and areas with existing infrastructure; education programs for developers and the public about project designs that minimize water quality impacts; and measures such as minimization of percent impervious area after development and minimization of directly connected impervious areas. Structural BMPs include: storage practices such as wet ponds and extended-detention outlet structures; filtration practices such as grassed swales, sand filters and filter strips; and infiltration practices such as infiltration basins and infiltration trenches. EPA recommends that you ensure the appropriate implementation of the structural BMPs by considering some or all of the following: pre-construction review of BMP designs; inspections during construction to verify BMPs are built as designed; post-construction inspection and maintenance of BMPs; and penalty provisions for the noncompliance with design, construction or operation and maintenance. Storm water technologies are constantly being improved, and EPA recommends that your requirements be responsive to these changes, developments or improvements in control technologies.

(6) *Pollution prevention/good housekeeping for municipal operations.* (i) You must develop and implement an operation and maintenance program that includes a training component and has the ultimate goal of preventing or reducing pollutant runoff from municipal operations. Using training materials that are available from EPA, your State, Tribe, or other organizations, your program must include employee training to prevent and reduce storm water pollution from activities such as park and open space maintenance, fleet and building maintenance, new construction and land disturbances, and storm water system maintenance.

(ii) Guidance: EPA recommends that, at a minimum, you consider the following in developing your program: maintenance activities, maintenance schedules, and long-term inspection procedures for structural and non-structural storm water controls to reduce floatables and other pollutants discharged from your separate storm sewers; controls for reducing or eliminating the discharge of pollutants from streets, roads, highways, municipal parking lots, maintenance and storage yards, fleet or maintenance shops with outdoor storage areas, salt/sand storage locations and snow disposal areas operated by you, and waste transfer stations; procedures for properly disposing of waste removed from the separate storm sewers and areas listed above (such as dredge spoil, accumulated sediments, floatables, and other debris); and ways to ensure that new flood management projects assess the impacts on water quality and examine existing projects for incorporating additional water quality protection devices or practices. Operation and maintenance should be an integral component of all storm water management programs. This measure is intended to improve the efficiency of these programs and require new programs where necessary. Properly developed and implemented operation and maintenance programs reduce the risk of water quality problems.

(c) If an existing qualifying local program requires you to implement one or more of the minimum control measures of paragraph (b) of this section, the NPDES permitting authority may include conditions in your NPDES permit that direct you to follow that qualifying program's requirements rather than the requirements of paragraph (b) of this section. A qualifying local program is a local, State or Tribal municipal storm water management program that imposes, at a minimum, the relevant requirements of paragraph (b) of this section.

(d)(1) In your permit application (either a notice of intent for coverage under a general permit or an individual permit application), you must identify and submit to your NPDES permitting authority the following information:

(i) The best management practices (BMPs) that you or another entity will implement for each of the storm water minimum control measures at paragraphs (b)(1) through (b)(6) of this section;

(ii) The measurable goals for each of the BMPs including, as appropriate, the months and years in which you will undertake required actions, including interim milestones and the frequency of the action; and

(iii) The person or persons responsible for implementing or coordinating your storm water management program.

(2) If you obtain coverage under a general permit, you are not required to meet any measurable goal(s) identified in your notice of intent in order to demonstrate compliance with the minimum control measures in paragraphs (b)(3) through (b)(6) of this section unless, prior to submitting your NOI, EPA or your State or Tribe has provided or issued a menu of BMPs that addresses each such minimum measure. Even if no regulatory authority issues the menu of BMPs, however, you still must comply with other requirements of the general permit, including good faith implementation of BMPs designed to comply with the minimum measures.

(3) Guidance: Either EPA or your State or Tribal permitting authority will provide a menu of BMPs. You may choose BMPs from the menu or select others that satisfy the minimum control measures.

(e)(1) You must comply with any more stringent effluent limitations in your permit, including permit requirements that modify, or are in addition to, the minimum control measures based on an approved total maximum daily load (TMDL) or equivalent analysis. The permitting authority may include such more stringent limitations based on a TMDL or equivalent analysis that determines such limitations are needed to protect water quality.

(2) Guidance: EPA strongly recommends that until the evaluation of the storm water program in §122.37, no additional requirements beyond the minimum control measures be imposed on regulated small MS4s without the agreement of the operator of the affected small MS4, except where an approved TMDL or equivalent analysis provides adequate information to develop more specific measures to protect water quality.

(f) You must comply with other applicable NPDES permit requirements, standards and conditions established in the individual or general permit, developed consistent with the provisions of §§122.41 through 122.49, as appropriate.

(g) *Evaluation and assessment*—(1) *Evaluation.* You must evaluate program compliance, the appropriateness of your identified best management practices, and progress towards achieving your identified measurable goals.

NOTE TO PARAGRAPH (g)(1): The NPDES permitting authority may determine monitoring requirements for you in accordance with State/Tribal monitoring plans appropriate to your watershed. Participation in a group monitoring program is encouraged.

(2) *Recordkeeping.* You must keep records required by the NPDES permit for at least 3 years. You must submit your records to the NPDES permitting authority only when specifically asked to do so. You must make your records, including a description of your storm water management program, available to the public at reasonable times during regular business hours (see §122.7 for confidentiality provision). (You may assess a reasonable charge for copying. You may require a member of the public to provide advance notice.)

(3) *Reporting.* Unless you are relying on another entity to satisfy your NPDES permit obligations under §122.35(a), you must submit annual reports to the NPDES permitting author-

ity for your first permit term. For subsequent permit terms, you must submit reports in year two and four unless the NPDES permitting authority requires more frequent reports. Your report must include:

(i) The status of compliance with permit conditions, an assessment of the appropriateness of your identified best management practices and progress towards achieving your identified measurable goals for each of the minimum control measures;

(ii) Results of information collected and analyzed, including monitoring data, if any, during the reporting period;

(iii) A summary of the storm water activities you plan to undertake during the next reporting cycle;

(iv) A change in any identified best management practices or measurable goals for any of the minimum control measures; and

(v) Notice that you are relying on another governmental entity to satisfy some of your permit obligations (if applicable).

[64 FR 68843, Dec. 8, 1999]

§122.35 As an operator of a regulated small MS4, may I share the responsibility to implement the minimum control measures with other entities?

(a) You may rely on another entity to satisfy your NPDES permit obligations to implement a minimum control measure if:

(1) The other entity, in fact, implements the control measure;

(2) The particular control measure, or component thereof, is at least as stringent as the corresponding NPDES permit requirement; and

(3) The other entity agrees to implement the control measure on your behalf. In the reports you must submit under §122.34(g)(3), you must also specify that you rely on another entity to satisfy some of your permit obligations. If you are relying on another governmental entity regulated under section 122 to satisfy all of your permit obligations, including your obligation to file periodic reports required by §122.34(g)(3), you must note that fact in your NOI, but you are not required to file the periodic reports. You remain

responsible for compliance with your permit obligations if the other entity fails to implement the control measure (or component thereof). Therefore, EPA encourages you to enter into a legally binding agreement with that entity if you want to minimize any uncertainty about compliance with your permit.

(b) In some cases, the NPDES permitting authority may recognize, either in your individual NPDES permit or in an NPDES general permit, that another governmental entity is responsible under an NPDES permit for implementing one or more of the minimum control measures for your small MS4 or that the permitting authority itself is responsible. Where the permitting authority does so, you are not required to include such minimum control measure(s) in your storm water management program. (For example, if a State or Tribe is subject to an NPDES permit that requires it to administer a program to control construction site runoff at the State or Tribal level and that program satisfies all of the requirements of § 122.34(b)(4), you could avoid responsibility for the construction measure, but would be responsible for the remaining minimum control measures.) Your permit may be reopened and modified to include the requirement to implement a minimum control measure if the entity fails to implement it.

[64 FR 68846, Dec. 8, 1999]

§ 122.36 As an operator of a regulated small MS4, what happens if I don't comply with the application or permit requirements in §§ 122.33 through 122.35?

NPDES permits are federally enforceable. Violators may be subject to the enforcement actions and penalties described in Clean Water Act sections 309 (b), (c), and (g) and 505, or under applicable State, Tribal, or local law. Compliance with a permit issued pursuant to section 402 of the Clean Water Act is deemed compliance, for purposes of sections 309 and 505, with sections 301, 302, 306, 307, and 403, except any standard imposed under section 307 for toxic pollutants injurious to human health. If you are covered as a co-permittee under an individual permit or

under a general permit by means of a joint Notice of Intent you remain subject to the enforcement actions and penalties for the failure to comply with the terms of the permit in your jurisdiction except as set forth in § 122.35(b).

[64 FR 68847, Dec. 8, 1999]

§ 122.37 Will the small MS4 storm water program regulations at §§ 122.32 through 122.36 and § 123.35 of this chapter change in the future?

EPA will evaluate the small MS4 regulations at §§ 122.32 through 122.36 and § 123.35 of this chapter after December 10, 2012 and make any necessary revisions. (EPA intends to conduct an enhanced research effort and compile a comprehensive evaluation of the NPDES MS4 storm water program. EPA will re-evaluate the regulations based on data from the NPDES MS4 storm water program, from research on receiving water impacts from storm water, and the effectiveness of best management practices (BMPs), as well as other relevant information sources.)

[64 FR 68847, Dec. 8, 1999]

Subpart C—Permit Conditions

§ 122.41 Conditions applicable to all permits (applicable to State programs, see § 123.25).

The following conditions apply to all NPDES permits. Additional conditions applicable to NPDES permits are in § 122.42. All conditions applicable to NPDES permits shall be incorporated into the permits either expressly or by reference. If incorporated by reference, a specific citation to these regulations (or the corresponding approved State regulations) must be given in the permit.

(a) *Duty to comply.* The permittee must comply with all conditions of this permit. Any permit noncompliance constitutes a violation of the Clean Water Act and is grounds for enforcement action; for permit termination, revocation and reissuance, or modification; or denial of a permit renewal application.

(1) The permittee shall comply with effluent standards or prohibitions established under section 307(a) of the Clean Water Act for toxic pollutants

and with standards for sewage sludge use or disposal established under section 405(d) of the CWA within the time provided in the regulations that establish these standards or prohibitions or standards for sewage sludge use or disposal, even if the permit has not yet been modified to incorporate the requirement.

(2) The Clean Water Act provides that any person who violates section 301, 302, 306, 307, 308, 318 or 405 of the Act, or any permit condition or limitation implementing any such sections in a permit issued under section 402, or any requirement imposed in a pretreatment program approved under sections 402(a)(3) or 402(b)(8) of the Act, is subject to a civil penalty not to exceed $25,000 per day for each violation. The Clean Water Act provides that any person who *negligently* violates sections 301, 302, 306, 307, 308, 318, or 405 of the Act, or any condition or limitation implementing any of such sections in a permit issued under section 402 of the Act, or any requirement imposed in a pretreatment program approved under section 402(a)(3) or 402(b)(8) of the Act, is subject to criminal penalties of $2,500 to $25,000 per day of violation, or imprisonment of not more than 1 year, or both. In the case of a second or subsequent conviction for a negligent violation, a person shall be subject to criminal penalties of not more than $50,000 per day of violation, or by imprisonment of not more than 2 years, or both. Any person who *knowingly* violates such sections, or such conditions or limitations is subject to criminal penalties of $5,000 to $50,000 per day of violation, or imprisonment for not more than 3 years, or both. In the case of a second or subsequent conviction for a knowing violation, a person shall be subject to criminal penalties of not more than $100,000 per day of violation, or imprisonment of not more than 6 years, or both. Any person who knowingly violates section 301, 302, 303, 306, 307, 308, 318 or 405 of the Act, or any permit condition or limitation implementing any of such sections in a permit issued under section 402 of the Act, and who knows at that time that he thereby places another person in imminent danger of death or serious bodily injury, shall, upon conviction, be sub-

ject to a fine of not more than $250,000 or imprisonment of not more than 15 years, or both. In the case of a second or subsequent conviction for a knowing endangerment violation, a person shall be subject to a fine of not more than $500,000 or by imprisonment of not more than 30 years, or both. An organization, as defined in section 309(c)(3)(B)(iii) of the CWA, shall, upon conviction of violating the imminent danger provision, be subject to a fine of not more than $1,000,000 and can be fined up to $2,000,000 for second or subsequent convictions.

(3) Any person may be assessed an administrative penalty by the Administrator for violating section 301, 302, 306, 307, 308, 318 or 405 of this Act, or any permit condition or limitation implementing any of such sections in a permit issued under section 402 of this Act. Administrative penalties for Class I violations are not to exceed $10,000 per violation, with the maximum amount of any Class I penalty assessed not to exceed $25,000. Penalties for Class II violations are not to exceed $10,000 per day for each day during which the violation continues, with the maximum amount of any Class II penalty not to exceed $125,000.

(b) *Duty to reapply.* If the permittee wishes to continue an activity regulated by this permit after the expiration date of this permit, the permittee must apply for and obtain a new permit.

(c) *Need to halt or reduce activity not a defense.* It shall not be a defense for a permittee in an enforcement action that it would have been necessary to halt or reduce the permitted activity in order to maintain compliance with the conditions of this permit.

(d) *Duty to mitigate.* The permittee shall take all reasonable steps to minimize or prevent any discharge or sludge use or disposal in violation of this permit which has a reasonable likelihood of adversely affecting human health or the environment.

(e) *Proper operation and maintenance.* The permittee shall at all times properly operate and maintain all facilities and systems of treatment and control (and related appurtenances) which are installed or used by the permittee to achieve compliance with the conditions of this permit. Proper operation and maintenance also includes adequate laboratory controls and appropriate quality assurance procedures. This provision requires the operation of back-up or auxiliary facilities or similar systems which are installed by a permittee only when the operation is necessary to achieve compliance with the conditions of the permit.

(f) *Permit actions.* This permit may be modified, revoked and reissued, or terminated for cause. The filing of a request by the permittee for a permit modification, revocation and reissuance, or termination, or a notification of planned changes or anticipated noncompliance does not stay any permit condition.

(g) *Property rights.* This permit does not convey any property rights of any sort, or any exclusive privilege.

(h) *Duty to provide information.* The permittee shall furnish to the Director, within a reasonable time, any information which the Director may request to determine whether cause exists for modifying, revoking and reissuing, or terminating this permit or to determine compliance with this permit. The permittee shall also furnish to the Director upon request, copies of records required to be kept by this permit.

(i) *Inspection and entry.* The permittee shall allow the Director, or an authorized representative (including an authorized contractor acting as a representative of the Administrator), upon presentation of credentials and other documents as may be required by law, to:

(1) Enter upon the permittee's premises where a regulated facility or activity is located or conducted, or where records must be kept under the conditions of this permit;

(2) Have access to and copy, at reasonable times, any records that must be kept under the conditions of this permit;

(3) Inspect at reasonable times any facilities, equipment (including monitoring and control equipment), practices, or operations regulated or required under this permit; and

(4) Sample or monitor at reasonable times, for the purposes of assuring permit compliance or as otherwise authorized by the Clean Water Act, any substances or parameters at any location.

(j) *Monitoring and records.* (1) Samples and measurements taken for the purpose of monitoring shall be representative of the monitored activity.

(2) Except for records of monitoring information required by this permit related to the permittee's sewage sludge use and disposal activities, which shall be retained for a period of at least five years (or longer as required by 40 CFR part 503), the permittee shall retain records of all monitoring information, including all calibration and maintenance records and all original strip chart recordings for continuous monitoring instrumentation, copies of all reports required by this permit, and records of all data used to complete the application for this permit, for a period of at least 3 years from the date of the sample, measurement, report or application. This period may be extended by request of the Director at any time.

(3) Records of monitoring information shall include:

(i) The date, exact place, and time of sampling or measurements;

(ii) The individual(s) who performed the sampling or measurements;

(iii) The date(s) analyses were performed;

(iv) The individual(s) who performed the analyses;

(v) The analytical techniques or methods used; and

(vi) The results of such analyses.

(4) Monitoring must be conducted according to test procedures approved under 40 CFR Part 136 unless another method is required under 40 CFR subchapters N or O.

(5) The Clean Water Act provides that any person who falsifies, tampers with, or knowingly renders inaccurate any monitoring device or method required to be maintained under this permit shall, upon conviction, be punished by a fine of not more than $10,000, or by imprisonment for not more than 2 years, or both. If a conviction of a person is for a violation committed after a first conviction of such person under this paragraph, punishment is a fine of not more than $20,000 per day of violation, or by imprisonment of not more than 4 years, or both.

(k) *Signatory requirement.* (1) All applications, reports, or information submitted to the Director shall be signed and certified. (See §122.22)

(2) The CWA provides that any person who knowingly makes any false statement, representation, or certification in any record or other document submitted or required to be maintained under this permit, including monitoring reports or reports of compliance or non-compliance shall, upon conviction, be punished by a fine of not more than $10,000 per violation, or by imprisonment for not more than 6 months per violation, or by both.

(1) *Reporting requirements*—(1) *Planned changes.* The permittee shall give notice to the Director as soon as possible of any planned physical alterations or additions to the permitted facility. Notice is required only when:

(i) The alteration or addition to a permitted facility may meet one of the criteria for determining whether a facility is a new source in §122.29(b); or

(ii) The alteration or addition could significantly change the nature or increase the quantity of pollutants discharged. This notification applies to pollutants which are subject neither to effluent limitations in the permit, nor to notification requirements under §122.42(a)(1).

(iii) The alteration or addition results in a significant change in the permittee's sludge use or disposal practices, and such alteration, addition, or change may justify the application of permit conditions that are different from or absent in the existing permit, including notification of additional use or disposal sites not reported during the permit application process or not reported pursuant to an approved land application plan;

(2) *Anticipated noncompliance.* The permittee shall give advance notice to the Director of any planned changes in the permitted facility or activity which may result in noncompliance with permit requirements.

(3) *Transfers.* This permit is not transferable to any person except after notice to the Director. The Director may require modification or revocation and reissuance of the permit to change the name of the permittee and incorporate such other requirements as may

be necessary under the Clean Water Act. (See §122.61; in some cases, modification or revocation and reissuance is mandatory.)

(4) *Monitoring reports.* Monitoring results shall be reported at the intervals specified elsewhere in this permit.

(i) Monitoring results must be reported on a Discharge Monitoring Report (DMR) or forms provided or specified by the Director for reporting results of monitoring of sludge use or disposal practices.

(ii) If the permittee monitors any pollutant more frequently than required by the permit using test procedures approved under 40 CFR Part 136, or another method required for an industry-specific waste stream under 40 CFR subchapters N or O, the results of such monitoring shall be included in the calculation and reporting of the data submitted in the DMR or sludge reporting form specified by the Director.

(iii) Calculations for all limitations which require averaging of measurements shall utilize an arithmetic mean unless otherwise specified by the Director in the permit.

(5) *Compliance schedules.* Reports of compliance or noncompliance with, or any progress reports on, interim and final requirements contained in any compliance schedule of this permit shall be submitted no later than 14 days following each schedule date.

(6) *Twenty-four hour reporting.* (i) The permittee shall report any noncompliance which may endanger health or the environment. Any information shall be provided orally within 24 hours from the time the permittee becomes aware of the circumstances. A written submission shall also be provided within 5 days of the time the permittee becomes aware of the circumstances. The written submission shall contain a description of the noncompliance and its cause; the period of noncompliance, including exact dates and times, and if the noncompliance has not been corrected, the anticipated time it is expected to continue; and steps taken or planned to reduce, eliminate, and prevent reoccurrence of the noncompliance.

(ii) The following shall be included as information which must be reported within 24 hours under this paragraph.

(A) Any unanticipated bypass which exceeds any effluent limitation in the permit. (See § 122.41(g).

(B) Any upset which exceeds any effluent limitation in the permit.

(C) Violation of a maximum daily discharge limitation for any of the pollutants listed by the Director in the permit to be reported within 24 hours. (See § 122.44(g).)

(iii) The Director may waive the written report on a case-by-case basis for reports under paragraph (l)(6)(ii) of this section if the oral report has been received within 24 hours.

(7) *Other noncompliance.* The permittee shall report all instances of noncompliance not reported under paragraphs (l) (4), (5), and (6) of this section, at the time monitoring reports are submitted. The reports shall contain the information listed in paragraph (l)(6) of this section.

(8) *Other information.* Where the permittee becomes aware that it failed to submit any relevant facts in a permit application, or submitted incorrect information in a permit application or in any report to the Director, it shall promptly submit such facts or information.

(m) *Bypass*—(1) *Definitions.* (i) *Bypass* means the intentional diversion of waste streams from any portion of a treatment facility.

(ii) *Severe property damage* means substantial physical damage to property, damage to the treatment facilities which causes them to become inoperable, or substantial and permanent loss of natural resources which can reasonably be expected to occur in the absence of a bypass. Severe property damage does not mean economic loss caused by delays in production.

(2) *Bypass not exceeding limitations.* The permittee may allow any bypass to occur which does not cause effluent limitations to be exceeded, but only if it also is for essential maintenance to assure efficient operation. These bypasses are not subject to the provisions of paragraphs (m)(3) and (m)(4) of this section.

(3) *Notice*—(i) *Anticipated bypass.* If the permittee knows in advance of the need for a bypass, it shall submit prior notice, if possible at least ten days before the date of the bypass.

(ii) *Unanticipated bypass.* The permittee shall submit notice of an unanticipated bypass as required in paragraph (l)(6) of this section (24-hour notice).

(4) *Prohibition of bypass.* (i) Bypass is prohibited, and the Director may take enforcement action against a permittee for bypass, unless:

(A) Bypass was unavoidable to prevent loss of life, personal injury, or severe property damage;

(B) There were no feasible alternatives to the bypass, such as the use of auxiliary treatment facilities, retention of untreated wastes, or maintenance during normal periods of equipment downtime. This condition is not satisfied if adequate back-up equipment should have been installed in the exercise of reasonable engineering judgment to prevent a bypass which occurred during normal periods of equipment downtime or preventive maintenance; and

(C) The permittee submitted notices as required under paragraph (m)(3) of this section.

(ii) The Director may approve an anticipated bypass, after considering its adverse effects, if the Director determines that it will meet the three conditions listed above in paragraph (m)(4)(i) of this section.

(n) *Upset*—(1) *Definition. Upset* means an exceptional incident in which there is unintentional and temporary noncompliance with technology based permit effluent limitations because of factors beyond the reasonable control of the permittee. An upset does not include noncompliance to the extent caused by operational error, improperly designed treatment facilities, inadequate treatment facilities, lack of preventive maintenance, or careless or improper operation.

(2) *Effect of an upset.* An upset constitutes an affirmative defense to an action brought for noncompliance with such technology based permit effluent limitations if the requirements of paragraph (n)(3) of this section are met. No determination made during administrative review of claims that noncompliance was caused by upset, and

before an action for noncompliance, is final administrative action subject to judicial review.

(3) *Conditions necessary for a demonstration of upset.* A permittee who wishes to establish the affirmative defense of upset shall demonstrate, through properly signed, contemporaneous operating logs, or other relevant evidence that:

(i) An upset occurred and that the permittee can identify the cause(s) of the upset;

(ii) The permitted facility was at the time being properly operated; and

(iii) The permittee submitted notice of the upset as required in paragraph (1)(6)(ii)(B) of this section (24 hour notice).

(iv) The permittee complied with any remedial measures required under paragraph (d) of this section.

(4) *Burden of proof.* In any enforcement proceeding the permittee seeking to establish the occurrence of an upset has the burden of proof.

(Clean Water Act (33 U.S.C. 1251 *et seq.*), Safe Drinking Water Act (42 U.S.C. 300f *et seq.*), Clean Air Act (42 U.S.C. 7401 *et seq.*), Resource Conservation and Recovery Act (42 U.S.C. 6901 *et seq.*))

[48 FR 14153, Apr. 1, 1983, as amended at 48 FR 39620, Sept. 1, 1983; 49 FR 38049, Sept. 26, 1984; 50 FR 4514, Jan. 31, 1985; 50 FR 6940, Feb. 19, 1985; 54 FR 255, Jan. 4, 1989; 54 FR 18783, May 2, 1989; 65 FR 30908, May 15, 2000; 72 FR 11211, Mar. 12, 2007]

§122.42 Additional conditions applicable to specified categories of NPDES permits (applicable to State NPDES programs, see §123.25).

The following conditions, in addition to those set forth in §122.41, apply to all NPDES permits within the categories specified below:

(a) *Existing manufacturing, commercial, mining, and silvicultural dischargers.* In addition to the reporting requirements under §122.41(1), all existing manufacturing, commercial, mining, and silvicultural dischargers must notify the Director as soon as they know or have reason to believe:

(1) That any activity has occurred or will occur which would result in the discharge, on a routine or frequent basis, of any toxic pollutant which is not limited in the permit, if that dis-

charge will exceed the highest of the following "notification levels":

(i) One hundred micrograms per liter (100 µg/l);

(ii) Two hundred micrograms per liter (200 µg/l) for acrolein and acrylonitrile; five hundred micrograms per liter (500 µg/l) for 2,4-dinitrophenol and for 2-methyl-4,6-dinitrophenol; and one milligram per liter (1 mg/l) for antimony;

(iii) Five (5) times the maximum concentration value reported for that pollutant in the permit application in accordance with §122.21(g)(7); or

(iv) The level established by the Director in accordance with §122.44(f).

(2) That any activity has occurred or will occur which would result in any discharge, on a non-routine or infrequent basis, of a toxic pollutant which is not limited in the permit, if that discharge will exceed the highest of the following "notification levels":

(i) Five hundred micrograms per liter (500 µg/l);

(ii) One milligram per liter (1 mg/l) for antimony;

(iii) Ten (10) times the maximum concentration value reported for that pollutant in the permit application in accordance with §122.21(g)(7).

(iv) The level established by the Director in accordance with §122.44(f).

(b) *Publicly owned treatment works.* All POTWs must provide adequate notice to the Director of the following:

(1) Any new introduction of pollutants into the POTW from an indirect discharger which would be subject to section 301 or 306 of CWA if it were directly discharging those pollutants; and

(2) Any substantial change in the volume or character of pollutants being introduced into that POTW by a source introducing pollutants into the POTW at the time of issuance of the permit.

(3) For purposes of this paragraph, adequate notice shall include information on (i) the quality and quantity of effluent introduced into the POTW, and (ii) any anticipated impact of the change on the quantity or quality of effluent to be discharged from the POTW.

(c) *Municipal separate storm sewer systems.* The operator of a large or medium municipal separate storm sewer system or a municipal separate storm sewer that has been designated by the Director under § 122.26(a)(1)(v) of this part must submit an annual report by the anniversary of the date of the issuance of the permit for such system. The report shall include:

(1) The status of implementing the components of the storm water management program that are established as permit conditions;

(2) Proposed changes to the storm water management programs that are established as permit condition. Such proposed changes shall be consistent with § 122.26(d)(2)(iii) of this part; and

(3) Revisions, if necessary, to the assessment of controls and the fiscal analysis reported in the permit application under § 122.26(d)(2)(iv) and (d)(2)(v) of this part;

(4) A summary of data, including monitoring data, that is accumulated throughout the reporting year;

(5) Annual expenditures and budget for year following each annual report;

(6) A summary describing the number and nature of enforcement actions, inspections, and public education programs;

(7) Identification of water quality improvements or degradation;

(d) *Storm water discharges.* The initial permits for discharges composed entirely of storm water issued pursuant to § 122.26(e)(7) of this part shall require compliance with the conditions of the permit as expeditiously as practicable, but in no event later than three years after the date of issuance of the permit.

(e) *Concentrated animal feeding operations (CAFOs).* Any permit issued to a CAFO must include the requirements in paragraphs (e)(1) through (e)(6) of this section.

(1) *Requirement to implement a nutrient management plan.* Any permit issued to a CAFO must include a requirement to implement a nutrient management plan that, at a minimum, contains best management practices necessary to meet the requirements of this paragraph and applicable effluent limitations and standards, including those specified in 40 CFR part 412. The nutrient management plan must, to the extent applicable:

(i) Ensure adequate storage of manure, litter, and process wastewater, including procedures to ensure proper operation and maintenance of the storage facilities;

(ii) Ensure proper management of mortalities (*i.e.*, dead animals) to ensure that they are not disposed of in a liquid manure, storm water, or process wastewater storage or treatment system that is not specifically designed to treat animal mortalities;

(iii) Ensure that clean water is diverted, as appropriate, from the production area;

(iv) Prevent direct contact of confined animals with waters of the United States;

(v) Ensure that chemicals and other contaminants handled on-site are not disposed of in any manure, litter, process wastewater, or storm water storage or treatment system unless specifically designed to treat such chemicals and other contaminants;

(vi) Identify appropriate site specific conservation practices to be implemented, including as appropriate buffers or equivalent practices, to control runoff of pollutants to waters of the United States;

(vii) Identify protocols for appropriate testing of manure, litter, process wastewater, and soil;

(viii) Establish protocols to land apply manure, litter or process wastewater in accordance with site specific nutrient management practices that ensure appropriate agricultural utilization of the nutrients in the manure, litter or process wastewater; and

(ix) Identify specific records that will be maintained to document the implementation and management of the minimum elements described in paragraphs (e)(1)(i) through (e)(1)(viii) of this section.

(2) *Recordkeeping requirements.* (i) The permittee must create, maintain for five years, and make available to the Director, upon request, the following records:

(A) All applicable records identified pursuant paragraph (e)(1)(ix) of this section;

(B) In addition, all CAFOs subject to 40 CFR part 412 must comply with record keeping requirements as specified in §412.37(b) and (c) and §412.47(b) and (c).

(ii) A copy of the CAFO's site-specific nutrient management plan must be maintained on site and made available to the Director upon request.

(3) *Requirements relating to transfer of manure or process wastewater to other persons.* Prior to transferring manure, litter or process wastewater to other persons, Large CAFOs must provide the recipient of the manure, litter or process wastewater with the most current nutrient analysis. The analysis provided must be consistent with the requirements of 40 CFR part 412. Large CAFOs must retain for five years records of the date, recipient name and address, and approximate amount of manure, litter or process wastewater transferred to another person.

(4) *Annual reporting requirements for CAFOs.* The permittee must submit an annual report to the Director. The annual report must include:

(i) The number and type of animals, whether in open confinement or housed under roof (beef cattle, broilers, layers, swine weighing 55 pounds or more, swine weighing less than 55 pounds, mature dairy cows, dairy heifers, veal calves, sheep and lambs, horses, ducks, turkeys, other);

(ii) Estimated amount of total manure, litter and process wastewater generated by the CAFO in the previous 12 months (tons/gallons);

(iii) Estimated amount of total manure, litter and process wastewater transferred to other person by the CAFO in the previous 12 months (tons/gallons);

(iv) Total number of acres for land application covered by the nutrient management plan developed in accordance with paragraph (e)(1) of this section;

(v) Total number of acres under control of the CAFO that were used for land application of manure, litter and process wastewater in the previous 12 months;

(vi) Summary of all manure, litter and process wastewater discharges from the production area that have occurred in the previous 12 months, including date, time, and approximate volume; and

(vii) A statement indicating whether the current version of the CAFO's nutrient management plan was developed or approved by a certified nutrient management planner; and

(viii) The actual crop(s) planted and actual yield(s) for each field, the actual nitrogen and phosphorus content of the manure, litter, and process wastewater, the results of calculations conducted in accordance with paragraphs (e)(5)(i)(B) and (e)(5)(ii)(D) of this section, and the amount of manure, litter, and process wastewater applied to each field during the previous 12 months; and, for any CAFO that implements a nutrient management plan that addresses rates of application in accordance with paragraph (e)(5)(ii) of this section, the results of any soil testing for nitrogen and phosphorus taken during the preceding 12 months, the data used in calculations conducted in accordance with paragraph (e)(5)(ii)(D) of this section, and the amount of any supplemental fertilizer applied during the previous 12 months.

(5) *Terms of the nutrient management plan.* Any permit issued to a CAFO must require compliance with the terms of the CAFO's site-specific nutrient management plan. The terms of the nutrient management plan are the information, protocols, best management practices, and other conditions in the nutrient management plan determined by the Director to be necessary to meet the requirements of paragraph (e)(1) of this section. The terms of the nutrient management plan, with respect to protocols for land application of manure, litter, or process wastewater required by paragraph (e)(1)(viii) of this section and, as applicable, 40 CFR 412.4(c), must include the fields available for land application; field-specific rates of application properly developed, as specified in paragraphs (e)(5)(i) through (ii) of this section, to ensure appropriate agricultural utilization of the nutrients in the manure, litter, or process wastewater; and any timing limitations identified in the nutrient management plan concerning land application on the fields available for land application. The terms must address rates of application using one of the following two approaches, unless

245

the Director specifies that only one of these approaches may be used:

(i) *Linear approach.* An approach that expresses rates of application as pounds of nitrogen and phosphorus, according to the following specifications:

(A) The terms include maximum application rates from manure, litter, and process wastewater for each year of permit coverage, for each crop identified in the nutrient management plan, in chemical forms determined to be acceptable to the Director, in pounds per acre, per year, for each field to be used for land application, and certain factors necessary to determine such rates. At a minimum, the factors that are terms must include: The outcome of the field-specific assessment of the potential for nitrogen and phosphorus transport from each field; the crops to be planted in each field or any other uses of a field such as pasture or fallow fields; the realistic yield goal for each crop or use identified for each field; the nitrogen and phosphorus recommendations from sources specified by the Director for each crop or use identified for each field; credits for all nitrogen in the field that will be plant available; consideration of multi-year phosphorus application; and accounting for all other additions of plant available nitrogen and phosphorus to the field. In addition, the terms include the form and source of manure, litter, and process wastewater to be land-applied; the timing and method of land application; and the methodology by which the nutrient management plan accounts for the amount of nitrogen and phosphorus in the manure, litter, and process wastewater to be applied.

(B) Large CAFOs that use this approach must calculate the maximum amount of manure, litter, and process wastewater to be land applied at least once each year using the results of the most recent representative manure, litter, and process wastewater tests for nitrogen and phosphorus taken within 12 months of the date of land application; or

(ii) *Narrative rate approach.* An approach that expresses rates of application as a narrative rate of application that results in the amount, in tons or gallons, of manure, litter, and process wastewater to be land applied, according to the following specifications:

(A) The terms include maximum amounts of nitrogen and phosphorus derived from all sources of nutrients, for each crop identified in the nutrient management plan, in chemical forms determined to be acceptable to the Director, in pounds per acre, for each field, and certain factors necessary to determine such amounts. At a minimum, the factors that are terms must include: the outcome of the field-specific assessment of the potential for nitrogen and phosphorus transport from each field; the crops to be planted in each field or any other uses such as pasture or fallow fields (including alternative crops identified in accordance with paragraph (e)(5)(ii)(B) of this section); the realistic yield goal for each crop or use identified for each field; and the nitrogen and phosphorus recommendations from sources specified by the Director for each crop or use identified for each field. In addition, the terms include the methodology by which the nutrient management plan accounts for the following factors when calculating the amounts of manure, litter, and process wastewater to be land applied: Results of soil tests conducted in accordance with protocols identified in the nutrient management plan, as required by paragraph (e)(1)(vii) of this section; credits for all nitrogen in the field that will be plant available; the amount of nitrogen and phosphorus in the manure, litter, and process wastewater to be applied; consideration of multi-year phosphorus application; accounting for all other additions of plant available nitrogen and phosphorus to the field; the form and source of manure, litter, and process wastewater; the timing and method of land application; and volatilization of nitrogen and mineralization of organic nitrogen.

(B) The terms of the nutrient management plan include alternative crops identified in the CAFO's nutrient management plan that are not in the planned crop rotation. Where a CAFO includes alternative crops in its nutrient management plan, the crops must be listed by field, in addition to the crops identified in the planned crop rotation for that field, and the nutrient management plan must include realistic crop yield goals and the nitrogen and phosphorus recommendations from sources specified by the Director for each crop. Maximum amounts of nitrogen and phosphorus from all sources of nutrients and the amounts of manure, litter, and process wastewater to be applied must be determined in accordance with the methodology described in paragraph (e)(5)(ii)(A) of this section.

(C) For CAFOs using this approach, the following projections must be included in the nutrient management plan submitted to the Director, but are not terms of the nutrient management plan: The CAFO's planned crop rotations for each field for the period of permit coverage; the projected amount of manure, litter, or process wastewater to be applied; projected credits for all nitrogen in the field that will be plant available; consideration of multi-year phosphorus application; accounting for all other additions of plant available nitrogen and phosphorus to the field; and the predicted form, source, and method of application of manure, litter, and process wastewater for each crop. Timing of application for each field, insofar as it concerns the calculation of rates of application, is not a term of the nutrient management plan.

(D) CAFOs that use this approach must calculate maximum amounts of manure, litter, and process wastewater to be land applied at least once each year using the methodology required in paragraph (e)(5)(ii)(A) of this section before land applying manure, litter, and process wastewater and must rely on the following data:

(1) A field-specific determination of soil levels of nitrogen and phosphorus, including, for nitrogen, a concurrent determination of nitrogen that will be plant available consistent with the methodology required by paragraph (e)(5)(ii)(A) of this section, and for phosphorus, the results of the most recent soil test conducted in accordance with soil testing requirements approved by the Director; and

(2) The results of most recent representative manure, litter, and process wastewater tests for nitrogen and phosphorus taken within 12 months of the date of land application, in order to determine the amount of nitrogen and phosphorus in the manure, litter, and process wastewater to be applied.

(6) *Changes to a nutrient management plan.* Any permit issued to a CAFO must require the following procedures to apply when a CAFO owner or operator makes changes to the CAFO's nutrient management plan previously submitted to the Director:

(i) The CAFO owner or operator must provide the Director with the most current version of the CAFO's nutrient management plan and identify changes from the previous version, except that the results of calculations made in accordance with the requirements of paragraphs (e)(5)(i)(B) and (e)(5)(ii)(D) of this section are not subject to the requirements of paragraph (e)(6) of this section.

(ii) The Director must review the revised nutrient management plan to ensure that it meets the requirements of this section and applicable effluent limitations and standards, including those specified in 40 CFR part 412, and must determine whether the changes to the nutrient management plan necessitate revision to the terms of the nutrient management plan incorporated into the permit issued to the CAFO. If revision to the terms of the nutrient management plan is not necessary, the Director must notify the CAFO owner or operator and upon such notification the CAFO may implement the revised nutrient management plan. If revision to the terms of the nutrient management plan is necessary, the Director must determine whether such changes are substantial changes as described in paragraph (e)(6)(iii) of this section.

(A) If the Director determines that the changes to the terms of the nutrient management plan are not substantial, the Director must make the revised nutrient management plan publicly available and include it in the permit record, revise the terms of the nutrient management plan incorporated into the permit, and notify the owner or operator and inform the public of any changes to the terms of the nutrient management plan that are incorporated into the permit.

(B) If the Director determines that the changes to the terms of the nutrient management plan are substantial, the Director must notify the public and make the proposed changes and the information submitted by the CAFO owner or operator available for public review and comment. The process for public comments, hearing requests, and the hearing process if a hearing is held must follow the procedures applicable to draft permits set forth in 40 CFR 124.11 through 124.13. The Director may establish, either by regulation or in the CAFO's permit, an appropriate period of time for the public to comment and request a hearing on the proposed changes that differs from the time period specified in 40 CFR 124.10. The Director must respond to all significant comments received during the comment period as provided in 40 CFR 124.17, and require the CAFO owner or operator to further revise the nutrient management plan if necessary, in order to approve the revision to the terms of the nutrient management plan incorporated into the CAFO's permit. Once the Director incorporates the revised terms of the nutrient management plan into the permit, the Director must notify the owner or operator and inform the public of the final decision concerning revisions to the terms and conditions of the permit.

(iii) Substantial changes to the terms of a nutrient management plan incorporated as terms and conditions of a permit include, but are not limited to:

(A) Addition of new land application areas not previously included in the CAFO's nutrient management plan. Except that if the land application area that is being added to the nutrient management plan is covered by terms of a nutrient management plan incorporated into an existing NPDES permit in accordance with the requirements of paragraph (e)(5) of this section, and the CAFO owner or operator applies manure, litter, or process wastewater on the newly added land application area in accordance with the existing field-specific permit terms applicable to the newly added land application area, such addition of new land would be a change to the new CAFO owner or operator's nutrient management plan but

not a substantial change for purposes of this section;

(B) Any changes to the field-specific maximum annual rates for land application, as set forth in paragraphs (e)(5)(i) of this section, and to the maximum amounts of nitrogen and phosphorus derived from all sources for each crop, as set forth in paragraph (e)(5)(ii) of this section;

(C) Addition of any crop or other uses not included in the terms of the CAFO's nutrient management plan and corresponding field-specific rates of application expressed in accordance with paragraph (e)(5) of this section; and

(D) Changes to site-specific components of the CAFO's nutrient management plan, where such changes are likely to increase the risk of nitrogen and phosphorus transport to waters of the U.S.

(iv) *For EPA-issued permits only.* Upon incorporation of the revised terms of the nutrient management plan into the permit, 40 CFR 124.19 specifies procedures for appeal of the permit decision. In addition to the procedures specified at 40 CFR 124.19, a person must have submitted comments or participated in the public hearing in order to appeal the permit decision.

[48 FR 14153, Apr. 1, 1983, as amended at 49 FR 38049, Sept. 26, 1984; 50 FR 4514, Jan. 31, 1985; 55 FR 48073, Nov. 16, 1990; 57 FR 60448, Dec. 18, 1992; 68 FR 7268, Feb. 12, 2003; 71 FR 6984, Feb. 10, 2006; 72 FR 40250, July 24, 2007; 73 FR 70483, Nov. 20, 2008]

§ 122.43 Establishing permit conditions (applicable to State programs, see § 123.25).

(a) In addition to conditions required in all permits (§§ 122.41 and 122.42), the Director shall establish conditions, as required on a case-by-case basis, to provide for and assure compliance with all applicable requirements of CWA and regulations. These shall include conditions under §§ 122.46 (duration of permits), 122.47(a) (schedules of compliance), 122.48 (monitoring), and for EPA permits only 122.47(b) (alternates schedule of compliance) and 122.49 (considerations under Federal law).

(b)(1) For a State issued permit, an applicable requirement is a State statutory or regulatory requirement which takes effect prior to final administrative disposition of a permit. For a permit issued by EPA, an applicable requirement is a statutory or regulatory requirement (including any interim final regulation) which takes effect prior to the issuance of the permit. Section 124.14 (reopening of comment period) provides a means for reopening EPA permit proceedings at the discretion of the Director where new requirements become effective during the permitting process and are of sufficient magnitude to make additional proceedings desirable. For State and EPA administered programs, an applicable requirement is also any requirement which takes effect prior to the modification or revocation and reissuance of a permit, to the extent allowed in §122.62.

(2) New or reissued permits, and to the extent allowed under §122.62 modified or revoked and reissued permits, shall incorporate each of the applicable requirements referenced in §§122.44 and 122.45.

(c) *Incorporation.* All permit conditions shall be incorporated either expressly or by reference. If incorporated by reference, a specific citation to the applicable regulations or requirements must be given in the permit.

[48 FR 14153, Apr. 1, 1983, as amended at 65 FR 30908, May 15, 2000]

§122.44 Establishing limitations, standards, and other permit conditions (applicable to State NPDES programs, see §123.25).

In addition to the conditions established under §122.43(a), each NPDES permit shall include conditions meeting the following requirements when applicable.

(a)(1) *Technology-based effluent limitations and standards* based on: effluent limitations and standards promulgated under section 301 of the CWA, or new source performance standards promulgated under section 306 of CWA, on case-by-case effluent limitations determined under section 402(a)(1) of CWA, or a combination of the three, in accordance with §125.3 of this chapter. For new sources or new dischargers, these technology based limitations and standards are subject to the provisions of §122.29(d) (protection period).

(2) *Monitoring waivers for certain guideline-listed pollutants.* (i) The Director may authorize a discharger subject to technology-based effluent limitations guidelines and standards in an NPDES permit to forego sampling of a pollutant found at 40 CFR Subchapter N of this chapter if the discharger has demonstrated through sampling and other technical factors that the pollutant is not present in the discharge or is present only at background levels from intake water and without any increase in the pollutant due to activities of the discharger.

(ii) This waiver is good only for the term of the permit and is not available during the term of the first permit issued to a discharger.

(iii) Any request for this waiver must be submitted when applying for a reissued permit or modification of a reissued permit. The request must demonstrate through sampling or other technical information, including information generated during an earlier permit term that the pollutant is not present in the discharge or is present only at background levels from intake water and without any increase in the pollutant due to activities of the discharger.

(iv) Any grant of the monitoring waiver must be included in the permit as an express permit condition and the reasons supporting the grant must be documented in the permit's fact sheet or statement of basis.

(v) This provision does not supersede certification processes and requirements already established in existing effluent limitations guidelines and standards.

(b)(1) *Other effluent limitations and standards* under sections 301, 302, 303, 307, 318 and 405 of CWA. If any applicable toxic effluent standard or prohibition (including any schedule of compliance specified in such effluent standard or prohibition) is promulgated under section 307(a) of CWA for a toxic pollutant and that standard or prohibition is more stringent than any limitation on the pollutant in the permit, the Director shall institute proceedings under these regulations to modify or revoke and reissue the permit to conform to the toxic effluent standard or prohibition. See also §122.41(a).

(2) *Standards for sewage sludge use or disposal* under section 405(d) of the CWA unless those standards have been included in a permit issued under the appropriate provisions of subtitle C of the Solid Waste Disposal Act, Part C of Safe Drinking Water Act, the Marine Protection, Research, and Sanctuaries Act of 1972, or the Clean Air Act, or under State permit programs approved by the Administrator. When there are no applicable standards for sewage sludge use or disposal, the permit may include requirements developed on a case-by-case basis to protect public health and the environment from any adverse effects which may occur from toxic pollutants in sewage sludge. If any applicable standard for sewage sludge use or disposal is promulgated under section 405(d) of the CWA and that standard is more stringent than any limitation on the pollutant or practice in the permit, the Director may initiate proceedings under these regulations to modify or revoke and reissue the permit to conform to the standard for sewage sludge use or disposal.

(3) Requirements applicable to cooling water intake structures under section 316(b) of the CWA, in accordance with part 125, subparts I, J, and N of this chapter.

(c) *Reopener clause:* For any permit issued to a treatment works treating domestic sewage (including "sludge-only facilities"), the Director shall include a reopener clause to incorporate any applicable standard for sewage sludge use or disposal promulgated under section 405(d) of the CWA. The Director may promptly modify or revoke and reissue any permit containing the reopener clause required by this paragraph if the standard for sewage sludge use or disposal is more stringent than any requirements for sludge use or disposal in the permit, or controls a pollutant or practice not limited in the permit.

(d) *Water quality standards and State requirements:* any requirements in addition to or more stringent than promulgated effluent limitations guidelines or standards under sections 301, 304, 306, 307, 318 and 405 of CWA necessary to:

(1) Achieve water quality standards established under section 303 of the CWA, including State narrative criteria for water quality.

(i) Limitations must control all pollutants or pollutant parameters (either conventional, nonconventional, or toxic pollutants) which the Director determines are or may be discharged at a level which will cause, have the reasonable potential to cause, or contribute to an excursion above any State water quality standard, including State narrative criteria for water quality.

(ii) When determining whether a discharge causes, has the reasonable potential to cause, or contributes to an in-stream excursion above a narrative or numeric criteria within a State water quality standard, the permitting authority shall use procedures which account for existing controls on point and nonpoint sources of pollution, the variability of the pollutant or pollutant parameter in the effluent, the sensitivity of the species to toxicity testing (when evaluating whole effluent toxicity), and where appropriate, the dilution of the effluent in the receiving water.

(iii) When the permitting authority determines, using the procedures in paragraph (d)(1)(ii) of this section, that a discharge causes, has the reasonable potential to cause, or contributes to an in-stream excursion above the allowable ambient concentration of a State numeric criteria within a State water quality standard for an individual pollutant, the permit must contain effluent limits for that pollutant.

(iv) When the permitting authority determines, using the procedures in paragraph (d)(1)(ii) of this section, that a discharge causes, has the reasonable potential to cause, or contributes to an in-stream excursion above the numeric criterion for whole effluent toxicity, the permit must contain effluent limits for whole effluent toxicity.

(v) Except as provided in this subparagraph, when the permitting authority determines, using the procedures in paragraph (d)(1)(ii) of this section, toxicity testing data, or other information, that a discharge causes, has the reasonable potential to cause, or contributes to an in-stream excursion above a narrative criterion within an applicable State water quality standard, the permit must contain effluent limits for whole effluent toxicity. Limits on whole effluent toxicity are not necessary where the permitting authority demonstrates in the fact sheet or statement of basis of the NPDES permit, using the procedures in paragraph (d)(1)(ii) of this section, that chemical-specific limits for the effluent are sufficient to attain and maintain applicable numeric and narrative State water quality standards.

(vi) Where a State has not established a water quality criterion for a specific chemical pollutant that is present in an effluent at a concentration that causes, has the reasonable potential to cause, or contributes to an excursion above a narrative criterion within an applicable State water quality standard, the permitting authority must establish effluent limits using one or more of the following options:

(A) Establish effluent limits using a calculated numeric water quality criterion for the pollutant which the permitting authority demonstrates will attain and maintain applicable narrative water quality criteria and will fully protect the designated use. Such a criterion may be derived using a proposed State criterion, or an explicit State policy or regulation interpreting its narrative water quality criterion, supplemented with other relevant information which may include: EPA's Water Quality Standards Handbook, October 1983, risk assessment data, exposure data, information about the pollutant from the Food and Drug Administration, and current EPA criteria documents; or

(B) Establish effluent limits on a case-by-case basis, using EPA's water quality criteria, published under section 304(a) of the CWA, supplemented where necessary by other relevant information; or

(C) Establish effluent limitations on an indicator parameter for the pollutant of concern, provided:

(1) The permit identifies which pollutants are intended to be controlled by the use of the effluent limitation;

(2) The fact sheet required by §124.56 sets forth the basis for the limit, including a finding that compliance with the effluent limit on the indicator parameter will result in controls on the pollutant of concern which are sufficient to attain and maintain applicable water quality standards;

(3) The permit requires all effluent and ambient monitoring necessary to show that during the term of the permit the limit on the indicator parameter continues to attain and maintain applicable water quality standards; and

(4) The permit contains a reopener clause allowing the permitting authority to modify or revoke and reissue the permit if the limits on the indicator parameter no longer attain and maintain applicable water quality standards.

(vii) When developing water quality-based effluent limits under this paragraph the permitting authority shall ensure that:

(A) The level of water quality to be achieved by limits on point sources established under this paragraph is derived from, and complies with all applicable water quality standards; and

(B) Effluent limits developed to protect a narrative water quality criterion, a numeric water quality criterion, or both, are consistent with the assumptions and requirements of any available wasteload allocation for the discharge prepared by the State and approved by EPA pursuant to 40 CFR 130.7.

(2) Attain or maintain a specified water quality through water quality related effluent limits established under section 302 of CWA;

(3) Conform to the conditions to a State certification under section 401 of the CWA that meets the requirements of §124.53 when EPA is the permitting authority. If a State certification is stayed by a court of competent jurisdiction or an appropriate State board or agency, EPA shall notify the State that the Agency will deem certification waived unless a finally effective State certification is received within sixty days from the date of the notice. If the State does not forward a finally effective certification within the sixty day period, EPA shall include conditions in the permit that may be necessary to meet EPA's obligation under section 301(b)(1)(C) of the CWA;

(4) Conform to applicable water quality requirements under section 401(a)(2)

of CWA when the discharge affects a State other than the certifying State;

(5) Incorporate any more stringent limitations, treatment standards, or schedule of compliance requirements established under Federal or State law or regulations in accordance with section 301(b)(1)(C) of CWA;

(6) Ensure consistency with the requirements of a Water Quality Management plan approved by EPA under section 208(b) of CWA;

(7) Incorporate section 403(c) criteria under part 125, subpart M, for ocean discharges;

(8) Incorporate alternative effluent limitations or standards where warranted by "fundamentally different factors," under 40 CFR part 125, subpart D;

(9) Incorporate any other appropriate requirements, conditions, or limitations (other than effluent limitations) into a new source permit to the extent allowed by the National Environmental Policy Act, 42 U.S.C. 4321 et seq. and section 511 of the CWA, when EPA is the permit issuing authority. (See § 122.29(c)).

(e) *Technology-based controls for toxic pollutants.* Limitations established under paragraphs (a), (b), or (d) of this section, to control pollutants meeting the criteria listed in paragraph (e)(1) of this section. Limitations will be established in accordance with paragraph (e)(2) of this section. An explanation of the development of these limitations shall be included in the fact sheet under § 124.56(b)(1)(i).

(1) Limitations must control all toxic pollutants which the Director determines (based on information reported in a permit application under § 122.21(g)(7) or in a notification under § 122.42(a)(1) or on other information) are or may be discharged at a level greater than the level which can be achieved by the technology-based treatment requirements appropriate to the permittee under § 125.3(c) of this chapter; or

(2) The requirement that the limitations control the pollutants meeting the criteria of paragraph (e)(1) of this section will be satisfied by:

(i) Limitations on those pollutants; or

(ii) Limitations on other pollutants which, in the judgment of the Director, will provide treatment of the pollutants under paragraph (e)(1) of this section to the levels required by § 125.3(c).

(f) *Notification level.* A "notification level" which exceeds the notification level of § 122.42(a)(1)(i), (ii) or (iii), upon a petition from the permittee or on the Director's initiative. This new notification level may not exceed the level which can be achieved by the technology-based treatment requirements appropriate to the permittee under § 125.3(c)

(g) *Twenty-four hour reporting.* Pollutants for which the permittee must report violations of maximum daily discharge limitations under § 122.41(1)(6)(ii)(C) (24-hour reporting) shall be listed in the permit. This list shall include any toxic pollutant or hazardous substance, or any pollutant specifically identified as the method to control a toxic pollutant or hazardous substance.

(h) *Durations* for permits, as set forth in § 122.46.

(i) *Monitoring requirements.* In addition to § 122.48, the following monitoring requirements:

(1) To assure compliance with permit limitations, requirements to monitor:

(i) The mass (or other measurement specified in the permit) for each pollutant limited in the permit;

(ii) The volume of effluent discharged from each outfall;

(iii) Other measurements as appropriate including pollutants in internal waste streams under § 122.45(i); pollutants in intake water for net limitations under § 122.45(f); frequency, rate of discharge, etc., for noncontinuous discharges under § 122.45(e); pollutants subject to notification requirements under § 122.42(a); and pollutants in sewage sludge or other monitoring as specified in 40 CFR part 503; or as determined to be necessary on a case-by-case basis pursuant to section 405(d)(4) of the CWA.

(iv) According to test procedures approved under 40 CFR Part 136 for the analyses of pollutants or another method is required under 40 CFR subchapters N or O. In the case of pollutants for which there are no approved methods under 40 CFR Part 136 or otherwise required under 40 CFR subchapters N or O, monitoring must be conducted according to a test procedure specified in the permit for such pollutants.

(2) Except as provided in paragraphs (i)(4) and (i)(5) of this section, requirements to report monitoring results shall be established on a case-by-case basis with a frequency dependent on the nature and effect of the discharge, but in no case less than once a year. For sewage sludge use or disposal practices, requirements to monitor and report results shall be established on a case-by-case basis with a frequency dependent on the nature and effect of the sewage sludge use or disposal practice; minimally this shall be as specified in 40 CFR part 503 (where applicable), but in no case less than once a year.

(3) Requirements to report monitoring results for storm water discharges associated with industrial activity which are subject to an effluent limitation guideline shall be established on a case-by-case basis with a frequency dependent on the nature and effect of the discharge, but in no case less than once a year.

(4) Requirements to report monitoring results for storm water discharges associated with industrial activity (other than those addressed in paragraph (i)(3) of this section) shall be established on a case-by-case basis with a frequency dependent on the nature and effect of the discharge. At a minimum, a permit for such a discharge must require:

(i) The discharger to conduct an annual inspection of the facility site to identify areas contributing to a storm water discharge associated with industrial activity and evaluate whether measures to reduce pollutant loadings identified in a storm water pollution prevention plan are adequate and properly implemented in accordance with the terms of the permit or whether additional control measures are needed;

(ii) The discharger to maintain for a period of three years a record summarizing the results of the inspection and a certification that the facility is in compliance with the plan and the permit, and identifying any incidents of non-compliance;

(iii) Such report and certification be signed in accordance with §122.22; and

(iv) Permits for storm water discharges associated with industrial activity from inactive mining operations may, where annual inspections are impracticable, require certification once every three years by a Registered Professional Engineer that the facility is in compliance with the permit, or alternative requirements.

(5) Permits which do not require the submittal of monitoring result reports at least annually shall require that the permittee report all instances of non-compliance not reported under §122.41(l) (1), (4), (5), and (6) at least annually.

(j) *Pretreatment program for POTWs.* Requirements for POTWs to:

(1) Identify, in terms of character and volume of pollutants, any Significant Industrial Users discharging into the POTW subject to Pretreatment Standards under section 307(b) of CWA and 40 CFR part 403.

(2)(i) Submit a local program when required by and in accordance with 40 CFR part 403 to assure compliance with pretreatment standards to the extent applicable under section 307(b). The local program shall be incorporated into the permit as described in 40 CFR part 403. The program must require all indirect dischargers to the POTW to comply with the reporting requirements of 40 CFR part 403.

(ii) Provide a written technical evaluation of the need to revise local limits under 40 CFR 403.5(c)(1), following permit issuance or reissuance.

(3) For POTWs which are "sludge-only facilities," a requirement to develop a pretreatment program under 40 CFR part 403 when the Director determines that a pretreatment program is necessary to assure compliance with Section 405(d) of the CWA.

(k) *Best management practices (BMPs)* to control or abate the discharge of pollutants when:

(1) Authorized under section 304(e) of the CWA for the control of toxic pollutants and hazardous substances from ancillary industrial activities;

(2) Authorized under section 402(p) of the CWA for the control of storm water discharges;

(3) Numeric effluent limitations are infeasible; or

(4) The practices are reasonably necessary to achieve effluent limitations and standards or to carry out the purposes and intent of the CWA.

NOTE TO PARAGRAPH (k)(4): Additional technical information on BMPs and the elements of BMPs is contained in the following documents: Guidance Manual for Developing Best Management Practices (BMPs), October 1993, EPA No. 833/B–93–004, NTIS No. PB 94–178324, ERIC No. W498); Storm Water Management for Construction Activities: Developing Pollution Prevention Plans and Best Management Practices, September 1992, EPA No. 832/R–92–005, NTIS No. PB 92–235951, ERIC No. N482); Storm Water Management for Construction Activities, Developing Pollution Prevention Plans and Best Management Practices: Summary Guidance, EPA No. 833/R–92–001, NTIS No. PB 93–223550; ERIC No. W139; Storm Water Management for Industrial Activities, Developing Pollution Prevention Plans and Best Management Practices, September 1992; EPA 832/R–92–006, NTIS No. PB 92–235969, ERIC No. N477; Storm Water Management for Industrial Activities, Developing Pollution Prevention Plans and Best Management Practices: Summary Guidance, EPA 833/R–92–002, NTIS No. PB 94–133782; ERIC No. W492. Copies of those documents (or directions on how to obtain them) can be obtained by contacting either the Office of Water Resource Center (using the EPA document number as a reference) at (202) 260–7786; or the Educational Resources Information Center (ERIC) (using the ERIC number as a reference) at (800) 276–0462. Updates of these documents or additional BMP documents may also be available. A list of EPA BMP guidance documents is available on the OWM Home Page at *http://www.epa.gov/owm*. In addition, States may have BMP guidance documents.

These EPA guidance documents are listed here only for informational purposes; they are not binding and EPA does not intend that these guidance documents have any mandatory, regulatory effect by virtue of their listing in this note.

(l) *Reissued permits.* (1) Except as provided in paragraph (l)(2) of this section when a permit is renewed or reissued, interim effluent limitations, standards or conditions must be at least as stringent as the final effluent limitations, standards, or conditions in the previous permit (unless the circumstances on which the previous permit was based have materially and substantially changed since the time the permit was issued and would constitute cause for permit modification or revocation and reissuance under § 122.62.)

(2) In the case of effluent limitations established on the basis of Section 402(a)(1)(B) of the CWA, a permit may not be renewed, reissued, or modified on the basis of effluent guidelines promulgated under section 304(b) subsequent to the original issuance of such permit, to contain effluent limitations which are less stringent than the comparable effluent limitations in the previous permit.

(i) Exceptions—A permit with respect to which paragraph (l)(2) of this section applies may be renewed, reissued, or modified to contain a less stringent effluent limitation applicable to a pollutant, if—

(A) Material and substantial alterations or additions to the permitted facility occurred after permit issuance which justify the application of a less stringent effluent limitation;

(B)(*1*) Information is available which was not available at the time of permit issuance (other than revised regulations, guidance, or test methods) and which would have justified the application of a less stringent effluent limitation at the time of permit issuance; or

(*2*) The Administrator determines that technical mistakes or mistaken interpretations of law were made in issuing the permit under section 402(a)(1)(b);

(C) A less stringent effluent limitation is necessary because of events over which the permittee has no control and for which there is no reasonably available remedy;

(D) The permittee has received a permit modification under section 301(c), 301(g), 301(h), 301(i), 301(k), 301(n), or 316(a); or

(E) The permittee has installed the treatment facilities required to meet the effluent limitations in the previous permit and has properly operated and maintained the facilities but has nevertheless been unable to achieve the previous effluent limitations, in which case the limitations in the reviewed, reissued, or modified permit may reflect the level of pollutant control actually achieved (but shall not be less stringent than required by effluent guidelines in effect at the time of permit renewal, reissuance, or modification).

(ii) *Limitations.* In no event may a permit with respect to which paragraph (l)(2) of this section applies be renewed, reissued, or modified to contain an effluent limitation which is less stringent than required by effluent guidelines in effect at the time the permit is renewed, reissued, or modified. In no event may such a permit to discharge into waters be renewed, issued, or modified to contain a less stringent effluent limitation if the implementation of such limitation would result in a violation of a water quality standard under section 303 applicable to such waters.

(m) *Privately owned treatment works.* For a privately owned treatment works, any conditions expressly applicable to any user, as a limited co-permittee, that may be necessary in the permit issued to the treatment works to ensure compliance with applicable requirements under this part. Alternatively, the Director may issue separate permits to the treatment works and to its users, or may require a separate permit application from any user. The Director's decision to issue a permit with no conditions applicable to any user, to impose conditions on one or more users, to issue separate permits, or to require separate applications, and the basis for that decision, shall be stated in the fact sheet for the draft permit for the treatment works.

(n) *Grants.* Any conditions imposed in grants made by the Administrator to POTWs under sections 201 and 204 of CWA which are reasonably necessary for the achievement of effluent limitations under section 301 of CWA.

(o) *Sewage sludge.* Requirements under section 405 of CWA governing the disposal of sewage sludge from publicly owned treatment works or any other treatment works treating domestic sewage for any use for which regulations have been established, in accordance with any applicable regulations.

(p) *Coast Guard.* When a permit is issued to a facility that may operate at certain times as a means of transportation over water, a condition that the discharge shall comply with any applicable regulations promulgated by the Secretary of the department in which the Coast Guard is operating, that establish specifications for safe transportation, handling, carriage, and storage of pollutants.

(q) *Navigation.* Any conditions that the Secretary of the Army considers necessary to ensure that navigation and anchorage will not be substantially impaired, in accordance with §124.59 of this chapter.

(r) *Great Lakes.* When a permit is issued to a facility that discharges into the Great Lakes System (as defined in 40 CFR 132.2), conditions promulgated by the State, Tribe, or EPA pursuant to 40 CFR part 132.

(s) *Qualifying State, Tribal, or local programs.* (1) For storm water discharges associated with small construction activity identified in §122.26(b)(15), the Director may include permit conditions that incorporate qualifying State, Tribal, or local erosion and sediment control program requirements by reference. Where a qualifying State, Tribal, or local program does not include one or more of the elements in this paragraph (s)(1), then the Director must include those elements as conditions in the permit. A qualifying State, Tribal, or local erosion and sediment control program is one that includes:

(i) Requirements for construction site operators to implement appropriate erosion and sediment control best management practices;

(ii) Requirements for construction site operators to control waste such as discarded building materials, concrete truck washout, chemicals, litter, and sanitary waste at the construction site that may cause adverse impacts to water quality;

(iii) Requirements for construction site operators to develop and implement a storm water pollution prevention plan. (A storm water pollution prevention plan includes site descriptions, descriptions of appropriate control measures, copies of approved State, Tribal or local requirements, maintenance procedures, inspection procedures, and identification of non-storm water discharges); and

(iv) Requirements to submit a site plan for review that incorporates consideration of potential water quality impacts.

(2) For storm water discharges from construction activity identified in § 122.26(b)(14)(x), the Director may include permit conditions that incorporate qualifying State, Tribal, or local erosion and sediment control program requirements by reference. A qualifying State, Tribal or local erosion and sediment control program is one that includes the elements listed in paragraph (s)(1) of this section and any additional requirements necessary to achieve the applicable technology-based standards of "best available technology" and "best conventional technology" based on the best professional judgment of the permit writer.

[48 FR 14153, Apr. 1, 1983]

EDITORIAL NOTE: For FEDERAL REGISTER citations affecting § 122.44, see the List of CFR Sections Affected, which appears in the Finding Aids section of the printed volume and at *www.fdsys.gov.*

§ 122.45 Calculating NPDES permit conditions (applicable to State NPDES programs, see § 123.25).

(a) *Outfalls and discharge points.* All permit effluent limitations, standards and prohibitions shall be established for each outfall or discharge point of the permitted facility, except as otherwise provided under § 122.44(k) (BMPs where limitations are infeasible) and paragraph (i) of this section (limitations on internal waste streams).

(b) *Production-based limitations.* (1) In the case of POTWs, permit effluent limitations, standards, or prohibitions shall be calculated based on design flow.

(2)(i) Except in the case of POTWs or as provided in paragraph (b)(2)(ii) of this section, calculation of any permit limitations, standards, or prohibitions which are based on production (or other measure of operation) shall be based not upon the designed production capacity but rather upon a reasonable measure of actual production of the facility. For new sources or new dischargers, actual production shall be estimated using projected production. The time period of the measure of production shall correspond to the time

period of the calculated permit limitations; for example, monthly production shall be used to calculate average monthly discharge limitations.

(ii)(A)(*1*) The Director may include a condition establishing alternate permit limitations, standards, or prohibitions based upon anticipated increased (not to exceed maximum production capability) or decreased production levels.

(*2*) For the automotive manufacturing industry only, the Regional Administrator shall, and the State Director may establish a condition under paragraph (b)(2)(ii)(A)(*1*) of this section if the applicant satisfactorily demonstrates to the Director at the time the application is submitted that its actual production, as indicated in paragraph (b)(2)(i) of this section, is substantially below maximum production capability and that there is a reasonable potential for an increase above actual production during the duration of the permit.

(B) If the Director establishes permit conditions under paragraph (b)(2)(ii)(A) of this section:

(*1*) The permit shall require the permittee to notify the Director at least two business days prior to a month in which the permittee expects to operate at a level higher than the lowest production level identified in the permit. The notice shall specify the anticipated level and the period during which the permittee expects to operate at the alternate level. If the notice covers more than one month, the notice shall specify the reasons for the anticipated production level increase. New notice of discharge at alternate levels is required to cover a period or production level not covered by prior notice or, if during two consecutive months otherwise covered by a notice, the production level at the permitted facility does not in fact meet the higher level designated in the notice.

(*2*) The permittee shall comply with the limitations, standards, or prohibitions that correspond to the lowest level of production specified in the permit, unless the permittee has notified the Director under paragraph (b)(2)(ii)(B)(*1*) of this section, in which case the permittee shall comply with the lower of the actual level of production during each month or the level specified in the notice.

(3) The permittee shall submit with the DMR the level of production that actually occurred during each month and the limitations, standards, or prohibitions applicable to that level of production.

(c) *Metals.* All permit effluent limitations, standards, or prohibitions for a metal shall be expressed in terms of "total recoverable metal" as defined in 40 CFR part 136 unless:

(1) An applicable effluent standard or limitation has been promulgated under the CWA and specifies the limitation for the metal in the dissolved or valent or total form; or

(2) In establishing permit limitations on a case-by-case basis under § 125.3, it is necessary to express the limitation on the metal in the dissolved or valent or total form to carry out the provisions of the CWA; or

(3) All approved analytical methods for the metal inherently measure only its dissolved form (e.g., hexavalent chromium).

(d) *Continuous discharges.* For continuous discharges all permit effluent limitations, standards, and prohibitions, including those necessary to achieve water quality standards, shall unless impracticable be stated as:

(1) Maximum daily and average monthly discharge limitations for all dischargers other than publicly owned treatment works; and

(2) Average weekly and average monthly discharge limitations for POTWs.

(e) *Non-continuous discharges.* Discharges which are not continuous, as defined in § 122.2, shall be particularly described and limited, considering the following factors, as appropriate:

(1) Frequency (for example, a batch discharge shall not occur more than once every 3 weeks);

(2) Total mass (for example, not to exceed 100 kilograms of zinc and 200 kilograms of chromium per batch discharge);

(3) Maximum rate of discharge of pollutants during the discharge (for example, not to exceed 2 kilograms of zinc per minute); and

(4) Prohibition or limitation of specified pollutants by mass, concentration, or other appropriate measure (for example, shall not contain at any time more than 0.1 mg/1 zinc or more than 250 grams (¼ kilogram) of zinc in any discharge).

(f) *Mass limitations.* (1) All pollutants limited in permits shall have limitations, standards or prohibitions expressed in terms of mass except:

(i) For pH, temperature, radiation, or other pollutants which cannot appropriately be expressed by mass;

(ii) When applicable standards and limitations are expressed in terms of other units of measurement; or

(iii) If in establishing permit limitations on a case-by-case basis under § 125.3, limitations expressed in terms of mass are infeasible because the mass of the pollutant discharged cannot be related to a measure of operation (for example, discharges of TSS from certain mining operations), and permit conditions ensure that dilution will not be used as a substitute for treatment.

(2) Pollutants limited in terms of mass additionally may be limited in terms of other units of measurement, and the permit shall require the permittee to comply with both limitations.

(g) *Pollutants in intake water.* (1) Upon request of the discharger, technology-based effluent limitations or standards shall be adjusted to reflect credit for pollutants in the discharger's intake water if:

(i) The applicable effluent limitations and standards contained in 40 CFR subchapter N specifically provide that they shall be applied on a net basis; or

(ii) The discharger demonstrates that the control system it proposes or uses to meet applicable technology-based limitations and standards would, if properly installed and operated, meet the limitations and standards in the absence of pollutants in the intake waters.

(2) Credit for generic pollutants such as biochemical oxygen demand (BOD) or total suspended solids (TSS) should not be granted unless the permittee demonstrates that the constituents of the generic measure in the effluent are

substantially similar to the constituents of the generic measure in the intake water or unless appropriate additional limits are placed on process water pollutants either at the outfall or elsewhere.

(3) Credit shall be granted only to the extent necessary to meet the applicable limitation or standard, up to a maximum value equal to the influent value. Additional monitoring may be necessary to determine eligibility for credits and compliance with permit limits.

(4) Credit shall be granted only if the discharger demonstrates that the intake water is drawn from the same body of water into which the discharge is made. The Director may waive this requirement if he finds that no environmental degradation will result.

(5) This section does not apply to the discharge of raw water clarifier sludge generated from the treatment of intake water.

(h) *Internal waste streams.* (1) When permit effluent limitations or standards imposed at the point of discharge are impractical or infeasible, effluent limitations or standards for discharges of pollutants may be imposed on internal waste streams before mixing with other waste streams or cooling water streams. In those instances, the monitoring required by § 122.48 shall also be applied to the internal waste streams.

(2) Limits on internal waste streams will be imposed only when the fact sheet under § 124.56 sets forth the exceptional circumstances which make such limitations necessary, such as when the final discharge point is inaccessible (for example, under 10 meters of water), the wastes at the point of discharge are so diluted as to make monitoring impracticable, or the interferences among pollutants at the point of discharge would make detection or analysis impracticable.

(i) *Disposal of pollutants into wells, into POTWs or by land application.* Permit limitations and standards shall be calculated as provided in § 122.50.

[48 FR 14153, Apr. 1, 1983, as amended at 49 FR 38049, Sept. 26, 1984; 50 FR 4514, Jan. 31, 1985; 54 FR 258, Jan. 4, 1989; 54 FR 18784, May 2, 1989; 65 FR 30909, May 15, 2000]

§ 122.46 **Duration of permits (applicable to State programs, see § 123.25).**

(a) NPDES permits shall be effective for a fixed term not to exceed 5 years.

(b) Except as provided in § 122.6, the term of a permit shall not be extended by modification beyond the maximum duration specified in this section.

(c) The Director may issue any permit for a duration that is less than the full allowable term under this section.

(d) A permit may be issued to expire on or after the statutory deadline set forth in section 301(b)(2) (A), (C), and (E), if the permit includes effluent limitations to meet the requirements of section 301(b)(2) (A), (C), (D), (E) and (F), whether or not applicable effluent limitations guidelines have been promulgated or approved.

(e) A determination that a particular discharger falls within a given industrial category for purposes of setting a permit expiration date under paragraph (d) of this section is not conclusive as to the discharger's inclusion in that industrial category for any other purposes, and does not prejudice any rights to challenge or change that inclusion at the time that a permit based on that determination is formulated.

[48 FR 14153, Apr. 1, 1983, as amended at 49 FR 31842, Aug. 8, 1984; 50 FR 6940, Feb. 19, 1985; 60 FR 33931, June 29, 1995]

§ 122.47 **Schedules of compliance.**

(a) *General (applicable to State programs, see § 123.25).* The permit may, when appropriate, specify a schedule of compliance leading to compliance with CWA and regulations.

(1) *Time for compliance.* Any schedules of compliance under this section shall require compliance as soon as possible, but not later than the applicable statutory deadline under the CWA.

(2) The first NPDES permit issued to a new source or a new discharger shall contain a schedule of compliance only when necessary to allow a reasonable opportunity to attain compliance with requirements issued or revised after commencement of construction but less than three years before commencement of the relevant discharge. For recommencing dischargers, a schedule of compliance shall be available only when necessary to allow a reasonable opportunity to attain compliance with

requirements issued or revised less than three years before recommencement of discharge.

(3) *Interim dates.* Except as provided in paragraph (b)(1)(ii) of this section, if a permit establishes a schedule of compliance which exceeds 1 year from the date of permit issuance, the schedule shall set forth interim requirements and the dates for their achievement.

(i) The time between interim dates shall not exceed 1 year, except that in the case of a schedule for compliance with standards for sewage sludge use and disposal, the time between interim dates shall not exceed six months.

(ii) If the time necessary for completion of any interim requirement (such as the construction of a control facility) is more than 1 year and is not readily divisible into stages for completion, the permit shall specify interim dates for the submission of reports of progress toward completion of the interim requirements and indicate a projected completion date.

NOTE: Examples of interim requirements include: (a) Submit a complete Step 1 construction grant (for POTWs); (b) let a contract for construction of required facilities; (c) commence construction of required facilities; (d) complete construction of required facilities.

(4) *Reporting.* The permit shall be written to require that no later than 14 days following each interim date and the final date of compliance, the permittee shall notify the Director in writing of its compliance or non-compliance with the interim or final requirements, or submit progress reports if paragraph (a)(3)(ii) is applicable.

(b) *Alternative schedules of compliance.* An NPDES permit applicant or permittee may cease conducting regulated activities (by terminating of direct discharge for NPDES sources) rather than continuing to operate and meet permit requirements as follows:

(1) If the permittee decides to cease conducting regulated activities at a given time within the term of a permit which has already been issued:

(i) The permit may be modified to contain a new or additional schedule leading to timely cessation of activities; or

(ii) The permittee shall cease conducting permitted activities before non-compliance with any interim or final compliance schedule requirement already specified in the permit.

(2) If the decision to cease conducting regulated activities is made before issuance of a permit whose term will include the termination date, the permit shall contain a schedule leading to termination which will ensure timely compliance with applicable requirements no later than the statutory deadline.

(3) If the permittee is undecided whether to cease conducting regulated activities, the Director may issue or modify a permit to contain two schedules as follows:

(i) Both schedules shall contain an identical interim deadline requiring a final decision on whether to cease conducting regulated activities no later than a date which ensures sufficient time to comply with applicable requirements in a timely manner if the decision is to continue conducting regulated activities;

(ii) One schedule shall lead to timely compliance with applicable requirements, no later than the statutory deadline;

(iii) The second schedule shall lead to cessation of regulated activities by a date which will ensure timely compliance with applicable requirements no later than the statutory deadline.

(iv) Each permit containing two schedules shall include a requirement that after the permittee has made a final decision under paragraph (b)(3)(i) of this section it shall follow the schedule leading to compliance if the decision is to continue conducting regulated activities, and follow the schedule leading to termination if the decision is to cease conducting regulated activities.

(4) The applicant's or permittee's decision to cease conducting regulated activities shall be evidenced by a firm public commitment satisfactory to the Director, such as a resolution of the board of directors of a corporation.

[48 FR 14153, Apr. 1, 1983, as amended at 49 FR 38050, Sept. 26, 1984; 50 FR 6940, Feb. 19, 1985; 54 FR 18784, May 2, 1989; 65 FR 30909, May 15, 2000]

§ 122.48 Requirements for recording and reporting of monitoring results (applicable to State programs, see § 123.25).

All permits shall specify:

(a) Requirements concerning the proper use, maintenance, and installation, when appropriate, of monitoring equipment or methods (including biological monitoring methods when appropriate);

(b) Required monitoring including type, intervals, and frequency sufficient to yield data which are representative of the monitored activity including, when appropriate, continuous monitoring;

(c) Applicable reporting requirements based upon the impact of the regulated activity and as specified in § 122.44. Reporting shall be no less frequent than specified in the above regulation.

[48 FR 14153, Apr. 1, 1983; 50 FR 6940, Feb. 19, 1985]

§ 122.49 Considerations under Federal law.

The following is a list of Federal laws that may apply to the issuance of permits under these rules. When any of these laws is applicable, its procedures must be followed. When the applicable law requires consideration or adoption of particular permit conditions or requires the denial of a permit, those requirements also must be followed.

(a) The *Wild and Scenic Rivers Act,* 16 U.S.C. 1273 *et seq.* section 7 of the Act prohibits the Regional Administrator from assisting by license or otherwise the construction of any water resources project that would have a direct, adverse effect on the values for which a national wild and scenic river was established.

(b) The *National Historic Preservation Act of 1966,* 16 U.S.C. 470 *et seq.* section 106 of the Act and implementing regulations (36 CFR part 800) require the Regional Administrator, before issuing a license, to adopt measures when feasible to mitigate potential adverse effects of the licensed activity and properties listed or eligible for listing in the National Register of Historic Places. The Act's requirements are to be implemented in cooperation with State Historic Preservation Officers and upon notice to, and when appropriate, in consultation with the Advisory Council on Historic Preservation.

(c) The *Endangered Species Act,* 16 U.S.C. 1531 *et seq.* section 7 of the Act and implementing regulations (50 CFR part 402) require the Regional Administrator to ensure, in consultation with the Secretary of the Interior or Commerce, that any action authorized by EPA is not likely to jeopardize the continued existence of any endangered or threatened species or adversely affect its critical habitat.

(d) The *Coastal Zone Management Act,* 16 U.S.C. 1451 *et seq.* section 307(c) of the Act and implementing regulations (15 CFR part 930) prohibit EPA from issuing a permit for an activity affecting land or water use in the coastal zone until the applicant certifies that the proposed activity complies with the State Coastal Zone Management program, and the State or its designated agency concurs with the certification (or the Secretary of Commerce overrides the State's nonconcurrence).

(e) The *Fish and Wildlife Coordination Act,* 16 U.S.C. 661 *et seq.,* requires that the Regional Administrator, before issuing a permit proposing or authorizing the impoundment (with certain exemptions), diversion, or other control or modification of any body of water, consult with the appropriate State agency exercising jurisdiction over wildlife resources to conserve those resources.

(f) *Executive orders.* [Reserved]

(g) The National Environmental Policy Act, 42 U.S.C. 4321 *et seq.,* may require preparation of an Environmental Impact Statement and consideration of EIS-related permit conditions (other than effluent limitations) as provided in § 122.29(c).

(Clean Water Act (33 U.S.C. 1251 *et seq.*), Safe Drinking Water Act (42 U.S.C. 300f *et seq.*), Clean Air Act (42 U.S.C. 7401 *et seq.*), Resource Conservation and Recovery Act (42 U.S.C. 6901 *et seq.*))

[48 FR 14153, Apr. 1, 1983, as amended at 48 FR 39620, Sept. 1, 1983; 49 FR 38050, Sept. 26, 1984]

§122.50 Disposal of pollutants into wells, into publicly owned treatment works or by land application (applicable to State NPDES programs, see §123.25).

(a) When part of a discharger's process wastewater is not being discharged into waters of the United States or contiguous zone because it is disposed into a well, into a POTW, or by land application thereby reducing the flow or level of pollutants being discharged into waters of the United States, applicable effluent standards and limitations for the discharge in an NPDES permit shall be adjusted to reflect the reduced raw waste resulting from such disposal. Effluent limitations and standards in the permit shall be calculated by one of the following methods:

(1) If none of the waste from a particular process is discharged into waters of the United States, and effluent limitations guidelines provide separate allocation for wastes from that process, all allocations for the process shall be eliminated from calculation of permit effluent limitations or standards.

(2) In all cases other than those described in paragraph (a)(1) of this section, effluent limitations shall be adjusted by multiplying the effluent limitation derived by applying effluent limitation guidelines to the total waste stream by the amount of wastewater flow to be treated and discharged into waters of the United States, and dividing the result by the total wastewater flow. Effluent limitations and standards so calculated may be further adjusted under part 125, subpart D to make them more or less stringent if discharges to wells, publicly owned treatment works, or by land application change the character or treatability of the pollutants being discharged to receiving waters. This method may be algebraically expressed as:

$$P = \frac{E \times N}{T}$$

where P is the permit effluent limitation, E is the limitation derived by applying effluent guidelines to the total wastestream, N is the wastewater flow to be treated and discharged

to waters of the United States, and T is the total wastewater flow.

(b) Paragraph (a) of this section does not apply to the extent that promulgated effluent limitations guidelines:

(1) Control concentrations of pollutants discharged but not mass; or

(2) Specify a different specific technique for adjusting effluent limitations to account for well injection, land application, or disposal into POTWs.

(c) Paragraph (a) of this section does not alter a discharger's obligation to meet any more stringent requirements established under §§122.41, 122.42, 122.43, and 122.44.

[48 FR 14153, Apr. 1, 1983, as amended at 49 FR 38050, Sept. 26, 1984]

Subpart D—Transfer, Modification, Revocation and Reissuance, and Termination of Permits

§122.61 Transfer of permits (applicable to State programs, see §123.25).

(a) *Transfers by modification.* Except as provided in paragraph (b) of this section, a permit may be transferred by the permittee to a new owner or operator only if the permit has been modified or revoked and reissued (under §122.62(b)(2)), or a minor modification made (under §122.63(d)), to identify the new permittee and incorporate such other requirements as may be necessary under CWA.

(b) *Automatic transfers.* As an alternative to transfers under paragraph (a) of this section, any NPDES permit may be automatically transferred to a new permittee if:

(1) The current permittee notifies the Director at least 30 days in advance of the proposed transfer date in paragraph (b)(2) of this section;

(2) The notice includes a written agreement between the existing and new permittees containing a specific date for transfer of permit responsibility, coverage, and liability between them; and

(3) The Director does not notify the existing permittee and the proposed new permittee of his or her intent to modify or revoke and reissue the permit. A modification under this subparagraph may also be a minor modification under §122.63. If this notice is

not received, the transfer is effective on the date specified in the agreement mentioned in paragraph (b)(2) of this section.

§ 122.62 Modification or revocation and reissuance of permits (applicable to State programs, see § 123.25).

When the Director receives any information (for example, inspects the facility, receives information submitted by the permittee as required in the permit (see § 122.41), receives a request for modification or revocation and reissuance under § 124.5, or conducts a review of the permit file) he or she may determine whether or not one or more of the causes listed in paragraphs (a) and (b) of this section for modification or revocation and reissuance or both exist. If cause exists, the Director may modify or revoke and reissue the permit accordingly, subject to the limitations of § 124.5(c), and may request an updated application if necessary. When a permit is modified, only the conditions subject to modification are reopened. If a permit is revoked and reissued, the entire permit is reopened and subject to revision and the permit is reissued for a new term. See § 124.5(c)(2). If cause does not exist under this section or § 122.63, the Director shall not modify or revoke and reissue the permit. If a permit modification satisfies the criteria in § 122.63 for "minor modifications" the permit may be modified without a draft permit or public review. Otherwise, a draft permit must be prepared and other procedures in part 124 (or procedures of an approved State program) followed.

(a) *Causes for modification.* The following are causes for modification but not revocation and reissuance of permits except when the permittee requests or agrees.

(1) *Alterations.* There are material and substantial alterations or additions to the permitted facility or activity (including a change or changes in the permittee's sludge use or disposal practice) which occurred after permit issuance which justify the application of permit conditions that are different or absent in the existing permit.

NOTE: Certain reconstruction activities may cause the new source provisions of § 122.29 to be applicable.

(2) *Information.* The Director has received new information. Permits may be modified during their terms for this cause only if the information was not available at the time of permit issuance (other than revised regulations, guidance, or test methods) and would have justified the application of different permit conditions at the time of issuance. For NPDES general permits (§ 122.28) this cause includes any information indicating that cumulative effects on the environment are unacceptable. For new source or new discharger NPDES permits §§ 122.21, 122.29), this cause shall include any significant information derived from effluent testing required under § 122.21(k)(5)(vi) or § 122.21(h)(4)(iii) after issuance of the permit.

(3) *New regulations.* The standards or regulations on which the permit was based have been changed by promulgation of amended standards or regulations or by judicial decision after the permit was issued. Permits may be modified during their terms for this cause only as follows:

(i) For promulgation of amended standards or regulations, when:

(A) The permit condition requested to be modified was based on a promulgated effluent limitation guideline, EPA approved or promulgated water quality standards, or the Secondary Treatment Regulations under part 133; and

(B) EPA has revised, withdrawn, or modified that portion of the regulation or effluent limitation guideline on which the permit condition was based, or has approved a State action with regard to a water quality standard on which the permit condition was based; and

(C) A permittee requests modification in accordance with § 124.5 within ninety (90) days after FEDERAL REGISTER notice of the action on which the request is based.

(ii) For judicial decisions, a court of competent jurisdiction has remanded and stayed EPA promulgated regulations or effluent limitation guidelines, if the remand and stay concern that portion of the regulations or guidelines on which the permit condition was based and a request is filed by the permittee in accordance with §124.5 within ninety (90) days of judicial remand.

(iii) For changes based upon modified State certifications of NPDES permits, see §124.55(b).

(4) *Compliance schedules.* The Director determines good cause exists for modification of a compliance schedule, such as an act of God, strike, flood, or materials shortage or other events over which the permittee has little or no control and for which there is no reasonably available remedy. However, in no case may an NPDES compliance schedule be modified to extend beyond an applicable CWA statutory deadline. See also §122.63(c) (minor modifications) and paragraph (a)(14) of this section (NPDES innovative technology).

(5) When the permittee has filed a request for a variance under CWA section 301(c), 301(g), 301(h), 301(i), 301(k), or 316(a) or for "fundamentally different factors" within the time specified in §122.21 or §125.27(a).

(6) *307(a) toxics.* When required to incorporate an applicable 307(a) toxic effluent standard or prohibition (see §122.44(b)).

(7) *Reopener.* When required by the "reopener" conditions in a permit, which are established in the permit under §122.44(b) (for CWA toxic effluent limitations and Standards for sewage sludge use or disposal, see also §122.44(c)) or 40 CFR 403.18(e) (Pretreatment program).

(8)(i) *Net limits.* Upon request of a permittee who qualifies for effluent limitations on a net basis under §122.45(g).

(ii) When a discharger is no longer eligible for net limitations, as provided in §122.45(g)(1)(ii).

(9) *Pretreatment.* As necessary under 40 CFR 403.8(e) (compliance schedule for development of pretreatment program).

(10) *Failure to notify.* Upon failure of an approved State to notify, as required by section 402(b)(3), another State whose waters may be affected by a discharge from the approved State.

(11) *Non-limited pollutants.* When the level of discharge of any pollutant which is not limited in the permit exceeds the level which can be achieved by the technology-based treatment requirements appropriate to the permittee under §125.3(c).

(12) *Notification levels.* To establish a "notification level" as provided in §122.44(f).

(13) *Compliance schedules.* To modify a schedule of compliance to reflect the time lost during construction of an innovative or alternative facility, in the case of a POTW which has received a grant under section 202(a)(3) of CWA for 100% of the costs to modify or replace facilities constructed with a grant for innovative and alternative wastewater technology under section 202(a)(2). In no case shall the compliance schedule be modified to extend beyond an applicable CWA statutory deadline for compliance.

(14) For a small MS4, to include an effluent limitation requiring implementation of a minimum control measure or measures as specified in §122.34(b) when:

(i) The permit does not include such measure(s) based upon the determination that another entity was responsible for implementation of the requirement(s); and

(ii) The other entity fails to implement measure(s) that satisfy the requirement(s).

(15) To correct technical mistakes, such as errors in calculation, or mistaken interpretations of law made in determining permit conditions.

(16) When the discharger has installed the treatment technology considered by the permit writer in setting effluent limitations imposed under section 402(a)(1) of the CWA and has properly operated and maintained the facilities but nevertheless has been unable to achieve those effluent limitations. In this case, the limitations in the modified permit may reflect the level of pollutant control actually achieved (but shall not be less stringent than required by a subsequently promulgated effluent limitations guideline).

(17) *Nutrient Management Plans.* The incorporation of the terms of a CAFO's nutrient management plan into the terms and conditions of a general permit when a CAFO obtains coverage under a general permit in accordance with §§122.23(h) and 122.28 is not a cause

for modification pursuant to the requirements of this section.

(18) *Land application plans.* When required by a permit condition to incorporate a land application plan for beneficial reuse of sewage sludge, to revise an existing land application plan, or to add a land application plan.

(b) *Causes for modification or revocation and reissuance.* The following are causes to modify or, alternatively, revoke and reissue a permit:

(1) Cause exists for termination under § 122.64, and the Director determines that modification or revocation and reissuance is appropriate.

(2) The Director has received notification (as required in the permit, see § 122.41(l)(3)) of a proposed transfer of the permit. A permit also may be modified to reflect a transfer after the effective date of an automatic transfer (§ 122.61(b)) but will not be revoked and reissued after the effective date of the transfer except upon the request of the new permittee.

[48 FR 14153, Apr. 1, 1983, as amended at 49 FR 25981, June 25, 1984; 49 FR 37009, Sept. 29, 1984; 49 FR 38050, Sept. 26, 1984; 50 FR 4514, Jan. 31, 1985; 51 FR 20431, June 4, 1986; 51 FR 26993, July 28, 1986; 54 FR 256, 258, Jan. 4, 1989; 54 FR 18784, May 2, 1989; 60 FR 33931, June 29, 1995; 64 FR 68847, Dec. 8, 1999; 65 FR 30909, May 15, 2000; 70 FR 60191, Oct. 14, 2005; 73 FR 70485, Nov. 20, 2008]

§ 122.63 Minor modifications of permits.

Upon the consent of the permittee, the Director may modify a permit to make the corrections or allowances for changes in the permitted activity listed in this section, without following the procedures of part 124. Any permit modification not processed as a minor modification under this section must be made for cause and with part 124 draft permit and public notice as required in § 122.62. Minor modifications may only:

(a) Correct typographical errors;

(b) Require more frequent monitoring or reporting by the permittee;

(c) Change an interim compliance date in a schedule of compliance, provided the new date is not more than 120 days after the date specified in the existing permit and does not interfere with attainment of the final compliance date requirement; or

(d) Allow for a change in ownership or operational control of a facility where the Director determines that no other change in the permit is necessary, provided that a written agreement containing a specific date for transfer of permit responsibility, coverage, and liability between the current and new permittees has been submitted to the Director.

(e)(1) Change the construction schedule for a discharger which is a new source. No such change shall affect a discharger's obligation to have all pollution control equipment installed and in operation prior to discharge under § 122.29.

(2) Delete a point source outfall when the discharge from that outfall is terminated and does not result in discharge of pollutants from other outfalls except in accordance with permit limits.

(f) [Reserved]

(g) Incorporate conditions of a POTW pretreatment program that has been approved in accordance with the procedures in 40 CFR 403.11 (or a modification thereto that has been approved in accordance with the procedures in 40 CFR 403.18) as enforceable conditions of the POTW's permits.

(h) Incorporate changes to the terms of a CAFO's nutrient management plan that have been revised in accordance with the requirements of § 122.42(e)(6).

[48 FR 14153, Apr. 1, 1983, as amended at 49 FR 38051, Sept. 26, 1984; 51 FR 20431, June 4, 1986; 53 FR 40616, Oct. 17, 1988; 60 FR 33931, June 29, 1995; 73 FR 70485, Nov. 20, 2008]

§ 122.64 Termination of permits (applicable to State programs, see § 123.25).

(a) The following are causes for terminating a permit during its term, or for denying a permit renewal application:

(1) Noncompliance by the permittee with any condition of the permit;

(2) The permittee's failure in the application or during the permit issuance process to disclose fully all relevant facts, or the permittee's misrepresentation of any relevant facts at any time;

(3) A determination that the permitted activity endangers human health or the environment and can only be regulated to acceptable levels

by permit modification or termination; or

(4) A change in any condition that requires either a temporary or permanent reduction or elimination of any discharge or sludge use or disposal practice controlled by the permit (for example, plant closure or termination of discharge by connection to a POTW).

(b) The Director shall follow the applicable procedures in part 124 or part 22 of this chapter, as appropriate (or State procedures equivalent to part 124) in terminating any NPDES permit under this section, except that if the entire discharge is permanently terminated by elimination of the flow or by connection to a POTW (but not by land application or disposal into a well), the Director may terminate the permit by notice to the permittee. Termination by notice shall be effective 30 days after notice is sent, unless the permittee objects within that time. If the permittee objects during that period, the Director shall follow part 124 of this chapter or applicable State procedures for termination. Expedited permit termination procedures are not available to permittees that are subject to pending State and/or Federal enforcement actions including citizen suits brought under State or Federal law. If requesting expedited permit termination procedures, a permittee must certify that it is not subject to any pending State or Federal enforcement actions including citizen suits brought under State or Federal law. State-authorized NPDES programs are not required to use part 22 of this chapter procedures for NPDES permit terminations.

[48 FR 14153, Apr. 1, 1983; 50 FR 6940, Feb. 19, 1985, as amended at 54 FR 18784, May 2, 1989; 65 FR 30909, May 15, 2000]

APPENDIX A TO PART 122—NPDES PRIMARY INDUSTRY CATEGORIES

Any permit issued after June 30, 1981 to dischargers in the following categories shall include effluent limitations and a compliance schedule to meet the requirements of section 301(b)(2)(A), (C), (D), (E) and (F) of CWA, whether or not applicable effluent limitations guidelines have been promulgated. See §§ 122.44 and 122.46.

Industry Category

Adhesives and sealants
Aluminum forming
Auto and other laundries
Battery manufacturing
Coal mining
Coil coating
Copper forming
Electrical and electronic components
Electroplating
Explosives manufacturing
Foundries
Gum and wood chemicals
Inorganic chemicals manufacturing
Iron and steel manufacturing
Leather tanning and finishing
Mechanical products manufacturing
Nonferrous metals manufacturing
Ore mining
Organic chemicals manufacturing
Paint and ink formulation
Pesticides
Petroleum refining
Pharmaceutical preparations
Photographic equipment and supplies
Plastics processing
Plastic and synthetic materials manufacturing
Porcelain enameling
Printing and publishing
Pulp and paper mills
Rubber processing
Soap and detergent manufacturing
Steam electric power plants
Textile mills
Timber products processing

APPENDIX B TO PART 122 [RESERVED]

APPENDIX C TO PART 122—CRITERIA FOR DETERMINING A CONCENTRATED AQUATIC ANIMAL PRODUCTION FACILITY (§ 122.24)

A hatchery, fish farm, or other facility is a concentrated aquatic animal production facility for purposes of § 122.24 if it contains, grows, or holds aquatic animals in either of the following categories:

(a) Cold water fish species or other cold water aquatic animals in ponds, raceways, or other similar structures which discharge at least 30 days per year but does not include:

(1) Facilities which produce less than 9,090 harvest weight kilograms (approximately 20,000 pounds) of aquatic animals per year; and

(2) Facilities which feed less than 2,272 kilograms (approximately 5,000 pounds) of food during the calendar month of maximum feeding.

(b) Warm water fish species or other warm water aquatic animals in ponds, raceways, or other similar structures which discharge at least 30 days per year, but does not include:

(1) Closed ponds which discharge only during periods of excess runoff; or

(2) Facilities which produce less than 45,454 harvest weight kilograms (approximately 100,000 pounds) of aquatic animals per year.

"Cold water aquatic animals" include, but are not limited to, the *Salmonidae* family of fish; e.g., trout and salmon.

"Warm water aquatic animals" include, but are not limited to, the *Ameiuride, Centrarchidae* and *Cyprinidae* families of fish; e.g., respectively, catfish, sunfish and minnows.

APPENDIX D TO PART 122—NPDES PERMIT APPLICATION TESTING REQUIREMENTS (§ 122.21)

TABLE I—TESTING REQUIREMENTS FOR ORGANIC TOXIC POLLUTANTS BY INDUSTRIAL CATEGORY FOR EXISTING DISCHARGERS

Industrial category	GC/MS Fraction [1]			
	Volatile	Acid	Base/neutral	Pesticide
Adhesives and Sealants	2	2	2	
Aluminum Forming	2	2	2	
Auto and Other Laundries	2	2	2	2
Battery Manufacturing	2		2	
Coal Mining	2	2	2	
Coil Coating	2	2	2	
Copper Forming	2	2	2	
Electric and Electronic Components	2	2	2	2
Electroplating	2	2	2	
Explosives Manufacturing		2	2	
Foundries	2	2	2	
Gum and Wood Chemicals	2	2	2	2
Inorganic Chemicals Manufacturing	2	2	2	
Iron and Steel Manufacturing	2	2	2	
Leather Tanning and Finishing	2	2	2	2
Mechanical Products Manufacturing	2	2	2	
Nonferrous Metals Manufacturing	2	2	2	2
Ore Mining	2	2	2	2
Organic Chemicals Manufacturing	2	2	2	2
Paint and Ink Formulation	2	2	2	2
Pesticides	2	2	2	2
Petroleum Refining	2	2	2	2
Pharmaceutical Preparations	2	2	2	
Photographic Equipment and Supplies	2	2	2	2
Plastic and Synthetic Materials Manufacturing	2	2	2	2
Plastic Processing	2			
Porcelain Enameling	2		2	
Printing and Publishing	2	2	2	2
Pulp and Paper Mills	2	2	2	2
Rubber Processing	2	2	2	
Soap and Detergent Manufacturing	2	2	2	

TABLE I—TESTING REQUIREMENTS FOR ORGANIC TOXIC POLLUTANTS BY INDUSTRIAL CATEGORY FOR EXISTING DISCHARGERS—Continued

Industrial category	GC/MS Fraction [1]			
	Volatile	Acid	Base/neutral	Pesticide
Steam Electric Power Plants	2	2	2	
Textile Mills	2	2	2	2
Timber Products Processing	2	2	2	2

[1] The toxic pollutants in each fraction are listed in Table II.
[2] Testing required.

TABLE II—ORGANIC TOXIC POLLUTANTS IN EACH OF FOUR FRACTIONS IN ANALYSIS BY GAS CHROMATOGRAPHY/MASS SPECTROSCOPY (GS/MS)

Volatiles

1V   acrolein
2V   acrylonitrile
3V   benzene
5V   bromoform
6V   carbon tetrachloride
7V   chlorobenzene
8V   chlorodibromomethane
9V   chloroethane
10V 2-chloroethylvinyl ether
11V chloroform
12V dichlorobromomethane
14V 1,1-dichloroethane
15V 1,2-dichloroethane
16V 1,1-dichloroethylene
17V 1,2-dichloropropane
18V 1,3-dichloropropylene
19V ethylbenzene
20V methyl bromide
21V methyl chloride
22V methylene chloride
23V 1,1,2,2-tetrachloroethane
24V tetrachloroethylene
25V toluene
26V 1,2-trans-dichloroethylene
27V 1,1,1-trichloroethane
28V 1,1,2-trichloroethane
29V trichloroethylene
31V vinyl chloride

Acid Compounds

1A   2-chlorophenol
2A   2,4-dichlorophenol
3A   2,4-dimethylphenol
4A   4,6-dinitro-o-cresol
5A   2,4-dinitrophenol
6A   2-nitrophenol
7A   4-nitrophenol
8A   p-chloro-m-cresol
9A   pentachlorophenol
10A phenol
11A 2,4,6-trichlorophenol

Base/Neutral

1B   acenaphthene

2B	acenaphthylene
3B	anthracene
4B	benzidine
5B	benzo(a)anthracene
6B	benzo(a)pyrene
7B	3,4-benzofluoranthene
8B	benzo(ghi)perylene
9B	benzo(k)fluoranthene
10B	bis(2-chloroethoxy)methane
11B	bis(2-chloroethyl)ether
12B	bis(2-chloroisopropyl)ether
13B	bis (2-ethylhexyl)phthalate
14B	4-bromophenyl phenyl ether
15B	butylbenzyl phthalate
16B	2-chloronaphthalene
17B	4-chlorophenyl phenyl ether
18B	chrysene
19B	dibenzo(a,h)anthracene
20B	1,2-dichlorobenzene
21B	1,3-dichlorobenzene
22B	1,4-dichlorobenzene
23B	3,3'-dichlorobenzidine
24B	diethyl phthalate
25B	dimethyl phthalate
26B	di-n-butyl phthalate
27B	2,4-dinitrotoluene
28B	2,6-dinitrotoluene
29B	di-n-octyl phthalate
30B	1,2-diphenylhydrazine (as azobenzene)
31B	fluroranthene
32B	fluorene
33B	hexachlorobenzene
34B	hexachlorobutadiene
35B	hexachlorocyclopentadiene
36B	hexachloroethane
37B	indeno(1,2,3-cd)pyrene
38B	isophorone
39B	napthalene
40B	nitrobenzene
41B	N-nitrosodimethylamine
42B	N-nitrosodi-n-propylamine
43B	N-nitrosodiphenylamine
44B	phenanthrene
45B	pyrene
46B	1,2,4-trichlorobenzene

Pesticides

1P	aldrin
2P	alpha-BHC
3P	beta-BHC
4P	gamma-BHC
5P	delta-BHC
6P	chlordane
7P	4,4'-DDT
8P	4,4'-DDE
9P	4,4'-DDD
10P	dieldrin
11P	alpha-endosulfan
12P	beta-endosulfan
13P	endosulfan sulfate
14P	endrin
15P	endrin aldehyde
16P	heptachlor
17P	heptachlor epoxide
18P	PCB-1242
19P	PCB-1254
20P	PCB-1221
21P	PCB-1232
22P	PCB-1248
23P	PCB-1260
24P	PCB-1016
25P	toxaphene

TABLE III—OTHER TOXIC POLLUTANTS (METALS AND CYANIDE) AND TOTAL PHENOLS

Antimony, Total
Arsenic, Total
Beryllium, Total
Cadmium, Total
Chromium, Total
Copper, Total
Lead, Total
Mercury, Total
Nickel, Total
Selenium, Total
Silver, Total
Thallium, Total
Zinc, Total
Cyanide, Total
Phenols, Total

TABLE IV—CONVENTIONAL AND NONCONVENTIONAL POLLUTANTS REQUIRED TO BE TESTED BY EXISTING DISCHARGERS IF EXPECTED TO BE PRESENT

Bromide
Chlorine, Total Residual
Color
Fecal Coliform
Fluoride
Nitrate-Nitrite
Nitrogen, Total Organic
Oil and Grease
Phosphorus, Total
Radioactivity
Sulfate
Sulfide
Sulfite
Surfactants
Aluminum, Total
Barium, Total
Boron, Total
Cobalt, Total
Iron, Total
Magnesium, Total
Molybdenum, Total
Manganese, Total
Tin, Total
Titanium, Total

TABLE V—TOXIC POLLUTANTS AND HAZARDOUS SUBSTANCES REQUIRED TO BE IDENTIFIED BY EXISTING DISCHARGERS IF EXPECTED TO BE PRESENT

Toxic Pollutants

Asbestos

Hazardous Substances

Acetaldehyde
Allyl alcohol
Allyl chloride
Amyl acetate

Aniline
Benzonitrile
Benzyl chloride
Butyl acetate
Butylamine
Captan
Carbaryl
Carbofuran
Carbon disulfide
Chlorpyrifos
Coumaphos
Cresol
Crotonaldehyde
Cyclohexane
2,4-D (2,4-Dichlorophenoxy acetic acid)
Diazinon
Dicamba
Dichlobenil
Dichlone
2,2-Dichloropropionic acid
Dichlorvos
Diethyl amine
Dimethyl amine
Dintrobenzene
Diquat
Disulfoton
Diuron
Epichlorohydrin
Ethion
Ethylene diamine
Ethylene dibromide
Formaldehyde
Furfural
Guthion
Isoprene
Isopropanolamine Dodecylbenzenesulfonate
Kelthane
Kepone
Malathion
Mercaptodimethur
Methoxychlor
Methyl mercaptan
Methyl methacrylate
Methyl parathion
Mevinphos
Mexacarbate
Monoethyl amine
Monomethyl amine
Naled
Napthenic acid
Nitrotoluene
Parathion
Phenolsulfanate
Phosgene
Propargite
Propylene oxide
Pyrethrins
Quinoline
Resorcinol
Strontium
Strychnine
Styrene
2,4,5-T (2,4,5-Trichlorophenoxy acetic acid)
TDE (Tetrachlorodiphenylethane)
2,4,5-TP　[2-(2,4,5-Trichlorophenoxy)　propanoic acid]
Trichlorofan

Triethanolamine dodecylbenzenesulfonate
Triethylamine
Trimethylamine
Uranium
Vanadium
Vinyl acetate
Xylene
Xylenol
Zirconium

[Note 1: The Environmental Protection Agency has suspended the requirements of § 122.21(g)(7)(ii)(A) and Table I of Appendix D as they apply to certain industrial categories. The suspensions are as follows:

a. At 46 FR 2046, Jan. 8, 1981, the Environmental Protection Agency suspended until further notice § 122.21(g)(7)(ii)(A) as it applies to coal mines.

b. At 46 FR 22585, Apr. 20, 1981, the Environmental Protection Agency suspended until further notice § 122.21(g)(7)(ii)(A) and the corresponding portions of Item V-C of the NPDES application Form 2c as they apply to:

1. Testing and reporting for all four organic fractions in the Greige Mills Subcategory of the Textile Mills industry (Subpart C—Low water use processing of 40 CFR part 410), and testing and reporting for the pesticide fraction in all other subcategories of this industrial category.

2. Testing and reporting for the volatile, base/neutral and pesticide fractions in the Base and Precious Metals Subcategory of the Ore Mining and Dressing industry (subpart B of 40 CFR part 440), and testing and reporting for all four fractions in all other subcategories of this industrial category.

3. Testing and reporting for all four GC/MS fractions in the Porcelain Enameling industry.

c. At 46 FR 35090, July 1, 1981, the Environmental Protection Agency suspended until further notice § 122.21(g)(7)(ii)(A) and the corresponding portions of Item V-C of the NPDES application Form 2c as they apply to:

1. Testing and reporting for the pesticide fraction in the Tall Oil Rosin Subcategory (subpart D) and Rosin-Based Derivatives Subcategory (subpart F) of the Gum and Wood Chemicals industry (40 CFR part 454), and testing and reporting for the pesticide and base/netural fractions in all other subcategories of this industrial category.

2. Testing and reporting for the pesticide fraction in the Leather Tanning and Finishing, Paint and Ink Formulation, and Photographic Supplies industrial categories.

3. Testing and reporting for the acid, base/neutral and pesticide fractions in the Petroleum Refining industrial category.

4. Testing and reporting for the pesticide fraction in the Papergrade Sulfite subcategories (subparts J and U) of the Pulp and Paper industry (40 CFR part 430); testing and reporting for the base/neutral and pesticide

fractions in the following subcategories: Deink (subpart Q), Dissolving Kraft (subpart F), and Paperboard from Waste Paper (subpart E); testing and reporting for the volatile, base/neutral and pesticide fractions in the following subcategories: BCT Bleached Kraft (subpart H), Semi-Chemical (subparts B and C), and Nonintegrated-Fine Papers (subpart R); and testing and reporting for the acid, base/neutral, and pesticide fractions in the following subcategories: Fine Bleached Kraft (subpart I), Dissolving Sulfite Pulp (subpart K), Groundwood-Fine Papers (subpart O), Market Bleached Kraft (subpart G), Tissue from Wastepaper (subpart T), and Nonintegrated-Tissue Papers (subpart S).

5. Testing and reporting for the base/neutral fraction in the Once-Through Cooling Water, Fly Ash and Bottom Ash Transport Water process wastestreams of the Steam Electric Power Plant industrial category.

This revision continues these suspensions.]*

For the duration of the suspensions, therefore, Table I effectively reads:

TABLE I—TESTING REQUIREMENTS FOR OR-GANIC TOXIC POLLUTANTS BY INDUSTRY CAT-EGORY

Industry category	GC/MS fraction [2]			
	Vola-tile	Acid	Neu-tral	Pes-ticide
Adhesives and sealants	([1])	([1])	([1])	
Aluminum forming	([1])	([1])	([1])	
Auto and other laundries ...	([1])	([1])	([1])	([1])
Battery manufacturing	([1])		([1])	
Coal mining				
Coil coating	([1])	([1])	([1])	
Copper forming	([1])	([1])	([1])	
Electric and electronic compounds	([1])	([1])	([1])	([1])
Electroplating	([1])	([1])	([1])	
Explosives manufacturing		([1])	([1])	
Foundries	([1])	([1])	([1])	
Gum and wood (all sub-parts except D and F)	([1])	([1])		
Subpart D—tall oil rosin	([1])	([1])	([1])	
Subpart F—rosin-based derivatives	([1])	([1])	([1])	
Inorganic chemicals manu-facturing	([1])	([1])	([1])	
Iron and steel manufac-turing	([1])	([1])	([1])	
Leather tanning and fin-ishing	([1])	([1])	([1])	
Mechanical products man-ufacturing	([1])	([1])	([1])	
Nonferrous metals manu-facturing	([1])	([1])	([1])	([1])
Ore mining (applies to the base and precious met-als/Subpart B)		([1])		
Organic chemicals manu-facturing	([1])	([1])	([1])	([1])
Paint and ink formulation ..	([1])	([1])	([1])	

TABLE I—TESTING REQUIREMENTS FOR OR-GANIC TOXIC POLLUTANTS BY INDUSTRY CAT-EGORY—Continued

Industry category	GC/MS fraction [2]			
	Vola-tile	Acid	Neu-tral	Pes-ticide
Pesticides	([1])	([1])	([1])	([1])
Petroleum refining	([1])			
Pharmaceutical prepara-tions	([1])	([1])	([1])	
Photographic equipment and supplies	([1])	([1])	([1])	
Plastic and synthetic mate-rials manufacturing	([1])	([1])	([1])	([1])
Plastic processing	([1])			
Porcelain enameling.				
Printing and publishing	([1])	([1])	([1])	([1])
Pulp and paperboard mills—see footnote [3].				
Rubber processing	([1])	([1])	([1])	
Soap and detergent manu-facturing	([1])	([1])	([1])	
Steam electric power plants	([1])	([1])		
Textile mills (Subpart C—Greige Mills are exempt from this table)	([1])	([1])	([1])	
Timber products proc-essing	([1])	([1])	([1])	([1])

[1] Testing required.
[2] The pollutants in each fraction are listed in Item V-C.
[3] Pulp and Paperboard Mills:

Subpart [3]	GS/MS fractions			
	VOA	Acid	Base/neu-tral	Pes-ticides
A	2	([1])	2	([1])
B	2	([1])	2	2
C	2	([1])	2	2
D	2	([1])	2	2
E	([1])	([1])	2	([1])
F	([1])	([1])	2	2
G	([1])	([1])	2	2
H	([1])	([1])	2	2
I	([1])	([1])	2	2
J	([1])	([1])	([1])	2
K	([1])	([1])	2	2
L	([1])	([1])	2	2
M	([1])	([1])	2	2
N	([1])	([1])	2	2
O	([1])	([1])	2	2
P	([1])	([1])	2	2
Q	([1])	([1])	2	([1])
R	2	([1])	2	2
S	([1])	([1])	2	([1])
T	([1])	([1])	2	([1])
U	([1])	([1])	([1])	2

[1] Must test.
[2] Do not test unless "reason to believe" it is discharged.
[3] Subparts are defined in 40 CFR Part 430.

*Editorial Note: The words "This revision" refer to the document published at 48 FR 14153, Apr. 1, 1983.

[48 FR 14153, Apr. 1, 1983, as amended at 49 FR 38050, Sept. 26, 1984; 50 FR 6940, Feb. 19, 1985]

APPENDIX E TO PART 122—RAINFALL ZONES OF THE UNITED STATES

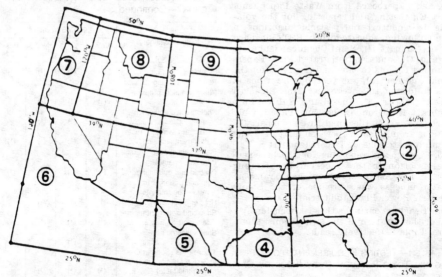

Not Shown: Alaska (Zone 7); Hawaii (Zone 7); Northern Mariana Islands (Zone 7); Guam (Zone 7); American Samoa (Zone 7); Trust Territory of the Pacific Islands (Zone 7); Puerto Rico (Zone 3) Virgin Islands (Zone 3).

Source: Methodology for Analysis of Detention Basins for Control of Urban Runoff Quality, prepared for U.S. Environmental Protection Agency, Office of Water, Nonpoint Source Division, Washington, DC, 1986.

[55 FR 48073, Nov. 16, 1990]

APPENDIX F TO PART 122—INCORPORATED PLACES WITH POPULATIONS GREATER THAN 250,000 ACCORDING TO THE 1990 DECENNIAL CENSUS BY THE BUREAU OF THE CENSUS

State	Incorporated Place
Alabama	Birmingham.
Arizona	Phoenix.
	Tucson.
California	Long Beach.
	Los Angeles.
	Oakland.
	Sacramento.
	San Diego.
	San Francisco.
	San Jose.
Colorado	Denver.
District of Columbia..	
Florida	Jacksonville.
	Miami.
	Tampa.
Georgia.	Atlanta.
Illinois	Chicago.
Indiana	Indianapolis.
Kansas	Wichita.
Kentucky	Louisville.
Louisiana	New Orleans.
Maryland	Baltimore.
Massachusetts	Boston.

State	Incorporated Place
Michigan	Detroit.
Minnesota	Minneapolis.
	St. Paul.
Missouri	Kansas City.
	St. Louis.
Nebraska	Omaha.
New Jersey	Newark.
New Mexico	Albuquerque.
New York	Buffalo.
	Bronx Borough.
	Brooklyn Borough.
	Manhattan Borough.
	Queens Borough.
	Staten Island Borough.
North Carolina	Charlotte.
Ohio	Cincinnati.
	Cleveland.
	Columbus.
	Toledo.
Oklahoma	Oklahoma City.
	Tulsa.
Oregon	Portland.
Pennsylvania	Philadelphia.
	Pittsburgh.
Tennessee	Memphis.
	Nashville/Davidson.
Texas	Austin.
	Dallas.
	El Paso.
	Fort Worth.

State	Incorporated Place
	Houston.
	San Antonio.
Virginia	Norfolk.
	Virginia Beach.
Washington	Seattle.
Wisconsin	Milwaukee.

[64 FR 68847, Dec. 8, 1999, as amended at 67 FR 47152, July 17, 2002]

APPENDIX G TO PART 122—INCOR-PORATED PLACES WITH POPULATIONS GREATER THAN 100,000 BUT LESS THAN 250,000 ACCORDING TO THE 1990 DECENNIAL CENSUS BY THE BUREAU OF THE CENSUS

State	Incorporated place
Alabama	Huntsville.
	Mobile.
	Montgomery.
Alaska	Anchorage.
Arizona	Mesa.
	Tempe.
Arkansas	Little Rock.
California	Anaheim.
	Bakersfield.
	Berkeley.
	Chula Vista.
	Concord.
	El Monte.
	Escondido.
	Fremont.
	Fresno.
	Fullerton.
	Garden Grove.
	Glendale.
	Hayward.
	Huntington Beach.
	Inglewood.
	Irvine.
	Modesto.
	Moreno Valley.
	Oceanside.
	Ontario.
	Orange.
Colorado	Aurora.
	Colorado Springs.
	Lakewood.
	Pueblo.
Connecticut	Bridgeport.
	Hartford.
	New Haven.
	Stamford.
	Waterbury.
Florida	Fort Lauderdale.
	Hialeah.
	Hollywood.
	Orlando.
	St. Petersburg.
	Tallahassee.
Georgia	Columbus.
	Macon.
	Savannah.
Idaho	Boise City.
Illinois	Peoria.
	Rockford.
Indiana	Evansville.
	Fort Wayne.
	Gary.

State	Incorporated place
	South Bend.
Iowa	Cedar Rapids.
	Davenport.
	Des Moines.
Kansas	Kansas City.
	Topeka.
Kentucky	Lexington-Fayette.
Louisiana	Baton Rouge.
	Shreveport.
Massachusetts	Springfield.
	Worcester.
Michigan	Ann Arbor.
	Flint.
	Grand Rapids.
	Lansing.
	Livonia.
	Sterling Heights.
	Warren.
Mississippi	Jackson.
Missouri	Independence.
	Springfield.
Nebraska	Lincoln.
Nevada	Las Vegas.
	Reno.
New Jersey	Elizabeth.
	Jersey City.
	Paterson.
New York	Albany.
	Rochester.
	Syracuse.
	Yonkers.
North Carolina	Durham.
	Greensboro.
	Raleigh.
	Winston-Salem.
Ohio	Akron.
	Dayton.
	Youngstown.
Oregon	Eugene.
Pennsylvania	Allentown.
	Erie.
Rhode Island	Providence.
South Carolina	Columbia.
Tennessee	Chattanooga.
	Knoxville.
Texas	Abilene.
	Amarillo.
	Arlington.
	Beaumont.
	Corpus Christi.
	Garland.
	Irving.
	Laredo.
	Lubbock.
	Mesquite.
	Pasadena.
	Plano.
	Waco.
Utah	Salt Lake City.
Virginia	Alexandria.
	Chesapeake.
	Hampton.
	Newport News.
	Portsmouth.
	Richmond.
	Roanoke.
Washington	Spokane.
	Tacoma.
Wisconsin	Madison.

[64 FR 68848, Dec. 8, 1999]

APPENDIX H TO PART 122—COUNTIES WITH UNINCORPORATED URBANIZED AREAS WITH A POPULATION OF 250,000 OR MORE ACCORDING TO THE 1990 DECENNIAL CENSUS BY THE BUREAU OF THE CENSUS

State	County	Unincorporated urbanized population
California	Los Angeles	886,780
	Sacramento	594,889
	San Diego	250,414
Delaware	New Castle	296,996
Florida	Dade	1,014,504
Georgia	DeKalb	448,686
Hawaii	Honolulu [1]	114,506
Maryland	Anne Arundel	344,654
	Baltimore	627,593
	Montgomery	599,028
	Prince George's	494,369
Texas	Harris	729,206
Utah	Salt Lake	270,989
Virginia	Fairfax	760,730
Washington	King	520,468

[1] County was previously listed in this appendix; however, population dropped to below 250,000 in the 1990 Census.

[64 FR 68848, Dec. 8, 1999]

APPENDIX I TO PART 122—COUNTIES WITH UNINCORPORATED URBANIZED AREAS GREATER THAN 100,000, BUT LESS THAN 250,000 ACCORDING TO THE 1990 DECENNIAL CENSUS BY THE BUREAU OF THE CENSUS

State	County	Unincorporated urbanized population
Alabama	Jefferson	78,608
Arizona	Pima	162,202
California	Alameda	115,082
	Contra Costa	131,082
	Kern	128,503
	Orange	223,081
	Riverside	166,509
	San Bernardino	162,202
Colorado	Arapahoe	103,248
Florida	Broward	142,329
	Escambia	167,463
	Hillsborough	398,593
	Lee	102,337
	Manatee	123,828
	Orange	378,611
	Palm Beach	360,553
	Pasco	148,907
	Pinellas	255,772
	Polk	121,528
	Sarasota	172,600
	Seminole	127,873
Georgia	Clayton	133,237
	Cobb	322,595
	Fulton	127,776
	Gwinnett	237,305
	Richmond	126,476
Kentucky	Jefferson	239,430
Louisiana	East Baton Rouge	102,539
	Parish	331,307

State	County	Unincorporated urbanized population
	Jefferson Parish.	
Maryland	Howard	157,972
North Carolina	Cumberland	146,827
Nevada	Clark	327,618
Oregon	Multnomah [1]	52,923
	Washington	116,687
South Carolina	Greenville	147,464
	Richland	130,589
Virginia	Arlington	170,936
	Chesterfield	174,488
	Henrico	201,367
	Prince William	157,131
Washington	Pierce	258,530
	Snohomish	157,218

[1] County was previously listed in this appendix; however, population dropped to below 100,000 in the 1990 Census.

[64 FR 68849, Dec. 8, 1999]

APPENDIX J TO PART 122—NPDES PERMIT TESTING REQUIREMENTS FOR PUBLICLY OWNED TREATMENT WORKS (§ 122.21(j))

TABLE 1A—EFFLUENT PARAMETERS FOR ALL POTWS

Biochemical oxygen demand (BOD-5 or CBOD-5)
Fecal coliform
Design Flow Rate
pH
Temperature
Total suspended solids

TABLE 1—EFFLUENT PARAMETERS FOR ALL POTWS WITH A FLOW EQUAL TO OR GREATER THAN 0.1 MGD

Ammonia (as N)
Chlorine (total residual, TRC)
Dissolved oxygen
Nitrate/Nitrite
Kjeldahl nitrogen
Oil and grease
Phosphorus
Total dissolved solids

TABLE 2—EFFLUENT PARAMETERS FOR SELECTED POTWS

Hardness
Metals (total recoverable), cyanide and total phenols
Antimony
Arsenic
Beryllium
Cadmium
Chromium
Copper
Lead
Mercury
Nickel
Selenium
Silver
Thallium

Zinc
Cyanide
Total phenolic compounds
Volatile organic compounds
Acrolein
Acrylonitrile
Benzene
Bromoform
Carbon tetrachloride
Chlorobenzene
Chlorodibromomethane
Chloroethane
2-chloroethylvinyl ether
Chloroform
Dichlorobromomethane
1,1-dichloroethane
1,2-dichloroethane
Trans-1,2-dichloroethylene
1,1-dichloroethylene
1,2-dichloropropane
1,3-dichloropropylene
Ethylbenzene
Methyl bromide
Methyl chloride
Methylene chloride
1,1,2,2-tetrachloroethane
Tetrachloroethylene
Toluene
1,1,1-trichloroethane
1,1,2-trichloroethane
Trichloroethylene
Vinyl chloride
Acid-extractable compounds
P-chloro-m-creso
2-chlorophenol
2,4-dichlorophenol
2,4-dimethylphenol
4,6-dinitro-o-cresol
2,4-dinitrophenol
2-nitrophenol
4-nitrophenol
Pentachlorophenol
Phenol
2,4,6-trichlorophenol
Base-neutral compounds
Acenaphthene
Acenaphthylene
Anthracene
Benzidine
Benzo(a)anthracene
Benzo(a)pyrene
3,4 benzofluoranthene
Benzo(ghi)perylene
Benzo(k)fluoranthene
Bis (2-chloroethoxy) methane
Bis (2-chloroethyl) ether
Bis (2-chloroisopropyl) ether
Bis (2-ethylhexyl) phthalate
4-bromophenyl phenyl ether
Butyl benzyl phthalate
2-chloronaphthalene
4-chlorophenyl phenyl ether
Chrysene
Di-n-butyl phthalate
Di-n-octyl phthalate
Dibenzo(a,h)anthracene
1,2-dichlorobenzene

1,3-dichlorobenzene
1,4-dichlorobenzene
3,3-dichlorobenzidine
Diethyl phthalate
Dimethyl phthalate
2,4-dinitrotoluene
2,6-dinitrotoluene
1,2-diphenylhydrazine
Fluoranthene
Fluorene
Hexachlorobenzene
Hexachlorobutadiene
Hexachlorocyclo-pentadiene
Hexachloroethane
Indeno(1,2,3-cd)pyrene
Isophorone
Naphthalene
Nitrobenzene
N-nitrosodi-n-propylamine
N-nitrosodimethylamine
N-nitrosodiphenylamine
Phenanthrene
Pyrene
1,2,4,-trichlorobenzene

[65 FR 42469, Aug. 4, 2000]

PART 123—STATE PROGRAM REQUIREMENTS

Subpart A—General

Sec.
123.1   Purpose and scope.
123.2   Definitions.
123.3   Coordination with other programs.

Subpart B—State Program Submissions

123.21   Elements of a program submission.
123.22   Program description.
123.23   Attorney General's statement.
123.24   Memorandum of Agreement with the Regional Administrator.
123.25   Requirements for permitting.
123.26   Requirements for compliance evaluation programs.
123.27   Requirements for enforcement authority.
123.28   Control of disposal of pollutants into wells.
123.29   Prohibition.
123.30   Judicial review of approval or denial of permits.
123.31   Requirements for eligibility of Indian Tribes.
123.32   Request by an Indian Tribe for a determination of eligibility.
123.33   Procedures for processing an Indian Tribe's application.
123.34   Provisions for Tribal criminal enforcement authority.
123.35   As the NPDES Permitting Authority for regulated small MS4s, what is my role?

123.36 Establishment of technical standards for concentrated animal feeding operations.

Subpart C—Transfer of Information and Permit Review

123.41 Sharing of information.
123.42 Receipt and use of Federal information.
123.43 Transmission of information to EPA.
123.44 EPA review of and objections to State permits.
123.45 Noncompliance and program reporting by the Director.
123.46 Individual control strategies.

Subpart D—Program Approval, Revision, and Withdrawal

123.61 Approval process.
123.62 Procedures for revision of State programs.
123.63 Criteria for withdrawal of State programs.
123.64 Procedures for withdrawal of State programs.

AUTHORITY: Clean Water Act, 33 U.S.C. 1251 *et seq.*

SOURCE: 48 FR 14178, Apr. 1, 1983, unless otherwise noted.

Subpart A—General

§ 123.1 Purpose and scope.

(a) This part specifies the procedures EPA will follow in approving, revising, and withdrawing State programs and the requirements State programs must meet to be approved by the Administrator under sections 318, 402, and 405(a) (National Pollutant Discharge Elimination System—NPDES) of the CWA. This part also specifies the procedures EPA will follow in approving, revising, and withdrawing State programs under section 405(f) (sludge management programs) of the CWA. The requirements that a State sewage sludge management program must meet for approval by the Administrator under section 405(f) are set out at 40 CFR part 501.

(b) These regulations are promulgated under the authority of sections 304(i), 101(e), 405, and 518(e) of the CWA, and implement the requirements of those sections.

(c) The Administrator will approve State programs which conform to the applicable requirements of this part. A State NPDES program will not be approved by the Administrator under sec-

tion 402 of CWA unless it has authority to control the discharges specified in sections 318 and 405(a) of CWA. Permit programs under sections 318 and 405(a) will not be approved independent of a section 402 program.

(d)(1) Upon approval of a State program, the Administrator shall suspend the issuance of Federal permits for those activities subject to the approved State program. After program approval EPA shall retain jurisdiction over any permits (including general permits) which it has issued unless arrangements have been made with the State in the Memorandum of Agreement for the State to assume responsibility for these permits. Retention of jurisdiction shall include the processing of any permit appeals, modification requests, or variance requests; the conduct of inspections, and the receipt and review of self-monitoring reports. If any permit appeal, modification request or variance request is not finally resolved when the federally issued permit expires, EPA may, with the consent of the State, retain jurisdiction until the matter is resolved.

(2) The procedures outlined in the preceding paragraph (d)(1) of this section for suspension of permitting authority and transfer of existing permits will also apply when EPA approves an Indian Tribe's application to operate a State program and a State was the authorized permitting authority under § 123.23(b) for activities within the scope of the newly approved program. The authorized State will retain jurisdiction over its existing permits as described in paragraph (d)(1) of this section absent a different arrangement stated in the Memorandum of Agreement executed between EPA and the Tribe.

(e) Upon submission of a complete program, EPA will conduct a public hearing, if interest is shown, and determine whether to approve or disapprove the program taking into consideration the requirements of this part, the CWA and any comments received.

(f) Any State program approved by the Administrator shall at all times be conducted in accordance with the requirements of this part.

(g)(1) Except as may be authorized pursuant to paragraph (g)(2) of this section or excluded by §122.3, the State program must prohibit all point source discharges of pollutants, all discharges into aquaculture projects, and all disposal of sewage sludge which results in any pollutant from such sludge entering into any waters of the United States within the State's jurisdiction except as authorized by a permit in effect under the State program or under section 402 of CWA. NPDES authority may be shared by two or more State agencies but each agency must have Statewide jurisdiction over a class of activities or discharges. When more than one agency is responsible for issuing permits, each agency must make a submission meeting the requirements of §123.21 before EPA will begin formal review.

(2) A State may seek approval of a partial or phased program in accordance with section 402(n) of the CWA.

(h) In many cases, States (other than Indian Tribes) will lack authority to regulate activities on Indian lands. This lack of authority does not impair that State's ability to obtain full program approval in accordance with this part, i.e., inability of a State to regulate activities on Indian lands does not constitute a partial program. EPA will administer the program on Indian lands if a State (or Indian Tribe) does not seek or have authority to regulate activities on Indian lands.

NOTE: States are advised to contact the United States Department of the Interior, Bureau of Indian Affairs, concerning authority over Indian lands.

(i) Nothing in this part precludes a State from:

(1) Adopting or enforcing requirements which are more stringent or more extensive than those required under this part;

(2) Operating a program with a greater scope of coverage than that required under this part. If an approved State program has greater scope of coverage than required by Federal law the additional coverage is not part of the Federally approved program.

NOTE: For example, if a State requires permits for discharges into publicly owned treatment works, these permits are not NPDES permits.

[48 FR 14178, Apr. 1, 1983, as amended at 54 FR 256, Jan. 4, 1989; 54 FR 18784, May 2, 1989; 58 FR 67981, Dec. 22, 1993; 59 FR 64343, Dec. 14, 1994; 63 FR 45122, Aug. 24, 1998]

§123.2 Definitions.

The definitions in part 122 apply to all subparts of this part.

[63 FR 45122, Aug. 24, 1998]

§123.3 Coordination with other programs.

Issuance of State permits under this part may be coordinated with issuance of RCRA, UIC, NPDES, and 404 permits whether they are controlled by the State, EPA, or the Corps of Engineers. See §124.4.

Subpart B—State Program Submissions

§123.21 Elements of a program submission.

(a) Any State that seeks to administer a program under this part shall submit to the Administrator at least three copies of a program submission. The submission shall contain the following:

(1) A letter from the Governor of the State (or in the case of an Indian Tribe in accordance with §123.33(b), the Tribal authority exercising powers substantially similar to those of a State Governor) requesting program approval;

(2) A complete program description, as required by §123.22, describing how the State intends to carry out its responsibilities under this part;

(3) An Attorney General's statement as required by §123.23;

(4) A Memorandum of Agreement with the Regional Administrator as required by §123.24;

(5) Copies of all applicable State statutes and regulations, including those governing State administrative procedures;

(b)(1) Within 30 days of receipt by EPA of a State program submission, EPA will notify the State whether its submission is complete. If EPA finds that a State's submission is complete, the statutory review period (i.e., the period of time allotted for formal EPA review of a proposed State program under CWA) shall be deemed to have begun on the date of receipt of the State's submission. If EPA finds that a State's submission is incomplete, the statutory review period shall not begin until all the necessary information is received by EPA.

(2) In the case of an Indian Tribe eligible under § 123.33(b), EPA shall take into consideration the contents of the Tribe's request submitted under § 123.32, in determining if the program submission required by § 123.21(a) is complete.

(c) If the State's submission is materially changed during the statutory review period, the statutory review period shall begin again upon receipt of the revised submission.

(d) The State and EPA may extend the statutory review period by agreement.

[48 FR 14178, Apr. 1, 1983; 50 FR 6941, Feb. 19, 1985, as amended at 58 FR 67981, Dec. 22, 1993; 59 FR 64343, Dec. 14, 1994]

§ 123.22 Program description.

Any State that seeks to administer a program under this part shall submit a description of the program it proposes to administer in lieu of the Federal program under State law or under an interstate compact. The program description shall include:

(a) A description in narrative form of the scope, structure, coverage and processes of the State program.

(b) A description (including organization charts) of the organization and structure of the State agency or agencies which will have responsibility for administering the program, including the information listed below. If more than one agency is responsible for administration of a program, each agency must have statewide jurisdiction over a class of activities. The responsibilities of each agency must be delineated, their procedures for coordination set forth, and an agency may be designated as a "lead agency" to facilitate communications between EPA and the State agencies having program responsibility. If the State proposes to administer a program of greater scope of coverage than is required by Federal law, the information provided under this paragraph shall indicate the resources dedicated to administering the Federally required portion of the program.

(1) A description of the State agency staff who will carry out the State program, including the number, occupations, and general duties of the employees. The State need not submit complete job descriptions for every employee carrying out the State program.

(2) An itemization of the estimated costs of establishing and administering the program for the first two years after approval, including cost of the personnel listed in paragraph (b)(1) of this section, cost of administrative support, and cost of technical support.

(3) An itemization of the sources and amounts of funding, including an estimate of Federal grant money, available to the State Director for the first two years after approval to meet the costs listed in paragraph (b)(2) of this section, identifying any restrictions or limitations upon this funding.

(c) A description of applicable State procedures, including permitting procedures and any State administrative or judicial review procedures;

(d) Copies of the permit form(s), application form(s), and reporting form(s) the State intends to employ in its program. Forms used by States need not be identical to the forms used by EPA but should require the same basic information, except that State NPDES programs are required to use standard Discharge Monitoring Reports (DMR). The State need not provide copies of uniform national forms it intends to use but should note its intention to use such forms.

NOTE: States are encouraged to use uniform national forms established by the Administrator. If uniform national forms are used, they may be modified to include the State Agency's name, address, logo, and other similar information, as appropriate, in place of EPA's.

(e) A complete description of the State's compliance tracking and enforcement program.

(f) In the case of Indian Tribes eligible under § 123.33(b), if a State has been authorized by EPA to issue permits on the Federal Indian reservation in accordance with § 123.23(b), a description of how responsibility for pending permit applications, existing permits, and supporting files will be transferred from the State to the eligible Indian

Tribe. To the maximum extent practicable, this should include a Memorandum of Agreement negotiated between the State and the Indian Tribe addressing the arrangements for such transfer.

[48 FR 14178, Apr. 1, 1983; 50 FR 6941, Feb. 19, 1985, as amended at 54 FR 18784, May 2, 1989; 58 FR 67981, Dec. 22, 1993; 59 FR 64343, Dec. 14, 1994; 63 FR 45122, Aug. 24, 1998]

§123.23 Attorney General's statement.

(a) Any State that seeks to administer a program under this part shall submit a statement from the State Attorney General (or the attorney for those State or interstate agencies which have independent legal counsel) that the laws of the State, or an interstate compact, provide adequate authority to carry out the program described under §123.22 and to meet the requirements of this part. This statement shall include citations to the specific statutes, administrative regulations, and, where appropriate, judicial decisions which demonstrate adequate authority. State statutes and regulations cited by the State Attorney General or independent legal counsel shall be in the form of lawfully adopted State statutes and regulations at the time the statement is signed and shall be fully effective by the time the program is approved. To qualify as "independent legal counsel" the attorney signing the statement required by this section must have full authority to independently represent the State agency in court on all matters pertaining to the State program.

NOTE: EPA will supply States with an Attorney General's statement format on request.

(b) If a State (which is not an Indian Tribe) seeks authority over activities on Indian lands, the statement shall contain an appropriate analysis of the State's authority.

(c) The Attorney General's statement shall certify that the State has adequate legal authority to issue and enforce general permits if the State seeks to implement the general permit program under §122.28.

[48 FR 14178, Apr. 1, 1983, as amended at 58 FR 67981, Dec. 22, 1993]

§123.24 Memorandum of Agreement with the Regional Administrator.

(a) Any State that seeks to administer a program under this part shall submit a Memorandum of Agreement. The Memorandum of Agreement shall be executed by the State Director and the Regional Administrator and shall become effective when approved by the Administrator. In addition to meeting the requirements of paragraph (b) of this section, the Memorandum of Agreement may include other terms, conditions, or agreements consistent with this part and relevant to the administration and enforcement of the State's regulatory program. The Administrator shall not approve any Memorandum of Agreement which contains provisions which restrict EPA's statutory oversight responsibility.

(b) The Memorandum of Agreement shall include the following:

(1)(i) Provisions for the prompt transfer from EPA to the State of pending permit applications and any other information relevant to program operation not already in the possession of the State Director (e.g., support files for permit issuance, compliance reports, etc.). If existing permits are transferred from EPA to the State for administration, the Memorandum of Agreement shall contain provisions specifying a procedure for transferring the administration of these permits. If a State lacks the authority to directly administer permits issued by the Federal government, a procedure may be established to transfer responsibility for these permits.

NOTE: For example, EPA and the State and the permittee could agree that the State would issue a permit(s) identical to the outstanding Federal permit which would simultaneously be terminated.

(ii) Where a State has been authorized by EPA to issue permits in accordance with §123.23(b) on the Federal Indian reservation of the Indian Tribe seeking program approval, provisions describing how the transfer of pending permit applications, permits, and any other information relevant to the program operation not already in the possession of the Indian Tribe (support files for permit issuance, compliance reports, etc.) will be accomplished.

(2) Provisions specifying classes and categories of permit applications, draft permits, and proposed permits that the State will send to the Regional Administrator for review, comment and, where applicable, objection.

(3) Provisions specifying the frequency and content of reports, documents and other information which the State is required to submit to EPA. The State shall allow EPA to routinely review State records, reports, and files relevant to the administration and enforcement of the approved program. State reports may be combined with grant reports where appropriate. These procedures shall implement the requirements of § 123.43.

(4) Provisions on the State's compliance monitoring and enforcement program, including:

(i) Provisions for coordination of compliance monitoring activities by the State and by EPA. These may specify the basis on which the Regional Administrator will select facilities or activities within the State for EPA inspection. The Regional Administrator will normally notify the State at least 7 days before any such inspection; and

(ii) Procedures to assure coordination of enforcement activities.

(5) When appropriate, provisions for joint processing of permits by the State and EPA for facilities or activities which require permits from both EPA and the State under different programs. (See § 124.4.)

NOTE: To promote efficiency and to avoid duplication and inconsistency, States are encouraged to enter into joint processing agreements with EPA for permit issuance. Likewise, States are encouraged (but not required) to consider steps to coordinate or consolidate their own permit programs and activities.

(6) Provisions for modification of the Memorandum of Agreement in accordance with this part.

(c) The Memorandum of Agreement, the annual program grant and the State/EPA Agreement should be consistent. If the State/EPA Agreement indicates that a change is needed in the Memorandum of Agreement, the Memorandum of Agreement may be amended through the procedures set forth in this part. The State/EPA Agreement may not override the Memorandum of Agreement.

NOTE: Detailed program priorities and specific arrangements for EPA support of the State program will change and are therefore more appropriately negotiated in the context of annual agreements rather than in the MOA. However, it may still be appropriate to specify in the MOA the basis for such detailed agreements, e.g., a provision in the MOA specifying that EPA will select facilities in the State for inspection annually as part of the State/EPA agreement.

(d) The Memorandum of Agreement shall also specify the extent to which EPA will waive its right to review, object to, or comment upon State-issued permits under section 402(d)(3), (e) or (f) of CWA. While the Regional Administrator and the State may agree to waive EPA review of certain "classes or categories" of permits, no waiver of review may be granted for the following classes or categories:

(1) Discharges into the territorial sea;

(2) Discharges which may affect the waters of a State other than the one in which the discharge originates;

(3) Discharges proposed to be regulated by general permits (see § 122.28);

(4) Discharges from publicly owned treatment works with a daily average discharge exceeding 1 million gallons per day;

(5) Discharges of uncontaminated cooling water with a daily average discharge exceeding 500 million gallons per day;

(6) Discharges from any major discharger or from any discharger within any of the 21 industrial categories listed in appendix A to part 122;

(7) Discharges from other sources with a daily average discharge exceeding 0.5 (one-half) million gallons per day, except that EPA review of permits for discharges of non-process wastewater may be waived regardless of flow.

(e) Whenever a waiver is granted under paragraph (d) of this section, the Memorandum of Agreement shall contain:

(1) A statement that the Regional Administrator retains the right to terminate the waiver as to future permit actions, in whole or in part, at any time by sending the State Director written notice of termination; and

(2) A statement that the State shall supply EPA with copies of final permits.

[48 FR 14178, Apr. 1, 1983; 50 FR 6941, Feb. 19, 1985, as amended at 54 FR 18784, May 2, 1989; 58 FR 67981, Dec. 22, 1993; 63 FR 45122, Aug. 24, 1998]

§ 123.25 Requirements for permitting.

(a) All State Programs under this part must have legal authority to implement each of the following provisions and must be administered in conformance with each, except that States are not precluded from omitting or modifying any provisions to impose more stringent requirements:

(1) § 122.4—(Prohibitions):

(2) § 122.5(a) and (b)—(Effect of permit);

(3) § 122.7(b) and (c)—(Confidential information);

(4) § 122.21 (a)-(b), (c)(2), (e)-(k), (m)-(p), (q), and (r)—(Application for a permit);

(5) § 122.22—(Signatories);

(6) § 122.23—(Concentrated animal feeding operations);

(7) § 122.24—(Concentrated aquatic animal production facilities);

(8) § 122.25—(Aquaculture projects);

(9) § 122.26—(Storm water discharges);

(10) § 122.27—(Silviculture);

(11) § 122.28—(General permits), *Provided* that States which do not seek to implement the general permit program under § 122.28 need not do so.

(12) Section 122.41 (a)(1) and (b) through (n)—(Applicable permit conditions) (Indian Tribes can satisfy enforcement authority requirements under § 123.34);

(13) § 122.42—(Conditions applicable to specified categories of permits);

(14) § 122.43—(Establishing permit conditions);

(15) § 122.44—(Establishing NPDES permit conditions);

(16) § 122.45—(Calculating permit conditions);

(17) § 122.46—(Duration);

(18) § 122.47(a)—(Schedules of compliance);

(19) § 122.48—(Monitoring requirements);

(20) § 122.50—(Disposal into wells);

(21) § 122.61—(Permit transfer);

(22) § 122.62—(Permit modification);

(23) § 122.64—(Permit termination);

(24) § 124.3(a)—(Application for a permit);

(25) § 124.5 (a), (c), (d), and (f)—(Modification of permits);

(26) § 124.6 (a), (c), (d), and (e)—(Draft permit);

(27) § 124.8—(Fact sheets);

(28) § 124.10 (a)(1)(ii), (a)(1)(iii), (a)(1)(v), (b), (c), (d), and (e)—(Public notice);

(29) § 124.11—(Public comments and requests for hearings);

(30) § 124.12(a)—(Public hearings); and

(31) § 124.17 (a) and (c)—(Response to comments);

(32) § 124.56—(Fact sheets);

(33) § 124.57(a)—(Public notice);

(34) § 124.59—(Comments from government agencies);

(35) § 124.62—(Decision on variances);

(36) Subparts A, B, D, H, I, J, and N of part 125 of this chapter;

(37) 40 CFR parts 129, 133, and subchapter N;

(38) For a Great Lakes State or Tribe (as defined in 40 CFR 132.2), 40 CFR part 132 (NPDES permitting implementation procedures only);

(39) § 122.30 (What are the objectives of the storm water regulations for small MS4s?);

(40) § 122.31 (For Indian Tribes only) (As a Tribe, what is my role under the NPDES storm water program?);

(41) § 122.32 (As an operator of a small MS4, am I regulated under the NPDES storm water program?);

(42) § 122.33 (If I am an operator of a regulated small MS4, how do I apply for an NPDES permit? When do I have to apply?);

(43) § 122.34 (As an operator of a regulated small MS4, what will my NPDES MS4 storm water permit require?);

(44) § 122.35 (As an operator of a regulated small MS4, may I share the responsibility to implement the minimum control measures with other entities?);

(45) § 122.36 (As an operator of a regulated small MS4, what happens if I don't comply with the application or permit requirements in §§ 122.33 through 122.35?); and

(46) For states that wish to receive electronic documents, 40 CFR Part 3— (Electronic reporting).

NOTE: Except for paragraph (a)(46) of this section, states need not implement provisions identical to the above listed provisions.

Implemented provisions must, however, establish requirements at least as stringent as the corresponding listed provisions. While States may impose more stringent requirements, they may not make one requirement more lenient as a tradeoff for making another requirement more stringent; for example, by requiring that public hearings be held prior to issuing any permit while reducing the amount of advance notice of such a hearing.

State programs may, if they have adequate legal authority, implement any of the provisions of parts 122 and 124. See, for example, § 122.5(d) (continuation of permits) and § 124.4 (consolidation of permit processing).

For example, a State may impose more stringent requirements in an NPDES program by omitting the upset provision of § 122.41 or by requiring more prompt notice of an upset.

(b) State NPDES programs shall have an approved continuing planning process under 40 CFR 130.5 and shall assure that the approved planning process is at all times consistent with the CWA.

(c) State NPDES programs shall ensure that any board or body which approves all or portions of permits shall not include as a member any person who receives, or has during the previous 2 years received, a significant portion of income directly or indirectly from permit holders or applicants for a permit.

(1) For the purposes of this paragraph:

(i) *Board or body* includes any individual, including the Director, who has or shares authority to approve all or portions of permits either in the first instance, as modified or reissued, or on appeal.

(ii) *Significant portion of income* means 10 percent or more of gross personal income for a calendar year, except that it means 50 percent or more of gross personal income for a calendar year if the recipient is over 60 years of age and is receiving that portion under retirement, pension, or similar arrangement.

(iii) *Permit holders or applicants for a permit* does not include any department or agency of a State government, such as a Department of Parks or a Department of Fish and Wildlife.

(iv) *Income* includes retirement benefits, consultant fees, and stock dividends.

(2) For the purposes of paragraph (c) of this section, income is not received "directly or indirectly from permit holders or applicants for a permit" when it is derived from mutual fund payments, or from other diversified investments for which the recipient does not know the identity of the primary sources of income.

[48 FR 14178, Apr. 1, 1983]

EDITORIAL NOTE: For FEDERAL REGISTER citations affecting § 123.25, see the List of CFR Sections Affected, which appears in the Finding Aids section of the printed volume and at *www.fdsys.gov*.

§ 123.26 Requirements for compliance evaluation programs.

(a) State programs shall have procedures for receipt, evaluation, retention and investigation for possible enforcement of all notices and reports required of permittees and other regulated persons (and for investigation for possible enforcement of failure to submit these notices and reports).

(b) State programs shall have inspection and surveillance procedures to determine, independent of information supplied by regulated persons, compliance or noncompliance with applicable program requirements. The State shall maintain:

(1) A program which is capable of making comprehensive surveys of all facilities and activities subject to the State Director's authority to identify persons subject to regulation who have failed to comply with permit application or other program requirements. Any compilation, index or inventory of such facilities and activities shall be made available to the Regional Administrator upon request;

(2) A program for periodic inspections of the facilities and activities subject to regulation. These inspections shall be conducted in a manner designed to:

(i) Determine compliance or noncompliance with issued permit conditions and other program requirements;

(ii) Verify the accuracy of information submitted by permittees and other regulated persons in reporting forms and other forms supplying monitoring data; and

(iii) Verify the adequacy of sampling, monitoring, and other methods used by permittees and other regulated persons to develop that information;

(3) A program for investigating information obtained regarding violations

of applicable program and permit requirements; and

(4) Procedures for receiving and ensuring proper consideration of information submitted by the Public about violations. Public effort in reporting violations shall be encouraged, and the State Director shall make available information on reporting procedures.

(c) The State Director and State officers engaged in compliance evaluation shall have authority to enter any site or premises subject to regulation or in which records relevant to program operation are kept in order to copy any records, inspect, monitor or otherwise investigate compliance with the State program including compliance with permit conditions and other program requirements. States whose law requires a search warrant before entry conform with this requirement.

(d) Investigatory inspections shall be conducted, samples shall be taken and other information shall be gathered in a manner (e.g., using proper "chain of custody" procedures) that will produce evidence admissible in an enforcement proceeding or in court.

(e) State NPDES compliance evaluation programs shall have procedures and ability for:

(1) Maintaining a comprehensive inventory of all sources covered by NPDES permits and a schedule of reports required to be submitted by permittees to the State agency;

(2) Initial screening (i.e., pre-enforcement evaluation) of all permit or grant-related compliance information to identify violations and to establish priorities for further substantive technical evaluation;

(3) When warranted, conducting a substantive technical evaluation following the initial screening of all permit or grant-related compliance information to determine the appropriate agency response;

(4) Maintaining a management information system which supports the compliance evaluation activities of this part; and

(5) Inspecting the facilities of all major dischargers at least annually.

[48 FR 14178, Apr. 1, 1983, as amended at 54 FR 18785, May 2, 1989; 63 FR 45122, Aug. 24, 1998]

§123.27 Requirements for enforcement authority.

(a) Any State agency administering a program shall have available the following remedies for violations of State program requirements:

(1) To restrain immediately and effectively any person by order or by suit in State court from engaging in any unauthorized activity which is endangering or causing damage to public health or the environment;

NOTE: This paragraph (a)(1) requires that States have a mechanism (e.g., an administrative cease and desist order or the ability to seek a temporary restraining order) to stop any unauthorized activity endangering public health or the environment.

(2) To sue in courts of competent jurisdiction to enjoin any threatened or continuing violation of any program requirement, including permit conditions, without the necessity of a prior revocation of the permit;

(3) To assess or sue to recover in court civil penalties and to seek criminal remedies, including fines, as follows:

(i) Civil penalties shall be recoverable for the violation of any NPDES permit condition; any NPDES filing requirement; any duty to allow or carry out inspection, entry or monitoring activities; or, any regulation or orders issued by the State Director. These penalties shall be assessable in at least the amount of $5,000 a day for each violation.

(ii) Criminal fines shall be recoverable against any person who willfully or negligently violates any applicable standards or limitations; any NPDES permit condition; or any NPDES filing requirement. These fines shall be assessable in at least the amount of $10,000 a day for each violation.

NOTE: States which provide the criminal remedies based on "criminal negligence," "gross negligence" or strict liability satisfy the requirement of paragraph (a)(3)(ii) of this section.

(iii) Criminal fines shall be recoverable against any person who knowingly makes any false statement, representation or certification in any NPDES form, in any notice or report required by an NPDES permit, or who knowingly renders inaccurate any monitoring device or method required to be maintained by the Director. These fines shall be recoverable in at least the amount of $5,000 for each instance of violation.

NOTE: In many States the State Director will be represented in State courts by the State Attorney General or other appropriate legal officer. Although the State Director need not appear in court actions he or she should have power to request that any of the above actions be brought.

(b)(1) The maximum civil penalty or criminal fine (as provided in paragraph (a)(3) of this section) shall be assessable for each instance of violation and, if the violation is continuous, shall be assessable up to the maximum amount for each day of violation.

(2) The burden of proof and degree of knowledge or intent required under State law for establishing violations under paragraph (a)(3) of this section, shall be no greater than the burden of proof or degree of knowledge or intent EPA must provide when it brings an action under the appropriate Act;

NOTE: For example, this requirement is not met if State law includes mental state as an element of proof for civil violations.

(c) A civil penalty assessed, sought, or agreed upon by the State Director under paragraph (a)(3) of this section shall be appropriate to the violation.

NOTE: To the extent that State judgments or settlements provide penalties in amounts which EPA believes to be substantially inadequate in comparison to the amounts which EPA would require under similar facts, EPA, when authorized by the applicable statute, may commence separate actions for penalties.

Procedures for assessment by the State of the cost of investigations, inspections, or monitoring surveys which lead to the establishment of violations;

In addition to the requirements of this paragraph, the State may have other enforcement remedies. The following enforcement options, while not mandatory, are highly recommended:

Procedures which enable the State to assess or to sue any persons responsible for unauthorized activities for any expenses incurred by the State in removing, correcting, or terminating any adverse effects upon human health and the environment resulting from the unauthorized activity, whether or not accidental;

Procedures which enable the State to sue for compensation for any loss or destruction of wildlife, fish or aquatic life, or their habitat, and for any other damages caused by unauthorized activity, either to the State or to any residents of the State who are directly aggrieved by the unauthorized activity, or both; and

Procedures for the administrative assessment of penalties by the Director.

(d) Any State administering a program shall provide for public participation in the State enforcement process by providing either:

(1) Authority which allows intervention as of right in any civil or administrative action to obtain remedies specified in paragraphs (a)(1), (2) or (3) of this section by any citizen having an interest which is or may be adversely affected; or

(2) Assurance that the State agency or enforcement authority will:

(i) Investigate and provide written responses to all citizen complaints submitted pursuant to the procedures specified in § 123.26(b)(4);

(ii) Not oppose intervention by any citizen when permissive intervention may be authorized by statute, rule, or regulation; and

(iii) Publish notice of and provide at least 30 days for public comment on any proposed settlement of a State enforcement action.

(e) Indian Tribes that cannot satisfy the criminal enforcement authority requirements of this section may still receive program approval if they meet the requirement for enforcement authority established under § 123.34.

(Clean Water Act (33 U.S.C. 1251 *et seq.*), Safe Drinking Water Act (42 U.S.C. 300f *et seq.*), Clean Air Act (42 U.S.C. 7401 *et seq.*), Resource Conservation and Recovery Act (42 U.S.C. 6901 *et seq.*))

[48 FR 14178, Apr. 1, 1983, as amended at 48 FR 39620, Sept. 1, 1983; 50 FR 6941, Feb. 19, 1985; 54 FR 258, Jan. 4, 1989; 58 FR 67981, Dec. 22, 1993]

§ 123.28 Control of disposal of pollutants into wells.

State law must provide authority to issue permits to control the disposal of pollutants into wells. Such authority shall enable the State to protect the public health and welfare and to prevent the pollution of ground and surface waters by prohibiting well discharges or by issuing permits for such discharges with appropriate permit terms and conditions. A program approved under section 1422 of SDWA satisfies the requirements of this section.

NOTE: States which are authorized to administer the NPDES permit program under section 402 of CWA are encouraged to rely on existing statutory authority, to the extent possible, in developing a State UIC program under section 1422 of SDWA. Section 402(b)(1)(D) of CWA requires that NPDES States have the authority "to issue permits which * * * control the disposal of pollutants into wells." In many instances, therefore, NPDES States will have existing statutory authority to regulate well disposal which satisfies the requirements of the UIC program. Note, however, that CWA excludes certain types of well injections from the definition of "pollutant." If the State's statutory authority contains a similar exclusion it may need to be modified to qualify for UIC program approval.

§123.29 Prohibition.

State permit programs shall provide that no permit shall be issued when the Regional Administrator has objected in writing under §123.44.

§123.30 Judicial review of approval or denial of permits.

All States that administer or seek to administer a program under this part shall provide an opportunity for judicial review in State Court of the final approval or denial of permits by the State that is sufficient to provide for, encourage, and assist public participation in the permitting process. A State will meet this standard if State law allows an opportunity for judicial review that is the same as that available to obtain judicial review in federal court of a federally-issued NPDES permit (see §509 of the Clean Water Act). A State will not meet this standard if it narrowly restricts the class of persons who may challenge the approval or denial of permits (for example, if only the permittee can obtain judicial review, if persons must demonstrate injury to a pecuniary interest in order to obtain judicial review, or if persons must have a property interest in close proximity to a discharge or surface waters in order to obtain judicial review.) This requirement does not apply to Indian Tribes.

[61 FR 20980, May 8, 1996]

§123.31 Requirements for eligibility of Indian Tribes.

(a) Consistent with section 518(e) of the CWA, 33 U.S.C. 1377(e), the Re-gional Administrator will treat an Indian Tribe as eligible to apply for NPDES program authority if it meets the following criteria:

(1) The Indian Tribe is recognized by the Secretary of the Interior.

(2) The Indian Tribe has a governing body carrying out substantial governmental duties and powers.

(3) The functions to be exercised by the Indian Tribe pertain to the management and protection of water resources which are held by an Indian Tribe, held by the United States in trust for the Indians, held by a member of an Indian Tribe if such property interest is subject to a trust restriction on alienation, or otherwise within the borders of an Indian reservation.

(4) The Indian Tribe is reasonably expected to be capable, in the Regional Administrator's judgment, of carrying out the functions to be exercised, in a manner consistent with the terms and purposes of the Act and applicable regulations, of an effective NPDES permit program.

(b) An Indian Tribe which the Regional Administrator determines meets the criteria described in paragraph (a) of this section must also satisfy the State program requirements described in this part for assumption of the State program.

[58 FR 67981, Dec. 22, 1993, as amended at 59 FR 64343, Dec. 14, 1994]

§123.32 Request by an Indian Tribe for a determination of eligibility.

An Indian Tribe may apply to the Regional Administrator for a determination that it qualifies pursuant to section 518 of the Act for purposes of seeking NPDES permit program approval. The application shall be concise and describe how the Indian Tribe will meet each of the requirements of §123.31. The application shall include the following information:

(a) A statement that the Tribe is recognized by the Secretary of the Interior;

(b) A descriptive statement demonstrating that the Tribal governing body is currently carrying out substantial governmental duties and powers over a defined area. This statement should:

(1) Describe the form of the Tribal government;

(2) Describe the types of governmental functions currently performed by the Tribal governing body, such as, but not limited to, the exercise of police powers affecting (or relating to) the health, safety, and welfare of the affected population; taxation; and the exercise of the power of eminent domain; and

(3) Identify the source of the Tribal government's authority to carry out the governmental functions currently being performed.

(c) A map or legal description of the area over which the Indian Tribe asserts authority under section 518(e)(2) of the Act; a statement by the Tribal Attorney General (or equivalent official authorized to represent the Tribe in all legal matters in court pertaining to the program for which it seeks approval) which describes the basis for the Tribe's assertion (including the nature or subject matter of the asserted regulatory authority); copies of those documents such as Tribal constitutions, by-laws, charters, executive orders, codes, ordinances, and/or resolutions which support the Tribe believes are relevant to its assertion under section 518(e)(2) of the Act; and a description of the location of the surface waters for which the Tribe proposes to establish an NPDES permit program.

(d) A narrative statement describing the capability of the Indian Tribe to administer an effective, environmentally sound NPDES permit program. The statement should include:

(1) A description of the Indian Tribe's previous management experience which may include the administration of programs and service authorized by the Indian Self-Determination and Education Assistance Act (25 U.S.C. 450 et seq.), the Indian Mineral Development Act (25 U.S.C. 2101 et seq.), or the Indian Sanitation Facility Construction Activity Act (42 U.S.C. 2004a);

(2) A list of existing environmental or public health programs administered by the Tribal governing body, and a copy of related Tribal laws, regulations, and policies;

(3) A description of the entity (or entities) which exercise the executive, legislative, and judicial functions of the Tribal government;

(4) A description of the existing, or proposed, agency of the Indian Tribe which will assume primary responsibility for establishing and administering an NPDES permit program (including a description of the relationship between the existing or proposed agency and its regulated entities);

(5) A description of the technical and administrative abilities of the staff to administer and manage an effective, environmentally sound NPDES permit program or a plan which proposes how the Tribe will acquire additional administrative and technical expertise. The plan must address how the Tribe will obtain the funds to acquire the administrative and technical expertise.

(e) The Regional Administrator may, at his or her discretion, request further documentation necessary to support a Tribe's eligibility.

(f) If the Administrator or his or her delegatee has previously determined that a Tribe has met the prerequisites that make it eligible to assume a role similar to that of a state as provided by statute under the Safe Drinking Water Act, the Clean Water Act, or the Clean Air Act, then that Tribe need provide only that information unique to the NPDES program which is requested by the Regional Administrator.

[58 FR 67982, Dec. 22, 1993, as amended at 59 FR 64343, Dec. 14, 1994]

§ 123.33 Procedures for processing an Indian Tribe's application.

(a) The Regional Administrator shall process an application of an Indian Tribe submitted pursuant to § 123.32 in a timely manner. He shall promptly notify the Indian Tribe of receipt of the application.

(b) The Regional Administrator shall follow the procedures described in 40 CFR part 123, subpart D in processing a Tribe's request to assume the NPDES program.

[58 FR 67982, Dec. 22, 1993, as amended at 59 FR 64343, Dec. 14, 1994]

§123.34 Provisions for Tribal criminal enforcement authority.

To the extent that an Indian Tribe is precluded from asserting criminal enforcement authority as required under §123.27, the Federal Government will exercise primary criminal enforcement responsibility. The Tribe, with the EPA Region, shall develop a procedure by which the Tribal agency will refer potential criminal violations to the Regional Administrator, as agreed to by the parties, in an appropriate and timely manner. This procedure shall encompass all circumstances in which the Tribe is incapable of exercising the enforcement requirements of §123.27. This agreement shall be incorporated into a joint or separate Memorandum of Agreement with the EPA Region, as appropriate.

[58 FR 67983, Dec. 22, 1993]

§123.35 As the NPDES Permitting Authority for regulated small MS4s, what is my role?

(a) You must comply with the requirements for all NPDES permitting authorities under Parts 122, 123, 124, and 125 of this chapter. (This section is meant only to supplement those requirements and discuss specific issues related to the small MS4 storm water program.)

(b) You must develop a process, as well as criteria, to designate small MS4s other than those described in §122.32(a)(1) of this chapter, as regulated small MS4s to be covered under the NPDES storm water discharge control program. This process must include the authority to designate a small MS4 waived under paragraph (d) of this section if circumstances change. EPA may make designations under this section if a State or Tribe fails to comply with the requirements listed in this paragraph. In making designations of small MS4s, you must:

(1)(i) Develop criteria to evaluate whether a storm water discharge results in or has the potential to result in exceedances of water quality standards, including impairment of designated uses, or other significant water quality impacts, including habitat and biological impacts.

(ii) Guidance: For determining other significant water quality impacts, EPA recommends a balanced consideration of the following designation criteria on a watershed or other local basis: discharge to sensitive waters, high growth or growth potential, high population density, contiguity to an urbanized area, significant contributor of pollutants to waters of the United States, and ineffective protection of water quality by other programs;

(2) Apply such criteria, at a minimum, to any small MS4 located outside of an urbanized area serving a jurisdiction with a population density of at least 1,000 people per square mile and a population of at least 10,000;

(3) Designate any small MS4 that meets your criteria by December 9, 2002. You may wait until December 8, 2004 to apply the designation criteria on a watershed basis if you have developed a comprehensive watershed plan. You may apply these criteria to make additional designations at any time, as appropriate; and

(4) Designate any small MS4 that contributes substantially to the pollutant loadings of a physically interconnected municipal separate storm sewer that is regulated by the NPDES storm water program.

(c) You must make a final determination within 180 days from receipt of a petition under §122.26(f) of this chapter (or analogous State or Tribal law). If you do not do so within that time period, EPA may make a determination on the petition.

(d) You must issue permits consistent with §§122.32 through 122.35 of this chapter to all regulated small MS4s. You may waive or phase in the requirements otherwise applicable to regulated small MS4s, as defined in §122.32(a)(1) of this chapter, under the following circumstances:

(1) You may waive permit coverage for each small MS4s in jurisdictions with a population under 1,000 within the urbanized area where all of the following criteria have been met:

(i) Its discharges are not contributing substantially to the pollutant loadings of a physically interconnected regulated MS4 (see paragraph (b)(4) of this section); and

(ii) If the small MS4 discharges any pollutant(s) that have been identified as a cause of impairment of any water body to which it discharges, storm water controls are not needed based on wasteload allocations that are part of an EPA approved or established "total maximum daily load" (TMDL) that address the pollutant(s) of concern.

(2) You may waive permit coverage for each small MS4 in jurisdictions with a population under 10,000 where all of the following criteria have been met:

(i) You have evaluated all waters of the U.S., including small streams, tributaries, lakes, and ponds, that receive a discharge from the MS4 eligible for such a waiver.

(ii) For all such waters, you have determined that storm water controls are not needed based on wasteload allocations that are part of an EPA approved or established TMDL that addresses the pollutant(s) of concern or, if a TMDL has not been developed or approved, an equivalent analysis that determines sources and allocations for the pollutant(s) of concern.

(iii) For the purpose of paragraph (d)(2)(ii) of this section, the pollutant(s) of concern include biochemical oxygen demand (BOD), sediment or a parameter that addresses sediment (such as total suspended solids, turbidity or siltation), pathogens, oil and grease, and any pollutant that has been identified as a cause of impairment of any water body that will receive a discharge from the MS4.

(iv) You have determined that current and future discharges from the MS4 do not have the potential to result in exceedances of water quality standards, including impairment of designated uses, or other significant water quality impacts, including habitat and biological impacts.

(v) Guidance: To help determine other significant water quality impacts, EPA recommends a balanced consideration of the following criteria on a watershed or other local basis: discharge to sensitive waters, high growth or growth potential, high population or commercial density, significant contributor of pollutants to waters of the United States, and ineffective protection of water quality by other programs.

(3) You may phase in permit coverage for small MS4s serving jurisdictions with a population under 10,000 on a schedule consistent with a State watershed permitting approach. Under this approach, you must develop and implement a schedule to phase in permit coverage for approximately 20 percent annually of all small MS4s that qualify for such phased-in coverage. Under this option, all regulated small MS4s are required to have coverage under an NPDES permit by no later than March 8, 2007. Your schedule for phasing in permit coverage for small MS4s must be approved by the Regional Administrator no later than December 10, 2001.

(4) If you choose to phase in permit coverage for small MS4s in jurisdictions with a population under 10,000, in accordance with paragraph (d)(3) of this section, you may also provide waivers in accordance with paragraphs (d)(1) and (d)(2) of this section pursuant to your approved schedule.

(5) If you do not have an approved schedule for phasing in permit coverage, you must make a determination whether to issue an NPDES permit or allow a waiver in accordance with paragraph (d)(1) or (d)(2) of this section, for each eligible MS4 by December 9, 2002.

(6) You must periodically review any waivers granted in accordance with paragraph (d)(2) of this section to determine whether any of the information required for granting the waiver has changed. At a minimum, you must conduct such a review once every five years. In addition, you must consider any petition to review any waiver when the petitioner provides evidence that the information required for granting the waiver has substantially changed.

(e) You must specify a time period of up to 5 years from the date of permit issuance for operators of regulated small MS4s to fully develop and implement their storm water program.

(f) You must include the requirements in §§ 122.33 through 122.35 of this chapter in any permit issued for regulated small MS4s or develop permit limits based on a permit application submitted by a regulated small MS4. (You may include conditions in a regulated small MS4 NPDES permit that direct the MS4 to follow an existing qualifying local program's requirements, as a way of complying with some or all of the requirements in §122.34(b) of this chapter. See §122.34(c) of this chapter. Qualifying local, State or Tribal program requirements must impose, at a minimum, the relevant requirements of §122.34(b) of this chapter.)

(g) If you issue a general permit to authorize storm water discharges from small MS4s, you must make available a menu of BMPs to assist regulated small MS4s in the design and implementation of municipal storm water management programs to implement the minimum measures specified in §122.34(b) of this chapter. EPA plans to develop a menu of BMPs that will apply in each State or Tribe that has not developed its own menu. Regardless of whether a menu of BMPs has been developed by EPA, EPA encourages State and Tribal permitting authorities to develop a menu of BMPs that is appropriate for local conditions. EPA also intends to provide guidance on developing BMPs and measurable goals and modify, update, and supplement such guidance based on the assessments of the NPDES MS4 storm water program and research to be conducted over the next thirteen years.

(h)(1) You must incorporate any additional measures necessary to ensure effective implementation of your State or Tribal storm water program for regulated small MS4s.

(2) Guidance: EPA recommends consideration of the following:

(i) You are encouraged to use a general permit for regulated small MS4s;

(ii) To the extent that your State or Tribe administers a dedicated funding source, you should play an active role in providing financial assistance to operators of regulated small MS4s;

(iii) You should support local programs by providing technical and programmatic assistance, conducting research projects, performing watershed monitoring, and providing adequate legal authority at the local level;

(iv) You are encouraged to coordinate and utilize the data collected under several programs including water quality management programs, TMDL programs, and water quality monitoring programs;

(v) Where appropriate, you may recognize existing responsibilities among governmental entities for the control measures in an NPDES small MS4 permit (see §122.35(b) of this chapter); and

(vi) You are encouraged to provide a brief (e.g., two page) reporting format to facilitate compiling and analyzing data from submitted reports under §122.34(g)(3) of this chapter. EPA intends to develop a model form for this purpose.

[64 FR 68850, Dec. 8, 1999]

§123.36 Establishment of technical standards for concentrated animal feeding operations.

If the State has not already established technical standards for nutrient management that are consistent with 40 CFR 412.4(c)(2), the Director shall establish such standards by the date specified in §123.62(e).

[68 FR 7269, Feb. 12, 2003]

Subpart C—Transfer of Information and Permit Review

§123.41 Sharing of information.

(a) Any information obtained or used in the administration of a State program shall be available to EPA upon request without restriction. If the information has been submitted to the State under a claim of confidentiality, the State must submit that claim to EPA when providing information under this section. Any information obtained from a State and subject to a claim of confidentiality will be treated in accordance with the regulations in 40 CFR part 2. If EPA obtains from a State information that is not claimed to be confidential, EPA may make that information available to the public without further notice.

(b) EPA shall furnish to States with approved programs the information in its files not submitted under a claim of confidentiality which the State needs to implement its approved program. EPA shall furnish to States with approved programs information submitted to EPA under a claim of confidentiality, which the State needs to implement its approved program, subject to the conditions in 40 CFR part 2.

§ 123.42   Receipt and use of Federal information.

Upon approving a State permit program, EPA will send to the State agency administering the permit program any relevant information which was collected by EPA. The Memorandum of Agreement under § 123.24 (or, in the case of a sewage sludge management program, § 501.14 of this chapter) will provide for the following, in such manner as the State Director and the Regional Administrator agree:

(a) Prompt transmission to the State Director from the Regional Administrator of copies of any pending permit applications or any other relevant information collected before the approval of the State permit program and not already in the possession of the State Director. When existing permits are transferred to the State Director (e.g., for purposes of compliance monitoring, enforcement or reissuance), relevant information includes support files for permit issuance, compliance reports and records of enforcement actions.

(b) Procedures to ensure that the State Director will not issue a permit on the basis of any application received from the Regional Administrator which the Regional Administrator identifies as incomplete or otherwise deficient until the State Director receives information sufficient to correct the deficiency.

[48 FR 14178, Apr. 1, 1983, as amended at 63 FR 45122, Aug. 24, 1998]

§ 123.43   Transmission of information to EPA.

(a) Each State agency administering a permit program shall transmit to the Regional Administrator copies of permit program forms and any other relevant information to the extent and in the manner agreed to by the State Director and Regional Administrator in the Memorandum of Agreement and not inconsistent with this part. Proposed permits shall be prepared by State agencies unless agreement to the contrary has been reached under § 123.44(j). The Memorandum of Agreement shall provide for the following:

(1) Prompt transmission to the Regional Administrator of a copy of all complete permit applications received by the State Director, except those for which permit review has been waived under § 123.24(d). The State shall supply EPA with copies of permit applications for which permit review has been waived whenever requested by EPA;

(2) Prompt transmission to the Regional Administrator of notice of every action taken by the State agency related to the consideration of any permit application or general permit, including a copy of each proposed or draft permit and any conditions, requirements, or documents which are related to the proposed or draft permit or which affect the authorization of the proposed permit, except those for which permit review has been waived under § 123.24(d). The State shall supply EPA with copies of notices for which permit review has been waived whenever requested by EPA; and

(3) Transmission to the Regional Administrator of a copy of every issued permit following issuance, along with any and all conditions, requirements, or documents which are related to or affect the authorization of the permit.

(b) If the State intends to waive any of the permit application requirements of § 122.21(j) or (q) of this chapter for a specific applicant, the Director must submit a written request to the Regional Administrator no less than 210 days prior to permit expiration. This request must include the State's justification for granting the waiver.

(c) The State program shall provide for transmission by the State Director to EPA of:

(1) Notices from publicly owned treatment works under § 122.42(b) and 40 CFR part 403, upon request of the Regional Administrator;

(2) A copy of any significant comments presented in writing pursuant to the public notice of a draft permit and a summary of any significant comments presented at any hearing on any draft permit, except those comments regarding permits for which permit review has been waived under § 123.24(d) and for which EPA has not otherwise requested receipt, if:

(i) The Regional Administrator requests this information; or

(ii) The proposed permit contains requirements significantly different from those contained in the tentative determination and draft permit; or

(iii) Significant comments objecting to the tentative determination and draft permit have been presented at the hearing or in writing pursuant to the public notice.

(d) Any State permit program shall keep such records and submit to the Administrator such information as the Administrator may reasonably require to ascertain whether the State program complies with the requirements of CWA or of this part.

[48 FR 14178, Apr. 1, 1983, as amended at 60 FR 33931, June 29, 1995; 64 FR 42470, Aug. 4, 1999]

§123.44 EPA review of and objections to State permits.

(a)(1) The Memorandum of Agreement shall provide a period of time (up to 90 days from receipt of proposed permits) to which the Regional Administrator may make general comments upon, objections to, or recommendations with respect to proposed permits. EPA reserves the right to take 90 days to supply specific grounds for objection, notwithstanding any shorter period specified in the Memorandum of Agreement, when a general objection is filed within the review period specified in the Memorandum of Agreement. The Regional Administrator shall send a copy of any comment, objection or recommendation to the permit applicant.

(2) In the case of general permits, EPA shall have 90 days from the date of receipt of the proposed general permit to comment upon, object to or make recommendations with respect to the proposed general permit, and is not bound by any shorter time limits set by the Memorandum of Agreement for general comments, objections or recommendations.

(b)(1) Within the period of time provided under the Memorandum of Agreement for making general comments upon, objections to or recommendations with respect to proposed permits, the Regional Administrator shall notify the State Director of any objection to issuance of a proposed permit (except as provided in paragraph (a)(2) of this section for proposed general permits). This notification shall set forth in writing the general nature of the objection.

(2) Within 90 days following receipt of a proposed permit to which he or she has objected under paragraph (b)(1) of this section, or in the case of general permits within 90 days after receipt of the proposed general permit, the Regional Administrator shall set forth in writing and transmit to the State Director:

(i) A statement of the reasons for the objection (including the section of CWA or regulations that support the objection), and

(ii) The actions that must be taken by the State Director to eliminate the objection (including the effluent limitations and conditions which the permit would include if it were issued by the Regional Administrator.)

NOTE: Paragraphs (a) and (b) of this section, in effect, modify any existing agreement between EPA and the State which provides less than 90 days for EPA to supply the specific grounds for an objection. However, when an agreement provides for an EPA review period of less than 90 days, EPA must file a general objection, in accordance with paragraph (b)(1) of this section within the time specified in the agreement. This general objection must be followed by a specific objection within the 90-day period. This modification to MOA's allows EPA to provide detailed information concerning acceptable permit conditions, as required by section 402(d) of CWA. To avoid possible confusion, MOA's should be changed to reflect this arrangement.

(c) The Regional Administrator's objection to the issuance of a proposed permit must be based upon one or more of the following grounds:

(1) The permit fails to apply, or to ensure compliance with, any applicable requirement of this part;

NOTE: For example, the Regional Administrator may object to a permit not requiring the achievement of required effluent limitations by applicable statutory deadlines.

(2) In the case of a proposed permit for which notification to the Administrator is required under section 402(b)(5) of CWA, the written recommendations of an affected State have not been accepted by the permitting State and the Regional Administrator finds the reasons for rejecting the recommendations are inadequate;

(3) The procedures followed in connection with formulation of the proposed permit failed in a material respect to comply with procedures required by CWA or by regulations thereunder or by the Memorandum of Agreement;

(4) Any finding made by the State Director in connection with the proposed permit misinterprets CWA or any guidelines or regulations under CWA, or misapplies them to the facts;

(5) Any provisions of the proposed permit relating to the maintenance of records, reporting, monitoring, sampling, or the provision of any other information by the permittee are inadequate, in the judgment of the Regional Administrator, to assure compliance with permit conditions, including effluent standards and limitations or standards for sewage sludge use and disposal required by CWA, by the guidelines and regulations issued under CWA, or by the proposed permit;

(6) In the case of any proposed permit with respect to which applicable effluent standards and limitations or standards for sewage sludge use and disposal under sections 301, 302, 306, 307, 318, 403, and 405 of CWA have not yet been promulgated by the Agency, the proposed permit, in the judgment of the Regional Administrator, fails to carry out the provisions of CWA or of any regulations issued under CWA; the provisions of this paragraph apply to determinations made pursuant to § 125.3(c)(2) in the absence of applicable guidelines, to best management practices under section 304(e) of CWA, which must be incorporated into permits as requirements under section 301, 306, 307, 318, 403 or 405, and to sewage sludge use and disposal requirements developed on a case-by-case basis pursuant to section 405(d) of CWA, as the case may be;

(7) Issuance of the proposed permit would in any other respect be outside the requirements of CWA, or regulations issued under CWA.

(8) The effluent limits of a permit fail to satisfy the requirements of 40 CFR 122.44(d).

(9) For a permit issued by a Great Lakes State or Tribe (as defined in 40 CFR 132.2), the permit does not satisfy the conditions promulgated by the State, Tribe, or EPA pursuant to 40 CFR part 132.

(d) Prior to notifying the State Director of an objection based upon any of the grounds set forth in paragraph (c) of this section, the Regional Administrator:

(1) Will consider all data transmitted pursuant to § 123.43 (or, in the case of a sewage sludge management program, § 501.21 of this chapter);

(2) May, if the information provided is inadequate to determine whether the proposed permit meets the guidelines and requirements of CWA, request the State Director to transmit to the Regional Administrator the complete record of the permit proceedings before the State, or any portions of the record that the Regional Administrator determines are necessary for review. If this request is made within 30 days of receipt of the State submittal under § 123.43 (or, in the case of a sewage sludge management program, § 501.21 of this chapter), it will constitute an interim objection to the issuance of the permit, and the full period of time specified in the Memorandum of Agreement for the Regional Administrator's review will recommence when the Regional Administrator has received such record or portions of the record; and

(3) May, in his or her discretion, and to the extent feasible within the period of time available under the Memorandum of Agreement, afford to interested persons an opportunity to comment on the basis for the objection;

(e) Within 90 days of receipt by the State Director of an objection by the Regional Administrator, the State or interstate agency or any interested person may request that a public hearing be held by the Regional Administrator on the objection. A public hearing in accordance with the procedures of § 124.12 (c) and (d) of this chapter (or, in the case of a sewage sludge management program, § 501.15(d)(7) of this chapter) will be held, and public notice provided in accordance with § 124.10 of this chapter, (or, in the case of a sewage sludge management program, § 501.15(d)(5) of this chapter), whenever requested by the State or the interstate agency which proposed the permit or if warranted by significant public interest based on requests received.

(f) A public hearing held under paragraph (e) of this section shall be conducted by the Regional Administrator, and, at the Regional Administrator's discretion, with the assistance of an EPA panel designated by the Regional Administrator, in an orderly and expeditious manner.

(g) Following the public hearing, the Regional Administrator shall reaffirm the original objection, modify the terms of the objection, or withdraw the objection, and shall notify the State of this decision.

(h)(1) If no public hearing is held under paragraph (e) of this section and the State does not resubmit a permit revised to meet the Regional Administrator's objection within 90 days of receipt of the objection, the Regional Administrator may issue the permit in accordance with parts 121, 122 and 124 of this chapter and any other guidelines and requirements of CWA.

(2) If a public hearing is held under paragraph (e) of this section, the Regional Administrator does not withdraw the objection, and the State does not resubmit a permit revised to meet the Regional Administrator's objection or modified objection within 30 days of the date of the Regional Administrator's notification under paragraph (g) of this section, the Regional Administrator may issue the permit in accordance with parts 121, 122 and 124 of this chapter and any other guidelines and requirements of CWA.

(3) Exclusive authority to issue the permit passes to EPA when the times set out in this paragraph expire.

(i) [Reserved]

(j) The Regional Administrator may agree, in the Memorandum of Agreement under §123.24 (or, in the case of a sewage sludge management program, §501.14 of this chapter), to review draft permits rather than proposed permits. In such a case, a proposed permit need not be prepared by the State and transmitted to the Regional Administrator for review in accordance with this section unless the State proposes to issue a permit which differs from the draft permit reviewed by the Regional Administrator, the Regional Adminis-

trator has objected to the draft permit, or there is significant public comment.

[48 FR 14178, Apr. 1, 1983, as amended at 54 FR 18785, May 2, 1989; 54 FR 23896, June 2, 1989; 60 FR 15386, Mar. 23, 1995; 63 FR 45122, Aug. 24, 1998; 65 FR 30910, May 15, 2000]

§123.45 Noncompliance and program reporting by the Director.

The Director shall prepare quarterly, semi-annual, and annual reports as detailed below. When the State is the permit-issuing authority, the State Director shall submit all reports required under this section to the Regional Administrator, and the EPA Region in turn shall submit the State reports to EPA Headquarters. When EPA is the permit-issuing authority, the Regional Administrator shall submit all reports required under this section to EPA Headquarters.

(a) *Quarterly reports.* The Director shall submit quarterly narrative reports for major permittees as follows:

(1) *Format.* The report shall use the following format:

(i) Provide a separate list of major NPDES permittees which shall be subcategorized as non-POTWs, POTWs, and Federal permittees.

(ii) Alphabetize each list by permittee name. When two or more permittees have the same name, the permittee with the lowest permit number shall be entered first.

(iii) For each permittee on the list, include the following information in the following order:

(A) The name, location, and permit number.

(B) A brief description and date of each instance of noncompliance for which paragraph (a)(2) of this section requires reporting. Each listing shall indicate each specific provision of paragraph (a)(2) (e.g., (ii)(A) thru (iii)(G)) which describes the reason for reporting the violation on the quarterly report.

(C) The date(s), and a brief description of the action(s) taken by the Director to ensure compliance.

(D) The status of the instance(s) of noncompliance and the date noncompliance was resolved.

(E) Any details which tend to explain or mitigate the instance(s) of noncompliance.

(2) *Instances of noncompliance by major dischargers to be reported*—(i) *General.* Instances of noncompliance, as defined in paragraphs (a)(2)(ii) and (iii) of this section, by major dischargers shall be reported in successive reports until the noncompliance is reported as resolved (i.e., the permittee is no longer violating the permit conditions reported as noncompliance in the QNCR). Once an instance of noncompliance is reported as resolved in the QNCR, it need not appear in subsequent reports.

(A) All reported violations must be listed on the QNCR for the reporting period when the violation occurred, even if the violation is resolved during that reporting period.

(B) All permittees under current enforcement orders (i.e., administrative and judicial orders and consent decrees) for previous instances of noncompliance must be listed in the QNCR until the orders have been satisfied in full and the permittee is in compliance with permit conditions. If the permittee is in compliance with the enforcement order, but has not achieved full compliance with permit conditions, the compliance status shall be reported as "resolved pending," but the permittee will continue to be listed on the QNCR.

(ii) *Category I noncompliance.* The following instances of noncompliance by major dischargers are Category I noncompliance:

(A) Violations of conditions in enforcement orders except compliance schedules and reports.

(B) Violations of compliance schedule milestones for starting construction, completing construction, and attaining final compliance by 90 days or more from the date of the milestone specified in an enforcement order or a permit.

(C) Violations of permit effluent limits that exceed the Appendix A "Criteria for Noncompliance Reporting in the NPDES Program".

(D) Failure to provide a compliance schedule report for final compliance or a monitoring report. This applies when the permittee has failed to submit a final compliance schedule progress report, pretreatment report, or a Discharge Monitoring Report within 30 days from the due date specified in an enforcement order or a permit.

(iii) *Category II noncompliance.* Category II noncompliance includes violations of permit conditions which the Agency believes to be of substantial concern and may not meet the Category I criteria. The following are instances of noncompliance which must be reported as Category II noncompliance unless the same violation meets the criteria for Category I noncompliance:

(A) (*1*) Violation of a permit limit;

(*2*) An unauthorized bypass;

(*3*) An unpermitted discharge; or

(*4*) A pass-through of pollutants which causes or has the potential to cause a water quality problem (e.g., fish kills, oil sheens) or health problems (e.g., beach closings, fishing bans, or other restrictions of beneficial uses).

(B) Failure of an approved POTW to implement its approved pretreatment program adequately including failure to enforce industrial pretreatment requirements on industrial users as required in the approved program.

(C) Violations of any compliance schedule milestones (except those milestones listed in paragraph (a)(2)(ii)(B) of this section) by 90 days or more from the date specified in an enforcement order or a permit.

(D) Failure of the permittee to provide reports (other than those reports listed in paragraph (a)(2)(ii)(D) of this section) within 30 days from the due date specified in an enforcement order or a permit.

(E) Instances when the required reports provided by the permittee are so deficient or incomplete as to cause misunderstanding by the Director and thus impede the review of the status of compliance.

(F) Violations of narrative requirements (e.g., requirements to develop Spill Prevention Control and Countermeasure Plans and requirements to implement Best Management Practices), which are of substantial concern to the regulatory agency.

(G) Any other violation or group of permit violations which the Director or Regional Administrator considers to be of substantial concern.

(b) *Semi-annual statistical summary report.* Summary information shall be provided twice a year on the number of major permittees with two or more violations of the same monthly average permit limitation in a six month period, including those otherwise reported under paragraph (a) of this section. This report shall be submitted at the same time, according to the Federal fiscal year calendar, as the first and third quarter QNCRs.

(c) *Annual reports for NPDES*—(1) *Annual noncompliance report.* Statistical reports shall be submitted by the Director on nonmajor NPDES permittees indicating the total number reviewed, the number of noncomplying nonmajor permittees, the number of enforcement actions, and number of permit modifications extending compliance deadlines. The statistical information shall be organized to follow the types of noncompliance listed in paragraph (a) of this section.

(2) A separate list of nonmajor discharges which are one or more years behind in construction phases of the compliance schedule shall also be submitted in alphabetical order by name and permit number.

(d) *Schedule*—(1) *For all quarterly reports.* On the last working day of May, August, November, and February, the State Director shall submit to the Regional Administrator information concerning noncompliance with NPDES permit requirements by major dischargers in the State in acordance with the following schedule. The Regional Administrator shall prepare and submit information for EPA-issued permits to EPA Headquarters in accordance with the same schedule:

QUARTERS COVERED BY REPORTS ON NONCOMPLIANCE BY MAJOR DISCHARGERS:
[Date for completion of reports]

January, February, and March	[1] May 31
April, May, and June	[1] August 31
July, August, and September	[1] November 30
October, November, and December.	[1] February 28

[1] Reports must be made available to the public for inspection and copying on this date.

(2) *For all annual reports.* The period for annual reports shall be for the calendar year ending December 31, with reports completed and available to the public no more than 60 days later.

APPENDIX A TO §123.45—CRITERIA FOR NON-COMPLIANCE REPORTING IN THE NPDES PROGRAM

This appendix describes the criteria for reporting violations of NPDES permit effluent limits in the quarterly noncompliance report (QNCR) as specified under §123.45(a)(2)(ii)(c). Any violation of an NPDES permit is a violation of the Clean Water Act (CWA) for which the permittee is liable. An agency's decision as to what enforcement action, if any, should be taken in such cases, will be based on an analysis of facts and legal requirements.

Violations of Permit Effluent Limits

Cases in which violations of permit effluent limits must be reported depend upon the magnitude and/or frequency of the violation. Effluent violations should be evaluated on a parameter-by-parameter and outfall-by-outfall basis. The criteria for reporting effluent violations are as follows:

a. Reporting Criteria for Violations of Monthly Average Permit Limits—Magnitude and Frequency

Violations of monthly average effluent limits which exceed or equal the product of the Technical Review Criteria (TRC) times the effluent limit, and occur two months in a six month period must be reported. TRCs are for two groups of pollutants.

Group I Pollutants—TRC=1.4
Group II Pollutants—TRC=1.2

b. Reporting Criteria for Chronic Violations of Monthly Average Limits

Chronic violations must be reported in the QNCR if the monthly average permit limits are exceeded any four months in a six-month period. These criteria apply to all Group I and Group II pollutants.

GROUP I POLLUTANTS—TRC=1.4

Oxygen Demand

Biochemical Oxygen Demand
Chemical Oxygen Demand
Total Oxygen Demands
Total Organic Carbon
Other

Solids

Total Suspended Solids (Residues)
Total Dissolved Solids (Residues)
Other

Nutrients

Inorganic Phosphorus Compounds
Inorganic Nitrogen Compounds
Other

Detergents and Oils

MBAS

NTA
Oil and Grease
Other detergents or algicides

Minerals

Calcium
Chloride
Fluoride
Magnesium
Sodium
Potassium
Sulfur
Sulfate
Total Alkalinity
Total Hardness
Other Minerals

Metals

Aluminum
Cobalt
Iron
Vanadium

GROUP II POLLUTANTS—TRC=1.2

METALS (ALL FORMS)

Other metals not specifically listed under
Group I

Inorganic

Cyanide
Total Residual Chlorine

Organics

All organics are Group II except those specifically listed under Group I.

(Approved by the Office of Management and
Budget under control number 2040–0082)

[48 FR 14178, Apr. 1, 1983, as amended at 50
FR 34653, Aug. 26, 1985; 54 FR 18785, May 2,
1989; 63 FR 45123, Aug. 24, 1998]

§ 123.46   Individual control strategies.

(a) Not later than February 4, 1989,
each State shall submit to the Regional Administrator for review, approval, and implementation an individual control strategy for each point
source identified by the State pursuant
to section 304(l)(1)(C) of the Act which
discharges to a water identified by the
State pursuant to section 304(l)(1)(B)
which will produce a reduction in the
discharge of toxic pollutants from the
point sources identified under section
304(l)(1)(C) through the establishment
of effluent limitations under section
402 of the CWA and water quality
standards under section 303(c)(2)(B) of
the CWA, which reduction is sufficient,
in combination with existing controls
on point and nonpoint sources of pollution, to achieve the applicable water
quality standard as soon as possible,
but not later than three years after the
date of establishment of such strategy.

(b) The Administrator shall approve
or disapprove the control strategies
submitted by any State pursuant to
paragraph (a) of this section, not later
than June 4, 1989. If a State fails to
submit control strategies in accordance with paragraph (a) of this section
or the Administrator does not approve
the control strategies submitted by
such State in accordance with paragraph (a), then, not later than June 4,
1990, the Administrator in cooperation
with such State and after notice and
opportunity for public comment, shall
implement the requirements of CWA
section 304(l)(1) in such State. In the
implementation of such requirements,
the Administrator shall, at a minimum, consider for listing under CWA
section 304(l)(1) any navigable waters
for which any person submits a petition to the Administrator for listing
not later than October 1, 1989.

(c) For the purposes of this section
the term individual control strategy,
as set forth in section 304(l) of the
CWA, means a final NPDES permit
with supporting documentation showing that effluent limits are consistent
with an approved wasteload allocation,
or other documentation which shows
that applicable water quality standards
will be met not later than three years
after the individual control strategy is
established. Where a State is unable to
issue a final permit on or before February 4, 1989, an individual control
strategy may be a draft permit with an
attached schedule (provided the State
meets the schedule for issuing the final
permit) indicating that the permit will
be issued on or before February 4, 1990.
If a point source is subject to section
304(l)(1)(C) of the CWA and is also subject to an on-site response action under
sections 104 or 106 of the Comprehensive Environmental Response, Compensation, and Liability Act of 1980
(CERCLA), (42 U.S.C. 9601 *et seq.*), an
individual control strategy may be the
decision document (which incorporates
the applicable or relevant and appropriate requirements under the CWA)
prepared under sections 104 or 106 of
CERCLA to address the release or

threatened release of hazardous substances to the environment.

(d) A petition submitted pursuant to section 304(l)(3) of the CWA must be submitted to the appropriate Regional Administrator. Petitions must identify a waterbody in sufficient detail so that EPA is able to determine the location and boundaries of the waterbody. The petition must also identify the list or lists for which the waterbody qualifies, and the petition must explain why the waterbody satisfies the criteria for listing under CWA section 304(l) and 40 CFR 130.10(d)(6).

(e) If the Regional Administrator disapproves one or more individual control strategies, or if a State fails to provide adequate public notice and an opportunity to comment on the ICSs, then, not later than June 4, 1989, the Regional Administrator shall give a notice of approval or disapproval of the individual control strategies submitted by each State pursuant to this section as follows:

(1) The notice of approval or disapproval given under this paragraph shall include the following:

(i) The name and address of the EPA office that reviews the State's submittals.

(ii) A brief description of the section 304(l) process.

(iii) A list of ICSs disapproved under this section and a finding that the ICSs will not meet all applicable review criteria under this section and section 304(l) of the CWA.

(iv) If the Regional Administrator determines that a State did not provide adequate public notice and an opportunity to comment on the waters, point sources, or ICSs prepared pursuant to section 304(l), or if the Regional Administrator chooses to exercise his or her discretion, a list of the ICSs approved under this section, and a finding that the ICSs satisfy all applicable review criteria.

(v) The location where interested persons may examine EPA's records of approval and disapproval.

(vi) The name, address, and telephone number of the person at the Regional Office from whom interested persons may obtain more information.

(vii) Notice that written petitions or comments are due within 120 days.

(2) The Regional Administrator shall provide the notice of approval or disapproval given under this paragraph to the appropriate State Director. The Regional Administrator shall publish a notice of availability, in a daily or weekly newspaper with State-wide circulation or in the FEDERAL REGISTER, for the notice of approval or disapproval. The Regional Administrator shall also provide written notice to each discharger identified under section 304(l)(1)(C), that EPA has listed the discharger under section 304(l)(1)(C).

(3) As soon as practicable but not later than June 4, 1990, the Regional Offices shall issue a response to petitions or comments received under section 304(l). The response to comments shall be given in the same manner as the notice described in paragraph (e) of this section except for the following changes:

(i) The lists of ICSs reflecting any changes made pursuant to comments or petitions received.

(ii) A brief description of the subsequent steps in the section 304(l) process.

(f) EPA shall review, and approve or disapprove, the individual control strategies prepared under section 304(l) of the CWA, using the applicable criteria set forth in section 304(l) of the CWA, and in 40 CFR part 122, including § 122.44(d). At any time after the Regional Administrator disapproves an ICS (or conditionally approves a draft permit as an ICS), the Regional Office may submit a written notification to the State that the Regional Office intends to issue the ICS. Upon mailing the notification, and notwithstanding any other regulation, exclusive authority to issue the permit passes to EPA.

[54 FR 256, Jan. 4, 1989, as amended at 54 FR 23896, June 2, 1989; 57 FR 33049, July 24, 1992]

Subpart D—Program Approval, Revision, and Withdrawal

§ 123.61   Approval process.

(a) After determining that a State program submission is complete, EPA shall publish notice of the State's application in the FEDERAL REGISTER, and in enough of the largest newspapers in the State to attract statewide attention, and shall mail notice to persons known to be interested in such matters, including all persons on appropriate State and EPA mailing lists and all permit holders and applicants within the State. The notice shall:

(1) Provide a comment period of not less than 45 days during which interested members of the public may express their views on the State program;

(2) Provide for a public hearing within the State to be held no less than 30 days after notice is published in the FEDERAL REGISTER;

(3) Indicate the cost of obtaining a copy of the State's submission;

(4) Indicate where and when the State's submission may be reviewed by the public;

(5) Indicate whom an interested member of the public should contact with any questions; and

(6) Briefly outline the fundamental aspects of the State's proposed program, and the process for EPA review and decision.

(b) Within 90 days of the receipt of a complete program submission under § 123.21 the Administrator shall approve or disapprove the program based on the requirements of this part and of CWA and taking into consideration all comments received. A responsiveness summary shall be prepared by the Regional Office which identifies the public participation activities conducted, describes the matters presented to the public, summarizes significant comments received and explains the Agency's response to these comments.

(c) If the Administrator approves the State's program he or she shall notify the State and publish notice in the FEDERAL REGISTER. The Regional Administrator shall suspend the issuance of permits by EPA as of the date of program approval.

(d) If the Administrator disapproves the State program he or she shall notify the State of the reasons for disapproval and of any revisions or modifications to the State program which are necessary to obtain approval.

[48 FR 14178, Apr. 1, 1983; 50 FR 6941, Feb. 19, 1985]

§ 123.62 Procedures for revision of State programs.

(a) Either EPA or the approved State may initiate program revision. Program revision may be necessary when the controlling Federal or State statutory or regulatory authority is modified or supplemented. The State shall keep EPA fully informed of any proposed modifications to its basic statutory or regulatory authority, its forms, procedures, or priorities. Grounds for program revision include cases where a State's existing approved program includes authority to issue NPDES permits for activities on a Federal Indian reservation and an Indian Tribe has subsequently been approved for assumption of the NPDES program under 40 CFR part 123 extending to those lands.

(b) Revision of a State program shall be accomplished as follows:

(1) The State shall submit a modified program description, Attorney General's statement, Memorandum of Agreement, or such other documents as EPA determines to be necessary under the circumstances.

(2) Whenever EPA determines that the proposed program revision is substantial, EPA shall issue public notice and provide an opportunity to comment for a period of at least 30 days. The public notice shall be mailed to interested persons and shall be published in the FEDERAL REGISTER and in enough of the largest newspapers in the State to provide Statewide coverage. The public notice shall summarize the proposed revisions and provide for the opportunity to request a public hearing. Such a hearing will be held if there is significant public interest based on requests received.

(3) The Administrator will approve or disapprove program revisions based on the requirements of this part (or, in the case of a sewage sludge management program, 40 CFR part 501) and of the CWA.

(4) A program revision shall become effective upon the approval of the Administrator. Notice of approval of any substantial revision shall be published in the FEDERAL REGISTER. Notice of approval of non-substantial program revisions may be given by a letter from the

Administrator to the State Governor or his designee.

(c) States with approved programs must notify EPA whenever they propose to transfer all or part of any program from the approved State agency to any other State agency, and must identify any new division of responsibilities among the agencies involved. The new agency is not authorized to administer the program until approved by the Administrator under paragraph (b) of this section. Organizational charts required under §123.22(b) (or, in the case of a sewage sludge management program, §501.12(b) of this chapter) must be revised and resubmitted.

(d) Whenever the Administrator has reason to believe that circumstances have changed with respect to a State program, he may request, and the State shall provide, a supplemental Attorney General's statement, program description, or such other documents or information as are necessary.

(e) *State NPDES programs only.* All new programs must comply with these regulations immediately upon approval. Any approved State section 402 permit program which requires revision to conform to this part shall be so revised within one year of the date of promulgation of these regulations, unless a State must amend or enact a statute in order to make the required revision in which case such revision shall take place within 2 years, except that revision of State programs to implement the requirements of 40 CFR part 403 (pretreatment) shall be accomplished as provided in 40 CFR 403.10. In addition, approved States shall submit, within 6 months, copies of their permit forms for EPA review and approval. Approved States shall also assure that permit applicants, other than POTWs, submit, as part of their application, the information required under §§124.4(d) and 122.21 (g) or (h), as appropriate.

(f) Revision of a State program by a Great Lakes State or Tribe (as defined in 40 CFR 132.2) to conform to section 118 of the CWA and 40 CFR part 132 shall be accomplished pursuant to 40 CFR part 132.

[48 FR 14178, Apr. 1, 1983, as amended at 49 FR 31842, Aug. 8, 1984; 50 FR 6941, Feb. 19, 1985; 53 FR 33007, Sept. 6, 1988; 58 FR 67983, Dec. 22, 1993; 60 FR 15386, Mar. 23, 1995; 63 FR 45123, Aug. 24, 1998]

§ 123.63 Criteria for withdrawal of State programs.

(a) In the case of a sewage sludge management program, references in this section to "this part" will be deemed to refer to 40 CFR part 501. The Administrator may withdraw program approval when a State program no longer complies with the requirements of this part, and the State fails to take corrective action. Such circumstances include the following:

(1) Where the State's legal authority no longer meets the requirements of this part, including:

(i) Failure of the State to promulgate or enact new authorities when necessary; or

(ii) Action by a State legislature or court striking down or limiting State authorities.

(2) Where the operation of the State program fails to comply with the requirements of this part, including:

(i) Failure to exercise control over activities required to be regulated under this part, including failure to issue permits;

(ii) Repeated issuance of permits which do not conform to the requirements of this part; or

(iii) Failure to comply with the public participation requirements of this part.

(3) Where the State's enforcement program fails to comply with the requirements of this part, including:

(i) Failure to act on violations of permits or other program requirements;

(ii) Failure to seek adequate enforcement penalties or to collect administrative fines when imposed; or

(iii) Failure to inspect and monitor activities subject to regulation.

(4) Where the State program fails to comply with the terms of the Memorandum of Agreement required under §123.24 (or, in the case of a sewage sludge management program, §501.14 of this chapter).

(5) Where the State fails to develop an adequate regulatory program for developing water quality-based effluent limits in NPDES permits.

(6) Where a Great Lakes State or Tribe (as defined in 40 CFR 132.2) fails to adequately incorporate the NPDES permitting implementation procedures promulgated by the State, Tribe, or EPA pursuant to 40 CFR part 132 into individual permits.

(b) [Reserved]

[48 FR 14178, Apr. 1, 1983; 50 FR 6941, Feb. 19, 1985, as amended at 54 FR 23897, June 2, 1989; 60 FR 15386, Mar. 23, 1995; 63 FR 45123, Aug. 24, 1998]

§ 123.64 Procedures for withdrawal of State programs.

(a) A State with a program approved under this part (or, in the case of a sewage sludge management program, 40 CFR part 501) may voluntarily transfer program responsibilities required by Federal law to EPA by taking the following actions, or in such other manner as may be agreed upon with the Administrator.

(1) The State shall give the Administrator 180 days notice of the proposed transfer and shall submit a plan for the orderly transfer of all relevant program information not in the possession of EPA (such as permits, permit files, compliance files, reports, permit applications) which are necessary for EPA to administer the program.

(2) Within 60 days of receiving the notice and transfer plan, the Administrator shall evaluate the State's transfer plan and shall identify any additional information needed by the Federal government for program administration and/or identify any other deficiencies in the plan.

(3) At least 30 days before the transfer is to occur the Administrator shall publish notice of the transfer in the FEDERAL REGISTER and in enough of the largest newspapers in the State to provide Statewide coverage, and shall mail notice to all permit holders, permit applicants, other regulated persons and other interested persons on appropriate EPA and State mailing lists.

(b) The following procedures apply when the Administrator orders the commencement of proceedings to determine whether to withdraw approval of a State program.

(1) *Order.* The Administrator may order the commencement of withdrawal proceedings on his or her own initiative or in response to a petition from an interested person alleging failure of the State to comply with the requirements of this part as set forth in § 123.63 (or, in the case of a sewage sludge management program, § 501.33 of this chapter). The Administrator will respond in writing to any petition to commence withdrawal proceedings. He may conduct an informal investigation of the allegations in the petition to determine whether cause exists to commence proceedings under this paragraph. The Administrator's order commencing proceedings under this paragraph will fix a time and place for the commencement of the hearing and will specify the allegations against the State which are to be considered at the hearing. Within 30 days the State must admit or deny these allegations in a written answer. The party seeking withdrawal of the State's program will have the burden of coming forward with the evidence in a hearing under this paragraph.

(2) *Definitions.* For purposes of this paragraph the definitions of "Act," "Administrative Law Judge," "Hearing Clerk," and "Presiding Officer" in 40 CFR 22.03 apply in addition to the following:

(i) *Party* means the petitioner, the State, the Agency, and any other person whose request to participate as a party is granted.

(ii) *Person* means the Agency, the State and any individual or organization having an interest in the subject matter of the proceeding.

(iii) *Petitioner* means any person whose petition for commencement of withdrawal proceedings has been granted by the Administrator.

(3) *Procedures.* (i) The following provisions of 40 CFR part 22 (Consolidated Rules of Practice) are applicable to proceedings under this paragraph:

(A) § 22.02—(use of number/gender);

(B) § 22.04(c)—(authorities of Presiding Officer);

(C) § 22.06—(filing/service of rulings and orders);

(D) § 22.09—(examination of filed documents);

(E) § 22.19(a), (b) and (c)—(prehearing conference);

(F) § 22.22—(evidence);

(G) § 22.23—(objections/offers of proof);

(H) § 22.25—(filing the transcript); and

(I) § 22.26—(findings/conclusions).

(ii) The following provisions are also applicable:

(A) *Computation and extension of time—(1) Computation.* In computing any period of time prescribed or allowed in these rules of practice, except as otherwise provided, the day of the event from which the designated period begins to run shall not be included. Saturdays, Sundays, and Federal legal holidays shall be included. When a stated time expires on a Saturday, Sunday, or legal holiday, the stated time period shall be extended to include the next business day.

(2) *Extensions of time.* The Administrator, Regional Administrator, or Presiding Officer, as appropriate, may grant an extension of time for the filing of any pleading, document, or motion (*i*) upon timely motion of a party to the proceeding, for good cause shown, and after consideration of prejudice to other parties, or (*ii*) upon his own motion. Such a motion by a party may only be made after notice to all other parties, unless the movant can show good cause why serving notice is impracticable. The motion shall be filed in advance of the date on which the pleading, document or motion is due to be filed, unless the failure of a party to make timely motion for extension of time was the result of excusable neglect.

(3) The time for commencement of the hearing shall not be extended beyond the date set in the Administrator's order without approval of the Administrator.

(B) *Ex parte discussion of proceedings.* At no time after the issuance of the order commencing proceedings shall the Administrator, the Regional Administrator, the Regional Judicial Officer, the Presiding Officer, or any other person who is likely to advise these officials in the decision on the case, discuss ex parte the merits of the proceeding with any interested person outside the Agency, with any Agency staff member who performs a prosecutorial or investigative function in such proceeding or a factually related proceeding, or with any representative of such person. Any ex parte memorandum or other communication addressed to the Administrator, the Regional Administrator, the Regional Judicial Officer, or the Presiding Officer during the pendency of the proceeding and relating to the merits thereof, by or on behalf of any party, shall be regarded as argument made in the proceeding and shall be served upon all other parties. The other parties shall be given an opportunity to reply to such memorandum or communication.

(C) *Intervention—(1) Motion.* A motion for leave to intervene in any proceeding conducted under these rules of practice must set forth the grounds for the proposed intervention, the position and interest of the movant and the likely impact that intervention will have on the expeditious progress of the proceeding. Any person already a party to the proceeding may file an answer to a motion to intervene, making specific reference to the factors set forth in the foregoing sentence and paragraph (b)(3)(ii)(C)(3) of this section, within ten (10) days after service of the motion for leave to intervene.

(2) However, motions to intervene must be filed within 15 days from the date the notice of the Administrator's order is first published.

(3) *Disposition.* Leave to intervene may be granted only if the movant demonstrates that (*i*) his presence in the proceeding would not unduly prolong or otherwise prejudice that adjudication of the rights of the original parties; (*ii*) the movant will be adversely affected by a final order; and (*iii*) the interests of the movant are not being adequately represented by the original parties. The intervenor shall become a full party to the proceeding upon the granting of leave to intervene.

(4) *Amicus curiae.* Persons not parties to the proceeding who wish to file briefs may so move. The motion shall identify the interest of the applicant and shall state the reasons why the proposed amicus brief is desirable. If the motion is granted, the Presiding Officer or Administrator shall issue an

order setting the time for filing such brief. An amicus curiae is eligible to participate in any briefing after his motion is granted, and shall be served with all briefs, reply briefs, motions, and orders relating to issues to be briefed.

(D) *Motions*—(1) *General.* All motions, except those made orally on the record during a hearing, shall (*i*) be in writing; (*ii*) state the grounds therefor with particularity; (*iii*) set forth the relief or order sought; and (*iv*) be accompanied by any affidavit, certificate, other evidence, or legal memorandum relied upon. Such motions shall be served as provided by paragraph (b)(4) of this section.

(2) *Response to motions.* A party's response to any written motion must be filed within ten (10) days after service of such motion, unless additional time is allowed for such response. The response shall be accompanied by any affidavit, certificate, other evidence, or legal memorandum relied upon. If no response is filed within the designated period, the parties may be deemed to have waived any objection to the granting of the motion. The Presiding Officer, Regional Administrator, or Administrator, as appropriate, may set a shorter time for response, or make such other orders concerning the disposition of motions as they deem appropriate.

(3) *Decision.* The Administrator shall rule on all motions filed or made after service of the recommended decision upon the parties. The Presiding Officer shall rule on all other motions. Oral argument on motions will be permitted where the Presiding Officer, Regional Administrator, or the Administrator considers it necessary or desirable.

(4) *Record of proceedings.* (i) The hearing shall be either stenographically reported verbatim or tape recorded, and thereupon transcribed by an official reporter designated by the Presiding Officer;

(ii) All orders issued by the Presiding Officer, transcripts of testimony, written statements of position, stipulations, exhibits, motions, briefs, and other written material of any kind submitted in the hearing shall be a part of the record and shall be available for inspection or copying in the Office of the Hearing Clerk, upon payment of costs. Inquiries may be made at the Office of the Administrative Law Judges, Hearing Clerk, 1200 Pennsylvania Ave., NW., Washington, DC 20460;

(iii) Upon notice to all parties the Presiding Officer may authorize corrections to the transcript which involves matters of substance;

(iv) An original and two (2) copies of all written submissions to the hearing shall be filed with the Hearing Clerk;

(v) A copy of each submission shall be served by the person making the submission upon the Presiding Officer and each party of record. Service under this paragraph shall take place by mail or personal delivery;

(vi) Every submission shall be accompanied by an acknowledgement of service by the person served or proof of service in the form of a statement of the date, time, and manner of service and the names of the persons served, certified by the person who made service, and;

(vii) The Hearing Clerk shall maintain and furnish to any person upon request, a list containing the name, service address, and telephone number of all parties and their attorneys or duly authorized representatives.

(5) *Participation by a person not a party.* A person who is not a party may, in the discretion of the Presiding Officer, be permitted to make a limited appearance by making oral or written statement of his/her position on the issues within such limits and on such conditions as may be fixed by the Presiding Officer, but he/she may not otherwise participate in the proceeding.

(6) *Rights of parties.* (i) All parties to the proceeding may:

(A) Appear by counsel or other representative in all hearing and pre-hearing proceedings;

(B) Agree to stipulations of facts which shall be made a part of the record.

(7) *Recommended decision.* (i) Within 30 days after the filing of proposed findings and conclusions, and reply briefs, the Presiding Officer shall evaluate the record before him/her, the proposed findings and conclusions and any briefs filed by the parties and shall prepare a recommended decision, and shall certify the entire record, including the

recommended decision, to the Administrator.

(ii) Copies of the recommended decision shall be served upon all parties.

(iii) Within 20 days after the certification and filing of the record and recommended decision, all parties may file with the Administrator exceptions to the recommended decision and a supporting brief.

(8) *Decision by Administrator.* (i) Within 60 days after the certification of the record and filing of the Presiding Officer's recommeded decision, the Administrator shall review the record before him and issue his own decision.

(ii) If the Administrator concludes that the State has administered the program in conformity with the appropriate Act and regulations his decision shall constitute "final agency action" within the meaning of 5 U.S.C. 704.

(iii) If the Administrator concludes that the State has not administered the program in conformity with the appropriate Act and regulations he shall list the deficiencies in the program and provide the State a reasonable time, not to exceed 90 days, to take such appropriate corrective action as the Administrator determines necessary.

(iv) Within the time prescribed by the Administrator the State shall take such appropriate corrective action as required by the Administrator and shall file with the Administrator and all parties a statement certified by the State Director that such appropriate corrective action has been taken.

(v) The Administrator may require a further showing in addition to the certified statement that corrective action has been taken.

(vi) If the State fails to take such appropriate corrective action and file a certified statement thereof within the time prescribed by the Administrator, the Administrator shall issue a supplementary order withdrawing approval of the State program. If the State takes such appropriate corrective action, the Administrator shall issue a supplementary order stating that approval of authority is not withdrawn.

(vii) The Administrator's supplementary order shall constitute final Agency action within the meaning of 5 U.S.C. 704.

(viii) Withdrawal of authorization under this section and the appropriate Act does not relieve any person from complying with the requirements of State law, nor does it affect the validity of actions by the State prior to withdrawal.

[48 FR 14178, Apr. 1, 1983; 50 FR 6941, Feb. 19, 1985, as amended at 57 FR 5335, Feb. 13, 1992; 63 FR 45123, Aug. 24, 1998]

PART 124—PROCEDURES FOR DECISIONMAKING

Subpart A—General Program Requirements

Subpart B—Specific Procedures Applicable to RCRA Permits

Subpart C—Specific Procedures Applicable to PSD Permits

AUTHORITY: Resource Conservation and Recovery Act, 42 U.S.C. 6901 *et seq.*; Safe Drinking Water Act, 42 U.S.C. 300f *et seq.*; Clean Water Act, 33 U.S.C. 1251 *et seq.*; Clean Air Act, 42 U.S.C. 7401 *et seq.*

SOURCE: 48 FR 14264, Apr. 1, 1983, unless otherwise noted.

Subpart A—General Program Requirements

§ 124.1 Purpose and scope.

(a) This part contains EPA procedures for issuing, modifying, revoking and reissuing, or terminating all RCRA, UIC, PSD and NPDES "permits" (including "sludge-only" permits issued pursuant to § 122.1(b)(2) of this chapter. The latter kinds of permits are governed by part 270. RCRA interim status and UIC authorization by rule are not "permits" and are covered by specific provisions in parts 144, subpart C, and 270. This part also does not apply to permits issued, modified, revoked and reissued or terminated by the Corps of Engineers. Those procedures are specified in 33 CFR parts 320–327. The procedures of this part also apply to denial of a permit for the active life of a RCRA hazardous waste management facility or unit under § 270.29.

(b) Part 124 is organized into five subparts. Subpart A contains general procedural requirements applicable to all permit programs covered by these provisions. Subparts B through D and Subpart G supplement these general provisions with requirements that apply to only one or more of the programs. Subpart A describes the steps EPA will follow in receiving permit applications, preparing draft permits, issuing public notice, inviting public comment and holding public hearings on draft permits. Subpart A also covers assembling an administrative record, responding to comments, issuing a final permit decision, and allowing for administrative appeal of the final permit decisions. Subpart B contains public participation requirements applicable to all RCRA hazardous waste management facilities. Subpart C contains definitions and specific procedural requirements for PSD permits. Subpart D contains specific procedural requirements for NPDES permits. Subpart G contains specific procedural requirements for RCRA standardized permits, which, in some instances, change how the General Program Requirements of subpart A apply in the context of the RCRA standardized permit.

(c) Part 124 offers an opportunity for public hearings (see § 124.12).

(d) This part is designed to allow permits for a given facility under two or more of the listed programs to be processed separately or together at the choice of the Regional Administrator. This allows EPA to combine the processing of permits only when appropriate, and not necessarily in all cases. The Regional Administrator may consolidate permit processing when the permit applications are submitted, when draft permits are prepared, or when final permit decisions are issued. This part also allows consolidated permits to be subject to a single public hearing under § 124.12. Permit applicants may recommend whether or not their applications should be consolidated in any given case.

(e) Certain procedural requirements set forth in part 124 must be adopted by States in order to gain EPA approval to operate RCRA, UIC, NPDES, and 404 permit programs. These requirements are listed in §§ 123.25 (NPDES), 145.11 (UIC), 233.26 (404), and 271.14 (RCRA) and signaled by the following words at the end of the appropriate part 124 section or paragraph heading: (*applicable to State programs see §§ 123.25 (NPDES), 145.11 (UIC), 233.26 (404), and 271.14 (RCRA)*). Part 124 does not apply to

PSD permits issued by an approved State.

(f) To coordinate decisionmaking when different permits will be issued by EPA and approved State programs, this part allows applications to be jointly processed, joint comment periods and hearings to be held, and final permits to be issued on a cooperative basis whenever EPA and a State agree to take such steps in general or in individual cases. These joint processing agreements may be provided in the Memorandum of Agreement developed under §§ 123.24 (NPDES), 145.24 (UIC), 233.24 (404), and 271.8 (RCRA).

[48 FR 14264, Apr. 1, 1983, as amended at 54 FR 9607, Mar. 7, 1989; 54 FR 18785, May 2, 1989; 65 FR 30910, May 15, 2000; 70 FR 53448, Sept. 8, 2005]

§ 124.2 Definitions.

(a) In addition to the definitions given in §§ 122.2 and 123.2 (NPDES), 501.2 (sludge management), 144.3 and 145.2 (UIC), 233.3 (404), and 270.2 and 271.2 (RCRA), the definitions below apply to this part, except for PSD permits which are governed by the definitions in § 124.41. Terms not defined in this section have the meaning given by the appropriate Act.

Administrator means the Administrator of the U.S. Environmental Protection Agency, or an authorized representative.

Application means the EPA standard national forms for applying for a permit, including any additions, revisions or modifications to the forms; or forms approved by EPA for use in "approved States," including any approved modifications or revisions. For RCRA, application also includes the information required by the Director under §§ 270.14 through 270.29 [contents of Part B of the RCRA application].

Appropriate Act and regulations means the Clean Water Act (CWA); the Solid Waste Disposal Act, as amended by the Resource Conservation Recovery Act (RCRA); or Safe Drinking Water Act (SDWA), whichever is applicable; and applicable regulations promulgated under those statutes. In the case of an "approved State program" appropriate Act and regulations includes program requirements.

CWA means the Clean Water Act (formerly referred to as the Federal Water Pollution Control Act of Federal Pollution Control Act Amendments of 1972) Public Law 92–500, as amended by Public Law 95–217 and Public Law 95–576; 33 U.S.C. 1251 *et seq.*

Director means the Regional Administrator, the State director or the Tribal director as the context requires, or an authorized representative. When there is no approved State or Tribal program, and there is an EPA administered program, *Director* means the Regional Administrator. When there is an approved State or Tribal program, "Director" normally means the State or Tribal director. In some circumstances, however, EPA retains the authority to take certain actions even when there is an approved State or Tribal program. (For example, when EPA has issued an NPDES permit prior to the approval of a State program, EPA may retain jurisdiction over that permit after program approval; see § 123.1) In such cases, the term "Director" means the Regional Administrator and not the State or Tribal director.

Draft permit means a document prepared under § 124.6 indicating the Director's tentative decision to issue or deny, modify, revoke and reissue, terminate, or reissue a "permit." A notice of intent to terminate a permit and a notice of intent to deny a permit as discussed in § 124.5, are types of "draft permits." A denial of a request for modification, revocation and reissuance or termination, as discussed in § 124.5, is not a "draft permit." A "proposal permit" is not a "draft permit."

Environmental Appeals Board shall mean the Board within the Agency described in § 1.25(e) of this title. The Administrator delegates authority to the Environmental Appeals Board to issue final decisions in RCRA, PSD, UIC, or NPDES permit appeals filed under this subpart, including informal appeals of denials of requests for modification, revocation and reissuance, or termination of permits under Section 124.5(b). An appeal directed to the Administrator, rather than to the Environmental Appeals Board, will not be considered. This delegation does not preclude the Environmental Appeals Board from referring an appeal or a motion under this subpart to the Administrator when the Environmental Appeals Board, in its discretion, deems it appropriate to do so. When an appeal or motion is referred to the Administrator by the Environmental Appeals Board, all parties shall be so notified and the rules in this subpart referring to the Environmental Appeals Board shall be interpreted as referring to the Administrator.

EPA ("EPA") means the United States "Environmental Protection Agency."

Facility or activity means any "HWM facility," UIC "injection well," NPDES "point source" or "treatment works treating domestic sewage" or State 404 dredge or fill activity, or any other facility or activity (including land or appurtenances thereto) that is subject to regulation under the RCRA, UIC, NPDES, or 404 programs.

Federal Indian reservation (in the case of NPDES) means all land within the limits of any Indian reservation under the jurisdiction of the United States Government, notwithstanding the issuance of any patent, and including rights-of-way running through the reservation.

General permit (NPDES and 404) means an NPDES or 404 "permit" authorizing a category of discharges or activities under the CWA within a geographical area. For NPDES, a general permit means a permit issued under § 122.28. For 404, a general permit means a permit issued under § 233.37.

Indian Tribe means (in the case of UIC) any Indian Tribe having a federally recognized governing body carrying out substantial governmental duties and powers over a defined area. For the NPDES program, the term "Indian Tribe" means any Indian Tribe, band, group, or community recognized by the Secretary of the Interior and exercising governmental authority over a Federal Indian reservation.

Interstate agency means an agency of two or more States established by or under an agreement or compact approved by the Congress, or any other agency of two or more States having substantial powers or duties pertaining to the control of pollution as determined and approved by the Administrator under the "appropriate Act and regulations."

Major facility means any RCRA, UIC, NPDES, or 404 "facility or activity" classified as such by the Regional Administrator, or, in the case of "approved State programs," the Regional Administrator in conjunction with the State Director.

Owner or operator means owner or operator of any "facility or activity" subject to regulation under the RCRA, UIC, NPDES, or 404 programs.

Permit means an authorization, license or equivalent control document issued by EPA or an "approved State" to implement the requirements of this part and parts 122, 123, 144, 145, 233, 270, and 271 of this chapter. "Permit" includes RCRA "permit by rule" (§270.60), RCRA standardized permit (§270.67), UIC area permit (§144.33), NPDES or 404 "general permit" (§§270.61, 144.34, and 233.38). Permit does not include RCRA interim status (§270.70), UIC authorization by rule (§144.21), or any permit which has not yet been the subject of final agency action, such as a "draft permit" or a "proposed permit."

Person means an individual, association, partnership, corporation, municipality, State, Federal, or Tribal agency, or an agency or employee thereof.

RCRA means the Solid Waste Disposal Act as amended by the Resource Conservation and Recovery Act of 1976 (Pub. L. 94–580, as amended by Pub. L. 95–609, 42 U.S.C. 6901 *et seq*).

Regional Administrator means the Regional Administrator of the appropriate Regional Office of the Environmental Protection Agency or the authorized representative of the Regional Administrator.

Schedule of compliance means a schedule of remedial measures included in a "permit," including an enforceable sequence of interim requirements (for example, actions, operations, or milestone events) leading to compliance with the "appropriate Act and regulations."

SDWA means the Safe Drinking Water Act (Pub. L. 95–523, as amended by Pub. L. 95–1900; 42 U.S.C. 300f *et seq*).

Section 404 program or State 404 program or 404 means an "approved State program" to regulate the "discharge of dredged material" and the "discharge of fill material" under section 404 of the Clean Water Act in "State regulated waters."

Site means the land or water area where any "facility or activity" is physically located or conducted, including adjacent land used in connection with the facility or activity.

Standardized permit means a RCRA permit authorizing management of hazardous waste issued under subpart G of this part and part 270, subpart J. The standardized permit may have two parts: A uniform portion issued in all cases and a supplemental portion issued at the Director's discretion.

State means one of the States of the United States, the District of Columbia, the Commonwealth of Puerto Rico, the Virgin Islands, Guam, American Samoa, the Trust Territory of the Pacific Islands (except in the case of RCRA), the Commonwealth of the Northern Mariana Islands, or an Indian Tribe that meets the statutory criteria which authorize EPA to treat the Tribe in a manner similar to that in which it treats a State (except in the case of RCRA).

State Director means the chief administrative officer of any State, interstate, or Tribal agency operating an approved program, or the delegated representative of the State director. If the responsibility is divided among two or more States, interstate, or Tribal agencies, "State Director" means the chief administrative officer of the State, interstate, or Tribal agency authorized to perform the particular procedure or function to which reference is made.

State Director means the chief administrative officer of any State or interstate agency operating an "approved program," or the delegated representative of the state Director. If responsibility is divided among two or more State or interstate agencies, "State Director" means the chief administrative officer of the State or interstate agency authorized to perform the particular procedure or function to which reference is made.

UIC means the Underground Injection Control program under Part C of the Safe Drinking Water Act, including an "approved program."

(b) For the purposes of part 124, the term *Director* means the State Director or Regional Administrator and is used when the accompanying provision is required of EPA-administered programs and of State programs under §§ 123.25 (NPDES), 145.11 (UIC), 233.26 (404), and 271.14 (RCRA). The term *Regional Administrator* is used when the accompanying provision applies exclusively to EPA-issued permits and is not applicable to State programs under these sections. While States are not required to implement these latter provisions, they are not precluded from doing so, notwithstanding use of the term "Regional Administrator."

[48 FR 14264, Apr. 1, 1983; 48 FR 30115, June 30, 1983, as amended at 49 FR 25981, June 25, 1984; 53 FR 37410, Sept. 26, 1988; 54 FR 18785, May 2, 1989; 57 FR 5335, Feb. 13, 1992; 57 FR 60129, Dec. 18, 1992; 58 FR 67983, Dec. 22, 1993; 59 FR 64343, Dec. 14, 1994; 65 FR 30910, May 15, 2000; 70 FR 53449, Sept. 8, 2005]

§ 124.3 Application for a permit.

(a) *Applicable to State programs, see §§ 123.25 (NPDES), 145.11 (UIC), 233.26 (404), and 271.14 (RCRA).* (1) Any person who requires a permit under the RCRA, UIC, NPDES, or PSD programs shall complete, sign, and submit to the Director an application for each permit required under §§ 270.1 (RCRA), 144.1 (UIC), 40 CFR 52.21 (PSD), and 122.1 (NPDES). Applications are not required for RCRA permits by rule (§ 270.60), underground injections authorized by rules (§§ 144.21 through 144.26), NPDES general permits (§ 122.28) and 404 general permits (§ 233.37).

(2) The Director shall not begin the processing of a permit until the applicant has fully complied with the application requirements for that permit. See §§ 270.10, 270.13 (RCRA), 144.31 (UIC), 40 CFR 52.21 (PSD), and 122.21 (NPDES).

(3) Permit applications (except for PSD permits) must comply with the signature and certification requirements of §§ 122.22 (NPDES), 144.32 (UIC), 233.6 (404), and 270.11 (RCRA).

(b) [Reserved]

(c) The Regional Administrator shall review for completeness every application for an EPA-issued permit. Each application for an EPA-issued permit submitted by a new HWM facility, a new UIC injection well, a major PSD stationary source or major PSD modification, or an NPDES new source or NPDES new discharger should be reviewed for completeness by the Regional Administrator within 30 days of its receipt. Each application for an EPA-issued permit submitted by an existing HWM facility (both Parts A and B of the application), existing injection well or existing NPDES source or sludge-only facility should be reviewed for completeness within 60 days of receipt. Upon completing the review, the Regional Administrator shall notify the applicant in writing whether the application is complete. If the application is incomplete, the Regional Administrator shall list the information necessary to make the application complete. When the application is for an existing HWM facility, an existing UIC injection well or an existing NPDES source or "sludge-only facility" the Regional Administrator shall specify in the notice of deficiency a date for submitting the necessary information. The Regional Administrator shall notify the applicant that the application is complete upon receiving this information. After the application is completed, the Regional Administrator may request additional information from an applicant but only when necessary to clarify, modify, or supplement previously submitted material. Requests for such additional information will not render an application incomplete.

(d) If an applicant fails or refuses to correct deficiencies in the application, the permit may be denied and appropriate enforcement actions may be taken under the applicable statutory provision including RCRA section 3008, SDWA sections 1423 and 1424, CAA section 167, and CWA sections 308, 309, 402(h), and 402(k).

(e) If the Regional Administrator decides that a site visit is necessary for any reason in conjunction with the processing of an application, he or she shall notify the applicant and a date shall be scheduled.

(f) The effective date of an application is the date on which the Regional

Administrator notifies the applicant that the application is complete as provided in paragraph (c) of this section.

(g) For each application from a major new HWM facility, major new UIC injection well, major NPDES new source, major NPDES new discharger, or a permit to be issued under provisions of §122.28(c), the Regional Administrator shall, no later than the effective date of the application, prepare and mail to the applicant a project decision schedule. (This paragraph does not apply to PSD permits.) The schedule shall specify target dates by which the Regional Administrator intends to:

(1) Prepare a draft permit;

(2) Give public notice;

(3) Complete the public comment period, including any public hearing; and

(4) Issue a final permit.

(Clean Water Act (33 U.S.C. 1251 et seq.), Safe Drinking Water Act (42 U.S.C. 300f et seq.), Clean Air Act (42 U.S.C. 7401 et seq.), Resource Conservation and Recovery Act (42 U.S.C. 6901 et seq.))

[48 FR 14264, Apr. 1, 1983, as amended at 48 FR 39620, Sept. 1, 1983; 54 FR 18785, May 2, 1989; 65 FR 30910, May 15, 2000]

§124.4 Consolidation of permit processing.

(a)(1) Whenever a facility or activity requires a permit under more than one statute covered by these regulations, processing of two or more applications for those permits may be consolidated. The first step in consolidation is to prepare each draft permit at the same time.

(2) Whenever draft permits are prepared at the same time, the statements of basis (required under §124.7 for EPA-issued permits only) or fact sheets (§124.8), administrative records (required under §124.9 for EPA-issued permits only), public comment periods (§124.10), and any public hearings (§124.12) on those permits should also be consolidated. The final permits may be issued together. They need not be issued together if in the judgment of the Regional Administrator or State Director(s), joint processing would result in unreasonable delay in the issuance of one or more permits.

(b) Whenever an existing facility or activity requires additional permits under one or more of the statutes covered by these regulations, the permitting authority may coordinate the expiration date(s) of the new permit(s) with the expiration date(s) of the existing permit(s) so that all permits expire simultaneously. Processing of the subsequent applications for renewal permits may then be consolidated.

(c) Processing of permit applications under paragraph (a) or (b) of this section may be consolidated as follows:

(1) The Director may consolidate permit processing at his or her discretion whenever a facility or activity requires all permits either from EPA or from an approved State.

(2) The Regional Administrator and the State Director(s) may agree to consolidate draft permits whenever a facility or activity requires permits from both EPA and an approved State.

(3) Permit applicants may recommend whether or not the processing of their applications should be consolidated.

(d) [Reserved]

(e) Except with the written consent of the permit applicant, the Regional Administrator shall not consolidate processing a PSD permit with any other permit under paragraph (a) or (b) of this section when to do so would delay issuance of the PSD permit more than one year from the effective date of the application under §124.3(f).

[48 FR 14264, Apr. 1, 1983, as amended at 65 FR 30910, May 15, 2000]

§124.5 Modification, revocation and reissuance, or termination of permits.

(a) (Applicable to State programs, see §§123.25 (NPDES), 145.11 (UIC), 233.26 (404), and 271.14 (RCRA).) Permits (other than PSD permits) may be modified, revoked and reissued, or terminated either at the request of any interested person (including the permittee) or upon the Director's initiative. However, permits may only be modified, revoked and reissued, or terminated for the reasons specified in §122.62 or §122.64 (NPDES), 144.39 or 144.40 (UIC), 233.14 or 233.15 (404), and 270.41 or 270.43 (RCRA). All requests shall be in writing and shall contain facts or reasons supporting the request.

(b) If the Director decides the request is not justified, he or she shall send the requester a brief written response giving a reason for the decision. Denials of requests for modification, revocation and reissuance, or termination are not subject to public notice, comment, or hearings. Denials by the Regional Administrator may be informally appealed to the Environmental Appeals Board by a letter briefly setting forth the relevant facts. The Environmental Appeals Board may direct the Regional Administrator to begin modification, revocation and reissuance, or termination proceedings under paragraph (c) of this section. The appeal shall be considered denied if the Environmental Appeals Board takes no action on the letter within 60 days after receiving it. This informal appeal is, under 5 U.S.C. 704, a prerequisite to seeking judicial review of EPA action in denying a request for modification, revocation and reissuance, or termination.

(c) (*Applicable to State programs, see 40 CFR 123.25 (NPDES), 145.11 (UIC), 233.26 (404), and 271.14 (RCRA)*). (1) If the Director tentatively decides to modify or revoke and reissue a permit under 40 CFR 122.62 (NPDES), 144.39 (UIC), 233.14 (404), or 270.41 (other than § 270.41(b)(3)) or § 270.42(c) (RCRA), he or she shall prepare a draft permit under § 124.6 incorporating the proposed changes. The Director may request additional information and, in the case of a modified permit, may require the submission of an updated application. In the case of revoked and reissued permits, other than under 40 CFR 270.41(b)(3), the Director shall require the submission of a new application. In the case of revoked and reissued permits under 40 CFR 270.41(b)(3), the Director and the permittee shall comply with the appropriate requirements in 40 CFR part 124, subpart G for RCRA standardized permits.

(2) In a permit modification under this section, only those conditions to be modified shall be reopened when a new draft permit is prepared. All other aspects of the existing permit shall remain in effect for the duration of the unmodified permit. When a permit is revoked and reissued under this section, the entire permit is reopened just as if the permit had expired and was being reissued. During any revocation and reissuance proceeding the permittee shall comply with all conditions of the existing permit until a new final permit is reissued.

(3) "Minor modifications" as defined in §§ 122.63 (NPDES), 144.41 (UIC), and 233.16 (404), and "Classes 1 and 2 modifications" as defined in § 270.42 (a) and (b) (RCRA) are not subject to the requirements of this section.

(d) (*Applicable to State programs, see §§ 123.25 (NPDES) of this chapter, 145.11 (UIC) of this chapter, and 271.14 (RCRA) of this chapter.*) (1) If the Director tentatively decides to terminate: A permit under § 144.40 (UIC) of this chapter, a permit under §§ 122.64(a) (NPDES) of this chapter or 270.43 (RCRA) of this chapter (for EPA-issued NPDES permits, only at the request of the permittee), or a permit under § 122.64(b) (NPDES) of this chapter where the permittee objects, he or she shall issue a notice of intent to terminate. A notice of intent to terminate is a type of draft permit which follows the same procedures as any draft permit prepared under § 124.6 of this chapter.

(2) For EPA-issued NPDES or RCRA permits, if the Director tentatively decides to terminate a permit under § 122.64(a) (NPDES) of this chapter, other than at the request of the permittee, or decides to conduct a hearing under section 3008 of RCRA in connection with the termination of a RCRA permit, he or she shall prepare a complaint under 40 CFR 22.13 and 22.44 of this chapter. Such termination of NPDES and RCRA permits shall be subject to the procedures of part 22 of this chapter.

(3) In the case of EPA-issued permits, a notice of intent to terminate or a complaint shall not be issued if the Regional Administrator and the permittee agree to termination in the course of transferring permit responsibility to an approved State under §§ 123.24(b)(1) (NPDES) of this chapter, 145.25(b)(1) (UIC) of this chapter, 271.8(b)(6) (RCRA) of this chapter, or 501.14(b)(1) (sludge) of this chapter. In addition, termination of an NPDES permit for cause pursuant to § 122.64 of this chapter may be accomplished by providing written notice to the permittee, unless the permittee objects.

(e) When EPA is the permitting authority, all draft permits (including notices of intent to terminate) prepared under this section shall be based on the administrative record as defined in §124.9.

(f) (*Applicable to State programs, see §233.26 (404).*) Any request by the permittee for modification to an existing 404 permit (other than a request for a minor modification as defined in §233.16 (404)) shall be treated as a permit application and shall be processed in accordance with all requirements of §124.3.

(g)(1) (Reserved for PSD Modification Provisions).

(2) PSD permits may be terminated only by rescission under §52.21(w) or by automatic expiration under §52.21(r). Applications for rescission shall be precessed under §52.21(w) and are not subject to this part.

[48 FR 14264, Apr. 1, 1983, as amended at 53 FR 37934, Sept. 28, 1988; 54 FR 18785, May 2, 1989; 57 FR 60129, Dec. 18, 1992; 65 FR 30910, May 15, 2000; 70 FR 53449, Sept. 8, 2005]

§124.6 Draft permits.

(a) (*Applicable to State programs, see §§123.25 (NPDES), 145.11 (UIC), 233.26 (404), and 271.14 (RCRA).*) Once an application is complete, the Director shall tentatively decide whether to prepare a draft permit (except in the case of State section 404 permits for which no draft permit is required under §233.39) or to deny the application.

(b) If the Director tentatively decides to deny the permit application, he or she shall issue a notice of intent to deny. A notice of intent to deny the permit application is a type of draft permit which follows the same procedures as any draft permit prepared under this section. See §124.6(e). If the Director's final decision (§124.15) is that the tentative decision to deny the permit application was incorrect, he or she shall withdraw the notice of intent to deny and proceed to prepare a draft permit under paragraph (d) of this section.

(c) (*Applicable to State programs, see §§123.25 (NPDES) and 233.26 (404).*) If the Director tentatively decides to issue an NPDES or 404 general permit, he or she shall prepare a draft general permit under paragraph (d) of this section.

(d) (*Applicable to State programs, see §§123.25 (NPDES), 145.11 (UIC), 233.26 (404), and 271.14 (RCRA).*) If the Director decides to prepare a draft permit, he or she shall prepare a draft permit that contains the following information:

(1) All conditions under §§122.41 and 122.43 (NPDES), 144.51 and 144.42 (UIC, 233.7 and 233.8 (404, or 270.30 and 270.32 (RCRA) (except for PSD permits)));

(2) All compliance schedules under §§122.47 (NPDES), 144.53 (UIC), 233.10 (404), or 270.33 (RCRA) (except for PSD permits);

(3) All monitoring requirements under §§122.48 (NPDES), 144.54 (UIC), 233.11 (404), or 270.31 (RCRA) (except for PSD permits); and

(4) For:

(i) RCRA permits, standards for treatment, storage, and/or disposal and other permit conditions under §270.30;

(ii) UIC permits, permit conditions under §144.52;

(iii) PSD permits, permit conditions under 40 CFR §52.21;

(iv) 404 permits, permit conditions under §§233.7 and 233.8;

(v) NPDES permits, effluent limitations, standards, prohibitions, standards for sewage sludge use or disposal, and conditions under §§122.41, 122.42, and 122.44, including when applicable any conditions certified by a State agency under §124.55, and all variances that are to be included under §124.63.

(e) (*Applicable to State programs, see §§123.25 (NPDES), 145.11 (UIC), 233.26 (404), and 271.14 (RCRA).*) All draft permits prepared by EPA under this section shall be accompanied by a statement of basis (§124.7) or fact sheet (§124.8), and shall be based on the administrative record (§124.9), publicly noticed (§124.10) and made available for public comment (§124.11). The Regional Administrator shall give notice of opportunity for a public hearing (§124.12), issue a final decision (§124.15) and respond to comments (§124.17). For RCRA, UIC or PSD permits, an appeal may be taken under §124.19 and, for NPDES permits, an appeal may be taken under §124.74. Draft permits prepared by a State shall be accompanied by a fact sheet if required under §124.8.

[48 FR 14264, Apr. 1, 1983, as amended at 54 FR 18785, May 2, 1989; 65 FR 30910, May 15, 2000]

§ 124.7 Statement of basis.

EPA shall prepare a statement of basis for every draft permit for which a fact sheet under § 124.8 is not prepared. The statement of basis shall briefly describe the derivation of the conditions of the draft permit and the reasons for them or, in the case of notices of intent to deny or terminate, reasons supporting the tentative decision. The statement of basis shall be sent to the applicant and, on request, to any other person.

§ 124.8 Fact sheet.

(*Applicable to State programs, see §§ 123.25 (NPDES), 145.11 (UIC), 233.26 (404), and 271.14 (RCRA).*)

(a) A fact sheet shall be prepared for every draft permit for a major HWM, UIC, 404, or NPDES facility or activity, for every Class I sludge management facility, for every 404 and NPDES general permit (§§ 237.37 and 122.28), for every NPDES draft permit that incorporates a variance or requires an explanation under § 124.56(b), for every draft permit that includes a sewage sludge land application plan under 40 CFR 501.15(a)(2)(ix), and for every draft permit which the Director finds is the subject of wide-spread public interest or raises major issues. The fact sheet shall briefly set forth the principal facts and the significant factual, legal, methodological and policy questions considered in preparing the draft permit. The Director shall send this fact sheet to the applicant and, on request, to any other person.

(b) The fact sheet shall include, when applicable:

(1) A brief description of the type of facility or activity which is the subject of the draft permit;

(2) The type and quantity of wastes, fluids, or pollutants which are proposed to be or are being treated, stored, disposed of, injected, emitted, or discharged.

(3) For a PSD permit, the degree of increment consumption expected to result from operation of the facility or activity.

(4) A brief summary of the basis for the draft permit conditions including references to applicable statutory or regulatory provisions and appropriate supporting references to the adminis-

trative record required by § 124.9 (for EPA-issued permits);

(5) Reasons why any requested variances or alternatives to required standards do or do not appear justified;

(6) A description of the procedures for reaching a final decision on the draft permit including:

(i) The beginning and ending dates of the comment period under § 124.10 and the address where comments will be received;

(ii) Procedures for requesting a hearing and the nature of that hearing; and

(iii) Any other procedures by which the public may participate in the final decision.

(7) Name and telephone number of a person to contact for additional information.

(8) For NPDES permits, provisions satisfying the requirements of § 124.56.

(9) Justification for waiver of any application requirements under § 122.21(j) or (q) of this chapter.

[48 FR 14264, Apr. 1, 1983, as amended at 54 FR 18786, May 2, 1989; 64 FR 42470, Aug. 4, 1999]

§ 124.9 Administrative record for draft permits when EPA is the permitting authority.

(a) The provisions of a draft permit prepared by EPA under § 124.6 shall be based on the administrative record defined in this section.

(b) For preparing a draft permit under § 124.6, the record shall consist of:

(1) The application, if required, and any supporting data furnished by the applicant;

(2) The draft permit or notice of intent to deny the application or to terminate the permit;

(3) The statement of basis (§ 124.7) or fact sheet (§ 124.8);

(4) All documents cited in the statement of basis or fact sheet; and

(5) Other documents contained in the supporting file for the draft permit.

(6) For NPDES new source draft permits only, any environmental assessment, environmental impact statement (EIS), finding of no significant impact, or environmental information document and any supplement to an EIS that may have been prepared. NPDES permits other than permits to new

sources as well as all RCRA, UIC and PSD permits are not subject to the environmental impact statement provisions of section 102(2)(C) of the National Environmental Policy Act, 42 U.S.C. 4321.

(c) Material readily available at the issuing Regional Office or published material that is generally available, and that is included in the administrative record under paragraphs (b) and (c) of this section, need not be physically included with the rest of the record as long as it is specifically referred to in the statement of basis or the fact sheet.

(d) This section applies to all draft permits when public notice was given after the effective date of these regulations.

§ 124.10 **Public notice of permit actions and public comment period.**

(a) *Scope.* (1) The Director shall give public notice that the following actions have occurred:

(i) A permit application has been tentatively denied under § 124.6(b);

(ii) (*Applicable to State programs, see §§ 123.25 (NPDES), 145.11 (UIC), 233.26 (404), and 271.14 (RCRA).*) A draft permit has been prepared under § 124.6(d);

(iii) (*Applicable to State programs, see §§ 123.25 (NPDES), 145.11 (UIC), 233.26 (404) and 271.14 (RCRA)).*) A hearing has been scheduled under § 124.12;

(iv) An appeal has been granted under § 124.19(c);

(v) (*Applicable to State programs, see § 233.26 (404).*) A State section 404 application has been received in cases when no draft permit will be prepared (see § 233.39); or

(vi) An NPDES new source determination has been made under § 122.29.

(2) No public notice is required when a request for permit modification, revocation and reissuance, or termination is denied under § 124.5(b). Written notice of that denial shall be given to the requester and to the permittee.

(3) Public notices may describe more than one permit or permit actions.

(b) *Timing (applicable to State programs, see §§ 123.25 (NPDES), 145.11 (UIC), 233.26 (404, and 271.14 (RCRA)).* (1) Public notice of the preparation of a draft permit (including a notice of intent to deny a permit application) re-quired under paragraph (a) of this section shall allow at least 30 days for public comment. For RCRA permits only, public notice shall allow at least 45 days for public comment. For EPA-issued permits, if the Regional Administrator determines under 40 CFR part 6, subpart F that an Environmental Impact Statement (EIS) shall be prepared for an NPDES new source, public notice of the draft permit shall not be given until after a draft EIS is issued.

(2) Public notice of a public hearing shall be given at least 30 days before the hearing. (Public notice of the hearing may be given at the same time as public notice of the draft permit and the two notices may be combined.)

(c) Methods (applicable to State programs, see 40 CFR 123.25 (NPDES), 145.11 (UIC), 233.23 (404), and 271.14 (RCRA)). Public notice of activities described in paragraph (a)(1) of this section shall be given by the following methods:

(1) By mailing a copy of a notice to the following persons (any person otherwise entitled to receive notice under this paragraph may waive his or her rights to receive notice for any classes and categories of permits);

(i) The applicant (except for NPDES and 404 general permits when there is no applicant);

(ii) Any other agency which the Director knows has issued or is required to issue a RCRA, UIC, PSD (or other permit under the Clean Air Act), NPDES, 404, sludge management permit, or ocean dumping permit under the Marine Research Protection and Sanctuaries Act for the same facility or activity (including EPA when the draft permit is prepared by the State);

(iii) Federal and State agencies with jurisdiction over fish, shellfish, and wildlife resources and over coastal zone management plans, the Advisory Council on Historic Preservation, State Historic Preservation Officers, including any affected States (Indian Tribes). (For purposes of this paragraph, and in the context of the Underground Injection Control Program only, the term State includes Indian Tribes treated as States.)

(iv) For NPDES and 404 permits only, any State agency responsible for plan development under CWA section

311

208(b)(2), 208(b)(4) or 303(e) and the U.S. Army Corps of Engineers, the U.S. Fish and Wildlife Service and the National Marine Fisheries Service;

(v) For NPDES permits only, any user identified in the permit application of a privately owned treatment works;

(vi) For 404 permits only, any reasonably ascertainable owner of property adjacent to the regulated facility or activity and the Regional Director of the Federal Aviation Administration if the discharge involves the construction of structures which may affect aircraft operations or for purposes associated with seaplane operations;

(vii) For PSD permits only, affected State and local air pollution control agencies, the chief executives of the city and county where the major stationary source or major modification would be located, any comprehensive regional land use planning agency and any State, Federal Land Manager, or Indian Governing Body whose lands may be affected by emissions from the regulated activity;

(viii) For Class I injection well UIC permits only, state and local oil and gas regulatory agencies and state agencies regulating mineral exploration and recovery;

(ix) Persons on a mailing list developed by:

(A) Including those who request in writing to be on the list;

(B) Soliciting persons for "area lists" from participants in past permit proceedings in that area; and

(C) Notifying the public of the opportunity to be put on the mailing list through periodic publication in the public press and in such publications as Regional and State funded newsletters, environmental bulletins, or State law journals. (The Director may update the mailing list from time to time by requesting written indication of continued interest from those listed. The Director may delete from the list the name of any person who fails to respond to such a request.)

(x)(A) To any unit of local government having jurisdiction over the area where the facility is proposed to be located; and (B) to each State agency having any authority under State law

with respect to the construction or operation of such facility.

(xi) For Class VI injection well UIC permits, mailing or e-mailing a notice to State and local oil and gas regulatory agencies and State agencies regulating mineral exploration and recovery, the Director of the Public Water Supply Supervision program in the State, and all agencies that oversee injection wells in the State.

(2)(i) For major permits, NPDES and 404 general permits, and permits that include sewage sludge land application plans under 40 CFR 501.15(a)(2)(ix), publication of a notice in a daily or weekly newspaper within the area affected by the facility or activity; and for EPA-issued NPDES general permits, in the FEDERAL REGISTER;

NOTE: The Director is encouraged to provide as much notice as possible of the NPDES or Section 404 draft general permit to the facilities or activities to be covered by the general permit.

(ii) For all RCRA permits, publication of a notice in a daily or weekly major local newspaper of general circulation and broadcast over local radio stations.

(3) When the program is being administered by an approved State, in a manner constituting legal notice to the public under State law; and

(4) Any other method reasonably calculated to give actual notice of the action in question to the persons potentially affected by it, including press releases or any other forum or medium to elicit public participation.

(d) *Contents (applicable to State programs, see §§ 123.25 (NPDES), 145.11 (UIC), 233.26 (404), and 271.14 (RCRA))*—

(1) *All public notices.* All public notices issued under this part shall contain the following minimum information:

(i) Name and address of the office processing the permit action for which notice is being given;

(ii) Name and address of the permittee or permit applicant and, if different, of the facility or activity regulated by the permit, except in the case of NPDES and 404 draft general permits under §§ 122.28 and 233.37;

(iii) A brief description of the business conducted at the facility or activity described in the permit application or the draft permit, for NPDES or 404 general permits when there is no application.

(iv) Name, address and telephone number of a person from whom interested persons may obtain further information, including copies of the draft permit or draft general permit, as the case may be, statement of basis or fact sheet, and the application; and

(v) A brief description of the comment procedures required by §§124.11 and 124.12 and the time and place of any hearing that will be held, including a statement of procedures to request a hearing (unless a hearing has already been scheduled) and other procedures by which the public may participate in the final permit decision.

(vi) For EPA-issued permits, the location of the administrative record required by §124.9, the times at which the record will be open for public inspection, and a statement that all data submitted by the applicant is available as part of the administrative record.

(vii) For NPDES permits only (including those for "sludge-only facilities"), a general description of the location of each existing or proposed discharge point and the name of the receiving water and the sludge use and disposal practice(s) and the location of each sludge treatment works treating domestic sewage and use or disposal sites known at the time of permit application. For EPA-issued NPDES permits only, if the discharge is from a new source, a statement as to whether an environmental impact statement will be or has been prepared.

(viii) For 404 permits only,

(A) The purpose of the proposed activity (including, in the case of fill material, activities intended to be conducted on the fill), a description of the type, composition, and quantity of materials to be discharged and means of conveyance; and any proposed conditions and limitations on the discharge;

(B) The name and water quality standards classification, if applicable, of the receiving waters into which the discharge is proposed, and a general description of the site of each proposed discharge and the portions of the site and the discharges which are within State regulated waters;

(C) A description of the anticipated environmental effects of activities conducted under the permit;

(D) References to applicable statutory or regulatory authority; and

(E) Any other available information which may assist the public in evaluating the likely impact of the proposed activity upon the integrity of the receiving water.

(ix) Requirements applicable to cooling water intake structures under section 316(b) of the CWA, in accordance with part 125, subparts I , J, and N of this chapter.

(x) Any additional information considered necessary or proper.

(2) *Public notices for hearings.* In addition to the general public notice described in paragraph (d)(1) of this section, the public notice of a hearing under §124.12 shall contain the following information:

(i) Reference to the date of previous public notices relating to the permit;

(ii) Date, time, and place of the hearing;

(iii) A brief description of the nature and purpose of the hearing, including the applicable rules and procedures; and

(iv) For 404 permits only, a summary of major issues raised to date during the public comment period.

(e) (*Applicable to State programs, see §§123.25 (NPDES), 145.11 (UIC), 233.26 (404), and 271.14 (RCRA).*) In addition to the general public notice described in paragraph (d)(1) of this section, all persons identified in paragraphs (c)(1) (i), (ii), (iii), and (iv) of this section shall be mailed a copy of the fact sheet or statement of basis (for EPA-issued permits), the permit application (if any) and the draft permit (if any).

[48 FR 14264, Apr. 1, 1983; 48 FR 30115, June 30, 1983, as amended at 53 FR 28147, July 26, 1988; 53 FR 37410, Sept. 26, 1988; 54 FR 258, Jan. 4, 1989; 54 FR 18786, May 2, 1989; 65 FR 30911, May 15, 2000; 66 FR 65338, Dec. 18, 2001; 69 FR 41683, July 9, 2004; 71 FR 35040, June 16, 2006; 75 FR 77286, Dec. 10, 2010]

§124.11 **Public comments and requests for public hearings.**

(*Applicable to State programs, see §§123.25 (NPDES), 145.11 (UIC), 233.26 (404), and 271.14 (RCRA).*) During the public comment period provided under

§ 124.10, any interested person may submit written comments on the draft permit or the permit application for 404 permits when no draft permit is required (see § 233.39) and may request a public hearing, if no hearing has already been scheduled. A request for a public hearing shall be in writing and shall state the nature of the issues proposed to be raised in the hearing. All comments shall be considered in making the final decision and shall be answered as provided in § 124.17.

§ 124.12 Public hearings.

(a) (*Applicable to State programs, see §§ 123.25 (NPDES), 145.11 (UIC), 233.26 (404), and 271.14 (RCRA).*) (1) The Director shall hold a public hearing whenever he or she finds, on the basis of requests, a significant degree of public interest in a draft permit(s);

(2) The Director may also hold a public hearing at his or her discretion, whenever, for instance, such a hearing might clarify one or more issues involved in the permit decision;

(3) For RCRA permits only, (i) the Director shall hold a public hearing whenever he or she receives written notice of opposition to a draft permit and a request for a hearing within 45 days of public notice under § 124.10(b)(1); (ii) whenever possible the Director shall schedule a hearing under this section at a location convenient to the nearest population center to the proposed facility;

(4) Public notice of the hearing shall be given as specified in § 124.10.

(b) Whenever a public hearing will be held and EPA is the permitting authority, the Regional Administrator shall designate a Presiding Officer for the hearing who shall be responsible for its scheduling and orderly conduct.

(c) Any person may submit oral or written statements and data concerning the draft permit. Reasonable limits may be set upon the time allowed for oral statements, and the submission of statements in writing may be required. The public comment period under § 124.10 shall automatically be extended to the close of any public hearing under this section. The hearing officer may also extend the comment period by so stating at the hearing.

(d) A tape recording or written transcript of the hearing shall be made available to the public.

[48 FR 14264, Apr. 1, 1983, as amended at 49 FR 17718, Apr. 24, 1984; 50 FR 6941, Feb. 19, 1985; 54 FR 258, Jan. 4, 1989; 65 FR 30911, May 15, 2000]

§ 124.13 Obligation to raise issues and provide information during the public comment period.

All persons, including applicants, who believe any condition of a draft permit is inappropriate or that the Director's tentative decision to deny an application, terminate a permit, or prepare a draft permit is inappropriate, must raise all reasonably ascertainable issues and submit all reasonably available arguments supporting their position by the close of the public comment period (including any public hearing) under § 124.10. Any supporting materials which are submitted shall be included in full and may not be incorporated by reference, unless they are already part of the administrative record in the same proceeding, or consist of State or Federal statutes and regulations, EPA documents of general applicability, or other generally available reference materials. Commenters shall make supporting materials not already included in the administrative record available to EPA as directed by the Regional Administrator. (A comment period longer than 30 days may be necessary to give commenters a reasonable opportunity to comply with the requirements of this section. Additional time shall be granted under § 124.10 to the extent that a commenter who requests additional time demonstrates the need for such time.)

[49 FR 38051, Sept. 26, 1984]

§ 124.14 Reopening of the public comment period.

(a)(1) The Regional Administrator may order the public comment period reopened if the procedures of this paragraph could expedite the decisionmaking process. When the public comment period is reopened under this paragraph, all persons, including applicants, who believe any condition of a draft permit is inappropriate or that the Regional Administrator's tentative decision to deny an application, terminate a permit, or prepare a draft permit is inappropriate, must submit all reasonably available factual grounds supporting their position, including all supporting material, by a date, not less than sixty days after public notice under paragraph (a)(2) of this section, set by the Regional Administrator. Thereafter, any person may file a written response to the material filed by any other person, by a date, not less than twenty days after the date set for filing of the material, set by the Regional Administrator.

(2) Public notice of any comment period under this paragraph shall identify the issues to which the requirements of § 124.14(a) shall apply.

(3) On his own motion or on the request of any person, the Regional Administrator may direct that the requirements of paragraph (a)(1) of this section shall apply during the initial comment period where it reasonably appears that issuance of the permit will be contested and that applying the requirements of paragraph (a)(1) of this section will substantially expedite the decisionmaking process. The notice of the draft permit shall state whenever this has been done.

(4) A comment period of longer than 60 days will often be necessary in complicated proceedings to give commenters a reasonable opportunity to comply with the requirements of this section. Commenters may request longer comment periods and they shall be granted under § 124.10 to the extent they appear necessary.

(b) If any data information or arguments submitted during the public comment period, including information or arguments required under § 124.13, appear to raise substantial new questions concerning a permit, the Regional Administrator may take one or more of the following actions:

(1) Prepare a new draft permit, appropriately modified, under § 124.6;

(2) Prepare a revised statement of basis under § 124.7, a fact sheet or revised fact sheet under § 124.8 and reopen the comment period under § 124.14; or

(3) Reopen or extend the comment period under § 124.10 to give interested persons an opportunity to comment on the information or arguments submitted.

(c) Comments filed during the reopened comment period shall be limited to the substantial new questions that caused its reopening. The public notice under § 124.10 shall define the scope of the reopening.

(d) [Reserved]

(e) Public notice of any of the above actions shall be issued under § 124.10.

[48 FR 14264, Apr. 1, 1983, as amended at 49 FR 38051, Sept. 26, 1984; 65 FR 30911, May 15, 2000]

§ 124.15 Issuance and effective date of permit.

(a) After the close of the public comment period under § 124.10 on a draft permit, the Regional Administrator shall issue a final permit decision (or a decision to deny a permit for the active life of a RCRA hazardous waste management facility or unit under § 270.29). The Regional Administrator shall notify the applicant and each person who has submitted written comments or requested notice of the final permit decision. This notice shall include reference to the procedures for appealing a decision on a RCRA, UIC, PSD, or NPDES permit under § 124.19 of this part. For the purposes of this section, a final permit decision means a final decision to issue, deny, modify, revoke and reissue, or terminate a permit.

(b) A final permit decision (or a decision to deny a permit for the active life of a RCRA hazardous waste management facility or unit under § 270.29) shall become effective 30 days after the service of notice of the decision unless:

(1) A later effective date is specified in the decision; or

(2) Review is requested on the permit under § 124.19.

(3) No comments requested a change in the draft permit, in which case the permit shall become effective immediately upon issuance.

[48 FR 14264, Apr. 1, 1983, as amended at 54 FR 9607, Mar. 7, 1989; 65 FR 30911, May 15, 2000]

§ 124.16 Stays of contested permit conditions.

(a) *Stays.* (1) If a request for review of a RCRA, UIC, or NPDES permit under § 124.19 of this part is filed, the effect of the contested permit conditions shall be stayed and shall not be subject to judicial review pending final agency action. Uncontested permit conditions shall be stayed only until the date specified in paragraph (a)(2)(i) of this section. (No stay of a PSD permit is available under this section.) If the permit involves a new facility or new injection well, new source, new discharger or a recommencing discharger, the applicant shall be without a permit for the proposed new facility, injection well, source or discharger pending final agency action. See also § 124.60.

(2)(i) Uncontested conditions which are not severable from those contested shall be stayed together with the contested conditions. The Regional Administrator shall identify the stayed provisions of permits for existing facilities, injection wells, and sources. All other provisions of the permit for the existing facility, injection well, or source become fully effective and enforceable 30 days after the date of the notification required in paragraph (a)(2)(ii) of this section.

(ii) The Regional Administrator shall, as soon as possible after receiving notification from the EAB of the filing of a petition for review, notify the EAB, the applicant, and all other interested parties of the uncontested (and severable) conditions of the final permit that will become fully effective enforceable obligations of the permit as of the date specified in paragraph (a)(2)(i) of this section . For NPDES permits only, the notice shall comply with the requirements of § 124.60(b).

(b) *Stays based on cross effects.* (1) A stay may be granted based on the grounds that an appeal to the Administrator under § 124.19 of one permit may result in changes to another EPA-issued permit only when each of the permits involved has been appealed to the Administrator and he or she has accepted each appeal.

(2) No stay of an EPA-issued RCRA, UIC, or NPDES permit shall be granted based on the staying of any State-issued permit except at the discretion of the Regional Administrator and only upon written request from the State Director.

(c) Any facility or activity holding an existing permit must:

(1) Comply with the conditions of that permit during any modification or revocation and reissuance proceeding under § 124.5; and

(2) To the extent conditions of any new permit are stayed under this section, comply with the conditions of the existing permit which correspond to the stayed conditions, unless compliance with the existing conditions would be technologically incompatible with compliance with other conditions of the new permit which have not been stayed.

[48 FR 14264, Apr. 1, 1983, as amended at 65 FR 30911, May 15, 2000]

§ 124.17   Response to comments.

(a) (*Applicable to State programs, see §§ 123.25 (NPDES), 145.11 (UIC), 233.26 (404), and 271.14 (RCRA).*) At the time that any final permit decision is issued under § 124.15, the Director shall issue a response to comments. States are only required to issue a response to comments when a final permit is issued. This response shall:

(1) Specify which provisions, if any, of the draft permit have been changed in the final permit decision, and the reasons for the change; and

(2) Briefly describe and respond to all significant comments on the draft permit or the permit application (for section 404 permits only) raised during the public comment period, or during any hearing.

(b) For EPA-issued permits, any documents cited in the response to comments shall be included in the administrative record for the final permit decision as defined in § 124.18. If new points are raised or new material supplied during the public comment period, EPA may document its response to those matters by adding new materials to the administrative record.

(c) (*Applicable to State programs, see §§ 123.25 (NPDES), 145.11 (UIC), 233.26 (404), and 271.14 (RCRA).*) The response to comments shall be available to the public.

§ 124.18   Administrative record for final permit when EPA is the permitting authority.

(a) The Regional Administrator shall base final permit decisions under

§124.15 on the administrative record defined in this section.

(b) The administrative record for any final permit shall consist of the administrative record for the draft permit and:

(1) All comments received during the public comment period provided under §124.10 (including any extension or reopening under §124.14);

(2) The tape or transcript of any hearing(s) held under §124.12;

(3) Any written materials submitted at such a hearing;

(4) The response to comments required by §124.17 and any new material placed in the record under that section;

(5) For NPDES new source permits only, final environmental impact statement and any supplement to the final EIS;

(6) Other documents contained in the supporting file for the permit; and

(7) The final permit.

(c) The additional documents required under paragraph (b) of this section should be added to the record as soon as possible after their receipt or publication by the Agency. The record shall be complete on the date the final permit is issued.

(d) This section applies to all final RCRA, UIC, PSD, and NPDES permits when the draft permit was subject to the administrative record requirements of §124.9 and to all NPDES permits when the draft permit was included in a public notice after October 12, 1979.

(e) Material readily available at the issuing Regional Office, or published materials which are generally available and which are included in the administrative record under the standards of this section or of §124.17 ("Response to comments"), need not be physically included in the same file as the rest of the record as long as it is specifically referred to in the statement of basis or fact sheet or in the response to comments.

§124.19 Appeal of RCRA, UIC, NPDES, and PSD Permits.

(a) Within 30 days after a RCRA, UIC, NPDES, or PSD final permit decision (or a decision under 270.29 of this chapter to deny a permit for the active life of a RCRA hazardous waste manage-

ment facility or unit) has been issued under §124.15 of this part, any person who filed comments on that draft permit or participated in the public hearing may petition the Environmental Appeals Board to review any condition of the permit decision. Persons affected by an NPDES general permit may not file a petition under this section or otherwise challenge the conditions of the general permit in further Agency proceedings. They may, instead, either challenge the general permit in court, or apply for an individual NPDES permit under §122.21 as authorized in §122.28 and then petition the Board for review as provided by this section. As provided in §122.28(b)(3), any interested person may also petition the Director to require an individual NPDES permit for any discharger eligible for authorization to discharge under an NPDES general permit. Any person who failed to file comments or failed to participate in the public hearing on the draft permit may petition for administrative review only to the extent of the changes from the draft to the final permit decision. The 30-day period within which a person may request review under this section begins with the service of notice of the Regional Administrator's action unless a later date is specified in that notice. The petition shall include a statement of the reasons supporting that review, including a demonstration that any issues being raised were raised during the public comment period (including any public hearing) to the extent required by these regulations and when appropriate, a showing that the condition in question is based on:

(1) A finding of fact or conclusion of law which is clearly erroneous, or

(2) An exercise of discretion or an important policy consideration which the Environmental Appeals Board should, in its discretion, review.

(b) The Environmental Appeals Board may also decide on its own initiative to review any condition of any RCRA, UIC, NPDES, or PSD permit decision issued under this part for which review is available under paragraph (a) of this section. The Environmental Appeals Board must act under this paragraph within 30 days of the service date of notice of the Regional Administrator's action.

(c) Within a reasonable time following the filing of the petition for review, the Environmental Appeals Board shall issue an order granting or denying the petition for review. To the extent review is denied, the conditions of the final permit decision become final agency action. Public notice of any grant of review by the Environmental Appeals Board under paragraph (a) or (b) of this section shall be given as provided in § 124.10. Public notice shall set forth a briefing schedule for the appeal and shall state that any interested person may file an amicus brief. Notice of denial of review shall be sent only to the person(s) requesting review.

(d) The Regional Administrator, at any time prior to the rendering of a decision under paragraph (c) of this section to grant or deny review of a permit decision, may, upon notification to the Board and any interested parties, withdraw the permit and prepare a new draft permit under § 124.6 addressing the portions so withdrawn. The new draft permit shall proceed through the same process of public comment and opportunity for a public hearing as would apply to any other draft permit subject to this part. Any portions of the permit which are not withdrawn and which are not stayed under § 124.16(a) continue to apply.

(e) A petition to the Environmental Appeals Board under paragraph (a) of this section is, under 5 U.S.C. 704, a prerequisite to the seeking of judicial review of the final agency action.

(f)(1) For purposes of judicial review under the appropriate Act, final agency action occurs when a final RCRA, UIC, NPDES, or PSD permit decision is issued by EPA and agency review procedures under this section are exhausted. A final permit decision shall be issued by the Regional Administrator:

(i) When the Environmental Appeals Board issues notice to the parties that review has been denied;

(ii) When the Environmental Appeals Board issues a decision on the merits of the appeal and the decision does not include a remand of the proceedings; or

(iii) Upon the completion of remand proceedings if the proceedings are remanded, unless the Environmental Appeals Board's remand order specifically provides that appeal of the remand decision will be required to exhaust administrative remedies.

(2) Notice of any final agency action regarding a PSD permit shall promptly be published in the FEDERAL REGISTER.

(g) Motions to reconsider a final order shall be filed within ten (10) days after service of the final order. Every such motion must set forth the matters claimed to have been erroneously decided and the nature of the alleged errors. Motions for reconsideration under this provision shall be directed to, and decided by, the Environmental Appeals Board. Motions for reconsideration directed to the administrator, rather than to the Environmental Appeals Board, will not be considered, except in cases that the Environmental Appeals Board has referred to the Administrator pursuant to § 124.2 and in which the Administrator has issued the final order. A motion for reconsideration shall not stay the effective date of the final order unless specifically so ordered by the Environmental Appeals Board.

[48 FR 14264, Apr. 1, 1983, as amended at 54 FR 9607, Mar. 7, 1989; 57 FR 5335, Feb. 13, 1992; 65 FR 30911, May 15, 2000]

§ 124.20   Computation of time.

(a) Any time period scheduled to begin on the occurrence of an act or event shall begin on the day after the act or event.

(b) Any time period scheduled to begin before the occurrence of an act or event shall be computed so that the period ends on the day before the act or event.

(c) If the final day of any time period falls on a weekend or legal holiday, the time period shall be extended to the next working day.

(d) Whenever a party or interested person has the right or is required to act within a prescribed period after the service of notice or other paper upon him or her by mail, 3 days shall be added to the prescribed time.

§ 124.21 Effective date of part 124.

(a) Part 124 of this chapter became effective for all permits except for RCRA permits on July 18, 1980. Part 124 of this chapter became effective for RCRA permits on November 19, 1980.

(b) EPA eliminated the previous requirement for NPDES permits to undergo an evidentiary hearing after permit issuance, and modified the procedures for termination of NPDES and RCRA permits, on June 14, 2000.

(c)(1) For any NPDES permit decision for which a request for evidentiary hearing was granted on or prior to June 13, 2000, the hearing and any subsequent proceedings (including any appeal to the Environmental Appeals Board) shall proceed pursuant to the procedures of this part as in effect on June 13, 2000.

(2) For any NPDES permit decision for which a request for evidentiary hearing was denied on or prior to June 13, 2000, but for which the Board has not yet completed proceedings under § 124.91, the appeal, and any hearing or other proceedings on remand if the Board so orders, shall proceed pursuant to the procedures of this part as in effect on June 13, 2000.

(3) For any NPDES permit decision for which a request for evidentiary hearing was filed on or prior to June 13, 2000 but was neither granted nor denied prior to that date, the Regional Administrator shall, no later than July 14, 2000, notify the requester that the request for evidentiary hearing is being returned without prejudice. Notwithstanding the time limit in § 124.19(a), the requester may file an appeal with the Board, in accordance with the other requirements of § 124.19(a), no later than August 13, 2000.

(4) A party to a proceeding otherwise subject to paragraph (c) (1) or (2) of this section may, no later than June 14, 2000, request that the evidentiary hearing process be suspended. The Regional Administrator shall inquire of all other parties whether they desire the evidentiary hearing to continue. If no party desires the hearing to continue, the Regional Administrator shall return the request for evidentiary hearing in the manner specified in paragraph (c)(3) of this section.

(d) For any proceeding to terminate an NPDES or RCRA permit commenced on or prior to June 13, 2000, the Regional Administrator shall follow the procedures of § 124.5(d) as in effect on June 13, 2000, and any formal hearing shall follow the procedures of subpart E of this part as in effect on the same date.

[65 FR 30911, May 15, 2000]

Subpart B—Specific Procedures Applicable to RCRA Permits

SOURCE: 60 FR 63431, Dec. 11, 1995, unless otherwise noted.

§ 124.31 Pre-application public meeting and notice.

(a) *Applicability.* The requirements of this section shall apply to all RCRA part B applications seeking initial permits for hazardous waste management units over which EPA has permit issuance authority. The requirements of this section shall also apply to RCRA part B applications seeking renewal of permits for such units, where the renewal application is proposing a significant change in facility operations. For the purposes of this section, a "significant change" is any change that would qualify as a class 3 permit modification under 40 CFR 270.42. For the purposes of this section only, "hazardous waste management units over which EPA has permit issuance authority" refers to hazardous waste management units for which the State where the units are located has not been authorized to issue RCRA permits pursuant to 40 CFR part 271. The requirements of this section shall also apply to hazardous waste management facilities for which facility owners or operators are seeking coverage under a RCRA standardized permit (see 40 part 270, subpart J), including renewal of a standardized permit for such units, where the renewal is proposing a significant change in facility operations, as defined at § 124.211(c). The requirements of this section do not apply to permit modifications under 40 CFR 270.42 or to applications that are submitted for the sole purpose of conducting post-closure activities or post-closure activities and corrective action at a facility.

(b) Prior to the submission of a part B RCRA permit application for a facility, or to the submission of a written Notice of Intent to be covered by a RCRA standardized permit (see 40 CFR part 270, subpart J), the applicant must hold at least one meeting with the public in order to solicit questions from the community and inform the community of proposed hazardous waste management activities. The applicant shall post a sign-in sheet or otherwise provide a voluntary opportunity for attendees to provide their names and addresses.

(c) The applicant shall submit a summary of the meeting, along with the list of attendees and their addresses developed under paragraph (b) of this section, and copies of any written comments or materials submitted at the meeting, to the permitting agency as a part of the part B application, in accordance with 40 CFR 270.14(b), or with the written Notice of Intent to be covered by a RCRA standardized permit (see 40 CFR part 270, subpart J).

(d) The applicant must provide public notice of the pre-application meeting at least 30 days prior to the meeting. The applicant must maintain, and provide to the permitting agency upon request, documentation of the notice.

(1) The applicant shall provide public notice in all of the following forms:

(i) *A newspaper advertisement.* The applicant shall publish a notice, fulfilling the requirements in paragraph (d)(2) of this section, in a newspaper of general circulation in the county or equivalent jurisdiction that hosts the proposed location of the facility. In addition, the Director shall instruct the applicant to publish the notice in newspapers of general circulation in adjacent counties or equivalent jurisdictions, where the Director determines that such publication is necessary to inform the affected public. The notice must be published as a display advertisement.

(ii) *A visible and accessible sign.* The applicant shall post a notice on a clearly marked sign at or near the facility, fulfilling the requirements in paragraph (d)(2) of this section. If the applicant places the sign on the facility property, then the sign must be large enough to be readable from the nearest point where the public would pass by the site.

(iii) *A broadcast media announcement.* The applicant shall broadcast a notice, fulfilling the requirements in paragraph (d)(2) of this section, at least once on at least one local radio station or television station. The applicant may employ another medium with prior approval of the Director.

(iv) *A notice to the permitting agency.* The applicant shall send a copy of the newspaper notice to the permitting agency and to the appropriate units of State and local government, in accordance with § 124.10(c)(1)(x).

(2) The notices required under paragraph (d)(1) of this section must include:

(i) The date, time, and location of the meeting;

(ii) A brief description of the purpose of the meeting;

(iii) A brief description of the facility and proposed operations, including the address or a map (e.g., a sketched or copied street map) of the facility location;

(iv) A statement encouraging people to contact the facility at least 72 hours before the meeting if they need special access to participate in the meeting; and

(v) The name, address, and telephone number of a contact person for the applicant.

[60 FR 63431, Dec. 11, 1995, as amended at 70 FR 53449, Sept. 8, 2005]

§ 124.32 Public notice requirements at the application stage.

(a) *Applicability.* The requirements of this section shall apply to all RCRA part B applications seeking initial permits for hazardous waste management units over which EPA has permit issuance authority. The requirements of this section shall also apply to RCRA part B applications seeking renewal of permits for such units under 40 CFR 270.51. For the purposes of this section only, "hazardous waste management units over which EPA has permit issuance authority" refers to hazardous waste management units for which the State where the units are located has not been authorized to issue RCRA permits pursuant to 40 CFR part

271. The requirements of this section do not apply to hazardous waste units for which facility owners or operators are seeking coverage under a RCRA standardized permit (see 40 CFR part 270, subpart J)). The requirements of this section also do not apply to permit modifications under 40 CFR 270.42 or permit applications submitted for the sole purpose of conducting post-closure activities or post-closure activities and corrective action at a facility.

(b) *Notification at application submittal.*

(1) The Director shall provide public notice as set forth in §124.10(c)(1)(ix), and notice to appropriate units of State and local government as set forth in §124.10(c)(1)(x), that a part B permit application has been submitted to the Agency and is available for review.

(2) The notice shall be published within a reasonable period of time after the application is received by the Director. The notice must include:

(i) The name and telephone number of the applicant's contact person;

(ii) The name and telephone number of the permitting agency's contact office, and a mailing address to which information, opinions, and inquiries may be directed throughout the permit review process;

(iii) An address to which people can write in order to be put on the facility mailing list;

(iv) The location where copies of the permit application and any supporting documents can be viewed and copied;

(v) A brief description of the facility and proposed operations, including the address or a map (e.g., a sketched or copied street map) of the facility location on the front page of the notice; and

(vi) The date that the application was submitted.

(c) Concurrent with the notice required under §124.32(b) of this subpart, the Director must place the permit application and any supporting documents in a location accessible to the public in the vicinity of the facility or at the permitting agency's office.

[60 FR 63431, Dec. 11, 1995, as amended at 70 FR 53449, Sept. 8, 2005]

§124.33 **Information repository.**

(a) *Applicability.* The requirements of this section apply to all applications seeking RCRA permits for hazardous waste management units over which EPA has permit issuance authority. For the purposes of this section only, "hazardous waste management units over which EPA has permit issuance authority" refers to hazardous waste management units for which the State where the units are located has not been authorized to issue RCRA permits pursuant to 40 CFR part 271.

(b) The Director may assess the need, on a case-by-case basis, for an information repository. When assessing the need for an information repository, the Director shall consider a variety of factors, including: the level of public interest; the type of facility; the presence of an existing repository; and the proximity to the nearest copy of the administrative record. If the Director determines, at any time after submittal of a permit application, that there is a need for a repository, then the Director shall notify the facility that it must establish and maintain an information repository. (See 40 CFR 270.30(m) for similar provisions relating to the information repository during the life of a permit).

(c) The information repository shall contain all documents, reports, data, and information deemed necessary by the Director to fulfill the purposes for which the repository is established. The Director shall have the discretion to limit the contents of the repository.

(d) The information repository shall be located and maintained at a site chosen by the facility. If the Director finds the site unsuitable for the purposes and persons for which it was established, due to problems with the location, hours of availability, access, or other relevant considerations, then the Director shall specify a more appropriate site.

(e) The Director shall specify requirements for informing the public about the information repository. At a minimum, the Director shall require the facility to provide a written notice about the information repository to all individuals on the facility mailing list.

(f) The facility owner/operator shall be responsible for maintaining and updating the repository with appropriate information throughout a time period specified by the Director. The Director may close the repository at his or her discretion, based on the factors in paragraph (b) of this section.

Subpart C—Specific Procedures Applicable to PSD Permits

§ 124.41 Definitions applicable to PSD permits.

Whenever PSD permits are processed under this part, the following terms shall have the following meanings:

Administrator, EPA, and *Regional Administrator* shall have the meanings set forth in § 124.2, except when EPA has delegated authority to administer those regulations to another agency under the applicable subsection of 40 CFR 52.21, the term *EPA* shall mean the delegate agency and the term *Regional Administrator* shall mean the chief administrative officer of the delegate agency.

Application means an application for a PSD permit.

Appropriate Act and Regulations means the Clean Air Act and applicable regulations promulgated under it.

Approved program means a State implementation plan providing for issuance of PSD permits which has been approved by EPA under the Clean Air Act and 40 CFR part 51. An *approved State* is one administering an *approved program. State Director* as used in § 124.4 means the person(s) responsible for issuing PSD permits under an approved program, or that person's delegated representative.

Construction has the meaning given in 40 CFR 52.21.

Director means the Regional Administrator.

Draft permit shall have the meaning set forth in § 124.2.

Facility or activity means a *major PSD stationary source* or *major PSD modification.*

Federal Land Manager has the meaning given in 40 CFR 52.21.

Indian Governing Body has the meaning given in 40 CFR 52.21.

Major PSD modification means a *major modification* as defined in 40 CFR 52.21.

Major PSD stationary source means a *major stationary source* as defined in 40 CFR 52.21(b)(1).

Owner or operator means the owner or operator of any facility or activity subject to regulation under 40 CFR 52.21 or by an approved State.

Permit or *PSD permit* means a permit issued under 40 CFR 52.21 or by an approved State.

Person includes an individual, corporation, partnership, association, State, municipality, political subdivision of a State, and any agency, department, or instrumentality of the United States and any officer, agent or employee thereof.

Regulated activity or *activity subject to regulation* means a *major PSD stationary source* or *major PSD modification.*

Site means the land or water area upon which a *major PSD stationary source* or *major PSD modification* is physically located or conducted, including but not limited to adjacent land used for utility systems; as repair, storage, shipping or processing areas; or otherwise in connection with the *major PSD stationary source* or *major PSD modification.*

State means a State, the District of Columbia, the Commonwealth of Puerto Rico, the Virgin Islands, Guam, and American Samoa and includes the Commonwealth of the Northern Mariana Islands.

§ 124.42 Additional procedures for PSD permits affecting Class I areas.

(a) The Regional Administrator shall provide notice of any permit application for a proposed major PSD stationary source or major PSD modification the emissions from which would affect a Class I area to the Federal Land Manager, and the Federal official charged with direct responsibility for management of any lands within such area. The Regional Administrator shall provide such notice promptly after receiving the application.

(b) Any demonstration which the Federal Land Manager wishes to present under 40 CFR 52.21(q)(3), and any variances sought by an owner or operator under § 52.21(q)(4) shall be submitted in writing, together with any necessary supporting analysis, by the end of the public comment period under §124.10 or §124.118. (40 CFR 52.21(q)(3) provides for denial of a PSD permit to a facility or activity when the Federal Land Manager demonstrates that its emissions would adversely affect a Class I area even though the applicable increments would not be exceeded. 40 CFR 52.21(q)(4) conversely authorizes EPA, with the concurrence of the Federal Land Manager and State responsible, to grant certain variances from the otherwise applicable emission limitations to a facility or activity whose emissions would affect a Class I area.)

(c) Variances authorized by 40 CFR 52.21 (q)(5) through (q)(7) shall be handled as specified in those paragraphs and shall not be subject to this part. Upon receiving appropriate documentation of a variance properly granted under any of these provisions, the Regional Administrator shall enter the variance in the administrative record. Any decisions later made in proceedings under this part concerning that permit shall be consistent with the conditions of that variance.

Subpart D—Specific Procedures Applicable to NPDES Permits

§ 124.51 Purpose and scope.

(a) This subpart sets forth additional requirements and procedures for decisionmaking for the NPDES program.

(b) Decisions on NPDES variance requests ordinarily will be made during the permit issuance process. Variances and other changes in permit conditions ordinarily will be decided through the same notice-and-comment and hearing procedures as the basic permit.

(c) As stated in 40 CFR 131.4, an Indian Tribe that meets the statutory criteria which authorize EPA to treat the Tribe in a manner similar to that in which it treats a State for purposes of the Water Quality Standards program is likewise qualified for such treatment for purposes of State certification of water quality standards pursuant to section 401(a)(1) of the Act and subpart D of this part.

[48 FR 14264, Apr. 1, 1983, as amended at 58 FR 67983, Dec. 22, 1993; 59 FR 64343, Dec. 14, 1994]

§ 124.52 Permits required on a case-by-case basis.

(a) Various sections of part 122, subpart B allow the Director to determine, on a case-by-case basis, that certain concentrated animal feeding operations (§122.23), concentrated aquatic animal production facilities (§122.24), storm water discharges (§122.26), and certain other facilities covered by general permits (§122.28) that do not generally require an individual permit may be required to obtain an individual permit because of their contributions to water pollution.

(b) Whenever the Regional Administrator decides that an individual permit is required under this section, except as provided in paragraph (c) of this section, the Regional Administrator shall notify the discharger in writing of that decision and the reasons for it, and shall send an application form with the notice. The discharger must apply for a permit under §122.21 within 60 days of notice, unless permission for a later date is granted by the Regional Administrator. The question whether the designation was proper will remain open for consideration during the public comment period under §124.11 and in any subsequent hearing.

(c) Prior to a case-by-case determination that an individual permit is required for a storm water discharge under this section (see §122.26(a)(1)(v), (c)(1)(v), and (a)(9)(iii) of this chapter), the Regional Administrator may require the discharger to submit a permit application or other information regarding the discharge under section 308 of the CWA. In requiring such information, the Regional Administrator shall notify the discharger in writing and shall send an application form with the notice. The discharger must apply for a permit within 180 days of notice, unless permission for a later date is granted by the Regional Administrator. The question whether the initial designation was proper will remain open for consideration during the public comment period under §124.11 and in any subsequent hearing.

[55 FR 48075, Nov. 16, 1990, as amended at 60 FR 17957, Apr. 7, 1995; 60 FR 19464, Apr. 18, 1995; 60 FR 40235, Aug. 7, 1995; 64 FR 68851, Dec. 8, 1999; 65 FR 30912, May 15, 2000]

§ 124.53 State certification.

(a) Under CWA section 401(a)(1), EPA may not issue a permit until a certification is granted or waived in accordance with that section by the State in which the discharge originates or will originate.

(b) Applications received without a State certification shall be forwarded by the Regional Administrator to the certifying State agency with a request that certification be granted or denied.

(c) If State certification has not been received by the time the draft permit is prepared, the Regional Administrator shall send the certifying State agency:

(1) A copy of a draft permit;

(2) A statement that EPA cannot issue or deny the permit until the certifying State agency has granted or denied certification under § 124.55, or waived its right to certify; and

(3) A statement that the State will be deemed to have waived its right to certify unless that right is exercised within a specified reasonable time not to exceed 60 days from the date the draft permit is mailed to the certifying State agency unless the Regional Administrator finds that unusual circumstances require a longer time.

(d) State certification shall be granted or denied within the reasonable time specified under paragraph (c)(3) of this section. The State shall send a notice of its action, including a copy of any certification, to the applicant and the Regional Administrator.

(e) State certification shall be in writing and shall include:

(1) Conditions which are necessary to assure compliance with the applicable provisions of CWA sections 208(e), 301, 302, 303, 306, and 307 and with appropriate requirements of State law;

(2) When the State certifies a draft permit instead of a permit application, any conditions more stringent than those in the draft permit which the State finds necessary to meet the requirements listed in paragraph (e)(1) of this section. For each more stringent condition, the certifying State agency shall cite the CWA or State law references upon which that condition is based. Failure to provide such a citation waives the right to certify with respect to that condition; and

(3) A statement of the extent to which each condition of the draft permit can be made less stringent without violating the requirements of State law, including water quality standards. Failure to provide this statement for any condition waives the right to certify or object to any less stringent condition which may be established during the EPA permit issuance process.

§ 124.54 Special provisions for State certification and concurrence on applications for section 301(h) variances.

(a) When an application for a permit incorporating a variance request under CWA section 301(h) is submitted to a State, the appropriate State official shall either:

(1) Deny the request for the CWA section 301(h) variance (and so notify the applicant and EPA) and, if the State is an approved NPDES State and the permit is due for reissuance, process the permit application under normal procedures; or

(2) Forward a certification meeting the requirements of § 124.53 to the Regional Administrator.

(b) When EPA issues a tentative decision on the request for a variance under CWA section 301(h), and no certification has been received under paragraph (a) of this section, the Regional Administrator shall forward the tentative decision to the State in accordance with § 124.53(b) specifying a reasonable time for State certification and concurrence. If the State fails to deny or grant certification and concurrence under paragraph (a) of this section within such reasonable time, certification shall be waived and the State shall be deemed to have concurred in the issuance of a CWA section 301(h) variance.

(c) Any certification provided by a State under paragraph (a)(2) of this section shall constitute the State's concurrence (as required by section 301(h)) in the issuance of the permit incorporating a section 301(h) variance subject to any conditions specified therein by the State. CWA section 301(h) certification and concurrence under this section will not be forwarded to the State by EPA for recertification after the permit issuance process; States must specify any conditions required by State law, including water quality standards, in the initial certification.

§ 124.55 Effect of State certification.

(a) When certification is required under CWA section 401(a)(1) no final permit shall be issued:

(1) If certification is denied, or

(2) Unless the final permit incorporates the requirements specified in the certification under § 124.53(e).

(b) If there is a change in the State law or regulation upon which a certification is based, or if a court of competent jurisdiction or appropriate State board or agency stays, vacates, or remands a certification, a State which has issued a certification under § 124.53 may issue a modified certification or notice of waiver and forward it to EPA. If the modified certification is received before final agency action on the permit, the permit shall be consistent with the more stringent conditions which are based upon State law identified in such certification. If the certification or notice of waiver is received after final agency action on the permit, the Regional Administrator may modify the permit on request of the permittee only to the extent necessary to delete any conditions based on a condition in a certification invalidated by a court of competent jurisdiction or by an appropriate State board or agency.

(c) A State may not condition or deny a certification on the grounds that State law allows a less stringent permit condition. The Regional Administrator shall disregard any such certification conditions, and shall consider those conditions or denials as waivers of certification.

(d) A condition in a draft permit may be changed during agency review in any manner consistent with a certification meeting the requirements of § 124.53(e). No such changes shall require EPA to submit the permit to the State for recertification.

(e) Review and appeals of limitations and conditions attributable to State certification shall be made through the applicable procedures of the State and may not be made through the procedures in this part.

(f) Nothing in this section shall affect EPA's obligation to comply with § 122.47. See CWA section 301(b)(1)(C).

[48 FR 14264, Apr. 1, 1983, as amended at 65 FR 30912, May 15, 2000]

§ 124.56 Fact sheets.

(*Applicable to State programs, see § 123.25 (NPDES).*) In addition to meeting the requirements of § 124.8, NPDES fact sheets shall contain the following:

(a) Any calculations or other necessary explanation of the derivation of specific effluent limitations and conditions or standards for sewage sludge use or disposal, including a citation to the applicable effluent limitation guideline, performance standard, or standard for sewage sludge use or disposal as required by § 122.44 and reasons why they are applicable or an explanation of how the alternate effluent limitations were developed.

(b)(1) When the draft permit contains any of the following conditions, an explanation of the reasons that such conditions are applicable:

(i) Limitations to control toxic pollutants under § 122.44(e) of this chapter;

(ii) Limitations on internal waste streams under § 122.45(i) of this chapter;

(iii) Limitations on indicator pollutants under § 125.3(g) of this chapter;

(iv) Limitations set on a case-by-case basis under § 125.3 (c)(2) or (c)(3) of this chapter, or pursuant to Section 405(d)(4) of the CWA;

(v) Limitations to meet the criteria for permit issuance under § 122.4(i) of this chapter, or

(vi) Waivers from monitoring requirements granted under § 122.44(a) of this chapter.

(2) For every permit to be issued to a treatment works owned by a person other than a State or municipality, an explanation of the Director's decision on regulation of users under § 122.44(m).

(c) When appropriate, a sketch or detailed description of the location of the discharge or regulated activity described in the application; and

(d) For EPA-issued NPDES permits, the requirements of any State certification under § 124.53.

(e) For permits that include a sewage sludge land application plan under 40 CFR 501.15(a)(2)(ix), a brief description of how each of the required elements of the land application plan are addressed in the permit.

[48 FR 14264, Apr. 1, 1983, as amended at 49 FR 38051, Sept. 26, 1984; 54 FR 18786, May 2, 1989; 65 FR 30912, May 15, 2000]

§ 124.57 Public notice.

(a) *Section 316(a) requests (applicable to State programs, see § 123.25).* In addition to the information required under § 124.10(d)(1), public notice of an NPDES draft permit for a discharge where a CWA section 316(a) request has been filed under § 122.21(l) shall include:

(1) A statement that the thermal component of the discharge is subject to effluent limitations under CWA section 301 or 306 and a brief description, including a quantitative statement, of the thermal effluent limitations proposed under section 301 or 306;

(2) A statement that a section 316(a) request has been filed and that alternative less stringent effluent limitations may be imposed on the thermal component of the discharge under section 316(a) and a brief description, including a quantitative statement, of the alternative effluent limitations, if any, included in the request; and

(3) If the applicant has filed an early screening request under § 125.72 for a section 316(a) variance, a statement that the applicant has submitted such a plan.

(b) [Reserved]

[48 FR 14264, Apr. 1, 1983; 50 FR 6941, Feb. 19, 1985, as amended at 65 FR 30912, May 15, 2000]

§ 124.58 [Reserved]

§ 124.59 Conditions requested by the Corps of Engineers and other government agencies.

(*Applicable to State programs, see § 123.25 (NPDES).*)

(a) If during the comment period for an NPDES draft permit, the District Engineer advises the Director in writing that anchorage and navigation of any of the waters of the United States would be substantially impaired by the granting of a permit, the permit shall be denied and the applicant so notified. If the District Engineer advised the Director that imposing specified conditions upon the permit is necessary to avoid any substantial impairment of anchorage or navigation, then the Di-

rector shall include the specified conditions in the permit. Review or appeal of denial of a permit or of conditions specified by the District Engineer shall be made through the applicable procedures of the Corps of Engineers, and may not be made through the procedures provided in this part. If the conditions are stayed by a court of competent jurisdiction or by applicable procedures of the Corps of Engineers, those conditions shall be considered stayed in the NPDES permit for the duration of that stay.

(b) If during the comment period the U.S. Fish and Wildlife Service, the National Marine Fisheries Service, or any other State or Federal agency with jurisdiction over fish, wildlife, or public health advises the Director in writing that the imposition of specified conditions upon the permit is necessary to avoid substantial impairment of fish, shellfish, or wildlife resources, the Director may include the specified conditions in the permit to the extent they are determined necessary to carry out the provisions of § 122.49 and of the CWA.

(c) In appropriate cases the Director may consult with one or more of the agencies referred to in this section before issuing a draft permit and may reflect their views in the statement of basis, the fact sheet, or the draft permit.

[48 FR 14264, Apr. 1, 1983, as amended at 54 FR 258, Jan. 4, 1989]

§ 124.60 Issuance and effective date and stays of NPDES permits.

In addition to the requirements of §§ 124.15, 124.16, and 124.19, the following provisions apply to NPDES permits:

(a) Notwithstanding the provisions of § 124.16(a)(1), if, for any offshore or coastal mobile exploratory drilling rig or coastal mobile developmental drilling rig which has never received a final effective permit to discharge at a "site," but which is not a "new discharger" or a "new source," the Regional Administrator finds that compliance with certain permit conditions may be necessary to avoid irreparable environmental harm during the administrative review, he or she may specify in the statement of basis or fact sheet that those conditions, even if contested, shall remain enforceable obligations of the discharger during administrative review.

(b)(1) As provided in §124.16(a), if an appeal of an initial permit decision is filed under §124.19, the force and effect of the contested conditions of the final permit shall be stayed until final agency action under §124.19(f). The Regional Administrator shall notify, in accordance with §124.16(a)(2)(ii), the discharger and all interested parties of the uncontested conditions of the final permit that are enforceable obligations of the discharger.

(2) When effluent limitations are contested, but the underlying control technology is not, the notice shall identify the installation of the technology in accordance with the permit compliance schedules (if uncontested) as an uncontested, enforceable obligation of the permit.

(3) When a combination of technologies is contested, but a portion of the combination is not contested, that portion shall be identified as uncontested if compatible with the combination of technologies proposed by the requester.

(4) Uncontested conditions, if inseverable from a contested condition, shall be considered contested.

(5) Uncontested conditions shall become enforceable 30 days after the date of notice under paragraph (b)(1) of this section.

(6) Uncontested conditions shall include:

(i) Preliminary design and engineering studies or other requirements necessary to achieve the final permit conditions which do not entail substantial expenditures;

(ii) Permit conditions which will have to be met regardless of the outcome of the appeal under §124.19;

(iii) When the discharger proposed a less stringent level of treatment than that contained in the final permit, any permit conditions appropriate to meet the levels proposed by the discharger, if the measures required to attain that less stringent level of treatment are consistent with the measures required to attain the limits proposed by any other party; and

(iv) Construction activities, such as segregation of waste streams or installation of equipment, which would partially meet the final permit conditions and could also be used to achieve the discharger's proposed alternative conditions.

(c) In addition to the requirements of §124.16(c)(2), when an appeal is filed under §124.19 on an application for a renewal of an existing permit and upon written request from the applicant, the Regional Administrator may delete requirements from the existing permit which unnecessarily duplicate uncontested provisions of the new permit.

[65 FR 30912, May 15, 2000]

§124.61 Final environmental impact statement.

No final NPDES permit for a new source shall be issued until at least 30 days after the date of issuance of a final environmental impact statement if one is required under 40 CFR 6.805.

§124.62 Decision on variances.

(*Applicable to State programs, see §123.25 (NPDES).*)

(a) The Director may grant or deny requests for the following variances (subject to EPA objection under §123.44 for State permits):

(1) Extensions under CWA section 301(i) based on delay in completion of a publicly owned treatment works;

(2) After consultation with the Regional Administrator, extensions under CWA section 301(k) based on the use of innovative technology; or

(3) Variances under CWA section 316(a) for thermal pollution.

(b) The State Director may deny, or forward to the Regional Administrator with a written concurrence, or submit to EPA without recommendation a completed request for:

(1) A variance based on the economic capability of the applicant under CWA section 301(c); or

(2) A variance based on water quality related effluent limitations under CWA section 302(b)(2).

(c) The Regional Administrator may deny, forward, or submit to the EPA Office Director for Water Enforcement and Permits with a recommendation for approval, a request for a variance listed in paragraph (b) of this section that is forwarded by the State Director, or that is submitted to the Regional Administrator by the requester where EPA is the permitting authority.

(d) The EPA Office Director for Water Enforcement and Permits may approve or deny any variance request submitted under paragraph (c) of this section. If the Office Director approves the variance, the Director may prepare a draft permit incorporating the variance. Any public notice of a draft permit for which a variance or modification has been approved or denied shall identify the applicable procedures for appealing that decision under § 124.64.

(e) The State Director may deny or forward to the Administrator (or his delegate) with a written concurrence a completed request for:

(1) A variance based on the presence of "fundamentally different factors" from those on which an effluent limitations guideline was based;

(2) A variance based upon certain water quality factors under CWA section 301(g).

(f) The Administrator (or his delegate) may grant or deny a request for a variance listed in paragraph (e) of this section that is forwarded by the State Director, or that is submitted to EPA by the requester where EPA is the permitting authority. If the Administrator (or his delegate) approves the variance, the State Director or Regional Administrator may prepare a draft permit incorporating the variance. Any public notice of a draft permit for which a variance or modification has been approved or denied shall identify the applicable procedures for appealing that decision under § 124.64.

[48 FR 14264, Apr. 1, 1983; 50 FR 6941, Feb. 19, 1985, as amended at 51 FR 16030, Apr. 30, 1986; 54 FR 256, 258, Jan. 4, 1989]

§ 124.63 Procedures for variances when EPA is the permitting authority.

(a) In States where EPA is the permit issuing authority and a request for a variance is filed as required by § 122.21, the request shall be processed as follows:

(1)(i) If, at the time, that a request for a variance based on the presence of fundamentally different factors or on section 301(g) of the CWA is submitted, the Regional Administrator has received an application under § 124.3 for issuance or renewal of that permit, but has not yet prepared a draft permit under § 124.6 covering the discharge in question, the Administrator (or his delegate) shall give notice of a tentative decision on the request at the time the notice of the draft permit is prepared as specified in § 124.10, unless this would significantly delay the processing of the permit. In that case the processing of the variance request may be separated from the permit in accordance with paragraph (a)(3) of this section, and the processing of the permit shall proceed without delay.

(ii) If, at the time, that a request for a variance under sections 301(c) or 302(b)(2) of the CWA is submitted, the Regional Administrator has received an application under § 124.3 for issuance or renewal of that permit, but has not yet prepared a draft permit under § 124.6 covering the discharge in question, the Regional Administrator, after obtaining any necessary concurrence of the EPA Deputy Assistant Administrator for Water Enforcement under § 124.62, shall give notice of a tentative decision on the request at the time notice of the draft permit is prepared as specified in § 124.10, unless this would significantly delay the processing of the permit. In that case the processing of the variance request may be separated from the permit in accordance with paragraph (a)(3) of this section, and the processing of the permit shall proceed without delay.

(2) If, at the time that a request for a variance is filed the Regional Administrator has given notice under § 124.10 of a draft permit covering the discharge in question, but that permit has not yet become final, administrative proceedings concerning that permit may be stayed and the Regional Administrator shall prepare a new draft permit including a tentative decision on the request, and the fact sheet required by § 124.8. However, if this will significantly delay the processing of the existing draft permit or the Regional Administrator, for other reasons, considers combining the variance request and the existing draft permit

inadvisable, the request may be separated from the permit in accordance with paragraph (a)(3) of this section, and the administrative disposition of the existing draft permit shall proceed without delay.

(3) If the permit has become final and no application under §124.3 concerning it is pending or if the variance request has been separated from a draft permit as described in paragraphs (a) (1) and (2) of this section, the Regional Administrator may prepare a new draft permit and give notice of it under §124.10. This draft permit shall be accompanied by the fact sheet required by §124.8 except that the only matters considered shall relate to the requested variance.

[48 FR 14264, Apr. 1, 1983, as amended at 51 FR 16030, Apr. 30, 1986]

§124.64 Appeals of variances.

(a) When a State issues a permit on which EPA has made a variance decision, separate appeals of the State permit and of the EPA variance decision are possible. If the owner or operator is challenging the same issues in both proceedings, the Regional Administrator will decide, in consultation with State officials, which case will be heard first.

(b) Variance decisions made by EPA may be appealed under the provisions of §124.19.

(c) *Stays for section 301(g) variances.* If an appeal is filed under §124.19 of a variance requested under CWA section 301(g), any otherwise applicable standards and limitations under CWA section 301 shall not be stayed unless:

(1) In the judgment of the Regional Administrator, the stay or the variance sought will not result in the discharge of pollutants in quantities which may reasonably be anticipated to pose an unacceptable risk to human health or the environment because of bioaccumulation, persistency in the environment, acute toxicity, chronic toxicity, or synergistic propensities; and

(2) In the judgment of the Regional Administrator, there is a substantial likelihood that the discharger will succeed on the merits of its appeal; and

(3) The discharger files a bond or other appropriate security which is required by the Regional Administrator

to assure timely compliance with the requirements from which a variance is sought in the event that the appeal is unsuccessful.

(d) Stays for variances other than section 301(g) variances are governed by §§124.16 and 124.60.

[48 FR 14264, Apr. 1, 1983, as amended at 65 FR 30912, May 15, 2000]

§124.65 [Reserved]

§124.66 Special procedures for decisions on thermal variances under section 316(a).

(a) The only issues connected with issuance of a particular permit on which EPA will make a final Agency decision before the final permit is issued under §§124.15 and 124.60 are whether alternative effluent limitations would be justified under CWA section 316(a) and whether cooling water intake structures will use the best available technology under section 316(b). Permit applicants who wish an early decision on these issues should request it and furnish supporting reasons at the time their permit applications are filed under §122.21. The Regional Administrator will then decide whether or not to make an early decision. If it is granted, both the early decision on CWA section 316 (a) or (b) issues and the grant of the balance of the permit shall be considered permit issuance under these regulations, and shall be subject to the same requirements of public notice and comment and the same opportunity for an appeal under §124.19.

(b) If the Regional Administrator, on review of the administrative record, determines that the information necessary to decide whether or not the CWA section 316(a) issue is not likely to be available in time for a decision on permit issuance, the Regional Administrator may issue a permit under §124.15 for a term up to 5 years. This permit shall require achievement of the effluent limitations initially proposed for the thermal component of the discharge no later than the date otherwise required by law. However, the permit shall also afford the permittee an opportunity to file a demonstration under CWA section 316(a) after conducting such studies as are required under 40 CFR part 125, subpart H. A new discharger may not exceed the thermal effluent limitation which is initially proposed unless and until its CWA section 316(a) variance request is finally approved.

(c) Any proceeding held under paragraph (a) of this section shall be publicly noticed as required by § 124.10 and shall be conducted at a time allowing the permittee to take necessary measures to meet the final compliance date in the event its request for modification of thermal limits is denied.

(d) Whenever the Regional Administrator defers the decision under CWA section 316(a), any decision under section 316(b) may be deferred.

[48 FR 14264, Apr. 1, 1983, as amended at 65 FR 30912, May 15, 2000]

Subparts E—F [Reserved]

Subpart G—Procedures for RCRA Standardized Permit

SOURCE: 70 FR 53449, Sept. 8, 2005, unless otherwise noted.

GENERAL INFORMATION ABOUT STANDARDIZED PERMITS

§ 124.200 What is a RCRA standardized permit?

The standardized permit is a special form of RCRA permit, that may consist of two parts: A uniform portion that the Director issues in all cases, and a supplemental portion that the Director issues at his or her discretion. We formally define the term "Standardized permit" in § 124.2.

(a) What comprises the uniform portion? The uniform portion of a standardized permit consists of terms and conditions, relevant to the unit(s) you are operating at your facility, that EPA has promulgated in 40 CFR part 267 (Standards for Owners and Operators of Hazardous Waste Facilities Operating under a Standardized Permit). If you intend to operate under the standardized permit, you must comply with these nationally applicable terms and conditions.

(b) What comprises the supplemental portion? The supplemental portion of a standardized permit consists of site-specific terms and conditions, beyond those of the uniform portion, that the Director may impose on your particular facility, as necessary to protect human health and the environment. If the Director issues you a supplemental portion, you must comply with the site-specific terms and conditions it imposes.

(1) When required under § 267.101, provisions to implement corrective action will be included in the supplemental portion.

(2) Unless otherwise specified, these supplemental permit terms and conditions apply to your facility in addition to the terms and conditions of the uniform portion of the standardized permit and not in place of any of those terms and conditions.

§ 124.201 Who is eligible for a standardized permit?

(a) You may be eligible for a standardized permit if:

(1) You generate hazardous waste and then store or non-thermally treat the hazardous waste on-site in containers, tanks, or containment buildings; or

(2) You receive hazardous waste generated off-site by a generator under the same ownership as the receiving facility, and then you store or non-thermally treat the hazardous waste in containers, tanks, or containment buildings.

(3) In either case, the Director will inform you of your eligibility when a decision is made on your permit.

(b) [Reserved]

APPLYING FOR A STANDARDIZED PERMIT

§ 124.202 How do I as a facility owner or operator apply for a standardized permit?

(a) You must follow the requirements in this subpart as well as those in § 124.31, 40 CFR 270.10, and 40 CFR part 270, subpart J.

(b) You must submit to the Director a written Notice of Intent to operate under the standardized permit. You must also include the information and certifications required under 40 CFR part 270, subpart J.

§124.203 How may I switch from my individual RCRA permit to a standardized permit?

Where all units in the RCRA permit are eligible for the standardized permit, you may request that your individual permit be revoked and reissued as a standardized permit, in accordance with §124.5. Where only some of the units in the RCRA permit are eligible for the standardized permit, you may request that your individual permit be modified to no longer include those units and issue a standardized permit for those units in accordance with §124.204.

ISSUING A STANDARDIZED PERMIT

§124.204 What must I do as the Director of the regulatory agency to prepare a draft standardized permit?

(a) You must review the Notice of Intent and supporting information submitted by the facility owner or operator.

(b) You must determine whether the facility is or is not eligible to operate under the standardized permit.

(1) If the facility is eligible for the standardized permit, you must propose terms and conditions, if any, to include in a supplemental portion. If you determine that these terms and conditions are necessary to protect human health and the environment and cannot be imposed, you must tentatively deny coverage under the standardized permit.

(2) If the facility is not eligible for the standardized permit, you must tentatively deny coverage under the standardized permit. Cause for ineligibility may include, but is not limited to, the following:

(i) Failure of owner or operator to submit all the information required under §270.275.

(ii) Information submitted that is required under §270.275 is determined to be inadequate.

(iii) Facility does not meet the eligibility requirements (activities are outside the scope of the standardized permit).

(iv) Demonstrated history of significant non-compliance with applicable requirements.

(v) Permit conditions cannot ensure protection of human health and the environment.

(c) You must prepare your draft permit decision within 120 days after receiving the Notice of Intent and supporting documents from a facility owner or operator. Your tentative determination under this section to deny or grant coverage under the standardized permit, including any proposed site-specific conditions in a supplemental portion, constitutes a draft permit decision. You are allowed a one time extension of 30 days to prepare the draft permit decision. When the use of the 30-day extension is anticipated, you should inform the permit applicant during the initial 120-day review period. Reasons for an extension may include, but is not limited to, needing to complete review of submissions with the Notice of Intent (e.g., closure plans, waste analysis plans, for facilities seeking to manage hazardous waste generated off-site).

(d) Many requirements in subpart A of this part apply to processing the standardized permit application and preparing your draft permit decision. For example, your draft permit decision must be accompanied by a statement of basis or fact sheet and must be based on the administrative record. In preparing your draft permit decision, the following provisions of subpart A of this part apply (subject to the following modifications):

(1) Section 124.1 Purpose and Scope. All paragraphs.

(2) Section 124.2 Definitions. All paragraphs.

(3) Section 124.3 Application for a permit. All paragraphs, except paragraphs (c), (d), (f), and (g) of this section apply.

(4) Section 124.4 Consolidation of permit processing. All paragraphs apply; however, in the context of the RCRA standardized permit, the reference to the public comment period is §124.208 instead of §124.10.

(5) Section 124.5 Modification, revocation and re-issuance, or termination of permits. Not applicable.

(6) Section 124.6 Draft permits. This section does not apply to the RCRA standardized permit; procedures in this subpart apply instead.

(7) Section 124.7 Statement of basis. The entire section applies.

(8) Section 124.8 Fact sheet. All paragraphs apply; however, in the context of the RCRA standardized permit, the reference to the public comment period is § 124.208 instead of § 124.10.

(9) Section 124.9 Administrative record for draft permits when EPA is the permitting authority. All paragraphs apply; however, in the context of the RCRA standardized permit, the reference to draft permits is § 24.204(c) instead of § 124.6.

(10) Section 124.10 Public notice of permit actions and public comment period. Only §§ 124.10(c)(1)(ix) and (c)(1)(x)(A) apply to the RCRA standardized permit. Most of § 124.10 does not apply to the RCRA standardized permit; §§ 124.207, 124.208, and 124.209 apply instead.

§ 124.205 What must I do as the Director of the regulatory agency to prepare a final standardized permit?

As Director of the regulatory agency, you must consider all comments received during the public comment period (see § 124.208) in making your final permit decision. In addition, many requirements in subpart A of this part apply to the public comment period, public hearings, and preparation of your final permit decision. In preparing a final permit decision, the following provisions of subpart A of this part apply (subject to the following modifications):

(a) Section 124.1 Purpose and Scope. All paragraphs.

(b) Section 124.2 Definitions. All paragraphs.

(c) Section 124.11 Public comments and requests for public hearings. This section does not apply to the RCRA standardized permit; the procedures in § 124.208 apply instead.

(d) Section 124.12 Public hearings. Paragraphs (b), (c), and (d) apply.

(e) Section 124.13 Obligation to raise issues and provide information during the public comment period. The entire section applies; however, in the context of the RCRA standardized permit, the reference to the public comment period is § 124.208 instead of § 124.10.

(f) Section 124.14 Reopening of the public comment period. All paragraphs apply; however, in the context of the RCRA standardized permit, use the following reference: in § 124.14(b)(1) use reference to § 124.204 instead of § 124.6; in § 124.14(b)(3) use reference to § 124.208 instead of § 124.10; in § 124.14(c) use reference to § 124.207 instead of § 124.10.

(g) Section 124.15 Issuance and effective date of permit. All paragraphs apply, however, in the context of the RCRA standardized permit, the reference to the public comment period is § 124.208 instead of § 124.10.

(h) Section 124.16 Stays of contested permit conditions. All paragraphs apply.

(i) Section 124.17 Response to comments. This section does not apply to the RCRA standardized permit; procedures in § 124.209 apply instead.

(j) Section 124.18 Administrative record for final permit when EPA is the permitting authority. All paragraphs apply, however, use reference to § 124.209 instead of § 124.17.

(k) Seciton124.19 Appeal of RCRA, UIC, NPDES, and PSD permits. All paragraphs apply.

(l) Section 124.20 Computation of time. All paragraphs apply.

§ 124.206 In what situations may I require a facility owner or operator to apply for an individual permit?

(a) Cases where you may determine that a facility is not eligible for the standardized permit include, but are not limited to, the following:

(1) The facility does not meet the criteria in § 124.201.

(2) The facility has a demonstrated history of significant non-compliance with regulations or permit conditions.

(3) The facility has a demonstrated history of submitting incomplete or deficient permit application information.

(4) The facility has submitted an incomplete or inadequate materials with the Notice of Intent.

(b) If you determine that a facility is not eligible for the standardized permit, you must inform the facility owner or operator that they must apply for an individual permit.

(c) You may require any facility that has a standardized permit to apply for and obtain an individual RCRA permit. Any interested person may petition you to take action under this paragraph. Cases where you may require an individual RCRA permit include, but are not limited to, the following:

(1) The facility is not in compliance with the terms and conditions of the standardized RCRA permit.

(2) Circumstances have changed since the time the facility owner or operator applied for the standardized permit, so that the facility's hazardous waste management practices are no longer appropriately controlled under the standardized permit.

(d) You may require any facility authorized by a standardized permit to apply for an individual RCRA permit only if you have notified the facility owner or operator in writing that an individual permit application is required. You must include in this notice a brief statement of the reasons for your decision, a statement setting a deadline for the owner or operator to file the application, and a statement that, on the effective date of the individual RCRA permit, the facility's standardized permit automatically terminates. You may grant additional time upon request from the facility owner or operator.

(e) When you issue an individual RCRA permit to an owner or operator otherwise subject to a standardized RCRA permit, the standardized permit for their facility will automatically cease to apply on the effective date of the individual permit.

OPPORTUNITIES FOR PUBLIC INVOLVEMENT IN THE STANDARDIZED PERMIT PROCESS

§ 124.207 What are the requirements for public notices?

(a) You, as the Director, must provide public notice of your draft permit decision and must provide an opportunity for the public to submit comments and request a hearing on that decision. You must provide the public notice to:

(1) The applicant;

(2) Any other agency which you know has issued or is required to issue a RCRA permit for the same facility or activity (including EPA when the draft permit is prepared by the State);

(3) Federal and State agencies with jurisdiction over fish, shellfish, and wildlife resources and over coastal zone management plans, the Advisory Council on Historic Preservation, State Historic Preservation Officers, including any affected States;

(4) To everyone on the facility mailing list developed according to the requirements in § 124.10(c)(1)(ix); and

(5) To any units of local government having jurisdiction over the area where the facility is proposed to be located and to each State agency having any authority under State law with respect to the construction or operation of the facility.

(b) You must issue the public notice according to the following methods:

(1) Publication in a daily or weekly major local newspaper of general circulation and broadcast over local radio stations;

(2) When the program is being administered by an approved State, in a manner constituting legal notice to the public under State law; and

(3) Any other method reasonably calculated to give actual notice of the draft permit decision to the persons potentially affected by it, including press releases or any other forum or medium to elicit public participation.

(c) You must include the following information in the public notice:

(1) The name and telephone number of the contact person at the facility.

(2) The name and telephone number of your contact office, and a mailing address to which people may direct comments, information, opinions, or inquiries.

(3) An address to which people may write to be put on the facility mailing list.

(4) The location where people may view and make copies of the draft standardized permit and the Notice of Intent and supporting documents.

(5) A brief description of the facility and proposed operations, including the address or a map (for example, a sketched or copied street map) of the facility location on the front page of the notice.

(6) The date that the facility owner or operator submitted the Notice of Intent and supporting documents.

(d) At the same time that you issue the public notice under this section, you must place the draft standardized permit (including both the uniform portion and the supplemental portion, if any), the Notice of Intent and supporting documents, and the statement of basis or fact sheet in a location accessible to the public in the vicinity of the facility or at your office.

§ 124.208 **What are the opportunities for public comments and hearings on draft permit decisions?**

(a) The public notice that you issue under § 124.207 must allow at least 45 days for people to submit written comments on your draft permit decision. This time is referred to as the public comment period. You must automatically extend the public comment period to the close of any public hearing under this section. The hearing officer may also extend the comment period by so stating at the hearing.

(b) During the public comment period, any interested person may submit written comments on the draft permit and may request a public hearing. If someone wants to request a public hearing, they must submit their request in writing to you. Their request must state the nature of the issues they propose to raise during the hearing.

(c) You must hold a public hearing whenever you receive a written notice of opposition to a standardized permit and a request for a hearing within the public comment period under paragraph (a) of this section. You may also hold a public hearing at your discretion, whenever, for instance, such a hearing might clarify one or more issues involved in the permit decision.

(d) Whenever possible, you must schedule a hearing under this section at a location convenient to the nearest population center to the facility. You must give public notice of the hearing at least 30 days before the date set for the hearing. (You may give the public notice of the hearing at the same time you provide public notice of the draft permit, and you may combine the two notices.)

(e) You must give public notice of the hearing according to the methods in § 124.207(a) and (b). The hearing must be conducted according to the procedures in § 124.12(b), (c), and (d).

(f) In their written comments and during the public hearing, if held, interested parties may provide comments on the draft permit decision. These comments may include, but are not limited to, the facility's eligibility for the standardized permit, the tentative supplemental conditions you proposed, and the need for additional supplemental conditions.

§ 124.209 **What are the requirements for responding to comments?**

(a) At the time you issue a final standardized permit, you must also respond to comments received during the public comment period on the draft permit. Your response must:

(1) Specify which additional conditions (*i.e.*, those in the supplemental portion), if any, you changed in the final permit, and the reasons for the change.

(2) Briefly describe and respond to all significant comments on the facility's ability to meet the general requirements (*i.e.*, those terms and conditions in the uniform portion) and on any additional conditions necessary to protect human health and the environment raised during the public comment period or during the hearing.

(3) Make the comments and responses accessible to the public.

(b) You may request additional information from the facility owner or operator or inspect the facility if you need additional information to adequately respond to significant comments or to make decisions about conditions you may need to add to the supplemental portion of the standardized permit.

(c) If you are the Director of an EPA permitting agency, you must include in the administrative record for your final permit decision any documents cited in the response to comments. If new points are raised or new material supplied during the public comment period, you may document your response to those matters by adding new materials to the administrative record.

§124.210 May I, as an interested party in the permit process, appeal a final standardized permit?

You may petition for administrative review of the Director's final permit decision, including his or her decision that the facility is eligible for the standardized permit, according to the procedures of §124.19. However, the terms and conditions of the uniform portion of the standardized permit are not subject to administrative review under this provision.

MAINTAINING A STANDARDIZED PERMIT

§124.211 What types of changes may I make to my standardized permit?

You may make both routine changes, routine changes with prior Agency approval, and significant changes. For the purposes of this section:

(a) "Routine changes" are any changes to the standardized permit that qualify as a class 1 permit modification (without prior Agency approval) under 40 CFR 270.42, appendix I, and

(b) "Routine changes with prior Agency approval" are for those changes to the standardized permit that would qualify as a class 1 modification with prior agency approval, or a class 2 permit modification under 40 CFR 270.42, appendix I; and

(c) "Significant changes" are any changes to the standardized permit that:

(1) Qualify as a class 3 permit modification under 40 CFR 270.42, appendix I;

(2) Are not explicitly identified in 40 CFR 270.42, appendix I; or

(3) Amend any terms or conditions in the supplemental portion of your standardized permit.

§124.212 What procedures must I follow to make routine changes?

(a) You can make routine changes to the standardized permit without obtaining approval from the Director. However, you must first determine whether the routine change you will make amends the information you submitted under 40 CFR 270.275 with your Notice of Intent to operate under the standardized permit.

(b) If the routine changes you make amend the information you submitted under 40 CFR 270.275 with your Notice of Intent to operate under the standardized permit, then before you make the routine changes you must:

(1) Submit to the Director the revised information pursuant to 40 CFR 270.275(a); and

(2) Provide notice of the changes to the facility mailing list and to state and local governments in accordance with the procedures in §124.10(c)(1)(ix) and (x).

§124.213 What procedures must I follow to make routine changes with prior approval?

(a) Routine changes to the standardized permit with prior Agency approval may only be made with the prior written approval of the Director.

(b) You must also follow the procedures in §124.212(b)(1)–(2).

§124.214 What procedures must I follow to make significant changes?

(a) You must first provide notice of and conduct a public meeting.

(1) Public Meeting. You must hold a meeting with the public to solicit questions from the community and inform the community of your proposed modifications to your hazardous waste management activities. You must post a sign-in sheet or otherwise provide a voluntary opportunity for people attending the meeting to provide their names and addresses.

(2) Public Notice. At least 30 days before you plan to hold the meeting, you must issue a public notice in accordance with the requirements of §124.31(d).

(b) After holding the public meeting, you must submit a modification request to the Director that:

(1) Describes the exact change(s) you want and whether they are changes to information you provided under 40 CFR 270.275 or to terms and conditions in the supplemental portion of your standardized permit;

(2) Explain why the modification is needed; and

(3) Includes a summary of the public meeting under paragraph (a) of this section, along with the list of

attendees and their addresses and copies of any written comments or materials they submitted at the meeting.

(c) Once the Director receives your modification request, he or she must make a tentative determination within 120 days to approve or disapprove your request. You are allowed a one time extension of 30 days to prepare the draft permit decision. When the use of the 30-day extension is anticipated, you should inform the permit applicant during the initial 120-day review period.

(d) After the Director makes this tentative determination, the procedures in § 124.205 and §§ 124.207 through 124.210 for processing an initial request for coverage under the standardized permit apply to making the final determination on the modification request.

PART 125—CRITERIA AND STANDARDS FOR THE NATIONAL POLLUTANT DISCHARGE ELIMINATION SYSTEM

AUTHORITY: The Clean Water Act, 33 U.S.C. 1251 *et seq.*, unless otherwise noted.

SOURCE: 44 FR 32948, June 7, 1979, unless otherwise noted.

Subpart A—Criteria and Standards for Imposing Technology-Based Treatment Requirements Under Sections 301(b) and 402 of the Act

§ 125.1 Purpose and scope.

This subpart establishes criteria and standards for the imposition of technology-based treatment requirements in permits under section 301(b) of the Act, including the application of EPA promulgated effluent limitations and case-by-case determinations of effluent limitations under section 402(a)(1) of the Act.

§ 125.2 Definitions.

For the purposes of this part, any reference to *the Act* shall mean the Clean

Water Act of 1977 (CWA). Unless otherwise noted, the definitions in parts 122, 123 and 124 apply to this part.

[45 FR 33512, May 19, 1980]

§ 125.3 Technology-based treatment requirements in permits.

(a) *General.* Technology-based treatment requirements under section 301(b) of the Act represent the minimum level of control that must be imposed in a permit issued under section 402 of the Act. (See §§ 122.41, 122.42 and 122.44 for a discussion of additional or more stringent effluent limitations and conditions.) Permits shall contain the following technology-based treatment requirements in accordance with the following statutory deadlines;

(1) For POTW's, effluent limitations based upon:

(i) Secondary treatment—from date of permit issuance; and

(ii) The best practicable waste treatment technology—not later than July 1, 1983; and

(2) For dischargers other than POTWs except as provided in § 122.29(d), effluent limitations requiring:

(i) The best practicable control technology currently available (BPT)—

(A) For effluent limitations promulgated under Section 304(b) after January 1, 1982 and requiring a level of control substantially greater or based on fundamentally different control technology than under permits for an industrial category issued before such date, compliance as expeditiously as practicable but in no case later than three years after the date such limitations are promulgated under section 304(b) and in no case later than March 31, 1989;

(B) For effluent limitations established on a case-by-case basis based on Best Professional Judgment (BPJ) under Section 402(a)(1)(B) of the Act in a permit issued after February 4, 1987, compliance as expeditiously as practicable but in no case later than three years after the date such limitations are established and in no case later than March 31, 1989;

(C) For all other BPT effluent limitations compliance is required from the date of permit issuance.

(ii) For conventional pollutants, the best conventional pollutant control technology (BCT)—

(A) For effluent limitations promulgated under section 304(b), as expeditiously as practicable but in no case later than three years after the date such limitations are promulgated under section 304(b), and in no case later than March 31, 1989.

(B) For effluent limitations established on a case-by-case (BPJ) basis under section 402(a)(1)(B) of the Act in a permit issued after February 4, 1987, compliance as expeditiously as practicable but in no case later than three years after the date such limitations are established and in no case later than March 31, 1989;

(iii) For all toxic pollutants referred to in Committee Print No. 95–30, House Committee on Public Works and Transportation, the best available technology economically achievable (BAT)—

(A) For effluent limitations established under section 304(b), as expeditiously as practicable but in no case later than three years after the date such limitations are promulgated under section 304(b), and in no case later than March 31, 1989.

(B) For permits issued on a case-by-case (BPJ) basis under section 402(a)(1)(B) of the Act after February 4, 1987 establishing BAT effluent limitations, compliance is required as expeditiously as practicable but in no case later than three years after the date such limitations are promulgated under section 304(b), and in no case later than March 31, 1989.

(iv) For all toxic pollutants other than those listed in Committee Print No. 95–30, effluent limitations based on BAT—

(A) For effluent limitations promulgated under section 304(b) compliance is required as expeditiously as practicable, but in no case later than three years after the date such limitations are promulgated under section 304(b) and in no case later than March 31, 1989.

(B) For permits issued on a case-by-case (BPJ) basis under Section 402(a)(1)(B) of the Act after February 4,

1987 establishing BAT effluent limitations, compliance is required as expeditiously as practicable but in no case later than 3 years after the date such limitations are established and in no case later than March 31, 1989.

(v) For all pollutants which are neither toxic nor conventional pollutants, effluent limitations based on BAT—

(A) For effluent limitations promulgated under section 304(b), compliance is required as expeditiously as practicable but in no case later than 3 years after the date such limitations are established and in no case later than March 31, 1989.

(B) For permits issued on a case-by-case (BPJ) basis under section 402(a)(1)(B) of the Act after February 4, 1987 establishing BAT effluent limitations compliance is required as expeditiously as practicable but in no case later than three years after the date such limitations are established and in no case later than March 31, 1989.

(b) *Statutory variances and extensions.* (1) The following variances from technology-based treatment requirements are authorized by the Act and may be applied for under §122.21;

(i) For POTW's, a section 301(h) marine discharge variance from secondary treatment (subpart G);

(ii) For dischargers other than POTW's;

(A) A section 301(c) economic variance from BAT (subpart E);

(B) A section 301(g) water quality related variance from BAT (subpart F); and

(C) A section 316(a) thermal variance from BPT, BCT and BAT (subpart H).

(2) The following extensions of deadlines for compliance with technology-based treatment requirements are authorized by the Act and may be applied for under §124.53:

(i) For POTW's a section 301(i) extension of the secondary treatment deadline (subpart J);

(ii) For dischargers other than POTW's:

(A) A section 301(i) extension of the BPT deadline (subpart J); and

(B) A section 301(k) extension of the BAT deadline (subpart C).

(c) *Methods of imposing technology-based treatment requirements in permits.* Technology-based treatment requirements may be imposed through one of the following three methods:

(1) Application of EPA-promulgated effluent limitations developed under section 304 of the Act to dischargers by category or subcategory. These effluent limitations are not applicable to the extent that they have been remanded or withdrawn. However, in the case of a court remand, determinations underlying effluent limitations shall be binding in permit issuance proceedings where those determinations are not required to be reexamined by a court remanding the regulations. In addition, dischargers may seek fundamentally different factors variances from these effluent limitations under §122.21 and subpart D of this part.

(2) On a case-by-case basis under section 402(a)(1) of the Act, to the extent that EPA-promulgated effluent limitations are inapplicable. The permit writer shall apply the appropriate factors listed in §125.3(d) and shall consider:

(i) The appropriate technology for the category or class of point sources of which the applicant is a member, based upon all available information; and

(ii) Any unique factors relating to the applicant.

[*Comment:* These factors must be considered in all cases, regardless of whether the permit is being issued by EPA or an approved State.]

(3) Through a combination of the methods in paragraphs (d) (1) and (2) of this section. Where promulgated effluent limitations guidelines only apply to certain aspects of the discharger's operation, or to certain pollutants, other aspects or activities are subject to regulation on a case-by-case basis in order to carry out the provisions of the Act.

(4) Limitations developed under paragraph (d)(2) of this section may be expressed, where appropriate, in terms of toxicity (e.g., "the LC_{50} for fat head minnow of the effluent from outfall 001 shall be greater than 25%"). *Provided,* That is shown that the limits reflect the appropriate requirements (for example, technology-based or water-quality-based standards) of the Act.

(d) In setting case-by-case limitations pursuant to §125.3(c), the permit writer must consider the following factors:

(1) *For BPT requirements:* (i) The total cost of application of technology in relation to the effluent reduction benefits to be achieved from such application;

(ii) The age of equipment and facilities involved;

(iii) The process employed;

(iv) The engineering aspects of the application of various types of control techniques;

(v) Process changes; and

(vi) Non-water quality environmental impact (including energy requirements).

(2) *For BCT requirements:* (i) The reasonableness of the relationship between the costs of attaining a reduction in effluent and the effluent reduction benefits derived;

(ii) The comparison of the cost and level of reduction of such pollutants from the discharge from publicly owned treatment works to the cost and level of reduction of such pollutants from a class or category of industrial sources;

(iii) The age of equipment and facilities involved;

(iv) The process employed;

(v) The engineering aspects of the application of various types of control techniques;

(vi) Process changes; and

(vii) Non-water quality environmental impact (including energy requirements).

(3) *For BAT requirements:* (i) The age of equipment and facilities involved;

(ii) The process employed;

(iii) The engineering aspects of the application of various types of control techniques;

(iv) Process changes;

(v) The cost of achieving such effluent reduction; and

(vi) Non-water quality environmental impact (including energy requirements).

(e) Technology-based treatment requirements are applied prior to or at the point of discharge.

(f) Technology-based treatment requirements cannot be satisfied through the use of "non-treatment" techniques such as flow augmentation and in-stream mechanical aerators. However, these techniques may be considered as a method of achieving water quality standards on a case-by-case basis when:

(1) The technology-based treatment requirements applicable to the discharge are not sufficient to achieve the standards;

(2) The discharger agrees to waive any opportunity to request a variance under section 301 (c), (g) or (h) of the Act; and

(3) The discharger demonstrates that such a technique is the preferred environmental and economic method to achieve the standards after consideration of alternatives such as advanced waste treatment, recycle and reuse, land disposal, changes in operating methods, and other available methods.

(g) Technology-based effluent limitations shall be established under this subpart for solids, sludges, filter backwash, and other pollutants removed in the course of treatment or control of wastewaters in the same manner as for other pollutants.

(h)(1) The Director may set a permit limit for a conventional pollutant at a level more stringent than the best conventional pollution control technology (BCT), or a limit for a nonconventional pollutant which shall not be subject to modification under section 301 (c) or (g) of the Act where:

(i) Effluent limitations guidelines specify the pollutant as an indicator for a toxic pollutant, or

(ii)(A) The limitation reflects BAT-level control of discharges of one or more toxic pollutants which are present in the waste stream, and a specific BAT limitation upon the toxic pollutant(s) is not feasible for economic or technical reasons;

(B) The permit identifies which toxic pollutants are intended to be controlled by use of the limitation; and

(C) The fact sheet required by § 124.56 sets forth the basis for the limitation, including a finding that compliance with the limitation will result in BAT-level control of the toxic pollutant discharges identified in paragraph (h)(1)(ii)(B) of this section, and a finding that it would be economically or technically infeasible to directly limit the toxic pollutant(s).

(2) The Director may set a permit limit for a conventional pollutant at a level more stringent than BCT when:

(i) Effluent limitations guidelines specify the pollutant as an indicator for a hazardous substance, or

(ii)(A) The limitation reflects BAT-level control of discharges (or an appropriate level determined under section 301(c) or (g) of the Act) of one or more hazardous substance(s) which are present in the waste stream, and a specific BAT (or other appropriate) limitation upon the hazardous substance(s) is not feasible for economic or technical reasons;

(B) The permit identifies which hazardous substances are intended to be controlled by use of the limitation; and

(C) The fact sheet required by § 124.56 sets forth the basis for the limitation, including a finding that compliance with the limitations will result in BAT-level (or other appropriate level) control of the hazardous substances discharges identified in paragraph (h)(2)(ii)(B) of this section, and a finding that it would be economically or technically infeasible to directly limit the hazardous substance(s).

(iii) Hazardous substances which are also toxic pollutants are subject to paragraph (h)(1) of this section.

(3) The Director may not set a more stringent limit under the preceding paragraphs if the method of treatment required to comply with the limit differs from that which would be required if the toxic pollutant(s) or hazardous substance(s) controlled by the limit were limited directly.

(4) Toxic pollutants identified under paragraph (h)(1) of this section remain subject to the requirements of § 122.42(a)(1) (notification of increased discharges of toxic pollutants above levels reported in the application form).

(Clean Water Act, Safe Drinking Water Act, Clean Air Act, Resource Conservation and Recovery Act: 42 U.S.C. 6905, 6912, 6925, 6927, 6974)

[44 FR 32948, June 7, 1979, as amended at 45 FR 33512, May 19, 1980; 48 FR 14293, Apr. 1, 1983; 49 FR 38052, Sept. 26, 1984; 50 FR 6941, Feb. 19, 1985; 54 FR 257, Jan. 4, 1989]

Subpart B—Criteria for Issuance of Permits to Aquaculture Projects

§ 125.10 Purpose and scope.

(a) These regulations establish guidelines under sections 318 and 402 of the Act for approval of any discharge of pollutants associated with an aquaculture project.

(b) The regulations authorize, on a selective basis, controlled discharges which would otherwise be unlawful under the Act in order to determine the feasibility of using pollutants to grow aquatic organisms which can be harvested and used beneficially. EPA policy is to encourage such projects, while at the same time protecting other beneficial uses of the waters.

(c) Permits issued for discharges into aquaculture projects under this subpart are NPDES permits and are subject to the applicable requirements of parts 122, 123 and 124. Any permit shall include such conditions (including monitoring and reporting requirements) as are necessary to comply with those parts. Technology-based effluent limitations need not be applied to discharges into the approved project except with respect to toxic pollutants.

§ 125.11 Criteria.

(a) No NPDES permit shall be issued to an aquaculture project unless:

(1) The Director determines that the aquaculture project:

(i) Is intended by the project operator to produce a crop which has significant direct or indirect commercial value (or is intended to be operated for research into possible production of such a crop); and

(ii) Does not occupy a designated project area which is larger than can be economically operated for the crop under cultivation or than is necessary for research purposes.

(2) The applicant has demonstrated, to the satisfaction of the Director, that the use of the pollutant to be discharged to the aquaculture project will result in an increased harvest of organisms under culture over what would naturally occur in the area;

(3) The applicant has demonstrated, to the satisfaction of the Director, that if the species to be cultivated in the aquaculture project is not indigenous

to the immediate geographical area, there will be minimal adverse effects on the flora and fauna indigenous to the area, and the total commercial value of the introduced species is at least equal to that of the displaced or affected indigenous flora and fauna;

(4) The Director determines that the crop will not have a significant potential for human health hazards resulting from its consumption;

(5) The Director determines that migration of pollutants from the designated project area to water outside of the aquaculture project will not cause or contribute to a violation of water quality standards or a violation of the applicable standards and limitations applicable to the supplier of the pollutant that would govern if the aquaculture project were itself a point source. The approval of an aquaculture project shall not result in the enlargement of a pre-existing mixing zone area beyond what had been designated by the State for the original discharge.

(b) No permit shall be issued for any aquaculture project in conflict with a plan or an amendment to a plan approved under section 208(b) of the Act.

(c) No permit shall be issued for any aquaculture project located in the territorial sea, the waters of the contiguous zone, or the oceans, except in conformity with guidelines issued under section 403(c) of the Act.

(d) Designated project areas shall not include a portion of a body of water large enough to expose a substantial portion of the indigenous biota to the conditions within the designated project area. For example, the designated project area shall not include the entire width of a watercourse, since all organisms indigenous to that watercourse might be subjected to discharges of pollutants that would, except for the provisions of section 318 of the Act, violate section 301 of the Act.

(e) Any modifications caused by the construction or creation of a reef, barrier or containment structure shall not unduly alter the tidal regimen of an estuary or interfere with migrations of unconfined aquatic species.

[*Comment:* Any modifications described in this paragraph which result in the discharge of dredged or fill material into navigable waters may be subject to the permit requirements of section 404 of the Act.]

(f) Any pollutants not required by or beneficial to the aquaculture crop shall not exceed applicable standards and limitations when entering the designated project area.

Subpart C [Reserved]

Subpart D—Criteria and Standards for Determining Fundamentally Different Factors Under Sections 301(b)(1)(A), 301(b)(2) (A) and (E) of the Act

§ 125.30 Purpose and scope.

(a) This subpart establishes the criteria and standards to be used in determining whether effluent limitations alternative to those required by promulgated EPA effluent limitations guidelines under sections 301 and 304 of the Act (hereinafter referred to as "national limits") should be imposed on a discharger because factors relating to the discharger's facilities, equipment, processes or other factors related to the discharger are fundamentally different from the factors considered by EPA in development of the national limits. This subpart applies to all national limitations promulgated under sections 301 and 304 of the Act, except for the BPT limits contained in 40 CFR 423.12 (steam electric generating point source category).

(b) In establishing national limits, EPA takes into account all the information it can collect, develop and solicit regarding the factors listed in sections 304(b) and 304(g) of the Act. In some cases, however, data which could affect these national limits as they apply to a particular discharge may not be available or may not be considered during their development. As a result, it may be necessary on a case-by-case basis to adjust the national limits, and make them either more or less stringent as they apply to certain dischargers within an industrial category or subcategory. This will only be done if data specific to that discharger indicates it presents factors fundamentally different from those considered by EPA in developing the limit at issue. Any interested person believing that factors relating to a discharger's facilities, equipment, processes or other facilities related to the discharger are fundamentally different from the factors considered during development of the national limits may request a fundamentally different factors variance under §122.21(1)(1). In addition, such a variance may be proposed by the Director in the draft permit.

(Secs. 301, 304, 306, 307, 308, and 501 of the Clean Water Act (the Federal Water Pollution Control Act Amendments of 1972, Pub. L. 92–500 as amended by the Clean Water Act of 1977, Pub. L. 95–217 (the "Act"); Clean Water Act, Safe Drinking Water Act, Clean Air Act, Resource Conservation and Recovery Act: 42 U.S.C. 6905, 6912, 6925, 6927, 6974)

[44 FR 32948, June 7, 1979, as amended at 45 FR 33512, May 19, 1980; 46 FR 9460, Jan. 28, 1981; 47 FR 52309, Nov. 19, 1982; 48 FR 14293, Apr. 1, 1983]

§125.31 Criteria.

(a) A request for the establishment of effluent limitations under this subpart (fundamentally different factors variance) shall be approved only if:

(1) There is an applicable national limit which is applied in the permit and specifically controls the pollutant for which alternative effluent limitations or standards have been requested; and

(2) Factors relating to the discharge controlled by the permit are fundamentally different from those considered by EPA in establishing the national limits; and

(3) The request for alternative effluent limitations or standards is made in accordance with the procedural requirements of part 124.

(b) A request for the establishment of effluent limitations less stringent than those required by national limits guidelines shall be approved only if:

(1) The alternative effluent limitation or standard requested is no less stringent than justified by the fundamental difference; and

(2) The alternative effluent limitation or standard will ensure compliance with sections 208(e) and 301(b)(1)(C) of the Act; and

(3) Compliance with the national limits (either by using the technologies upon which the national limits are

based or by other control alternatives) would result in:

(i) A removal cost wholly out of proportion to the removal cost considered during development of the national limits; or

(ii) A non-water quality environmental impact (including energy requirements) fundamentally more adverse than the impact considered during development of the national limits.

(c) A request for alternative limits more stringent than required by national limits shall be approved only if:

(1) The alternative effluent limitation or standard requested is no more stringent than justified by the fundamental difference; and

(2) Compliance with the alternative effluent limitation or standard would not result in:

(i) A removal cost wholly out of proportion to the removal cost considered during development of the national limits; or

(ii) A non-water quality environmental impact (including energy requirements) fundamentally more adverse than the impact considered during development of the national limits.

(d) Factors which may be considered fundamentally different are:

(1) The nature or quality of pollutants contained in the raw waste load of the applicant's process wastewater;

[Comment: (1) In determining whether factors concerning the discharger are fundamentally different, EPA will consider, where relevant, the applicable development document for the national limits, associated technical and economic data collected for use in developing each respective national limit, records of legal proceedings, and written and printed documentation including records of communication, etc., relevant to the development of respective national limits which are kept on public file by EPA.

(2) Waste stream(s) associated with a discharger's process wastewater which were not considered in the development of the national limits will not ordinarily be treated as fundamentally different under paragraph (a) of this section. Instead, national limits should be applied to the other streams, and the unique stream(s) should be subject to limitations based on section 402(a)(1) of the Act. See §125.2(c)(2).]

(2) The volume of the discharger's process wastewater and effluent discharged;

(3) Non-water quality environmental impact of control and treatment of the discharger's raw waste load;

(4) Energy requirements of the application of control and treatment technology;

(5) Age, size, land availability, and configuration as they relate to the discharger's equipment or facilities; processes employed; process changes; and engineering aspects of the application of control technology;

(6) Cost of compliance with required control technololgy.

(e) A variance request or portion of such a request under this section shall not be granted on any of the following grounds:

(1) The infeasibility of installing the required waste treatment equipment within the time the Act allows.

[*Comment:* Under this section a variance request may be approved if it is based on factors which relate to the discharger's ability ultimately to achieve national limits but not if it is based on factors which merely affect the discharger's ability to meet the statutory deadlines of sections 301 and 307 of the Act such as labor difficulties, construction schedules, or unavailability of equipment.]

(2) The assertion that the national limits cannot be achieved with the appropriate waste treatment facilities installed, if such assertion is not based on factor(s) listed in paragraph (d) of this section;

[*Comment:* Review of the Administrator's action in promulgating national limits is available only through the judicial review procedures set forth in section 509(b) of the Act.]

(3) The discharger's ability to pay for the required waste treatment; or

(4) The impact of a discharge on local receiving water quality.

(f) Nothing in this section shall be construed to impair the right of any State or locality under section 510 of the Act to impose more stringent limitations than those required by Federal law.

§ 125.32   Method of application.

(a) A written request for a variance under this subpart D shall be submitted in duplicate to the Director in accordance with §§ 122.21(m)(1) and 124.3 of this chapter.

(b) The burden is on the person requesting the variance to explain that:

(1) Factor(s) listed in § 125.31(b) regarding the discharger's facility are fundamentally different from the factors EPA considered in establishing the national limits. The requester should refer to all relevant material and information, such as the published guideline regulations development document, all associated technical and economic data collected for use in developing each national limit, all records of legal proceedings, and all written and printed documentation including records of communication, etc., relevant to the regulations which are kept on public file by the EPA;

(2) The alternative limitations requested are justified by the fundamental difference alleged in paragraph (b)(1) of this section; and

(3) The appropriate requirements of § 125.31 have been met.

[44 FR 32948, June 7, 1979, as amended at 65 FR 30913, May 15, 2000]

Subpart E—Criteria for Granting Economic Variances From Best Available Technology Economically Achievable Under Section 301(c) of the Act [Reserved]

Subpart F—Criteria for Granting Water Quality Related Variances Under Section 301(g) of the Act [Reserved]

Subpart G—Criteria for Modifying the Secondary Treatment Requirements Under Section 301(h) of the Clean Water Act

AUTHORITY: Clean Water Act, as amended by the Clean Water Act of 1977, 33 U.S.C. 1251 *et seq.*, unless otherwise noted.

SOURCE: 59 FR 40658, Aug. 9, 1994, unless otherwise noted.

§ 125.56   Scope and purpose.

This subpart establishes the criteria to be applied by EPA in acting on section 301(h) requests for modifications to the secondary treatment requirements. It also establishes special permit conditions which must be included in any permit incorporating a section 301(h) modification of the secondary treatment requirements ("section 301(h) modified permit").

§125.57 Law governing issuance of a section 301(h) modified permit.

(a) Section 301(h) of the Clean Water Act provides that:

Administrator, with the concurrence of the State, may issue a permit under section 402 which modifies the requirements of paragraph (b)(1)(B) of this section with respect to the discharge of any pollutant from a publicly owned treatment works into marine waters, if the applicant demonstrates to the satisfaction of the Administrator that—

(1) There is an applicable water quality standard specific to the pollutant for which the modification is requested, which has been identified under section 304(a)(6) of this Act;

(2) The discharge of pollutants in accordance with such modified requirements will not interfere, alone or in combination with pollutants from other sources, with the attainment or maintenance of that water quality which assures protection of public water supplies and protection and propagation of a balanced indigenous population of shellfish, fish, and wildlife, and allows recreational activities, in and on the water;

(3) The applicant has established a system for monitoring the impact of such discharge on a representative sample of aquatic biota, to the extent practicable, and the scope of such monitoring is limited to include only those scientific investigations which are necessary to study the effects of the proposed discharge;

(4) Such modified requirements will not result in any additional requirements on any other point or nonpoint source;

(5) All applicable pretreatment requirements for sources introducing waste into such treatment works will be enforced;

(6) In the case of any treatment works serving a population of 50,000 or more, with respect to any toxic pollutant introduced into such works by an industrial discharger for which pollutant there is no applicable pretreatment requirement in effect, sources introducing waste into such works are in compliance with all applicable pretreatment requirements, the applicant will enforce such requirements, and the applicant has in effect a pretreatment program which, in combination with the treatment of discharges from such works, removes the same amount of such pollutant as would be removed if such works were to apply secondary treatment to discharges and if such works

had no pretreatment program with respect to such pollutant;

(7) To the extent practicable, the applicant has established a schedule of activities designed to eliminate the entrance of toxic pollutants from nonindustrial sources into such treatment works;

(8) There will be no new or substantially increased discharges from the point source of the pollutant to which the modification applies above that volume of discharge specified in the permit;

(9) The applicant at the time such modification becomes effective will be discharging effluent which has received at least primary or equivalent treatment and which meets the criteria established under section 304(a)(1) of this Act after initial mixing in the waters surrounding or adjacent to the point at which such effluent is discharged.

For the purposes of this section, the phrase "the discharge of any pollutant into marine waters" refers to a discharge into deep waters of the territorial sea or the waters of the contiguous zone, or into saline estuarine waters where there is strong tidal movement and other hydrological and geological characteristics which the Administrator determines necessary to allow compliance with paragraph (2) of this section, and section 101(a)(2) of this Act. For the purposes of paragraph (9), "primary or equivalent treatment" means treatment by screening, sedimentation, and skimming adequate to remove at least 30 percent of the biological oxygen demanding material and of the suspended solids in the treatment works influent, and disinfection, where appropriate. A municipality which applies secondary treatment shall be eligible to receive a permit pursuant to this subsection which modifies the requirements of paragraph (b)(1)(B) of this section with respect to the discharge of any pollutant from any treatment works owned by such municipality into marine waters. No permit issued under this subsection shall authorize the discharge of sewage sludge into marine waters. In order for a permit to be issued under this subsection for the discharge of a pollutant into marine waters, such marine waters must exhibit characteristics assuring that water providing dilution does not contain significant amounts of previously discharged effluent from such treatment works. No permit issued under this subsection shall authorize the discharge of any pollutant into saline estuarine waters which at the time of application do not support a balanced indigenous population of shellfish, fish, and wildlife, or allow recreation in and on the waters or which exhibit ambient water quality below applicable water quality standards adopted for the protection of public water supplies, shellfish, fish, and wildlife or recreational activities or such other standards necessary to assure support and protection of such uses. The prohibition contained in the preceding sentence shall apply without regard to the presence or absence of a causal relationship between such characteristics and the applicant's current or proposed discharge. Notwithstanding any other provisions of this subsection, no permit may be issued under this subsection for discharge of a pollutant into the New York Bight Apex consisting of the ocean waters of the Atlantic Ocean westward of 73 degrees 30 minutes west longitude and northward of 40 degrees 10 minutes north latitude.

(b) Section 301(j)(1) of the Clean Water Act provides that:

Any application filed under this section for a modification of the provisions of—

(A) subsection (b)(1)(B) under subsection (h) of this section shall be filed not later than the 365th day which begins after the date of enactment of the Municipal Wastewater Treatment Construction Grant Amendments of 1981, except that a publicly owned treatment works which prior to December 31, 1982, had a contractual arrangement to use a portion of the capacity of an ocean outfall operated by another publicly owned treatment works which has applied for or received modification under subsection (h) may apply for a modification of subsection (h) in its own right not later than 30 days after the date of the enactment of the Water Quality Act of 1987.

(c) Section 22(e) of the Municipal Wastewater Treatment Construction Grant Amendments of 1981, Public Law 97–117, provides that:

The amendments made by this section shall take effect on the date of enactment of this Act except that no applicant, other than the city of Avalon, California, who applies after the date of enactment of this Act for a permit pursuant to subsection (h) of section 301 of the Federal Water Pollution Control Act which modifies the requirements of subsection (b)(1)(B) of section 301 of such Act shall receive such permit during the one-year period which begins on the date of enactment of this Act.

(d) Section 303(b)(2) of the Water Quality Act, Public Law 100–4, provides that:

Section 301(h)(3) shall only apply to modifications and renewals of modifications which are tentatively or finally approved after the date of the enactment of this Act.

(e) Section 303(g) of the Water Quality Act provides that:

The amendments made to sections 301(h) and (h)(2), as well as provisions of (h)(6) and (h)(9), shall not apply to an application for a permit under section 301(h) of the Federal Water Pollution Control Act which has been tentatively or finally approved by the Administrator before the date of the enactment of this Act; except that such amendments shall apply to all renewals of such permits after such date of enactment.

§ 125.58 Definitions.

For the purpose of this subpart:

(a) *Administrator* means the EPA Administrator or a person designated by the EPA Administrator.

(b) *Altered discharge* means any discharge other than a current discharge or improved discharge, as defined in this regulation.

(c) *Applicant* means an applicant for a new or renewed section 301(h) modified permit. Large applicants have populations contributing to their POTWs equal to or more than 50,000 people or average dry weather flows of 5.0 million gallons per day (mgd) or more; small applicants have contributing populations of less than 50,000 people and average dry weather flows of less than 5.0 mgd. For the purposes of this definition the contributing population and flows shall be based on projections for the end of the five-year permit term. Average dry weather flows shall be the average daily total discharge flows for the maximum month of the dry weather season.

(d) *Application* means a final application previously submitted in accordance with the June 15, 1979, section 301(h) regulations (44 FR 34784); an application submitted between December 29, 1981, and December 29, 1982; or a section 301(h) renewal application submitted in accordance with these regulations. It does not include a preliminary application submitted in accordance with the June 15, 1979, section 301(h) regulations.

(e) *Application questionnaire* means EPA's "Applicant Questionnaire for Modification of Secondary Treatment Requirements," published as an appendix to this subpart.

(f) *Balanced indigenous population* means an ecological community which:

(1) Exhibits characteristics similar to those of nearby, healthy communities existing under comparable but unpolluted environmental conditions; or

(2) May reasonably be expected to become re-established in the polluted water body segment from adjacent waters if sources of pollution were removed.

(g) *Categorical pretreatment standard* means a standard promulgated by EPA under 40 CFR Chapter I, Subchapter N.

(h) *Current discharge* means the volume, composition, and location of an applicant's discharge at the time of permit application.

(i) *Improved discharge* means the volume, composition, and location of an applicant's discharge following:

(1) Construction of planned outfall improvements, including, without limitation, outfall relocation, outfall repair, or diffuser modification; or

(2) Construction of planned treatment system improvements to treatment levels or discharge characteristics; or

(3) Implementation of a planned program to improve operation and maintenance of an existing treatment system or to eliminate or control the introduction of pollutants into the applicant's treatment works.

(j) *Industrial discharger* or *industrial source* means any source of non-domestic pollutants regulated under section 307(b) or (c) of the Clean Water Act which discharges into a POTW.

(k) *Modified discharge* means the volume, composition, and location of the discharge proposed by the applicant for which a modification under section 301(h) of the Act is requested. A modified discharge may be a current discharge, improved discharge, or altered discharge.

(l) *New York Bight Apex* means the ocean waters of the Atlantic Ocean westward of 73 degrees 30 minutes west longitude and northward of 40 degrees 10 minutes north latitude.

(m) *Nonindustrial source* means any source of pollutants which is not an industrial source.

(n) *Ocean waters* means those coastal waters landward of the baseline of the territorial seas, the deep waters of the territorial seas, or the waters of the contiguous zone. The term "ocean waters" excludes saline estuarine waters.

(o) *Permittee* means an NPDES permittee with an effective section 301(h) modified permit.

(p) *Pesticides* means demeton, guthion, malathion, mirex, methoxychlor, and parathion.

(q) *Pretreatment* means the reduction of the amount of pollutants, the elimination of pollutants, or the alteration of the nature of pollutant properties in wastewater prior to or in lieu of discharging or otherwise introducing such pollutants into a POTW. The reduction or alteration may be obtained by physical, chemical, or biological processes, process changes, or by other means, except as prohibited by 40 CFR part 403.

(r) *Primary or equivalent treatment* for the purposes of this subpart means treatment by screening, sedimentation, and skimming adequate to remove at least 30 percent of the biochemical oxygen demanding material and of the suspended solids in the treatment works influent, and disinfection, where appropriate.

(s) *Public water supplies* means water distributed from a public water system.

(t) *Public water system* means a system for the provision to the public of piped water for human consumption, if such system has at least fifteen (15) service connections or regularly serves at least twenty-five (25) individuals. This term includes: (1) Any collection, treatment, storage, and distribution facilities under the control of the operator of the system and used primarily in connection with the system, and (2) Any collection or pretreatment storage facilities not under the control of the operator of the system which are used primarily in connection with the system.

(u) *Publicly owned treatment works* or *POTW* means a treatment works, as defined in section 212(2) of the Act, which is owned by a State, municipality, or intermunicipal or interstate agency.

(v) *Saline estuarine waters* means those semi-enclosed coastal waters which have a free connection to the territorial sea, undergo net seaward exchange with ocean waters, and have salinities comparable to those of the ocean. Generally, these waters are near the mouth of estuaries and have cross-sectional annual mean salinities greater than twenty-five (25) parts per thousand.

(w) *Secondary removal equivalency* means that the amount of a toxic pollutant removed by the combination of the applicant's own treatment of its influent and pretreatment by its industrial users is equal to or greater than the amount of the toxic pollutant that would be removed if the applicant were to apply secondary treatment to its discharge where the discharge has not undergone pretreatment by the applicant's industrial users.

(x) *Secondary treatment* means the term as defined in 40 CFR part 133.

(y) *Shellfish, fish, and wildlife* means any biological population or community that might be adversely affected by the applicant's modified discharge.

(z) *Stressed waters* means those ocean waters for which an applicant can demonstrate to the satisfaction of the Administrator, that the absence of a balanced indigenous population is caused solely by human perturbations other than the applicant's modified discharge.

(aa) *Toxic pollutants* means those substances listed in 40 CFR 401.15.

(bb) *Water quality criteria* means scientific data and guidance developed and periodically updated by EPA under section 304(a)(1) of the Clean Water Act, which are applicable to marine waters.

(cc) *Water quality standards* means applicable water quality standards which have been approved, left in effect, or promulgated under section 303 of the Clean Water Act.

(dd) *Zone of initial dilution* (ZID) means the region of initial mixing surrounding or adjacent to the end of the outfall pipe or diffuser ports, provided that the ZID may not be larger than allowed by mixing zone restrictions in applicable water quality standards.

§ 125.59 General.

(a) *Basis for application.* An application under this subpart shall be based on a current, improved, or altered discharge into ocean waters or saline estuarine waters.

(b) *Prohibitions.* No section 301(h) modified permit shall be issued:

(1) Where such issuance would not assure compliance with all applicable requirements of this subpart and part 122;

(2) For the discharge of sewage sludge;

(3) Where such issuance would conflict with applicable provisions of State, local, or other Federal laws or Executive Orders. This includes compliance with the Coastal Zone Management Act of 1972, as amended, 16 U.S.C. 1451 *et seq.*; the Endangered Species Act of 1973, as amended, 16 U.S.C. 1531 *et seq.*; and Title III of the Marine Protection, Research and Sanctuaries Act, as amended, 16 U.S.C. 1431 *et seq.*;

(4) Where the discharge of any pollutant enters into saline estuarine waters which at the time of application do not support a balanced indigenous population of shellfish, fish, and wildlife, or allow recreation in and on the waters or which exhibit ambient water quality below applicable water quality standards adopted for the protection of public water supplies, shellfish, fish, and wildlife or recreational activities or such other standards necessary to assure support and protection of such uses. The prohibition contained in the preceding sentence shall apply without regard to the presence or absence of a causal relationship between such characteristics and the applicant's current or proposed discharge; or

(5) Where the discharge of any pollutant is into the New York Bight Apex.

(c) *Applications.* Each applicant for a modified permit under this subpart shall submit an application to EPA signed in compliance with 40 CFR part 122, subpart B, which shall contain:

(1) A signed, completed NPDES Application Standard form A, parts I, II, III;

(2) A completed Application Questionnaire;

(3) The certification in accordance with 40 CFR 122.22(d);

(4) In addition to the requirements of § 125.59(c) (1) through (3), applicants for permit renewal shall support continuation of the modification by supplying to EPA the results of studies and monitoring performed in accordance with § 125.63 during the life of the permit. Upon a demonstration meeting the statutory criteria and requirements of

this subpart, the permit may be re-newed under the applicable procedures of 40 CFR part 124.

(d) *Revisions to applications.* (1) POTWs which submitted applications in accordance with the June 15, 1979, regulations (44 FR 34784) may revise their applications one time following a tentative decision to propose changes to treatment levels and/or outfall and diffuser location and design in accord-ance with §125.59(f)(2)(i); and

(2) Other applicants may revise their applications one time following a ten-tative decision to propose changes to treatment levels and/or outfall and dif-fuser location and design in accordance with §125.59(f)(2)(i). Revisions by such applicants which propose downgrading treatment levels and/or outfall and dif-fuser location and design must be justi-fied on the basis of substantial changes in circumstances beyond the appli-cant's control since the time of appli-cation submission.

(3) Applicants authorized or re-quested to submit additional informa-tion under §125.59(g) may submit a re-vised application in accordance with §125.59(f)(2)(ii) where such additional information supports changes in pro-posed treatment levels and/or outfall location and diffuser design. The oppor-tunity for such revision shall be in ad-dition to the one-time revision allowed under §125.59(d) (1) and (2).

(4) POTWs which revise their applica-tions must:

(i) Modify their NPDES form and Ap-plication Questionnaire as needed to ensure that the information filed with their application is correct and com-plete;

(ii) Provide additional analysis and data as needed to demonstrate compli-ance with this subpart;

(iii) Obtain new State determinations under §§125.61(b)(2) and 125.64(b); and

(iv) Provide the certification de-scribed in paragraph (c)(3) of this sec-tion.

(5) Applications for permit renewal may not be revised.

(e) *Submittal of additional information to demonstrate compliance with §§125.60 and 125.65.* (1) On or before the deadline established in paragraph (f)(3) of this section, applicants shall submit a let-ter of intent to demonstrate compli-ance with §§125.60 and 125.65. The letter of intent is subject to approval by the Administrator based on the require-ments of this paragraph and paragraph (f)(3) of this section. The letter of in-tent shall consist of the following:

(i) For compliance with §125.60: (A) A description of the proposed treatment system which upgrades treatment to satisfy the requirements of §125.60.

(B) A project plan, including a sched-ule for data collection and for achiev-ing compliance with §125.60. The project plan shall include dates for de-sign and construction of necessary fa-cilities, submittal of influent/effluent data, and submittal of any other infor-mation necessary to demonstrate com-pliance with §125.60. The Administrator will review the project plan and may require revisions prior to authorizing submission of the additional informa-tion.

(ii) For compliance with §125.65: (A) A determination of what approach will be used to achieve compliance with §125.65.

(B) A project plan for achieving com-pliance. The project plan shall include any necessary data collection activi-ties, submittal of additional informa-tion, and/or development of appropriate pretreatment limits to demonstrate compliance with §125.65. The Adminis-trator will review the project plan and may require revisions prior to submis-sion of the additional information.

(iii) POTWs which submit additional information must:

(A) Modify their NPDES form and Application Questionnaire as needed to ensure that the information filed with their application is correct and com-plete;

(B) Obtain new State determinations under §§125.61(b)(2) and 125.64(b); and

(C) Provide the certification de-scribed in paragraph (c)(3) of this sec-tion.

(2) The information required under this paragraph must be submitted in accordance with the schedules in §125.59(f)(3)(ii). If the applicant does not meet these schedules for compli-ance, EPA may deny the application on that basis.

(f) *Deadlines and distribution—*(1) *Ap-plications.*(i) The application for an original 301(h) permit for POTWs which

349

directly discharges effluent into saline waters shall be submitted to the appropriate EPA Regional Administrator no later than December 29, 1982.

(ii) The application for renewal of a 301(h) modified permit shall be submitted no less than 180 days prior to the expiration of the existing permit, unless permission for a later date has been granted by the Administrator. (The Administrator shall not grant permission for applications to be submitted later than the expiration date of the existing permit.)

(iii) A copy of the application shall be provided to the State and interstate agency(s) authorized to provide certification/concurrence under §§ 124.53 through 124.55 on or before the date the application is submitted to EPA.

(2) *Revisions to Applications.* (i) Applicants desiring to revise their applications under § 125.59 (d)(1) or (d)(2) must:

(A) Submit to the appropriate Regional Administrator a letter of intent to revise their application either within 45 days of the date of EPA's tentative decision on their original application or within 45 days of November 26, 1982, whichever is later. Following receipt by EPA of a letter of intent, further EPA proceedings on the tentative decision under 40 CFR part 124 will be stayed.

(B) Submit the revised application as described for new applications in § 125.59(f)(1) either within one year of the date of EPA's tentative decision on their original application or within one year of November 26, 1982, if a tentative decision has already been made, whichever is later.

(ii) Applicants desiring to revise their applications under § 125.59(d)(3) must submit the revised application as described for new applications in § 125.59(f)(1) concurrent with submission of the additional information under § 125.59(g).

(3) Deadline for additional information to demonstrate compliance with §§ 125.60 and 125.65.

(i) A letter of intent required under § 125.59(e)(1) must be submitted by the following dates: for permittees with 301(h) modifications or for applicants to which a tentative or final decision has been issued, November 7, 1994; for all others, within 90 days after the Administrator issues a tentative decision on an application. Following receipt by EPA of a letter of intent containing the information required in § 125.59(e)(1), further EPA proceedings on the tentative decision under 40 CFR part 124 will be stayed.

(ii) The project plan submitted under § 125.59(e)(1) shall ensure that the applicant meets all the requirements of §§ 125.60 and 125.65 by the following deadlines:

(A) By August 9, 1996 for applicants that are not grandfathered under § 125.59(j).

(B) At the time of permit renewal or by August 9, 1996, whichever is later, for applicants that are grandfathered under § 125.59(j).

(4) *State determination deadline.* State determinations, as required by §§ 125.61(b)(2) and 125.64(b) shall be filed by the applicant with the appropriate Regional Administrator no later than 90 days after submission of the revision to the application or additional information to EPA. Extensions to this deadline may be provided by EPA upon request. However, EPA will not begin review of the revision to the application or additional information until a favorable State determination is received by EPA. Failure to provide the State determination within the timeframe required by this paragraph (f)(4) is a basis for denial of the application.

(g)(1) The Administrator may authorize or request an applicant to submit additional information by a specified date not to exceed one year from the date of authorization or request.

(2) Applicants seeking authorization to submit additional information on current/modified discharge characteristics, water quality, biological conditions or oceanographic characteristics must:

(i) Demonstrate that they made a diligent effort to provide such information with their application and were unable to do so, and

(ii) Submit a plan of study, including a schedule, for data collection and submittal of the additional information. EPA will review the plan of study and may require revisions prior to authorizing submission of the additional information.

(h) *Tentative decisions on section 301(h) modifications.* The Administrator shall grant a tentative approval or a tentative denial of a section 301(h) modified permit application. To qualify for a tentative approval, the applicant shall demonstrate to the satisfaction of the Administrator that it is using good faith means to come into compliance with all the requirements of this subpart and that it will meet all such requirements based on a schedule approved by the Administrator. For compliance with §§125.60 and 125.65, such schedule shall be in accordance with §125.59(f)(3)(ii).

(i) *Decisions on section 301(h) modifications.* (1) The decision to grant or deny a section 301(h) modification shall be made by the Administrator and shall be based on the applicant's demonstration that it has met all the requirements of §§125.59 through 125.68.

(2) No section 301(h) modified permit shall be issued until the appropriate State certification/concurrence is granted or waived pursuant to §124.54 or if the State denies certification/concurrence pursuant to §124.54.

(3) In the case of a modification issued to an applicant in a State administering an approved permit program under 40 CFR part 123, the State Director may:

(i) Revoke an existing permit as of the effective date of the EPA issued section 301(h) modified permit; and

(ii) Cosign the section 301(h) modified permit if the Director has indicated an intent to do so in the written concurrence.

(4) Any section 301(h) modified permit shall:

(i) Be issued in accordance with the procedures set forth in 40 CFR part 124, except that, because section 301(h) permits may be issued only by EPA, the terms "Administrator or a person designated by the Administrator" shall be substituted for the term "Director" as appropriate; and

(ii) Contain all applicable terms and conditions set forth in 40 CFR part 122 and §125.68.

(5) Appeals of section 301(h) determinations shall be governed by the procedures in 40 CFR part 124.

(j) *Grandfathering provision.* Applicants that received tentative or final approval for a section 301(h) modified permit prior to February 4, 1987, are not subject to §125.60, the water quality criteria provisions of §125.62(a)(1), or §125.65 until the time of permit renewal. In addition, if permit renewal will occur prior to August 9, 1996, applicants may have additional time to come into compliance with §§125.60 and 125.65, as determined appropriate by EPA on a case-by-case basis. Such additional time, however, shall not extend beyond August 9, 1996. This paragraph does not apply to any application that was initially tentatively approved, but as to which EPA withdrew its tentative approval or issued a tentative denial prior to February 4, 1987.

§125.60 Primary or equivalent treatment requirements.

(a) The applicant shall demonstrate that, at the time its modification becomes effective, it will be discharging effluent that has received at least primary or equivalent treatment.

(b) The applicant shall perform monitoring to ensure, based on the monthly average results of the monitoring, that the effluent it discharges has received primary or equivalent treatment.

(c)(1) An applicant may request that the demonstration of compliance with the requirement under paragraph (b) of this section to provide 30 percent removal of BOD be allowed on an averaging basis different from monthly (e.g., quarterly), subject to the demonstrations provided in paragraphs (c)(1)(i), (ii) and (iii) of this section. The Administrator may approve such requests if the applicant demonstrates to the Administrator's satisfaction that:

(i) The applicant's POTW is adequately designed and well operated;

(ii) The applicant will be able to meet all requirements under section 301(h) of the CWA and these subpart G regulations with the averaging basis selected; and

(iii) The applicant cannot achieve 30 percent removal on a monthly average basis because of circumstances beyond the applicant's control. Circumstances beyond the applicant's control may include seasonally dilute influent BOD concentrations due to relatively high

351

(although nonexcessive) inflow and in-filtration; relatively high soluble to in-soluble BOD ratios on a fluctuating basis; or cold climates resulting in cold influent. Circumstances beyond the ap-plicant's control shall not include less concentrated wastewater due to exces-sive inflow and infiltration (I&I). The determination of whether the less con-centrated wastewater is the result of excessive I&I will be based on the defi-nition of excessive I&I in 40 CFR 35.2005(b)(16) plus the additional cri-terion that inflow is nonexcessive if the total flow to the POTW (i.e., waste-water plus inflow plus infiltration) is less than 275 gallons per capita per day.

(2) In no event shall averaging on a less frequent basis than annually be al-lowed.

[59 FR 40658, Aug. 9, 1994, as amended at 61 FR 45833, Aug. 29, 1996]

§ 125.61 Existence of and compliance with applicable water quality standards.

(a) There must exist a water quality standard or standards applicable to the pollutant(s) for which a section 301(h) modified permit is requested, includ-ing:

(1) Water quality standards for bio-chemical oxygen demand or dissolved oxygen;

(2) Water quality standards for sus-pended solids, turbidity, light trans-mission, light scattering, or mainte-nance of the euphotic zone; and

(3) Water quality standards for pH.

(b) The applicant must: (1) Dem-onstrate that the modified discharge will comply with the above water qual-ity standard(s); and

(2) Provide a determination signed by the State or interstate agency(s) au-thorized to provide certification under §§ 124.53 and 124.54 that the proposed modified discharge will comply with applicable provisions of State law in-cluding water quality standards. This determination shall include a discus-sion of the basis for the conclusion reached.

§ 125.62 Attainment or maintenance of water quality which assures protec-tion of public water supplies; assures the protection and propaga-tion of a balanced indigenous popu-lation of shellfish, fish, and wildlife; and allows recreational activities.

(a) *Physical characteristics of dis-charge.* (1) At the time the 301(h) modi-fication becomes effective, the appli-cant's outfall and diffuser must be lo-cated and designed to provide adequate initial dilution, dispersion, and trans-port of wastewater such that the dis-charge does not exceed at and beyond the zone of initial dilution:

(i) All applicable water quality standards; and

(ii) All applicable EPA water quality criteria for pollutants for which there is no applicable EPA-approved water quality standard that directly cor-responds to the EPA water quality cri-terion for the pollutant.

(iii) For purposes of paragraph (a)(1)(ii) of this section, a State water quality standard "directly cor-responds" to an EPA water quality cri-terion only if:

(A) The State water quality standard addresses the same pollutant as the EPA water quality criterion and

(B) The State water quality standard specifies a numeric criterion for that pollutant or State objective method-ology for deriving such a numeric cri-terion.

(iv) The evaluation of compliance with paragraphs (a)(1) (i) and (ii) of this section shall be based upon conditions reflecting periods of maximum strati-fication and during other periods when discharge characteristics, water qual-ity, biological seasons, or oceano-graphic conditions indicate more crit-ical situations may exist.

(2) The evaluation under paragraph (a)(1)(ii) of this section as to compli-ance with applicable section 304(a)(1) water quality criteria shall be based on the following:

(i) *For aquatic life criteria:* The pollut-ant concentrations that must not be exceeded are the numeric ambient val-ues, if any, specified in the EPA sec-tion 304(a)(1) water quality criteria documents as the concentrations at which acute and chronic toxicity to aquatic life occurs or that are otherwise identified as the criteria to protect aquatic life.

(ii) *For human health criteria for carcinogens:* (A) For a known or suspected carcinogen, the Administrator shall determine the pollutant concentration that shall not be exceeded. To make this determination, the Administrator shall first determine a level of risk associated with the pollutant that is acceptable for purposes of this section. The Administrator shall then use the information in the section 304(a)(1) water quality criterion document, supplemented by all other relevant information, to determine the specific pollutant concentration that corresponds to the identified risk level.

(B) For purposes of paragraph (a)(2)(ii)(A) of this section, an acceptable risk level will be a single level that has been consistently used, as determined by the Administrator, as the basis of the State's EPA-approved water quality standards for carcinogenic pollutants. Alternatively, the Administrator may consider a State's recommendation to use a risk level that has been otherwise adopted or formally proposed by the State. The State recommendation must demonstrate, to the satisfaction of the Administrator, that the recommended level is sufficiently protective of human health in light of the exposure and uncertainty factors associated with the estimate of the actual risk posed by the applicant's discharge. The State must include with its demonstration a showing that the risk level selected is based on the best information available and that the State has held a public hearing to review the selection of the risk level, in accordance with provisions of State law and public participation requirements of 40 CFR part 25. If the Administrator neither determines that there is a consistently used single risk level nor accepts a risk level recommended by the State, then the Administrator shall otherwise determine an acceptable risk level based on all relevant information.

(iii) *For human health criteria for noncarcinogens:* For noncarcinogenic pollutants, the pollutant concentrations that must not be exceeded are the numeric ambient values, if any, specified in the EPA section 304(a)(1) water quality criteria documents as protective against the potential toxicity of the contaminant through ingestion of contaminated aquatic organisms.

(3) The requirements of paragraphs (a)(1) and (a)(2) of this section apply in addition to, and do not waive or substitute for, the requirements of §125.61.

(b) *Impact of discharge on public water supplies.* (1) The applicant's modified discharge must allow for the attainment or maintenance of water quality which assures protection of public water supplies.

(2) The applicant's modified discharge must not:

(i) Prevent a planned or existing public water supply from being used, or from continuing to be used, as a public water supply; or

(ii) Have the effect of requiring treatment over and above that which would be necessary in the absence of such discharge in order to comply with local and EPA drinking water standards.

(c) *Biological impact of discharge.* (1) The applicant's modified discharge must allow for the attainment or maintenance of water quality which assures protection and propagation of a balanced indigenous population of shellfish, fish, and wildlife.

(2) A balanced indigenous population of shellfish, fish, and wildlife must exist:

(i) Immediately beyond the zone of initial dilution of the applicant's modified discharge; and

(ii) In all other areas beyond the zone of initial dilution where marine life is actually or potentially affected by the applicant's modified discharge.

(3) Conditions within the zone of initial dilution must not contribute to extreme adverse biological impacts, including, but not limited to, the destruction of distinctive habitats of limited distribution, the presence of disease epicenter, or the stimulation of phytoplankton blooms which have adverse effects beyond the zone of initial dilution.

(4) In addition, for modified discharges into saline estuarine water:

(i) Benthic populations within the zone of initial dilution must not differ

353

substantially from the balanced indigenous populations which exist immediately beyond the boundary of the zone of initial dilution;

(ii) The discharge must not interfere with estuarine migratory pathways within the zone of initial dilution; and

(iii) The discharge must not result in the accumulation of toxic pollutants or pesticides at levels which exert adverse effects on the biota within the zone of initial dilution.

(d) *Impact of discharge on recreational activities.* (1) The applicant's modified discharge must allow for the attainment or maintenance of water quality which allows for recreational activities beyond the zone of initial dilution, including, without limitation, swimming, diving, boating, fishing, and picnicking, and sports activities along shorelines and beaches.

(2) There must be no Federal, State, or local restrictions on recreational activities within the vicinity of the applicant's modified outfall unless such restrictions are routinely imposed around sewage outfalls. This exception shall not apply where the restriction would be lifted or modified, in whole or in part, if the applicant were discharging a secondary treatment effluent.

(e) *Additional requirements for applications based on improved or altered discharges.* An application for a section 301(h) modified permit on the basis of an improved or altered discharge must include:

(1) A demonstration that such improvements or alterations have been thoroughly planned and studied and can be completed or implemented expeditiously;

(2) Detailed analyses projecting changes in average and maximum monthly flow rates and composition of the applicant's discharge which are expected to result from proposed improvements or alterations;

(3) The assessments required by paragraphs (a) through (d) of this section based on its current discharge; and

(4) A detailed analysis of how the applicant's planned improvements or alterations will comply with the requirements of paragraphs (a) through (d) of this section.

(f) *Stressed waters.* An applicant must demonstrate compliance with paragraphs (a) through (e) of this section not only on the basis of the applicant's own modified discharge, but also taking into account the applicant's modified discharge in combination with pollutants from other sources. However, if an applicant which discharges into ocean waters believes that its failure to meet the requirements of paragraphs (a) through (e) of this section is entirely attributable to conditions resulting from human perturbations other than its modified discharge (including, without limitation, other municipal or industrial discharges, nonpoint source runoff, and the applicant's previous discharges), the applicant need not demonstrate compliance with those requirements if it demonstrates, to the satisfaction of the Administrator, that its modified discharge does not or will not:

(1) Contribute to, increase, or perpetuate such stressed conditions;

(2) Contribute to further degradation of the biota or water quality if the level of human perturbation from other sources increases; and

(3) Retard the recovery of the biota or water quality if the level of human perturbation from other sources decreases.

§ 125.63 Establishment of a monitoring program.

(a) *General requirements.* (1) The applicant must:

(i) Have a monitoring program that is:

(A) Designed to provide data to evaluate the impact of the modified discharge on the marine biota, demonstrate compliance with applicable water quality standards or water quality criteria, as applicable, and measure toxic substances in the discharge, and

(B) Limited to include only those scientific investigations necessary to study the effects of the proposed discharge;

(ii) Describe the sampling techniques, schedules and locations (including appropriate control sites), analytical techniques, quality control and verification procedures to be used in the monitoring program;

(iii) Demonstrate that it has the resources necessary to implement the program upon issuance of the modified permit and to carry it out for the life of the modified permit; and

(iv) Determine the frequency and extent of the monitoring program taking into consideration the applicant's rate of discharge, quantities of toxic pollutants discharged, and potentially significant impacts on receiving water quality, marine biota, and designated water uses.

(2) The Administrator may require revision of the proposed monitoring program before issuing a modified permit and during the term of any modified permit.

(b) *Biological monitoring program.* The biological monitoring program for both small and large applicants shall provide data adequate to evaluate the impact of the modified discharge on the marine biota.

(1) Biological monitoring shall include to the extent practicable:

(i) Periodic surveys of the biological communities and populations which are most likely affected by the discharge to enable comparisons with baseline conditions described in the application and verified by sampling at the control stations/reference sites during the periodic surveys;

(ii) Periodic determinations of the accumulation of toxic pollutants and pesticides in organisms and examination of adverse effects, such as disease, growth abnormalities, physiological stress, or death;

(iii) Sampling of sediments in areas of solids deposition in the vicinity of the ZID, in other areas of expected impact, and at appropriate reference sites to support the water quality and biological surveys and to measure the accumulation of toxic pollutants and pesticides; and

(iv) Where the discharge would affect commercial or recreational fisheries, periodic assessments of the conditions and productivity of fisheries.

(2) Small applicants are not subject to the requirements of paragraph (b)(1) (ii) through (iv) of this section if they discharge at depths greater than 10 meters and can demonstrate through a suspended solids deposition analysis that there will be negligible seabed ac-

cumulation in the vicinity of the modified discharge.

(3) For applicants seeking a section 301(h) modified permit based on:

(i) A current discharge, biological monitoring shall be designed to demonstrate ongoing compliance with the requirements of § 125.62(c);

(ii) An improved discharge or altered discharge other than outfall relocation, biological monitoring shall provide baseline data on the current impact of the discharge and data which demonstrate, upon completion of improvements or alterations, that the requirements of § 125.62(c) are met; or

(iii) An improved or altered discharge involving outfall relocation, the biological monitoring shall:

(A) Include the current discharge site until such discharge ceases; and

(B) Provide baseline data at the relocation site to demonstrate the impact of the discharge and to provide the basis for demonstrating that requirements of § 125.62(c) will be met.

(c) *Water quality monitoring program.* The water quality monitoring program shall to the extent practicable:

(1) Provide adequate data for evaluating compliance with water quality standards or water quality criteria, as applicable under § 125.62(a)(1);

(2) Measure the presence of toxic pollutants which have been identified or reasonably may be expected to be present in the discharge.

(d) *Effluent monitoring program.* (1) In addition to the requirements of 40 CFR part 122, to the extent practicable, monitoring of the POTW effluent shall provide quantitative and qualitative data which measure toxic substances and pesticides in the effluent and the effectiveness of the toxic control program.

(2) The permit shall require the collection of data on a frequency specified in the permit to provide adequate data for evaluating compliance with the percent removal efficiency requirements under § 125.60.

§ 125.64 Effect of the discharge on other point and nonpoint sources.

(a) No modified discharge may result in any additional pollution control requirements on any other point or nonpoint source.

(b) The applicant shall obtain a determination from the State or interstate agency(s) having authority to establish wasteload allocations indicating whether the applicant's discharge will result in an additional treatment pollution control, or other requirement on any other point or nonpoint sources. The State determination shall include a discussion of the basis for its conclusion.

§ 125.65 Urban area pretreatment program.

(a) *Scope and applicability.* (1) The requirements of this section apply to each POTW serving a population of 50,000 or more that has one or more toxic pollutants introduced into the POTW by one or more industrial dischargers and that seeks a section 301(h) modification.

(2) The requirements of this section apply in addition to any applicable requirements of 40 CFR part 403, and do not waive or substitute for the part 403 requirements in any way.

(b) *Toxic pollutant control.* (1) As to each toxic pollutant introduced by an industrial discharger, each POTW subject to the requirements of this section shall demonstrate that it either:

(i) Has an applicable pretreatment requirement in effect in accordance with paragraph (c) of this section; or

(ii) Has in effect a program that achieves secondary removal equivalency in accordance with paragraph (d) of this section.

(2) Each applicant shall demonstrate that industrial sources introducing waste into the applicant's treatment works are in compliance with all applicable pretreatment requirements, including numerical standards set by local limits, and that it will enforce those requirements.

(c) *Applicable pretreatment requirement.* (1) An applicable pretreatment requirement under paragraph (b)(1)(i) of this section with respect to a toxic pollutant shall consist of the following:

(i) As to a toxic pollutant introduced into the applicant's treatment works by an industrial discharger for which there is no applicable categorical pretreatment standard for the toxic pollutant, a local limit or limits on the toxic pollutant as necessary to satisfy the requirements of 40 CFR part 403; and

(ii) As to a toxic pollutant introduced into the applicant's treatment works by an industrial discharger that is subject to a categorical pretreatment standard for the toxic pollutant, the categorical standard and a local limit or limits as necessary to satisfy the requirements of 40 CFR part 403;

(iii) As to a toxic pollutant introduced into the applicant's treatment works by an industrial discharger for which there is no applicable categorical pretreatment standard for the toxic pollutant, and the 40 CFR part 403 analysis on the toxic pollutant shows that no local limit is necessary, the applicant shall demonstrate to EPA on an annual basis during the term of the permit through continued monitoring and appropriate technical review that a local limit is not necessary, and, where appropriate, require industrial management practices plans and other pollution prevention activities to reduce or control the discharge of each such pollutant by industrial dischargers to the POTW. If such monitoring and technical review of data indicate that a local limit is needed, the POTW shall establish and implement a local limit.

(2) Any local limits developed to meet the requirements of paragraphs (b)(1)(i) and (c)(1) of this section shall be:

(i) Consistent with all applicable requirements of 40 CFR part 403 and

(ii) Subject to approval by the Administrator as part of the 301(h) application review. The Administrator may require such local limits to be revised as necessary to meet the requirements of this section or 40 CFR part 403.

(d) *Secondary removal equivalency.* An applicant shall demonstrate that it achieves secondary removal equivalency through the use of a secondary treatment pilot (demonstration) plant at the applicant's facility which provides an empirical determination of the amount of a toxic pollutant removed by the application of secondary treatment to the applicant's influent where the applicant's influent has not been pretreated. Alternatively, an applicant may make this determination using influent that has received industrial pretreatment, notwithstanding the definition of secondary removal equivalency in §125.58(w). The NPDES permit shall include effluent limits based on the data from the secondary equivalency demonstration when those limits are more stringent than effluent limits based on State water quality standards or water quality criteria, if applicable, or are otherwise required to assure that all applicable environmental protection criteria are met. Once such effluent limits are established in the NPDES permit, the POTW may either establish local limits or perform additional treatment at the POTW or a combination of the two to achieve the permit limit.

§125.66 Toxics control program.

(a) *Chemical analysis.* (1) The applicant shall submit at the time of application a chemical analysis of its current discharge for all toxic pollutants and pesticides as defined in §125.58(aa) and (p). The analysis shall be performed on two 24-hour composite samples (one dry weather and one wet weather). Applicants may supplement or substitute chemical analyses if composition of the supplemental or substitute samples typifies that which occurs during dry and wet weather conditions.

(2) Unless required by the State, this requirement shall not apply to any small section 301(h) applicant which certifies that there are no known or suspected sources of toxic pollutants or pesticides and documents the certification with an industrial user survey as described by 40 CFR 403.8(f)(2).

(b) *Identification of sources.* The applicant shall submit at the time of application an analysis of the known or suspected sources of toxic pollutants or pesticides identified in §125.66(a). The applicant shall to the extent practicable categorize the sources according to industrial and nonindustrial types.

(c) *Industrial pretreatment requirements.* (1) An applicant that has known or suspected industrial sources of toxic pollutants shall have an approved pretreatment program in accordance with 40 CFR part 403.

(2) This requirement shall not apply to any applicant which has no known or suspected industrial sources of toxic pollutants or pesticides and so certifies to the Administrator.

(3) The pretreatment program submitted by the applicant under this section shall be subject to revision as required by the Administrator prior to issuing or renewing any section 301(h) modified permit and during the term of any such permit.

(4) Implementation of all existing pretreatment requirements and authorities must be maintained through the period of development of any additional pretreatment requirements that may be necessary to comply with the requirements of this subpart.

(d) *Nonindustrial source control program.* (1) The applicant shall submit a proposed public education program designed to minimize the entrance of nonindustrial toxic pollutants and pesticides into its POTW(s) which shall be implemented no later than 18 months after issuance of a 301(h) modified permit.

(2) The applicant shall also develop and implement additional nonindustrial source control programs on the earliest possible schedule. This requirement shall not apply to a small applicant which certifies that there are no known or suspected water quality, sediment accumulation, or biological problems related to toxic pollutants or pesticides in its discharge.

(3) The applicant's nonindustrial source control programs under paragraph (d)(2) of this section shall include the following schedules which are to be implemented no later than 18 months after issuance of a section 301(h) modified permit:

(i) A schedule of activities for identifying nonindustrial sources of toxic pollutants and pesticides; and

(ii) A schedule for the development and implementation of control programs, to the extent practicable, for nonindustrial sources of toxic pollutants and pesticides.

(4) Each proposed nonindustrial source control program and/or schedule submitted by the applicant under this section shall be subject to revision as determined by the Administrator prior to issuing or renewing any section 301(h) modified permit and during the term of any such permit.

§ 125.67 Increase in effluent volume or amount of pollutants discharged.

(a) No modified discharge may result in any new or substantially increased discharges of the pollutant to which the modification applies above the discharge specified in the section 301(h) modified permit.

(b) Where pollutant discharges are attributable in part to combined sewer overflows, the applicant shall minimize existing overflows and prevent increases in the amount of pollutants discharged.

(c) The applicant shall provide projections of effluent volume and mass loadings for any pollutants to which the modification applies in 5-year increments for the design life of its facility.

§ 125.68 Special conditions for section 301(h) modified permits.

Each section 301(h) modified permit issued shall contain, in addition to all applicable terms and conditions required by 40 CFR part 122, the following:

(a) Effluent limitations and mass loadings which will assure compliance with the requirements of this subpart;

(b) A schedule or schedules of compliance for:

(1) Pretreatment program development required by § 125.66(c);

(2) Nonindustrial toxics control program required by § 125.66(d); and

(3) Control of combined sewer overflows required by § 125.67.

(c) Monitoring program requirements that include:

(1) Biomonitoring requirements of § 125.63(b);

(2) Water quality requirements of § 125.63(c);

(3) Effluent monitoring requirements of §§ 125.60(b), 125.62(c) and (d), and 125.63(d).

(d) Reporting requirements that include the results of the monitoring programs required by paragraph (c) of this section at such frequency as prescribed in the approved monitoring program.

APPENDIX TO SUBPART G OF PART 125— APPLICANT QUESTIONNAIRE FOR MODIFICATION OF SECONDARY TREATMENT REQUIREMENTS

OMB Control Number 2040–0088 Expires on 2/28/96 Public reporting burden for this collection of information is estimated to average 1,295 - 19,552 hours per response, for small and large applicants, respectively. The reporting burden includes time for reviewing instructions, gathering data, including monitoring and toxics control activities, and completing and reviewing the questionnaire. Send comments regarding the burden estimate or any other aspect of this collection, including suggestions for reducing the burden, to Chief, Information Policy Branch, U.S. Environmental Protection Agency, 1200 Pennsylvania Ave., NW. (2136), Washington, DC 20460 and Office of Management and Budget, Office of Information and Regulatory Affairs, Attn: Desk Officer for EPA, Washington, DC 20503.

I. INTRODUCTION

1. This questionnaire is to be submitted by both small and large applicants for modification of secondary treatment requirements under section 301(h) of the Clean Water Act (CWA). A small applicant is defined as a POTW that has a contributing population to its wastewater treatment facility of less than 50,000 and a projected average dry weather flow of less than 5.0 million gallons per day (mgd, 0.22 cubic meters/sec) [40 CFR 125.58(c)]. A large applicant is defined as a POTW that has a population contributing to its wastewater treatment facility of at least 50,000 or a projected average dry weather flow of its discharge of at least 5.0 million gallons per day (mgd, 0.22 cubic meters/sec) [40 CFR 125.58(c)]. The questionnaire is in two sections, a general information and basic requirements section (part II) and a technical evaluation section (part III). Satisfactory completion by small and large dischargers of the appropriate questions of this questionnaire is necessary to enable EPA to determine whether the applicant's modified discharge meets the criteria of section 301(h) and EPA regulations (40 CFR part 125, subpart G).

2. Most small applicants should be able to complete the questionnaire using available information. However, small POTWs with low initial dilution discharging into shallow waters or waters with poor dispersion and transport characteristics, discharging near distinctive and susceptible biological habitats, or discharging substantial quantities of toxics should anticipate the need to collect additional information and/or conduct additional analyses to demonstrate compliance

with section 301(h) criteria. If there are questions in this regard, applicants should contact the appropriate EPA Regional Office for guidance.

3. Guidance for responding to this questionnaire is provided by the newly amended section 301(h) technical support document. Where available information is incomplete and the applicant needs to collect additional data during the period it is preparing the application or a letter of intent, EPA encourages the applicant to consult with EPA prior to data collection and submission. Such consultation, particularly if the applicant provides a project plan, will help ensure that the proper data are gathered in the most efficient matter.

4. The notation (L) means large applicants must respond to the question, and (S) means small applicants must respond.

II. GENERAL INFORMATION AND BASIC DATA
REQUIREMENTS

A. Treatment System Description

1. (L,S) On which of the following are you basing your application: a current discharge, improved discharge, or altered discharge, as defined in 40 CFR 125.58? [40 CFR 125.59(a)]

2. (L,S) Description of the Treatment/Outfall System [40 CFR 125.62(a) and 125.62(e)]

a. Provide detailed descriptions and diagrams of the treatment system and outfall configuration which you propose to satisfy the requirements of section 301(h) and 40 CFR part 125, subpart G. What is the total discharge design flow upon which this application is based?

b. Provide a map showing the geographic location of proposed outfall(s) (i.e., discharge). What is the latitude and longitude of the proposed outfall(s)?

c. For a modification based on an improved or altered discharge, provide a description and diagram of your current treatment system and outfall configuration. Include the current outfall's latitude and longitude, if different from the proposed outfall.

3. (L,S) Primary or equivalent treatment requirements [40 CFR 125.60]

a. Provide data to demonstrate that your effluent meets at least primary or equivalent treatment requirements as defined in 40 CFR 125.58(r) [40 CFR 125.60]

b. If your effluent does not meet the primary or equivalent treatment requirements, when do you plan to meet them? Provide a detailed schedule, including design, construction, start-up and full operation, with your application. This requirement must be met by the effective date of the new section 301(h) modified permit.

4. (L,S) Effluent Limitations and Characteristics [40 CFR 125.61(b) and 125.62(e)(2)]

a. Identify the final effluent limitations for five-day biochemical oxygen demand

(BOD_5), suspended solids, and pH upon which your application for a modification is based:

—BOD_5 _____ mg/L
—Suspended solids _____ mg/L
—pH _____ (range)

b. Provide data on the following effluent characteristics for your current discharge as well as for the modified discharge if different from the current discharge:

Flow (m^3/sec):
—minimum
—average dry weather
—average wet weather
—maximum
—annual average

BOD_5 (mg/L) for the following plant flows:
—minimum
—average dry weather
—average wet weather
—maximum
—annual average

Suspended solids (mg/L) for the following plant flows:
—minimum
—average dry weather
—average wet weather
—maximum
—annual average

Toxic pollutants and pesticides (ug/L):
—list each toxic pollutant and pesticide
—list each 304(a)(1) criteria and toxic pollutant and pesticide

pH:
—minimum
—maximum

Dissolved oxygen (mg/L, prior to chlorination) for the following plant flows:
—minimum
—average dry weather
—average wet weather
—maximum
—annual average

Immediate dissolved oxygen demand (mg/L).

5. (L,S) Effluent Volume and Mass Emissions [40 CFR 125.62(e)(2) and 125.67]

a. Provide detailed analyses showing projections of effluent volume (annual average, m^3/sec) and mass loadings (mt/yr) of BOD_5 and suspended solids for the design life of your treatment facility in five-year increments. If the application is based upon an improved or altered discharge, the projections must be provided with and without the proposed improvements or alterations.

b. Provide projections for the end of your five-year permit term for 1) the treatment facility contributing population and 2) the average daily total discharge flow for the maximum month of the dry weather season.

6. (L,S) Average Daily Industrial Flow (m^3/sec). Provide or estimate the average daily industrial inflow to your treatment facility

for the same time increments as in question II.A.5 above. [40 CFR 125.66]

7. (L,S) Combined Sewer Overflows [40 CFR 125.67(b)]

a. Does (will) your treatment and collection system include combined sewer overflows?

b. If yes, provide a description of your plan for minimizing combined sewer overflows to the receiving water.

8. (L,S) Outfall/Diffuser Design. Provide the following data for your current discharge as well as for the modified discharge, if different from the current discharge: [40 CFR 125.62(a)(1)]

—Diameter and length of the outfall(s) (meters)

—Diameter and length of the diffuser(s) (meters)

—Angle(s) of port orientation(s) from horizontal (degrees)

—Port diameter(s) (meters)

—Orifice contraction coefficient(s), if known

—Vertical distance from mean lower low water (or mean low water) surface and outfall port(s) centerline (meters)

—Number of ports

—Port spacing (meters)

—Design flow rate for each port, if multiple ports are used (m³/sec)

B. Receiving Water Description

1. (L,S) Are you applying for a modification based on a discharge to the ocean [40 CFR 125.58(n)] or to a saline estuary [40 CFR 125.58(v)]? [40 CFR 125.59(a)].

2. (L,S) Is your current discharge or modified discharge to stressed waters as defined in 40 CFR 125.58(z)? If yes, what are the pollution sources contributing to the stress? [40 CFR 125.59(b)(4) and 125.62(f)].

3. (L,S) Provide a description and data on the seasonal circulation patterns in the vicinity of your current and modified discharge(s). [40 CFR 125.62(a)].

4. (L) Oceanographic conditions in the vicinity of the current and proposed modified discharge(s). Provide data on the following: [40 CFR 125.62(a)].

—Lowest ten percentile current speed (m/sec)

—Predominant current speed (m/sec) and direction (true) during the four seasons

—Period(s) of maximum stratification (months)

—Period(s) of natural upwelling events (duration and frequency, months)

—Density profiles during period(s) of maximum stratification

5. (L,S) Do the receiving waters for your discharge contain significant amounts of effluent previously discharged from the treatment works for which you are applying for a section 301(h) modified permit? [40 CFR 125.57(a)(9)]

6. Ambient water quality conditions during the period(s) of maximum stratification: at the zone of initial dilution (ZID) boundary, at other areas of potential impact, and at control stations. [40 CFR 125.62(a)]

a. (L) Provide profiles (with depth) on the following for the current discharge location and for the modified discharge location, if different from the current discharge:

—BOD_5 (mg/L)

—Dissolved oxygen (mg/L)

—Suspended solids (mg/L)

—pH

—Temperature (°C)

—Salinity (ppt)

—Transparency (turbidity, percent light transmittance)

—Other significant variables (e.g., nutrients, 304(a)(1) criteria and toxic pollutants and pesticides, fecal coliform bacteria)

b. (S) Provide available data on the following in the vicinity of the current discharge location and for the modified discharge location, if different from the current discharge: [40 CFR 125.61(b)(1)]

—Dissolved oxygen (mg/L)

—Suspended solids (mg/L)

—pH

—Temperature (°C)

—Salinity (ppt)

—Transparency (turbidity, percent light transmittance)

—Other significant variables (e.g., nutrients, 304(a)(1) criteria and toxic pollutants and pesticides, fecal coliform bacteria)

c. (L,S)Are there other periods when receiving water quality conditions may be more critical than the period(s) of maximum stratification? If so, describe these and other critical periods and data requested in 6.a. for the other critical period(s). [40 CFR 125.62(a)(1)].

7. (L) Provide data on steady state sediment dissolved oxygen demand and dissolved oxygen demand due to resuspension of sediments in the vicinity of your current and modified discharge(s) (mg/L/day).

C. Biological Conditions

1. (L) Provide a detailed description of representative biological communities (e.g., plankton, macrobenthos, demersal fish, etc.) in the vicinity of your current and modified discharge(s): within the ZID, at the ZID boundary, at other areas of potential discharge-related impact, and at reference (control) sites. Community characteristics to be described shall include (but not be limited to) species composition; abundance; dominance and diversity; spatial/temporal distribution; growth and reproduction; disease frequency; trophic structure and productivity patterns; presence of opportunistic species; bioaccumulation of toxic materials; and the occurrence of mass mortalities.

2. (L,S)a. Are distinctive habitats of limited distribution (such as kelp beds or coral reefs) located in areas potentially affected by the modified discharge? [40 CFR 125.62(c)]

b. If yes, provide information on type, extent, and location of habitats.

3. (L,S)a. Are commercial or recreational fisheries located in areas potentially affected by the discharge? [40 CFR 125.62 (c) and (d)]

b. If yes, provide information on types, location, and value of fisheries.

D. State and Federal Laws [40 CFR 125.61 and 125.62(a)(1)]

1. (L,S) Are there water quality standards applicable to the following pollutants for which a modification is requested:

—Biochemical oxygen demand or dissolved oxygen?

—Suspended solids, turbidity, light transmission, light scattering, or maintenance of the euphotic zone?

—pH of the receiving water?

2. (L,S) If yes, what is the water use classification for your discharge area? What are the applicable standards for your discharge area for each of the parameters for which a modification is requested? Provide a copy of all applicable water quality standards or a citation to where they can be found.

3. (L,S) Will the modified discharge: [40 CFR 125.59(b)(3)].

—Be consistent with applicable State coastal zone management program(s) approved under the Coastal Zone Management Act as amended, 16 U.S.C. 1451 et seq.? [See 16 U.S.C. 1456(c)(3)(A)]

—Be located in a marine sanctuary designated under Title III of the Marine Protection, Research, and Sanctuaries Act (MPRSA) as amended, 16 U.S.C. 1431 et seq., or in an estuarine sanctuary designated under the Coastal Zone Management Act as amended, 16 U.S.C. 1461? If located in a marine sanctuary designated under Title III of the MPRSA, attach a copy of any certification or permit required under regulations governing such marine sanctuary. [See 16 U.S.C. 1432(f)(2)]

—Be consistent with the Endangered Species Act as amended, 16 U.S.C. 1531 et seq.? Provide the names of any threatened or endangered species that inhabit or obtain nutrients from waters that may be affected by the modified discharge. Identify any critical habitat that may be affected by the modified discharge and evaluate whether the modified discharge will affect threatened or endangered species or modify a critical habitat. [See 16 U.S.C. 1536(a)(2)].

4. (L,S) Are you aware of any State or Federal laws or regulations (other than the Clean Water Act or the three statutes identified in item 3 above) or an Executive Order which is applicable to your discharge? If yes, provide sufficient information to demonstrate that your modified discharge will comply with such law(s), regulation(s), or order(s). [40 CFR 125.59 (b)(3)].

III. TECHNICAL EVALUATION

A. Physical Characteristics of Discharge [40 CFR 125.62(a)]

1. (L,S) What is the critical initial dilution for your current and modified discharge(s) during (1) the period(s) of maximum stratification? and (2) any other critical period(s) of discharge volume/composition, water quality, biological seasons, or oceanographic conditions?

2. (L,S) What are the dimensions of the zone of initial dilution for your modified discharge(s)?

3. (L) What are the effects of ambient currents and stratification on dispersion and transport of the discharge plume/wastefield?

4. (S) Will there be significant sedimentation of suspended solids in the vicinity of the modified discharge?

5. (L) Sedimentation of suspended solids

a. What fraction of the modified discharge's suspended solids will accumulate within the vicinity of the modified discharge?

b. What are the calculated area(s) and rate(s) of sediment accumulation within the vicinity of the modified discharge(s) (g/m²/yr)?

c. What is the fate of settleable solids transported beyond the calculated sediment accumulation area?

B. Compliance with Applicable Water Quality Standards and CWA § 304(a)(1) water quality criteria [40 CFR 125.61(b) and 125.62(a)]

1. (L,S) What is the concentration of dissolved oxygen immediately following initial dilution for the period(s) of maximum stratification and any other critical period(s) of discharge volume/composition, water quality, biological seasons, or oceanographic conditions?

2. (L,S) What is the farfield dissolved oxygen depression and resulting concentration due to BOD exertion of the wastefield during the period(s) of maximum stratification and any other critical period(s)?

3. (L) What are the dissolved oxygen depressions and resulting concentrations near the bottom due to steady sediment demand and resuspension of sediments?

4. (L,S) What is the increase in receiving water suspended solids concentration immediately following initial dilution of the modified discharge(s)?

5. (L) What is the change in receiving water pH immediately following initial dilution of the modified discharge(s)?

6. (L,S) Does (will) the modified discharge comply with applicable water quality standards for:

—Dissolved oxygen?
—Suspended solids or surrogate standards?
—pH?

7. (L,S) Provide data to demonstrate that all applicable State water quality standards, and all applicable water quality criteria established under Section 304(a)(1) of the Clean Water Act for which there are no directly corresponding numerical applicable water quality standards approved by EPA, are met at and beyond the boundary of the ZID under critical environmental and treatment plant conditions in the waters surrounding or adjacent to the point at which your effluent is discharged. [40 CFR 125.62(a)(1)]

8. (L,S) Provide the determination required by 40 CFR 125.61(b)(2) for compliance with all applicable provisions of State law, including water quality standards or, if the determination has not yet been received, a copy of a letter to the appropriate agency(s) requesting the required determination.

C. Impact on Public Water Supplies [40 CFR 125.62(b)]

1. (L,S) Is there a planned or existing public water supply (desalinization facility) intake in the vicinity of the current or modified discharge?

2. (L,S) If yes:

a. What is the location of the intake(s) (latitude and longitude)?

b. Will the modified discharge(s) prevent the use of intake(s) for public water supply?

c. Will the modified discharge(s) cause increased treatment requirements for public water supply(s) to meet local, State, and EPA drinking water standards?

D. Biological Impact of Discharge [40 CFR 125.62(c)]

1. (L,S) Does (will) a balanced indigenous population of shellfish, fish, and wildlife exist:

—Immediately beyond the ZID of the current and modified discharge(s)?

—In all other areas beyond the ZID where marine life is actually or potentially affected by the current and modified discharge(s)?

2. (L,S) Have distinctive habitats of limited distribution been impacted adversely by the current discharge and will such habitats be impacted adversely by the modified discharge?

3. (L,S) Have commercial or recreational fisheries been impacted adversely by the current discharge (e.g., warnings, restrictions, closures, or mass mortalities) or will they be impacted adversely by the modified discharge?

4. (L,S*) Does the current or modified discharge cause the following within or beyond the ZID: [40 CFR 125.62(c)(3)]

—Mass mortality of fishes or invertebrates due to oxygen depletion, high concentrations of toxics, or other conditions?

—An increased incidence of disease in marine organisms?

—An abnormal body burden of any toxic material in marine organisms?

—Any other extreme, adverse biological impacts?

5. (L,S) For discharges into saline estuarine waters: [40 CFR 125.62 (c)(4)]

—Does or will the current or modified discharge cause substantial differences in the benthic population within the ZID and beyond the ZID?

—Does or will the current or modified discharge interfere with migratory pathways within the ZID?

—Does or will the current or modified discharge result in bioaccumulation of toxic pollutants or pesticides at levels which exert adverse effects on the biota within the ZID?

No section (h) modified permit shall be issued where the discharge enters into stressed saline estuarine waters as stated in 40 CFR 125.59(b)(4).

6. (L,S) For improved discharges, will the proposed improved discharge(s) comply with the requirements of 40 CFR 125.62(a) through 125.62(d)? [40 CFR 125.62(e)]

7. (L,S) For altered discharge(s), will the altered discharge(s) comply with the requirements of 40 CFR 125.62(a) through 125.62(d)? [40 CFR 125.62(e)]

8. (L,S) If your current discharge is to stressed ocean waters, does or will your current or modified discharge: [40 CFR 125.62(f)]

—Contribute to, increase, or perpetuate such stressed condition?

—Contribute to further degradation of the biota or water quality if the level of human perturbation from other sources increases?

—Retard the recovery of the biota or water quality if human perturbation from other sources decreases?

E. Impacts of Discharge on Recreational Activities [40 CFR 125.62(d)]

1. (L,S) Describe the existing or potential recreational activities likely to be affected by the modified discharge(s) beyond the zone of initial dilution.

2. (L,S) What are the existing and potential impacts of the modified discharge(s) on recreational activities? Your answer should include, but not be limited to, a discussion of fecal coliform bacteria.

3. (L,S) Are there any Federal, State, or local restrictions on recreational activities in the vicinity of the modified discharge(s)? If yes, describe the restrictions and provide citations to available references.

4. (L,S) If recreational restrictions exist, would such restrictions be lifted or modified

if you were discharging a secondary treatment effluent?

F. Establishment of a Monitoring Program [40 CFR 125.63]

1. (L,S) Describe the biological, water quality, and effluent monitoring programs which you propose to meet the criteria of 40 CFR 125.63. Only those scientific investigations that are necessary to study the effects of the proposed discharge should be included in the scope of the 301(h) monitoring program [40 CFR 125.63(a)(1)(i)(B)].

2. (L,S) Describe the sampling techniques, schedules, and locations, analytical techniques, quality control and verification procedures to be used.

3. (L,S) Describe the personnel and financial resources available to implement the monitoring programs upon issuance of a modified permit and to carry it out for the life of the modified permit.

G. Effect of Discharge on Other Point and Nonpoint Sources [40 CFR 125.64]

1. (L,S) Does (will) your modified discharge(s) cause additional treatment or control requirements for any other point or nonpoint pollution source(s)?

2. (L,S) Provide the determination required by 40 CFR 125.64(b) or, if the determination has not yet been received, a copy of a letter to the appropriate agency(s) requesting the required determination.

H. Toxics Control Program and Urban Area Pretreatment Program [40 CFR 125.65 and 125.66]

1. a. (L,S) Do you have any known or suspected industrial sources of toxic pollutants or pesticides?

b. (L,S) If no, provide the certification required by 40 CFR 125.66(a)(2) for small dischargers, and required by 40 CFR 125.66(c)(2) for large dischargers.

c. (L,S*) Provide the results of wet and dry weather effluent analyses for toxic pollutants and pesticides as required by 40 CFR 125.66(a)(1). (* to the extent practicable)

d. (L,S*) Provide an analysis of known or suspected industrial sources of toxic pollutants and pesticides identified in (1)(c) above as required by 40 CFR 125.66(b). (* to the extent practicable)

2. (S)a. Are there any known or suspected water quality, sediment accumulation, or biological problems related to toxic pollutants or pesticides from your modified discharge(s)?

(S)b. If no, provide the certification required by 40 CFR 125.66(d)(2) together with available supporting data.

(S)c. If yes, provide a schedule for development and implementation of nonindustrial toxics control programs to meet the requirements of 40 CFR 126.66(d)(3).

(L)d. Provide a schedule for development and implementation of a nonindustrial toxics control program to meet the requirements of 40 CFR 125.66(d)(3).

3. (L,S) Describe the public education program you propose to minimize the entrance of nonindustrial toxic pollutants and pesticides into your treatment system. [40 CFR 125.66(d)(1)]

4. (L,S) Do you have an approved industrial pretreatment program?

a. If yes, provide the date of EPA approval.

b. If no, and if required by 40 CFR part 403 to have an industrial pretreatment program, provide a proposed schedule for development and implementation of your industrial pretreatment program to meet the requirements of 40 CFR part 403.

5. Urban area pretreatment requirement [40 CFR 125.65] Dischargers serving a population of 50,000 or more must respond.

a. Provide data on all toxic pollutants introduced into the treatment works from industrial sources (categorical and noncategorical).

b. Note whether applicable pretreatment requirements are in effect for each toxic pollutant. Are the industrial sources introducing such toxic pollutants in compliance with all of their pretreatment requirements? Are these pretreatment requirements being enforced? [40 CFR 125.65(b)(2)]

c. If applicable pretreatment requirements do not exist for each toxic pollutant in the POTW effluent introduced by industrial sources,

—provide a description and a schedule for your development and implementation of applicable pretreatment requirements [40 CFR 125.65(c)], or

—describe how you propose to demonstrate secondary removal equivalency for each of those toxic pollutants, including a schedule for compliance, by using a secondary treatment pilot plant. [40 CFR 125.65(d)]

Subpart H—Criteria for Determining Alternative Effluent Limitations Under Section 316(a) of the Act

§125.70 Purpose and scope.

Section 316(a) of the Act provides that:

"With respect to any point source otherwise subject to the provisions of section 301 or section 306 of this Act, whenever the owner or operator of any such source, after opportunity for public hearing, can demonstrate to the satisfaction of the Administrator (or, if appropriate, the State) that any effluent limitation proposed for the control of the thermal component of any discharge from such source will require effluent limitations more stringent than necessary to assure the projection [sic] and propagation of a balanced, indigenous population of shellfish, fish and wildlife in and on the body of water into which the discharge is to be made, the Administrator (or, if appropriate, the State) may impose an effluent limitation under such sections on such plant, with respect to the thermal component of such discharge (taking into account the interaction of such thermal component with other pollutants), that will assure the protection and propagation of a balanced indigenous population of shellfish, fish and wildlife in and on that body of water."

This subpart describes the factors, criteria and standards for the establishment of alternative thermal effluent limitations under section 316(a) of the Act in permits issued under section 402(a) of the Act.

§ 125.71 Definitions.

For the purpose of this subpart:

(a) *Alternative effluent limitations* means all effluent limitations or standards of performance for the control of the thermal component of any discharge which are established under section 316(a) and this subpart.

(b) *Representative important species* means species which are representative, in terms of their biological needs, of a balanced, indigenous community of shellfish, fish and wildlife in the body of water into which a discharge of heat is made.

(c) The term *balanced, indigenous community* is synonymous with the term *balanced, indigenous population* in the Act and means a biotic community typically characterized by diversity, the capacity to sustain itself through cyclic seasonal changes, presence of necessary food chain species and by a lack of domination by pollution tolerant species. Such a community may include historically non-native species introduced in connection with a program of wildlife management and species whose presence or abundance results from substantial, irreversible environmental modifications. Normally, however, such a community will not include species whose presence or abundance is attributable to the introduction of pollutants that will be eliminated by compliance by all sources with section 301(b)(2) of the Act; and

may not include species whose presence or abundance is attributable to alternative effluent limitations imposed pursuant to section 316(a).

§ 125.72 Early screening of applications for section 316(a) variances.

(a) Any initial application for a section 316(a) variance shall include the following early screening information:

(1) A description of the alternative effluent limitation requested;

(2) A general description of the method by which the discharger proposes to demonstrate that the otherwise applicable thermal discharge effluent limitations are more stringent than necessary;

(3) A general description of the type of data, studies, experiments and other information which the discharger intends to submit for the demonstration; and

(4) Such data and information as may be available to assist the Director in selecting the appropriate representative important species.

(b) After submitting the early screening information under paragraph (a) of this section, the discharger shall consult with the Director at the earliest practicable time (but not later than 30 days after the application is filed) to discuss the discharger's early screening information. Within 60 days after the application is filed, the discharger shall submit for the Director's approval a detailed plan of study which the discharger will undertake to support its section 316(a) demonstration. The discharger shall specify the nature and extent of the following type of information to be included in the plan of study: Biological, hydrographical and meteorological data; physical monitoring data; engineering or diffusion models; laboratory studies; representative important species; and other relevant information. In selecting representative important species, special consideration shall be given to species mentioned in applicable water quality standards. After the discharger submits its detailed plan of study, the Director shall either approve the plan or specify any necessary revisions to the plan. The discharger shall provide any additional information or studies which the Director subsequently determines necessary to support the demonstration, including such studies or inspections as may be necessary to select representative important species. The discharger may provide any additional information or studies which the discharger feels are appropriate to support the demonstration.

(c) Any application for the renewal of a section 316(a) variance shall include only such information described in paragraphs (a) and (b) of this section as the Director requests within 60 days after receipt of the permit application.

(d) The Director shall promptly notify the Secretary of Commerce and the Secretary of the Interior, and any affected State of the filing of the request and shall consider any timely recommendations they submit.

(e) In making the demonstration the discharger shall consider any information or guidance published by EPA to assist in making such demonstrations.

(f) If an applicant desires a ruling on a section 316(a) application before the ruling on any other necessary permit terms and conditions, (as provided by §124.65), it shall so request upon filing its application under paragraph (a) of this section. This request shall be granted or denied at the discretion of the Director.

NOTE: At the expiration of the permit, any discharger holding a section 316(a) variance should be prepared to support the continuation of the variance with studies based on the discharger's actual operation experience.

[44 FR 32948, June 7, 1979, as amended at 45 FR 33513, May 19, 1980; 65 FR 30913, May 15, 2000]

§125.73 Criteria and standards for the determination of alternative effluent limitations under section 316(a).

(a) Thermal discharge effluent limitations or standards established in permits may be less stringent than those required by applicable standards and limitations if the discharger demonstrates to the satisfaction of the director that such effluent limitations are more stringent than necessary to assure the protection and propagation of a balanced, indigenous community of shellfish, fish and wildlife in and on the body of water into which the discharge is made. This demonstration must show that the alternative effluent limitation desired by the discharger, considering the cumulative impact of its thermal discharge together with all other significant impacts on the species affected, will assure the protection and propagation of a balanced indigenous community of shellfish, fish and wildlife in and on the body of water into which the discharge is to be made.

(b) In determining whether or not the protection and propagation of the affected species will be assured, the Director may consider any information contained or referenced in any applicable thermal water quality criteria and thermal water quality information published by the Administrator under section 304(a) of the Act, or any other information he deems relevant.

(c) (1) Existing dischargers may base their demonstration upon the absence of prior appreciable harm in lieu of predictive studies. Any such demonstrations shall show:

(i) That no appreciable harm has resulted from the normal component of the discharge (taking into account the interaction of such thermal component with other pollutants and the additive effect of other thermal sources to a balanced, indigenous community of shellfish, fish and wildlife in and on the body of water into which the discharge has been made; or

(ii) That despite the occurrence of such previous harm, the desired alternative effluent limitations (or appropriate modifications thereof) will nevertheless assure the protection and propagation of a balanced, indigenous community of shellfish, fish and wildlife in and on the body of water into which the discharge is made.

(2) In determining whether or not prior appreciable harm has occurred, the Director shall consider the length of time in which the applicant has been discharging and the nature of the discharge.

Subpart I—Requirements Applicable to Cooling Water Intake Structures for New Facilities Under Section 316(b) of the Act

SOURCE: 66 FR 65338, Dec. 18, 2001, unless otherwise noted.

§ 125.80 What are the purpose and scope of this subpart?

(a) This subpart establishes requirements that apply to the location, design, construction, and capacity of cooling water intake structures at new facilities. The purpose of these requirements is to establish the best technology available for minimizing adverse environmental impact associated with the use of cooling water intake structures. These requirements are implemented through National Pollutant Discharge Elimination System (NPDES) permits issued under section 402 of the Clean Water Act (CWA).

(b) This subpart implements section 316(b) of the CWA for new facilities. Section 316(b) of the CWA provides that any standard established pursuant to sections 301 or 306 of the CWA and applicable to a point source shall require that the location, design, construction, and capacity of cooling water intake structures reflect the best technology available for minimizing adverse environmental impact.

(c) New facilities that do not meet the threshold requirements regarding amount of water withdrawn or percentage of water withdrawn for cooling water purposes in § 125.81(a) must meet requirements determined on a case-by-case, best professional judgement (BPJ) basis.

(d) Nothing in this subpart shall be construed to preclude or deny the right of any State or political subdivision of a State or any interstate agency under section 510 of the CWA to adopt or enforce any requirement with respect to control or abatement of pollution that is more stringent than those required by Federal law.

§ 125.81 Who is subject to this subpart?

(a) This subpart applies to a new facility if it:

(1) Is a point source that uses or proposes to use a cooling water intake structure;

(2) Has at least one cooling water intake structure that uses at least 25 percent of the water it withdraws for cooling purposes as specified in paragraph (c) of this section; and

(3) Has a design intake flow greater than two (2) million gallons per day (MGD).

(b) Use of a cooling water intake structure includes obtaining cooling water by any sort of contract or arrangement with an independent supplier (or multiple suppliers) of cooling water if the supplier or suppliers withdraw(s) water from waters of the United States. Use of cooling water does not include obtaining cooling water from a public water system or the use of treated effluent that otherwise would be discharged to a water of the U.S. This provision is intended to prevent circumvention of these requirements by creating arrangements to receive cooling water from an entity that is not itself a point source.

(c) The threshold requirement that at least 25 percent of water withdrawn be used for cooling purposes must be measured on an average monthly basis. A new facility meets the 25 percent cooling water threshold if, based on the new facility's design, any monthly average over a year for the percentage of cooling water withdrawn is expected to equal or exceed 25 percent of the total water withdrawn.

(d) This subpart does not apply to facilities that employ cooling water intake structures in the offshore and coastal subcategories of the oil and gas extraction point source category as defined under 40 CFR 435.10 and 40 CFR 435.40.

§ 125.82 When must I comply with this subpart?

You must comply with this subpart when an NPDES permit containing requirements consistent with this subpart is issued to you.

§ 125.83 What special definitions apply to this subpart?

The following special definitions apply to this subpart:

Annual mean flow means the average of daily flows over a calendar year. Historical data (up to 10 years) must be used where available.

Closed-cycle recirculating system means a system designed, using minimized makeup and blowdown flows, to withdraw water from a natural or other water source to support contact and/or noncontact cooling uses within a facility. The water is usually sent to a cooling canal or channel, lake, pond, or tower to allow waste heat to be dissipated to the atmosphere and then is returned to the system. (Some facilities divert the waste heat to other process operations.) New source water (make-up water) is added to the system to replenish losses that have occurred due to blowdown, drift, and evaporation.

Cooling water means water used for contact or noncontact cooling, including water used for equipment cooling, evaporative cooling tower makeup, and dilution of effluent heat content. The intended use of the cooling water is to absorb waste heat rejected from the process or processes used, or from auxiliary operations on the facility's premises. Cooling water that is used in a manufacturing process either before or after it is used for cooling is considered process water for the purposes of calculating the percentage of a new facility's intake flow that is used for cooling purposes in § 125.81(c).

Cooling water intake structure means the total physical structure and any associated constructed waterways used to withdraw cooling water from waters of the U.S. The cooling water intake structure extends from the point at which water is withdrawn from the surface water source up to, and including, the intake pumps.

Design intake flow means the value assigned (during the facility's design) to the total volume of water withdrawn from a source water body over a specific time period.

Design intake velocity means the value assigned (during the design of a cooling water intake structure) to the average speed at which intake water passes through the open area of the intake screen (or other device) against which organisms might be impinged or through which they might be entrained.

Entrainment means the incorporation of all life stages of fish and shellfish with intake water flow entering and passing through a cooling water intake structure and into a cooling water system.

Estuary means a semi-enclosed body of water that has a free connection with open seas and within which the seawater is measurably diluted with fresh water derived from land drainage. The salinity of an estuary exceeds 0.5 parts per thousand (by mass) but is typically less than 30 parts per thousand (by mass).

Existing facility means any facility that is not a new facility.

Freshwater river or stream means a lotic (free-flowing) system that does not receive significant inflows of water from oceans or bays due to tidal action. For the purposes of this rule, a flow-through reservoir with a retention time of 7 days or less will be considered a freshwater river or stream.

Hydraulic zone of influence means that portion of the source waterbody hydraulically affected by the cooling water intake structure withdrawal of water.

Impingement means the entrapment of all life stages of fish and shellfish on the outer part of an intake structure or against a screening device during periods of intake water withdrawal.

Lake or *reservoir* means any inland body of open water with some minimum surface area free of rooted vegetation and with an average hydraulic retention time of more than 7 days. Lakes or reservoirs might be natural water bodies or impounded streams, usually fresh, surrounded by land or by land and a man-made retainer (e.g., a dam). Lakes or reservoirs might be fed by rivers, streams, springs, and/or local precipitation. Flow-through reservoirs with an average hydraulic retention time of 7 days or less should be considered a freshwater river or stream.

Maximize means to increase to the greatest amount, extent, or degree reasonably possible.

Minimize means to reduce to the smallest amount, extent, or degree reasonably possible.

Natural thermal stratification means the naturally-occurring division of a waterbody into horizontal layers of differing densities as a result of variations in temperature at different depths.

New facility means any building, structure, facility, or installation that meets the definition of a "new source" or "new discharger" in 40 CFR 122.2 and 122.29(b)(1), (2), and (4) and is a greenfield or stand-alone facility; commences construction after January 17, 2002; and uses either a newly constructed cooling water intake structure, or an existing cooling water intake structure whose design capacity is increased to accommodate the intake of additional cooling water. New facilities include only "greenfield" and "stand-alone" facilities. A greenfield facility is a facility that is constructed at a site at which no other source is located, or that totally replaces the process or production equipment at an existing facility (*see* 40 CFR 122.29(b)(1)(i) and (ii)). A stand-alone facility is a new, separate facility that is constructed on property where an existing facility is located and whose processes are substantially independent of the existing facility at the same site (*see* 40 CFR 122.29(b)(1)(iii)). New facility does not include new units that are added to a facility for purposes of the same general industrial operation (for example, a new peaking unit at an electrical generating station).

(1) Examples of "new facilities" include, but are not limited to: the following scenarios:

(i) A new facility is constructed on a site that has never been used for industrial or commercial activity. It has a new cooling water intake structure for its own use.

(ii) A facility is demolished and another facility is constructed in its place. The newly-constructed facility uses the original facility's cooling water intake structure, but modifies it to increase the design capacity to accommodate the intake of additional cooling water.

(iii) A facility is constructed on the same property as an existing facility, but is a separate and independent industrial operation. The cooling water intake structure used by the original facility is modified by constructing a new intake bay for the use of the newly constructed facility or is otherwise modified to increase the intake capacity for the new facility.

(2) Examples of facilities that would not be considered a "new facility" include, but are not limited to, the following scenarios:

(i) A facility in commercial or industrial operation is modified and either continues to use its original cooling water intake structure or uses a new or modified cooling water intake structure.

(ii) A facility has an existing intake structure. Another facility (a separate and independent industrial operation), is constructed on the same property and connects to the facility's cooling water intake structure behind the intake pumps, and the design capacity of the cooling water intake structure has not been increased. This facility would not be considered a "new facility" even if routine maintenance or repairs that do not increase the design capacity were performed on the intake structure.

Ocean means marine open coastal waters with a salinity greater than or equal to 30 parts per thousand (by mass).

Source water means the water body (waters of the U.S.) from which the cooling water is withdrawn.

Thermocline means the middle layer of a thermally stratified lake or reservoir. In this layer, there is a rapid decrease in temperatures.

Tidal excursion means the horizontal distance along the estuary or tidal river that a particle moves during one tidal cycle of ebb and flow.

Tidal river means the most seaward reach of a river or stream where the salinity is typically less than or equal to 0.5 parts per thousand (by mass) at a time of annual low flow and whose surface elevation responds to the effects of coastal lunar tides.

[66 FR 65338, Dec. 18, 2001, as amended at 68 FR 36754, June 19, 2003]

§ 125.84 As an owner or operator of a new facility, what must I do to comply with this subpart?

(a)(1) The owner or operator of a new facility must comply with either:

(i) Track I in paragraph (b) or (c) of this section; or

(ii) Track II in paragraph (d) of this section.

(2) In addition to meeting the requirements in paragraph (b), (c), or (d) of this section, the owner or operator of a new facility may be required to comply with paragraph (e) of this section.

(b) *Track I requirements for new facilities that withdraw equal to or greater than 10 MGD.* You must comply with all of the following requirements:

(1) You must reduce your intake flow, at a minimum, to a level commensurate with that which can be attained by a closed-cycle recirculating cooling water system;

(2) You must design and construct each cooling water intake structure at your facility to a maximum through-screen design intake velocity of 0.5 ft/s;

(3) You must design and construct your cooling water intake structure such that the total design intake flow from all cooling water intake structures at your facility meets the following requirements:

(i) For cooling water intake structures located in a freshwater river or stream, the total design intake flow must be no greater than five (5) percent of the source water annual mean flow;

(ii) For cooling water intake structures located in a lake or reservoir, the total design intake flow must not disrupt the natural thermal stratification or turnover pattern (where present) of the source water except in cases where the disruption is determined to be beneficial to the management of fisheries for fish and shellfish by any fishery management agency(ies);

(iii) For cooling water intake structures located in an estuary or tidal river, the total design intake flow over one tidal cycle of ebb and flow must be no greater than one (1) percent of the volume of the water column within the area centered about the opening of the intake with a diameter defined by the distance of one tidal excursion at the mean low water level;

(4) You must select and implement design and construction technologies or operational measures for minimizing impingement mortality of fish and shellfish if:

(i) There are threatened or endangered or otherwise protected federal, state, or tribal species, or critical habitat for these species, within the hydraulic zone of influence of the cooling water intake structure; or

(ii) Based on information submitted by any fishery management agency(ies) or other relevant information, there are migratory and/or sport or commercial species of impingement concern to the Director that pass through the hydraulic zone of influence of the cooling water intake structure; or

(iii) It is determined by the Director, based on information submitted by any fishery management agency(ies) or other relevant information, that the proposed facility, after meeting the technology-based performance requirements in paragraphs (b)(1), (2), and (3) of this section, would still contribute unacceptable stress to the protected species, critical habitat of those species, or species of concern;

(5) You must select and implement design and construction technologies or operational measures for minimizing entrainment of entrainable life stages of fish and shellfish if:

(i) There are threatened or endangered or otherwise protected federal, state, or tribal species, or critical habitat for these species, within the hydraulic zone of influence of the cooling water intake structure; or

(ii) Based on information submitted by any fishery management agency(ies) or other relevant information, there are or would be undesirable cumulative stressors affecting entrainable life stages of species of concern to the Director and the Director determines that the proposed facility, after meeting the technology-based performance requirements in paragraphs (b)(1), (2), and (3) of this section, would still contribute unacceptable stress to the protected species , critical habitat of those species, or these species of concern;

(6) You must submit the application information required in 40 CFR 122.21(r) and §125.86(b);

(7) You must implement the monitoring requirements specified in §125.87;

(8) You must implement the record-keeping requirements specified in § 125.88.

(c) *Track I requirements for new facilities that withdraw equal to or greater than 2 MGD and less than 10 MGD and that choose not to comply with paragraph (b) of this section.* You must comply with all the following requirements:

(1) You must design and construct each cooling water intake structure at your facility to a maximum through-screen design intake velocity of 0.5 ft/s;

(2) You must design and construct your cooling water intake structure such that the total design intake flow from all cooling water intake structures at your facility meets the following requirements:

(i) For cooling water intake structures located in a freshwater river or stream, the total design intake flow must be no greater than five (5) percent of the source water annual mean flow;

(ii) For cooling water intake structures located in a lake or reservoir, the total design intake flow must not disrupt the natural thermal stratification or turnover pattern (where present) of the source water except in cases where the disruption is determined to be beneficial to the management of fisheries for fish and shellfish by any fishery management agency(ies);

(iii) For cooling water intake structures located in an estuary or tidal river, the total design intake flow over one tidal cycle of ebb and flow must be no greater than one (1) percent of the volume of the water column within the area centered about the opening of the intake with a diameter defined by the distance of one tidal excursion at the mean low water level;

(3) You must select and implement design and construction technologies or operational measures for minimizing impingement mortality of fish and shellfish if:

(i) There are threatened or endangered or otherwise protected federal, state, or tribal species, or critical habitat for these species, within the hydraulic zone of influence of the cooling water intake structure; or

(ii) Based on information submitted by any fishery management agency(ies) or other relevant information, there are migratory and/or sport or commercial species of impingement concern to the Director that pass through the hydraulic zone of influence of the cooling water intake structure; or

(iii) It is determined by the Director, based on information submitted by any fishery management agency(ies) or other relevant information, that the proposed facility, after meeting the technology-based performance requirements in paragraphs (c)(1) and (2) of this section, would still contribute unacceptable stress to the protected species, critical habitat of those species, or species of concern;

(4) You must select and implement design and construction technologies or operational measures for minimizing entrainment of entrainable life stages of fish and shellfish;

(5) You must submit the application information required in 40 CFR 122.21(r) and § 125.86(b)(2), (3), and (4);

(6) You must implement the monitoring requirements specified in § 125.87;

(7) You must implement the record-keeping requirements specified in § 125.88.

(d) *Track II.* The owner or operator of a new facility that chooses to comply under Track II must comply with the following requirements:

(1) You must demonstrate to the Director that the technologies employed will reduce the level of adverse environmental impact from your cooling water intake structures to a comparable level to that which you would achieve were you to implement the requirements of paragraphs (b)(1) and (2) of this section. This demonstration must include a showing that the impacts to fish and shellfish, including important forage and predator species, within the watershed will be comparable to those which would result if you were to implement the requirements of paragraphs (b)(1) and (2) of this section. This showing may include consideration of impacts other than impingement mortality and entrainment, including measures that will result in increases in fish and shellfish, but it must demonstrate comparable performance for species that the Director identifies as species of concern. In identifying such species, the Director may consider information provided by

any fishery management agency(ies) along with data and information from other sources.

(2) You must design and construct your cooling water intake structure such that the total design intake flow from all cooling water intake structures at your facility meet the following requirements:

(i) For cooling water intake structures located in a freshwater river or stream, the total design intake flow must be no greater than five (5) percent of the source water annual mean flow;

(ii) For cooling water intake structures located in a lake or reservoir, the total design intake flow must not disrupt the natural thermal stratification or turnover pattern (where present) of the source water except in cases where the disruption is determined to be beneficial to the management of fisheries for fish and shellfish by any fishery management agency(ies);

(iii) For cooling water intake structures located in an estuary or tidal river, the total design intake flow over one tidal cycle of ebb and flow must be no greater than one (1) percent of the volume of the water column within the area centered about the opening of the intake with a diameter defined by the distance of one tidal excursion at the mean low water level.

(3) You must submit the application information required in 40 CFR 122.21(r) and §125.86(c).

(4) You must implement the monitoring requirements specified in §125.87.

(5) You must implement the recordkeeping requirements specified in §125.88.

(e) You must comply with any more stringent requirements relating to the location, design, construction, and capacity of a cooling water intake structure or monitoring requirements at a new facility that the Director deems are reasonably necessary to comply with any provision of state law, including compliance with applicable state water quality standards (including designated uses, criteria, and antidegradation requirements).

[66 FR 65338, Dec. 18, 2001, as amended at 68 FR 36754, June 19, 2003]

§125.85 May alternative requirements be authorized?

(a) Any interested person may request that alternative requirements less stringent than those specified in §125.84(a) through (e) be imposed in the permit. The Director may establish alternative requirements less stringent than the requirements of §125.84(a) through (e) only if:

(1) There is an applicable requirement under §125.84(a) through (e);

(2) The Director determines that data specific to the facility indicate that compliance with the requirement at issue would result in compliance costs wholly out of proportion to the costs EPA considered in establishing the requirement at issue or would result in significant adverse impacts on local air quality, significant adverse impacts on local water resources other than impingement or entrainment, or significant adverse impacts on local energy markets;

(3) The alternative requirement requested is no less stringent than justified by the wholly out of proportion cost or the significant adverse impacts on local air quality, significant adverse impacts on local water resources other than impingement or entrainment, or significant adverse impacts on local energy markets; and

(4) The alternative requirement will ensure compliance with other applicable provisions of the Clean Water Act and any applicable requirement of state law.

(b) The burden is on the person requesting the alternative requirement to demonstrate that alternative requirements should be authorized.

[66 FR 65338, Dec. 18, 2001, as amended at 68 FR 36755, June 19, 2003]

§125.86 As an owner or operator of a new facility, what must I collect and submit when I apply for my new or reissued NPDES permit?

(a)(1) As an owner or operator of a new facility, you must submit to the Director a statement that you intend to comply with either:

(i) The Track I requirements for new facilities that withdraw equal to or greater than 10 MGD in §125.84(b);

(ii) The Track I requirements for new facilities that withdraw equal to or greater than 2 MGD and less than 10 MGD in § 125.84(c);

(iii) The requirements for Track II in § 125.84 (d).

(2) You must also submit the application information required by 40 CFR 122.21(r) and the information required in either paragraph (b) of this section for Track I or paragraph (c) of this section for Track II when you apply for a new or reissued NPDES permit in accordance with 40 CFR 122.21.

(b) *Track I application requirements.* To demonstrate compliance with Track I requirements in § 125.84(b) or (c), you must collect and submit to the Director the information in paragraphs (b)(1) through (4) of this section.

(1) *Flow reduction information.* If you must comply with the flow reduction requirements in § 125.84(b)(1), you must submit the following information to the Director to demonstrate that you have reduced your flow to a level commensurate with that which can be attained by a closed-cycle recirculating cooling water system:

(i) A narrative description of your system that has been designed to reduce your intake flow to a level commensurate with that which can be attained by a closed-cycle recirculating cooling water system and any engineering calculations, including documentation demonstrating that your make-up and blowdown flows have been minimized; and

(ii) If the flow reduction requirement is met entirely, or in part, by reusing or recycling water withdrawn for cooling purposes in subsequent industrial processes, you must provide documentation that the amount of cooling water that is not reused or recycled has been minimized.

(2) *Velocity information.* You must submit the following information to the Director to demonstrate that you are complying with the requirement to meet a maximum through-screen design intake velocity of no more than 0.5 ft/s at each cooling water intake structure as required in § 125.84(b)(2) and (c)(1):

(i) A narrative description of the design, structure, equipment, and operation used to meet the velocity requirement; and

(ii) Design calculations showing that the velocity requirement will be met at minimum ambient source water surface elevations (based on best professional judgement using available hydrological data) and maximum head loss across the screens or other device.

(3) *Source waterbody flow information.* You must submit to the Director the following information to demonstrate that your cooling water intake structure meets the flow requirements in § 125.84(b)(3) and (c)(2):

(i) If your cooling water intake structure is located in a freshwater river or stream, you must provide the annual mean flow and any supporting documentation and engineering calculations to show that your cooling water intake structure meets the flow requirements;

(ii) If your cooling water intake structure is located in an estuary or tidal river, you must provide the mean low water tidal excursion distance and any supporting documentation and engineering calculations to show that your cooling water intake structure facility meets the flow requirements; and

(iii) If your cooling water intake structure is located in a lake or reservoir, you must provide a narrative description of the water body thermal stratification, and any supporting documentation and engineering calculations to show that the natural thermal stratification and turnover pattern will not be disrupted by the total design intake flow. In cases where the disruption is determined to be beneficial to the management of fisheries for fish and shellfish you must provide supporting documentation and include a written concurrence from any fisheries management agency(ies) with responsibility for fisheries potentially affected by your cooling water intake structure(s).

(4) *Design and Construction Technology Plan.* To comply with § 125.84(b)(4) and (5), or (c)(3) and (c)(4), you must submit to the Director the following information in a Design and Construction Technology Plan:

(i) Information to demonstrate whether or not you meet the criteria in § 125.84(b)(4) and (b)(5), or (c)(3) and (c)(4);

(ii) Delineation of the hydraulic zone of influence for your cooling water intake structure;

(iii) New facilities required to install design and construction technologies and/or operational measures must develop a plan explaining the technologies and measures you have selected based on information collected for the Source Water Biological Baseline Characterization required by 40 CFR 122.21(r)(3). (Examples of appropriate technologies include, but are not limited to, wedgewire screens, fine mesh screens, fish handling and return systems, barrier nets, aquatic filter barrier systems, etc. Examples of appropriate operational measures include, but are not limited to, seasonal shutdowns or reductions in flow, continuous operations of screens, etc.) The plan must contain the following information:

(A) A narrative description of the design and operation of the design and construction technologies, including fish-handling and return systems, that you will use to maximize the survival of those species expected to be most susceptible to impingement. Provide species-specific information that demonstrates the efficacy of the technology;

(B) A narrative description of the design and operation of the design and construction technologies that you will use to minimize entrainment of those species expected to be most susceptible to entrainment. Provide species-specific information that demonstrates the efficacy of the technology; and

(C) Design calculations, drawings, and estimates to support the descriptions provided in paragraphs (b)(4)(iii)(A) and (B) of this section.

(c) *Application requirements for Track II.* If you have chosen to comply with the requirements of Track II in §125.84(d) you must collect and submit the following information:

(1) *Source waterbody flow information.* You must submit to the Director the following information to demonstrate that your cooling water intake structure meets the source water body requirements in §125.84(d)(2):

(i) If your cooling water intake structure is located in a freshwater river or stream, you must provide the annual mean flow and any supporting documentation and engineering calculations to show that your cooling water intake structure meets the flow requirements;

(ii) If your cooling water intake structure is located in an estuary or tidal river, you must provide the mean low water tidal excursion distance and any supporting documentation and engineering calculations to show that your cooling water intake structure facility meets the flow requirements; and

(iii) If your cooling water intake structure is located in a lake or reservoir, you must provide a narrative description of the water body thermal stratification, and any supporting documentation and engineering calculations to show that the natural thermal stratification and thermal or turnover pattern will not be disrupted by the total design intake flow. In cases where the disruption is determined to be beneficial to the management of fisheries for fish and shellfish you must provide supporting documentation and include a written concurrence from any fisheries management agency(ies) with responsibility for fisheries potentially affected by your cooling water intake structure(s).

(2) *Track II Comprehensive Demonstration Study.* You must perform and submit the results of a Comprehensive Demonstration Study (Study). This information is required to characterize the source water baseline in the vicinity of the cooling water intake structure(s), characterize operation of the cooling water intake(s), and to confirm that the technology(ies) proposed and/or implemented at your cooling water intake structure reduce the impacts to fish and shellfish to levels comparable to those you would achieve were you to implement the requirements in §125.84(b)(1)and (2) of Track I. To meet the "comparable level" requirement, you must demonstrate that:

(i) You have reduced both impingement mortality and entrainment of all life stages of fish and shellfish to 90 percent or greater of the reduction that would be achieved through §125.84(b)(1) and (2); or

(ii) If your demonstration includes consideration of impacts other than impingement mortality and entrainment, that the measures taken will maintain the fish and shellfish in the waterbody at a substantially similar level to that which would be achieved through § 125.84(b)(1) and (2); and

(iii) You must develop and submit a plan to the Director containing a proposal for how information will be collected to support the study. The plan must include:

(A) A description of the proposed and/or implemented technology(ies) to be evaluated in the Study;

(B) A list and description of any historical studies characterizing the physical and biological conditions in the vicinity of the proposed or actual intakes and their relevancy to the proposed Study. If you propose to rely on existing source water body data, it must be no more than 5 years old, you must demonstrate that the existing data are sufficient to develop a scientifically valid estimate of potential impingement and entrainment impacts, and provide documentation showing that the data were collected using appropriate quality assurance/quality control procedures;

(C) Any public participation or consultation with Federal or State agencies undertaken in developing the plan; and

(D) A sampling plan for data that will be collected using actual field studies in the source water body. The sampling plan must document all methods and quality assurance procedures for sampling, and data analysis. The sampling and data analysis methods you propose must be appropriate for a quantitative survey and based on consideration of methods used in other studies performed in the source water body. The sampling plan must include a description of the study area (including the area of influence of the cooling water intake structure and at least 100 meters beyond); taxonomic identification of the sampled or evaluated biological assemblages (including all life stages of fish and shellfish); and sampling and data analysis methods; and

(iv) You must submit documentation of the results of the Study to the Director. Documentation of the results of the Study must include:

(A) *Source Water Biological Study.* The Source Water Biological Study must include:

(1) A taxonomic identification and characterization of aquatic biological resources including: a summary of historical and contemporary aquatic biological resources; determination and description of the target populations of concern (those species of fish and shellfish and all life stages that are most susceptible to impingement and entrainment); and a description of the abundance and temporal/spatial characterization of the target populations based on the collection of multiple years of data to capture the seasonal and daily activities (e.g., spawning, feeding and water column migration) of all life stages of fish and shellfish found in the vicinity of the cooling water intake structure;

(2) An identification of all threatened or endangered species that might be susceptible to impingement and entrainment by the proposed cooling water intake structure(s); and

(3) A description of additional chemical, water quality, and other anthropogenic stresses on the source waterbody.

(B) *Evaluation of potential cooling water intake structure effects.* This evaluation will include:

(1) Calculations of the reduction in impingement mortality and entrainment of all life stages of fish and shellfish that would need to be achieved by the technologies you have selected to implement to meet requirements under Track II. To do this, you must determine the reduction in impingement mortality and entrainment that would be achieved by implementing the requirements of § 125.84(b)(1) and (2) of Track I at your site.

(2) An engineering estimate of efficacy for the proposed and/or implemented technologies used to minimize impingement mortality and entrainment of all life stages of fish and shellfish and maximize survival of impinged life stages of fish and shellfish. You must demonstrate that the technologies reduce impingement mortality and entrainment of all life stages of fish and shellfish to a comparable level to that which you would

achieve were you to implement the requirements in §125.84(b)(1) and (2) of Track I. The efficacy projection must include a site-specific evaluation of technology(ies) suitability for reducing impingement mortality and entrainment based on the results of the Source Water Biological Study in paragraph (c)(2)(iv)(A) of this section. Efficacy estimates may be determined based on case studies that have been conducted in the vicinity of the cooling water intake structure and/or site-specific technology prototype studies.

(C) *Evaluation of proposed restoration measures.* If you propose to use restoration measures to maintain the fish and shellfish as allowed in §125.84(d)(1)(i), you must provide the following information to the Director:

(1) Information and data to show that you have coordinated with the appropriate fishery management agency(ies); and

(2) A plan that provides a list of the measures you plan to implement and how you will demonstrate and continue to ensure that your restoration measures will maintain the fish and shellfish in the waterbody to a substantially similar level to that which would be achieved through §125.84(b)(1) and (2).

(D) *Verification monitoring plan.* You must include in the Study the following:

(1) A plan to conduct, at a minimum, two years of monitoring to verify the full-scale performance of the proposed or implemented technologies, operational measures. The verification study must begin at the start of operations of the cooling water intake structure and continue for a sufficient period of time to demonstrate that the facility is reducing the level of impingement and entrainment to the level documented in paragraph (c)(2)(iv)(B) of this section. The plan must describe the frequency of monitoring and the parameters to be monitored. The Director will use the verification monitoring to confirm that you are meeting the level of impingement mortality and entrainment reduction required in §125.84(d), and that the operation of the technology has been optimized.

(2) A plan to conduct monitoring to verify that the restoration measures will maintain the fish and shellfish in the waterbody to a substantially similar level as that which would be achieved through §125.84(b)(1) and (2).

§125.87 As an owner or operator of a new facility, must I perform monitoring?

As an owner or operator of a new facility, you will be required to perform monitoring to demonstrate your compliance with the requirements specified in §125.84.

(a) *Biological monitoring.* You must monitor both impingement and entrainment of the commercial, recreational, and forage base fish and shellfish species identified in either the Source Water Baseline Biological Characterization data required by 40 CFR 122.21(r)(3) or the Comprehensive Demonstration Study required by §125.86(c)(2), depending on whether you chose to comply with Track I or Track II. The monitoring methods used must be consistent with those used for the Source Water Baseline Biological Characterization data required in 40 CFR 122.21(r)(3) or the Comprehensive Demonstration Study required by §125.86(c)(2). You must follow the monitoring frequencies identified below for at least two (2) years after the initial permit issuance. After that time, the Director may approve a request for less frequent sampling in the remaining years of the permit term and when the permit is reissued, if supporting data show that less frequent monitoring would still allow for the detection of any seasonal and daily variations in the species and numbers of individuals that are impinged or entrained.

(1) *Impingement sampling.* You must collect samples to monitor impingement rates (simple enumeration) for each species over a 24-hour period and no less than once per month when the cooling water intake structure is in operation.

(2) *Entrainment sampling.* You must collect samples to monitor entrainment rates (simple enumeration) for each species over a 24-hour period and

no less than biweekly during the primary period of reproduction, larval recruitment, and peak abundance identified during the Source Water Baseline Biological Characterization required by 40 CFR 122.21(r)(3) or the Comprehensive Demonstration Study required in § 125.86(c)(2). You must collect samples only when the cooling water intake structure is in operation.

(b) *Velocity monitoring.* If your facility uses surface intake screen systems, you must monitor head loss across the screens and correlate the measured value with the design intake velocity. The head loss across the intake screen must be measured at the minimum ambient source water surface elevation (best professional judgment based on available hydrological data). The maximum head loss across the screen for each cooling water intake structure must be used to determine compliance with the velocity requirement in § 125.84(b)(2) or (c)(1). If your facility uses devices other than surface intake screens, you must monitor velocity at the point of entry through the device. You must monitor head loss or velocity during initial facility startup, and thereafter, at the frequency specified in your NPDES permit, but no less than once per quarter.

(c) *Visual or remote inspections.* You must either conduct visual inspections or employ remote monitoring devices during the period the cooling water intake structure is in operation. You must conduct visual inspections at least weekly to ensure that any design and construction technologies required in § 125.84(b)(4) and (5), or (c)(3) and (4) are maintained and operated to ensure that they will continue to function as designed. Alternatively, you must inspect via remote monitoring devices to ensure that the impingement and entrainment technologies are functioning as designed.

§ 125.88 As an owner or operator of a new facility, must I keep records and report?

As an owner or operator of a new facility you are required to keep records and report information and data to the Director as follows:

(a) You must keep records of all the data used to complete the permit application and show compliance with the requirements, any supplemental information developed under § 125.86, and any compliance monitoring data submitted under § 125.87, for a period of at least three (3) years from the date of permit issuance. The Director may require that these records be kept for a longer period.

(b) You must provide the following to the Director in a yearly status report:

(1) Biological monitoring records for each cooling water intake structure as required by § 125.87(a);

(2) Velocity and head loss monitoring records for each cooling water intake structure as required by § 125.87(b); and

(3) Records of visual or remote inspections as required in § 125.87(c).

§ 125.89 As the Director, what must I do to comply with the requirements of this subpart?

(a) *Permit application.* As the Director, you must review materials submitted by the applicant under 40 CFR 122.21(r)(3) and § 125.86 at the time of the initial permit application and before each permit renewal or reissuance.

(1) After receiving the initial permit application from the owner or operator of a new facility, the Director must determine applicable standards in § 125.84 to apply to the new facility. In addition, the Director must review materials to determine compliance with the applicable standards.

(2) For each subsequent permit renewal, the Director must review the application materials and monitoring data to determine whether requirements, or additional requirements, for design and construction technologies or operational measures should be included in the permit.

(3) For Track II facilities, the Director may review the information collection proposal plan required by § 125.86(c)(2)(iii). The facility may initiate sampling and data collection activities prior to receiving comment from the Director.

(b) *Permitting requirements.* Section 316(b) requirements are implemented for a facility through an NPDES permit. As the Director, you must determine, based on the information submitted by the new facility in its permit application, the appropriate requirements and conditions to include in the permit based on the track (Track I or Track II) the new facility has chosen to comply with. The following requirements must be included in each permit:

(1) *Cooling water intake structure requirements.* At a minimum, the permit conditions must include the performance standards that implement the requirements of §125.84(b)(1), (2), (3), (4) and (5); §125.84(c)(1), (2), (3) and (4); or §125.84(d)(1) and (2). In determining compliance with proportional flow requirement in §§125.84(b)(3)(ii); (c)(2)(ii); and (d)(2)(ii), the director must consider anthropogenic factors (those not considered "natural") unrelated to the new facility's cooling water intake structure that can influence the occurrence and location of a thermocline. These include source water inflows, other water withdrawals, managed water uses, wastewater discharges, and flow/level management practices (e.g., some reservoirs release water from below the surface, close to the deepest areas).

(i) For a facility that chooses Track I, you must review the Design and Construction Technology Plan required in §125.86(b)(4) to evaluate the suitability and feasibility of the technology proposed to minimize impingement mortality and entrainment of all life stages of fish and shellfish. In the first permit issued, you must put a condition requiring the facility to reduce impingement mortality and entrainment commensurate with the implementation of the technologies in the permit. Under subsequent permits, the Director must review the performance of the technologies implemented and require additional or different design and construction technologies, if needed to minimize impingement mortality and entrainment of all life stages of fish and shellfish. In addition, you must consider whether more stringent conditions are reasonably necessary in accordance with §125.84(e).

(ii) For a facility that chooses Track II, you must review the information submitted with the Comprehensive Demonstration Study information required in §125.86(c)(2), evaluate the suitability of the proposed design and construction technologies and oper-ational measures to determine whether they will reduce both impingement mortality and entrainment of all life stages of fish and shellfish to 90 percent or greater of the reduction that could be achieved through Track I. If you determine that restoration measures are appropriate at the new facility for consideration of impacts other than impingement mortality and entrainment, you must review the Evaluation of Proposed Restoration Measures and evaluate whether the proposed measures will maintain the fish and shellfish in the waterbody at a substantially similar level to that which would be achieved through §125.84(b)(1) and (2). In addition, you must review the Verification Monitoring Plan in §125.86(c)(2)(iv)(D) and require that the proposed monitoring begin at the start of operations of the cooling water intake structure and continue for a sufficient period of time to demonstrate that the technologies, operational measures and restoration measures meet the requirements in §125.84(d)(1). Under subsequent permits, the Director must review the performance of the additional and/or different technologies or measures used and determine that they reduce the level of adverse environmental impact from the cooling water intake structures to a comparable level that the facility would achieve were it to implement the requirements of §125.84(b)(1) and (2).

(2) *Monitoring conditions.* At a minimum, the permit must require the permittee to perform the monitoring required in §125.87. You may modify the monitoring program when the permit is reissued and during the term of the permit based on changes in physical or biological conditions in the vicinity of the cooling water intake structure. The Director may require continued monitoring based on the results of the Verification Monitoring Plan in §125.86(c)(2)(iv)(D).

(3) *Record keeping and reporting.* At a minimum, the permit must require the permittee to report and keep records as required by §125.88.

[66 FR 65338, Dec. 18, 2001]

Subpart J—Requirements Applicable to Cooling Water Intake Structures for Phase II Existing Facilities Under Section 316(b) of the Act

SOURCE: 69 FR 41683, July 9, 2004, unless otherwise noted.

§ 125.90 What are the purpose and scope of this subpart?

(a) This subpart establishes requirements that apply to the location, design, construction, and capacity of cooling water intake structures at existing facilities that are subject to this subpart (*i.e.*, Phase II existing facilities). The purpose of these requirements is to establish the best technology available for minimizing adverse environmental impact associated with the use of cooling water intake structures. These requirements are implemented through National Pollutant Discharge Elimination System (NPDES) permits issued under section 402 of the Clean Water Act (CWA).

(b) Existing facilities that are not subject to requirements under this or another subpart of this part must meet requirements under section 316(b) of the CWA determined by the Director on a case-by-case, best professional judgment (BPJ) basis.

(c) *Alternative regulatory requirements.* Notwithstanding any other provision of this subpart, if a State demonstrates to the Administrator that it has adopted alternative regulatory requirements in its NPDES program that will result in environmental performance within a watershed that is comparable to the reductions of impingement mortality and entrainment that would otherwise be achieved under § 125.94, the Administrator must approve such alternative regulatory requirements.

(d) Nothing in this subpart shall be construed to preclude or deny the right of any State or political subdivision of a State or any interstate agency under section 510 of the CWA to adopt or enforce any requirement with respect to control or abatement of pollution that is not less stringent than those required by Federal law.

EFFECTIVE DATE NOTE: At 72 FR 37109, July 9, 2007, § 125.90(a), (c), and (d) were suspended.

§ 125.91 What is a "Phase II Existing Facility"?

(a) An existing facility, as defined in § 125.93, is a Phase II existing facility subject to this subpart if it meets each of the following criteria:

(1) It is a point source.

(2) It uses or proposes to use cooling water intake structures with a total design intake flow of 50 million gallons per day (MGD) or more to withdraw cooling water from waters of the United States;

(3) As its primary activity, the facility both generates and transmits electric power, or generates electric power but sells it to another entity for transmission; and

(4) It uses at least 25 percent of water withdrawn exclusively for cooling purposes, measured on an average annual basis.

(b) In the case of a Phase II existing facility that is co-located with a manufacturing facility, only that portion of the combined cooling water intake flow that is used by the Phase II facility to generate electricity for sale to another entity will be considered for purposes of determining whether the 50 MGD and 25 percent criteria in paragraphs (a)(2) and (4) of this section have been exceeded.

(c) Use of a cooling water intake structure includes obtaining cooling water by any sort of contract or arrangement with one or more independent suppliers of cooling water if the supplier withdraws water from waters of the United States but is not itself a Phase II existing facility, except as provided in paragraph (d) of this section. This provision is intended to prevent circumvention of these requirements by creating arrangements to receive cooling water from an entity that is not itself a Phase II existing facility.

(d) Notwithstanding paragraph (c) of this section, obtaining cooling water from a public water system or using treated effluent as cooling water does not constitute use of a cooling water intake structure for purposes of this subpart.

EFFECTIVE DATE NOTE: At 72 FR 37109, July 9, 2007, § 125.91 was suspended.

§125.92 [Reserved]

§125.93 What special definitions apply to this subpart?

In addition to the definitions provided in §122.3 of this chapter, the following special definitions apply to this subpart:

Adaptive management method is a type of project management method where a facility chooses an approach to meeting the project goal, monitors the effectiveness of that approach, and then based on monitoring and any other relevant information, makes any adjustments necessary to ensure continued progress toward the project's goal. This cycle of activity is repeated as necessary to reach the project's goal.

Annual mean flow means the average of daily flows over a calendar year.

All life stages means eggs, larvae, juveniles, and adults.

Calculation baseline means an estimate of impingement mortality and entrainment that would occur at your site assuming that: the cooling water system has been designed as a once-through system; the opening of the cooling water intake structure is located at, and the face of the standard ⅜-inch mesh traveling screen is oriented parallel to, the shoreline near the surface of the source waterbody; and the baseline practices, procedures, and structural configuration are those that your facility would maintain in the absence of any structural or operational controls, including flow or velocity reductions, implemented in whole or in part for the purposes of reducing impingement mortality and entrainment. You may also choose to use the current level of impingement mortality and entrainment as the calculation baseline. The calculation baseline may be estimated using: historical impingement mortality and entrainment data from your facility or from another facility with comparable design, operational, and environmental conditions; current biological data collected in the waterbody in the vicinity of your cooling water intake structure; or current impingement mortality and entrainment data collected at your facility. You may request that the calculation baseline be modified to be based on a location of the opening of the cooling water intake structure at a depth other than at or near the surface if you can demonstrate to the Director that the other depth would correspond to a higher baseline level of impingement mortality and/or entrainment.

Capacity utilization rate means the ratio between the average annual net generation of power by the facility (in MWh) and the total net capability of the facility to generate power (in MW) multiplied by the number of hours during a year. In cases where a facility has more than one intake structure, and each intake structure provides cooling water exclusively to one or more generating units, the capacity utilization rate may be calculated separately for each intake structure, based on the capacity utilization of the units it services. Applicable requirements under this subpart would then be determined separately for each intake structure. The average annual net generation should be measured over a five year period (if available) of representative operating conditions, unless the facility makes a binding commitment to maintain capacity utilization below 15 percent for the life of the permit, in which case the rate may be based on this commitment. For purposes of this subpart, the capacity utilization rate applies to only that portion of the facility that generates electricity for transmission or sale using a thermal cycle employing the steam water system as the thermodynamic medium.

Closed-cycle recirculating system means a system designed, using minimized make-up and blowdown flows, to withdraw water from a natural or other water source to support contact and/or noncontact cooling uses within a facility. The water is usually sent to a cooling canal or channel, lake, pond, or tower to allow waste heat to be dissipated to the atmosphere and then is returned to the system. (Some facilities divert the waste heat to other process operations.) New source water (make-up water) is added to the system to replenish losses that have occurred due to blowdown, drift, and evaporation.

Cooling water means water used for contact or noncontact cooling, including water used for equipment cooling, evaporative cooling tower makeup, and dilution of effluent heat content. The intended use of the cooling water is to absorb waste heat rejected from the process or processes used, or from auxiliary operations on the facility's premises. Cooling water that is used in a manufacturing process either before or after it is used for cooling is considered process water for the purposes of calculating the percentage of a facility's intake flow that is used for cooling purposes in § 125.91(a)(4).

Cooling water intake structure means the total physical structure and any associated constructed waterways used to withdraw cooling water from waters of the U.S. The cooling water intake structure extends from the point at which water is withdrawn from the surface water source up to, and including, the intake pumps.

Design and construction technology means any physical configuration of the cooling water intake structure, or a technology that is placed in the water body in front of the cooling water intake structure, to reduce impingement mortality and/or entrainment. Design and construction technologies include, but are not limited to, location of the intake structure, intake screen systems, passive intake systems, fish diversion and/or avoidance systems, and fish handling and return systems. Restoration measures are not design and construction technologies for purposes of this definition.

Design intake flow means the value assigned (during the cooling water intake structure design) to the total volume of water withdrawn from a source waterbody over a specific time period.

Design intake velocity means the value assigned (during the design of a cooling water intake structure) to the average speed at which intake water passes through the open area of the intake screen (or other device) against which organisms might be impinged or through which they might be entrained.

Diel means daily and refers to variation in organism abundance and density over a 24-hour period due to the influence of water movement, physical or chemical changes, and changes in light intensity.

Entrainment means the incorporation of any life stages of fish and shellfish

with intake water flow entering and passing through a cooling water intake structure and into a cooling water system.

Estuary means a semi-enclosed body of water that has a free connection with open seas and within which the seawater is measurably diluted with fresh water derived from land drainage. The salinity of an estuary exceeds 0.5 parts per thousand (by mass) but is typically less than 30 parts per thousand (by mass).

Existing facility means any facility that commenced construction as described in 40 CFR 122.29(b)(4) on or before January 17, 2002 or July 17, 2006 for an offshore oil and gas extraction facility); and any modification of, or any addition of a unit at such a facility that does not meet the definition of a new facility at § 125.83.

Freshwater river or stream means a lotic (free-flowing) system that does not receive significant inflows of water from oceans or bays due to tidal action. For the purposes of this rule, a flow-through reservoir with a retention time of 7 days or less will be considered a freshwater river or stream.

Impingement means the entrapment of any life stages of fish and shellfish on the outer part of an intake structure or against a screening device during periods of intake water withdrawal.

Lake or reservoir means any inland body of open water with some minimum surface area free of rooted vegetation and with an average hydraulic retention time of more than 7 days. Lakes or reservoirs might be natural water bodies or impounded streams, usually fresh, surrounded by land or by land and a man-made retainer (e.g., a dam). Lakes or reservoirs might be fed by rivers, streams, springs, and/or local precipitation.

Moribund means dying; close to death.

Natural thermal stratification means the naturally occurring and/or existing division of a waterbody into horizontal layers of differing densities as a result of variations in temperature at different depths.

Ocean means marine open coastal waters with a salinity greater than or equal to 30 parts per thousand (by mass).

Once-through cooling water system means a system designed to withdraw water from a natural or other water source, use it at the facility to support contact and/or noncontact cooling uses, and then discharge it to a waterbody without recirculation. Once-through cooling systems sometimes employ canals/channels, ponds, or non-recirculating cooling towers to dissipate waste heat from the water before it is discharged.

Operational measure means a modification to any operation at a facility that serves to minimize impact to fish and shellfish from the cooling water intake structure. Examples of operational measures include, but are not limited to: reductions in cooling water intake flow through the use of variable speed pumps and seasonal flow reductions or shutdowns; and more frequent rotation of traveling screens.

Phase II existing facility means any existing facility that meets the criteria specified in §125.91.

Source water means the waters of the U.S. from which the cooling water is withdrawn.

Supplier means an entity, other than the regulated facility, that owns and operates its own cooling water intake structure and directly withdraws water from waters of the United States. The supplier sells the cooling water to other facilities for their use, but may also use a portion of the water itself. An entity that provides potable water to residential populations (e.g., public water system) is not a supplier for purposes of this subpart.

Thermocline means the middle layer of a thermally stratified lake or a reservoir. In this layer, there is a rapid change in temperatures between the top and bottom of the layer.

Tidal river means the most seaward reach of a river or stream where the salinity is typically less than or equal to 0.5 parts per thousand (by mass) at a time of annual low flow and whose surface elevation responds to the effects of coastal lunar tides.

[44 FR 32948, June 7, 1979, as amended at 71 FR 35040, June 16, 2006]

EFFECTIVE DATE NOTE: At 72 FR 37109, July 9, 2007, §125.93 was suspended.

§125.94 How will requirements reflecting best technology available for minimizing adverse environmental impact be established for my Phase II existing facility?

(a) *Compliance alternatives.* You must select and implement one of the following five alternatives for establishing best technology available for minimizing adverse environmental impact at your facility:

(1)(i) You may demonstrate to the Director that you have reduced, or will reduce, your flow commensurate with a closed-cycle recirculating system. In this case, you are deemed to have met the applicable performance standards and will *not* be required to demonstrate further that your facility meets the impingement mortality and entrainment performance standards specified in paragraph (b) of this section. In addition, you are not subject to the requirements in §§125.95, 125.96, 125.97, or 125.98. However, you may still be subject to any more stringent requirements established under paragraph (e) of this section; or

(ii) You may demonstrate to the Director that you have reduced, or will reduce, your maximum through-screen design intake velocity to 0.5 ft/s or less. In this case, you are deemed to have met the impingement mortality performance standards and will not be required to demonstrate further that your facility meets the performance standards for impingement mortality specified in paragraph (b) of this section and you are not subject to the requirements in §§125.95, 125.96, 125.97, or 125.98 as they apply to impingement mortality. However, you are still subject to any applicable requirements for entrainment reduction and may still be subject to any more stringent requirements established under paragraph (e) of this section.

(2) You may demonstrate to the Director that your existing design and construction technologies, operational measures, and/or restoration measures meet the performance standards specified in paragraph (b) of this section and/or the restoration requirements in paragraph (c) of this section.

(3) You may demonstrate to the Director that you have selected, and will

install and properly operate and maintain, design and construction technologies, operational measures, and/or restoration measures that will, in combination with any existing design and construction technologies, operational measures, and/or restoration measures, meet the performance standards specified in paragraph (b) of this section and/or the restoration requirements in paragraph (c) of this section;

(4) You may demonstrate to the Director that you have installed, or will install, and properly operate and maintain an approved design and construction technology in accordance with § 125.99(a) or (b); or

(5) You may demonstrate to the Director that you have selected, installed, and are properly operating and maintaining, or will install and properly operate and maintain design and construction technologies, operational measures, and/or restoration measures that the Director has determined to be the best technology available to minimize adverse environmental impact for your facility in accordance with paragraphs (a)(5)(i) or (ii) of this section.

(i) If the Director determines that data specific to your facility demonstrate that the costs of compliance under alternatives in paragraphs (a)(2) through (4) of this section would be significantly greater than the costs considered by the Administrator for a facility like yours in establishing the applicable performance standards in paragraph (b) of this section, the Director must make a site-specific determination of the best technology available for minimizing adverse environmental impact. This determination must be based on reliable, scientifically valid cost and performance data submitted by you and any other information that the Director deems appropriate. The Director must establish site-specific alternative requirements based on new and/or existing design and construction technologies, operational measures, and/or restoration measures that achieve an efficacy that is, in the judgment of the Director, as close as practicable to the applicable performance standards in paragraph (b) of this section, without resulting in costs that are significantly greater than the costs considered by the Ad-

ministrator for a facility like yours in establishing the applicable performance standards. The Director's site-specific determination may conclude that design and construction technologies, operational measures, and/or restoration measures in addition to those already in place are not justified because of the significantly greater costs. To calculate the costs considered by the Administrator for a facility like yours in establishing the applicable performance standards you must:

(A) Determine which technology the Administrator modeled as the most appropriate compliance technology for your facility;

(B) Using the Administrator's costing equations, calculate the annualized capital and net operation and maintenance (O&M) costs for a facility with your design intake flow using this technology;

(C) Determine the annualized net revenue loss associated with net construction downtime that the Administrator modeled for your facility to install this technology;

(D) Determine the annualized pilot study costs that the Administrator modeled for your facility to test and optimize this technology;

(E) Sum the cost items in paragraphs (a)(5)(i)(B), (C), and (D) of this section; and

(F) Determine if the performance standards that form the basis of these estimates (*i.e.*, impingement mortality reduction only or impingement mortality and entrainment reduction) are applicable to your facility, and if necessary, adjust the estimates to correspond to the applicable performance standards.

(ii) If the Director determines that data specific to your facility demonstrate that the costs of compliance under alternatives in paragraphs (a)(2) through (4) of this section would be significantly greater than the benefits of complying with the applicable performance standards at your facility, the Director must make a site-specific determination of best technology available for minimizing adverse environmental impact. This determination must be based on reliable, scientifically valid cost and performance data submitted by you and any other information the

Director deems appropriate. The Director must establish site-specific alternative requirements based on new and/or existing design and construction technologies, operational measures, and/or restoration measures that achieve an efficacy that, in the judgment of the Director, is as close as practicable to the applicable performance standards in paragraph (b) of this section without resulting in costs that are significantly greater than the benefits at your facility. The Director's site-specific determination may conclude that design and construction technologies, operational measures, and/or restoration measures in addition to those already in place are not justified because the costs would be significantly greater than the benefits at your facility.

(b) *National performance standards*—(1) *Impingement mortality performance standards.* If you choose compliance alternatives in paragraphs (a)(2), (a)(3), or (a)(4) of this section, you must reduce impingement mortality for all life stages of fish and shellfish by 80 to 95 percent from the calculation baseline.

(2) *Entrainment performance standards.* If you choose compliance alternatives in paragraphs (a)(1)(ii), (a)(2), (a)(3), or (a)(4) of this section, you must also reduce entrainment of all life stages of fish and shellfish by 60 to 90 percent from the calculation baseline if:

(i) Your facility has a capacity utilization rate of 15 percent or greater, and

(ii)(A) Your facility uses cooling water withdrawn from a tidal river, estuary, ocean, or one of the Great Lakes; or

(B) Your facility uses cooling water withdrawn from a freshwater river or stream and the design intake flow of your cooling water intake structures is greater than five percent of the mean annual flow.

(3) *Additional performance standards for facilities withdrawing from a lake (other than one of the Great Lakes) or a reservoir.* If your facility withdraws cooling water from a lake (other than one of the Great Lakes) or a reservoir and you propose to increase the design intake flow of cooling water intake structures it uses, your increased design intake flow must not disrupt the natural thermal stratification or turnover pattern (where present) of the source water, except in cases where the disruption does not adversely affect the management of fisheries. In determining whether any such disruption does not adversely affect the management of fisheries, you must consult with Federal, State, or Tribal fish and wildlife management agencies).

(4) *Use of performance standards for site-specific determinations of best technology available.* The performance standards in paragraphs (b)(1) through (3) of this section must also be used for determining eligibility for site-specific determinations of best technology available for minimizing adverse environmental impact and establishing site specific requirements that achieve an efficacy as close as practicable to the applicable performance standards without resulting in costs that are significantly greater than those considered by the Administrator for a facility like yours in establishing the performance standards or costs that are significantly greater than the benefits at your facility, pursuant to §125.94(a)(5).

(c) *Requirements for restoration measures.* With the approval of the Director, you may implement and adaptively manage restoration measures that produce and result in increases of fish and shellfish in your facility's watershed in place of or as a supplement to installing design and control technologies and/or adopting operational measures that reduce impingement mortality and entrainment. You must demonstrate to the Director that:

(1) You have evaluated the use of design and construction technologies and operational measures for your facility and determined that the use of restoration measures is appropriate because meeting the applicable performance standards or site-specific requirements through the use of design and construction technologies and/or operational measures alone is less feasible, less cost-effective, or less environmentally desirable than meeting the standards or requirements in whole or in part through the use of restoration measures; and

(2) The restoration measures you will implement, alone or in combination with design and construction technologies and/or operational measures,

will produce ecological benefits (fish and shellfish), including maintenance or protection of community structure and function in your facility's waterbody or watershed, at a level that is substantially similar to the level you would achieve by meeting the applicable performance standards under paragraph (b) of this section, or that satisfies alternative site-specific requirements established pursuant to paragraph (a)(5) of this section.

(d)(1) *Compliance using a technology installation and operation plan or restoration plan.* If you choose one of the compliance alternatives in paragraphs (a)(2), (3), (4), or (5) of this section, you may request that compliance with the requirements of §125.94(b) during the first permit containing requirements consistent with this subpart be determined based on whether you have complied with the construction, operational, maintenance, monitoring, and adaptive management requirements of a Technology Installation and Operation Plan developed in accordance with §125.95(b)(4)(ii) (for any design and construction technologies and/or operational measures) and/or a Restoration Plan developed in accordance with §125.95(b)(5) (for any restoration measures). The Technology Installation and Operation Plan must be designed to meet applicable performance standards in paragraph (b) of this section or alternative site-specific requirements developed pursuant to paragraph (a)(5) of this section. The Restoration Plan must be designed to achieve compliance with the applicable requirements in paragraph (c) of this section.

(2) During subsequent permit terms, if you selected and installed design and construction technologies and/or operational measures and have been in compliance with the construction, operational, maintenance, monitoring, and adaptive management requirements of your Technology Installation and Operation Plan during the preceding permit term, you may request that compliance with the requirements of §125.94 during the following permit term be determined based on whether you remain in compliance with your Technology Installation and Operation Plan, revised in accordance with your adaptive management plan in

§125.95(b)(4)(ii)(C) if applicable performance standards are not being met. Each request and approval of a Technology Installation and Operation Plan shall be limited to one permit term.

(3) During subsequent permit terms, if you selected and installed restoration measures and have been in compliance with the construction, operational, maintenance, monitoring, and adaptive management requirements in your Restoration Plan during the preceding permit term, you may request that compliance with the requirements of this section during the following permit term be determined based on whether you remain in compliance with your Restoration Plan, revised in accordance with your adaptive management plan in §125.95(b)(5)(v) if applicable performance standards are not being met. Each request and approval of a Restoration Plan shall be limited to one permit term.

(e) *More stringent standards.* The Director may establish more stringent requirements as best technology available for minimizing adverse environmental impact if the Director determines that your compliance with the applicable requirements of this section would not meet the requirements of applicable State and Tribal law, or other Federal law.

(f) *Nuclear facilities.* If you demonstrate to the Director based on consultation with the Nuclear Regulatory Commission that compliance with this subpart would result in a conflict with a safety requirement established by the Commission, the Director must make a site-specific determination of best technology available for minimizing adverse environmental impact that would not result in a conflict with the Nuclear Regulatory Commission's safety requirement.

EFFECTIVE DATE NOTE: At 72 FR 37109, July 9, 2007, §125.94 was suspended.

§ 125.95 As an owner or operator of a Phase II existing facility, what must I collect and submit when I apply for my reissued NPDES permit?

(a)(1) You must submit to the Director the Proposal for Information Collection required in paragraph (b)(1) of this section prior to the start of information collection activities;

(2) You must submit to the Director the information required in 40 CFR 122.21(r)(2), (r)(3) and (r)(5) and any applicable portions of the Comprehensive Demonstration Study (Study), except for the Proposal for Information Collection required by paragraph (b)(1) of this section; and

(i) You must submit your NPDES permit application in accordance with the time frames specified in 40 CFR 122.21(d)(2).

(ii) If your existing permit expires before July 9, 2008, you may request that the Director establish a schedule for you to submit the information required by this section as expeditiously as practicable, but not later than January 7, 2008. Between the time your existing permit expires and the time an NPDES permit containing requirements consistent with this subpart is issued to your facility, the best technology available to minimize adverse environmental impact will continue to be determined based on the Director's best professional judgment.

(3) In subsequent permit terms, the Director may approve a request to reduce the information required to be submitted in your permit application on the cooling water intake structure(s) and the source waterbody, if conditions at your facility and in the waterbody remain substantially unchanged since your previous application. You must submit your request for reduced cooling water intake structure and waterbody application information to the Director at least one year prior to the expiration of the permit. Your request must identify each required information item in §122.21(r) and this section that you determine has not substantially changed since the previous permit application and the basis for your determination.

(b) *Comprehensive Demonstration Study.* The purpose of the Comprehensive Demonstration Study (The Study) is to characterize impingement mortality and entrainment, to describe the operation of your cooling water intake structures, and to confirm that the technologies, operational measures, and/or restoration measures you have selected and installed, or will install, at your facility meet the applicable requirements of §125.94. All facilities ex-

cept those that have met the applicable requirements in accordance with §§125.94(a)(1)(i), 125.94(a)(1)(ii), and 125.94(a)(4) must submit all applicable portions of the Comprehensive Demonstration Study to the Director in accordance with paragraph (a) of this section. Facilities that meet the requirements in §125.94(a)(1)(i) by reducing their flow commensurate with a closed-cycle, recirculating system are not required to submit a Comprehensive Demonstration Study. Facilities that meet the requirements in §125.94(a)(1)(ii) by reducing their design intake velocity to 0.5 ft/sec or less are required to submit a Study only for the entrainment requirements, if applicable. Facilities that meet the requirements in §125.94(a)(4) and have installed and properly operate and maintain an approved design and construction technology (in accordance with §125.99) are required to submit only the Technology Installation and Operation Plan in paragraph (b)(4) of this section and the Verification Monitoring Plan in paragraph (b)(7) of this section. Facilities that are required to meet only impingement mortality performance standards in §125.94(b)(1) are required to submit only a Study for the impingement mortality reduction requirements. The Comprehensive Demonstration Study must include:

(1) *Proposal For Information Collection.* You must submit to the Director for review and comment a description of the information you will use to support your Study. The Proposal for Information must be submitted prior to the start of information collection activities, but you may initiate such activities prior to receiving comment from the Director. The proposal must include:

(i) A description of the proposed and/or implemented technologies, operational measures, and/or restoration measures to be evaluated in the Study;

(ii) A list and description of any historical studies characterizing impingement mortality and entrainment and/or the physical and biological conditions in the vicinity of the cooling water intake structures and their relevance to this proposed Study. If you propose to use existing data, you must demonstrate the extent to which the data are representative of current conditions and that the data were collected using appropriate quality assurance/quality control procedures;

(iii) A summary of any past or ongoing consultations with appropriate Federal, State, and Tribal fish and wildlife agencies that are relevant to this Study and a copy of written comments received as a result of such consultations; and

(iv) A sampling plan for any new field studies you propose to conduct in order to ensure that you have sufficient data to develop a scientifically valid estimate of impingement mortality and entrainment at your site. The sampling plan must document all methods and quality assurance/quality control procedures for sampling and data analysis. The sampling and data analysis methods you propose must be appropriate for a quantitative survey and include consideration of the methods used in other studies performed in the source waterbody. The sampling plan must include a description of the study area (including the area of influence of the cooling water intake structure(s)), and provide a taxonomic identification of the sampled or evaluated biological assemblages (including all life stages of fish and shellfish).

(2) *Source waterbody flow information.* You must submit to the Director the following source waterbody flow information:

(i) If your cooling water intake structure is located in a freshwater river or stream, you must provide the annual mean flow of the waterbody and any supporting documentation and engineering calculations to support your analysis of whether your design intake flow is greater than five percent of the mean annual flow of the river or stream for purposes of determining applicable performance standards under paragraph (b) of this section. Representative historical data (from a period of time up to 10 years, if available) must be used; and

(ii) If your cooling water intake structure is located in a lake (other than one of the Great Lakes) or a reservoir and you propose to increase its design intake flow, you must provide a description of the thermal stratification in the waterbody, and any sup-

porting documentation and engineering calculations to show that the total design intake flow after the increase will not disrupt the natural thermal stratification and turnover pattern in a way that adversely impacts fisheries, including the results of any consultations with Federal, State, or Tribal fish and wildlife management agencies.

(3) *Impingement Mortality and/or Entrainment Characterization Study.* You must submit to the Director an Impingement Mortality and/or Entrainment Characterization Study whose purpose is to provide information to support the development of a calculation baseline for evaluating impingement mortality and entrainment and to characterize current impingement mortality and entrainment. The Impingement Mortality and/or Entrainment Characterization Study must include the following, in sufficient detail to support development of the other elements of the Comprehensive Demonstration Study:

(i) Taxonomic identifications of all life stages of fish, shellfish, and any species protected under Federal, State, or Tribal Law (including threatened or endangered species) that are in the vicinity of the cooling water intake structure(s) and are susceptible to impingement and entrainment;

(ii) A characterization of all life stages of fish, shellfish, and any species protected under Federal, State, or Tribal Law (including threatened or endangered species) identified pursuant to paragraph (b)(3)(i) of this section, including a description of the abundance and temporal and spatial characteristics in the vicinity of the cooling water intake structure(s), based on sufficient data to characterize annual, seasonal, and diel variations in impingement mortality and entrainment (*e.g.*, related to climate and weather differences, spawning, feeding and water column migration). These may include historical data that are representative of the current operation of your facility and of biological conditions at the site;

(iii) Documentation of the current impingement mortality and entrainment of all life stages of fish, shellfish, and any species protected under Federal, State, or Tribal Law (including threatened or endangered species) identified pursuant to paragraph (b)(3)(i) of this section and an estimate of impingement mortality and entrainment to be used as the calculation baseline. The documentation may include historical data that are representative of the current operation of your facility and of biological conditions at the site. Impingement mortality and entrainment samples to support the calculations required in paragraphs (b)(4)(i)(C) and (b)(5)(iii) of this section must be collected during periods of representative operational flows for the cooling water intake structure and the flows associated with the samples must be documented;

(4) *Technology and compliance assessment information*—(i) *Design and Construction Technology Plan.* If you choose to use design and construction technologies and/or operational measures, in whole or in part to meet the requirements of §125.94(a)(2) or (3), you must submit a Design and Construction Technology Plan to the Director for review and approval. In the plan, you must provide the capacity utilization rate for your facility (or for individual intake structures where applicable, in accordance with §125.93) and provide supporting data (including the average annual net generation of the facility (in MWh) measured over a five year period (if available) of representative operating conditions and the total net capacity of the facility (in MW)) and underlying calculations. The plan must explain the technologies and/or operational measures you have in place and/or have selected to meet the requirements in §125.94. (Examples of potentially appropriate technologies may include, but are not limited to, wedgewire screens, fine mesh screens, fish handling and return systems, barrier nets, aquatic filter barrier systems, vertical and/or lateral relocation of the cooling water intake structure, and enlargement of the cooling water intake structure opening to reduce velocity. Examples of potentially appropriate operational measures may include, but are not limited to, seasonal shutdowns, reductions in flow, and continuous or more frequent rotation of traveling screens.) The plan must contain the following information:

(A) A narrative description of the design and operation of all design and construction technologies and/or operational measures (existing and proposed), including fish handling and return systems, that you have in place or will use to meet the requirements to reduce impingement mortality of those species expected to be most susceptible to impingement, and information that demonstrates the efficacy of the technologies and/or operational measures for those species;

(B) A narrative description of the design and operation of all design and construction technologies and/or operational measures (existing and proposed) that you have in place or will use to meet the requirements to reduce entrainment of those species expected to be the most susceptible to entrainment, if applicable, and information that demonstrates the efficacy of the technologies and/or operational measures for those species;

(C) Calculations of the reduction in impingement mortality and entrainment of all life stages of fish and shellfish that would be achieved by the technologies and/or operational measures you have selected based on the Impingement Mortality and/or Entrainment Characterization Study in paragraph (b)(3) of this section. In determining compliance with any requirements to reduce impingement mortality or entrainment, you must assess the total reduction in impingement mortality and entrainment against the calculation baseline determined in accordance with paragraph (b)(3) of this section. Reductions in impingement mortality and entrainment from this calculation baseline as a result of any design and construction technologies and/or operational measures already implemented at your facility should be added to the reductions expected to be achieved by any additional design and/or construction technologies and operational measures that will be implemented, and any increases in fish and shellfish within the waterbody attributable to your restoration measures. Facilities that recirculate a portion of their flow, but do not reduce flow sufficiently to satisfy the compliance option in §125.94(a)(1)(i) may take into account the reduction in impingement mortality and entrainment associated with the reduction in flow when determining the net reduction associated with existing design and construction technologies and/or operational measures. This estimate must include a site-specific evaluation of the suitability of the technologies and/or operational measures based on the species that are found at the site, and may be determined based on representative studies (*i.e.*, studies that have been conducted at a similar facility's cooling water intake structures located in the same waterbody type with similar biological characteristics) and/or site-specific technology prototype or pilot studies; and

(D) Design and engineering calculations, drawings, and estimates prepared by a qualified professional to support the descriptions required by paragraphs (b)(4)(i)(A) and (B) of this section.

(ii) *Technology Installation and Operation Plan.* If you choose the compliance alternative in § 125.94(a)(2), (3), (4), or (5) and use design and construction technologies and/or operational measures in whole or in part to comply with the applicable requirements of § 125.94, you must submit the following information with your application for review and approval by the Director:

(A) A schedule for the installation and maintenance of any new design and construction technologies. Any downtime of generating units to accommodate installation and/or maintenance of these technologies should be scheduled to coincide with otherwise necessary downtime (*e.g.*, for repair, overhaul, or routine maintenance of the generating units) to the extent practicable. Where additional downtime is required, you may coordinate scheduling of this downtime with the North American Electric Reliability Council and/or other generators in your area to ensure that impacts to reliability and supply are minimized;

(B) List of operational and other parameters to be monitored, and the location and frequency that you will monitor them;

(C) List of activities you will undertake to ensure to the degree practicable the efficacy of installed design and construction technologies and

operational measures, and your schedule for implementing them;

(D) A schedule and methodology for assessing the efficacy of any installed design and construction technologies and operational measures in meeting applicable performance standards or site-specific requirements, including an adaptive management plan for revising design and construction technologies, operational measures, operation and maintenance requirements, and/or monitoring requirements if your assessment indicates that applicable performance standards or site-specific requirements are not being met; and

(E) If you choose the compliance alternative in § 125.94(a)(4), documentation that the appropriate site conditions in § 125.99(a) or (b) exist at your facility.

(5) *Restoration Plan.* If you propose to use restoration measures, in whole or in part, to meet the applicable requirements in § 125.94, you must submit the following information with your application for review and approval by the Director. You must address species of concern identified in consultation with Federal, State, and Tribal fish and wildlife management agencies with responsibility for fisheries and wildlife potentially affected by your cooling water intake structure(s).

(i) A demonstration to the Director that you have evaluated the use of design and construction technologies and/or operational measures for your facility and an explanation of how you determined that restoration would be more feasible, cost-effective, or environmentally desirable;

(ii) A narrative description of the design and operation of all restoration measures (existing and proposed) that you have in place or will use to produce fish and shellfish;

(iii) Quantification of the ecological benefits of the proposed restoration measures. You must use information from the Impingement Mortality and/or Entrainment Characterization Study required in paragraph (b)(3) of this section, and any other available and appropriate information, to estimate the reduction in fish and shellfish impingement mortality and/or entrainment that would be necessary for your facility to comply with § 125.94(c)(2).

You must then calculate the production of fish and shellfish that you will achieve with the restoration measures you will or have already installed. You must include a discussion of the nature and magnitude of uncertainty associated with the performance of these restoration measures. You must also include a discussion of the time frame within which these ecological benefits are expected to accrue;

(iv) Design calculations, drawings, and estimates to document that your proposed restoration measures in combination with design and construction technologies and/or operational measures, or alone, will meet the requirements of §125.94(c)(2). If the restoration measures address the same fish and shellfish species identified in the Impingement Mortality and/or Entrainment Characterization Study (in-kind restoration), you must demonstrate that the restoration measures will produce a level of these fish and shellfish substantially similar to that which would result from meeting applicable performance standards in §125.94(b), or that they will satisfy site-specific requirements established pursuant to §125.94(a)(5). If the restoration measures address fish and shellfish species different from those identified in the Impingement Mortality and/or Entrainment Characterization Study (out-of-kind restoration), you must demonstrate that the restoration measures produce ecological benefits substantially similar to or greater than those that would be realized through in-kind restoration. Such a demonstration should be based on a watershed approach to restoration planning and consider applicable multi-agency watershed restoration plans, site-specific peer-reviewed ecological studies, and/or consultation with appropriate Federal, State, and Tribal fish and wildlife management agencies.

(v) A plan utilizing an adaptive management method for implementing, maintaining, and demonstrating the efficacy of the restoration measures you have selected and for determining the extent to which the restoration measures, or the restoration measures in combination with design and construction technologies and operational

measures, have met the applicable requirements of §125.94(c)(2). The plan must include:

(A) A monitoring plan that includes a list of the restoration parameters that will be monitored, the frequency at which you will monitor them, and success criteria for each parameter;

(B) A list of activities you will undertake to ensure the efficacy of the restoration measures, a description of the linkages between these activities and the items in paragraph (b)(5)(v)(A) of this section, and an implementation schedule; and

(C) A process for revising the Restoration Plan as new information, including monitoring data, becomes available, if the applicable requirements under §125.94(c)(2) are not being met.

(vi) A summary of any past or ongoing consultation with appropriate Federal, State, and Tribal fish and wildlife management agencies on your use of restoration measures including a copy of any written comments received as a result of such consultations;

(vii) If requested by the Director, a peer review of the items you submit for the Restoration Plan. You must choose the peer reviewers in consultation with the Director who may consult with EPA and Federal, State, and Tribal fish and wildlife management agencies with responsibility for fish and wildlife potentially affected by your cooling water intake structure(s). Peer reviewers must have appropriate qualifications (*e.g.*, in the fields of geology, engineering, and/or biology, etc.) depending upon the materials to be reviewed; and

(viii) A description of the information to be included in a bi-annual status report to the Director.

(6) *Information to support site-specific determination of best technology available for minimizing adverse environmental impact.* If you have requested a site-specific determination of best technology available for minimizing adverse environmental impact pursuant to §125.94(a)(5)(i) because of costs significantly greater than those considered by the Administrator for a facility like yours in establishing the applicable performance standards of §125.94(b),

you are required to provide to the Director the information specified in paragraphs (b)(6)(i) and (b)(6)(iii) of this section. If you have requested a site-specific determination of best technology available for minimizing adverse environmental impact pursuant to § 125.94(a)(5)(ii) because of costs significantly greater than the benefits of meeting the applicable performance standards of § 125.94(b) at your facility, you must provide the information specified in paragraphs (b)(6)(i), (b)(6)(ii), and (b)(6)(iii) of this section:

(i) *Comprehensive Cost Evaluation Study.* You must perform and submit the results of a Comprehensive Cost Evaluation Study, that includes:

(A) Engineering cost estimates in sufficient detail to document the costs of implementing design and construction technologies, operational measures, and/or restoration measures at your facility that would be needed to meet the applicable performance standards of § 125.94(b);

(B) A demonstration that the costs documented in paragraph (b)(6)(i)(A) of this section significantly exceed either those considered by the Administrator for a facility like yours in establishing the applicable performance standards or the benefits of meeting the applicable performance standards at your facility; and

(C) Engineering cost estimates in sufficient detail to document the costs of implementing the design and construction technologies, operational measures, and/or restoration measures in your Site-Specific Technology Plan developed in accordance with paragraph (b)(6)(iii) of this section.

(ii) *Benefits Valuation Study.* If you are seeking a site-specific determination of best technology available for minimizing adverse environmental impact because of costs significantly greater than the benefits of meeting the applicable performance standards of § 125.94(b) at your facility, you must use a comprehensive methodology to fully value the impacts of impingement mortality and entrainment at your site and the benefits achievable by meeting the applicable performance standards. In addition to the valuation estimates, the benefit study must include the following:

(A) A description of the methodology(ies) used to value commercial, recreational, and ecological benefits (including any non-use benefits, if applicable);

(B) Documentation of the basis for any assumptions and quantitative estimates. If you plan to use an entrainment survival rate other than zero, you must submit a determination of entrainment survival at your facility based on a study approved by the Director;

(C) An analysis of the effects of significant sources of uncertainty on the results of the study; and

(D) If requested by the Director, a peer review of the items you submit in the Benefits Valuation Study. You must choose the peer reviewers in consultation with the Director who may consult with EPA and Federal, State, and Tribal fish and wildlife management agencies with responsibility for fish and wildlife potentially affected by your cooling water intake structure. Peer reviewers must have appropriate qualifications depending upon the materials to be reviewed.

(E) A narrative description of any non-monetized benefits that would be realized at your site if you were to meet the applicable performance standards and a qualitative assessment of their magnitude and significance.

(iii) *Site-Specific Technology Plan.* Based on the results of the Comprehensive Cost Evaluation Study required by paragraph (b)(6)(i) of this section, and the Benefits Valuation Study required by paragraph (b)(6)(ii) of this section, if applicable, you must submit a Site-Specific Technology Plan to the Director for review and approval. The plan must contain the following information:

(A) A narrative description of the design and operation of all existing and proposed design and construction technologies, operational measures, and/or restoration measures that you have selected in accordance with § 125.94(a)(5);

(B) An engineering estimate of the efficacy of the proposed and/or implemented design and construction technologies or operational measures, and/or restoration measures. This estimate must include a site-specific evaluation of the suitability of the technologies or operational measures for reducing impingement mortality and/or entrainment (as applicable) of all life stages of fish and shellfish based on representative studies (*e.g.*, studies that have been conducted at cooling water intake structures located in the same waterbody type with similar biological characteristics) and, if applicable, site-specific technology prototype or pilot studies. If restoration measures will be used, you must provide a Restoration Plan that includes the elements described in paragraph (b)(5) of this section.

(C) A demonstration that the proposed and/or implemented design and construction technologies, operational measures, and/or restoration measures achieve an efficacy that is as close as practicable to the applicable performance standards of § 125.94(b) without resulting in costs significantly greater than either the costs considered by the Administrator for a facility like yours in establishing the applicable performance standards, or as appropriate, the benefits of complying with the applicable performance standards at your facility;

(D) Design and engineering calculations, drawings, and estimates prepared by a qualified professional to support the elements of the Plan.

(7) *Verification Monitoring Plan.* If you comply using compliance alternatives in § 125.94(a)(2), (3), (4), or (5) using design and construction technologies and/or operational measures, you must submit a plan to conduct, at a minimum, two years of monitoring to verify the full-scale performance of the proposed or already implemented technologies and/or operational measures. The verification study must begin once the design and construction technologies and/or operational measures are installed and continue for a period of time that is sufficient to demonstrate to the Director whether the facility is meeting the applicable performance standards in § 125.94(b) or site-specific requirements developed pursuant to § 125.94(a)(5). The plan must provide the following:

(i) Description of the frequency and duration of monitoring, the parameters to be monitored, and the basis for determining the parameters and the frequency and duration for monitoring. The parameters selected and duration and frequency of monitoring must be consistent with any methodology for assessing success in meeting applicable performance standards in your Technology Installation and Operation Plan as required by paragraph (b)(4)(ii) of this section.

(ii) A proposal on how naturally moribund fish and shellfish that enter the cooling water intake structure would be identified and taken into account in assessing success in meeting the performance standards in § 125.94(b).

(iii) A description of the information to be included in a bi-annual status report to the Director.

[69 FR 41683, July 9, 2004, as amended at 69 FR 47210, Aug. 4, 2004]

EFFECTIVE DATE NOTE: At 72 FR 37109, July 9, 2007, § 125.95 was suspended.

§ 125.96 As an owner or operator of a Phase II existing facility, what monitoring must I perform?

As an owner or operator of a Phase II existing facility, you must perform monitoring, as applicable, in accordance with the Technology Installation and Operation Plan required by § 125.95(b)(4)(ii), the Restoration Plan required by § 125.95(b)(5), the Verification Monitoring Plan required by § 125.95(b)(7), and any additional monitoring specified by the Director to demonstrate compliance with the applicable requirements of § 125.94.

EFFECTIVE DATE NOTE: At 72 FR 37109, July 9, 2007, § 125.96 was suspended.

§ 125.97 As an owner or operator of a Phase II existing facility, what records must I keep and what information must I report?

As an owner or operator of a Phase II existing facility you are required to keep records and report information and data to the Director as follows:

(a) You must keep records of all the data used to complete the permit application and show compliance with the requirements of § 125.94, any supplemental information developed under § 125.95, and any compliance monitoring data submitted under § 125.96, for a period of at least three (3) years from date of permit issuance. The Director may require that these records be kept for a longer period.

(b) You must submit a status report to the Director for review every two years that includes appropriate monitoring data and other information as specified by the Director in accordance with § 125.98(b)(5).

EFFECTIVE DATE NOTE: At 72 FR 37109, July 9, 2007, § 125.97 was suspended.

§ 125.98 **As the Director, what must I do to comply with the requirements of this subpart?**

(a) *Permit application.* As the Director, you must review materials submitted by the applicant under 40 CFR 122.21(r) and § 125.95 before each permit renewal or reissuance.

(1) You must review and comment on the Proposal for Information Collection submitted by the facility in accordance with § 125.95(a)(1). You are encouraged to provide comments expeditiously so that the permit applicant can make responsive modifications to its information gathering activities. If a facility submits a request in accordance with § 125.95(a)(2)(ii) for an alternate schedule for submitting the information required in § 125.95, you must approve a schedule that is as expeditious as practicable, but does not extend beyond January 7, 2008. If a facility submits a request in accordance with § 125.95(a)(3) to reduce the information about their cooling water intake structures and the source waterbody required to be submitted in their permit application (other than with the first permit application after September 7, 2004), you must approve the request within 60 days if conditions at the facility and in the waterbody remain substantially unchanged since the previous application.

(2) After receiving the permit application from the owner or operator of a Phase II existing facility, you must determine which of the requirements specified in § 125.94 apply to the facility. In addition, you must review materials to determine compliance with the applicable requirements.

(3) At each permit renewal, you must review the application materials and monitoring data to determine whether new or revised requirements for design and construction technologies, operational measures, or restoration measures should be included in the permit to meet the applicable performance standards in § 125.94(b) or alternative site-specific requirements established pursuant to § 125.94(a)(5).

(b) *Permitting requirements.* Section 316(b) requirements are implemented for a facility through an NPDES permit. As the Director, you must consider the information submitted by the Phase II existing facility in its permit application, and determine the appropriate requirements and conditions to include in the permit based on the compliance alternatives in § 125.94(a). The following requirements must be included in each permit:

(1) *Cooling water intake structure requirements.* The permit conditions must include the requirements that implement the applicable provisions of § 125.94. You must evaluate the performance of the design and construction technologies, operational measures, and/or restoration measures proposed and implemented by the facility and require additional or different design and construction technologies, operational measure, and/or restoration measures, and/or improved operation and maintenance of existing technologies and measures, if needed to meet the applicable performance standards, restoration requirements, or alternative site-specific requirements. In determining compliance with the performance standards for facilities proposing to increase withdrawals of cooling water from a lake (other than a Great Lake) or a reservoir in § 125.94(b)(3), you must consider anthropogenic factors (those not considered "natural") unrelated to the Phase II existing facility's cooling water intake structures that can influence the occurrence and location of a thermocline. These include source water inflows, other water withdrawals, managed water uses, wastewater discharges, and flow/level management practices (*e.g.,* some reservoirs release water from deeper bottom layers). As the Director, you must coordinate with appropriate Federal, State, or Tribal fish and wildlife management agencies to determine if any disruption of the natural thermal stratification resulting from the

proposed increased withdrawal of cooling water does not adversely affect the management of fisheries. Specifically:

(i) You must review and approve the Design and Construction Technology Plan required in §125.95(b)(4) to evaluate the suitability and feasibility of the design and construction technology and/or operational measures proposed to meet the performance standards in §125.94(b) or site-specific requirements developed pursuant to §125.94(a)(5).

(ii) If the facility proposes restoration measures in accordance with §125.94(c), you must review and approve the Restoration Plan required under §125.95(b)(5) to determine whether the proposed measures, alone or in combination with design and construction technologies and/or operational measures, will meet the requirements under §125.94(c).

(iii) In each reissued permit, you must include a condition in the permit requiring the facility to reduce impingement mortality and entrainment (or to increase fish production, if applicable) commensurate with the efficacy at the facility of the installed design and construction technologies, operational measures, and/or restoration measures.

(iv) If the facility implements design and construction technologies and/or operational measures and requests that compliance with the requirements in §125.94 be measured for the first permit term (or subsequent permit terms, if applicable) employing the Technology Installation and Operation Plan in accordance with §125.95(b)(4)(ii), you must review the Technology Installation and Operation Plan to ensure it meets the requirements of §125.95(b)(4)(ii). If the Technology Installation and Operation Plan meets the requirements of §125.95(b)(4)(ii), you must approve the Technology Installation and Operation Plan and require the facility to meet the terms of the plan including any revision to the plan that may be necessary if applicable performance standards or alternative site-specific requirements are not being met. If the facility implements restoration measures and requests that compliance with the requirements in §125.94 be measured for the first permit term (or subsequent

permit terms, if applicable) employing a Restoration Plan in accordance with §125.95(b)(5), you must review the Restoration Plan to ensure it meets the requirements of §125.95(b)(5). If the Restoration Plan meets the requirements of §125.95(b)(5), you must approve the plan and require the facility to meet the terms of the plan including any revision to the plan that may be necessary if applicable performance standards or site-specific requirements are not being met. In determining whether to approve a Technology Installation and Operation Plan or Restoration Plan, you must evaluate whether the design and construction technologies, operational measures, and/or restoration measures the facility has installed, or proposes to install, can reasonably be expected to meet the applicable performance standards in §125.94(b), restoration requirements in §125.94(c)(2), and/or alternative site-specific requirements established pursuant to §125.94(a)(5), and whether the Technology Installation and Operation Plan and/or Restoration Plan complies with the applicable requirements of §125.95(b). In reviewing the Technology Installation and Operation Plan, you must approve any reasonable scheduling provisions that are designed to ensure that impacts to energy reliability and supply are minimized, in accordance with §125.95(b)(4)(ii)(A). If the facility does not request that compliance with the requirements in §125.94 be measured employing a Technology Installation and Operation Plan and/or Restoration Plan, or the facility has not been in compliance with the terms of its current Technology Installation and Operation Plan and/or Restoration Plan during the preceding permit term, you must require the facility to comply with the applicable performance standards in §125.94(b), restoration requirement in §125.94(c)(2), and/or alternative site-specific requirements developed pursuant to §125.94(a)(5). In considering a permit application, you must review the performance of the design and construction measures, operational measures, and/or restoration measures implemented and require additional or different design and construction technologies, operational measures, and/or restoration measures, ·

and/or improved operation and maintenance of existing technologies and measures, if needed to meet the applicable performance standards, restoration requirements, and/or alternative site-specific requirements.

(v) You must review and approve the proposed Verification Monitoring Plan submitted under § 125.95(b)(7) (for design and construction technologies) and/or monitoring provisions of the Restoration Plan submitted under § 125.95(b)(5)(v) and require that the monitoring continue for a sufficient period of time to demonstrate whether the design and construction technology, operational measures, and/or restoration measures meet the applicable performance standards in § 125.94(b), restoration requirements in 125.94(c)(2) and/or site-specific requirements established pursuant to § 125.94(a)(5).

(vi) If a facility requests requirements based on a site-specific determination of best technology available for minimizing adverse environmental impact, you must review the application materials submitted under § 125.95(b)(6) and any other information you may have, including quantitative and qualitative benefits, that would be relevant to a determination of whether alternative requirements are appropriate for the facility. If a facility submits a study to support entrainment survival at the facility, you must review and approve the results of that study. If you determine that alternative requirements are appropriate, you must make a site-specific determination of best technology available for minimizing adverse environmental impact in accordance with § 125.94(a)(5). You, as the Director, may request revisions to the information submitted by the facility in accordance with § 125.95(b)(6) if it does not provide an adequate basis for you to make this determination. Any alternative site-specific requirements established based on new and/or existing design and construction technologies, operational measures, and/or restoration measures, must achieve an efficacy that is, in your judgement, as close as practicable to the applicable performance standards of § 125.94(b) without resulting in costs that are significantly greater than the costs considered by the Administrator for a like facility in establishing the applicable performance standards in § 125.94(b), determined in accordance with § 125.94(a)(5)(i)(A) through (F), or the benefits of complying with the applicable performance standards at the facility; and

(vii) You must review the proposed methods for assessing success in meeting applicable performance standards and/or restoration requirements submitted by the facility under § 125.95(b)(4)(ii)(D) and/or (b)(5)(v)(A), evaluate those and other available methods, and specify how assessment of success in meeting the performance standards and/or restoration requirements must be determined including the averaging period for determining the percent reduction in impingement mortality and entrainment and/or the production of fish and shellfish. Compliance for facilities who request that compliance be measured employing a Technology Installation and Operation Plan and/or Restoration Plan must be determined in accordance with § 125.98(b)(1)(iv).

(2) *Monitoring conditions.* You must require the facility to perform monitoring in accordance with the Technology Installation and Operation Plan in § 125.95(b)(4)(ii), the Restoration Plan required by § 125.95(b)(5), if applicable, and the Verification Monitoring Plan required by § 125.95(b)(7). In determining any additional applicable monitoring requirements in accordance with § 125.96, you must consider the monitoring facility's Verification Monitoring, Technology Installation and Operation, and/or Restoration Plans, as appropriate. You may modify the monitoring program based on changes in physical or biological conditions in the vicinity of the cooling water intake structure.

(3) *Recordkeeping and reporting.* At a minimum, the permit must require the facility to report and keep records specified in § 125.97.

(4) *Design and construction technology approval*—(i) For a facility that chooses to demonstrate that it has installed and properly operate and maintain a design and construction technology approved in accordance with § 125.99, the Director must review and approve the

information submitted in the Technology Installation and Operation Plan in §125.95(b)(4)(ii) and determine if it meets the criteria in §125.99.

(ii) If a person requests approval of a technology under §125.99(b), the Director must review and approve the information submitted and determine its suitability for widespread use at facilities with similar site conditions in its jurisdiction with minimal study. As the Director, you must evaluate the adequacy of the technology when installed in accordance with the required design criteria and site conditions to consistently meet the performance standards in §125.94. You, as the Director, may only approve a technology following public notice and consideration of comment regarding such approval.

(5) *Bi-annual status report.* You must specify monitoring data and other information to be included in a status report every two years. The other information may include operation and maintenance records, summaries of adaptive management activities, or any other information that is relevant to determining compliance with the terms of the facility's Technology Operation and Installation Plan and/or Restoration Plan.

EFFECTIVE DATE NOTE: At 72 FR 37109, July 9, 2007, §125.98 was suspended.

§125.99 What are approved design and construction technologies?

(a) The following technologies constitute approved design and construction technologies for purposes of §125.94(a)(4):

(1) Submerged cylindrical wedge-wire screen technology, if you meet the following conditions:

(i) Your cooling water intake structure is located in a freshwater river or stream;

(ii) Your cooling water intake structure is situated such that sufficient ambient counter currents exist to promote cleaning of the screen face;

(iii)Your maximum through-screen design intake velocity is 0.5 ft/s or less;

(iv) The slot size is appropriate for the size of eggs, larvae, and juveniles of all fish and shellfish to be protected at the site; and

(v) Your entire main condenser cooling water flow is directed through the technology. Small flows totaling less than 2 MGD for auxiliary plant cooling uses are excluded from this provision.

(2) A technology that has been approved in accordance with the process described in paragraph (b) of this section.

(b) You or any other interested person may submit a request to the Director that a technology be approved in accordance with the compliance alternative in §125.94(a)(4) after providing the public with notice and an opportunity to comment on the request for approval of the technology. If the Director approves the technology, it may be used by all facilities with similar site conditions under the Director's jurisdiction. Requests for approval of a technology must be submitted to the Director and include the following information:

(1) A detailed description of the technology;

(2) A list of design criteria for the technology and site characteristics and conditions that each facility must have in order to ensure that the technology can consistently meet the appropriate impingement mortality and entrainment performance standards in §125.94(b); and

(3) Information and data sufficient to demonstrate that facilities under the jurisdiction of the Director can meet the applicable impingement mortality and entrainment performance standards in §125.94(b) if the applicable design criteria and site characteristics and conditions are present at the facility.

EFFECTIVE DATE NOTE. At 72 FR 37109, July 9, 2007, §125.99 was suspended.

Subpart K [Reserved]

Subpart L—Criteria and Standards for Imposing Conditions for the Disposal of Sewage Sludge Under Section 405 of the Act [Reserved]

Subpart M—Ocean Discharge Criteria

SOURCE: 45 FR 65953, Oct. 3, 1980, unless otherwise noted.

§ 125.120 Scope and purpose.

This subpart establishes guidelines for issuance of National Pollutant Discharge Elimination System (NPDES) permits for the discharge of pollutants from a point source into the territorial seas, the contiguous zone, and the oceans.

§ 125.121 Definitions.

(a) *Irreparable harm* means significant undesirable effects occurring after the date of permit issuance which will not be reversed after cessation or modification of the discharge.

(b) *Marine environment* means that territorial seas, the contiguous zone and the oceans.

(c) *Mixing zone* means the zone extending from the sea's surface to seabed and extending laterally to a distance of 100 meters in all directions from the discharge point(s) or to the boundary of the zone of initial dilution as calculated by a plume model approved by the director, whichever is greater, unless the director determines that the more restrictive mixing zone or another definition of the mixing zone is more appropriate for a specific discharge.

(d) *No reasonable alternatives* means:

(1) No land-based disposal sites, discharge point(s) within internal waters, or approved ocean dumping sites within a reasonable distance of the site of the proposed discharge the use of which would not cause unwarranted economic impacts on the discharger, or, notwithstanding the availability of such sites,

(2) On-site disposal is environmentally preferable to other alternative means of disposal after consideration of:

(i) The relative environmental harm of disposal on-site, in disposal sites located on land, from discharge point(s) within internal waters, or in approved ocean dumping sites, and

(ii) The risk to the environment and human safety posed by the transportation of the pollutants.

(e) *Unreasonable degradation of the marine environment* means: (1) Significant adverse changes in ecosystem diversity, productivity and stability of the biological community within the area of discharge and surrounding biological communities,

(2) Threat to human health through direct exposure to pollutants or through consumption of exposed aquatic organisms, or

(3) Loss of esthetic, recreational, scientific or economic values which is unreasonable in relation to the benefit derived from the discharge.

§ 125.122 Determination of unreasonable degradation of the marine environment.

(a) The director shall determine whether a discharge will cause unreasonable degradation of the marine environment based on consideration of:

(1) The quantities, composition and potential for bioaccumulation or persistence of the pollutants to be discharged;

(2) The potential transport of such pollutants by biological, physical or chemical processes;

(3) The composition and vulnerability of the biological communities which may be exposed to such pollutants, including the presence of unique species or communities of species, the presence of species identified as endangered or threatened pursuant to the Endangered Species Act, or the presence of those species critical to the structure or function of the ecosystem, such as those important for the food chain;

(4) The importance of the receiving water area to the surrounding biological community, including the presence of spawning sites, nursery/forage areas, migratory pathways, or areas necessary for other functions or critical stages in the life cycle of an organism.

(5) The existence of special aquatic sites including, but not limited to marine sanctuaries and refuges, parks, national and historic monuments, national seashores, wilderness areas and coral reefs;

(6) The potential impacts on human health through direct and indirect pathways;

(7) Existing or potential recreational and commercial fishing, including finfishing and shellfishing;

(8) Any applicable requirements of an approved Coastal Zone Management plan;

(9) Such other factors relating to the effects of the discharge as may be appropriate;

(10) Marine water quality criteria developed pursuant to section 304(a)(1).

(b) Discharges in compliance with section 301(g), 301(h), or 316(a) variance requirements or State water quality standards shall be presumed not to cause unreasonable degradation of the marine environment, for any specific pollutants or conditions specified in the variance or the standard.

§ 125.123 Permit requirements.

(a) If the director on the basis of available information including that supplied by the applicant pursuant to § 125.124 determines prior to permit issuance that the discharge will not cause unreasonable degradation of the marine environment after application of any necessary conditions specified in § 125.123(d), he may issue an NPDES permit containing such conditions.

(b) If the director, on the basis of available information including that supplied by the applicant pursuant to § 125.124 determines prior to permit issuance that the discharge will cause unreasonable degradation of the marine environment after application of all possible permit conditions specified in § 125.123(d), he may not issue an NPDES permit which authorizes the discharge of pollutants.

(c) If the director has insufficient information to determine prior to permit issuance that there will be no unreasonable degradation of the marine environment pursuant to § 125.122, there shall be no discharge of pollutants into the marine environment unless the director on the basis of available information, including that supplied by the applicant pursuant to § 125.124 determines that:

(1) Such discharge will not cause irreparable harm to the marine environment during the period in which monitoring is undertaken, and

(2) There are no reasonable alternatives to the on-site disposal of these materials, and

(3) The discharge will be in compliance with all permit conditions established pursuant to paragraph (d) of this section.

(d) All permits which authorize the discharge of pollutants pursuant to paragraph (c) of this section shall:

(1) Require that a discharge of pollutants will: (i) Following dilution as measured at the boundary of the mixing zone not exceed the limiting permissible concentration for the liquid and suspended particulate phases of the waste material as described in § 227.27(a) (2) and (3), § 227.27(b), and § 227.27(c) of the Ocean Dumping Criteria; and (ii) not exceed the limiting permissible concentration for the solid phase of the waste material or cause an accumulation of toxic materials in the human food chain as described in § 227.27 (b) and (d) of the Ocean Dumping Criteria;

(2) Specify a monitoring program, which is sufficient to assess the impact of the discharge on water, sediment, and biological quality including, where appropriate, analysis of the bioaccumulative and/or persistent impact on aquatic life of the discharge;

(3) Contain any other conditions, such as performance of liquid or suspended particulate phase bioaccumulation tests, seasonal restrictions on discharge, process modifications, dispersion of pollutants, or schedule of compliance for existing discharges, which are determined to be necessary because of local environmental conditions, and

(4) Contain the following clause: In addition to any other grounds specified herein, this permit shall be modified or revoked at any time if, on the basis of any new data, the director determines that continued discharges may cause unreasonable degradation of the marine environment.

§ 125.124 Information required to be submitted by applicant.

The applicant is responsible for providing information which the director may request to make the determination required by this subpart. The director may require the following information as well as any other pertinent information:

(a) An analysis of the chemical constituents of any discharge;

(b) Appropriate bioassays necessary to determine the limiting permissible concentrations for the discharge;

(c) An analysis of initial dilution;

(d) Available process modifications which will reduce the quantities of pollutants which will be discharged;

(e) Analysis of the location where pollutants are sought to be discharged, including the biological community and the physical description of the discharge facility;

(f) Evaluation of available alternatives to the discharge of the pollutants including an evaluation of the possibility of land-based disposal or disposal in an approved ocean dumping site.

Subpart N—Requirements Applicable to Cooling Water Intake Structures for New Offshore Oil and Gas Extraction Facilities Under Section 316(b) of the Act

SOURCE: 71 FR 35040, June 16, 2006, unless otherwise noted.

§ 125.130 What are the purpose and scope of this subpart?

(a) This subpart establishes requirements that apply to the location, design, construction, and capacity of cooling water intake structures at new offshore oil and gas extraction facilities. The purpose of these requirements is to establish the best technology available for minimizing adverse environmental impact associated with the use of cooling water intake structures at these facilities. These requirements are implemented through National Pollutant Discharge Elimination System (NPDES) permits issued under section 402 of the Clean Water Act (CWA).

(b) This subpart implements section 316(b) of the CWA for new offshore oil and gas extraction facilities. Section 316(b) of the CWA provides that any standard established pursuant to sections 301 or 306 of the CWA and applicable to a point source shall require that the location, design, construction, and capacity of cooling water intake structures reflect the best technology available for minimizing adverse environmental impact.

(c) New offshore oil and gas extraction facilities that do not meet the threshold requirements regarding amount of water withdrawn or percentage of water withdrawn for cooling water purposes in § 125.131(a) must meet requirements determined by the Direc-

tor on a case-by-case, best professional judgement (BPJ) basis.

(d) Nothing in this subpart shall be construed to preclude or deny the right of any State or political subdivision of a State or any interstate agency under section 510 of the CWA to adopt or enforce any requirement with respect to control or abatement of pollution that is more stringent than those required by Federal law.

§ 125.131 Who is subject to this subpart?

(a) This subpart applies to a new offshore oil and gas extraction facility if it meets all of the following criteria:

(1) It is a point source that uses or proposes to use a cooling water intake structure;

(2) It has at least one cooling water intake structure that uses at least 25 percent of the water it withdraws for cooling purposes as specified in paragraph (c) of this section; and

(3) It has a design intake flow greater than two (2) million gallons per day (MGD).

(b) Use of a cooling water intake structure includes obtaining cooling water by any sort of contract or arrangement with an independent supplier (or multiple suppliers) of cooling water if the supplier or suppliers withdraw(s) water from waters of the United States. Use of cooling water does not include obtaining cooling water from a public water system or the use of treated effluent that otherwise would be discharged to a water of the U.S.

(c) The threshold requirement that at least 25 percent of water withdrawn be used for cooling purposes must be measured on an average monthly basis. A new offshore oil and gas extraction facility meets the 25 percent cooling water threshold if, based on the new facility's design, any monthly average over a year for the percentage of cooling water withdrawn is expected to equal or exceed 25 percent of the total water withdrawn.

(d) Neither this subpart nor Subpart I of this part applies to seafood processing vessels or offshore liquefied natural gas import terminals that are new facilities as defined in 40 CFR 125.83. Seafood processing vessels and offshore liquefied natural gas import terminals must meet requirements established by the Director on a case-by-case, best professional judgment (BPJ) basis.

§125.132 When must I comply with this subpart?

You must comply with this subpart when an NPDES permit containing requirements consistent with this subpart is issued to you.

§125.133 What special definitions apply to this subpart?

In addition to the definitions set forth at 40 CFR 125.83, the following special definitions apply to this subpart:

Cooling water means water used for contact or noncontact cooling, including water used for equipment cooling, evaporative cooling tower makeup, and dilution of effluent heat content. The intended use of the cooling water is to absorb waste heat rejected from the process or processes used, or from auxiliary operations on the facility's premises. Cooling water that is used in another industrial process either before or after it is used for cooling is considered process water rather than cooling water for the purposes of calculating the percentage of a new offshore oil and gas extraction facility's intake flow that is used for cooling purposes in §125.131(c).

Fixed facility means a bottom founded offshore oil and gas extraction facility permanently attached to the seabed or subsoil of the outer continental shelf (e.g., platforms, guyed towers, articulated gravity platforms) or a buoyant facility securely and substantially moored so that it cannot be moved without a special effort (e.g., tension leg platforms, permanently moored semi-submersibles) and which is not intended to be moved during the production life of the well. This definition does not include mobile offshore drilling units (MODUs) (e.g., drill ships, temporarily moored semi-submersibles, jack-ups, submersibles, tender-assisted rigs, and drill barges).

Minimum ambient source water surface elevation means the mean low tidal water level for estuaries or oceans. The mean low tidal water level is the average height of the low water over at least 19 years.

New offshore oil and gas extraction facility means any building, structure, facility, or installation that: meets the definition of a "new facility" at 40 CFR 125.83; and is regulated by the Offshore or Coastal Subcategories of the Oil and Gas Extraction Point Source Category Effluent Guidelines in 40 CFR 435.10 or 40 CFR 435.40; but only if it commences construction after July 17, 2006.

Offshore liquefied natural gas (LNG) import terminal means any facility located in waters defined in 40 CFR 435.10 or 40 CFR 435.40 that liquefies, regasifies, transfers, or stores liquefied natural gas.

Sea chest means the underwater compartment or cavity within the facility or vessel hull or pontoon through which sea water is drawn in (for cooling and other purposes) or discharged.

Seafood processing vessel means any offshore or nearshore, floating, mobile, facility engaged in the processing of fresh, frozen, canned, smoked, salted or pickled seafood, seafood paste, mince, or meal.

§125.134 As an owner or operator of a new offshore oil and gas extraction facility, what must I do to comply with this subpart?

(a)(1) The owner or operator of a new offshore oil and gas extraction facility must comply with:

(i) Track I in paragraph (b) or Track II in paragraph (c) of this section, if it is a fixed facility; or

(ii) Track I in paragraph (b) of this section, if it is *not* a fixed facility.

(2) In addition to meeting the requirements in paragraph (b) or (c) of this section, the owner or operator of a new offshore oil and gas extraction facility may be required to comply with paragraph (d) of this section.

(b) *Track I requirements for new offshore oil and gas extraction facilities.* (1)(i) New offshore oil and gas extraction facilities that *do not* employ sea chests as cooling water intake structures and are fixed facilities must comply with all of the requirements in paragraphs (b)(2) through (8) of this section.

(ii) New offshore oil and gas extraction facilities that employ sea chests as cooling water intake structures and are fixed facilities must comply with the requirements in paragraphs (b)(2), (3), (4), (6), (7), and (8) of this section.

(iii) New offshore oil and gas extraction facilities that are *not* fixed facilities must comply with the requirements in paragraphs (b)(2), (4), (6), (7), and (8) of this section.

(2) You must design and construct each cooling water intake structure at your facility to a maximum through-screen design intake velocity of 0.5 ft/s;

(3) For cooling water intake structures located in an estuary or tidal river, the total design intake flow over one tidal cycle of ebb and flow must be no greater than one (1) percent of the volume of the water column within the area centered about the opening of the intake with a diameter defined by the distance of one tidal excursion at the mean low water level;

(4) You must select and implement design and construction technologies or operational measures for minimizing impingement mortality of fish and shellfish if the Director determines that:

(i) There are threatened or endangered or otherwise protected federal, state, or tribal species, or critical habitat for these species, within the hydraulic zone of influence of the cooling water intake structure; or

(ii) Based on information submitted by any fishery management agency(ies) or other relevant information, there are migratory and/or sport or commercial species of impingement concern to the Director that pass through the hydraulic zone of influence of the cooling water intake structure; or

(iii) Based on information submitted by any fishery management agency(ies) or other relevant information, that the proposed facility, after meeting the technology-based performance requirements in paragraphs (b)(2) and (5) of this section, would still contribute unacceptable stress to the protected species, critical habitat of those species, or species of concern;

(5) You must select and implement design and construction technologies or operational measures for minimizing entrainment of entrainable life stages of fish and shellfish;

(6) You must submit the applicable application information required in 40 CFR 122.21(r) and § 125.136(b). If you are a fixed facility you must submit the information required in 40 CFR 122.21(r)(2) (except (r)(2)(iv)), (3), and (4) and § 125.136(b) of this subpart as part of your application. If you are a not a fixed facility, you must only submit the information required in 40 CFR 122.21(r)(2)(iv), (r)(3) (except (r)(3)(ii)) and § 125.136(b) as part of your application.

(7) You must implement the monitoring requirements specified in § 125.137; and

(8) You must implement the recordkeeping requirements specified in § 125.138.

(c) *Track II requirements for new offshore oil and gas extraction facilities.* The owner or operator of a new offshore oil and gas extraction facility that is a fixed facility and chooses to comply under Track II must comply with the following requirements:

(1) You must demonstrate to the Director that the technologies employed will reduce the level of adverse environmental impact from your cooling water intake structures to a comparable level to that which you would achieve were you to implement the applicable requirements of paragraph (b)(2) and, if your facility is a fixed facility without a sea chest, also paragraph (b)(5) of this section. This demonstration must include a showing that the impacts to fish and shellfish, including important forage and predator species, will be comparable to those which would result if you were to implement the requirements of paragraph (b)(2) and, if your facility is a fixed facility without a sea chest, also paragraph (b)(5) of this section. In identifying such species, the Director may consider information provided by any fishery management agency(ies) along with data and information from other sources;

(2) For cooling water intake structures located in an estuary or tidal river, the total design intake flow over one tidal cycle of ebb and flow must be no greater than one (1) percent of the volume of the water column within the area centered about the opening of the intake with a diameter defined by the

distance of one tidal excursion at the mean low water level;

(3) You must submit the applicable information required in 40 CFR 122.21(r)(2) (except (r)(2)(iv)), (3) and (4) and §125.136(c);

(4) You must implement the monitoring requirements specified in §125.137;

(5) You must implement the record-keeping requirements specified in §125.138.

(d) You must comply with any more stringent requirements relating to the location, design, construction, and capacity of a cooling water intake structure or monitoring requirements at a new offshore oil and gas extraction facility that the Director deems are reasonably necessary to comply with any provision of federal or state law, including compliance with applicable state water quality standards (including designated uses, criteria, and antidegradation requirements).

§125.135 May alternative requirements be authorized?

(a) Any interested person may request that alternative requirements less stringent than those specified in §125.134(a) through (d) be imposed in the permit. The Director may establish alternative requirements less stringent than the requirements of §125.134(a) through (d) only if:

(1) There is an applicable requirement under §125.134(a) through (d);

(2) The Director determines that data specific to the facility indicate that compliance with the requirement at issue would result in compliance costs wholly out of proportion to the costs EPA considered in establishing the requirement at issue or would result in significant adverse impacts on local water resources other than impingement or entrainment, or significant adverse impacts on energy markets;

(3) The alternative requirement requested is no less stringent than justified by the wholly out of proportion cost or the significant adverse impacts on local water resources other than impingement or entrainment, or significant adverse impacts on energy markets; and

(4) The alternative requirement will ensure compliance with other applicable provisions of the Clean Water Act and any applicable requirement of federal or state law.

(b) The burden is on the person requesting the alternative requirement to demonstrate that alternative requirements should be authorized.

§125.136 As an owner or operator of a new offshore oil and gas extraction facility, what must I collect and submit when I apply for my new or reissued NPDES permit?

(a)(1) As an owner or operator of a new offshore oil and gas extraction facility, you must submit to the Director a statement that you intend to comply with either:

(i) The Track I requirements for new offshore oil and gas extraction facilities in §125.134(b); or

(ii) If you are a fixed facility, you may choose to comply with the Track II requirements in §125.134(c).

(2) You must also submit the application information required by 40 CFR 122.21(r) and the information required in either paragraph (b) of this section for Track I or, if you are a fixed facility that chooses to comply under Track II, paragraph (c) of this section when you apply for a new or reissued NPDES permit in accordance with 40 CFR 122.21.

(b) *Track I application requirements.* To demonstrate compliance with Track I requirements in §125.134(b), you must collect and submit to the Director the information in paragraphs (b)(1) through (3) of this section.

(1) *Velocity information.* You must submit the following information to the Director to demonstrate that you are complying with the requirement to meet a maximum through-screen design intake velocity of no more than 0.5 ft/s at each cooling water intake structure as required in §125.134(b)(2):

(i) A narrative description of the design, structure, equipment, and operation used to meet the velocity requirement; and

(ii) Design calculations showing that the velocity requirement will be met at minimum ambient source water surface elevations (based on best professional judgment using available hydrological data) and maximum head loss across the screens or other device.

(2) *Source waterbody flow information.* If you are a fixed facility and your cooling water intake structure is located in an estuary or tidal river, you must provide the mean low water tidal excursion distance and any supporting documentation and engineering calculations to show that your cooling water intake structure facility meets the flow requirements in § 125.134(b)(3).

(3) *Design and Construction Technology Plan.* To comply with § 125.134(b)(4) and/or (5), if applicable, you must submit to the Director the following information in a Design and Construction Technology Plan:

(i) If the Director determines that additional impingement requirements should be included in your permit:

(A) Information to demonstrate whether or not you meet the criteria in § 125.134(b)(4);

(B) Delineation of the hydraulic zone of influence for your cooling water intake structure;

(ii) New offshore oil and gas extraction facilities required to install design and construction technologies and/or operational measures must develop a plan explaining the technologies and measures you have selected. (Examples of appropriate technologies include, but are not limited to, increased opening to cooling water intake structure to decrease design intake velocity, wedgewire screens, fixed screens, velocity caps, location of cooling water intake opening in waterbody, etc. Examples of appropriate operational measures include, but are not limited to, seasonal shutdowns or reductions in flow, continuous operations of screens, etc.) The plan must contain the following information, if applicable:

(A) A narrative description of the design and operation of the design and construction technologies, including fish-handling and return systems, that you will use to maximize the survival of those species expected to be most susceptible to impingement. Provide species-specific information that demonstrates the efficacy of the technology;

(B) To demonstrate compliance with § 125.134(b)(5), if applicable, a narrative description of the design and operation of the design and construction technologies that you will use to minimize

entrainment of those species expected to be the most susceptible to entrainment. Provide species-specific information that demonstrates the efficacy of the technology; and

(C) Design calculations, drawings, and estimates to support the descriptions provided in paragraphs (b)(3)(ii)(A) and (B) of this section.

(c) *Application requirements for Track II.* If you are a fixed facility and have chosen to comply with the requirements of Track II in § 125.134(c) you must collect and submit the following information:

(1) *Source waterbody flow information.* If your cooling water intake structure is located in an estuary or tidal river, you must provide the mean low water tidal excursion distance and any supporting documentation and engineering calculations to show that your cooling water intake structure facility meets the flow requirements in § 125.134(c)(2);

(2) *Track II Comprehensive Demonstration Study.* You must perform and submit the results of a Comprehensive Demonstration Study (Study). This information is required to characterize the source water baseline in the vicinity of the cooling water intake structure(s), characterize operation of the cooling water intake(s), and to confirm that the technology(ies) proposed and/or implemented at your cooling water intake structure reduce the impacts to fish and shellfish to levels comparable to those you would achieve were you to implement the applicable requirements in § 125.134(b).

(i) To meet the "comparable level" requirement, you must demonstrate that:

(A) You have reduced impingement mortality of all life stages of fish and shellfish to 90 percent or greater of the reduction that would be achieved through the applicable requirements in § 125.134(b)(2); and

(B) If you are a facility without sea chests, you have minimized entrainment of entrainable life stages of fish and shellfish to 90 percent or greater of the reduction that would have been achieved through the applicable requirements in § 125.134(b)(5);

(ii) You must develop and submit a plan to the Director containing a proposal for how information will be collected to support the study. The plan must include:

(A) A description of the proposed and/or implemented technology(ies) to be evaluated in the Study;

(B) A list and description of any historical studies characterizing the physical and biological conditions in the vicinity of the proposed or actual intakes and their relevancy to the proposed Study. If you propose to rely on existing source water body data, it must be no more than 5 years old, you must demonstrate that the existing data are sufficient to develop a scientifically valid estimate of potential impingement mortality and (if applicable) entrainment impacts, and provide documentation showing that the data were collected using appropriate quality assurance/quality control procedures;

(C) Any public participation or consultation with Federal or State agencies undertaken in developing the plan; and

(D) A sampling plan for data that will be collected using actual field studies in the source water body. The sampling plan must document all methods and quality assurance procedures for sampling and data analysis. The sampling and data analysis methods you propose must be appropriate for a quantitative survey and based on consideration of methods used in other studies performed in the source water body. The sampling plan must include a description of the study area (including the area of influence of the cooling water intake structure and at least 100 meters beyond); taxonomic identification of the sampled or evaluated biological assemblages (including all life stages of fish and shellfish); and sampling and data analysis methods; and

(iii) You must submit documentation of the results of the Study to the Director. Documentation of the results of the Study must include:

(A) *Source Water Biological Study*. The Source Water Biological Study must include:

(1) A taxonomic identification and characterization of aquatic biological resources including: A summary of historical and contemporary aquatic biological resources; determination and description of the target populations of concern (those species of fish and shellfish and all life stages that are most susceptible to impingement and entrainment); and a description of the abundance and temporal/spatial characterization of the target populations based on the collection of multiple years of data to capture the seasonal and daily activities (e.g., spawning, feeding and water column migration) of all life stages of fish and shellfish found in the vicinity of the cooling water intake structure;

(2) An identification of all threatened or endangered species that might be susceptible to impingement and entrainment by the proposed cooling water intake structure(s); and

(3) A description of additional chemical, water quality, and other anthropogenic stresses on the source waterbody.

(B) *Evaluation of potential cooling water intake structure effects*. This evaluation must include:

(1) Calculations of the reduction in impingement mortality and, (if applicable), entrainment of all life stages of fish and shellfish that would need to be achieved by the technologies you have selected to implement to meet requirements under Track II. To do this, you must determine the reduction in impingement mortality and entrainment that would be achieved by implementing the requirements of §125.134(b)(2) and, for facilities without sea chests, §125.134(b)(5) of Track I at your site.

(2) An engineering estimate of efficacy for the proposed and/or implemented technologies used to minimize impingement mortality and (if applicable) entrainment of all life stages of fish and shellfish and maximize survival of impinged life stages of fish and shellfish. You must demonstrate that the technologies reduce impingement mortality and (if applicable) entrainment of all life stages of fish and shellfish to a comparable level to that which you would achieve were you to implement the requirements in §125.134(b)(2) and, for facilities without sea chests, §125.134(b)(5) of Track I. The efficacy projection must include a site-specific evaluation of technology(ies)

403

suitability for reducing impingement mortality and (if applicable) entrainment based on the results of the Source Water Biological Study in paragraph (c)(2)(iii)(A) of this section. Efficacy estimates may be determined based on case studies that have been conducted in the vicinity of the cooling water intake structure and/or site-specific technology prototype studies.

(C) *Verification monitoring plan.* You must include in the Study a plan to conduct, at a minimum, two years of monitoring to verify the full-scale performance of the proposed or implemented technologies and/or operational measures. The verification study must begin at the start of operations of the cooling water intake structure and continue for a sufficient period of time to demonstrate that the facility is reducing the level of impingement mortality and (if applicable) entrainment to the level documented in paragraph (c)(2)(iii)(B) of this section. The plan must describe the frequency of monitoring and the parameters to be monitored. The Director will use the verification monitoring to confirm that you are meeting the level of impingement mortality and entrainment reduction required in § 125.134(c), and that the operation of the technology has been optimized.

§ 125.137 As an owner or operator of a new offshore oil and gas extraction facility, must I perform monitoring?

As an owner or operator of a new offshore oil and gas extraction facility, you will be required to perform monitoring to demonstrate your compliance with the requirements specified in § 125.134 or alternative requirements under § 125.135.

(a) *Biological monitoring.* (1)(i) Fixed facilities without sea chests that choose to comply with the Track I requirements in § 125.134(b)(1)(i) must monitor for entrainment. These facilities are not required to monitor for impingement, unless the Director determines that the information would be necessary to evaluate the need for or compliance with additional requirements in accordance with § 125.134(b)(4) or more stringent requirements in accordance with § 125.134(d).

(ii) Fixed facilities with sea chests that choose to comply with Track I requirements are not required to perform biological monitoring unless the Director determines that the information would be necessary to evaluate the need for or compliance with additional requirements in accordance with § 125.134(b)(4) or more stringent requirements in accordance with § 125.134(d).

(iii) Facilities that are not fixed facilities are not required to perform biological monitoring unless the Director determines that the information would be necessary to evaluate the need for or compliance with additional requirements in accordance with § 125.134(b)(4) or more stringent requirements in accordance with § 125.134(d).

(iv) Fixed facilities with sea chests that choose to comply with Track II requirements in accordance with § 125.134(c), must monitor for impingement only. Fixed facilities without sea chests that choose to comply with Track II requirements, must monitor for both impingement and entrainment.

(2) Monitoring must characterize the impingement rates and (if applicable) entrainment rates) of commercial, recreational, and forage base fish and shellfish species identified in the Source Water Baseline Biological Characterization data required by 40 CFR 122.21(r)(4), identified in the Comprehensive Demonstration Study required by § 125.136(c)(2), or as specified by the Director.

(3) The monitoring methods used must be consistent with those used for the Source Water Baseline Biological Characterization data required in 40 CFR 122.21(r)(4), those used by the Comprehensive Demonstration Study required by § 125.136(c)(2), or as specified by the Director. You must follow the monitoring frequencies identified below for at least two (2) years after the initial permit issuance. After that time, the Director may approve a request for less frequent sampling in the remaining years of the permit term and when the permit is reissued, if supporting data show that less frequent monitoring would still allow for the detection of any seasonal variations in the species and numbers of individuals that are impinged or entrained.

(4) *Impingement sampling.* You must collect samples to monitor impingement rates (simple enumeration) for each species over a 24-hour period and no less than once per month when the cooling water intake structure is in operation.

(5) *Entrainment sampling.* If your facility is subject to the requirements of §125.134(b)(1)(i), or if your facility is subject to §125.134(c) and is a fixed facility without a sea chest, you must collect samples to monitor entrainment rates (simple enumeration) for each species over a 24-hour period and no less than biweekly during the primary period of reproduction, larval recruitment, and peak abundance identified during the Source Water Baseline Biological Characterization required by 40 CFR 122.21(r)(4) or the Comprehensive Demonstration Study required in §125.136(c)(2). You must collect samples only when the cooling water intake structure is in operation.

(b) *Velocity monitoring.* If your facility uses a surface intake screen systems, you must monitor head loss across the screens and correlate the measured value with the design intake velocity. The head loss across the intake screen must be measured at the minimum ambient source water surface elevation (best professional judgment based on available hydrological data). The maximum head loss across the screen for each cooling water intake structure must be used to determine compliance with the velocity requirement in §125.134(b)(2). If your facility uses devices other than surface intake screens, you must monitor velocity at the point of entry through the device. You must monitor head loss or velocity during initial facility startup, and thereafter, at the frequency specified in your NPDES permit, but no less than once per quarter.

(c) *Visual or remote inspections.* You must either conduct visual inspections or employ remote monitoring devices during the period the cooling water intake structure is in operation. You must conduct visual inspections at least weekly to ensure that any design and construction technologies required in §125.134(b)(4), (b)(5), (c), and/or (d) are maintained and operated to ensure that they will continue to function as designed. Alternatively, you must inspect via remote monitoring devices to ensure that the impingement and entrainment technologies are functioning as designed.

§ **125.138 As an owner or operator of a new offshore oil and gas extraction facility, must I keep records and report?**

As an owner or operator of a new offshore oil and gas extraction facility you are required to keep records and report information and data to the Director as follows:

(a) You must keep records of all the data used to complete the permit application and show compliance with the requirements, any supplemental information developed under §125.136, and any compliance monitoring data submitted under §125.137, for a period of at least three (3) years from the date of permit issuance. The Director may require that these records be kept for a longer period.

(b) You must provide the following to the Director in a yearly status report:

(1) For fixed facilities, biological monitoring records for each cooling water intake structure as required by §125.137(a);

(2) Velocity and head loss monitoring records for each cooling water intake structure as required by §125.137(b); and

(3) Records of visual or remote inspections as required in §125.137(c).

§ **125.139 As the Director, what must I do to comply with the requirements of this subpart?**

(a) *Permit application.* As the Director, you must review materials submitted by the applicant under 40 CFR 122.21(r), §125.135, and §125.136 at the time of the initial permit application and before each permit renewal or reissuance.

(1) After receiving the initial permit application from the owner or operator of a new offshore oil and gas extraction facility, the Director must determine applicable standards in §125.134 or §125.135 to apply to the new offshore oil and gas extraction facility. In addition, the Director must review materials to determine compliance with the applicable standards.

(2) For each subsequent permit renewal, the Director must review the application materials and monitoring data to determine whether requirements, or additional requirements, for design and construction technologies or operational measures should be included in the permit.

(3) For Track II facilities, the Director may review the information collection proposal plan required by § 125.136(c)(2)(ii). The facility may initiate sampling and data collection activities prior to receiving comment from the Director.

(b) *Permitting requirements.* Section 316(b) requirements are implemented for a facility through an NPDES permit. As the Director, you must determine, based on the information submitted by the new offshore oil and gas extraction facility in its permit application, the appropriate requirements and conditions to include in the permit based on the track (Track I or Track II), or alternative requirements in accordance with § 125.135, the new offshore oil and gas extraction facility has chosen to comply with. The following requirements must be included in each permit:

(1) *Cooling water intake structure requirements.* At a minimum, the permit conditions must include the performance standards that implement the applicable requirements of § 125.134(b)(2), (3), (4) and (5); § 125.134(c)(1) and (2); or § 125.135.

(i) For a facility that chooses Track I, you must review the Design and Construction Technology Plan required in § 125.136(b)(3) to evaluate the suitability and feasibility of the technology proposed to minimize impingement mortality and (if applicable) entrainment of all life stages of fish and shellfish. In the first permit issued, you must include a condition requiring the facility to reduce impingement mortality and/ or entrainment commensurate with the implementation of the technologies in the permit. Under subsequent permits, the Director must review the performance of the technologies implemented and require additional or different design and construction technologies, if needed to minimize impingement mortality and/or entrainment of all life stages of fish and shellfish. In addition,

you must consider whether more stringent conditions are reasonably necessary in accordance with § 125.134(d).

(ii) For a fixed facility that chooses Track II, you must review the information submitted with the Comprehensive Demonstration Study information required in § 125.136(c)(2), evaluate the suitability of the proposed design and construction technology and/or operational measures to determine whether they will reduce both impingement mortality and/or entrainment of all life stages of fish and shellfish to 90 percent or greater of the reduction that could be achieved through Track I. In addition, you must review the Verification Monitoring Plan in § 125.136(c)(2)(iii)(C) and require that the proposed monitoring begin at the start of operations of the cooling water intake structure and continue for a sufficient period of time to demonstrate that the technologies and operational measures meet the requirements in § 125.134(c)(1). Under subsequent permits, the Director must review the performance of the additional and /or different technologies or measures used and determine that they reduce the level of adverse environmental impact from the cooling water intake structures to a comparable level that the facility would achieve were it to implement the requirements of § 125.134(b)(2) and, if applicable, § 125.134(b)(5).

(iii) If a facility requests alternative requirements in accordance with § 125.135, you must determine if data specific to the facility meet the requirements in § 125.135(a) and include in the permit requirements that are no less stringent than justified by the wholly out of proportion cost or the significant adverse impacts on local water resources other than impingement or entrainment, or significant adverse impacts on energy markets.

(2) *Monitoring conditions.* At a minimum, the permit must require the permittee to perform the monitoring required in § 125.137. You may modify the monitoring program when the permit is reissued and during the term of the permit based on changes in physical or biological conditions in the vicinity of the cooling water intake structure. The Director may require continued

monitoring based on the results of monitoring done pursuant to the Verification Monitoring Plan in §125.136(c)(2)(iii)(C).

(3) *Record keeping and reporting.* At a minimum, the permit must require the permittee to report and keep records as required by §125.138.

PART 129—TOXIC POLLUTANT EFFLUENT STANDARDS

Subpart A—Toxic Pollutant Effluent Standards and Prohibitions

Sec.
129.1 Scope and purpose.
129.2 Definitions.
129.3 Abbreviations.
129.4 Toxic pollutants.
129.5 Compliance.
129.6 Adjustment of effluent standard for presence of toxic pollutant in the intake water.
129.7 Requirement and procedure for establishing a more stringent effluent limitation.
129.8 Compliance date.
129.9–129.99 [Reserved]
129.100 Aldrin/dieldrin.
129.101 DDT, DDD and DDE.
129.102 Endrin.
129.103 Toxaphene.
129.104 Benzidine.
129.105 Polychlorinated biphenyls (PCBs).

AUTHORITY: Secs. 307, 308, 501, Federal Water Pollution Control Act Amendments of 1972 (Pub. L. 92–500, 86 Stat. 816, (33 U.S.C. 1251 *et seq.*)).

SOURCE: 42 FR 2613, Jan. 12, 1977, unless otherwise noted.

Subpart A—Toxic Pollutant Effluent Standards and Prohibitions

§ 129.1 Scope and purpose.

(a) The provisions of this subpart apply to owners or operators of specified facilities discharging into navigable waters.

(b) The effluent standards or prohibitions for toxic pollutants established in this subpart shall be applicable to the sources and pollutants hereinafter set forth, and may be incorporated in any NPDES permit, modification or renewal thereof, in accordance with the provisions of this subpart.

(c) The provisions of 40 CFR parts 124 and 125 shall apply to any NPDES permit proceedings for any point source discharge containing any toxic pollutant for which a standard or prohibition is established under this part.

§ 129.2 Definitions.

All terms not defined herein shall have the meaning given them in the Act or in 40 CFR part 124 or 125. As used in this part, the term:

(a) *Act* means the Federal Water Pollution Control Act, as amended (Pub. L. 92–500, 86 Stat. 816 *et seq.*, 33 U.S.C. 1251 *et seq.*). Specific references to sections within the Act will be according to Pub. L. 92–500 notation.

(b) *Administrator* means the Administrator of the Environmental Protection Agency or any employee of the Agency to whom the Administrator may by order delegate the authority to carry out his functions under section 307(a) of the Act, or any person who shall by operation of law be authorized to carry out such functions.

(c) *Effluent standard* means, for purposes of section 307, the equivalent of *effluent limitation* as that term is defined in section 502(11) of the Act with the exception that it does not include a schedule of compliance.

(d) *Prohibited* means that the constituent shall be absent in any discharge subject to these standards, as determined by any analytical method.

(e) *Permit* means a permit for the discharge of pollutants into navigable waters under the National Pollutant Discharge Elimination System established by section 402 of the Act and implemented in regulations in 40 CFR parts 124 and 125.

(f) *Working day* means the hours during a calendar day in which a facility discharges effluents subject to this part.

(g) *Ambient water criterion* means that concentration of a toxic pollutant in a navigable water that, based upon available data, will not result in adverse impact on important aquatic life, or on consumers of such aquatic life, after exposure of that aquatic life for periods of time exceeding 96 hours and continuing at least through one reproductive cycle; and will not result in a significant risk of adverse health effects in a large human population based on available information such as mammalian laboratory toxicity data, epidemiological studies of human occupational exposures, or human exposure data, or any other relevant data.

(h) *New source* means any source discharging a toxic pollutant, the construction of which is commenced after proposal of an effluent standard or prohibition applicable to such source if such effluent standard or prohibition is thereafter promulgated in accordance with section 307.

(i) *Existing source* means any source which is not a new source as defined above.

(j) *Source* means any building, structure, facility, or installation from which there is or may be the discharge of toxic pollutants designated as such by the Administration under section 307(a)(1) of the Act.

(k) *Owner or operator* means any person who owns, leases, operates, controls, or supervises a source as defined above.

(l) *Construction* means any placement, assembly, or installation of facilities or equipment (including contractual obligations to purchase such facilities or equipment) at the premises where such equipment will be used, including preparation work at such premises.

(m) *Manufacturer* means any establishment engaged in the mechanical or chemical transformation of materials or substances into new products including but not limited to the blending of materials such as pesticidal products, resins, or liquors.

(n) *Process wastes* means any designated toxic pollutant, whether in wastewater or otherwise present, which is inherent to or unavoidably resulting from any manufacturing process, including that which comes into direct contact with or results from the production or use of any raw material, intermediate product, finished product, by-product or waste product and is discharged into the navigable waters.

(o) *Air emissions* means the release or discharge of a toxic pollutant by an owner or operator into the ambient air either (1) by means of a stack or (2) as a fugitive dust, mist or vapor as a result inherent to the manufacturing or formulating process.

(p) *Fugitive dust, mist or vapor* means dust, mist or vapor containing a toxic pollutant regulated under this part which is emitted from any source other than through a stack.

(q) *Stack* means any chimney, flue, conduit, or duct arranged to conduct emissions to the ambient air.

(r) *Ten year 24-hour rainfall event* means the maximum precipitation event with a probable recurrence interval of once in 10 years as defined by the National Weather Service in Technical Paper No. 40, *Rainfall Frequency Atlas of the United States*, May 1961, and subsequent amendments or equivalent regional or State rainfall probability information developed therefrom.

(s) *State Director* means the chief administrative officer of a State or interstate water pollution control agency operating an approved HPDES permit program. In the event responsibility for water pollution control and enforcement is divided among two or more State or interstate agencies, the term *State Director* means the administrative officer authorized to perform the particular procedure to which reference is made.

§ 129.3 Abbreviations.

The abbreviations used in this part represent the following terms:

lb=pound (or pounds)
g=gram
µg/l=micrograms per liter (1 one-millionth gram/liter)
kg=kilogram(s)
kkg=1000 kilogram(s)

§ 129.4 Toxic pollutants.

The following are the pollutants subject to regulation under the provisions of this subpart:

(a) Aldrin/Dieldrin—*Aldrin* means the compound aldrin as identified by the chemical name, 1,2,3,4,10,10-hexachloro-1,4,4a,5,8,8a-hexahydro-1,4 -endo-5,8-exo-dimethanonaphthalene; "Dieldrin" means the compound the dieldrin as identified by the chemical name 1,2,3,4,10,10-hexachloro-6,7-epoxy-1,4,4a,5,6,7,8,8a-octahydro-1,4-endo-5,8-exo-dimethanonaphthalene.

(b) DDT—*DDT* means the compounds DDT, DDD, and DDE as identified by the chemical names:(DDT)-1,1,1-

trichloro-2,2-bis(p-chlorophenyl) ethane and someo,p'-isomers; (DDD) or (TDE)-1,1-dichloro-2,2-bis(p-chlorophenyl) ethane and some o,p'-isomers; (DDE)-1,1-dichloro-2,2-bis(p-chlorophenyl) ethylene.

(c) Endrin—*Endrin* means the compound endrin as identified by the chemical name 1,2,3,4,10,10-hexachloro-6,7-epoxy-1,4,4a,5,6,7,8,8a-octahydro-1,4-endo-5,8-endodimethanonaphthalene.

(d) Toxaphene—*Toxaphene* means a material consisting of technical grade chlorinated camphene having the approximate formula of $C_{10} H_{10} Cl_8$ and normally containing 67–69 percent chlorine by weight.

(e) Benzidine—*Benzidine* means the compound benzidine and its salts as identified by the chemical name 4,4'-diaminobiphenyl.

(f) Polychlorinated Biphenyls (PCBs) *polychlorinated biphenyls* (PCBs) means a mixture of compounds composed of the biphenyl molecule which has been chlorinated to varying degrees.

[42 FR 2613, Jan. 12, 1977, as amended at 42 FR 2620, Jan. 12, 1977; 42 FR 6555, Feb. 2, 1977]

§129.5 Compliance.

(a)(1) Within 60 days from the date of promulgation of any toxic pollutant effluent standard or prohibition each owner or operator with a discharge subject to that standard or prohibition must notify the Regional Administrator (or State Director, if appropriate) of such discharge. Such notification shall include such information and follow such procedures as the Regional Administrator (or State Director, if appropriate) may require.

(2) Any owner or operator who does not have a discharge subject to any toxic pollutant effluent standard at the time of such promulgation but who thereafter commences or intends to commence any activity which would result in such a discharge shall first notify the Regional Administrator (or State Director, if appropriate) in the manner herein provided at least 60 days prior to any such discharge.

(b) Upon receipt of any application for issuance or reissuance of a permit or for a modification of an existing permit for a discharge subject to a toxic pollutant effluent standard or prohibition the permitting authority shall proceed thereon in accordance with 40 CFR part 124 or 125, whichever is applicable.

(c)(1) Every permit which contains limitations based upon a toxic pollutant effluent standard or prohibition under this part is subject to revision following the completion of any proceeding revising such toxic pollutant effluent standard or prohibition regardless of the duration specified on the permit.

(2) For purposes of this section, all toxic pollutants for which standards are set under this part are deemed to be injurious to human health within the meaning of section 402(k) of the Act unless otherwise specified in the standard established for any particular pollutant.

(d)(1) Upon the compliance date for any section 307(a) toxic pollutant effluent standard or prohibition, each owner or operator of a discharge subject to such standard or prohibition shall comply with such monitoring, sampling, recording, and reporting conditions as the Regional Administrator (or State Director, if appropriate) may require for that discharge. Notice of such conditions shall be provided in writing to the owner or operator.

(2) In addition to any conditions required pursuant to paragraph (d)(1) of this section and to the extent not required in conditions contained in NPDES permits, within 60 days following the close of each calendar year each owner or operator of a discharge subject to any toxic standard or prohibition shall report to the Regional Administrator (or State Director, if appropriate) concerning the compliance of such discharges. Such report shall include, as a minimum, information concerning (i) relevant identification of the discharger such as name, location of facility, discharge points, receiving waters, and the industrial process or operation emitting the toxic pollutant; (ii) relevant conditions (pursuant to paragraph (d)(1) of this section or to an NPDES permit) as to flow, section 307(a) toxic pollutant concentrations, and section 307(a) toxic pollutant mass emission rate; (iii) compliance by the discharger with such conditions.

(3) When samples collected for analysis are composited, such samples shall

be composited in proportion to the flow at time of collection and preserved in compliance with requirements of the Regional Administrator (or State Director, if appropriate), but shall include at least five samples, collected at approximately equal intervals throughout the working day.

(e)(1) Nothing in these regulations shall preclude a Regional Administrator from requiring in any permit a more stringent effluent limitation or standard pursuant to section 301(b)(1)(C) of the Act and implemented in 40 CFR 125.11 and other related provisions of 40 CFR part 125.

(2) Nothing in these regulations shall preclude the Director of a State Water Pollution Control Agency or interstate agency operating a National Pollutant Discharge Elimination System Program which has been approved by the Administrator pursuant to section 402 of the Act from requiring in any permit a more stringent effluent limitation or standard pursuant to section 301(b)(1)(C) of the Act and implemented in 40 CFR 124.42 and other related provisions of 40 CFR part 124.

(f) Any owner or operator of a facility which discharges a toxic pollutant to the navigable waters and to a publicly owned treatment system shall limit the summation of the mass emissions from both discharges to the less restrictive standard, either the direct discharge standard or the pretreatment standard; but in no case will this paragraph allow a discharge to the navigable waters greater than the toxic pollutant effluent standard established for a direct discharge to the navigable waters.

(g) In any permit hearing or other administrative proceeding relating to the implementation or enforcement of these standards, or any modification thereof, or in any judicial proceeding other than a petition for review of these standards pursuant to section 509(b)(1)(C) of the Act, the parties thereto may not contest the validity of any national standards established in this part, or the ambient water criterion established herein for any toxic pollutant.

§ 129.6 Adjustment of effluent standard for presence of toxic pollutant in the intake water.

(a) Upon the request of the owner or operator of a facility discharging a pollutant subject to a toxic pollutant effluent standard or prohibition, the Regional Administrator (or State Director, if appropriate) shall give credit, and shall adjust the effluent standard(s) in such permit to reflect credit for the toxic pollutant(s) in the owner's or operator's water supply if (1) the source of the owner's or operator's water supply is the same body of water into which the discharge is made and if (2) it is demonstrated to the Regional Administrator (or State Director, if appropriate) that the toxic pollutant(s) present in the owner's or operator's intake water will not be removed by any wastewater treatment systems whose design capacity and operation were such as to reduce toxic pollutants to the levels required by the applicable toxic pollutant effluent standards in the absence of the toxic pollutant in the intake water.

(b) Effluent limitations established pursuant to this section shall be calculated on the basis of the amount of section 307(a) toxic pollutant(s) present in the water after any water supply treatment steps have been performed by or for the owner or operator.

(c) Any permit which includes toxic pollutant effluent limitations established pursuant to this section shall also contain conditions requiring the permittee to conduct additional monitoring in the manner and locations determined by the Regional Administrator (or State Director, if appropriate) for those toxic pollutants for which the toxic pollutant effluent standards have been adjusted.

§ 129.7 Requirement and procedure for establishing a more stringent effluent limitation.

(a) *In exceptional cases:* (1) Where the Regional Administrator (or State Director, if appropriate) determines that the ambient water criterion established in these standards is not being met or will not be met in the receiving water as a result of one or more discharges at levels allowed by these standards, and

410

(2) Where he further determines that this is resulting in or may cause or contribute to significant adverse effects on aquatic or other organisms usually or potentially present, or on human health, he may issue to an owner or operator a permit or a permit modification containing a toxic pollutant effluent limitation at a more stringent level than that required by the standard set forth in these regulations. Any such action shall be taken pursuant to the procedural provisions of 40 CFR parts 124 and 125, as appropriate. In any proceeding in connection with such action the burden of proof and of going forward with evidence with regard to such more stringent effluent limitation shall be upon the Regional Administrator (or State Director, if appropriate) as the proponent of such more stringent effluent limitation.

(3) Evidence in such proceeding shall include at a minimum: An analysis using data and other information to demonstrate receiving water concentrations of the specified toxic pollutant, projections of the anticipated effects of the proposed modification on such receiving water concentrations, and the hydrologic and hydrographic characteristics of the receiving waters including the occurrence of dispersion of the effluent. Detailed specifications for presenting relevant information by any interested party may be prescribed in guidance documents published from time to time, whose availability will be announced in the FEDERAL REGISTER.

(b) Any effluent limitation in an NPDES permit which a State proposes to issue which is more stringent than the toxic pollutant effluent standards promulgated by the Administrator is subject to review by the Administrator under section 402(d) of the Act. The Administrator may approve or disapprove such limitation(s) or specify another limitation(s) upon review of any record of any proceedings held in connection with the permit issuance or modification and any other evidence available to him. If he takes no action within ninety days of his receipt of the notification of the action of the permit issuing authority and any record thereof, the action of the State permit issuing authority shall be deemed to be approved.

§129.8 Compliance date.

(a) The effluent standards or prohibitions set forth herein shall be complied with not later than one year after promulgation unless an earlier date is established by the Administrator for an industrial subcategory in the promulgation of the standards or prohibitions.

(b) Toxic pollutant effluent standards or prohibitions set forth herein shall become enforceable under sections 307(d) and 309 of the Act on the date established in paragraph (a) of this section regardless of proceedings in connection with the issuance of any NPDES permit or application therefor, or modification or renewal thereof.

§§129.9–129.99 [Reserved]

§129.100 Aldrin/dieldrin.

(a) *Specialized definitions.* (1) *Aldrin/ Dieldrin manufacturer* means a manufacturer, excluding any source which is exclusively an aldrin/dieldrin formulator, who produces, prepares or processes technical aldrin or dieldrin or who uses aldrin or dieldrin as a material in the production, preparation or processing of another synthetic organic substance.

(2) *Aldrin/Dieldrin formulator* means a person who produces, prepares or processes a formulated product comprising a mixture of either aldrin or dieldrin and inert materials or other diluents, into a product intended for application in any use registered under the Federal Insecticide, Fungicide and Rodenticide Act, as amended (7 U.S.C. 135, *et seq.*).

(3) The ambient water criterion for aldrin/dieldrin in navigable waters is 0.003 μg/l.

(b) *Aldrin/dieldrin manufacturer—(1) Applicability.* (i) These standards or prohibitions apply to:

(A) All discharges of process wastes; and

(B) All discharges from the manufacturing areas, loading and unloading areas, storage areas and other areas which are subject to direct contamination by aldrin/dieldrin as a result of the manufacturing process, including but not limited to:

(*1*) Stormwater and other runoff except as hereinafter provided in paragraph (b)(1)(ii) of this section; and

411

(2) Water used for routine cleanup or cleanup of spills.

(ii) These standards do not apply to stormwater runoff or other discharges from areas subject to contamination solely by fallout from air emissions of aldrin/dieldrin; or to stormwater runoff that exceeds that from the ten year 24-hour rainfall event.

(2) *Analytical method acceptable.* Environmental Protection Agency method specified in 40 CFR part 136, except that a 1-liter sample size is required to increase the analytical sensitivity.

(3) *Effluent standard*—(i) *Existing sources.* Aldrin or dieldrin is prohibited in any discharge from any aldrin/dieldrin manufacturer.

(ii) *New Sources.* Aldrin or dieldrin is prohibited in any discharge from any aldrin/dieldrin manufacturer.

(c) *Aldrin/dieldrin formulator*—(1) *Applicability.* (i) These standards or prohibitions apply to:

(A) All discharges of process wastes; and

(B) All discharges from the formulating areas, loading and unloading areas, storage areas and other areas which are subject to direct contamination by aldrin/dieldrin as a result of the formulating process, including but not limited to:

(*1*) Stormwater and other runoff except as hereinafter provided in paragraph (c)(1)(ii) of this section; and

(*2*) Water used for routine cleanup or cleanup of spills.

(ii) These standards do not apply to stormwater runoff or other discharges from areas subject to contamination solely by fallout from air emissions of aldrin/dieldrin; or to stormwater runoff that exceeds that from the ten year 24-hour rainfall event.

(2) *Analytical method acceptable.* Environmental Protection Agency method specified in 40 CFR part 136, except that a 1-liter sample size is required to increase the analytical sensitivity.

(3) *Effluent standard*—(i) *Existing sources.* Aldrin or dieldrin is prohibited in any discharge from any aldrin/dieldrin formulator.

(ii) *New sources.* Aldrin or dieldrin is prohibited in any discharge from any aldrin/dieldrin formulator.

§ 129.101 DDT, DDD and DDE.

(a) *Specialized definitions.* (1) *DDT Manufacturer* means a manufacturer, excluding any source which is exclusively a DDT formulator, who produces, prepares or processes technical DDT, or who uses DDT as a material in the production, preparation or processing of another synthetic organic substance.

(2) *DDT formulator* means a person who produces, prepares or processes a formulated product comprising a mixture of DDT and inert materials or other diluents into a product intended for application in any use registered under the Federal Insecticide, Fungicide and Rodenticide Act, as amended (7 U.S.C. 135, et seq.).

(3) The ambient water criterion for DDT in navigable waters is 0.001 µg/l.

(b) *DDT manufacturer*—(1) *Applicability.* (i) These standards or prohibitions apply to:

(A) All discharges of process wastes; and

(B) All discharges from the manufacturing areas, loading and unloading areas, storage areas and other areas which are subject to direct contamination by DDT as a result of the manufacturing process, including but not limited to:

(*1*) Stormwater and other runoff except as hereinafter provided in paragraph (b)(1)(ii) of this section; and

(*2*) Water used for routine cleanup or cleanup of spills.

(ii) These standards do not apply to stormwater runoff or other discharges from areas subject to contamination solely by fallout from air emissions of DDT; or to stormwater runoff that exceeds that from the ten year 24-hour rainfall event.

(2) *Analytical method acceptable.* Environmental Protection Agency method specified in 40 CFR part 136, except that a 1-liter sample size is required to increase the analytical sensitivity.

(3) *Effluent standard*—(i) *Existing sources.* DDT is prohibited in any discharge from any DDT manufacturer.

(ii) *New sources.* DDT is prohibited in any discharge from any DDT manufacturer.

(c) *DDT formulator*—(1) *Applicability.* (i) These standards or prohibitions apply to:

(A) All discharges of process wastes; and

(B) All discharges from the formulating areas, loading and unloading areas, storage areas and other areas which are subject to direct contamination by DDT as a result of the formulating process, including but not limited to:

(1) Stormwater and other runoff except as hereinafter provided in paragraph (c)(1)(ii) of this section; and

(2) Water used for routine cleanup or cleanup of spills.

(ii) These standards do not apply to stormwater runoff or other discharges from areas subject to contamination solely by fallout from air emissions of DDT; or to stormwater runoff that exceeds that from the ten year 24-hour rainfall event.

(2) *Analytical method acceptable.* Environmental Protection Agency method specified in 40 CFR part 136, except that a 1-liter sample size is required to increase the analytical sensitivity.

(3) *Effluent standard*—(i) *Existing sources.* DDT is prohibited in any discharge from any DDT formulator.

(ii) *New Sources.* DDT is prohibited in any discharge from any DDT formulator.

§129.102 Endrin.

(a) *Specialized definitions.* (1) *Endrin Manufacturer* means a manufacturer, excluding any source which is exclusively an endrin formulator, who produces, prepares or processes technical endrin or who uses endrin as a material in the production, preparation or processing of another synthetic organic substance.

(2) *Endrin Formulator* means a person who produces, prepares or processes a formulated product comprising a mixture of endrin and inert materials or other diluents into a product intended for application in any use registered under the Federal Insecticide, Fungicide and Rodenticide Act, as amended (7 U.S.C. 135 *et seq.*).

(3) The ambient water criterion for endrin in navigable waters is 0.004 µg/l.

(b) *Endrin manufacturer*—(1) *Applicability.* (i) These standards or prohibitions apply to:

(A) All discharges of process wastes; and

(B) All discharges from the manufacturing areas, loading and unloading areas, storage areas and other areas which are subject to direct contamination by endrin as a result of the manufacturing process, including but not limited to:

(1) Stormwater and other runoff except as hereinafter provided in paragraph (b)(1)(ii) of this section; and

(2) Water used for routine cleanup or cleanup of spills.

(ii) These standards do not apply to stormwater runoff or other discharges from areas subject to contamination solely by fallout from air emissions of endrin; or to stormwater runoff that exceeds that from the ten year 24-hour rainfall event.

(2) *Analytical method acceptable*—Environmental Protection Agency method specified in 40 CFR part 136.

(3) *Effluent standard*—(i) *Existing sources.* Discharges from an endrin manufacturer shall not contain endrin concentrations exceeding an average per working day of 1.5 µg/l calculated over any calendar month; and shall not exceed a monthly average daily loading of 0.0006 kg/kkg of endrin produced; and shall not exceed 7.5 µg/l in a sample(s) representing any working day.

(ii) *New sources.* Discharges from an endrin manufacturer shall not contain endrin concentrations exceeding an average per working day of 0.1 µg/l calculated over any calendar month; and shall not exceed a monthly average daily loading of 0.00004 kg/kkg of endrin produced; and shall not exceed 0.5 µg/l in a sample(s) representing any working day.

(iii) *Mass emission standard during shutdown of production.* In computing the allowable monthly average daily loading figure required under the preceding paragraphs (b)(3) (i) and (ii) of this section, for any calendar month for which there is no endrin being manufactured at any plant or facility which normally contributes to the discharge which is subject to these standards, the applicable production value shall be deemed to be the average monthly production level for the most recent preceding 360 days of actual operation of the plant or facility.

(c) *Endrin formulator*—(1) *Applicability*. (i) These standards or prohibitions apply to:

(A) All discharges of process wastes; and

(B) All discharges from the formulating areas, loading and unloading areas, storage areas and other areas which are subject to direct contamination by endrin as a result of the formulating process, including but not limited to: (*1*) Stormwater and other runoff except as hereinafter provided in paragraph (c)(1)(ii) of this section; and (*2*) water used for routine cleanup or cleanup of spills.

(ii) These standards do not apply to stormwater runoff or other discharges from areas subject to contamination solely by fallout from air emissions of endrin; or to storm-water runoff that exceeds that from the ten year 24-hour rainfall event.

(2) *Analytical method acceptable*—Environmental Protection Agency method specified in 40 CFR part 136, except that a 1-liter sample size is required to increase the analytical sensitivity.

(3) *Effluent standard*—(i) *Existing sources*. Endrin is prohibited in any discharge from any endrin formulator.

(ii) New sources—Endrin is prohibited in any discharge from any endrin formulator.

(d) The standards set forth in this section shall apply to the total combined weight or concentration of endrin, excluding any associated element or compound.

§ 129.103 Toxaphene.

(a) *Specialized definitions.* (1) *Toxaphene manufacturer* means a manufacturer, excluding any source which is exclusively a toxaphene formulator, who produces, prepares or processes toxaphene or who uses toxaphene as a material in the production, preparation or processing of another synthetic organic substance.

(2) *Toxaphene formulator* means a person who produces, prepares or processes a formulated product comprising a mixture of toxaphene and inert materials or other diluents into a product intended for application in any use registered under the Federal Insecticide, Fungicide and Rodenticide Act, as amended (7 U.S.C. 135, *et seq.*).

(3) The ambient water criterion for toxaphene in navigable waters is 0.005 µg/l.

(b) *Toxaphene manufacturer*—(1) *Applicability*. (i) These standards or prohibitions apply to:

(A) All discharges of process wastes; and

(B) All discharges from the manufacturing areas, loading and unloading areas, storage areas and other areas which are subject to direct contamination by toxaphene as a result of the manufacturing process, including but not limited to: (*1*) Stormwater and other runoff except as hereinafter provided in paragraph (b)(1)(ii) of this section; and (*2*) water used for routine cleanup or cleanup of spills.

(ii) These standards do not apply to stormwater runoff or other discharges from areas subject to contamination solely by fallout from air emissions of toxaphene; or to stormwater runoff that exceeds that from the ten year 24-hour rainfall event.

(2) *Analytical method acceptable*—Environmental Protection Agency method specified in 40 CFR part 136.

(3) *Effluent standard*—(i) *Existing sources*. Discharges from a toxaphene manufacturer shall not contain toxaphene concentrations exceeding an average per working day of 1.5 µg/l calculated over any calendar month; and shall not exceed a monthly average daily loading of 0.00003 kg/kkg of toxaphene produced, and shall not exceed 7.5 µg/l in a sample(s) representing any working day.

(ii) *New sources*. Discharges from a toxaphene manufacturer shall not contain toxaphene concentrations exceeding an average per working day of 0.1 µg/l calculated over any calendar month; and shall not exceed a monthly average daily loading of 0.000002 kg/kkg of toxaphene produced, and shall not exceed 0.5 µ/l in a sample(s) representing any working day.

(iii) *Mass emission during shutdown of production*. In computing the allowable monthly average daily loading figure required under the preceding paragraphs (b)(3)(i) and (ii) of this section, for any calendar month for which there is no toxaphene being manufactured at any plant or facility which normally contributes to the discharge which is subject to these standards, the applicable production value shall be deemed to be the average monthly production level for the most recent preceding 360 days of actual operation of the plant or facility.

(c) *Toxaphene formulator*—(1) *Applicability.* (i) These standards or prohibitions apply to:

(A) All discharges of process wastes; and

(B) All discharges from the formulating areas, loading and unloading areas, storage areas and other areas which are subject to direct contamination by toxaphene as a result of the formulating process, including but not limited to: (*1*) Stormwater and other runoff except as hereinafter provided in paragraph (c)(1)(ii) of this section; and (*2*) water used for routine cleanup or cleanup of spills.

(ii) These standards do not apply to stormwater runoff or other discharges from areas subject to contamination solely by fallout from air emissions of toxaphene; or to stormwater runoff that exceeds that from the ten year 24-hour rainfall event.

(2) *Analytical method acceptable*—Environmental Protection Agency method specified in 40 CFR part 136, except that a 1-liter sample size is required to increase the analytical sensitivity.

(3) *Effluent standards*—(i) *Existing sources.* Toxaphene is prohibited in any discharge from any toxaphene formulator.

(ii) *New sources.* Toxaphene is prohibited in any discharge from any toxaphene formulator.

(d) The standards set forth in this section shall apply to the total combined weight or concentration of toxaphene, excluding any associated element or compound.

§129.104 Benzidine.

(a) *Specialized definitions.* (1) *Benzidine Manufacturer* means a manufacturer who produces benzidine or who produces benzidine as an intermediate product in the manufacture of dyes commonly used for textile, leather and paper dyeing.

(2) *Benzidine-Based Dye Applicator* means an owner or operator who uses benzidine-based dyes in the dyeing of textiles, leather or paper.

(3) The ambient water criterion for benzidine in navigable waters is 0.1 µg/l.

(b) *Benzidine manufacturer*—(1) *Applicability.* (i) These standards apply to:

(A) All discharges into the navigable waters of process wastes, and

(B) All discharges into the navigable waters of wastes containing benzidine from the manufacturing areas, loading and unloading areas, storage areas, and other areas subject to direct contamination by benzidine or benzidine-containing product as a result of the manufacturing process, including but not limited to:

(*1*) Stormwater and other runoff except as hereinafter provided in paragraph (b)(1)(ii) of this section, and

(*2*) Water used for routine cleanup or cleanup of spills.

(ii) These standards do not apply to stormwater runoff or other discharges from areas subject to contamination solely by fallout from air emissions of benzidine; or to stormwater runoff that exceeds that from the ten year 24-hour rainfall event.

(2) *Analytical method acceptable*—Environmental Protection Agency method specified in 40 CFR part 136.

(3) *Effluent standards*—(i) *Existing sources.* Discharges from a benzidine manufacturer shall not contain benzidine concentrations exceeding an average per working day of 10 µg/l calculated over any calendar month, and shall not exceed a monthly average daily loading of 0.130 kg/kkg of benzidine produced, and shall not exceed 50 µg/l in a sample(s) representing any working day.

(ii) *New sources.* Discharges from a benzidine manufacturer shall not contain benzidine concentrations exceeding an average per working day of 10 µg/l calculated over any calendar month, and shall not exceed a monthly average daily loading of 0.130 kg/kkg of benzidine produced, and shall not exceed 50 µg/l in a sample(s) representing any working day.

(4) The standards set forth in this paragraph (b) shall apply to the total combined weight or concentration of benzidine, excluding any associated element or compound.

(c) *Benzidine-based dye applicators*—(1) *Applicability.* (i) These standards apply to:

(A) All discharges into the navigable waters of process wastes, and

(B) All discharges into the navigable waters of wastes containing benzidine from the manufacturing areas, loading and unloading areas, storage areas, and other areas subject to direct contamination by benzidine or benzidine-containing product as a result of the manufacturing process, including but not limited to:

(1) Stormwater and other runoff except as hereinafter provided in paragraph (c)(1)(ii) of this section, and

(2) Water used for routine cleanup or cleanup of spills.

(ii) These standards do not apply to stormwater runoff or other discharges from areas subject to contamination solely by fallout from air emissions of benzidine; or to stormwater that exceeds that from the ten year 24-hour rainfall event.

(2) *Analytical method acceptable.* (i) Environmental Protection Agency method specified in 40 CFR part 136; or

(ii) Mass balance monitoring approach which requires the calculation of the benzidine concentration by dividing the total benzidine contained in dyes used during a working day (as certified in writing by the manufacturer) by the total quantity of water discharged during the working day.

[*Comment:* The Regional Administrator (or State Director, if appropriate) shall rely entirely upon the method specified in 40 CFR part 136 in analyses performed by him for enforcement purposes.]

(3) *Effluent standards*—(i) *Existing sources.* Discharges from benzidine-based dye applicators shall not contain benzidine concentrations exceeding an average per working day of 10 µg/l calculated over any calendar month; and shall not exceed 25 µg/l in a sample(s) or calculation(s) representing any working day.

(ii) *New sources.* Discharges from benzidine-based dye applicators shall not contain benzidine concentrations exceeding an average per working day of 10 µg/l calculated over any calendar month; and shall not exceed 25 µg/l in a sample(s) or calculation(s) representing any working day.

(4) The standards set forth in this paragraph (c) shall apply to the total combined concentrations of benzidine, excluding any associated element or compound.

[42 FR 2620, Jan. 12, 1977]

§ 129.105   Polychlorinated   biphenyls (PCBs).

(a) *Specialized definitions.* (1) *PCB Manufacturer* means a manufacturer who produces polychlorinated biphenyls.

(2) *Electrical capacitor manufacturer* means a manufacturer who produces or assembles electrical capacitors in which PCB or PCB-containing compounds are part of the dielectric.

(3) *Electrical transformer manufacturer* means a manufacturer who produces or assembles electrical transformers in which PCB or PCB-containing compounds are part of the dielectric.

(4) The ambient water criterion for PCBs in navigable waters is 0.001 µg/l.

(b) *PCB manufacturer*—(1) *Applicability.* (i) These standards or prohibitions apply to:

(A) All discharges of process wastes;

(B) All discharges from the manufacturing or incinerator areas, loading and unloading areas, storage areas, and other areas which are subject to direct contamination by PCBs as a result of the manufacturing process, including but not limited to:

(1) Stormwater and other runoff except as hereinafter provided in paragraph (b)(1)(ii) of this section; and

(2) Water used for routine cleanup or cleanup of spills.

(ii) These standards do not apply to stormwater runoff or other discharges from areas subject to contamination solely by fallout from air emissions of PCBs; or to stormwater runoff that exceeds that from the ten-year 24-hour rainfall event.

(2) Analytical Method Acceptable—Environmental Protection Agency method specified in 40 CFR part 136 except that a 1-liter sample size is required to increase analytical sensitivity.

(3) *Effluent standards*—(i) *Existing sources.* PCBs are prohibited in any discharge from any PCB manufacturer;

(ii) *New sources.* PCBs are prohibited in any discharge from any PCB manufacturer.

(c) *Electrical capacitor manufacturer—* (1) *Applicability.* (i) These standards or prohibitions apply to:

(A) All discharges of process wastes; and

(B) All discharges from the manufacturing or incineration areas, loading and unloading areas, storage areas and other areas which are subject to direct contamination by PCBs as a result of the manufacturing process, including but not limited to:

(1) Stormwater and other runoff except as hereinafter provided in paragraph (c)(1)(ii) of this section; and

(2) Water used for routine cleanup or cleanup of spills.

(ii) These standards do not apply to stormwater runoff or other discharges from areas subject to contamination solely by fallout from air emissions of PCBs; or to stormwater runoff that exceeds that from the ten-year 24-hour rainfall event.

(2) *Analytical method acceptable.* Environmental Protection Agency method specified in 40 CFR part 136, except that a 1-liter sample size is required to increase analytical sensitivity.

(3) *Effluent standards*—(i) *Existing sources.* PCBs are prohibited in any discharge from any electrical capacitor manufacturer;

(ii) *New sources.* PCBs are prohibited in any discharge from any electrical capacitor manufacturer.

(d) *Electrical transformer manufacturer*—(1) *Applicability.* (i) These standards or prohibitions apply to:

(A) All discharges of process wastes; and

(B) All discharges from the manufacturing or incineration areas, loading and unloading areas, storage areas, and other areas which are subject to direct contamination by PCBs as a result of the manufacturing process, including but not limited to:

(1) Stormwater and other runoff except as hereinafter provided in paragraph (d)(1)(ii) of this section; and

(2) Water used for routine cleanup or cleanup of spills.

(ii) These standards do not apply to stormwater runoff or other discharges from areas subject to contamination

solely by fallout from air emissions of PCBs; or to stormwater runoff that exceeds that from the ten-year 24-hour rainfall event.

(2) *Analytical method acceptable.* Environmental Protection Agency method specified in 40 CFR part 136, except that a 1-liter sample size is required to increase analytical sensitivity.

(3) *Effluent standards*—(i) *Existing sources.* PCBs are prohibited in any discharge from any electrical transformer manufacturer;

(ii) *New sources.* PCBs are prohibited in any discharge from any electrical transformer manufacturer.

(e) *Adjustment of effluent standard for presence of PCBs in intake water.* Whenever a facility which is subject to these standards has PCBs in its effluent which result from the presence of PCBs in its intake waters, the owner may apply to the Regional Administrator (or State Director, if appropriate), for a credit pursuant to the provisions of § 129.6, where the source of the water supply is the same body of water into which the discharge is made. The requirement of paragraph (1) of § 129.6(a), relating to the source of the water supply, shall be waived, and such facility shall be eligible to apply for a credit under § 129.6, upon a showing by the owner or operator of such facility to the Regional Administrator (or State Director, if appropriate) that the concentration of PCBs in the intake water supply of such facility does not exceed the concentration of PCBs in the receiving water body to which the plant discharges its effluent.

[42 FR 6555, Feb. 2, 1977]

PART 130—WATER QUALITY PLANNING AND MANAGEMENT

130.11   Program management.
130.12   Coordination with other programs.
130.15   Processing application for Indian tribes.

AUTHORITY: 33 U.S.C. 1251 *et seq.*

SOURCE: 50 FR 1779, Jan. 11, 1985, unless otherwise noted.

§ 130.0   Program summary and purpose.

(a) This subpart establishes policies and program requirements for water quality planning, management and implementation under sections 106, 205(j), non-construction management 205(g), 208, 303 and 305 of the Clean Water Act. The Water Quality Management (WQM) process described in the Act and in this regulation provides the authority for a consistent national approach for maintaining, improving and protecting water quality while allowing States to implement the most effective individual programs. The process is implemented jointly by EPA, the States, interstate agencies, and areawide, local and regional planning organizations. This regulation explains the requirements of the Act, describes the relationships between the several components of the WQM process and outlines the roles of the major participants in the process. The components of the WQM process are discussed below.

(b) Water quality standards (WQS) are the State's goals for individual water bodies and provide the legal basis for control decisions under the Act. Water quality monitoring activities provide the chemical, physical and biological data needed to determine the present quality of a State's waters and to identify the sources of pollutants in those waters. The primary assessment of the quality of a State's water is contained in its biennial Report to Congress required by section 305(b) of the Act.

(c) This report and other assessments of water quality are used in the State's WQM plans to identify priority water quality problems. These plans also contain the results of the State's analyses and management decisions which are necessary to control specific sources of pollution. The plans recommend control measures and designated management agencies (DMAs) to attain the goals established in the State's water quality standards.

(d) These control measures are implemented by issuing permits, building publicly-owned treatment works (POTWs), instituting best management practices for nonpoint sources of pollution and other means. After control measures are in place, the State evaluates the extent of the resulting improvements in water quality, conducts additional data gathering and planning to determine needed modifications in control measures and again institutes control measures.

(e) This process is a dynamic one, in which requirements and emphases vary over time. At present, States have completed WQM plans which are generally comprehensive in geographic and programmatic scope. Technology based controls are being implemented for most point sources of pollution. However, WQS have not been attained in many water bodies and are threatened in others.

(f) Present continuing planning requirements serve to identify these critical water bodies, develop plans for achieving higher levels of abatement and specify additional control measures. Consequently, this regulation reflects a programmatic emphasis on concentrating planning and abatement activities on priority water quality issues and geographic areas. EPA will focus its grant funds on activities designed to address these priorities. Annual work programs negotiated between EPA and State and interstate agencies will reflect this emphasis.

§ 130.1   Applicability.

(a) This subpart applies to all State, eligible Indian Tribe, interstate, areawide and regional and local CWA water quality planning and management activities undertaken on or after February 11, 1985 including all updates and continuing certifications for approved Water Quality Management (WQM) plans developed under sections 208 and 303 of the Act.

(b) Planning and management activities undertaken prior to February 11, 1985 are governed by the requirements

418

of the regulations in effect at the time of the last grant award.

[50 FR 1779, Jan. 11, 1985, as amended at 54 FR 14359, Apr. 11, 1989; 59 FR 13817, Mar. 23, 1994]

§130.2 Definitions.

(a) *The Act.* The Clean Water Act, as amended, 33 U.S.C. 1251 *et seq.*

(b) *Indian Tribe.* Any Indian Tribe, band, group, or community recognized by the Secretary of the Interior and exercising governmental authority over a Federal Indian reservation.

(c) *Pollution.* The man-made or man-induced alteration of the chemical, physical, biological, and radiological integrity of water.

(d) *Water quality standards (WQS).* Provisions of State or Federal law which consist of a designated use or uses for the waters of the United States and water quality criteria for such waters based upon such uses. Water quality standards are to protect the public health or welfare, enhance the quality of water and serve the purposes of the Act.

(e) *Load or loading.* An amount of matter or thermal energy that is introduced into a receiving water; to introduce matter or thermal energy into a receiving water. Loading may be either man-caused (pollutant loading) or natural (natural background loading).

(f) *Loading capacity.* The greatest amount of loading that a water can receive without violating water quality standards.

(g) *Load allocation (LA).* The portion of a receiving water's loading capacity that is attributed either to one of its existing or future nonpoint sources of pollution or to natural background sources. Load allocations are best estimates of the loading, which may range from reasonably accurate estimates to gross allotments, depending on the availability of data and appropriate techniques for predicting the loading. Wherever possible, natural and nonpoint source loads should be distinguished.

(h) *Wasteload allocation (WLA).* The portion of a receiving water's loading capacity that is allocated to one of its existing or future point sources of pollution. WLAs constitute a type of water quality-based effluent limitation.

(i) *Total maximum daily load (TMDL).* The sum of the individual WLAs for point sources and LAs for nonpoint sources and natural background. If a receiving water has only one point source discharger, the TMDL is the sum of that point source WLA plus the LAs for any nonpoint sources of pollution and natural background sources, tributaries, or adjacent segments. TMDLs can be expressed in terms of either mass per time, toxicity, or other appropriate measure. If Best Management Practices (BMPs) or other nonpoint source pollution controls make more stringent load allocations practicable, then wasteload allocations can be made less stringent. Thus, the TMDL process provides for nonpoint source control tradeoffs.

(j) *Water quality limited segment.* Any segment where it is known that water quality does not meet applicable water quality standards, and/or is not expected to meet applicable water quality standards, even after the application of the technology-based effluent limitations required by sections 301(b) and 306 of the Act.

(k) *Water quality management (WQM) plan.* A State or areawide waste treatment management plan developed and updated in accordance with the provisions of sections 205(j), 208 and 303 of the Act and this regulation.

(l) *Areawide agency.* An agency designated under section 208 of the Act, which has responsibilities for WQM planning within a specified area of a State.

(m) *Best Management Practice (BMP).* Methods, measures or practices selected by an agency to meet its nonpoint source control needs. BMPs include but are not limited to structural and nonstructural controls and operation and maintenance procedures. BMPs can be applied before, during and after pollution-producing activities to reduce or eliminate the introduction of pollutants into receiving waters.

(n) *Designated management agency (DMA).* An agency identified by a WQM plan and designated by the Governor to implement specific control recommendations.

[50 FR 1779, Jan. 11, 1985, as amended at 54 FR 14359, Apr. 11, 1989]

§ 130.3 Water quality standards.

A water quality standard (WQS) defines the water quality goals of a water body, or portion thereof, by designating the use or uses to be made of the water and by setting criteria necessary to protect the uses. States and EPA adopt WQS to protect public health or welfare, enhance the quality of water and serve the purposes of the Clean Water Act (CWA). *Serve the purposes of Act* (as defined in sections 101(a)(2) and 303(c) of the Act) means that WQS should, wherever attainable, provide water quality for the protection and propagation of fish, shellfish and wildlife and for recreation in and on the water and take into consideration their use and value for public water supplies, propagation of fish, shellfish, wildlife, recreation in and on the water, and agricultural, industrial and other purposes including navigation.

Such standards serve the dual purposes of establishing the water quality goals for a specific water body and serving as the regulatory basis for establishment of water quality-based treatment controls and strategies beyond the technology-based level of treatment required by sections 301(b) and 306 of the Act. States shall review and revise WQS in accordance with applicable regulations and, as appropriate, update their Water Quality Management (WQM) plans to reflect such revisions. Specific WQS requirements are found in 40 CFR part 131.

§ 130.4 Water quality monitoring.

(a) In accordance with section 106(e)(1), States must establish appropriate monitoring methods and procedures (including biological monitoring) necessary to compile and analyze data on the quality of waters of the United States and, to the extent practicable, ground-waters. This requirement need not be met by Indian Tribes. However, any monitoring and/or analysis activities undertaken by a Tribe must be performed in accordance with EPA's quality assurance/quality control guidance.

(b) The State's water monitoring program shall include collection and analysis of physical, chemical and biological data and quality assurance and control programs to assure scientifically valid data. The uses of these data include determining abatement and control priorities; developing and reviewing water quality standards, total maximum daily loads, wasteload allocations and load allocations; assessing compliance with National Pollutant Discharge Elimination System (NPDES) permits by dischargers; reporting information to the public through the section 305(b) report and reviewing site-specific monitoring efforts.

[50 FR 1779, Jan. 11, 1985, as amended at 54 FR 14359, Apr. 11, 1989]

§ 130.5 Continuing planning process.

(a) *General.* Each State shall establish and maintain a continuing planning process (CPP) as described under section 303(e)(3)(A)–(H) of the Act. Each State is responsible for managing its water quality program to implement the processes specified in the continuing planning process. EPA is responsible for periodically reviewing the adequacy of the State's CPP.

(b) *Content.* The State may determine the format of its CPP as long as the mininum requirements of the CWA and this regulation are met. The following processes must be described in each State CPP, and the State may include other processes at its discretion.

(1) The process for developing effluent limitations and schedules of compliance at least as stringent as those required by sections 301(b) (1) and (2), 306 and 307, and at least as stringent as any requirements contained in applicable water quality standards in effect under authority of section 303 of the Act.

(2) The process for incorporating elements of any applicable areawide waste treatment plans under section 208, and applicable basin plans under section 209 of the Act.

(3) The process for developing total maximum daily loads (TMDLs) and individual water quality based effluent

limitations for pollutants in accordance with section 303(d) of the Act and §130.7(a) of this regulation.

(4) The process for updating and maintaining Water Quality Management (WQM) plans, including schedules for revision.

(5) The process for assuring adequate authority for intergovernmental cooperation in the implementation of the State WQM program.

(6) The process for establishing and assuring adequate implementation of new or revised water quality standards, including schedules of compliance, under section 303(c) of the Act.

(7) The process for assuring adequate controls over the disposition of all residual waste from any water treatment processing.

(8) The process for developing an inventory and ranking, in order of priority of needs for construction of waste treatment works required to meet the applicable requirements of sections 301 and 302 of the Act.

(9) The process for determining the priority of permit issuance.

(c) *Regional Administrator review.* The Regional Administrator shall review approved State CPPs from time to time to ensure that the planning processes are consistent with the Act and this regulation. The Regional Administrator shall not approve any permit program under Title IV of the Act for any State which does not have an approved continuing planning process.

§130.6 Water quality management plans.

(a) *Water quality management (WQM) plans.* WQM plans consist of initial plans produced in accordance with sections 208 and 303(e) of the Act and certified and approved updates to those plans. Continuing water quality planning shall be based upon WQM plans and water quality problems identified in the latest 305(b) reports. State water quality planning should focus annually on priority issues and geographic areas and on the development of water quality controls leading to implementation measures. Water quality planning directed at the removal of conditions placed on previously certified and approved WQM plans should focus on re-moval of conditions which will lead to control decisions.

(b) *Use of WQM plans.* WQM plans are used to direct implementation. WQM plans draw upon the water quality assessments to identify priority point and nonpoint water quality problems, consider alternative solutions and recommend control measures, including the financial and institutional measures necessary for implementing recommended solutions. State annual work programs shall be based upon the priority issues identified in the State WQM plan.

(c) *WQM plan elements.* Sections 205(j), 208 and 303 of the Act specify water quality planning requirements. The following plan elements shall be included in the WQM plan or referenced as part of the WQM plan if contained in separate documents when they are needed to address water quality problems.

(1) *Total maximum daily loads.* TMDLs in accordance with sections 303(d) and (e)(3)(C) of the Act and §130.7 of this part.

(2) *Effluent limitations.* Effluent limitations including water quality based effluent limitations and schedules of compliance in accordance with section 303(e)(3)(A) of the Act and §130.5 of this part.

(3) *Municipal and industrial waste treatment.* Identification of anticipated municipal and industrial waste treatment works, including facilities for treatment of stormwater-induced combined sewer overflows; programs to provide necessary financial arrangements for such works; establishment of construction priorities and schedules for initiation and completion of such treatment works including an identification of open space and recreation opportunities from improved water quality in accordance with section 208(b)(2) (A) and (B) of the Act.

(4) *Nonpoint source management and control.* (i) The plan shall describe the regulatory and non-regulatory programs, activities and Best Management Practices (BMPs) which the agency has selected as the means to control nonpoint source pollution where necessary to protect or achieve approved water uses. Economic, institutional,

and technical factors shall be considered in a continuing process of identifying control needs and evaluating and modifying the BMPs as necessary to achieve water quality goals.

(ii) Regulatory programs shall be identified where they are determined to be necessary by the State to attain or maintain an approved water use or where non-regulatory approaches are inappropriate in accomplishing that objective.

(iii) BMPs shall be identified for the nonpoint sources identified in section 208(b)(2)(F)–(K) of the Act and other nonpoint sources as follows:

(A) *Residual waste.* Identification of a process to control the disposition of all residual waste in the area which could affect water quality in accordance with section 208(b)(2)(J) of the Act.

(B) *Land disposal.* Identification of a process to control the disposal of pollutants on land or in subsurface excavations to protect ground and surface water quality in accordance with section 208(b)(2)(K) of the Act.

(C) *Agricultural and silvicultural.* Identification of procedures to control agricultural and silvicultural sources of pollution in accordance with section 208(b)(2)(F) of the Act.

(D) *Mines.* Identification of procedures to control mine-related sources of pollution in accordance with section 208(b)(2)(G) of the Act.

(E) *Construction.* Identification of procedures to control construction related sources of pollution in accordance with section 208(b)(2)(H) of the Act.

(F) *Saltwater intrusion.* Identification of procedures to control saltwater intrusion in accordance with section 208(b)(2)(I) of the Act.

(G) *Urban stormwater.* Identification of BMPs for urban stormwater control to achieve water quality goals and fiscal analysis of the necessary capital and operations and maintenance expenditures in accordance with section 208(b)(2)(A) of the Act.

(iv) The nonpoint source plan elements outlined in § 130.6(c) (4)(iii)(A)(G) of this regulation shall be the basis of water quality activities implemented through agreements or memoranda of understanding between EPA and other departments, agencies or instrumental-

ities of the United States in accordance with section 304(k) of the Act.

(5) *Management agencies.* Identification of agencies necessary to carry out the plan and provision for adequate authority for intergovernmental cooperation in accordance with sections 208(b)(2)(D) and 303(e)(3)(E) of the Act. Management agencies must demonstrate the legal, institutional, managerial and financial capability and specific activities necessary to carry out their responsibilities in accordance with section 208(c)(2)(A) through (I) of the Act.

(6) *Implementation measures.* Identification of implementation measures necessary to carry out the plan, including financing, the time needed to carry out the plan, and the economic, social and environmental impact of carrying out the plan in accordance with section 208(b)(2)(E).

(7) *Dredge or fill program.* Identification and development of programs for the control of dredge or fill material in accordance with section 208(b)(4)(B) of the Act.

(8) *Basin plans.* Identification of any relationship to applicable basin plans developed under section 209 of the Act.

(9) *Ground water.* Identification and development of programs for control of ground-water pollution including the provisions of section 208(b)(2)(K) of the Act. States are not required to develop ground-water WQM plan elements beyond the requirements of section 208(b)(2)(K) of the Act, but may develop a ground-water plan element if they determine it is necessary to address a ground-water quality problem. If a State chooses to develop a ground-water plan element, it should describe the essentials of a State program and should include, but is not limited to:

(i) Overall goals, policies and legislative authorities for protection of ground-water.

(ii) Monitoring and resource assessment programs in accordance with section 106(e)(1) of the Act.

(iii) Programs to control sources of contamination of ground-water including Federal programs delegated to the State and additional programs authorized in State statutes.

(iv) Procedures for coordination of ground-water protection programs among State agencies and with local and Federal agencies.

(v) Procedures for program management and administration including provision of program financing, training and technical assistance, public participation, and emergency management.

(d) *Indian Tribes*. An Indian Tribe is eligible for the purposes of this rule and the Clean Water Act assistance programs under 40 CFR part 35, subparts A and H if:

(1) The Indian Tribe has a governing body carrying out substantial governmental duties and powers;

(2) The functions to be exercised by the Indian Tribe pertain to the management and protection of water resources which are held by an Indian Tribe, held by the United States in trust for Indians, held by a member of an Indian Tribe if such property interest is subject to a trust restriction on alienation, or otherwise within the borders of an Indian reservation; and

(3) The Indian Tribe is reasonably expected to be capable, in the Regional Administrator's judgment, of carrying out the functions to be exercised in a manner consistent with the terms and purposes of the Clean Water Act and applicable regulations.

(e) *Update and certification*. State and/ or areawide agency WQM plans shall be updated as needed to reflect changing water quality conditions, results of implementation actions, new requirements or to remove conditions in prior conditional or partial plan approvals. Regional Administrators may require that State WQM plans be updated as needed. State Continuing Planning Processes (CPPs) shall specify the process and schedule used to revise WQM plans. The State shall ensure that State and areawide WQM plans together include all necessary plan elements and that such plans are consistent with one another. The Governor or the Governor's designee shall certify by letter to the Regional Administrator for EPA approval that WQM plan updates are consistent with all other parts of the plan. The certification may be contained in the annual State work program.

(f) *Consistency*. Construction grant and permit decisions must be made in accordance with certified and approved WQM plans as described in §§ 130.12(a) and 130.12(b).

[50 FR 1779, Jan. 11, 1985, as amended at 54 FR 14360, Apr. 11, 1989; 59 FR 13818, Mar. 23, 1994]

§ 130.7 Total maximum daily loads (TMDL) and individual water quality-based effluent limitations.

(a) *General*. The process for identifying water quality limited segments still requiring wasteload allocations, load allocations and total maximum daily loads (WLAs/LAs and TMDLs), setting priorities for developing these loads; establishing these loads for segments identified, including water quality monitoring, modeling, data analysis, calculation methods, and list of pollutants to be regulated; submitting the State's list of segments identified, priority ranking, and loads established (WLAs/LAs/TMDLs) to EPA for approval; incorporating the approved loads into the State's WQM plans and NPDES permits; and involving the public, affected dischargers, designated areawide agencies, and local governments in this process shall be clearly described in the State Continuing Planning Process (CPP).

(b) Identification and priority setting for water quality-limited segments still requiring TMDLs.

(1) Each State shall identify those water quality-limited segments still requiring TMDLs within its boundaries for which:

(i) Technology-based effluent limitations required by sections 301(b), 306, 307, or other sections of the Act;

(ii) More stringent effluent limitations (including prohibitions) required by either State or local authority preserved by section 510 of the Act, or Federal authority (law, regulation, or treaty); and

(iii) Other pollution control requirements (e.g., best management practices) required by local, State, or Federal authority are not stringent enough to implement any water quality standards (WQS) applicable to such waters.

(2) Each State shall also identify on the same list developed under paragraph (b)(1) of this section those water

quality-limited segments still requiring TMDLs or parts thereof within its boundaries for which controls on thermal discharges under section 301 or State or local requirements are not stringent enough to assure protection and propagation of a balanced indigenous population of shellfish, fish and wildlife.

(3) For the purposes of listing waters under § 130.7(b), the term "water quality standard applicable to such waters" and "applicable water quality standards" refer to those water quality standards established under section 303 of the Act, including numeric criteria, narrative criteria, waterbody uses, and antidegradation requirements.

(4) The list required under §§ 130.7(b)(1) and 130.7(b)(2) of this section shall include a priority ranking for all listed water quality-limited segments still requiring TMDLs, taking into account the severity of the pollution and the uses to be made of such waters and shall identify the pollutants causing or expected to cause violations of the applicable water quality standards. The priority ranking shall specifically include the identification of waters targeted for TMDL development in the next two years.

(5) Each State shall assemble and evaluate all existing and readily available water quality-related data and information to develop the list required by §§ 130.7(b)(1) and 130.7(b)(2). At a minimum "all existing and readily available water quality-related data and information" includes but is not limited to all of the existing and readily available data and information about the following categories of waters:

(i) Waters identified by the State in its most recent section 305(b) report as "partially meeting" or "not meeting" designated uses or as "threatened";

(ii) Waters for which dilution calculations or predictive models indicate nonattainment of applicable water quality standards;

(iii) Waters for which water quality problems have been reported by local, state, or federal agencies; members of the public; or academic institutions. These organizations and groups should be actively solicited for research they may be conducting or reporting. For example, university researchers, the United States Department of Agriculture, the National Oceanic and Atmospheric Administration, the United States Geological Survey, and the United States Fish and Wildlife Service are good sources of field data; and

(iv) Waters identified by the State as impaired or threatened in a nonpoint assessment submitted to EPA under section 319 of the CWA or in any updates of the assessment.

(6) Each State shall provide documentation to the Regional Administrator to support the State's determination to list or not to list its waters as required by §§ 130.7(b)(1) and 130.7(b)(2). This documentation shall be submitted to the Regional Administrator together with the list required by §§ 130.7(b)(1) and 130.7(b)(2) and shall include at a minimum:

(i) A description of the methodology used to develop the list; and

(ii) A description of the data and information used to identify waters, including a description of the data and information used by the State as required by § 130.7(b)(5); and

(iii) A rationale for any decision to not use any existing and readily available data and information for any one of the categories of waters as described in § 130.7(b)(5); and

(iv) Any other reasonable information requested by the Regional Administrator. Upon request by the Regional Administrator, each State must demonstrate good cause for not including a water or waters on the list. Good cause includes, but is not limited to, more recent or accurate data; more sophisticated water quality modeling; flaws in the original analysis that led to the water being listed in the categories in § 130.7(b)(5); or changes in conditions, e.g., new control equipment, or elimination of discharges.

(c) Development of TMDLs and individual water quality based effluent limitations.

(1) Each State shall establish TMDLs for the water quality limited segments identified in paragraph (b)(1) of this section, and in accordance with the priority ranking. For pollutants other than heat, TMDLs shall be established

at levels necessary to attain and maintain the applicable narrative and numerical WQS with seasonal variations and a margin of safety which takes into account any lack of knowledge concerning the relationship between effluent limitations and water quality. Determinations of TMDLs shall take into account critical conditions for stream flow, loading, and water quality parameters.

(i) TMDLs may be established using a pollutant-by-pollutant or biomonitoring approach. In many cases both techniques may be needed. Site-specific information should be used wherever possible.

(ii) TMDLs shall be established for all pollutants preventing or expected to prevent attainment of water quality standards as identified pursuant to paragraph (b)(1) of this section. Calculations to establish TMDLs shall be subject to public review as defined in the State CPP.

(2) Each State shall estimate for the water quality limited segments still requiring TMDLs identified in paragraph (b)(2) of this section, the total maximum daily thermal load which cannot be exceeded in order to assure protection and propagation of a balanced, indigenous population of shellfish, fish and wildlife. Such estimates shall take into account the normal water temperatures, flow rates, seasonal variations, existing sources of heat input, and the dissipative capacity of the identified waters or parts thereof. Such estimates shall include a calculation of the maximum heat input that can be made into each such part and shall include a margin of safety which takes into account any lack of knowledge concerning the development of thermal water quality criteria for protection and propagation of a balanced, indigenous population of shellfish, fish and wildlife in the identified waters or parts thereof.

(d) *Submission and EPA approval.* (1) Each State shall submit biennially to the Regional Administrator beginning in 1992 the list of waters, pollutants causing impairment, and the priority ranking including waters targeted for TMDL development within the next two years as required under paragraph (b) of this section. For the 1992 biennial

submission, these lists are due no later than October 22, 1992. Thereafter, each State shall submit to EPA lists required under paragraph (b) of this section on April 1 of every even-numbered year. For the year 2000 submission, a State must submit a list required under paragraph (b) of this section only if a court order or consent decree, or commitment in a settlement agreement dated prior to January 1, 2000, expressly requires EPA to take action related to that State's year 2000 list. For the year 2002 submission, a State must submit a list required under paragraph (b) of this section by October 1, 2002, unless a court order, consent decree or commitment in a settlement agreement expressly requires EPA to take an action related to that State's 2002 list prior to October 1, 2002, in which case, the State must submit a list by April 1, 2002. The list of waters may be submitted as part of the State's biennial water quality report required by §130.8 of this part and section 305(b) of the CWA or submitted under separate cover. All WLAs/LAs and TMDLs established under paragraph (c) for water quality limited segments shall continue to be submitted to EPA for review and approval. Schedules for submission of TMDLs shall be determined by the Regional Administrator and the State.

(2) The Regional Administrator shall either approve or disapprove such listing and loadings not later than 30 days after the date of submission. The Regional Administrator shall approve a list developed under §130.7(b) that is submitted after the effective date of this rule only if it meets the requirements of §130.7(b). If the Regional Administrator approves such listing and loadings, the State shall incorporate them into its current WQM plan. If the Regional Administrator disapproves such listing and loadings, he shall, not later than 30 days after the date of such disapproval, identify such waters in such State and establish such loads for such waters as determined necessary to implement applicable WQS. The Regional Administrator shall promptly issue a public notice seeking comment on such listing and loadings. After considering public comment and

making any revisions he deems appropriate, the Regional Administrator shall transmit the listing and loads to the State, which shall incorporate them into its current WQM plan.

(e) For the specific purpose of developing information and as resources allow, each State shall identify all segments within its boundaries which it has not identified under paragraph (b) of this section and estimate for such waters the TMDLs with seasonal variations and margins of safety, for those pollutants which the Regional Administrator identifies under section 304(a)(2) as suitable for such calculation and for thermal discharges, at a level that would assure protection and propagation of a balanced indigenous population of fish, shellfish and wildlife. However, there is no requirement for such loads to be submitted to EPA for approval, and establishing TMDLs for those waters identified in paragraph (b) of this section shall be given higher priority.

[50 FR 1779, Jan. 11, 1985, as amended at 57 FR 33049, July 24, 1992; 65 FR 17170, Mar. 31, 2000; 66 FR 53048, Oct. 18, 2001]

§ 130.8 Water quality report.

(a) Each State shall prepare and submit biennially to the Regional Administrator a water quality report in accordance with section 305(b) of the Act. The water quality report serves as the primary assessment of State water quality. Based upon the water quality data and problems identified in the 305(b) report, States develop water quality management (WQM) plan elements to help direct all subsequent control activities. Water quality problems identified in the 305(b) report should be analyzed through water quality management planning leading to the development of alternative controls and procedures for problems identified in the latest 305(b) report. States may also use the 305(b) report to describe ground-water quality and to guide development of ground-water plans and programs. Water quality problems identified in the 305(b) report should be emphasized and reflected in the State's WQM plan and annual work program under sections 106 and 205(j) of the Clean Water Act.

(b) Each such report shall include but is not limited to the following:

(1) A description of the water quality of all waters of the United States and the extent to which the quality of waters provides for the protection and propagation of a balanced population of shellfish, fish, and wildlife and allows recreational activities in and on the water.

(2) An estimate of the extent to which CWA control programs have improved water quality or will improve water quality for the purposes of paragraph (b)(1) of this section, and recommendations for future actions necessary and identifications of waters needing action.

(3) An estimate of the environmental, economic and social costs and benefits needed to achieve the objectives of the CWA and an estimate of the date of such achievement.

(4) A description of the nature and extent of nonpoint source pollution and recommendations of programs needed to control each category of nonpoint sources, including an estimate of implementation costs.

(5) An assessment of the water quality of all publicly owned lakes, including the status and trends of such water quality as specified in section 314(a)(1) of the Clean Water Act.

(c) States may include a description of the nature and extent of ground-water pollution and recommendations of State plans or programs needed to maintain or improve ground-water quality.

(d) In the years in which it is prepared the biennial section 305(b) report satisfies the requirement for the annual water quality report under section 205(j). In years when the 305(b) report is not required, the State may satisfy the annual section 205(j) report requirement by certifying that the most recently submitted section 305(b) report is current or by supplying an update of the sections of the most recently submitted section 305(b) report which require updating.

[50 FR 1779, Jan.11, 1985, as amended at 57 FR 33050, July 24, 1992]

§130.9 Designation and de-designation.

(a) *Designation.* Areawide planning agencies may be designated by the Governor in accordance with section 208(a) (2) and (3) of the Act or may self-designate in accordance with section 208(a)(4) of the Act. Such designations shall subject to EPA approval in accordance with section 208(a)(7) of the Act.

(b) *De-designation.* The Governor may modify or withdraw the planning designation of a designated planning agency other than an Indian tribal organization self-designated §130.6(c)(2) if:

(1) The areawide agency requests such cancellation; or

(2) The areawide agency fails to meet its planning requirements as specified in grant agreements, contracts or memoranda of understanding; or

(3) The areawide agency no longer has the resources or the commitment to continue water quality planning activities within the designated boundaries.

(c) *Impact of de-designation.* Once an areawide planning agency's designation has been withdrawn the State agency shall assume direct responsibility for continued water quality planning and oversight of implementation within the area.

(d) *Designated management agencies (DMA).* In accordance with section 208(c)(1) of the Act, management agencies shall be designated by the Governor in consultation with the designated planning agency. EPA shall approve such designations unless the DMA lacks the legal, financial and managerial authority required under section 208(c)(2) of the Act. Designated management agencies shall carry out responsibilities specified in Water Quality Management (WQM) plans. Areawide planning agencies shall monitor DMA activities in their area and recommend necessary plan changes during the WQM plan update. Where there is no designated areawide planning agency, States shall monitor DMA activities and make any necessary changes during the WQM plan update.

§130.10 State submittals to EPA.

(a) The following must be submitted regularly by the States to EPA:

(1) The section 305(b) report, in FY 84 and every two years thereafter, and the annual section 205(j) certification or update of the 305(b) water quality report; (Approved by OMB under the control number 2040–0071)

(2) The annual State work program(s) under sections 106 and 205(j) of the Act; and (Approved by OMB under the control number 2010–0004)

(3) Revisions or additions to water quality standards (WQS) (303(c)). (Approved by OMB under 2040–0049)

(b) The Act also requires that each State initially submit to EPA and revise as necessary the following:

(1) Continuing planning process (CPP) (303(e));

(2) Identification of water quality-limited waters still requiring TMDLs (section 303(d)), pollutants, and the priority ranking including waters targeted for TMDL development within the next two years as required under §130.7(b) in accordance with the schedule set for in §130.7(d)(1).

(Approved by the Office of Management and Budget under control number 2040–0071)

(3) Total maximum daily loads (TMDLs) (303(d)); and

(4) Water quality management (WQM) plan and certified and approved WQM plan updates (208, 303(e)). (Paragraph (b)(1), (4) approved by OMB under the control number 2010–0004).

(c) The form and content of required State submittals to EPA may be tailored to reflect the organization and needs of the State, as long as the requirements and purposes of the Act, this part and, where applicable, 40 CFR parts 29, 30, 33 and 35, subparts A and J are met. The need for revision and schedule of submittals shall be agreed to annually with EPA as the States annual work program is developed.

(d) Not later than February 4, 1989, each State shall submit to EPA for review, approval, and implementation—

(1) A list of those waters within the State which after the application of effluent limitations required under section 301(b)(2) of the CWA cannot reasonably be anticipated to attain or maintain (i) water quality standards for such waters reviewed, revised, or adopted in accordance with section 303(c)(2)(B) of the CWA, due to toxic pollutants, or (ii) that water quality

which shall assure protection of public health, public water supplies, agricultural and industrial uses, and the protection and propagation of a balanced population of shellfish, fish and wildlife, and allow recreational activities in and on the water;

(2) A list of all navigable waters in such State for which the State does not expect the applicable standard under section 303 of the CWA will be achieved after the requirements of sections 301(b), 306, and 307(b) are met, due entirely or substantially to discharges from point sources of any toxic pollutants listed pursuant to section 307(a);

(3) For each segment of navigable waters included on such lists, a determination of the specific point source discharging any such toxic pollutant which is believed to be preventing or impairing such water quality and the amount of each such toxic pollutant discharged by each such source.

(Approved by the Office of Management and Budget under control number 2040–0152)

(4) For the purposes of listing waters under § 130.10(d)(2), *applicable standard* means a numeric criterion for a priority pollutant promulgated as part of a state water quality standard. Where a state numeric criterion for a priority pollutant is not promulgated as part of a state water quality standard, for the purposes of listing waters "applicable standard" means the state narrative water quality criterion to control a priority pollutant (e.g., no toxics in toxic amounts) interpreted on a chemical-by-chemical basis by applying a proposed state cirterion, an explicit state policy or regulation, or an EPA national water quality criterion, supplemented with other relevant information.

(5) If a water meets either of the two conditions listed below the water must be listed under § 130.10(d)(2) on the grounds that the applicable standard is not achieved or expected to be achieved due entirely or substantially to discharges from point sources.

(i) Existing or additional water quality-based limits on one or more point sources would result in the achievement of an applicable water quality standard for a toxic pollutant; or

(ii) The discharge of a toxic pollutant from one or more point sources, regardless of any nonpoint source contribution of the same pollutant, is sufficient to cause or is expected to cause an excursion above the applicable water quality standard for the toxic pollutant.

(6) Each state shall assemble and evaluate all existing and readily available water quality-related data and information and each state shall develop the lists required by paragraphs (d)(1), (2), and (3) of this section based upon this data and information. At a minimum, all existing and readily available water quality-related data and information includes, but is not limited to, all of the existing and readily available data about the following categories of waters in the state:

(i) Waters where fishing or shellfish bans and/or advisories are currently in effect or are anticipated.

(ii) Waters where there have been repeated fishkills or where abnormalities (cancers, lesions, tumors, etc.) have been observed in fish or other aquatic life during the last ten years.

(iii) Waters where there are restrictions on water sports or recreational contact.

(iv) Waters identified by the state in its most recent state section 305(b) report as either "partially achieving" or "not achieving" designated uses.

(v) Waters identified by the states under section 303(d) of the CWA as waters needing water quality-based controls.

(vi) Waters identified by the state as priority waterbodies. (State Water Quality Management plans often include priority waterbody lists which are those waters that most need water pollution control decisions to achieve water quality standards or goals.)

(vii) Waters where ambient data indicate potential or actual exceedances of water quality criteria due to toxic pollutants from an industry classified as a primary industry in appendix A of 40 CFR part 122.

(viii) Waters for which effluent toxicity test results indicate possible or actual exceedances of state water quality standards, including narrative "free from" water quality criteria or EPA water quality criteria where state criteria are not available.

(ix) Waters with primary industrial major dischargers where dilution analyses indicate exceedances of state narrative or numeric water quality criteria (or EPA water quality criteria where state standards are not available) for toxic pollutants, ammonia, or chlorine. These dilution analyses must be based on estimates of discharge levels derived from effluent guidelines development documents, NPDES permits or permit application data (e.g., Form 2C), Discharge Monitoring Reports (DMRs), or other available information.

(x) Waters with POTW dischargers requiring local pretreatment programs where dilution analyses indicate exceedances of state water quality criteria (or EPA water quality criteria where state water quality criteria are not available) for toxic pollutants, ammonia, or chlorine. These dilution analyses must be based upon data from NPDES permits or permit applications (e.g., Form 2C), Discharge Monitoring Reports (DMRs), or other available information.

(xi) Waters with facilities not included in the previous two categories such as major POTWs, and industrial minor dischargers where dilution analyses indicate exceedances of numeric or narrative state water quality criteria (or EPA water quality criteria where state water quality criteria are not available) for toxic pollutants, ammonia, or chlorine. These dilution analyses must be based upon estimates of discharge levels derived from effluent guideline development documents, NPDES permits or permit application data, Discharge Monitoring Reports (DMRs), or other available information.

(xii) Waters classified for uses that will not support the "fishable/swimmable" goals of the Clean Water Act.

(xiii) Waters where ambient toxicity or adverse water quality conditions have been reported by local, state, EPA or other Federal Agencies, the private sector, public interest groups, or universities. These organizations and groups should be actively solicited for research they may be conducting or reporting. For example, university researchers, the United States Department of Agriculture, the National Oce-

anic and Atmospheric Administration, the United States Geological Survey, and the United States Fish and Wildlife Service are good sources of field data and research.

(xiv) Waters identified by the state as impaired in its most recent Clean Lake Assessments conducted under section 314 of the Clean Water Act.

(xv) Waters identified as impaired by nonpoint sources in the *America's Clean Water: The States' Nonpoint Source Assessments* 1985 (Association of State and Interstate Water Pollution Control Administrators (ASIWPCA)) or waters identified as impaired or threatened in a nonpoint source assessment submitted by the state to EPA under section 319 of the Clean Water Act.

(xvi) Surface waters impaired by pollutants from hazardous waste sites on the National Priority List prepared under section 105(8)(A) of CERCLA.

(7) Each state shall provide documentation to the Regional Administrator to support the state's determination to list or not to list waters as required by paragraphs (d)(1), (d)(2) and (d)(3) of this section. This documentation shall be submitted to the Regional Administrator together with the lists required by paragraphs (d)(1), (d)(2), and (d)(3) of this section and shall include as a minimum:

(i) A description of the methodology used to develop each list;

(ii) A description of the data and information used to identify waters and sources including a description of the data and information used by the state as required by paragraph (d)(6) of this section;

(iii) A rationale for any decision not to use any one of the categories of existing and readily available data required by paragraph (d)(6) of this section; and

(iv) Any other information requested by the Regional Administrator that is reasonable or necessary to determine the adequacy of a state's lists. Upon request by the Regional Administrator, each state must demonstrate good cause for not including a water or waters on one or more lists. Good cause includes, but is not limited to, more recent or accurate data; more accurate water quality modeling; flaws in the original analysis that led to the water being identified in a category in §130.10(d)(6); or changes in conditions, e.g., new control equipment, or elimination of discharges.

(8) The Regional Administrator shall approve or disapprove each list required by paragraphs (d)(1), (d)(2), and (d)(3) of this section no later than June 4, 1989. The Regional Administrator shall approve each list required under paragraphs (d)(1), (d)(2), and (d)(3) of this section only if it meets the regulatory requirements for listing under paragraphs (d)(1), (d)(2), and (d)(3) of this section and if the state has met all the requirements of paragraphs (d)(6) and (d)(7) of this section.

(9) If a state fails to submit lists in accordance with paragraph (d) of this section or the Regional Administrator does not approve the lists submitted by such state in accordance with this paragraph, then not later than June 4, 1990, the Regional Administrator, in cooperation with such state, shall implement the requirements of CWA section 304(l) (1) and (2) in such state.

(10) If the Regional Administrator disapproves a state's decision with respect to one or more of the waters required under paragraph (d) (1), (2), or (3) of this section, or one or more of the individual control strategies required pursuant to section 304(l)(1)(D), then not later than June 4, 1989, the Regional Administrator shall distribute the notice of approval or disapproval given under this paragraph to the appropriate state Director. The Regional Administrator shall also publish a notice of availability, in a daily or weekly newspaper with state-wide circulation or in the FEDERAL REGISTER, for the notice of approval or disapproval. The Regional Administrator shall also provide written notice to each discharger identified under section 304(l)(1)(C), that EPA has listed the discharger under section 304(l)(1)(C). The notice of approval and disapproval shall include the following:

(i) The name and address of the EPA office that reviews the state's submittals.

(ii) A brief description of the section 304(l) process.

(iii) A list of waters, point sources and pollutants disapproved under this paragraph.

(iv) If the Regional Administrator determines that a state did not provide adequate public notice and an opportunity to comment on the lists prepared under this section, or if the Regional Administrator chooses to exercise his or her discretion, a list of waters, point sources, or pollutants approved under this paragraph.

(v) The name, address, and telephone number of the person at the Regional Office from whom interested persons may obtain more information.

(vi) Notice that written petitions or comments are due within 120 days.

(11) As soon as practicable, but not later than June 4, 1990, the Regional Office shall issue a response to petitions or comments received under paragraph (d)(10) of this section. Notice shall be given in the same manner as notice described in paragraph (d)(10) of this section, except for the following changes to the notice of approvals and disapprovals:

(i) The lists of waters, point sources and pollutants must reflect any changes made pursuant to comments or petitions received.

(ii) A brief description of the subsequent steps in the section 304(l) process shall be included.

[50 FR 1779, Jan. 11, 1985, as amended at 54 FR 258, Jan. 4, 1989; 54 FR 23897, June 2, 1989; 57 FR 33050, July 24, 1992]

§ 130.11 Program management.

(a) State agencies may apply for grants under sections 106, 205(j) and 205(g) to carry out water quality planning and management activities. Interstate agencies may apply for grants under section 106 to carry out water quality planning and management activities. Local or regional planning organizations may request 106 and 205(j) funds from a State for planning and management activities. Grant administrative requirements for these funds appear in 40 CFR parts 25, 29, 30, 33 and 35, subparts A and J.

(b) Grants under section 106 may be used to fund a wide range of activities, including but not limited to assessments of water quality, revision of water quality standards (WQS), development of alternative approaches to control pollution, implementation and enforcement of control measures and development or implementation of ground water programs. Grants under section 205(j) may be used to fund water quality management (WQM) planning activities but may not be used to fund implementation of control measures (see part 35, subpart A). Section 205(g) funds are used primarily to manage the wastewater treatment works construction grants program pursuant to the provisions of 40 CFR part 35, subpart J. A State may also use part of the 205(g) funds to administer approved permit programs under sections 402 and 404, to administer a statewide waste treatment management program under section 208(b)(4) and to manage waste treatment construction grants for small communities.

(c) Grant work programs for water quality planning and management shall describe geographic and functional priorities for use of grant funds in a manner which will facilitate EPA review of the grant application and subsequent evaluation of work accomplished with the grant funds. A State's 305(b) Report, WQM plan and other water quality assessments shall identify the State's priority water quality problems and areas. The WQM plan shall contain an analysis of alternative control measures and recommendations to control specific problems. Work programs shall specify the activities to be carried out during the period of the grant; the cost of specific activities; the outputs, for example, permits issued, intensive surveys, wasteload allocations, to be produced by each activity; and where applicable, schedules indicating when activities are to be completed.

(d) State work programs under sections 106, 205(j) and 205(g) shall be coordinated in a manner which indicates the funding from these grants dedicated to major functions, such as permitting, enforcement, monitoring, planning and standards, nonpoint source implementation, management of construction grants, operation and maintenance of treatment works, ground-water, emergency response and program management. States shall also describe how the activities funded by these grants are used in a coordinated manner to address the priority water quality problems identified in the State's water quality assessment under section 305(b).

(e) EPA, States, areawide agencies, interstate agencies, local and Regional governments, and designated management agencies (DMAs) are joint participants in the water pollution control program. States may enter into contractual arrangements or intergovernmental agreements with other agencies concerning the performance of water quality planning and management tasks. Such arrangements shall reflect the capabilities of the respective agencies and shall efficiently utilize available funds and funding eligibilities to meet Federal requirements commensurate with State and local priorities. State work programs under section 205(j) shall be developed jointly with local, Regional and other comprehensive planning organizations.

§130.12 **Coordination with other programs.**

(a) Relationship to the National Pollutant Discharge Elimination System (NPDES) program. In accordance with section 208(e) of the Act, no NPDES permit may be issued which is in conflict with an approved Water Quality Management (WQM) plan. Where a State has assumed responsibility for the administration of the permit program under section 402, it shall assure consistency with the WQM plan.

(b) Relationship to the municipal construction grants program. In accordance with sections 205(j), 216 and 303(e)(3)(H) of the Act, each State shall develop a system for setting priorities for funding construction of municipal wastewater treatment facilities under section 201 of the Act. The State, or the agency to which the State has delegated WQM planning functions, shall review each facility plan in its area for consistency with the approved WQM plan. Under section 208(d) of the Act, after a waste treatment management agency has been designated and a WQM plan approved, section 201 construction grant funds may be awarded only to those agencies for construction of treatment works in conformity with the approved WQM plan.

(c) Relationship to Federal activities—Each department, agency or instrumentality of the executive, legislative and judicial branches of the Federal Government having jurisdiction over any property or facility or engaged in any activity resulting, or which may result, in the discharge or runoff of pollutants shall comply with all Federal, State, interstate and local requirements, administrative authority, and process and sanctions respecting the control and abatement of water pollution in the same manner and extent as any non-governmental entity in accordance with section 313 of the CWA.

§ 130.15 Processing application for Indian tribes.

The Regional Administrator shall process an application of an Indian Tribe submitted under § 130.6(d) in a timely manner. He shall promptly notify the Indian Tribe of receipt of the application.

[54 FR 14360, Apr. 11, 1989, as amended at 59 FR 13818, Mar. 23, 1994]

PART 131—WATER QUALITY STANDARDS

Subpart A—General Provisions

AUTHORITY: 33 U.S.C. 1251 et seq.

SOURCE: 48 FR 51405, Nov. 8, 1983, unless otherwise noted.

Subpart A—General Provisions

§ 131.1   Scope.

This part describes the requirements and procedures for developing, reviewing, revising, and approving water quality standards by the States as authorized by section 303(c) of the Clean Water Act. Additional specific procedures for developing, reviewing, revising, and approving water quality standards for Great Lakes States or Great Lakes Tribes (as defined in 40 CFR 132.2) to conform to section 118 of the Clean Water Act and 40 CFR part 132, are provided in 40 CFR part 132.

[60 FR 15386, Mar. 23, 1995]

§ 131.2   Purpose.

A water quality standard defines the water quality goals of a water body, or portion thereof, by designating the use or uses to be made of the water and by setting criteria necessary to protect the uses. States adopt water quality standards to protect public health or welfare, enhance the quality of water and serve the purposes of the Clean

Water Act (the Act). "Serve the purposes of the Act" (as defined in sections 101(a)(2) and 303(c) of the Act) means that water quality standards should, wherever attainable, provide water quality for the protection and propagation of fish, shellfish and wildlife and for recreation in and on the water and take into consideration their use and value of public water supplies, propagation of fish, shellfish, and wildlife, recreation in and on the water, and agricultural, industrial, and other purposes including navigation.

Such standards serve the dual purposes of establishing the water quality goals for a specific water body and serve as the regulatory basis for the establishment of water-quality-based treatment controls and strategies beyond the technology-based levels of treatment required by sections 301(b) and 306 of the Act.

§131.3 Definitions.

(a) *The Act* means the Clean Water Act (Pub. L. 92–500, as amended (33 U.S.C. 1251 *et seq.*)).

(b) *Criteria* are elements of State water quality standards, expressed as constituent concentrations, levels, or narrative statements, representing a quality of water that supports a particular use. When criteria are met, water quality will generally protect the designated use.

(c) *Section 304(a) criteria* are developed by EPA under authority of section 304(a) of the Act based on the latest scientific information on the relationship that the effect of a constituent concentration has on particular aquatic species and/or human health. This information is issued periodically to the States as guidance for use in developing criteria.

(d) *Toxic pollutants* are those pollutants listed by the Administrator under section 307(a) of the Act.

(e) *Existing uses* are those uses actually attained in the water body on or after November 28, 1975, whether or not they are included in the water quality standards.

(f) *Designated uses* are those uses specified in water quality standards for each water body or segment whether or not they are being attained.

(g) *Use attainability analysis* is a structured scientific assessment of the factors affecting the attainment of the use which may include physical, chemical, biological, and economic factors as described in § 131.10(g).

(h) *Water quality limited segment* means any segment where it is known that water quality does not meet applicable water quality standards, and/or is not expected to meet applicable water quality standards, even after the application of the technology-bases effluent limitations required by sections 301(b) and 306 of the Act.

(i) *Water quality standards* are provisions of State or Federal law which consist of a designated use or uses for the waters of the United States and water quality criteria for such waters based upon such uses. Water quality standards are to protect the public health or welfare, enhance the quality of water and serve the purposes of the Act.

(j) *States* include: The 50 States, the District of Columbia, Guam, the Commonwealth of Puerto Rico, Virgin Islands, American Samoa, the Trust Territory of the Pacific Islands, the Commonwealth of the Northern Mariana Islands, and Indian Tribes that EPA determines to be eligible for purposes of water quality standards program. .

(k) *Federal Indian Reservation, Indian Reservation,* or *Reservation* means all land within the limits of any Indian reservation under the jurisdiction of the United States Government, notwithstanding the issuance of any patent, and including rights-of-way running through the reservation."

(l) *Indian Tribe* or *Tribe* means any Indian Tribe, band, group, or community recognized by the Secretary of the Interior and exercising governmental authority over a Federal Indian reservation.

[48 FR 51405, Nov. 8, 1983, as amended at 56 FR 64893, Dec. 12, 1991; 59 FR 64344, Dec. 14, 1994]

§131.4 State authority.

(a) States (as defined in §131.3) are responsible for reviewing, establishing, and revising water quality standards. As recognized by section 510 of the Clean Water Act, States may develop water quality standards more stringent than required by this regulation. Consistent with section 101(g) and 518(a) of the Clean Water Act, water quality standards shall not be construed to supersede or abrogate rights to quantities of water.

(b) States (as defined in § 131.3) may issue certifications pursuant to the requirements of Clean Water Act section 401. Revisions adopted by States shall be applicable for use in issuing State certifications consistent with the provisions of § 131.21(c).

(c) Where EPA determines that a Tribe is eligible to the same extent as a State for purposes of water quality standards, the Tribe likewise is eligible to the same extent as a State for purposes of certifications conducted under Clean Water Act section 401.

[56 FR 64893, Dec. 12, 1991, as amended at 59 FR 64344, Dec. 14, 1994]

§ 131.5 EPA authority.

(a) Under section 303(c) of the Act, EPA is to review and to approve or disapprove State-adopted water quality standards. The review involves a determination of:

(1) Whether the State has adopted water uses which are consistent with the requirements of the Clean Water Act;

(2) Whether the State has adopted criteria that protect the designated water uses;

(3) Whether the State has followed its legal procedures for revising or adopting standards;

(4) Whether the State standards which do not include the uses specified in section 101(a)(2) of the Act are based upon appropriate technical and scientific data and analyses, and

(5) Whether the State submission meets the requirements included in § 131.6 of this part and, for Great Lakes States or Great Lakes Tribes (as defined in 40 CFR 132.2) to conform to section 118 of the Act, the requirements of 40 CFR part 132.

(b) If EPA determines that the State's or Tribe's water quality standards are consistent with the factors listed in paragraphs (a)(1) through (a)(5) of this section, EPA approves the standards. EPA must disapprove the State's or Tribe's water quality standards and promulgate Federal standards

under section 303(c)(4), and for Great Lakes States or Great Lakes Tribes under section 118(c)(2)(C) of the Act, if State or Tribal adopted standards are not consistent with the factors listed in paragraphs (a)(1) through (a)(5) of this section. EPA may also promulgate a new or revised standard when necessary to meet the requirements of the Act.

(c) Section 401 of the Clean Water Act authorizes EPA to issue certifications pursuant to the requirements of section 401 in any case where a State or interstate agency has no authority for issuing such certifications.

[48 FR 51405, Nov. 8, 1983, as amended at 56 FR 64894, Dec. 12, 1991; 60 FR 15387, Mar. 23, 1995]

§ 131.6 Minimum requirements for water quality standards submission.

The following elements must be included in each State's water quality standards submitted to EPA for review:

(a) Use designations consistent with the provisions of sections 101(a)(2) and 303(c)(2) of the Act.

(b) Methods used and analyses conducted to support water quality standards revisions.

(c) Water quality criteria sufficient to protect the designated uses.

(d) An antidegradation policy consistent with § 131.12.

(e) Certification by the State Attorney General or other appropriate legal authority within the State that the water quality standards were duly adopted pursuant to State law.

(f) General information which will aid the Agency in determining the adequacy of the scientific basis of the standards which do not include the uses specified in section 101(a)(2) of the Act as well as information on general policies applicable to State standards which may affect their application and implementation.

§ 131.7 Dispute resolution mechanism.

(a) Where disputes between States and Indian Tribes arise as a result of differing water quality standards on common bodies of water, the lead EPA Regional Administrator, as determined based upon OMB circular A-105, shall be responsible for acting in accordance with the provisions of this section.

(b) The Regional Administrator shall attempt to resolve such disputes where:

(1) The difference in water quality standards results in unreasonable consequences;

(2) The dispute is between a State (as defined in §131.3(j) but exclusive of all Indian Tribes) and a Tribe which EPA has determined is eligible to the same extent as a State for purposes of water quality standards;

(3) A reasonable effort to resolve the dispute without EPA involvement has been made;

(4) The requested relief is consistent with the provisions of the Clean Water Act and other relevant law;

(5) The differing State and Tribal water quality standards have been adopted pursuant to State and Tribal law and approved by EPA; and

(6) A valid written request has been submitted by either the Tribe or the State.

(c) Either a State or a Tribe may request EPA to resolve any dispute which satisfies the criteria of paragraph (b) of this section. Written requests for EPA involvement should be submitted to the lead Regional Administrator and must include:

(1) A concise statement of the unreasonable consequences that are alleged to have arisen because of differing water quality standards;

(2) A concise description of the actions which have been taken to resolve the dispute without EPA involvement;

(3) A concise indication of the water quality standards provision which has resulted in the alleged unreasonable consequences;

(4) Factual data to support the alleged unreasonable consequences; and

(5) A statement of the relief sought from the alleged unreasonable consequences.

(d) Where, in the Regional Administrator's judgment, EPA involvement is appropriate based on the factors of paragraph (b) of this section, the Regional Administrator shall, within 30 days, notify the parties in writing that he/she is initiating an EPA dispute resolution action and solicit their written response. The Regional Administrator shall also make reasonable efforts to ensure that other interested individuals or groups have notice of this action. Such efforts shall include but not be limited to the following:

(1) Written notice to responsible Tribal and State Agencies, and other affected Federal agencies,

(2) Notice to the specific individual or entity that is alleging that an unreasonable consequence is resulting from differing standards having been adopted on a common body of water,

(3) Public notice in local newspapers, radio, and television, as appropriate,

(4) Publication in trade journal newsletters, and

(5) Other means as appropriate.

(e) If in accordance with applicable State and Tribal law an Indian Tribe and State have entered into an agreement that resolves the dispute or establishes a mechanism for resolving a dispute, EPA shall defer to this agreement where it is consistent with the Clean Water Act and where it has been approved by EPA.

(f) EPA dispute resolution actions shall be consistent with one or a combination of the following options:

(1) *Mediation.* The Regional Administrator may appoint a mediator to mediate the dispute. Mediators shall be EPA employees, employees from other Federal agencies, or other individuals with appropriate qualifications.

(i) Where the State and Tribe agree to participate in the dispute resolution process, mediation with the intent to establish Tribal-State agreements, consistent with Clean Water Act section 518(d), shall normally be pursued as a first effort.

(ii) Mediators shall act as neutral facilitators whose function is to encourage communication and negotiation between all parties to the dispute.

(iii) Mediators may establish advisory panels, to consist in part of representatives from the affected parties, to study the problem and recommend an appropriate solution.

(iv) The procedure and schedule for mediation of individual disputes shall be determined by the mediator in consultation with the parties.

(v) If formal public hearings are held in connection with the actions taken under this paragraph, Agency requirements at 40 CFR 25.5 shall be followed.

(2) *Arbitration.* Where the parties to the dispute agree to participate in the dispute resolution process, the Regional Administrator may appoint an arbitrator or arbitration panel to arbitrate the dispute. Arbitrators and panel members shall be EPA employees, employees from other Federal agencies, or other individuals with appropriate qualifications. The Regional administrator shall select as arbitrators and arbitration panel members individuals who are agreeable to all parties, are knowledgeable concerning the requirements of the water quality standards program, have a basic understanding of the political and economic interests of Tribes and States involved, and are expected to fulfill the duties fairly and impartially.

(i) The arbitrator or arbitration panel shall conduct one or more private or public meetings with the parties and actively solicit information pertaining to the effects of differing water quality permit requirements on upstream and downstream dischargers, comparative risks to public health and the environment, economic impacts, present and historical water uses, the quality of the waters subject to such standards, and other factors relevant to the dispute, such as whether proposed water quality criteria are more stringent than necessary to support designated uses, more stringent than natural background water quality or whether designated uses are reasonable given natural background water quality.

(ii) Following consideration of relevant factors as defined in paragraph (f)(2)(i) of this section, the arbitrator or arbitration panel shall have the authority and responsibility to provide all parties and the Regional Administrator with a written recommendation for resolution of the dispute. Arbitration panel recommendations shall, in general, be reached by majority vote. However, where the parties agree to binding arbitration, or where required by the Regional Administrator, recommendations of such arbitration panels may be unanimous decisions. Where binding or non-binding arbitration panels cannot reach a unanimous rec-

ommendation after a reasonable period of time, the Regional Administrator may direct the panel to issue a nonbinding decision by majority vote.

(iii) The arbitrator or arbitration panel members may consult with EPA's Office of General Counsel on legal issues, but otherwise shall have no *ex parte* communications pertaining to the dispute. Federal employees who are arbitrators or arbitration panel members shall be neutral and shall not be predisposed for or against the position of any disputing party based on any Federal Trust responsibilities which their employers may have with respect to the Tribe. In addition, arbitrators or arbitration panel members who are Federal employees shall act independently from the normal hierarchy within their agency.

(iv) The parties are not obligated to abide by the arbitrator's or arbitration panel's recommendation unless they voluntarily entered into a binding agreement to do so.

(v) If a party to the dispute believes that the arbitrator or arbitration panel has recommended an action contrary to or inconsistent with the Clean Water Act, the party may appeal the arbitrator's recommendation to the Regional Administrator. The request for appeal must be in writing and must include a description of the statutory basis for altering the arbitrator's recommendation.

(vi) The procedure and schedule for arbitration of individual disputes shall be determined by the arbitrator or arbitration panel in consultation with parties.

(vii) If formal public hearings are held in connection with the actions taken under this paragraph, Agency requirements at 40 CFR 25.5 shall be followed.

(3) *Dispute resolution default procedure.* Where one or more parties (as defined in paragraph (g) of this section) refuse to participate in either the mediation or arbitration dispute resolution processes, the Regional Administrator may appoint a single official or panel to review available information pertaining to the dispute and to issue a written recommendation for resolving the dispute. Review officials shall be EPA employees, employees from other

Federal agencies, or other individuals with appropriate qualifications. Review panels shall include appropriate members to be selected by the Regional Administrator in consultation with the participating parties. Recommendations of such review officials or panels shall, to the extent possible given the lack of participation by one or more parties, be reached in a manner identical to that for arbitration of disputes specified in paragraphs (f)(2)(i) through (f)(2)(vii) of this section.

(g) *Definitions.* For the purposes of this section:

(1) *Dispute Resolution Mechanism* means the EPA mechanism established pursuant to the requirements of Clean Water Act section 518(e) for resolving unreasonable consequences that arise as a result of differing water quality standards that may be set by States and Indian Tribes located on common bodies of water.

(2) *Parties* to a State-Tribal dispute include the State and the Tribe and may, at the discretion of the Regional Administrator, include an NPDES permittee, citizen, citizen group, or other affected entity.

[56 FR 64894, Dec. 12, 1991, as amended at 59 FR 64344, Dec. 14, 1994]

§131.8 Requirements for Indian Tribes to administer a water quality standards program.

(a) The Regional Administrator, as determined based on OMB Circular A–105, may accept and approve a tribal application for purposes of administering a water quality standards program if the Tribe meets the following criteria:

(1) The Indian Tribe is recognized by the Secretary of the Interior and meets the definitions in §131.3 (k) and (l),

(2) The Indian Tribe has a governing body carrying out substantial governmental duties and powers,

(3) The water quality standards program to be administered by the Indian Tribe pertains to the management and protection of water resources which are within the borders of the Indian reservation and held by the Indian Tribe, within the borders of the Indian reservation and held by the United States in trust for Indians, within the borders of the Indian reservation and held by a member of the Indian Tribe if such property interest is subject to a trust restriction on alienation, or otherwise within the borders of the Indian reservation, and

(4) The Indian Tribe is reasonably expected to be capable, in the Regional Administrator's judgment, of carrying out the functions of an effective water quality standards program in a manner consistent with the terms and purposes of the Act and applicable regulations.

(b) Requests by Indian Tribes for administration of a water quality standards program should be submitted to the lead EPA Regional Administrator. The application shall include the following information:

(1) A statement that the Tribe is recognized by the Secretary of the Interior.

(2) A descriptive statement demonstrating that the Tribal governing body is currently carrying out substantial governmental duties and powers over a defined area. The statement should:

(i) Describe the form of the Tribal government;

(ii) Describe the types of governmental functions currently performed by the Tribal governing body such as, but not limited to, the exercise of police powers affecting (or relating to) the health, safety, and welfare of the affected population, taxation, and the exercise of the power of eminent domain; and

(iii) Identify the source of the Tribal government's authority to carry out the governmental functions currently being performed.

(3) A descriptive statement of the Indian Tribe's authority to regulate water quality. The statement should include:

(i) A map or legal description of the area over which the Indian Tribe asserts authority to regulate surface water quality;

(ii) A statement by the Tribe's legal counsel (or equivalent official) which describes the basis for the Tribes assertion of authority and which may include a copy of documents such as Tribal constitutions, by-laws, charters, executive orders, codes, ordinances, and/or resolutions which support the Tribe's assertion of authority; and

(iii) An identification of the surface waters for which the Tribe proposes to establish water quality standards.

(4) A narrative statement describing the capability of the Indian Tribe to administer an effective water quality standards program. The narrative statement should include:

(i) A description of the Indian Tribe's previous management experience which may include the administration of programs and services authorized by the Indian Self-Determination and Education Assistance Act (25 U.S.C. 450 *et seq.*), the Indian Mineral Development Act (25 U.S.C. 2101 *et seq.*), or the Indian Sanitation Facility Construction Activity Act (42 U.S.C. 2004a);

(ii) A list of existing environmental or public health programs administered by the Tribal governing body and copies of related Tribal laws, policies, and regulations;

(iii) A description of the entity (or entities) which exercise the executive, legislative, and judicial functions of the Tribal government;

(iv) A description of the existing, or proposed, agency of the Indian Tribe which will assume primary responsibility for establishing, reviewing, implementing and revising water quality standards;

(v) A description of the technical and administrative capabilities of the staff to administer and manage an effective water quality standards program or a plan which proposes how the Tribe will acquire additional administrative and technical expertise. The plan must address how the Tribe will obtain the funds to acquire the administrative and technical expertise.

(5) Additional documentation required by the Regional Administrator which, in the judgment of the Regional Administrator, is necessary to support a Tribal application.

(6) Where the Tribe has previously qualified for eligibility or "treatment as a state" under a Clean Water Act or Safe Drinking Water Act program, the Tribe need only provide the required information which has not been submitted in a previous application.

(c) Procedure for processing an Indian Tribe's application.

(1) The Regional Administrator shall process an application of an Indian

Tribe submitted pursuant to § 131.8(b) in a timely manner. He shall promptly notify the Indian Tribe of receipt of the application.

(2) Within 30 days after receipt of the Indian Tribe's application the Regional Administrator shall provide appropriate notice. Notice shall:

(i) Include information on the substance and basis of the Tribe's assertion of authority to regulate the quality of reservation waters; and

(ii) Be provided to all appropriate governmental entities.

(3) The Regional Administrator shall provide 30 days for comments to be submitted on the Tribal application. Comments shall be limited to the Tribe's assertion of authority.

(4) If a Tribe's asserted authority is subject to a competing or conflicting claim, the Regional Administrator, after due consideration, and in consideration of other comments received, shall determine whether the Tribe has adequately demonstrated that it meets the requirements of § 131.8(a)(3).

(5) Where the Regional Administrator determines that a Tribe meets the requirements of this section, he shall promptly provide written notification to the Indian Tribe that the Tribe is authorized to administer the Water Quality Standards program.

[56 FR 64895, Dec. 12, 1991, as amended at 59 FR 64344, Dec. 14, 1994]

Subpart B—Establishment of Water Quality Standards

§ 131.10 Designation of uses.

(a) Each State must specify appropriate water uses to be achieved and protected. The classification of the waters of the State must take into consideration the use and value of water for public water supplies, protection and propagation of fish, shellfish and wildlife, recreation in and on the water, agricultural, industrial, and other purposes including navigation. In no case shall a State adopt waste transport or waste assimilation as a designated use for any waters of the United States.

(b) In designating uses of a water body and the appropriate criteria for those uses, the State shall take into consideration the water quality standards of downstream waters and shall ensure that its water quality standards provide for the attainment and maintenance of the water quality standards of downstream waters.

(c) States may adopt sub-categories of a use and set the appropriate criteria to reflect varying needs of such sub-categories of uses, for instance, to differentiate between cold water and warm water fisheries.

(d) At a minimum, uses are deemed attainable if they can be achieved by the imposition of effluent limits required under sections 301(b) and 306 of the Act and cost-effective and reasonable best management practices for nonpoint source control.

(e) Prior to adding or removing any use, or establishing sub-categories of a use, the State shall provide notice and an opportunity for a public hearing under §131.20(b) of this regulation.

(f) States may adopt seasonal uses as an alternative to reclassifying a water body or segment thereof to uses requiring less stringent water quality criteria. If seasonal uses are adopted, water quality criteria should be adjusted to reflect the seasonal uses, however, such criteria shall not preclude the attainment and maintenance of a more protective use in another season.

(g) States may remove a designated use which is *not* an existing use, as defined in §131.3, or establish sub-categories of a use if the State can demonstrate that attaining the designated use is not feasible because:

(1) Naturally occurring pollutant concentrations prevent the attainment of the use; or

(2) Natural, ephemeral, intermittent or low flow conditions or water levels prevent the attainment of the use, unless these conditions may be compensated for by the discharge of sufficient volume of effluent discharges without violating State water conservation requirements to enable uses to be met; or

(3) Human caused conditions or sources of pollution prevent the attainment of the use and cannot be remedied or would cause more environmental damage to correct than to leave in place; or

(4) Dams, diversions or other types of hydrologic modifications preclude the attainment of the use, and it is not feasible to restore the water body to its original condition or to operate such modification in a way that would result in the attainment of the use; or

(5) Physical conditions related to the natural features of the water body, such as the lack of a proper substrate, cover, flow, depth, pools, riffles, and the like, unrelated to water quality, preclude attainment of aquatic life protection uses; or

(6) Controls more stringent than those required by sections 301(b) and 306 of the Act would result in substantial and widespread economic and social impact.

(h) States may not remove designated uses if:

(1) They are existing uses, as defined in §131.3, unless a use requiring more stringent criteria is added; or

(2) Such uses will be attained by implementing effluent limits required under sections 301(b) and 306 of the Act and by implementing cost-effective and reasonable best management practices for nonpoint source control.

(i) Where existing water quality standards specify designated uses less than those which are presently being attained, the State shall revise its standards to reflect the uses actually being attained.

(j) A State must conduct a use attainability analysis as described in §131.3(g) whenever:

(1) The State designates or has designated uses that do not include the uses specified in section 101(a)(2) of the Act, or

(2) The State wishes to remove a designated use that is specified in section 101(a)(2) of the Act or to adopt subcategories of uses specified in section 101(a)(2) of the Act which require less stringent criteria.

(k) A State is not required to conduct a use attainability analysis under this regulation whenever designating uses which include those specified in section 101(a)(2) of the Act.

§131.11 Criteria.

(a) *Inclusion of pollutants:* (1) States must adopt those water quality criteria that protect the designated use.

Such criteria must be based on sound scientific rationale and must contain sufficient parameters or constituents to protect the designated use. For waters with multiple use designations, the criteria shall support the most sensitive use.

(2) *Toxic pollutants.* States must review water quality data and information on discharges to identify specific water bodies where toxic pollutants may be adversely affecting water quality or the attainment of the designated water use or where the levels of toxic pollutants are at a level to warrant concern and must adopt criteria for such toxic pollutants applicable to the water body sufficient to protect the designated use. Where a State adopts narrative criteria for toxic pollutants to protect designated uses, the State must provide information identifying the method by which the State intends to regulate point source discharges of toxic pollutants on water quality limited segments based on such narrative criteria. Such information may be included as part of the standards or may be included in documents generated by the State in response to the Water Quality Planning and Management Regulations (40 CFR part 35).

(b) Form of criteria: In establishing criteria, States should:

(1) Establish numerical values based on:

(i) 304(a) Guidance; or

(ii) 304(a) Guidance modified to reflect site-specific conditions; or

(iii) Other scientifically defensible methods;

(2) Establish narrative criteria or criteria based upon biomonitoring methods where numerical criteria cannot be established or to supplement numerical criteria.

§ 131.12   Antidegradation policy.

(a) The State shall develop and adopt a statewide antidegradation policy and identify the methods for implementing such policy pursuant to this subpart. The antidegradation policy and implementation methods shall, at a minimum, be consistent with the following:

(1) Existing instream water uses and the level of water quality necessary to protect the existing uses shall be maintained and protected.

(2) Where the quality of the waters exceed levels necessary to support propagation of fish, shellfish, and wildlife and recreation in and on the water, that quality shall be maintained and protected unless the State finds, after full satisfaction of the intergovernmental coordination and public participation provisions of the State's continuing planning process, that allowing lower water quality is necessary to accommodate important economic or social development in the area in which the waters are located. In allowing such degradation or lower water quality, the State shall assure water quality adequate to protect existing uses fully. Further, the State shall assure that there shall be achieved the highest statutory and regulatory requirements for all new and existing point sources and all cost-effective and reasonable best management practices for nonpoint source control.

(3) Where high quality waters constitute an outstanding National resource, such as waters of National and State parks and wildlife refuges and waters of exceptional recreational or ecological significance, that water quality shall be maintained and protected.

(4) In those cases where potential water quality impairment associated with a thermal discharge is involved, the antidegradation policy and implementing method shall be consistent with section 316 of the Act.

§ 131.13   General policies.

States may, at their discretion, include in their State standards, policies generally affecting their application and implementation, such as mixing zones, low flows and variances. Such policies are subject to EPA review and approval.

Subpart C—Procedures for Review and Revision of Water Quality Standards

§ 131.20   State review and revision of water quality standards.

(a) *State review.* The State shall from time to time, but at least once every three years, hold public hearings for the purpose of reviewing applicable water quality standards and, as appropriate, modifying and adopting standards. Any water body segment with water quality standards that do not include the uses specified in section 101(a)(2) of the Act shall be re-examined every three years to determine if any new information has become available. If such new information indicates that the uses specified in section 101(a)(2) of the Act are attainable, the State shall revise its standards accordingly. Procedures States establish for identifying and reviewing water bodies for review should be incorporated into their Continuing Planning Process.

(b) *Public participation.* The State shall hold a public hearing for the purpose of reviewing water quality standards, in accordance with provisions of State law, EPA's water quality management regulation (40 CFR 130.3(b)(6)) and public participation regulation (40 CFR part 25). The proposed water quality standards revision and supporting analyses shall be made available to the public prior to the hearing.

(c) *Submittal to EPA.* The State shall submit the results of the review, any supporting analysis for the use attainability analysis, the methodologies used for site-specific criteria development, any general policies applicable to water quality standards and any revisions of the standards to the Regional Administrator for review and approval, within 30 days of the final

State action to adopt and certify the revised standard, or if no revisions are made as a result of the review, within 30 days of the completion of the review.

§131.21 EPA review and approval of water quality standards.

(a) After the State submits its officially adopted revisions, the Regional Administrator shall either:

(1) Notify the State within 60 days that the revisions are approved, or

(2) Notify the State within 90 days that the revisions are disapproved. Such notification of disapproval shall specify the changes needed to assure compliance with the requirements of the Act and this regulation, and shall explain why the State standard is not in compliance with such requirements. Any new or revised State standard must be accompanied by some type of supporting analysis.

(b) The Regional Administrator's approval or disapproval of a State water quality standard shall be based on the requirements of the Act as described in §§131.5 and 131.6, and, with respect to Great Lakes States or Tribes (as defined in 40 CFR 132.2), 40 CFR part 132.

(c) *How do I determine which water quality standards are applicable for purposes of the Act?* You may determine which water quality standards are applicable water quality standards for purposes of the Act from the following table:

If—	Then—	Unless or until—	In which case—
(1) A State or authorized Tribe has adopted a water quality standard that is effective under State or Tribal law and has been submitted to EPA before May 30, 2000...	...the State or Tribe's water quality standard is the applicable water quality standard for purposes of the Act...	...EPA has promulgated a more stringent water quality standard for the State or Tribe that is in effect...	...the EPA-promulgated water quality standard is the applicable water quality standard for purposes of the Act until EPA withdraws the Federal water quality standard.
(2) A State or authorized Tribe adopts a water quality standard that goes into effect under State or Tribal law on or *after* May 30, 2000...	...once EPA approves that water quality standard, it becomes the applicable water quality standard for purposes of the Act...	...EPA has promulgated a more stringent water quality standard for the State or Tribe that is in effect...	...the EPA promulgated water quality standard is the applicable water quality standard for purposes of the Act until EPA withdraws the Federal water quality standard.

(d) *When do I use the applicable water quality standards identified in paragraph (c) above?* Applicable water quality standards for purposes of the Act are the minimum standards which must be

used when the CWA and regulations implementing the CWA refer to water quality standards, for example, in identifying impaired waters and calculating TMDLs under section 303(d), developing NPDES permit limitations under section 301(b)(1)(C), evaluating proposed discharges of dredged or fill material under section 404, and in issuing certifications under section 401 of the Act.

(e) *For how long does an applicable water quality standard for purposes of the Act remain the applicable water quality standard for purposes of the Act?* A State or authorized Tribe's applicable water quality standard for purposes of the Act remains the applicable standard until EPA approves a change, deletion, or addition to that water quality standard, or until EPA promulgates a more stringent water quality standard.

(f) *How can I find out what the applicable standards are for purposes of the Act?* In each Regional office, EPA maintains a docket system for the States and authorized Tribes in that Region, available to the public, identifying the applicable water quality standards for purposes of the Act.

[48 FR 51405, Nov. 8, 1983, as amended at 60 FR 15387, Mar. 23, 1995; 65 FR 24653, Apr. 27, 2000]

§ 131.22 EPA promulgation of water quality standards.

(a) If the State does not adopt the changes specified by the Regional Administrator within 90 days after notification of the Regional Administrator's disapproval, the Administrator shall promptly propose and promulgate such standard.

(b) The Administrator may also propose and promulgate a regulation, applicable to one or more States, setting forth a new or revised standard upon determining such a standard is necessary to meet the requirements of the Act.

(c) In promulgating water quality standards, the Administrator is subject to the same policies, procedures, analyses, and public participation requirements established for States in these regulations.

Subpart D—Federally Promulgated Water Quality Standards

§ 131.31 Arizona.

(a) [Reserved]

(b) The following waters have, in addition to the uses designated by the State, the designated use of fish consumption as defined in R18–11–101 (which is available from the Arizona Department of Environmental Quality, Water Quality Division, 3033 North Central Ave., Phoenix, AZ 85012):

COLORADO MAIN STEM RIVER BASIN:
    Hualapai Wash
MIDDLE GILA RIVER BASIN:
    Agua Fria River (Camelback Road to Avondale WWTP)
    Galena Gulch
    Gila River (Felix Road to the Salt River)
    Queen Creek (Headwaters to the Superior WWTP)
    Queen Creek (Below Potts Canyon)
SAN PEDRO RIVER BASIN:
    Copper Creek
SANTA CRUZ RIVER BASIN:
    Agua Caliente Wash
    Nogales Wash
    Sonoita Creek (Above the town of Patagonia)
    Tanque Verde Creek
    Tinaja Wash
    Davidson Canyon
UPPER GILA RIVER BASIN
    Chase Creek

(c) To implement the requirements of R18–11–108.A.5 with respect to effects of mercury on wildlife, EPA (or the State with the approval of EPA) shall implement a monitoring program to assess attainment of the water quality standard.

(Sec. 303, Federal Water Pollution Control Act, as amended, 33 U.S.C. 1313, 86 Stat. 816 *et seq.*, Pub. L. 92–500; Clean Water Act, Pub. L. 92–500, as amended; 33 U.S.C. 1251 *et seq.*)

[41 FR 25000, June 22, 1976; 41 FR 48737, Nov. 5, 1976. Redesignated and amended at 42 FR 56740, Oct. 28, 1977. Further redesignated and amended at 48 FR 51408, Nov. 8, 1983; 61 FR 20693, May 7, 1996; 68 FR 62744, Nov. 6, 2003]

§ 131.32 [Reserved]

§ 131.33 Idaho.

(a) *Temperature criteria for bull trout.*
(1) Except for those streams or portions

of streams located in Indian country, or as may be modified by the Regional Administrator, EPA Region X, pursuant to paragraph (a)(3) of this section, a temperature criterion of 10 °C, expressed as an average of daily maximum temperatures over a seven-day period, applies to the waterbodies identified in paragraph (a)(2) of this section during the months of June, July, August and September.

(2) The following waters are protected for bull trout spawning and rearing:

(i) BOISE-MORE BASIN: Devils Creek, East Fork Sheep Creek, Sheep Creek.

(ii) BROWNLEE RESERVOIR BASIN: Crooked River, Indian Creek.

(iii) CLEARWATER BASIN: Big Canyon Creek, Cougar Creek, Feather Creek, Laguna Creek, Lolo Creek, Orofino Creek, Talapus Creek, West Fork Potlatch River.

(iv) COEUR D'ALENE LAKE BASIN: Cougar Creek, Fernan Creek, Kid Creek, Mica Creek, South Fork Mica Creek, Squaw Creek, Turner Creek.

(v) HELLS CANYON BASIN: Dry Creek, East Fork Sheep Creek, Getta Creek, Granite Creek, Kurry Creek, Little Granite Creek, Sheep Creek.

(vi) LEMHI BASIN: Adams Creek, Alder Creek, Basin Creek, Bear Valley Creek, Big Eightmile Creek, Big Springs Creek, Big Timber Creek, Bray Creek, Bull Creek, Cabin Creek, Canyon Creek, Carol Creek, Chamberlain Creek, Clear Creek, Climb Creek, Cooper Creek, Dairy Creek, Deer Creek, Deer Park Creek, East Fork Hayden Creek, Eighteenmile Creek, Falls Creek, Ferry Creek, Ford Creek, Geertson Creek, Grove Creek, Hawley Creek, Hayden Creek, Kadletz Creek, Kenney Creek, Kirtley Creek, Lake Creek, Lee Creek, Lemhi River (above Big Eightmile Creek), Little Eightmile Creek, Little Mill Creek, Little Timber Creek, Middle Fork Little Timber Creek, Milk Creek, Mill Creek, Mogg Creek, North Fork Kirtley Creek, North Fork Little Timber Creek, Paradise Creek, Patterson Creek, Payne Creek, Poison Creek, Prospect Creek, Rocky Creek, Short Creek, Squaw Creek, Squirrel Creek, Tobias Creek, Trail Creek, West Fork Hayden Creek, Wright Creek.

(vii) LITTLE LOST BASIN: Badger Creek, Barney Creek, Bear Canyon, Bear Creek, Bell Mountain Creek, Big Creek, Bird Canyon, Black Creek, Buck Canyon, Bull Creek, Cedar Run Creek, Chicken Creek, Coal Creek, Corral Creek, Deep Creek, Dry Creek, Dry Creek Canal, Firbox Creek, Garfield Creek, Hawley Canyon, Hawley Creek, Horse Creek, Horse Lake Creek, Iron Creek, Jackson Creek, Little Lost River (above Badger Creek), Mahogany Creek, Main Fork Sawmill Creek, Massacre Creek, Meadow Creek, Mill Creek, Moffett Creek, Moonshine Creek, Quigley Creek, Red Rock Creek, Sands Creek, Sawmill Creek, Slide Creek, Smithie Fork, Squaw Creek, Summerhouse Canyon, Summit Creek, Timber Creek, Warm Creek, Wet Creek, Williams Creek.

(viii) LITTLE SALMON BASIN: Bascum Canyon, Boulder Creek, Brown Creek, Campbell Ditch, Castle Creek, Copper Creek, Granite Fork Lake Fork Rapid River, Hard Creek, Hazard Creek, Lake Fork Rapid River, Little Salmon River (above Hazard Creek), Paradise Creek, Pony Creek, Rapid River, Squirrel Creek, Trail Creek, West Fork Rapid River.

(ix) LOCHSA BASIN: Apgar Creek, Badger Creek, Bald Mountain Creek, Beaver Creek, Big Flat Creek, Big Stew Creek, Boulder Creek, Brushy Fork, Cabin Creek, Castle Creek, Chain Creek, Cliff Creek, Coolwater Creek, Cooperation Creek, Crab Creek, Crooked Fork Lochsa River, Dan Creek, Deadman Creek, Doe Creek, Dutch Creek, Eagle Creek, East Fork Papoose Creek, East Fork Split Creek, East Fork Squaw Creek, Eel Creek, Fern Creek, Fire Creek, Fish Creek, Fish Lake Creek, Fox Creek, Gass Creek, Gold Creek, Ham Creek, Handy Creek, Hard Creek, Haskell Creek, Heather Creek, Hellgate Creek, Holly Creek, Hopeful Creek, Hungery Creek, Indian Grave Creek, Jay Creek, Kerr Creek, Kube Creek, Lochsa River, Lone Knob Creek, Lottie Creek, Macaroni Creek, Maud Creek, Middle Fork Clearwater River, No-see-um Creek, North Fork Spruce Creek, North Fork Storm Creek, Nut Creek, Otter Slide Creek, Pack Creek, Papoose Creek, Parachute Creek, Pass Creek, Pedro Creek, Pell Creek, Pete King Creek, Placer Creek,

Polar Creek, Postoffice Creek, Queen Creek, Robin Creek, Rock Creek, Rye Patch Creek, Sardine Creek, Shoot Creek, Shotgun Creek, Skookum Creek, Snowshoe Creek, South Fork Spruce Creek, South Fork Storm Creek, Split Creek, Sponge Creek, Spring Creek, Spruce Creek, Squaw Creek, Storm Creek, Tick Creek, Tomcat Creek, Tumble Creek, Twin Creek, Wag Creek, Walde Creek, Walton Creek, Warm Springs Creek, Weir Creek, Wendover Creek, West Fork Boulder Creek, West Fork Papoose Creek, West Fork Squaw Creek, West Fork Wendover Creek, White Sands Creek, Willow Creek.

(x) LOWER CLARK FORK BASIN: Cascade Creek, East Fork, East Fork Creek, East Forkast Fork Creek, Gold Creek, Johnson Creek, Lightning Creek, Mosquito Creek, Porcupine Creek, Rattle Creek, Spring Creek, Twin Creek, Wellington Creek.

(xi) LOWER KOOTENAI BASIN: Ball Creek, Boundary Creek, Brush Creek, Cabin Creek, Caribou Creek, Cascade Creek, Cooks Creek, Cow Creek, Curley Creek, Deep Creek, Grass Creek, Jim Creek, Lime Creek, Long Canyon Creek, Mack Creek, Mission Creek, Myrtle Creek, Peak Creek, Snow Creek, Trout Creek.

(xii) LOWER MIDDLE FORK SALMON BASIN: Acorn Creek, Alpine Creek, Anvil Creek, Arrastra Creek, Bar Creek, Beagle Creek, Beaver Creek, Belvidere Creek, Big Creek, Birdseye Creek, Boulder Creek, Brush Creek, Buck Creek, Bull Creek, Cabin Creek, Camas Creek, Canyon Creek, Castle Creek, Clark Creek, Coin Creek, Corner Creek, Coxey Creek, Crooked Creek, Doe Creek, Duck Creek, East Fork Holy Terror Creek, Fawn Creek, Flume Creek, Fly Creek, Forge Creek, Furnace Creek, Garden Creek, Government Creek, Grouse Creek, Hammer Creek, Hand Creek, Holy Terror Creek, J Fell Creek, Jacobs Ladder Creek, Lewis Creek, Liberty Creek, Lick Creek, Lime Creek, Little Jacket Creek, Little Marble Creek, Little White Goat Creek, Little Woodtick Creek, Logan Creek, Lookout Creek, Loon Creek, Martindale Creek, Meadow Creek, Middle Fork Smith Creek, Monumental Creek, Moore Creek, Mulligan Creek, North Fork Smith Creek, Norton

Creek, Placer Creek, Pole Creek, Rams Creek, Range Creek, Routson Creek, Rush Creek, Sawlog Creek, Sheep Creek, Sheldon Creek, Shellrock Creek, Ship Island Creek, Shovel Creek, Silver Creek, Smith Creek, Snowslide Creek, Soldier Creek, South Fork Camas Creek, South Fork Chamberlain Creek, South Fork Holy Terror Creek, South Fork Norton Creek, South Fork Rush Creek, South Fork Sheep Creek, Spider Creek, Spletts Creek, Telephone Creek, Trail Creek, Two Point Creek, West Fork Beaver Creek, West Fork Camas Creek, West Fork Monumental Creek, West Fork Rush Creek, White Goat Creek, Wilson Creek.

(xiii) LOWER NORTH FORK CLEARWATER BASIN: Adair Creek, Badger Creek, Bathtub Creek, Beaver Creek, Black Creek, Brush Creek, Buck Creek, Butte Creek, Canyon Creek, Caribou Creek, Crimper Creek, Dip Creek, Dog Creek, Elmer Creek, Falls Creek, Fern Creek, Goat Creek, Isabella Creek, John Creek, Jug Creek, Jungle Creek, Lightning Creek, Little Lost Lake Creek, Little North Fork Clearwater River, Lost Lake Creek, Lund Creek, Montana Creek, Mowitch Creek, Papoose Creek, Pitchfork Creek, Rocky Run, Rutledge Creek, Spotted Louis Creek, Triple Creek, Twin Creek, West Fork Montana Creek, Willow Creek.

(xiv) LOWER SALMON BASIN: Bear Gulch, Berg Creek, East Fork John Day Creek, Elkhorn Creek, Fiddle Creek, French Creek, Hurley Creek, John Day Creek, Kelly Creek, Klip Creek, Lake Creek, Little Slate Creek, Little Van Buren Creek, No Business Creek, North Creek, North Fork Slate Creek, North Fork White Bird Creek, Partridge Creek, Slate Creek, Slide Creek, South Fork John Day Creek, South Fork White Bird Creek, Warm Springs Creek.

(xv) LOWER SELWAY BASIN: Anderson Creek, Bailey Creek, Browns Spring Creek, Buck Lake Creek, Butte Creek, Butter Creek, Cabin Creek, Cedar Creek, Chain Creek, Chute Creek, Dent Creek, Disgrace Creek, Double Creek, East Fork Meadow Creek, East Fork Moose Creek, Elbow Creek, Fivemile Creek, Fourmile Creek, Gate Creek, Gedney Creek, Goddard Creek, Horse Creek, Indian Hill Creek, Little Boulder Creek, Little

Schwar Creek, Matteson Creek, Meadow Creek, Monument Creek, Moose Creek, Moss Creek, Newsome Creek, North Fork Moose Creek, Rhoda Creek, Saddle Creek, Schwar Creek, Shake Creek, Spook Creek, Spur Creek, Tamarack Creek, West Fork Anderson Creek, West Fork Gedney Creek, West Moose Creek, Wounded Doe Creek.

(xvi) MIDDLE FORK CLEARWATER BASIN: Baldy Creek, Big Cedar Creek, Browns Spring Creek, Clear Creek, Middle Fork Clear Creek, Pine Knob Creek, South Fork Clear Creek.

(xvii) MIDDLE FORK PAYETTE BASIN: Bull Creek, Middle Fork Payette River (above Fool Creek), Oxtail Creek, Silver Creek, Sixteen-to-one Creek.

(xviii) MIDDLE SALMON-CHAMBERLAIN BASIN: Arrow Creek, Bargamin Creek, Bat Creek, Bay Creek, Bear Creek, Bend Creek, Big Elkhorn Creek, Big Harrington Creek, Big Mallard Creek, Big Squaw Creek, Bleak Creek, Bronco Creek, Broomtail Creek, Brown Creek, Cayuse Creek, Center Creek, Chamberlain Creek, Cliff Creek, Colt Creek, Corn Creek, Crooked Creek, Deer Creek, Dennis Creek, Disappointment Creek, Dismal Creek, Dog Creek, East Fork Fall Creek, East Fork Horse Creek, East Fork Noble Creek, Fall Creek, Filly Creek, Fish Creek, Flossie Creek, Game Creek, Gap Creek, Ginger Creek, Green Creek, Grouse Creek, Guard Creek, Hamilton Creek, Horse Creek, Hot Springs Creek, Hotzel Creek, Hungry Creek, Iodine Creek, Jack Creek, Jersey Creek, Kitchen Creek, Lake Creek, Little Horse Creek, Little Lodgepole Creek, Little Mallard Creek, Lodgepole Creek, Mayflower Creek, McCalla Creek, Meadow Creek, Moose Creek, Moose Jaw Creek, Mule Creek, Mustang Creek, No Name Creek, Owl Creek, Poet Creek, Pole Creek, Porcupine Creek, Prospector Creek, Pup Creek, Queen Creek, Rainey Creek, Ranch Creek, Rattlesnake Creek, Red Top Creek, Reynolds Creek, Rim Creek, Ring Creek, Rock Creek, Root Creek, Runaway Creek, Sabe Creek, Saddle Creek, Salt Creek, Schissler Creek, Sheep Creek, Short Creek, Shovel Creek, Skull Creek, Slaughter Creek, Slide Creek, South Fork Cottonwood Creek, South Fork Chamberlain Creek, South Fork Kitchen Creek, South Fork Salmon River, Spread Creek, Spring Creek, Starvation Creek, Steamboat Creek, Steep Creek, Stud Creek, Warren Creek, Webfoot Creek, West Fork Chamberlain Creek, West Fork Rattlesnake Creek, West Horse Creek, Whimstick Creek, Wind River, Woods Fork Horse Creek.

(xix) MIDDLE SALMON-PANTHER BASIN: Allen Creek, Arnett Creek, Beaver Creek, Big Deer Creek, Blackbird Creek, Boulder Creek, Cabin Creek, Camp Creek, Carmen Creek, Clear Creek, Colson Creek, Copper Creek, Corral Creek, Cougar Creek, Cow Creek, Deadhorse Creek, Deep Creek, East Boulder Creek, Elkhorn Creek, Fawn Creek, Fourth Of July Creek, Freeman Creek, Homet Creek, Hughes Creek, Hull Creek, Indian Creek, Iron Creek, Jackass Creek, Jefferson Creek, Jesse Creek, Lake Creek, Little Deep Creek, Little Hat Creek, Little Sheep Creek, McConn Creek, McKim Creek, Mink Creek, Moccasin Creek, Moose Creek, Moyer Creek, Musgrove Creek, Napias Creek, North Fork Hughes Creek, North Fork Iron Creek, North Fork Salmon River, North Fork Williams Creek, Opal Creek, Otter Creek, Owl Creek, Panther Creek, Park Creek, Phelan Creek, Pine Creek, Pony Creek, Porphyry Creek, Pruvan Creek, Rabbit Creek, Rancherio Creek, Rapps Creek, Salt Creek, Salzer Creek, Saw Pit Creek, Sharkey Creek, Sheep Creek, South Fork Cabin Creek, South Fork Iron Creek, South Fork Moyer Creek, South Fork Phelan Creek, South Fork Sheep Creek, South Fork Williams Creek, Spring Creek, Squaw Creek, Trail Creek, Twelvemile Creek, Twin Creek, Weasel Creek, West Fork Blackbird Creek, West Fork Iron Creek, Williams Creek, Woodtick Creek.

(xx) MOYIE BASIN: Brass Creek, Bussard Creek, Copper Creek, Deer Creek, Faro Creek, Keno Creek, Kreist Creek, Line Creek, McDougal Creek, Mill Creek, Moyie River (above Skin Creek), Placer Creek, Rutledge Creek, Skin Creek, Spruce Creek, West Branch Deer Creek.

(xxi) NORTH AND MIDDLE FORK BOISE BASIN: Abby Creek, Arrastra Creek, Bald Mountain Creek, Ballentyne Creek, Banner Creek,

Bayhouse Creek, Bear Creek, Bear River, Big Gulch, Big Silver Creek, Billy Creek, Blackwarrior Creek, Bow Creek, Browns Creek, Buck Creek, Cabin Creek, Cahhah Creek, Camp Gulch, China Fork, Coma Creek, Corbus Creek, Cow Creek, Crooked River, Cub Creek, Decker Creek, Dutch Creek, Dutch Frank Creek, East Fork Roaring River, East Fork Swanholm Creek, East Fork Yuba River, Flint Creek, Flytrip Creek, Gotch Creek, Graham Creek, Granite Creek, Grays Creek, Greylock Creek, Grouse Creek, Hot Creek, Hungarian Creek, Joe Daley Creek, Johnson Creek, Kid Creek, King Creek, La Mayne Creek, Leggit Creek, Lightning Creek, Little Queens River, Little Silver Creek, Louise Creek, Lynx Creek, Mattingly Creek, McKay Creek, McLeod Creek, McPhearson Creek, Middle Fork Boise River (above Roaring River), Middle Fork Corbus Creek, Middle Fork Roaring River, Mill Creek, Misfire Creek, Montezuma Creek, North Fork Boise River (above Bear River), Phifer Creek, Pikes Fork, Quartz Gulch, Queens River, Rabbit Creek, Right Creek, Roaring River, Robin Creek, Rock Creek, Rockey Creek, Sawmill Creek, Scenic Creek, Scotch Creek, Scott Creek, Shorip Creek, Smith Creek, Snow Creek, Snowslide Creek, South Fork Corbus Creek, South Fork Cub Creek, Spout Creek, Steamboat Creek, Steel Creek, Steppe Creek, Swanholm Creek, Timpa Creek, Trail Creek, Trapper Creek, Tripod Creek, West Fork Creek, West Warrior Creek, Willow Creek, Yuba River.

(xxii) NORTH FORK PAYETTE BASIN: Gold Fork River, North Fork Gold Fork River, Pearsol Creek.

(xxiii) AHSIMEROI BASIN: Baby Creek, Bear Creek, Big Creek, Big Gulch, Burnt Creek, Christian Gulch, Dead Cat Canyon, Ditch Creek, Donkey Creek, Doublespring Creek, Dry Canyon, Dry Gulch, East Fork Burnt Creek, East Fork Morgan Creek, East Fork Pahsimeroi River, East Fork Patterson Creek, Elkhorn Creek, Falls Creek, Goldberg Creek, Hillside Creek, Inyo Creek, Long Creek, Mahogany Creek, Mill Creek, Morgan Creek, Morse Creek, Mulkey Gulch, North Fork Big Creek, North Fork Morgan Creek, Pahsimeroi River (above Big Creek), Patterson Creek, Rock Spring

Canyon, Short Creek, Snowslide Creek, South Fork Big Creek, Spring Gulch, Squaw Creek, Stinking Creek, Tater Creek, West Fork Burnt Creek, West Fork North Fork Big Creek.

(xxiv) PAYETTE BASIN: Squaw Creek, Third Fork Squaw Creek.

(xxv) PEND OREILLE LAKE BASIN: Branch North Gold Creek, Cheer Creek, Chloride Gulch, Dry Gulch, Dyree Creek, Flume Creek, Gold Creek, Granite Creek, Grouse Creek, Kick Bush Gulch, North Fork Grouse Creek, North Gold Creek, Plank Creek, Rapid Lightning Creek, South Fork Grouse Creek, Strong Creek, Thor Creek, Trestle Creek, West Branch Pack River, West Gold Creek, Wylie Creek, Zuni Creek.

(xxvi) PRIEST BASIN: Abandon Creek, Athol Creek, Bath Creek, Bear Creek, Bench Creek, Blacktail Creek, Bog Creek, Boulder Creek, Bugle Creek, Canyon Creek, Caribou Creek, Cedar Creek, Chicopee Creek, Deadman Creek, East Fork Trapper Creek, East River, Fedar Creek, Floss Creek, Gold Creek, Granite Creek, Horton Creek, Hughes Fork, Indian Creek, Jackson Creek, Jost Creek, Kalispell Creek, Kent Creek, Keokee Creek, Lime Creek, Lion Creek, Lost Creek, Lucky Creek, Malcom Creek, Middle Fork East River, Muskegon Creek, North Fork Granite Creek, North Fork Indian Creek, Packer Creek, Rock Creek, Ruby Creek, South Fork Granite Creek, South Fork Indian Creek, South Fork Lion Creek, Squaw Creek, Tango Creek, Tarlac Creek, The Thorofare, Trapper Creek, Two Mouth Creek, Uleda Creek, Priest R. (above Priest Lake), Zero Creek.

(xxvii) SOUTH FORK BOISE BASIN: Badger Creek, Bear Creek, Bear Gulch, Big Smoky Creek, Big Water Gulch, Boardman Creek, Burnt Log Creek, Cayuse Creek, Corral Creek, Cow Creek, Edna Creek, Elk Creek, Emma Creek, Feather River, Fern Gulch, Grape Creek, Gunsight Creek, Haypress Creek, Heather Creek, Helen Creek, Johnson Creek, Lincoln Creek, Little Cayuse Creek, Little Rattlesnake Creek, Little Skeleton Creek, Little Smoky Creek, Loggy Creek, Mule Creek, North Fork Ross Fork, Pinto Creek, Rattlesnake Creek, Ross Fork, Russel Gulch, Salt Creek, Shake Creek,

Skeleton Creek, Slater Creek, Smokey Dome Canyon, South Fork Ross Fork, Three Forks Creek, Tipton Creek, Vienna Creek, Weeks Gulch, West Fork Big Smoky Creek, West Fork Salt Creek, West Fork Skeleton Creek, Willow Creek.

(xxviii) SOUTH FORK CLEARWATER BASIN: American River, Baker Gulch, Baldy Creek, Bear Creek, Beaver Creek, Big Canyon Creek, Big Elk Creek, Blanco Creek, Boundary Creek, Box Sing Creek, Boyer Creek, Cartwright Creek, Cole Creek, Crooked River, Dawson Creek, Deer Creek, Ditch Creek, East Fork American River, East Fork Crooked River, Elk Creek, Fivemile Creek, Flint Creek, Fourmile Creek, Fox Creek, French Gulch, Galena Creek, Gospel Creek, Hagen Creek, Hays Creek, Johns Creek, Jungle Creek, Kirks Fork American River, Little Elk Creek, Little Moose Creek, Little Siegel Creek, Loon Creek, Mackey Creek, Meadow Creek, Melton Creek, Middle Fork Red River, Mill Creek, Monroe Creek, Moores Creek, Moores Lake Creek, Moose Butte Creek, Morgan Creek, Mule Creek, Newsome Creek, Nuggett Creek, Otterson Creek, Pat Brennan Creek, Pilot Creek, Quartz Creek, Queen Creek, Rabbit Creek, Rainbow Gulch, Red River, Relief Creek, Ryan Creek, Sally Ann Creek, Sawmill Creek, Schooner Creek, Schwartz Creek, Sharmon Creek, Siegel Creek, Silver Creek, Sixmile Creek, Sixtysix Creek, Snoose Creek, Sourdough Creek, South Fork Red River, Square Mountain Creek, Swale Creek, Swift Creek, Taylor Creek, Tenmile Creek, Trail Creek, Trapper Creek, Trout Creek, Twentymile Creek, Twin Lakes Creek, Umatilla Creek, West Fork Big Elk Creek, West Fork Crooked River, West Fork Gospel Creek, West Fork Newsome Creek, West Fork Red River, West Fork Twentymile Creek, Whiskey Creek, Whitaker Creek, Williams Creek.

(xxix) SOUTH FORK PAYETTE BASIN: Archie Creek, Ash Creek, Baron Creek, Basin Creek, Bear Creek, Beaver Creek, Big Spruce Creek, Bitter Creek, Blacks Creek, Blue Jay Creek, Burn Creek, Bush Creek, Camp Creek, Canyon Creek, Casner Creek, Cat Creek, Chapman Creek, Charters Creek, Clear Creek, Coski Creek, Cup Creek, Dead Man Creek, Deadwood River, Deer Creek, East Fork Deadwood Creek, East Fork Warm Springs Creek, Eby Creek, Elkhorn Creek, Emma Creek, Fall Creek, Fence Creek, Fern Creek, Fivemile Creek, Fox Creek, Garney Creek, Gates Creek, Goat Creek, Grandjem Creek, Grouse Creek, Habit Creek, Helende Creek, Horse Creek, Huckleberry Creek, Jackson Creek, Kettle Creek, Kirkham Creek, Lake Creek, Lick Creek, Little Tenmile Creek, Logging Gulch, Long Creek, MacDonald Creek, Meadow Creek, Middle Fork Warm Springs Creek, Miller Creek, Monument Creek, Moulding Creek, Ninemile Creek, No Man Creek, No Name Creek, North Fork Baron Creek, North Fork Canyon Creek, North Fork Deer Creek, North Fork Whitehawk Creek, O'Keefe Creek, Packsaddle Creek, Park Creek, Pass Creek, Pinchot Creek, Pine Creek, Pitchfork Creek, Pole Creek, Richards Creek, Road Fork Rock Creek, Rock Creek, Rough Creek, Scott Creek, Silver Creek, Sixmile Creek, Smith Creek, Smokey Creek, South Fork Beaver Creek, South Fork Canyon Creek, South Fork Clear Creek, South Fork Payette River (above Rock Creek), South Fork Scott Creek, South Fork Warm Spring Creek, Spring Creek, Steep Creek, Stratton Creek, Topnotch Creek, Trail Creek, Wapiti Creek, Warm Spring Creek, Warm Springs Creek, Whangdoodle Creek, Whitehawk Creek, Wild Buck Creek, Wills Gulch, Wilson Creek, Wolf Creek.

(xxx) SOUTH FORK SALMON BASIN: Alez Creek, Back Creek, Bear Creek, Bishop Creek, Blackmare Creek, Blue Lake Creek, Buck Creek, Buckhorn Bar Creek, Buckhorn Creek, Burgdorf Creek, Burntlog Creek, Cabin Creek, Calf Creek, Camp Creek, Cane Creek, Caton Creek, Cinnabar Creek, Cliff Creek, Cly Creek, Cougar Creek, Cow Creek, Cox Creek, Curtis Creek, Deep Creek, Dollar Creek, Dutch Creek, East Fork South Fork Salmon River, East Fork Zena Creek, Elk Creek, Enos Creek, Falls Creek, Fernan Creek, Fiddle Creek, Fitsum Creek, Flat Creek, Fourmile Creek, Goat Creek, Grimmet Creek, Grouse Creek, Halfway Creek, Hanson Creek, Hays

Creek, Holdover Creek, Hum Creek, Indian Creek, Jeanette Creek, Johnson Creek, Josephine Creek, Jungle Creek, Knee Creek, Krassel Creek, Lake Creek, Landmark Creek, Lick Creek, Little Buckhorn Creek, Little Indian Creek, Lodgepole Creek, Loon Creek, Maverick Creek, Meadow Creek, Middle Fork Elk Creek, Missouri Creek, Moose Creek, Mormon Creek, Nasty Creek, Nethker Creek, Nick Creek, No Mans Creek, North Fork Bear Creek, North Fork Buckhorn Creek, North Fork Camp Creek, North Fork Dollar Creek, North Fork Fitsum Creek, North Fork Lake Fork, North Fork Lick Creek, North Fork Riordan Creek, North Fork Six-bit Creek, Oompaul Creek, Paradise Creek, Park Creek, Peanut Creek, Pepper Creek, Phoebe Creek, Piah Creek, Pid Creek, Pilot Creek, Pony Creek, Porcupine Creek, Porphyry Creek, Prince Creek, Profile Creek, Quartz Creek, Reeves Creek, Rice Creek, Riordan Creek, Roaring Creek, Ruby Creek, Rustican Creek, Ryan Creek, Salt Creek, Sand Creek, Secesh River, Sheep Creek, Silver Creek, Sister Creek, Six-Bit Creek, South Fork Bear Creek, South Fork Blackmare Creek, South Fork Buckhorn Creek, South Fork Cougar Creek, South Fork Elk Creek, South Fork Fitsum Creek, South Fork Fourmile Creek, South Fork Salmon River, South Fork Threemile Creek, Split Creek, Steep Creek, Sugar Creek, Summit Creek, Tamarack Creek, Teepee Creek, Threemile Creek, Trail Creek, Trapper Creek, Trout Creek, Tsum Creek, Twobit Creek, Tyndall Creek, Vein Creek, Victor Creek, Wardenhoff Creek, Warm Lake Creek, Warm Spring Creek, West Fork Buckhorn Creek, West Fork Elk Creek, West Fork Enos Creek, West Fork Zena Creek, Whangdoodle Creek, Willow Basket Creek, Willow Creek, Zena Creek.

(xxxi) ST. JOE R. BASIN: Bad Bear Creek, Bean Creek, Bear Creek, Beaver Creek, Bedrock Creek, Berge Creek, Bird Creek, Blue Grouse Creek, Boulder Creek, Broadaxe Creek, Bruin Creek, California Creek, Cherry Creek, Clear Creek, Color Creek, Copper Creek, Dolly Creek, Dump Creek, Eagle Creek, East Fork Bluff Creek, East Fork Gold Creek, Emerald Creek, Fishhook Creek, Float Creek, Fly Creek, Fuzzy Creek, Gold Creek, Heller Creek, Indian Creek, Kelley Creek, Malin Creek, Marble Creek, Medicine Creek, Mica Creek, Mill Creek, Mosquito Creek, North Fork Bean Creek, North Fork Saint Joe River, North Fork Simmons Creek, Nugget Creek, Packsaddle Creek, Periwinkle Creek, Prospector Creek, Quartz Creek, Red Cross Creek, Red Ives Creek, Ruby Creek, Saint Joe River (above Siwash Creek), Setzer Creek, Sherlock Creek, Simmons Creek, Siwash Creek, Skookum Creek, Thomas Creek, Thorn Creek, Three Lakes Creek, Timber Creek, Tinear Creek, Trout Creek, Tumbledown Creek, Wahoo Creek, Washout Creek, Wilson Creek, Yankee Bar Creek.

(xxxii) UPPER COEUR D'ALENE BASIN: Brown Creek, Falls Creek, Graham Creek.

(xxxiii) UPPER KOOTENAI BASIN: Halverson Cr, North Callahan Creek, South Callahan Creek, West Fork Keeler Creek

(xxxiv) UPPER MIDDLE FORK SALMON BASIN: Asher Creek, Automatic Creek, Ayers Creek, Baldwin Creek, Banner Creek, Bear Creek, Bear Valley Creek, Bearskin Creek, Beaver Creek, Bernard Creek, Big Chief Creek, Big Cottonwood Creek, Birch Creek, Blue Lake Creek, Blue Moon Creek, Boundary Creek, Bridge Creek, Browning Creek, Buck Creek, Burn Creek, Cabin Creek, Cache Creek, Camp Creek, Canyon Creek, Cap Creek, Cape Horn Creek, Casner Creek, Castle Fork, Casto Creek, Cat Creek, Chokebore Creek, Chuck Creek, Cliff Creek, Cold Creek, Collie Creek, Colt Creek, Cook Creek, Corley Creek, Cornish Creek, Cottonwood Creek, Cougar Creek, Crystal Creek, Cub Creek, Cultus Creek, Dagger Creek, Deer Creek, Deer Horn Creek, Doe Creek, Dry Creek, Duffield Creek, Dynamite Creek, Eagle Creek, East Fork Elk Creek, East Fork Indian Creek, East Fork Mayfield Creek, Elk Creek, Elkhorn Creek, Endoah Creek, Fall Creek, Fawn Creek, Feltham Creek, Fir Creek, Flat Creek, Float Creek, Foresight Creek, Forty-five Creek, Forty-four Creek, Fox Creek, Full Moon Creek, Fuse Creek, Grays Creek, Grenade Creek, Grouse Creek, Gun Creek, Half Moon Creek, Hogback Creek, Honeymoon Creek, Hot Creek, Ibex Creek, Indian Creek, Jose Creek,

Kelly Creek, Kerr Creek, Knapp Creek, Kwiskwis Creek, Lime Creek, Lincoln Creek, Little Beaver Creek, Little Cottonwood Creek, Little East Fork Elk Creek, Little Indian Creek, Little Loon Creek, Little Pistol Creek, Lola Creek, Loon Creek, Lucinda Creek, Lucky Creek, Luger Creek, Mace Creek, Mack Creek, Marble Creek, Marlin Creek, Marsh Creek, Mayfield Creek, McHoney Creek, McKee Creek, Merino Creek, Middle Fork Elkhorn Creek, Middle Fork Indian Creek, Middle Fork Salmon River (above Soldier Creek), Mine Creek, Mink Creek, Moonshine Creek, Mowitch Creek, Muskeg Creek, Mystery Creek, Nelson Creek, New Creek, No Name Creek, North Fork Elk Creek, North Fork Elkhorn Creek, North Fork Sheep Creek, North Fork Sulphur Creek, Papoose Creek, Parker Creek, Patrol Creek, Phillips Creek, Pierson Creek, Pinyon Creek, Pioneer Creek, Pistol Creek, Placer Creek, Poker Creek, Pole Creek, Popgun Creek, Porter Creek, Prospect Creek, Rabbit Creek, Rams Horn Creek, Range Creek, Rapid River, Rat Creek, Remington Creek, Rock Creek, Rush Creek, Sack Creek, Safety Creek, Salt Creek, Savage Creek, Scratch Creek, Seafoam Creek, Shady Creek, Shake Creek, Sheep Creek, Sheep Trail Creek, Shell Creek, Shrapnel Creek, Siah Creek, Silver Creek, Slide Creek, Snowshoe Creek, Soldier Creek, South Fork Cottonwood Creek, South Fork Sheep Creek, Spike Creek, Springfield Creek, Squaw Creek, Sulphur Creek, Sunnyside Creek, Swamp Creek, Tennessee Creek, Thatcher Creek, Thicket Creek, Thirty-two Creek, Tomahawk Creek, Trail Creek, Trapper Creek, Trigger Creek, Twenty-two Creek, Vader Creek, Vanity Creek, Velvet Creek, Walker Creek, Wampum Creek, Warm Spring Creek, West Fork Elk Creek, West Fork Little Loon Creek, West Fork Mayfield Creek, White Creek, Wickiup Creek, Winchester Creek, Winnemucca Creek, Wyoming Creek.

(xxxv) UPPER NORTH FORK CLEARWATER BASIN: Adams Creek, Avalanche Creek, Bacon Creek, Ball Creek, Barn Creek, Barnard Creek, Barren Creek, Bear Creek, Beaver Dam Creek, Bedrock Creek, Bill Creek, Bostonian Creek, Boundary Creek, Burn Creek, Butter Creek, Camp George

Creek, Canyon Creek, Cayuse Creek, Chamberlain Creek, Clayton Creek, Cliff Creek, Coffee Creek, Cold Springs Creek, Collins Creek, Colt Creek, Cool Creek, Copper Creek, Corral Creek, Cougar Creek, Craig Creek, Crater Creek, Cub Creek, Davis Creek, Deadwood Creek, Deer Creek, Dill Creek, Drift Creek, Elizabeth Creek, Fall Creek, Fire Creek, Fix Creek, Flame Creek, Fly Creek, Fourth of July Creek, Fro Creek, Frog Creek, Frost Creek, Gilfillian Creek, Goose Creek, Grass Creek, Gravey Creek, Grizzly Creek, Hanson Creek, Heather Creek, Henry Creek, Hidden Creek, Howard Creek, Independence Creek, Jam Creek, Japanese Creek, Johnagan Creek, Johnny Creek, Junction Creek, Kelly Creek, Kid Lake Creek, Kodiak Creek, Lake Creek, Laundry Creek, Lightning Creek, Little Moose Creek, Little Weitas Creek, Liz Creek, Long Creek, Marten Creek, Meadow Creek, Middle Creek, Middle North Fork Kelly Creek, Mill Creek, Mire Creek, Monroe Creek, Moose Creek, Negro Creek, Nettle Creek, Niagra Gulch, North Fork Clearwater River (Fourth of July Creek), Nub Creek, Osier Creek, Perry Creek, Pete Ott Creek, Placer Creek, Polar Creek, Post Creek, Potato Creek, Quartz Creek, Rapid Creek, Rawhide Creek, Roaring Creek, Rock Creek, Rocky Ridge Creek, Ruby Creek, Saddle Creek, Salix Creek, Scurry Creek, Seat Creek, Short Creek, Shot Creek, Siam Creek, Silver Creek, Skull Creek, Slide Creek, Smith Creek, Snow Creek, South Fork Kelly Creek, Spud Creek, Spy Creek, Stolen Creek, Stove Creek, Sugar Creek, Swamp Creek, Tinear Creek, Tinkle Creek, Toboggan Creek, Trail Creek, Vanderbilt Gulch, Wall Creek, Weitas Creek, Williams Creek, Windy Creek, Wolf Creek, Young Creek.

(xxxvi) UPPER SALMON BASIN: Alder Creek, Alpine Creek, Alta Creek, Alturas Lake Creek, Anderson Creek, Aspen Creek, Basin Creek, Bayhorse Creek, Bear Creek, Beaver Creek, Big Boulder Creek, Block Creek, Blowfly Creek, Blue Creek, Boundary Creek, Bowery Creek, Broken Ridge Creek, Bruno Creek, Buckskin Creek, Cabin Creek, Camp Creek, Cash Creek, Challis Creek, Chamberlain Creek,

Champion Creek, Cherry Creek, Cinnabar Creek, Cleveland Creek, Coal Creek, Crooked Creek, Darling Creek, Deadwood Creek, Decker Creek, Deer Creek, Dry Creek, Duffy Creek, East Basin Creek, East Fork Salmon River, East Fork Valley Creek, East Pass Creek, Eddy Creek, Eightmile Creek, Elevenmile Creek, Elk Creek, Ellis Creek, Estes Creek, First Creek, Fisher Creek, Fishhook Creek, Fivemile Creek, Fourth of July Creek, Frenchman Creek, Garden Creek, Germania Creek, Goat Creek, Gold Creek, Gooseberry Creek, Greylock Creek, Hay Creek, Hell Roaring Creek, Herd Creek, Huckleberry Creek, Iron Creek, Job Creek, Jordan Creek, Juliette Creek, Kelly Creek, Kinnikinic Creek, Lick Creek, Lightning Creek, Little Basin Creek, Little Beaver Creek, Little Boulder Creek, Little West Fork Morgan Creek, Lodgepole Creek, Lone Pine Creek, Lost Creek, MacRae Creek, Martin Creek, McKay Creek, Meadow Creek, Mill Creek, Morgan Creek, Muley Creek, Ninemile Creek, Noho Creek, Pack Creek, Park Creek, Pat Hughes Creek, Pig Creek, Pole Creek, Pork Creek, Prospect Creek, Rainbow Creek, Redfish Lake Creek, Road Creek, Rough Creek, Sage Creek, Sagebrush Creek, Salmon River (Redfish Lake Creek), Sawmill Creek, Second Creek, Sevenmile Creek, Sheep Creek, Short Creek, Sixmile Creek, Slate Creek, Smiley Creek, South Fork East Fork Salmon River, Squaw Creek, Stanley Creek, Stephens Creek, Summit Creek, Sunday Creek, Swimm Creek, Taylor Creek, Tenmile Creek, Tennel Creek, Thompson Creek, Three Cabins Creek, Trail Creek, Trap Creek, Trealor Creek, Twelvemile Creek, Twin Creek, Valley Creek, Van Horn Creek, Vat Creek, Warm Spring Creek, Warm Springs Creek, Washington Creek, West Beaver Creek, West Fork Creek, West Fork East Fork Salmon River, West Fork Herd Creek, West Fork Morgan Creek, West Fork Yankee Fork, West Pass Creek, Wickiup Creek, Williams Creek, Willow Creek, Yankee Fork.

(xxxvii) UPPER SELWAY BASIN: Basin Creek, Bear Creek, Burn Creek, Camp Creek, Canyon Creek, Cliff Creek, Comb Creek, Cooper Creek, Cub Creek, Deep Creek, Eagle Creek, Elk Creek, Fall Creek, Fox Creek, Goat Creek, Gold Pan Creek, Granite Creek, Grass Gulch, Haystack Creek, Hells Half Acre Creek, Indian Creek, Kim Creek, Lake Creek, Langdon Gulch, Little Clearwater River, Lodge Creek, Lunch Creek, Mist Creek, Paloma Creek, Paradise Creek, Peach Creek, Pettibone Creek, Running Creek, Saddle Gulch, Schofield Creek, Selway River (above Pettibone Creek), South Fork Running Creek, South Fork Saddle Gulch, South Fork Surprise Creek, Spruce Creek, Squaw Creek, Stripe Creek, Surprise Creek, Set Creek, Tepee Creek, Thirteen Creek, Three Lakes Creek, Triple Creek, Wahoo Creek, White Cap Creek, Wilkerson Creek, Witter Creek.

(xxxviii) WEISER BASIN: Anderson Creek, Bull Corral Creek, Dewey Creek, East Fork Weiser River, Little Weiser River, above Anderson Creek, Sheep Creek, Wolf Creek.

(3) Procedures for site specific modification of listed waterbodies or temperature criteria for bull trout.

(i) The Regional Administrator may, in his discretion, determine that the temperature criteria in paragraph (a)(1) of this section shall not apply to a specific waterbody or portion thereof listed in paragraph (a)(2) of this section. Any such determination shall be made consistent with § 131.11 and shall be based on a finding that bull trout spawning and rearing is not an existing use in such waterbody or portion thereof.

(ii) The Regional Administrator may, in his discretion, raise the temperature criteria in paragraph (a)(1) of this section as they pertain to a specific waterbody or portion thereof listed in paragraph (a)(2) of this section. Any such determination shall be made consistent with § 131.11, and shall be based on a finding that bull trout would be fully supported at the higher temperature criteria.

(iii) For any determination made under paragraphs (a)(3)(i) or (a)(3)(ii) of this section, the Regional Administrator shall, prior to making such a determination, provide for public notice of and comment on a proposed determination. For any such proposed determination, the Regional Administrator shall prepare and make available to the public a technical support document addressing each waterbody or portion thereof that would be deleted or modified and the justification for each proposed determination. This document shall be made available to the public not later than the date of public notice.

(iv) The Regional Administrator shall maintain and make available to the public an updated list of determinations made pursuant to paragraphs (a)(3)(i) and (a)(3)(ii) of this section as well as the technical support documents for each determination.

(v) Nothing in this paragraph (a)(3) shall limit the Administrator's authority to modify the temperature criteria in paragraph (a)(1) of this section or the list of waterbodies in paragraph (a)(2) of this section through rulemaking.

(b) [Reserved]

(c) *Excluded waters.* Lakes, ponds, pools, streams, and springs outside public lands but located wholly and entirely upon a person's land are not protected specifically or generally for any beneficial use, unless such waters are designated in Idaho 16.01.02.110. through 160., or, although not so designated, are waters of the United States as defined at 40 CFR 122.2.

[62 FR 41183, July 31, 1997, as amended at 67 FR 11248, Mar. 13, 2002; 73 FR 65739, Nov. 5, 2008]

§131.34 Kansas.

(a) In addition to the State-adopted use designations, the following water body segment in Kansas is designated for an expected aquatic life use:

Stream segment name	HUC8	Segment #	Designated use
Basin: Missouri			
Subbasin: Independence-Sugar			
Whiskey Creek	10240011	235	Expected Aquatic Life.

(b) In addition to the State-adopted use designations, the following water body segments and lakes in Kansas are designated for recreation uses as specified in the following table:

Stream segment name	HUC8	Segment #	Designated use
Basin: Cimarron			
Subbasin: Upper Cimarron-Bluff			
Big Sandy Creek	11040008	6	Primary Contact Recreation
Gyp Creek	11040008	25	Secondary Contact Recreation
Indian Creek	11040008	14	Secondary Contact Recreation
Kiger Creek	11040008	8	Secondary Contact Recreation
Stink Creek	11040008	17	Secondary Contact Recreation
Two Mile Creek	11040008	15	Secondary Contact Recreation
Subbasin: Lower Cimarron-Eagle Chief			
Anderson Creek	11050001	39	Primary Contact Recreation
Basin: Kansas/Lower Republican			
Subbasin: Middle Republican			
Antelope Creek	10250016	66	Secondary Contact Recreation
Ash Creek	10250016	65	Secondary Contact Recreation
Bean Creek	10250016	76	Secondary Contact Recreation
Cora Creek	10250016	51	Secondary Contact Recreation
Crow Creek (Crystal Creek)	10250016	52	Secondary Contact Recreation
Korb Creek	10250016	72	Primary Contact Recreation
Long Branch	10250016	68	Secondary Contact Recreation
Lost Creek	10250016	53	Primary Contact Recreation
Louisa Creek	10250016	61	Secondary Contact Recreation
Norway Creek	10250016	73	Secondary Contact Recreation
Oak Creek	10250016	75	Secondary Contact Recreation
Rebecca Creek	10250016	39	Secondary Contact Recreation
Spring Creek	10250016	71	Secondary Contact Recreation
Spring Creek	10250016	78	Secondary Contact Recreation

Stream segment name	HUC8	Segment #	Designated use
Taylor Creek	10250016	74	Secondary Contact Recreation
Walnut Creek	10250016	40	Primary Contact Recreation
Walnut Creek	10250016	46	Secondary Contact Recreation
White Rock Creek, North Branch	10250016	60	Secondary Contact Recreation
Wolf Creek	10250016	67	Secondary Contact Recreation
Subbasin: Lower Republican			
Cool Creek	10250017	50	Secondary Contact Recreation
Elm Creek, West Branch	10250017	59	Secondary Contact Recreation
Gar Creek	10250017	12	Primary Contact Recreation
Mud Creek	10250017	63	Secondary Contact Recreation
Turkey Creek	10250017	51	Secondary Contact Recreation
Subbasin: Upper Kansas			
Dry Creek	10270101	19	Primary Contact Recreation
Humbolt Creek	10270101	10	Primary Contact Recreation
Kitten Creek	10270101	14	Primary Contact Recreation
Little Arkansas Creek	10270101	13	Primary Contact Recreation
Little Kitten Creek	10270101	16	Primary Contact Recreation
Mulberry Creek	10270101	20	Secondary Contact Recreation
Subbasin: Middle Kansas			
Adams Creek	10270102	53	Secondary Contact Recreation
Bartlett Creek	10270102	55	Secondary Contact Recreation
Big Elm Creek	10270102	90	Secondary Contact Recreation
Blackjack Creek	10270102	64	Secondary Contact Recreation
Blacksmith Creek	10270102	102	Secondary Contact Recreation
Bourbonais Creek	10270102	63	Primary Contact Recreation
Brush Creek	10270102	57	Primary Contact Recreation
Coal Creek	10270102	46	Secondary Contact Recreation
Coryell Creek	10270102	94	Secondary Contact Recreation
Cow Creek	10270102	45	Secondary Contact Recreation
Crow Creek	10270102	86	Primary Contact Recreation
Darnells Creek	10270102	51	Secondary Contact Recreation
Dog Creek	10270102	78	Secondary Contact Recreation
Doyle Creek	10270102	69	Primary Contact Recreation
Dry Creek	10270102	79	Primary Contact Recreation
Dutch Creek	10270102	92	Secondary Contact Recreation
Elm Creek	10270102	98	Primary Contact Recreation
Elm Creek	10270102	103	Secondary Contact Recreation
Elm Slough	10270102	58	Secondary Contact Recreation
Emmons Creek	10270102	66	Secondary Contact Recreation
French Creek	10270102	19	Primary Contact Recreation
Gilson Creek	10270102	47	Secondary Contact Recreation
Hendricks Creek	10270102	73	Primary Contact Recreation
Hise Creek	10270102	43	Secondary Contact Recreation
Indian Creek	10270102	20	Secondary Contact Recreation
James Creek	10270102	87	Secondary Contact Recreation
Jim Creek	10270102	52	Secondary Contact Recreation
Johnson Creek	10270102	84	Secondary Contact Recreation
Kuenzli Creek	10270102	82	Secondary Contact Recreation
Little Cross Creek	10270102	61	Secondary Contact Recreation
Little Muddy Creek	10270102	99	Primary Contact Recreation
Loire Creek	10270102	80	Primary Contact Recreation
Lost Creek	10270102	60	Secondary Contact Recreation
Messhoss Creek	10270102	96	Primary Contact Recreation
Mud Creek	10270102	44	Secondary Contact Recreation
Mud Creek	10270102	56	Secondary Contact Recreation
Muddy Creek, West Fork	10270102	93	Secondary Contact Recreation
Mulberry Creek	10270102	42	Secondary Contact Recreation
Mulberry Creek	10270102	77	Secondary Contact Recreation
Nehring Creek	10270102	81	Primary Contact Recreation
Paw Paw Creek	10270102	75	Secondary Contact Recreation
Pleasant Hill Run Creek	10270102	23	Primary Contact Recreation
Pomeroy Creek	10270102	59	Secondary Contact Recreation
Post Creek	10270102	101	Secondary Contact Recreation
Pretty Creek	10270102	74	Secondary Contact Recreation
Rock Creek	10270102	21	Primary Contact Recreation
Rock Creek, East Fork	10270102	22	Secondary Contact Recreation
Ross Creek	10270102	35	Secondary Contact Recreation
Salt Creek	10270102	88	Secondary Contact Recreation

Stream segment name	HUC8	Segment #	Designated use
Sand Creek	10270102	65	Secondary Contact Recreation
Shunganunga Creek, South Branch	10270102	106	Primary Contact Recreation
Snake Creek	10270102	95	Secondary Contact Recreation
Snokomo Creek	10270102	85	Secondary Contact Recreation
Spring Creek	10270102	48	Secondary Contact Recreation
Spring Creek	10270102	54	Primary Contact Recreation
Spring Creek	10270102	76	Secondary Contact Recreation
Spring Creek	10270102	105	Secondary Contact Recreation
Sullivan Creek	10270102	89	Primary Contact Recreation
Tecumseh Creek	10270102	107	Secondary Contact Recreation
Turkey Creek	10270102	71	Primary Contact Recreation
Unnamed Stream	10270102	8	Primary Contact Recreation
Vassar Creek	10270102	100	Secondary Contact Recreation
Vermillion Creek	10270102	15	Primary Contact Recreation
Walnut Creek	10270102	91	Secondary Contact Recreation
Wells Creek	10270102	68	Secondary Contact Recreation
Whetstone Creek	10270102	104	Secondary Contact Recreation
Wilson Creek	10270102	50	Primary Contact Recreation
Wolf Creek	10270102	49	Primary Contact Recreation

Subbasin: Delaware

Stream segment name	HUC8	Segment #	Designated use
Banner Creek	10270103	45	Secondary Contact Recreation
Barnes Creek	10270103	39	Secondary Contact Recreation
Bills Creek	10270103	47	Secondary Contact Recreation
Brush Creek	10270103	44	Secondary Contact Recreation
Brush Creek	10270103	54	Primary Contact Recreation
Burr Oak Branch	10270103	8	Primary Contact Recreation
Catamount Creek	10270103	49	Primary Contact Recreation
Cedar Creek, North	10270103	46	Primary Contact Recreation
Claywell Creek	10270103	56	Primary Contact Recreation
Clear Creek	10270103	19	Primary Contact Recreation
Coal Creek	10270103	50	Primary Contact Recreation
Grasshopper Creek	10270103	18	Primary Contact Recreation
Grasshopper Creek	10270103	20	Primary Contact Recreation
Gregg Creek	10270103	24	Primary Contact Recreation
Honey Creek	10270103	55	Secondary Contact Recreation
Little Grasshopper Creek	10270103	16	Secondary Contact Recreation
Little Wild Horse Creek	10270103	57	Primary Contact Recreation
Mission Creek	10270103	40	Primary Contact Recreation
Nebo Creek	10270103	48	Secondary Contact Recreation
Negro Creek	10270103	43	Secondary Contact Recreation
Otter Creek	10270103	41	Secondary Contact Recreation
Plum Creek	10270103	36	Secondary Contact Recreation
Rock Creek	10270103	34	Primary Contact Recreation
Rock Creek	10270103	53	Primary Contact Recreation
Spring Creek	10270103	42	Primary Contact Recreation
Squaw Creek	10270103	38	Secondary Contact Recreation
Straight Creek	10270103	28	Secondary Contact Recreation
Tick Creek	10270103	52	Primary Contact Recreation
Unnamed Stream	10270103	31	Secondary Contact Recreation
Walnut Creek	10270103	51	Primary Contact Recreation
Wolfley Creek	10270103	27	Secondary Contact Recreation

Subbasin: Lower Kansas

Stream segment name	HUC8	Segment #	Designated use
Baldwin Creek	10270104	69	Secondary Contact Recreation
Brush Creek	10270104	49	Secondary Contact Recreation
Brush Creek, West	10270104	46	Secondary Contact Recreation
Buttermilk Creek	10270104	44	Secondary Contact Recreation
Camp Creek	10270104	41	Secondary Contact Recreation
Camp Creek	10270104	74	Secondary Contact Recreation
Captain Creek	10270104	72	Primary Contact Recreation
Chicken Creek	10270104	79	Secondary Contact Recreation
Clear Creek	10270104	383	Primary Contact Recreation
Cow Creek	10270104	58	Secondary Contact Recreation
Crooked Creek	10270104	10	Primary Contact Recreation
Crooked Creek	10270104	12	Primary Contact Recreation
Dawson Creek	10270104	45	Secondary Contact Recreation
Elk Creek	10270104	68	Primary Contact Recreation
Full Creek	10270104	52	Primary Contact Recreation
Hanson Creek	10270104	437	Secondary Contact Recreation
Hog Creek	10270104	54	Secondary Contact Recreation
Howard Creek	10270104	43	Secondary Contact Recreation

Stream segment name	HUC8	Segment #	Designated use
Hulls Branch	10270104	42	Secondary Contact Recreation
Indian Creek	10270104	48	Secondary Contact Recreation
Jarbalo Creek	10270104	51	Secondary Contact Recreation
Kent Creek	10270104	73	Secondary Contact Recreation
Kill Creek	10270104	37	Primary Contact Recreation
Little Cedar Creek	10270104	76	Primary Contact Recreation
Little Mill Creek	10270104	78	Primary Contact Recreation
Little Turkey Creek	10270104	62	Primary Contact Recreation
Little Wakarusa Creek	10270104	71	Primary Contact Recreation
Mission Creek, East	10270104	61	Secondary Contact Recreation
Ninemile Creek	10270104	15	Secondary Contact Recreation
Ninemile Creek	10270104	17	Primary Contact Recreation
Oakley Creek	10270104	56	Secondary Contact Recreation
Plum Creek	10270104	50	Secondary Contact Recreation
Prairie Creek	10270104	47	Secondary Contact Recreation
Rock Creek	10270104	35	Primary Contact Recreation
Scatter Creek	10270104	13	Secondary Contact Recreation
Spoon Creek	10270104	75	Secondary Contact Recreation
Stone Horse Creek	10270104	57	Secondary Contact Recreation
Stranger Creek	10270104	7	Primary Contact Recreation
Stranger Creek	10270104	8	Primary Contact Recreation
Stranger Creek	10270104	9	Primary Contact Recreation
Tonganoxie Creek	10270104	14	Primary Contact Recreation
Tooley Creek	10270104	379	Secondary Contact Recreation
Turkey Creek	10270104	77	Primary Contact Recreation
Unnamed Stream	10270104	11	Primary Contact Recreation
Unnamed Stream	10270104	16	Secondary Contact Recreation
Wakarusa River, Middle Branch	10270104	64	Secondary Contact Recreation
Wakarusa River, South Branch	10270104	63	Primary Contact Recreation
Washington Creek	10270104	36	Primary Contact Recreation
Yankee Tank Creek	10270104	70	Primary Contact Recreation

Subbasin: Lower Big Blue

Stream segment name	HUC8	Segment #	Designated use
Ackerman Creek	10270205	49	Secondary Contact Recreation
Black Vermillion River, Clear Fork	10270205	9	Primary Contact Recreation
Black Vermillion River, North Fork	10270205	15	Secondary Contact Recreation
Black Vermillion River, South Fork	10270205	12	Secondary Contact Recreation
Bluff Creek	10270205	573	Primary Contact Recreation
Bommer Creek	10270205	40	Secondary Contact Recreation
Busksnort Creek	10270205	566	Secondary Contact Recreation
Carter Creek	10270205	59	Secondary Contact Recreation
Cedar Creek	10270205	56	Secondary Contact Recreation
Corndodger Creek	10270205	52	Primary Contact Recreation
De Shazer Creek	10270205	55	Secondary Contact Recreation
Deadman Creek	10270205	60	Secondary Contact Recreation
Deer Creek	10270205	36	Secondary Contact Recreation
Dog Walk Creek	10270205	53	Secondary Contact Recreation
Dutch Creek	10270205	44	Primary Contact Recreation
Elm Creek	10270205	46	Secondary Contact Recreation
Elm Creek, North	10270205	41	Secondary Contact Recreation
Fancy Creek, North Fork	10270205	61	Secondary Contact Recreation
Fancy Creek, West	10270205	29	Primary Contact Recreation
Game Fork	10270205	54	Secondary Contact Recreation
Hop Creek	10270205	43	Secondary Contact Recreation
Indian Creek	10270205	37	Secondary Contact Recreation
Jim Creek	10270205	57	Secondary Contact Recreation
Johnson Fork	10270205	51	Secondary Contact Recreation
Kearney Branch	10270205	58	Secondary Contact Recreation
Lily Creek	10270205	39	Secondary Contact Recreation
Little Indian Creek	10270205	35	Secondary Contact Recreation
Little Timber Creek	10270205	48	Primary Contact Recreation
Meadow Creek	10270205	34	Secondary Contact Recreation
Mission Creek	10270205	22	Primary Contact Recreation
Murdock Creek	10270205	42	Secondary Contact Recreation
Otter Creek	10270205	67	Secondary Contact Recreation
Otter Creek, North	10270205	62	Primary Contact Recreation
Perkins Creek	10270205	47	Secondary Contact Recreation
Phiel Creek	10270205	68	Primary Contact Recreation
Raemer Creek	10270205	33	Primary Contact Recreation
Robidoux Creek	10270205	16	Primary Contact Recreation
Schell Creek	10270205	45	Primary Contact Recreation
School Branch	10270205	63	Secondary Contact Recreation
Scotch Creek	10270205	38	Secondary Contact Recreation

Stream segment name	HUC8	Segment #	Designated use
Spring Creek	10270205	19	Primary Contact Recreation
Spring Creek	10270205	65	Primary Contact Recreation
Timber Creek	10270205	64	Primary Contact Recreation
Weyer Creek	10270205	50	Secondary Contact Recreation

Subbasin: Upper Little Blue

Dry Creek	10270206	41	Secondary Contact Recreation

Subbasin: Lower Little Blue

Ash Creek	10270207	36	Secondary Contact Recreation
Beaver Creek	10270207	38	Secondary Contact Recreation
Bolling Creek	10270207	42	Secondary Contact Recreation
Bowman Creek	10270207	21	Secondary Contact Recreation
Buffalo Creek	10270207	32	Secondary Contact Recreation
Camp Creek	10270207	35	Secondary Contact Recreation
Camp Creek	10270207	44	Primary Contact Recreation
Cedar Creek	10270207	40	Secondary Contact Recreation
Cherry Creek	10270207	25	Secondary Contact Recreation
Coon Creek	10270207	23	Primary Contact Recreation
Fawn Creek	10270207	45	Secondary Contact Recreation
Gray Branch	10270207	27	Secondary Contact Recreation
Humphrey Branch	10270207	24	Secondary Contact Recreation
Iowa Creek	10270207	34	Secondary Contact Recreation
Jones Creek	10270207	29	Secondary Contact Recreation
Joy Creek	10270207	13	Secondary Contact Recreation
Lane Branch	10270207	39	Secondary Contact Recreation
Malone Creek	10270207	37	Secondary Contact Recreation
Melvin Creek	10270207	33	Secondary Contact Recreation
Mercer Creek	10270207	43	Primary Contact Recreation
Mill Creek, South Fork	10270207	31	Secondary Contact Recreation
Myer Creek	10270207	26	Secondary Contact Recreation
Riddle Creek	10270207	17	Secondary Contact Recreation
Rose Creek	10270207	12	Secondary Contact Recreation
Salt Creek	10270207	19	Primary Contact Recreation
School Creek	10270207	49	Primary Contact Recreation
Silver Creek	10270207	28	Primary Contact Recreation
Spring Creek	10270207	15	Secondary Contact Recreation
Spring Creek	10270207	30	Secondary Contact Recreation
Walnut Creek	10270207	41	Primary Contact Recreation

Basin: Lower Arkansas
Subbasin: Rattlesnake

Spring Creek	11030009	7	Secondary Contact Recreation

Subbasin: Gar-Peace

Gar Creek	11030010	8	Primary Contact Recreation

Subbasin: Cow

Blood Creek	11030011	15	Secondary Contact Recreation
Deception Creek	11030011	13	Secondary Contact Recreation
Dry Creek	11030011	22	Primary Contact Recreation
Jarvis Creek	11030011	19	Primary Contact Recreation
Little Cheyenne Creek	11030011	7	Primary Contact Recreation
Little Cow Creek	11030011	2	Primary Contact Recreation
Lost Creek	11030011	17	Secondary Contact Recreation
Owl Creek	11030011	18	Primary Contact Recreation
Plum Creek	11030011	4	Secondary Contact Recreation
Salt Creek	11030011	21	Primary Contact Recreation
Spring Creek	11030011	20	Secondary Contact Recreation

Subbasin: Little Arkansas

Beaver Creek	11030012	26	Primary Contact Recreation
Bull Creek	11030012	24	Primary Contact Recreation
Dry Creek	11030012	22	Secondary Contact Recreation
Dry Turkey Creek	11030012	13	Primary Contact Recreation
Emma Creek	11030012	6	Primary Contact Recreation
Emma Creek	11030012	7	Primary Contact Recreation
Emma Creek, West	11030012	8	Primary Contact Recreation

Stream segment name	HUC8	Segment #	Designated use
Gooseberry Creek	11030012	17	Primary Contact Recreation
Horse Creek	11030012	19	Primary Contact Recreation
Jester Creek	11030012	2	Primary Contact Recreation
Jester Creek, East Fork	11030012	18	Primary Contact Recreation
Kisiwa Creek	11030012	15	Secondary Contact Recreation
Lone Tree Creek	11030012	20	Secondary Contact Recreation
Mud Creek	11030012	16	Primary Contact Recreation
Running Turkey Creek	11030012	25	Secondary Contact Recreation
Salt Creek	11030012	21	Primary Contact Recreation
Sun Creek	11030012	11	Primary Contact Recreation
Turkey Creek	11030012	12	Secondary Contact Recreation
Subbasin: Middle Arkansas—Slate			
Antelope Creek	11030013	25	Primary Contact Recreation
Badger Creek	11030013	31	Primary Contact Recreation
Beaver Creek	11030013	29	Primary Contact Recreation
Beaver Creek	11030013	33	Primary Contact Recreation
Big Slough	11030013	11	Primary Contact Recreation
Big Slough, South Fork	11030013	35	Primary Contact Recreation
Bitter Creek	11030013	28	Primary Contact Recreation
Dry Creek	11030013	15	Primary Contact Recreation
Dry Creek	11030013	16	Primary Contact Recreation
Gypsum Creek	11030013	5	Primary Contact Recreation
Hargis Creek	11030013	24	Primary Contact Recreation
Lost Creek	11030013	23	Primary Contact Recreation
Negro Creek	11030013	20	Primary Contact Recreation
Oak Creek	11030013	26	Secondary Contact Recreation
Salt Creek	11030013	22	Primary Contact Recreation
Spring Creek	11030013	19	Primary Contact Recreation
Spring Creek	11030013	21	Primary Contact Recreation
Spring Creek	11030013	27	Primary Contact Recreation
Spring Creek	11030013	34	Primary Contact Recreation
Spring Creek	11030013	37	Primary Contact Recreation
Winser Creek	11030013	32	Primary Contact Recreation
Subbasin: North Fork Ninnescah			
Crow Creek	11030014	11	Primary Contact Recreation
Dooleyville Creek	11030014	8	Primary Contact Recreation
Goose Creek	11030014	10	Primary Contact Recreation
Ninnescah River, North Fork	11030014	1	Primary Contact Recreation
Ninnescah River, North Fork	11030014	5	Primary Contact Recreation
Ninnescah River, North Fork	11030014	6	Primary Contact Recreation
Red Rock Creek	11030014	12	Primary Contact Recreation
Rock Creek	11030014	13	Primary Contact Recreation
Silver Creek	11030014	7	Primary Contact Recreation
Spring Creek	11030014	14	Primary Contact Recreation
Wolf Creek	11030014	9	Primary Contact Recreation
Subbasin: South Fork Ninnescah			
Coon Creek	11030015	9	Primary Contact Recreation
Coon Creek	11030015	17	Primary Contact Recreation
Hunter Creek	11030015	14	Primary Contact Recreation
Mead Creek	11030015	10	Primary Contact Recreation
Mod Creek	11030015	19	Primary Contact Recreation
Natrona Creek	11030015	K38	Primary Contact Recreation
Negro Creek	11030015	13	Primary Contact Recreation
Nester Creek	11030015	15	Primary Contact Recreation
Ninnescah River, West Branch South Fork	11030015	5	Primary Contact Recreation
Painter Creek	11030015	7	Primary Contact Recreation
Pat Creek	11030015	11	Primary Contact Recreation
Petyt Creek	11030015	12	Primary Contact Recreation
Sand Creek	11030015	18	Primary Contact Recreation
Spring Creek	11030015	8	Primary Contact Recreation
Wild Run Creek	11030015	16	Primary Contact Recreation
Subbasin: Ninnescah			
Afton Creek	11030016	5	Primary Contact Recreation
Clearwater Creek	11030016	4	Primary Contact Recreation
Clearwater Creek	11030016	7	Primary Contact Recreation
Dry Creek	11030016	16	Primary Contact Recreation
Elm Creek	11030016	10	Primary Contact Recreation

Stream segment name	HUC8	Segment #	Designated use
Garvey Creek	11030016	11	Primary Contact Recreation
Sand Creek	11030016	14	Primary Contact Recreation
Silver Creek	11030016	12	Primary Contact Recreation
Spring Creek	11030016	2	Primary Contact Recreation
Spring Creek	11030016	15	Primary Contact Recreation
Turtle Creek	11030016	13	Primary Contact Recreation
Subbasin: Kaw Lake			
Blue Branch	11060001	30	Primary Contact Recreation
Bullington Creek	11060001	28	Primary Contact Recreation
Cedar Creek	11060001	32	Primary Contact Recreation
Chilocco Creek	11060001	19	Primary Contact Recreation
Crabb Creek	11060001	29	Primary Contact Recreation
Ferguson Creek	11060001	38	Primary Contact Recreation
Franklin Creek	11060001	35	Primary Contact Recreation
Gardners Branch	11060001	39	Primary Contact Recreation
Goose Creek	11060001	34	Primary Contact Recreation
Myers Creek	11060001	24	Primary Contact Recreation
Otter Creek	11060001	20	Primary Contact Recreation
Pebble Creek	11060001	26	Primary Contact Recreation
Plum Creek	11060001	33	Primary Contact Recreation
Riley Creek	11060001	37	Primary Contact Recreation
School Creek	11060001	31	Primary Contact Recreation
Shellrock Creek	11060001	22	Primary Contact Recreation
Silver Creek	11060001	17	Primary Contact Recreation
Snake Creek	11060001	25	Primary Contact Recreation
Spring Creek	11060001	21	Primary Contact Recreation
Turkey Creek	11060001	27	Primary Contact Recreation
Wagoner Creek	11060001	36	Primary Contact Recreation
Subbasin: Upper Salt Fork Arkansas			
Ash Creek	11060002	20	Primary Contact Recreation
Big Sandy Creek	11060002	5	Primary Contact Recreation
Cave Creek	11060002	28	Primary Contact Recreation
Deadman Creek	11060002	22	Primary Contact Recreation
Dog Creek	11060002	29	Primary Contact Recreation
Hackberry Creek	11060002	23	Primary Contact Recreation
Indian Creek	11060002	9	Primary Contact Recreation
Inman Creek	11060002	21	Primary Contact Recreation
Mustang Creek	11060002	31	Primary Contact Recreation
Nescatunga Creek, East Branch	11060002	27	Primary Contact Recreation
Red Creek	11060002	16	Primary Contact Recreation
Spring Creek	11060002	24	Primary Contact Recreation
Wildcat Creek	11060002	12	Primary Contact Recreation
Yellowstone Creek	11060002	17	Primary Contact Recreation
Subbasin: Medicine Lodge			
Amber Creek	11060003	12	Primary Contact Recreation
Antelope Creek	11060003	22	Primary Contact Recreation
Bear Creek	11060003	13	Secondary Contact Recreation
Bitter Creek	11060003	18	Secondary Contact Recreation
Cedar Creek	11060003	20	Primary Contact Recreation
Cottonwood Creek	11060003	16	Primary Contact Recreation
Crooked Creek	11060003	11	Primary Contact Recreation
Litle Mule Creek	11060003	9	Primary Contact Recreation
Dry Creek	11060003	21	Secondary Contact Recreation
Elm Creek, East Branch South	11060003	10	Primary Contact Recreation
Elm Creek, North Branch	11060003	4	Primary Contact Recreation
Elm Creek, South Branch	11060003	5	Primary Contact Recreation
Little Bear Creek	11060003	19	Primary Contact Recreation
Medicine Lodge River, North Branch	11060003	24	Secondary Contact Recreation
Mulberry Creek	11060003	14	Primary Contact Recreation
Otter Creek	11060003	25	Secondary Contact Recreation
Puckett Creek	11060003	15	Primary Contact Recreation
Sand Creek	11060003	17	Primary Contact Recreation
Soldier Creek	11060003	27	Secondary Contact Recreation
Stink Creek	11060003	28	Primary Contact Recreation
Turkey Creek	11060003	7	Primary Contact Recreation
Wilson Slough	11060003	23	Primary Contact Recreation

Stream segment name	HUC8	Segment #	Designated use
Subbasin: Lower Salt Fork Arkansas			
Camp Creek	11060004	68	Primary Contact Recreation
Cooper Creek	11060004	71	Primary Contact Recreation
Crooked Creek	11060004	24	Primary Contact Recreation
Little Sandy Creek	11060004	39	Primary Contact Recreation
Little Sandy Creek, East Branch	11060004	65	Primary Contact Recreation
Osage Creek	11060004	17	Primary Contact Recreation
Plum Creek	11060004	70	Primary Contact Recreation
Pond Creek	11060004	18	Primary Contact Recreation
Rush Creek	11060004	69	Primary Contact Recreation
Salty Creek	11060004	40	Primary Contact Recreation
Sandy Creek	11060004	37	Primary Contact Recreation
Sandy Creek, West	11060004	56	Primary Contact Recreation
Spring Creek	11060004	66	Primary Contact Recreation
Unnamed Stream	11060004	25	Primary Contact Recreation
Subbasin: Chikaskia			
Allen Creek	11060005	40	Primary Contact Recreation
Baehr Creek	11060005	22	Primary Contact Recreation
Beaver Creek	11060005	28	Primary Contact Recreation
Beaver Creek	11060005	46	Primary Contact Recreation
Big Spring Creek	11060005	34	Primary Contact Recreation
Bitter Creek	11060005	4	Primary Contact Recreation
Bitter Creek, East	11060005	16	Primary Contact Recreation
Blue Stem Creek	11060005	48	Primary Contact Recreation
Chicken Creek	11060005	36	Primary Contact Recreation
Copper Creek	11060005	42	Primary Contact Recreation
Dry Creek	11060005	17	Primary Contact Recreation
Duck Creek	11060005	32	Primary Contact Recreation
Fall Creek	11060005	14	Primary Contact Recreation
Fall Creek, East Branch	11060005	27	Primary Contact Recreation
Goose Creek	11060005	38	Primary Contact Recreation
Kemp Creek	11060005	49	Primary Contact Recreation
Long Creek	11060005	529	Primary Contact Recreation
Meridian Creek	11060005	20	Primary Contact Recreation
Prairie Creek	11060005	512	Primary Contact Recreation
Prairie Creek, East	11060005	516	Primary Contact Recreation
Prairie Creek, West	11060005	527	Primary Contact Recreation
Red Creek	11060005	43	Primary Contact Recreation
Rock Creek	11060005	23	Primary Contact Recreation
Rodgers Branch	11060005	26	Primary Contact Recreation
Rose Bud Creek	11060005	44	Primary Contact Recreation
Rush Creek	11060005	45	Primary Contact Recreation
Sand Creek	11060005	11	Primary Contact Recreation
Sand Creek, East	11060005	12	Primary Contact Recreation
Sandy Creek	11060005	30	Primary Contact Recreation
Shoo Fly Creek, East	11060005	19	Secondary Contact Recreation
Shore Creek	11060005	35	Primary Contact Recreation
Silver Creek	11060005	29	Primary Contact Recreation
Skunk Creek	11060005	39	Primary Contact Recreation
Spring Branch	11060005	21	Primary Contact Recreation
Wild Horse Creek	11060005	41	Primary Contact Recreation
Wildcat Creek	11060005	24	Primary Contact Recreation
Basin: Marais Des Cygnes			
Subbasin: Upper Marais Des Cygnes			
Appanoose Creek	10290101	16	Primary Contact Recreation
Appanoose Creek, East	10290101	89	Primary Contact Recreation
Batch Creek	10290101	86	Primary Contact Recreation
Blue Creek	10290101	81	Primary Contact Recreation
Bradshaw Creek	10290101	75	Primary Contact Recreation
Cedar Creek	10290101	66	Primary Contact Recreation
Cherry Creek	10290101	74	Primary Contact Recreation
Chicken Creek	10290101	70	Primary Contact Recreation
Chicken Creek	10290101	93	Primary Contact Recreation
Coal Creek	10290101	48	Primary Contact Recreation
Dry Creek	10290101	57	Primary Contact Recreation
Dry Creek	10290101	95	Primary Contact Recreation
Duck Creek	10290101	41	Primary Contact Recreation
Eightmile Creek	10290101	13	Primary Contact Recreation
Frog Creek	10290101	42	Primary Contact Recreation

Stream segment name	HUC8	Segment #	Designated use
Hard Fish Creek	10290101	47	Primary Contact Recreation
Hickory Creek	10290101	8	Primary Contact Recreation
Hill Creek	10290101	71	Primary Contact Recreation
Iantha Creek	10290101	62	Primary Contact Recreation
Jersey Creek	10290101	76	Primary Contact Recreation
Kenoma Creek	10290101	64	Primary Contact Recreation
Little Rock Creek	10290101	73	Primary Contact Recreation
Long Creek	10290101	K36	Primary Contact Recreation
Locust Creek	10290101	69	Primary Contact Recreation
Middle Creek	10290101	50	Primary Contact Recreation
Mosquito Creek	10290101	52	Primary Contact Recreation
Mud Creek	10290101	49	Primary Contact Recreation
Mud Creek	10290101	78	Primary Contact Recreation
Mud Creek	10290101	91	Primary Contact Recreation
Mute Creek	10290101	92	Primary Contact Recreation
Ottawa Creek	10290101	K25	Primary Contact Recreation
Plum Creek	10290101	2	Primary Contact Recreation
Plum Creek	10290101	79	Primary Contact Recreation
Popcorn Creek	10290101	87	Primary Contact Recreation
Pottawatomie Creek, North Fork	10290101	65	Primary Contact Recreation
Pottawatomie Creek, South Fork	10290101	67	Primary Contact Recreation
Rock Creek	10290101	43	Primary Contact Recreation
Rock Creek	10290101	97	Primary Contact Recreation
Sac Branch, South Fork	10290101	54	Secondary Contact Recreation
Sac Creek	10290101	60	Primary Contact Recreation
Salt Creek	10290101	29	Primary Contact Recreation
Sand Creek	10290101	82	Primary Contact Recreation
Smith Creek	10290101	77	Primary Contact Recreation
Spring Creek	10290101	84	Primary Contact Recreation
Switzler Creek	10290101	80	Primary Contact Recreation
Tauy Creek	10290101	11	Primary Contact Recreation
Tauy Creek, West Fork	10290101	K26	Primary Contact Recreation
Tequa Creek	10290101	44	Primary Contact Recreation
Tequa Creek, East Branch	10290101	46	Primary Contact Recreation
Tequa Creek, South Branch	10290101	45	Primary Contact Recreation
Thomas Creek	10290101	72	Secondary Contact Recreation
Turkey Creek	10290101	4	Primary Contact Recreation
Turkey Creek	10290101	6	Primary Contact Recreation
Unnamed Stream	10290101	5	Primary Contact Recreation
Walnut Creek	10290101	90	Primary Contact Recreation
West Fork Eight Mile Creek	10290101	88	Primary Contact Recreation
Willow Creek	10290101	94	Primary Contact Recreation
Wilson Creek	10290101	83	Primary Contact Recreation
Wolf Creek	10290101	96	Primary Contact Recreation

Subbasin: Lower Marais Des Cygnes

Buck Creek	10290102	44	Primary Contact Recreation
Bull Creek	10290102	26	Secondary Contact Recreation
Davis Creek	10290102	38	Primary Contact Recreation
Dorsey Creek	10290102	22	Primary Contact Recreation
Elm Branch	10290102	48	Primary Contact Recreation
Elm Branch	10290102	53	Primary Contact Recreation
Elm Creek	10290102	40	Primary Contact Recreation
Hushpuckney Creek	10290102	37	Primary Contact Recreation
Jake Branch	10290102	54	Secondary Contact Recreation
Jordan Branch	10290102	36	Primary Contact Recreation
Little Bull Creek	10290102	51	Primary Contact Recreation
Little Sugar Creek	10290102	33	Primary Contact Recreation
Little Sugar Creek, North Fork	10290102	43	Primary Contact Recreation
Martin Creek	10290102	26	Primary Contact Recreation
Middle Creek	10290102	13	Primary Contact Recreation
Middle Creek	10290102	30	Primary Contact Recreation
Mound Creek	10290102	35	Primary Contact Recreation
Richland Creek	10290102	41	Primary Contact Recreation
Rock Creek	10290102	27	Primary Contact Recreation
Smith Branch	10290102	47	Primary Contact Recreation
Spring Creek	10290102	50	Primary Contact Recreation
Sugar Creek	10290102	42	Primary Contact Recreation
Turkey Creek	10290102	45	Primary Contact Recreation
Walnut Creek	10290102	14	Primary Contact Recreation
Walnut Creek	10290102	34	Primary Contact Recreation
Walnut Creek	10290102	52	Primary Contact Recreation
Wea Creek, North	10290102	21	Primary Contact Recreation

Stream segment name	HUC8	Segment #	Designated use
Wea Creek, South	10290102	18	Primary Contact Recreation
Wea Creek, South	10290102	19	Primary Contact Recreation
Wea Creek, South	10290102	20	Primary Contact Recreation
Subbasin: Little Osage			
Clever Creek	10290103	7	Primary Contact Recreation
Elk Creek	10290103	11	Primary Contact Recreation
Fish Creek	10290103	8	Primary Contact Recreation
Indian Creek	10290103	12	Primary Contact Recreation
Irish Creek	10290103	9	Primary Contact Recreation
Laberdie Creek, East	10290103	13	Primary Contact Recreation
Limestone Creek	10290103	5	Primary Contact Recreation
Lost Creek	10290103	10	Primary Contact Recreation
Reagan Branch	10290103	6	Primary Contact Recreation
Subbasin: Marmaton			
Buck Run	10290104	46	Primary Contact Recreation
Bunion Creek	10290104	39	Primary Contact Recreation
Cedar Creek	10290104	41	Primary Contact Recreation
Drywood Creek, Moores Branch	10290104	17	Primary Contact Recreation
Drywood Creek, West Fork	10290104	19	Primary Contact Recreation
Elm Creek	10290104	15	Secondary Contact Recreation
Hinton Creek	10290104	38	Primary Contact Recreation
Lath Branch	10290104	42	Primary Contact Recreation
Little Mill Creek	10290104	34	Primary Contact Recreation
Mill Creek	10290104	6	Primary Contact Recreation
Owl Creek	10290104	45	Primary Contact Recreation
Paint Creek	10290104	13	Primary Contact Recreation
Paint Creek	10290104	14	Primary Contact Recreation
Prong Creek	10290104	44	Secondary Contact Recreation
Robinson Branch	10290104	40	Primary Contact Recreation
Shiloh Creek	10290104	36	Primary Contact Recreation
Sweet Branch	10290104	30	Primary Contact Recreation
Tennyson Creek	10290104	31	Primary Contact Recreation
Turkey Creek	10290104	33	Primary Contact Recreation
Walnut Creek	10290104	32	Primary Contact Recreation
Walnut Creek	10290104	47	Primary Contact Recreation
Wolfpen Creek	10290104	37	Primary Contact Recreation
Wolverine Creek	10290104	35	Primary Contact Recreation
Subbasin: South Grand			
Harless Creek	10290108	67	Primary Contact Recreation
Poney Creek	10290108	48	Primary Contact Recreation
Basin: Missouri			
Subbasin: Tarkio-Wolf			
Cold Ryan Branch	10240005	70	Primary Contact Recreation
Coon Creek	10240005	71	Primary Contact Recreation
Halling Creek	10240005	68	Primary Contact Recreation
Mill Creek	10240005	52	Primary Contact Recreation
Rittenhouse Branch	10240005	69	Primary Contact Recreation
Spring Creek	10240005	65	Primary Contact Recreation
Striker Branch	10240005	72	Primary Contact Recreation
Wolf River, Middle Fork	10240005	67	Primary Contact Recreation
Wolf River, North Fork	10240005	66	Primary Contact Recreation
Wolf River, South Fork	10240005	57	Primary Contact Recreation
Unnamed Stream	10240005	55	Primary Contact Recreation
Subbasin: South Fork Big Nemaha			
Burger Creek	10240007	24	Secondary Contact Recreation
Deer Creek	10240007	18	Primary Contact Recreation
Fisher Creek	10240007	28	Primary Contact Recreation
Illinois Creek	10240007	30	Primary Contact Recreation
Rattlesnake Creek	10240007	27	Primary Contact Recreation
Rock Creek	10240007	20	Primary Contact Recreation
Tennessee Creek	10240007	29	Primary Contact Recreation
Turkey Creek	10240007	4	Primary Contact Recreation
Turkey Creek	10240007	5	Primary Contact Recreation
Wildcat Creek	10240007	23	Primary Contact Recreation

Stream segment name	HUC8	Segment #	Designated use
Wildcat Creek	10240007	22	Primary Contact Recreation
Wolf Pen Creek	10240007	25	Primary Contact Recreation

Subbasin: Big Nemaha

Noharts Creek	10240008	42	Primary Contact Recreation
Pedee Creek	10240008	41	Primary Contact Recreation
Pony Creek	10240008	38	Primary Contact Recreation
Roys Creek	10240008	40	Primary Contact Recreation

Subbasin: Independence—Sugar

Brush Creek	10240011	26	Primary Contact Recreation
Deer Creek	10240011	32	Primary Contact Recreation
Fivemile Creek	10240011	35	Primary Contact Recreation
Independence Creek, North Branch	10240011	29	Primary Contact Recreation
Jordan Creek	10240011	30	Primary Contact Recreation
Owl Creek	10240011	33	Primary Contact Recreation
Rock Creek	10240011	21	Primary Contact Recreation
Salt Creek	10240011	34	Primary Contact Recreation
Smith Creek	10240011	28	Primary Contact Recreation
Three Mile Creek	10240011	36	Primary Contact Recreation
Walnut Creek	10240011	23	Primary Contact Recreation
Walnut Creek	10240011	25	Primary Contact Recreation
White Clay Creek	10240011	31	Primary Contact Recreation
White Clay Creek	10240011	9031	Primary Contact Recreation
Whiskey Creek	10240011	235	Primary Contact Recreation
Whiskey Creek	10240011	9235	Primary Contact Recreation

Subbasin: Lower Missouri—Crooked

Brush Creek	10300101	54	Primary Contact Recreation
Camp Branch	10300101	56	Primary Contact Recreation
Coffee Creek	10300101	57	Primary Contact Recreation
Dyke Branch	10300101	55	Primary Contact Recreation
Indian Creek	10300101	32	Primary Contact Recreation
Negro Creek	10300101	58	Primary Contact Recreation
Tomahawk Creek	10300101	53	Primary Contact Recreation

Basin: Neosho
Subbasin: Neosho Headwaters

Allen Creek	11070201	5	Primary Contact Recreation
Badger Creek	11070201	45	Primary Contact Recreation
Big John Creek	11070201	37	Primary Contact Recreation
Bluff Creek	11070201	8	Primary Contact Recreation
Crooked Creek	11070201	35	Primary Contact Recreation
Dows Creek	11070201	3	Primary Contact Recreation
Dows Creek	11070201	4	Primary Contact Recreation
Eagle Creek	11070201	25	Primary Contact Recreation
Eagle Creek, South	11070201	47	Primary Contact Recreation
East Creek	11070201	39	Primary Contact Recreation
Elm Creek	11070201	36	Primary Contact Recreation
Fourmile Creek	11070201	24	Primary Contact Recreation
Fourmile Creek	11070201	48	Primary Contact Recreation
Haun Creek	11070201	29	Primary Contact Recreation
Horse Creek	11070201	33	Primary Contact Recreation
Kahola Creek	11070201	43	Primary Contact Recreation
Lairds Creek	11070201	30	Primary Contact Recreation
Lanos Creek	11070201	21	Primary Contact Recreation
Lebo Creek	11070201	51	Primary Contact Recreation
Munkers Creek, East Branch	11070201	31	Primary Contact Recreation
Munkers Creek, Middle Branch	11070201	32	Primary Contact Recreation
Neosho River, East Fork	11070201	18	Primary Contact Recreation
Neosho River, West Fork	11070201	28	Primary Contact Recreation
Parkers Creek	11070201	27	Primary Contact Recreation
Plum Creek	11070201	50	Primary Contact Recreation
Plumb Creek	11070201	49	Primary Contact Recreation
Rock Creek	11070201	7	Primary Contact Recreation
Rock Creek	11070201	9	Primary Contact Recreation
Rock Creek, East Branch	11070201	34	Primary Contact Recreation
Spring Creek	11070201	40	Primary Contact Recreation
Stillman Creek	11070201	44	Primary Contact Recreation
Taylor Creek	11070201	46	Primary Contact Recreation

Stream segment name	HUC8	Segment #	Designated use
Walker Branch	11070201	42	Primary Contact Recreation
Wolf Creek	11070201	41	Primary Contact Recreation
Wrights Creek	11070201	38	Primary Contact Recreation

Subbasin: Upper Cottonwood

Antelope Creek	11070202	19	Primary Contact Recreation
Bills Creek	11070202	30	Primary Contact Recreation
Bruno Creek	11070202	27	Primary Contact Recreation
Catlin Creek	11070202	20	Primary Contact Recreation
Clear Creek	11070202	5	Primary Contact Recreation
Clear Creek, East Branch	11070202	24	Primary Contact Recreation
Coon Creek	11070202	32	Primary Contact Recreation
Cottonwood River, South	11070202	17	Primary Contact Recreation
Cottonwood River, South	11070202	18	Primary Contact Recreation
Doyle Creek	11070202	21	Primary Contact Recreation
French Creek	11070202	16	Primary Contact Recreation
Mud Creek	11070202	6	Primary Contact Recreation
Perry Creek	11070202	23	Primary Contact Recreation
Spring Branch	11070202	26	Primary Contact Recreation
Spring Creek	11070202	28	Primary Contact Recreation
Spring Creek	11070202	29	Primary Contact Recreation
Stony Brook	11070202	25	Primary Contact Recreation
Turkey Creek	11070202	31	Primary Contact Recreation

Subbasin: Lower Cottonwood

Beaver Creek	11070203	29	Primary Contact Recreation
Bloody Creek	11070203	40	Primary Contact Recreation
Buck Creek	11070203	39	Primary Contact Recreation
Buckeye Creek	11070203	44	Primary Contact Recreation
Bull Creek	11070203	26	Primary Contact Recreation
Camp Creek	11070203	14	Primary Contact Recreation
Coal Creek	11070203	43	Primary Contact Recreation
Collett Creek	11070203	21	Primary Contact Recreation
Corn Creek	11070203	47	Primary Contact Recreation
Coyne Branch	11070203	33	Primary Contact Recreation
Crocker Creek	11070203	46	Primary Contact Recreation
Dodds Creek	11070203	15	Primary Contact Recreation
Fox Creek	11070203	19	Primary Contact Recreation
French Creek	11070203	32	Primary Contact Recreation
Gannon Creek	11070203	24	Primary Contact Recreation
Gould Creek	11070203	36	Primary Contact Recreation
Holmes Creek	11070203	35	Primary Contact Recreation
Jacob Creek	11070203	28	Primary Contact Recreation
Kirk Creek	11070203	48	Primary Contact Recreation
Little Cedar Creek	11070203	11	Primary Contact Recreation
Little Cedar Creek	11070203	45	Primary Contact Recreation
Middle Creek	11070203	5	Primary Contact Recreation
Mile-and-a-half Creek	11070203	13	Secondary Contact Recreation
Moon Creek	11070203	31	Primary Contact Recreation
Mulvane Creek	11070203	22	Primary Contact Recreation
Peyton Creek	11070203	25	Primary Contact Recreation
Phenis Creek	11070203	30	Primary Contact Recreation
Pickett Creek	11070203	18	Primary Contact Recreation
Prather Creek	11070203	23	Primary Contact Recreation
Rock Creek	11070203	37	Primary Contact Recreation
Schaffer Creek	11070203	17	Primary Contact Recreation
School Creek	11070203	16	Primary Contact Recreation
Sharpes Creek	11070203	38	Primary Contact Recreation
Silver Creek	11070203	34	Primary Contact Recreation
Spring Creek	11070203	41	Primary Contact Recreation
Stout Run	11070203	27	Primary Contact Recreation
Stribby Creek	11070203	20	Primary Contact Recreation

Subbasin: Upper Neosho

Badger Creek	11070204	42	Primary Contact Recreation
Big Creek, North	11070204	16	Primary Contact Recreation
Big Creek, South	11070204	17	Primary Contact Recreation
Bloody Run	11070204	25	Primary Contact Recreation
Carlyle Creek	11070204	47	Primary Contact Recreation
Charles Branch	11070204	27	Primary Contact Recreation
Cherry Creek	11070204	20	Primary Contact Recreation

Stream segment name	HUC8	Segment #	Designated use
Coal Creek	11070204	4	Primary Contact Recreation
Cottonwood Creek	11070204	48	Primary Contact Recreation
Crooked Creek	11070204	44	Primary Contact Recreation
Draw Creek	11070204	34	Primary Contact Recreation
Goose Creek	11070204	29	Primary Contact Recreation
Long Creek	11070204	12	Primary Contact Recreation
Martin Creek	11070204	49	Primary Contact Recreation
Mud Creek	11070204	26	Primary Contact Recreation
Mud Creek	11070204	31	Primary Contact Recreation
Onion Creek	11070204	24	Primary Contact Recreation
Owl Creek	11070204	19	Primary Contact Recreation
Owl Creek	11070204	21	Primary Contact Recreation
Plum Creek	11070204	22	Primary Contact Recreation
Rock Creek	11070204	7	Primary Contact Recreation
Rock Creek	11070204	23	Primary Contact Recreation
Rock Creek	11070204	15	Primary Contact Recreation
School Creek	11070204	38	Primary Contact Recreation
Scott Creek	11070204	40	Primary Contact Recreation
Slack Creek	11070204	30	Primary Contact Recreation
Spring Creek	11070204	46	Primary Contact Recreation
Sutton Creek	11070204	35	Primary Contact Recreation
Turkey Branch	11070204	28	Primary Contact Recreation
Turkey Creek	11070204	18	Primary Contact Recreation
Turkey Creek	11070204	32	Primary Contact Recreation
Twiss Creek	11070204	45	Primary Contact Recreation
Varvel Creek	11070204	43	Primary Contact Recreation
Village Creek	11070204	33	Primary Contact Recreation
Wolf Creek	11070204	37	Primary Contact Recreation

Subbasin: Middle Neosho

Stream segment name	HUC8	Segment #	Designated use
Bachelor Creek	11070205	40	Primary Contact Recreation
Canville Creek	11070205	16	Primary Contact Recreation
Center Creek	11070205	25	Primary Contact Recreation
Cherry Creek	11070205	4	Primary Contact Recreation
Deer Creek	11070205	27	Primary Contact Recreation
Denny Branch	11070205	31	Primary Contact Recreation
Elk Creek	11070205	19	Primary Contact Recreation
Elm Creek	11070205	43	Primary Contact Recreation
Flat Rock Creek	11070205	12	Primary Contact Recreation
Flat Rock Creek	11070205	14	Primary Contact Recreation
Fourmile Creek	11070205	49	Primary Contact Recreation
Grindstone Creek	11070205	42	Primary Contact Recreation
Hickory Creek	11070205	10	Primary Contact Recreation
Lake Creek	11070205	24	Primary Contact Recreation
Lightning Creek	11070205	6	Primary Contact Recreation
Lightning Creek	11070205	8	Primary Contact Recreation
Limestone Creek	11070205	7	Primary Contact Recreation
Little Cherry Creek	11070205	32	Primary Contact Recreation
Little Elk Creek	11070205	47	Primary Contact Recreation
Little Fly Creek	11070205	26	Secondary Contact Recreation
Little Labette Creek	11070205	23	Primary Contact Recreation
Little Walnut Creek	11070205	46	Primary Contact Recreation
Litup Creek	11070205	36	Primary Contact Recreation
Mulberry Creek	11070205	35	Primary Contact Recreation
Murphy Creek	11070205	41	Primary Contact Recreation
Ogeese Creek	11070205	38	Primary Contact Recreation
Pecan Creek	11070205	45	Primary Contact Recreation
Plum Creek	11070205	34	Primary Contact Recreation
Rock Creek	11070205	48	Primary Contact Recreation
Spring Creek	11070205	30	Primary Contact Recreation
Stink Branch	11070205	37	Primary Contact Recreation
Thunderbolt Creek	11070205	44	Primary Contact Recreation
Tolen Creek	11070205	39	Primary Contact Recreation
Town Creek	11070205	28	Primary Contact Recreation
Turkey Creek	11070205	29	Primary Contact Recreation
Walnut Creek	11070205	13	Primary Contact Recreation
Wolf Creek	11070205	33	Primary Contact Recreation

Subbasin: Lake O' the Cherokees

Stream segment name	HUC8	Segment #	Designated use
Fourmile Creek	11070206	18	Primary Contact Recreation
Tar Creek	11070206	19	Primary Contact Recreation

Stream segment name	HUC8	Segment #	Designated use
Subbasin: Spring			
Little Shawnee Creek	11070207	22	Primary Contact Recreation
Long Branch	11070207	21	Primary Contact Recreation
Shawnee Creek	11070207	17	Primary Contact Recreation
Taylor Branch	11070207	25	Primary Contact Recreation
Willow Creek	11070207	20	Primary Contact Recreation
Basin: Smoky Hill/Saline			
Subbasin: Middle Smoky Hill			
Ash Creek	10260006	37	Primary Contact Recreation
Big Timber Creek	10260006	24	Primary Contact Recreation
Big Timber Creek	10260006	25	Primary Contact Recreation
Big Timber Creek	10260006	27	Primary Contact Recreation
Blood Creek	10260006	35	Secondary Contact Recreation
Buck Creek	10260006	29	Primary Contact Recreation
Buffalo Creek	10260006	6	Primary Contact Recreation
Clear Creek	10260006	42	Primary Contact Recreation
Coal Creek	10260006	34	Primary Contact Recreation
Cow Creek	10260006	38	Primary Contact Recreation
Eagle Creek	10260006	30	Primary Contact Recreation
Fossil Creek	10260006	13	Primary Contact Recreation
Goose Creek	10260006	39	Primary Contact Recreation
Landon Creek	10260006	31	Primary Contact Recreation
Loss Creek	10260006	44	Primary Contact Recreation
Mud Creek	10260006	47	Primary Contact Recreation
Oxide Creek	10260006	45	Primary Contact Recreation
Sellens Creek	10260006	32	Primary Contact Recreation
Shelter Creek	10260006	43	Primary Contact Recreation
Skunk Creek	10260006	48	Primary Contact Recreation
Spring Creek	10260006	41	Primary Contact Recreation
Timber Creek	10260006	26	Primary Contact Recreation
Turkey Creek	10260006	46	Primary Contact Recreation
Unnamed Stream	10260006	20	Primary Contact Recreation
Unnamed Stream	10260006	23	Primary Contact Recreation
Unnamed Stream	10260006	28	Primary Contact Recreation
Wilson Creek	10260006	40	Primary Contact Recreation
Wolf Creek	10260006	36	Primary Contact Recreation
Subbasin: Lower Smoky Hill			
Basket Creek	10260008	40	Primary Contact Recreation
Battle Creek	10260008	23	Primary Contact Recreation
Carry Creek	10260008	32	Primary Contact Recreation
Carry Creek	10260008	35	Primary Contact Recreation
Chapman Creek, West	10260008	5	Primary Contact Recreation
Dry Creek	10260008	36	Primary Contact Recreation
Dry Creek, East	10260008	43	Primary Contact Recreation
Hobbs Creek	10260008	48	Primary Contact Recreation
Holland Creek	10260008	25	Primary Contact Recreation
Holland Creek, East	10260008	27	Primary Contact Recreation
Holland Creek, West	10260008	26	Primary Contact Recreation
Kentucky Creek	10260008	17	Secondary Contact Recreation
Kentucky Creek, West	10260008	54	Primary Contact Recreation
Lone Tree Creek	10260008	41	Primary Contact Recreation
Lyon Creek, West Branch	10260008	34	Primary Contact Recreation
Mcallister Creek	10260008	49	Primary Contact Recreation
Middle Branch	10260008	58	Primary Contact Recreation
Mud Creek	10260008	8	Primary Contact Recreation
Otter Creek	10260008	42	Primary Contact Recreation
Paint Creek	10260008	52	Secondary Contact Recreation
Pewee Creek	10260008	56	Primary Contact Recreation
Sand Creek	10260008	46	Primary Contact Recreation
Sharps Creek	10260008	16	Secondary Contact Recreation
Spring Creek	10260008	45	Primary Contact Recreation
Stag Creek	10260008	19	Primary Contact Recreation
Turkey Creek	10260008	28	Primary Contact Recreation
Turkey Creek	10260008	30	Primary Contact Recreation
Turkey Creek, East	10260008	50	Primary Contact Recreation
Turkey Creek, West Branch	10260008	62	Primary Contact Recreation
Unnamed Stream	10260008	K3	Primary Contact Recreation
Unnamed Stream	10260008	K4	Primary Contact Recreation
Unnamed Stream	10260008	K24	Primary Contact Recreation

Stream segment name	HUC8	Segment #	Designated use
Wiley Creek	10260008	47	Primary Contact Recreation

Subbasin: Upper Saline

Cedar Creek	10260009	30	Secondary Contact Recreation
Chalk Creek	10260009	26	Primary Contact Recreation
Coyote Creek	10260009	23	Primary Contact Recreation
Eagle Creek	10260009	6	Primary Contact Recreation
Happy Creek	10260009	25	Primary Contact Recreation
Paradise Creek	10260009	5	Primary Contact Recreation
Salt Creek	10260009	20	Primary Contact Recreation
Spring Creek, East	10260009	10	Primary Contact Recreation
Sweetwater Creek	10260009	29	Primary Contact Recreation
Trego Creek	10260009	19	Primary Contact Recreation
Unnamed Stream	10260009	13	Primary Contact Recreation
Wild Horse Creek	10260009	27	Primary Contact Recreation

Subbasin: Lower Saline

Bacon Creek	10260010	7	Primary Contact Recreation
Blue Stem Creek	10260010	33	Primary Contact Recreation
Coon Creek	10260010	31	Primary Contact Recreation
Dry Creek	10260010	29	Secondary Contact Recreation
Eff Creek	10260010	23	Primary Contact Recreation
Elkhorn Creek	10260010	17	Primary Contact Recreation
Elkhorn Creek, West	10260010	38	Primary Contact Recreation
Fourmile Creek	10260010	30	Primary Contact Recreation
Lost Creek	10260010	34	Secondary Contact Recreation
Owl Creek	10260010	18	Primary Contact Recreation
Owl Creek	10260010	39	Primary Contact Recreation
Ralston Creek	10260010	28	Primary Contact Recreation
Shaw Creek	10260010	41	Primary Contact Recreation
Spillman Creek	10260010	6	Primary Contact Recreation
Spillman Creek, North Branch	10260010	8	Primary Contact Recreation
Spring Creek	10260010	16	Primary Contact Recreation
Spring Creek	10260010	19	Primary Contact Recreation
Spring Creek	10260010	20	Primary Contact Recreation
Spring Creek	10260010	24	Primary Contact Recreation
Spring Creek	10260010	26	Primary Contact Recreation
Spring Creek	10260010	27	Primary Contact Recreation
Table Rock Creek	10260010	40	Primary Contact Recreation
Trail Creek	10260010	32	Secondary Contact Recreation
Twelvemile Creek	10260010	36	Primary Contact Recreation
Twin Creek, West	10260010	37	Secondary Contact Recreation
West Spring Creek	10260010	25	Primary Contact Recreation
Wolf Creek	10260010	10	Primary Contact Recreation
Wolf Creek, East Fork	10260010	11	Primary Contact Recreation
Wolf Creek, West Fork	10260010	12	Primary Contact Recreation
Yauger Creek	10260010	35	Primary Contact Recreation

Basin: Solomon
Subbasin: Upper North Fork Solomon

Ash Creek	10260011	24	Primary Contact Recreation
Beaver Creek	10260011	23	Primary Contact Recreation
Big Timber Creek	10260011	8	Primary Contact Recreation
Bow Creek	10260011	15	Primary Contact Recreation
Cactus Creek	10260011	28	Primary Contact Recreation
Crooked Creek	10260011	6	Primary Contact Recreation
Elk Creek	10260011	12	Primary Contact Recreation
Elk Creek, East	10260011	25	Primary Contact Recreation
Game Creek	10260011	10	Primary Contact Recreation
Game Creek	10260011	27	Primary Contact Recreation
Lost Creek	10260011	20	Primary Contact Recreation
Sand Creek	10260011	26	Primary Contact Recreation
Scull Creek	10260011	21	Primary Contact Recreation
Spring Creek	10260011	19	Primary Contact Recreation
Wolf Creek	10260011	22	Primary Contact Recreation

Subbasin: Lower North Fork Solomon

Beaver Creek	10260012	10	Primary Contact Recreation
Beaver Creek, East Branch	10260012	11	Primary Contact Recreation
Beaver Creek, Middle	10260012	12	Primary Contact Recreation

465

Stream segment name	HUC8	Segment #	Designated use
Beaver Creek, Middle	10260012	13	Primary Contact Recreation
Beaver Creek, West	10260012	14	Secondary Contact Recreation
Big Creek	10260012	26	Primary Contact Recreation
Boughton Creek	10260012	34	Primary Contact Recreation
Buck Creek	10260012	43	Secondary Contact Recreation
Cedar Creek	10260012	16	Primary Contact Recreation
Cedar Creek	10260012	18	Primary Contact Recreation
Cedar Creek, East	10260012	17	Primary Contact Recreation
Cedar Creek, East Middle	10260012	37	Primary Contact Recreation
Cedar Creek, Middle	10260012	19	Secondary Contact Recreation
Deer Creek	10260012	23	Primary Contact Recreation
Deer Creek	10260012	25	Primary Contact Recreation
Deer Creek	10260012	27	Primary Contact Recreation
Deer Creek	10260012	29	Primary Contact Recreation
Deer Creek	10260012	31	Primary Contact Recreation
Dry Creek	10260012	42	Primary Contact Recreation
Glen Rock Creek	10260012	41	Primary Contact Recreation
Lawrence Creek	10260012	44	Primary Contact Recreation
Lindley Creek	10260012	45	Primary Contact Recreation
Little Oak Creek	10260012	3	Primary Contact Recreation
Medicine Creek	10260012	33	Primary Contact Recreation
Oak Creek	10260012	2	Primary Contact Recreation
Oak Creek	10260012	4	Primary Contact Recreation
Oak Creek, East	10260012	40	Primary Contact Recreation
Oak Creek, West	10260012	39	Secondary Contact Recreation
Plotner Creek	10260012	30	Primary Contact Recreation
Plum Creek	10260012	20	Primary Contact Recreation
Spring Creek	10260012	8	Secondary Contact Recreation
Spring Creek	10260012	28	Secondary Contact Recreation
Starvation Creek	10260012	38	Primary Contact Recreation
Turner Creek	10260012	24	Primary Contact Recreation

Subbasin: Upper South Fork Solomon

Spring Creek	10260013	5	Primary Contact Recreation

Subbasin: Lower South Fork Solomon

Ash Creek	10260014	22	Primary Contact Recreation
Boxelder Creek	10260014	14	Primary Contact Recreation
Carr Creek	10260014	21	Primary Contact Recreation
Covert Creek	10260014	19	Primary Contact Recreation
Crooked Creek	10260014	27	Primary Contact Recreation
Dibble Creek	10260014	23	Primary Contact Recreation
Elm Creek	10260014	15	Primary Contact Recreation
Jim Creek	10260014	25	Primary Contact Recreation
Kill Creek	10260014	18	Primary Contact Recreation
Kill Creek, East	10260014	28	Primary Contact Recreation
Lost Creek	10260014	13	Primary Contact Recreation
Lucky Creek	10260014	26	Primary Contact Recreation
Medicine Creek	10260014	16	Primary Contact Recreation
Medicine Creek	10260014	17	Primary Contact Recreation
Robbers Roost Creek	10260014	24	Primary Contact Recreation
Twin Creek	10260014	20	Primary Contact Recreation
Twin Creek, East	10260014	29	Primary Contact Recreation

Subbasin: Solomon River

Cow Creek	10260015	28	Primary Contact Recreation
Fifth Creek	10260015	45	Secondary Contact Recreation
Granite Creek	10260015	24	Secondary Contact Recreation
Leban Creek	10260015	41	Secondary Contact Recreation
Mill Creek	10260015	38	Secondary Contact Recreation
Mulberry Creek	10260015	36	Secondary Contact Recreation
Pipe Creek	10260015	9	Primary Contact Recreation
Walnut Creek	10260015	26	Secondary Contact Recreation

Basin: Upper Arkansas
Subbasin: Buckner

Buckner Creek, South Fork	11030006	6	Primary Contact Recreation
Duck Creek	11030006	8	Secondary Contact Recreation
Elm Creek	11030006	5	Primary Contact Recreation
Saw Log Creek	11030006	3	Primary Contact Recreation

Stream segment name	HUC8	Segment #	Designated use
Saw Log Creek	11030006	4	Secondary Contact Recreation

Subbasin: Lower Walnut Creek

Stream segment name	HUC8	Segment #	Designated use
Alexander Dry Creek	11030008	7	Secondary Contact Recreation
Bazine Creek	11030008	9	Secondary Contact Recreation
Boot Creek	11030008	15	Secondary Contact Recreation
Dry Creek	11030008	14	Secondary Contact Recreation
Dry Walnut Creek	11030008	13	Secondary Contact Recreation
Otter Creek	11030008	12	Primary Contact Recreation
Sand Creek	11030008	3	Secondary Contact Recreation
Sandy Creek	11030008	11	Secondary Contact Recreation
Walnut Creek	11030008	1	Primary Contact Recreation
Walnut Creek	11030008	2	Primary Contact Recreation
Walnut Creek	11030008	4	Primary Contact Recreation

Basin: Upper Republican
Subbasin: South Fork Republican

Stream segment name	HUC8	Segment #	Designated use
Big Timber Creek	10250003	61	Secondary Contact Recreation

Subbasin: Beaver

Stream segment name	HUC8	Segment #	Designated use
Beaver Creek	10250014	2	Secondary Contact Recreation

Basin: Verdigris
Subbasin: Upper Verdigris

Stream segment name	HUC8	Segment #	Designated use
Bachelor Creek	11070101	21	Primary Contact Recreation
Bernard Creek	11070101	24	Secondary Contact Recreation
Big Cedar Creek	11070101	39	Primary Contact Recreation
Brazil Creek	11070101	31	Primary Contact Recreation
Buffalo Creek	11070101	2	Primary Contact Recreation
Buffalo Creek, West	11070101	34	Primary Contact Recreation
Cedar Creek	11070101	32	Primary Contact Recreation
Chetopa Creek	11070101	22	Primary Contact Recreation
Crooked Creek	11070101	38	Primary Contact Recreation
Dry Creek	11070101	27	Primary Contact Recreation
Elder Branch	11070101	37	Primary Contact Recreation
Fancy Creek	11070101	28	Primary Contact Recreation
Greenhall Creek	11070101	26	Primary Contact Recreation
Holderman Creek	11070101	47	Primary Contact Recreation
Homer Creek	11070101	20	Primary Contact Recreation
Kelly Branch	11070101	42	Primary Contact Recreation
Kuntz Branch	11070101	29	Primary Contact Recreation
Little Sandy Creek	11070101	33	Primary Contact Recreation
Long Creek	11070101	45	Primary Contact Recreation
Miller Creek	11070101	30	Primary Contact Recreation
Moon Branch	11070101	43	Primary Contact Recreation
Onion Creek	11070101	23	Primary Contact Recreation
Rock Creek	11070101	14	Primary Contact Recreation
Ross Branch	11070101	35	Primary Contact Recreation
Sandy Creek	11070101	4	Primary Contact Recreation
Shaw Creek	11070101	40	Primary Contact Recreation
Slate Creek	11070101	25	Primary Contact Recreation
Snake Creek	11070101	36	Primary Contact Recreation
Tate Branch Creek	11070101	44	Primary Contact Recreation
Van Horn Creek	11070101	46	Primary Contact Recreation
Verdigris River, Bernard Branch	11070101	16	Primary Contact Recreation
Verdigris River, North Branch	11070101	13	Primary Contact Recreation
Verdigris River, North Branch	11070101	15	Primary Contact Recreation
Walnut Creek	11070101	19	Primary Contact Recreation
West Creek	11070101	17	Primary Contact Recreation
Wolf Creek	11070101	41	Primary Contact Recreation

Subbasin: Fall

Stream segment name	HUC8	Segment #	Designated use
Battle Creek	11070102	18	Primary Contact Recreation
Burnt Creek	11070102	24	Primary Contact Recreation
Clear Creek	11070102	37	Primary Contact Recreation
Coon Creek	11070102	25	Primary Contact Recreation
Coon Creek	11070102	36	Primary Contact Recreation
Crain Creek	11070102	32	Primary Contact Recreation
Honey Creek	11070102	26	Primary Contact Recreation

Stream segment name	HUC8	Segment #	Designated use
Indian Creek	11070102	15	Primary Contact Recreation
Ivanpah Creek	11070102	19	Primary Contact Recreation
Kitty Creek	11070102	27	Primary Contact Recreation
Little Indian Creek	11070102	34	Primary Contact Recreation
Little Salt Creek	11070102	35	Primary Contact Recreation
Oleson Creek	11070102	21	Primary Contact Recreation
Otis Creek	11070102	20	Primary Contact Recreation
Plum Creek	11070102	30	Primary Contact Recreation
Rainbow Creek, East	11070102	17	Primary Contact Recreation
Salt Creek	11070102	14	Primary Contact Recreation
Salt Creek	11070102	38	Primary Contact Recreation
Silver Creek	11070102	33	Primary Contact Recreation
Snake Creek	11070102	31	Primary Contact Recreation
Spring Creek	11070102	12	Primary Contact Recreation
Swing Creek	11070102	989	Primary Contact Recreation
Tadpole Creek	11070102	29	Primary Contact Recreation
Watson Branch	11070102	23	Primary Contact Recreation
Subbasin: Middle Verdigris			
Big Creek	11070103	21	Primary Contact Recreation
Biscuit Creek	11070103	53	Primary Contact Recreation
Bluff Run	11070103	54	Primary Contact Recreation
Choteau Creek	11070103	63	Primary Contact Recreation
Claymore Creek	11070103	50	Primary Contact Recreation
Deadman Creek	11070103	57	Primary Contact Recreation
Deer Creek	11070103	51	Primary Contact Recreation
Drum Creek	11070103	34	Primary Contact Recreation
Dry Creek	11070103	37	Primary Contact Recreation
Fawn Creek	11070103	56	Primary Contact Recreation
Mud Creek	11070103	59	Primary Contact Recreation
Onion Creek	11070103	39	Primary Contact Recreation
Potato Creek	11070103	31	Primary Contact Recreation
Prior Creek	11070103	62	Primary Contact Recreation
Pumpkin Creek	11070103	28	Primary Contact Recreation
Richland Creek	11070103	49	Primary Contact Recreation
Rock Creek	11070103	58	Primary Contact Recreation
Rock Creek	11070103	61	Primary Contact Recreation
Snow Creek	11070103	25	Primary Contact Recreation
Spring Creek	11070103	55	Primary Contact Recreation
Sycamore Creek	11070103	52	Primary Contact Recreation
Wildcat Creek	11070103	60	Primary Contact Recreation
Subbasin: Elk			
Bachelor Creek	11070104	25	Primary Contact Recreation
Bloody Run	11070104	26	Primary Contact Recreation
Bull Creek	11070104	33	Primary Contact Recreation
Card Creek	11070104	19	Primary Contact Recreation
Chetopa Creek	11070104	18	Primary Contact Recreation
Clear Creek	11070104	30	Primary Contact Recreation
Clear Creek	11070104	32	Primary Contact Recreation
Coffey Branch	11070104	20	Primary Contact Recreation
Duck Creek	11070104	3	Primary Contact Recreation
Elk River, Mound Branch	11070104	15	Primary Contact Recreation
Elk River, South Branch	11070104	38	Primary Contact Recreation
Elk River, Rowe Branch	11070104	39	Primary Contact Recreation
Elm Branch	11070104	23	Primary Contact Recreation
Hickory Creek	11070104	28	Primary Contact Recreation
Hitchen Creek	11070104	7	Primary Contact Recreation
Hitchen Creek, East	11070104	35	Primary Contact Recreation
Little Duck Creek	11070104	24	Primary Contact Recreation
Little Hitchen Creek	11070104	37	Primary Contact Recreation
Painterhood Creek	11070104	5	Primary Contact Recreation
Painterhood Creek, East	11070104	36	Primary Contact Recreation
Pan Creek	11070104	27	Primary Contact Recreation
Pawpaw Creek	11070104	11	Primary Contact Recreation
Racket Creek	11070104	21	Primary Contact Recreation
Rock Creek	11070104	13	Primary Contact Recreation
Salt Creek	11070104	17	Primary Contact Recreation
Salt Creek, South	11070104	29	Primary Contact Recreation
Skull Creek	11070104	31	Primary Contact Recreation
Snake Creek	11070104	34	Primary Contact Recreation
Sycamore Creek	11070104	22	Primary Contact Recreation

Stream segment name	HUC8	Segment #	Designated use
Wildcat Creek	11070104	16	Primary Contact Recreation

Subbasin: Caney			
Bachelor Creek	11070106	47	Primary Contact Recreation
Bee Creek	11070106	9	Primary Contact Recreation
California Creek	11070106	48	Primary Contact Recreation
Caney Creek	11070106	12	Primary Contact Recreation
Caney River, East Fork	11070106	52	Primary Contact Recreation
Caney Creek, North	11070106	11	Primary Contact Recreation
Cedar Creek	11070106	30	Primary Contact Recreation
Cedar Creek	11070106	32	Primary Contact Recreation
Cheyenne Creek	11070106	40	Primary Contact Recreation
Coon Creek	11070106	36	Primary Contact Recreation
Corum Creek	11070106	51	Primary Contact Recreation
Cotton Creek	11070106	38	Primary Contact Recreation
Cotton Creek, North Fork	11070106	37	Primary Contact Recreation
Dry Creek	11070106	29	Primary Contact Recreation
Fly Creek	11070106	46	Primary Contact Recreation
Illinois Creek	11070106	39	Primary Contact Recreation
Jim Creek	11070106	49	Primary Contact Recreation
Lake Creek	11070106	34	Primary Contact Recreation
Otter Creek	11070106	33	Primary Contact Recreation
Pool Creek	11070106	43	Primary Contact Recreation
Possum Trot Creek	11070106	74	Primary Contact Recreation
Rock Creek	11070106	28	Primary Contact Recreation
Spring Creek	11070106	44	Primary Contact Recreation
Spring Creek	11070106	53	Primary Contact Recreation
Squaw Creek	11070106	42	Primary Contact Recreation
Sycamore Creek	11070106	31	Primary Contact Recreation
Turkey Creek	11070106	45	Primary Contact Recreation
Union Creek	11070106	41	Primary Contact Recreation
Wolf Creek	11070106	35	Primary Contact Recreation
Wolf Creek	11070106	50	Primary Contact Recreation

Basin: Walnut			
Subbasin: Upper Walnut River			
Badger Creek	11030017	36	Primary Contact Recreation
Bemis Creek	11030017	8	Primary Contact Recreation
Cole Creek	11030017	15	Primary Contact Recreation
Constant Creek	11030017	41	Primary Contact Recreation
Dry Creek	11030017	27	Primary Contact Recreation
Dry Creek	11030017	32	Primary Contact Recreation
Durechen Creek	11030017	12	Primary Contact Recreation
Elm Creek	11030017	43	Primary Contact Recreation
Fourmile Creek	11030017	20	Primary Contact Recreation
Gilmore Branch	11030017	39	Primary Contact Recreation
Gypsum Creek	11030017	30	Primary Contact Recreation
Henry Creek	11030017	33	Primary Contact Recreation
Lower Branch	11030017	42	Primary Contact Recreation
Prairie Creek	11030017	35	Primary Contact Recreation
Rock Creek	11030017	37	Primary Contact Recreation
Sand Creek	11030017	29	Primary Contact Recreation
Satchel Creek	11030017	10	Primary Contact Recreation
School Branch	11030017	45	Primary Contact Recreation
Sutton Creek	11030017	40	Primary Contact Recreation
Walnut Creek	11030017	44	Primary Contact Recreation
Whitewater Creek	11030017	34	Primary Contact Recreation
Whitewater Creek, East Branch	11030017	31	Primary Contact Recreation
Whitewater River, East Branch	11030017	22	Primary Contact Recreation
Whitewater River, West Branch	11030017	24	Primary Contact Recreation
Whitewater River, West Branch	11030017	25	Primary Contact Recreation
Wildcat Creek	11030017	26	Primary Contact Recreation
Wildcat Creek, West	11030017	28	Primary Contact Recreation

Subbasin: Lower Walnut River			
Black Crook Creek	11030018	18	Primary Contact Recreation
Cedar Creek	11030018	19	Secondary Contact Recreation
Chigger Creek	11030018	21	Primary Contact Recreation
Crooked Creek	11030018	31	Primary Contact Recreation
Durham Creek	11030018	23	Primary Contact Recreation
Dutch Creek	11030018	2	Primary Contact Recreation

Stream segment name	HUC8	Segment #	Designated use
Dutch Creek	11030018	4	Primary Contact Recreation
Eightmile Creek	11030018	30	Primary Contact Recreation
Foos Creek	11030018	26	Primary Contact Recreation
Hickory Creek	11030018	12	Primary Contact Recreation
Honey Creek	11030018	33	Primary Contact Recreation
Little Dutch Creek	11030018	27	Primary Contact Recreation
Lower Dutch Creek	11030018	20	Primary Contact Recreation
Plum Creek	11030018	36	Primary Contact Recreation
Polecat Creek	11030018	17	Primary Contact Recreation
Posey Creek	11030018	37	Primary Contact Recreation
Richland Creek	11030018	25	Primary Contact Recreation
Rock Creek, North Branch	11030018	35	Primary Contact Recreation
Sanford Creek	11030018	29	Primary Contact Recreation
Spring Branch	11030018	32	Primary Contact Recreation
Stalter Branch	11030018	24	Primary Contact Recreation
Stewart Creek	11030018	28	Primary Contact Recreation
Swisher Branch	11030018	22	Primary Contact Recreation

Total = 1186

Lake name	County	Designated use
Basin: Cimarron		
Subbasin: Upper Cimarron (HUC 11040002)		
Moss Lake East	MORTON	Primary Contact Recreation
Moss Lake West	MORTON	Primary Contact Recreation
Subbasin: North Fork Cimarron (HUC 11040006)		
Russell Lake	STEVENS	Primary Contact Recreation
Subbasin: Upper Cimarron-Bluff (HUC 11040008)		
Clark State Fishing Lake	CLARK	Primary Contact Recreation
Saint Jacob's Well	CLARK	Primary Contact Recreation
Basin: Kansas/Lower Republican		
Subbasin: Middle Republican (HUC 10250016)		
Lake Jewell	JEWELL	Primary Contact Recreation
Subbasin: Lower Republican (HUC 10250017)		
Belleville City Lake	REPUBLIC	Primary Contact Recreation
Wakefield Lake	CLAY	Primary Contact Recreation
Subbasin: Middle Kansas (HUC 10270102)		
Alma City Reservoir	WABAUNSEE	Primary Contact Recreation
Cedar Crest Pond	SHAWNEE	Primary Contact Recreation
Central Park Lake	SHAWNEE	Primary Contact Recreation
Gage Park Lake	SHAWNEE	Primary Contact Recreation
Jeffrey Energy Center Lakes	POTTAWATOMIE	Primary Contact Recreation
Subbasin: Delaware (HUC 10270103)		
Atchison County Park Lake	ATCHISON	Primary Contact Recreation
Little Lake	BROWN	Primary Contact Recreation
Subbasin: Lower Kansas (HUC 10270104)		
Douglas County State Lake	DOUGLAS	Primary Contact Recreation
Lenexa Lake	JOHNSON	Primary Contact Recreation
Mahaffie Farmstead Pond	JOHNSON	Primary Contact Recreation
Pierson Park Lake	WYANDOTTE	Primary Contact Recreation
Waterworks Lakes	JOHNSON	Primary Contact Recreation
Subbasin: Lower Big Blue (HUC 10270205)		
Lake Idlewild	MARSHALL	Primary Contact Recreation

Lake name	County	Designated use

Subbasin: Lower Little Blue (HUC 10270207)

Washington County State Fishing Lake	WASHINGTON	Primary Contact Recreation

Basin: Lower Arkansas
Subbasin: Rattlesnake (HUC 11030009)

Kiowa County State Fishing Lake	KIOWA	Primary Contact Recreation

Subbasin: Cow (HUC 11030011)

Barton Lake	BARTON	Primary Contact Recreation
Sterling City Lake	RICE	Primary Contact Recreation

Subbasin: Little Arkansas (HUC 11030012)

Dillon Park Lakes #1	RENO	Primary Contact Recreation
Dillon Park Lake #2	RENO	Primary Contact Recreation
Newton City Park Lake	HARVEY	Primary Contact Recreation

Subbasin: Middle Arkansas-Slate (HUC 11030013)

Belaire Lake	SEDGWICK	Primary Contact Recreation
Buffalo Park Lake	SEDGWICK	Primary Contact Recreation
Emery Park	SEDGWICK	Primary Contact Recreation
Harrison Park Lake	SEDGWICK	Primary Contact Recreation
Riggs Park Lake	SEDGWICK	Primary Contact Recreation

Subbasin: South Fork Ninnescah (HUC 11030015)

Lemon Park Lake	PRATT	Primary Contact Recreation

Subbasin: Medicine Lodge (HUC 11060003)

Barber County State Fishing Lake	BARBER	Primary Contact Recreation

Subbasin: Lower Salt Fork Arkansas (HUC 11060004)

Hargis Lake	BARBER	Primary Contact Recreation

Basin: Marais Des Cygnes
Subbasin: Upper Marais Des Cygnes (HUC 10290101)

Allen City Lake	LYON	Primary Contact Recreation
Cedar Creek Lake	ANDERSON	Primary Contact Recreation
Crystal Lake	ANDERSON	Primary Contact Recreation
Lyon County State Fishing Lake	LYON	Primary Contact Recreation
Osage City Reservoir	OSAGE	Primary Contact Recreation
Waterworks Impoundment	ANDERSON	Primary Contact Recreation

Subbasin: Lower Marais Des Cygnes (HUC 10290102)

Edgerton City Lake	JOHNSON	Primary Contact Recreation
Edgerton South Lake	JOHNSON	Primary Contact Recreation
Lake LaCygne	LINN	Primary Contact Recreation
Louisburg State Fishing Lake	MIAMI	Primary Contact Recreation
Miami County State Fishing Lake	MIAMI	Primary Contact Recreation
Paola City Lake	MIAMI	Primary Contact Recreation
Pleasanton Lake #1	LINN	Primary Contact Recreation
Pleasanton Lake #2	LINN	Primary Contact Recreation
Spring Hill City Lake	JOHNSON	Primary Contact Recreation

Subbasin: Marmaton (HUC 10290104)

Gunn Park Lake, East	BOURBON	Primary Contact Recreation
Gunn Park Lake, West	BOURBON	Primary Contact Recreation
Rock Creek Lake	BOURBON	Primary Contact Recreation

Basin: Missouri
Subbasin: South Fork Big Nemaha (HUC 10240007)

Pony Creek Lake	NEMAHA	Primary Contact Recreation
Sabetha City Lake	NEMAHA	Primary Contact Recreation

Lake name	County	Designated use
Subbasin: Independence-Sugar (HUC 10240011)		
Atchison City Lakes	ATCHISON	Primary Contact Recreation
Big Eleven Lake	WYANDOTTE	Primary Contact Recreation
Doniphan Fair Association Lake	DONIPHAN	Primary Contact Recreation
Jerrys Lake	LEAVENWORTH	Primary Contact Recreation
Lansing City Lake	LEAVENWORTH	Primary Contact Recreation
South Park Lake	LEAVENWORTH	Primary Contact Recreation
Subbasin: Lower Missouri-Crooked (HUC 10300101)		
Prairie View Park	JOHNSON	Primary Contact Recreation
South Park Lake	JOHNSON	Primary Contact Recreation
Stanley Rural Water District Lake #2	JOHNSON	Primary Contact Recreation
Stohl Park Lake	JOHNSON	Primary Contact Recreation
Basin: Neosho		
Subbasin: Lower Cottonwood (HUC 11070203)		
Peter Pan Pond	LYON	Primary Contact Recreation
Subbasin: Upper Neosho (HUC 11070204)		
Chanute City (Santa Fe) Lake	NEOSHO	Primary Contact Recreation
Leonard's Lake	WOODSON	Primary Contact Recreation
Subbasin: Middle Neosho (HUC 11070205)		
Altamont City Lake #1	LABETTE	Primary Contact Recreation
Bartlett City Lake	LABETTE	Primary Contact Recreation
Harmon Wildlife Area Lakes	LABETTE	Primary Contact Recreation
Mined Land Wildlife Area Lakes	CHEROKEE	Primary Contact Recreation
Timber Lake	NEOSHO	Primary Contact Recreation
Subbasin: Spring (HUC 11070207)		
Empire Lake	CHEROKEE	Primary Contact Recreation
Frontenac City Park	CRAWFORD	Primary Contact Recreation
Mined Land Wildlife Area Lakes	CRAWFORD	Primary Contact Recreation
Pittsburg College Lake	CRAWFORD	Primary Contact Recreation
Playters Lake	CRAWFORD	Primary Contact Recreation
Basin: Smoky Hill/Saline		
Subbasin: Lower Smoky Hill (HUC 10260008)		
Herington City Park Lake	DICKINSON	Primary Contact Recreation
Herington Reservoir	DICKINSON	Primary Contact Recreation
Basin: Solomon		
Subbasin: Lower North Fork Solomon (HUC 10260012)		
Francis Wachs Wildlife Area Lakes	SMITH	Primary Contact Recreation
Subbasin: Solomon River (HUC 10260015)		
Jewell County State Fishing Lake	JEWELL	Primary Contact Recreation
Ottawa County State Fishing Lake	OTTAWA	Primary Contact Recreation
Basin: Upper Arkansas		
Subbasin: Middle Arkansas-Lake McKinney (HUC 11030001)		
Lake McKinney	KEARNY	Primary Contact Recreation
Subbasin: Arkansas-Dodge City (HUC 11030003)		
Lake Charles	FORD	Primary Contact Recreation
Subbasin: Pawnee (HUC 11030005)		
Concannon State Fishing Lake	FINNEY	Primary Contact Recreation
Finney County Game Refuge Lakes	FINNEY	Primary Contact Recreation
Subbasin: Buckner (HUC 11030006)		
Ford County Lake	FORD	Primary Contact Recreation

Lake name	County	Designated use
Hain State Fishing Lake	FORD	Primary Contact Recreation
Subbasin: Upper Walnut Creek (HUC 11030007)		
Goodman State Fishing Lake	NESS	Primary Contact Recreation
Subbasin: Lower Walnut Creek (HUC 11030008)		
Memorial Park Lake	BARTON	Primary Contact Recreation
Stone Lake	BARTON	Primary Contact Recreation
Basin: Verdigris		
Subbasin: Upper Verdigris (HUC 11070101)		
Quarry Lake	WILSON	Primary Contact Recreation
Thayer New City Lake	NEOSHO	Primary Contact Recreation
Subbasin: Middle Verdigris (HUC 11070103)		
La Claire Lake	MONTGOMERY	Primary Contact Recreation
Pfister Park Lakes	MONTGOMERY	Primary Contact Recreation
Subbasin: Caney (HUC 11070106)		
Caney City Lake	CHAUTAUQUA	Primary Contact Recreation
Basin: Walnut		
Subbasin: Lower Walnut River (HUC 11030018)		
Butler County State Fishing Lake	BUTLER	Primary Contact Recreation
Winfield Park Lagoon	COWLEY	Primary Contact Recreation
Total = 100		

(c) Water quality standard variances. (1) The Regional Administrator, EPA Region 7, is authorized to grant variances from the water quality standards in paragraphs (a) and (b) of this section where the requirements of this paragraph (c) are met. A water quality standard variance applies only to the permittee requesting the variance and only to the pollutant or pollutants specified in the variance; the underlying water quality standard otherwise remains in effect.

(2) A water quality standard variance shall not be granted if:

(i) Standards will be attained by implementing effluent limitations required under sections 301(b) and 306 of the CWA and by the permittee implementing reasonable best management practices for nonpoint source control; or

(ii) The variance would likely jeopardize the continued existence of any threatened or endangered species listed under section 4 of the Endangered Species Act or result in the destruction or adverse modification of such species' critical habitat.

(3) Subject to paragraph (c)(2) of this section, a water quality standards variance may be granted if the applicant demonstrates to EPA that attaining the water quality standard is not feasible because:

(i) Naturally occurring pollutant concentrations prevent the attainment of the use; or

(ii) Natural, ephemeral, intermittent or low flow conditions or water levels prevent the attainment of the use, unless these conditions may be compensated for by the discharge of sufficient volume of effluent discharges without violating State water conservation requirements to enable uses to be met; or

(iii) Human caused conditions or sources of pollution prevent the attainment of the use and cannot be remedied or would cause more environmental damage to correct than to leave in place; or

(iv) Dams, diversions or other types of hydrologic modifications preclude the attainment of the use, and it is not feasible to restore the water body to its original condition or to operate such modification in a way which would result in the attainment of the use; or

(v) Physical conditions related to the natural features of the water body, such as the lack of a proper substrate, cover, flow, depth, pools, riffles, and the like unrelated to water quality, preclude attainment of aquatic life protection uses; or

(vi) Controls more stringent than those required by sections 301(b) and 306 of the CWA would result in substantial and widespread economic and social impact.

(4) Procedures. An applicant for a water quality standards variance shall submit a request to the Regional Administrator of EPA Region 7. The application shall include all relevant information showing that the requirements for a variance have been satisfied. The burden is on the applicant to demonstrate to EPA's satisfaction that the designated use is unattainable for one of the reasons specified in paragraph (c)(3) of this section. If the Regional Administrator preliminarily determines that grounds exist for granting a variance, he shall provide public notice of the proposed variance and provide an opportunity for public comment. Any activities required as a condition of the Regional Administrator's granting of a variance shall be included as conditions of the NPDES permit for the applicant. These terms and conditions shall be incorporated into the applicant's NPDES permit through the permit reissuance process or through a modification of the permit pursuant to the applicable permit modification provisions of Kansas' NPDES program

(5) A variance may not exceed 3 years or the term of the NPDES permit, whichever is less. A variance may be renewed if the applicant reapplies and demonstrates that the use in question is still not attainable. Renewal of the variance may be denied if the applicant did not comply with the conditions of the original variance, or otherwise does not meet the requirements of this section.

[68 FR 40442, July 7, 2003]

§ 131.35 Colville Confederated Tribes Indian Reservation.

The water quality standards applicable to the waters within the Colville Indian Reservation, located in the State of Washington.

(a) *Background.* (1) It is the purpose of these Federal water quality standards to prescribe minimum water quality requirements for the surface waters located within the exterior boundaries of the Colville Indian Reservation to ensure compliance with section 303(c) of the Clean Water Act.

(2) The Colville Confederated Tribes have a primary interest in the protection, control, conservation, and utilization of the water resources of the Colville Indian Reservation. Water quality standards have been enacted into tribal law by the Colville Business Council of the Confederated Tribes of the Colville Reservation, as the Colville Water Quality Standards Act, CTC Title 33 (Resolution No. 1984-526 (August 6, 1984) as amended by Resolution No. 1985-20 (January 18, 1985)).

(b) *Territory covered.* The provisions of these water quality standards shall apply to all surface waters within the exterior boundaries of the Colville Indian Reservation.

(c) *Applicability, Administration and Amendment.* (1) The water quality standards in this section shall be used by the Regional Administrator for establishing any water quality based National Pollutant Discharge Elimination System Permit (NPDES) for point sources on the Colville Confederated Tribes Reservation.

(2) In conjunction with the issuance of section 402 or section 404 permits, the Regional Administrator may designate mixing zones in the waters of the United States on the reservation on a case-by-case basis. The size of such mixing zones and the in-zone water quality in such mixing zones shall be consistent with the applicable procedures and guidelines in EPA's Water Quality Standards Handbook and the Technical Support Document for Water Quality Based Toxics Control.

(3) Amendments to the section at the request of the Tribe shall proceed in the following manner.

(i) The requested amendment shall first be duly approved by the Confederated Tribes of the Colville Reservation (and so certified by the Tribes

Legal Counsel) and submitted to the Regional Administrator.

(ii) The requested amendment shall be reviewed by EPA (and by the State of Washington, if the action would affect a boundary water).

(iii) If deemed in compliance with the Clean Water Act, EPA will propose and promulgate an appropriate change to this section.

(4) Amendment of this section at EPA's initiative will follow consultation with the Tribe and other appropriate entities. Such amendments will then follow normal EPA rulemaking procedures.

(5) All other applicable provisions of this part 131 shall apply on the Colville Confederated Tribes Reservation. Special attention should be paid to §§131.6, 131.10, 131.11 and 131.20 for any amendment to these standards to be initiated by the Tribe.

(6) All numeric criteria contained in this section apply at all in-stream flow rates greater than or equal to the flow rate calculated as the minimum 7-consecutive day average flow with a recurrence frequency of once in ten years (7Q10); narrative criteria (§131.35(e)(3)) apply regardless of flow. The 7Q10 low flow shall be calculated using methods recommended by the U.S. Geological Survey.

(d) *Definitions.* (1) *Acute toxicity* means a deleterious response (e.g., mortality, disorientation, immobilization) to a stimulus observed in 96 hours or less.

(2) *Background conditions* means the biological, chemical, and physical conditions of a water body, upstream from the point or non-point source discharge under consideration. Background sampling location in an enforcement action will be upstream from the point of discharge, but not upstream from other inflows. If several discharges to any water body exist, and an enforcement action is being taken for possible violations to the standards, background sampling will be undertaken immediately upstream from each discharge.

(3) *Ceremonial and Religious water use* means activities involving traditional Native American spiritual practices which involve, among other things, primary (direct) contact with water.

(4) *Chronic toxicity* means the lowest concentration of a constituent causing observable effects (*i.e.,* considering lethality, growth, reduced reproduction, etc.) over a relatively long period of time, usually a 28-day test period for small fish test species.

(5) *Council* or *Tribal Council* means the Colville Business Council of the Colville Confederated Tribes.

(6) *Geometric mean* means the *nth* root of a product of *n* factors.

(7) *Mean retention time* means the time obtained by dividing a reservoir's mean annual minimum total storage by the non-zero 30-day, ten-year low-flow from the reservoir.

(8) *Mixing zone* or *dilution zone* means a limited area or volume of water where initial dilution of a discharge takes place; and where numeric water quality criteria can be exceeded but acutely toxic conditions are prevented from occurring.

(9) *pH* means the negative logarithm of the hydrogen ion concentration.

(10) *Primary contact recreation* means activities where a person would have direct contact with water to the point of complete submergence, including but not limited to skin diving, swimming, and water skiing.

(11) *Regional Administrator* means the Administrator of EPA's Region X.

(12) *Reservation* means all land within the limits of the Colville Indian Reservation, established on July 2, 1872 by Executive Order, presently containing 1,389,000 acres more or less, and under the jurisdiction of the United States government, notwithstanding the issuance of any patent, and including rights-of-way running through the reservation.

(13) *Secondary contact recreation* means activities where a person's water contact would be limited to the extent that bacterial infections of eyes, ears, respiratory, or digestive systems or urogenital areas would normally be avoided (such as wading or fishing).

(14) *Surface water* means all water above the surface of the ground within the exterior boundaries of the Colville Indian Reservation including but not limited to lakes, ponds, reservoirs, artificial impoundments, streams, rivers, springs, seeps and wetlands.

(15) *Temperature* means water temperature expressed in Centigrade degrees (C).

(16) *Total dissolved solids* (TDS) means the total filterable residue that passes through a standard glass fiber filter disk and remains after evaporation and drying to a constant weight at 180 degrees C. it is considered to be a measure of the dissolved salt content of the water.

(17) *Toxicity* means acute and/or chronic toxicity.

(18) *Tribe* or *Tribes* means the Colville Confederated Tribes.

(19) *Turbidity* means the clarity of water expressed as nephelometric turbidity units (NTU) and measured with a calibrated turbidimeter.

(20) *Wildlife habitat* means the waters and surrounding land areas of the Reservation used by fish, other aquatic life and wildlife at any stage of their life history or activity.

(e) *General considerations.* The following general guidelines shall apply to the water quality standards and classifications set forth in the use designation Sections.

(1) *Classification boundaries.* At the boundary between waters of different classifications, the water quality standards for the higher classification shall prevail.

(2) *Antidegradation policy.* This antidegradation policy shall be applicable to all surface waters of the Reservation.

(i) Existing in-stream water uses and the level of water quality necessary to protect the existing uses shall be maintained and protected.

(ii) Where the quality of the waters exceeds levels necessary to support propagation of fish, shellfish, and wildlife and recreation in and on the water, that quality shall be maintained and protected unless the Regional Administrator finds, after full satisfaction of the inter-governmental coordination and public participation provisions of the Tribes' continuing planning process, that allowing lower water quality is necessary to accommodate important economic or social development in the area in which the waters are located. In allowing such degradation or lower water quality, the Regional Administrator shall assure water quality adequate to protect existing uses fully. Further, the Regional Administrator shall assure that there shall be achieved the highest statutory and regulatory requirements for all new and existing point sources and all cost-effective and reasonable best management practices for nonpoint source control.

(iii) Where high quality waters are identified as constituting an outstanding national or reservation resource, such as waters within areas designated as unique water quality management areas and waters otherwise of exceptional recreational or ecological significance, and are designated as special resource waters, that water quality shall be maintained and protected.

(iv) In those cases where potential water quality impairment associated with a thermal discharge is involved, this antidegradation policy's implementing method shall be consistent with section 316 of the Clean Water Act.

(3) *Aesthetic qualities.* All waters within the Reservation, including those within mixing zones, shall be free from substances, attributable to wastewater discharges or other pollutant sources, that:

(i) Settle to form objectionable deposits;

(ii) Float as debris, scum, oil, or other matter forming nuisances;

(iii) Produce objectionable color, odor, taste, or turbidity;

(iv) Cause injury to, are toxic to, or produce adverse physiological responses in humans, animals, or plants; or

(v) produce undesirable or nuisance aquatic life.

(4) *Analytical methods.* (i) The analytical testing methods used to measure or otherwise evaluate compliance with water quality standards shall to the extent practicable, be in accordance with the "Guidelines Establishing Test Procedures for the Analysis of Pollutants" (40 CFR part 136). When a testing method is not available for a particular substance, the most recent edition of "Standard Methods for the Examination of Water and Wastewater" (published by the American Public Health Association, American Water Works

Association, and the Water Pollution Control Federation) and other or superseding methods published and/or approved by EPA shall be used.

(f) *General water use and criteria classes.* The following criteria shall apply to the various classes of surface waters on the Colville Indian Reservation:

(1) *Class I (Extraordinary)*—(i) *Designated uses.* The designated uses include, but are not limited to, the following:

(A) Water supply (domestic, industrial, agricultural).

(B) Stock watering.

(C) Fish and shellfish: Salmonid migration, rearing, spawning, and harvesting; other fish migration, rearing, spawning, and harvesting.

(D) Wildlife habitat.

(E) Ceremonial and religious water use.

(F) Recreation (primary contact recreation, sport fishing, boating and aesthetic enjoyment).

(G) Commerce and navigation.

(ii) *Water quality criteria.* (A) Bacteriological Criteria. The geometric mean of the enterococci bacteria densities in samples taken over a 30 day period shall not exceed 8 per 100 milliliters, nor shall any single sample exceed an enterococci density of 35 per 100 milliliters. These limits are calculated as the geometric mean of the collected samples approximately equally spaced over a thirty day period.

(B) Dissolved oxygen—The dissolved oxygen shall exceed 9.5 mg/l.

(C) Total dissolved gas—concentrations shall not exceed 110 percent of the saturation value for gases at the existing atmospheric and hydrostatic pressures at any point of sample collection.

(D) Temperature—shall not exceed 16.0 degrees C due to human activities. Temperature increases shall not, at any time, exceed t=23/(T+5).

(*1*) When natural conditions exceed 16.0 degrees C, no temperature increase will be allowed which will raise the receiving water by greater than 0.3 degrees C.

(*2*) For purposes hereof, "t" represents the permissive temperature change across the dilution zone; and "T" represents the highest existing temperature in this water classification outside of any dilution zone.

(*3*) Provided that temperature increase resulting from nonpoint source activities shall not exceed 2.8 degrees C, and the maximum water temperature shall not exceed 10.3 degrees C.

(E) pH shall be within the range of 6.5 to 8.5 with a human-caused variation of less than 0.2 units.

(F) Turbidity shall not exceed 5 NTU over background turbidity when the background turbidity is 50 NTU or less, or have more than a 10 percent increase in turbidity when the background turbidity is more than 50 NTU.

(G) Toxic, radioactive, nonconventional, or deleterious material concentrations shall be less than those of public health significance, or which may cause acute or chronic toxic conditions to the aquatic biota, or which may adversely affect designated water uses.

(2) *Class II (Excellent)*—(i) *Designated uses.* The designated uses include but are not limited to, the following:

(A) Water supply (domestic, industrial, agricultural).

(B) Stock watering.

(C) Fish and shellfish: Salmonid migration, rearing, spawning, and harvesting; other fish migration, rearing, spawning, and harvesting; crayfish rearing, spawning, and harvesting.

(D) Wildlife habitat.

(E) Ceremonial and religious water use.

(F) Recreation (primary contact recreation, sport fishing, boating and aesthetic enjoyment).

(G) Commerce and navigation.

(ii) *Water quality criteria.* (A) Bacteriological Criteria—The geometric mean of the enterococci bacteria densities in samples taken over a 30 day period shall not exceed 16/100 ml, nor shall any single sample exceed an enterococci density of 75 per 100 milliliters. These limits are calculated as the geometric mean of the collected samples approximately equally spaced over a thirty day period.

(B) Dissolved oxygen—The dissolved oxygen shall exceed 8.0 mg/l.

(C) Total dissolved gas—concentrations shall not exceed 110 percent of the saturation value for gases at the existing atmospheric and hydrostatic pressures at any point of sample collection.

(D) Temperature-shall not exceed 18.0 degrees C due to human activities. Temperature increases shall not, at any time, exceed t=28/(T+7).

(1) When natural conditions exceed 18 degrees C no temperature increase will be allowed which will raise the receiving water temperature by greater than 0.3 degrees C.

(2) For purposes hereof, "t" represents the permissive temperature change across the dilution zone; and "T" represents the highest existing temperature in this water classification outside of any dilution zone.

(3) Provided that temperature increase resulting from non-point source activities shall not exceed 2.8 degrees C, and the maximum water temperature shall not exceed 18.3 degrees C.

(E) pH shall be within the range of 6.5 to 8.5 with a human-caused variation of less than 0.5 units.

(F) Turbidity shall not exceed 5 NTU over background turbidity when the background turbidity is 50 NTU or less, or have more than a 10 percent increase in turbidity when the background turbidity is more than 50 NTU.

(G) Toxic, radioactive, nonconventional, or deleterious material concentrations shall be less than those of public health significance, or which may cause acute or chronic toxic conditions to the aquatic biota, or which may adversely affect designated water uses.

(3) *Class III (Good)*—(i) *Designated uses.* The designated uses include but are not limited to, the following:

(A) Water supply (industrial, agricultural).

(B) Stock watering.

(C) Fish and shellfish: Salmonid migration, rearing, spawning, and harvesting; other fish migration, rearing, spawning, and harvesting; crayfish rearing, spawning, and harvesting.

(D) Wildlife habitat.

(E) Recreation (secondary contact recreation, sport fishing, boating and aesthetic enjoyment).

(F) Commerce and navigation.

(ii) *Water quality criteria.* (A) Bacteriological Criteria—The geometric mean of the enterococci bacteria densities in samples taken over a 30 day period shall not exceed 33/100 ml, nor shall any single sample exceed an enterococci density of 150 per 100 milliliters. These limits are calculated as the geometric mean of the collected samples approximately equally spaced over a thirty day period.

(B) Dissolved oxygen.

	Early life stages [1,2]	Other life stages
7 day mean	9.5 (6.5)	[3] NA
1 day minimum [4]	8.0 (5.0)	6.5

[1] These are water column concentrations recommended to achieve the required intergravel dissolved oxygen concentrations shown in parentheses. The 3 mg/L differential is discussed in the dissolved oxygen criteria document (EPA 440/5–86–003, April 1986). For species that have early life stages exposed directly to the water column, the figures in parentheses apply.

[2] Includes all embryonic and larval stages and all juvenile forms to 30-days following hatching.

[3] NA (not applicable)

[4] All minima should be considered as instantaneous concentrations to be achieved at all times.

(C) Total dissolved gas concentrations shall not exceed 110 percent of the saturation value for gases at the existing atmospheric and hydrostatic pressures at any point of sample collection.

(D) Temperature shall not exceed 21.0 degrees C due to human activities. Temperature increases shall not, at any time, exceed t=34/(T+9).

(1) When natural conditions exceed 21.0 degrees C no temperature increase will be allowed which will raise the receiving water temperature by greater than 0.3 degrees C.

(2) For purposes hereof, "t" represents the permissive temperature change across the dilution zone; and "T" represents the highest existing temperature in this water classification outside of any dilution zone.

(3) Provided that temperature increase resulting from nonpoint source activities shall not exceed 2.8 degrees C, and the maximum water temperature shall not exceed 21.3 degrees C.

(E) pH shall be within the range of 6.5 to 8.5 with a human-caused variation of less than 0.5 units.

(F) Turbidity shall not exceed 10 NTU over background turbidity when the background turbidity is 50 NTU or less, or have more than a 20 percent increase in turbidity when the background turbidity is more than 50 NTU.

(G) Toxic, radioactive, nonconventional, or deleterious material concentrations shall be less than those of public health significance, or which may cause acute or chronic toxic conditions to the aquatic biota, or which may adversely affect designated water uses.

(4) *Class IV (Fair)*—(i) *Designated uses.* The designated uses include but are not limited to, the following:

(A) Water supply (industrial).

(B) Stock watering.

(C) Fish (salmonid and other fish migration).

(D) Recreation (secondary contact recreation, sport fishing, boating and aesthetic enjoyment).

(E) Commerce and navigation.

(ii) *Water quality criteria.* (A) Dissolved oxygen.

	During periods of salmonid and other fish migration	During all other time periods
30 day mean	6.5	5.5
7 day mean	[1]NA	[1]NA
7 day mean minimum	5.0	4.0
1 day minimum[2]	4.0	3.0

[1] NA (not applicable).
[2] All minima should be considered as instantaneous concentrations to be achieved at all times.

(B) Total dissolved gas—concentrations shall not exceed 110 percent of the saturation value for gases at the existing atmospheric and hydrostatic pressures at any point of sample collection.

(C) Temperature shall not exceed 22.0 degrees C due to human activities. Temperature increases shall not, at any time, exceed t=20/(T+2).

(*1*) When natural conditions exceed 22.0 degrees C, no temperature increase will be allowed which will raise the receiving water temperature by greater than 0.3 degrees C.

(*2*) For purposes hereof, "t" represents the permissive temperature change across the dilution zone; and "T" represents the highest existing temperature in this water classification outside of any dilution zone.

(D) pH shall be within the range of 6.5 to 9.0 with a human-caused variation of less than 0.5 units.

(E) Turbidity shall not exceed 10 NTU over background turbidity when the background turbidity is 50 NTU or less,

or have more than a 20 percent increase in turbidity when the background turbidity is more than 50 NTU.

(F) Toxic, radioactive, nonconventional, or deleterious material concentrations shall be less than those of public health significance, or which may cause acute or chronic toxic conditions to the aquatic biota, or which may adversely affect designated water uses.

(5) *Lake Class*—(i) *Designated uses.* The designated uses include but are not limited to, the following:

(A) Water supply (domestic, industrial, agricultural).

(B) Stock watering.

(C) Fish and shellfish: Salmonid migration, rearing, spawning, and harvesting; other fish migration, rearing, spawning, and harvesting; crayfish rearing, spawning, and harvesting.

(D) Wildlife habitat.

(E) Ceremonial and religious water use.

(F) Recreation (primary contact recreation, sport fishing, boating and aesthetic enjoyment).

(G) Commerce and navigation.

(ii) *Water quality criteria.* (A) Bacteriological Criteria. The geometric mean of the enterococci bacteria densities in samples taken over a 30 day period shall not exceed 33/100 ml, nor shall any single sample exceed an enterococci density of 150 per 100 milliliters. These limits are calculated as the geometric mean of the collected samples approximately equally spaced over a thirty day period.

(B) Dissolved oxygen—no measurable decrease from natural conditions.

(C) Total dissolved gas concentrations shall not exceed 110 percent of the saturation value for gases at the existing atmospheric and hydrostatic pressures at any point of sample collection.

(D) Temperature—no measurable change from natural conditions.

(E) pH—no measurable change from natural conditions.

(F) Turbidity shall not exceed 5 NTU over natural conditions.

(G) Toxic, radioactive, nonconventional, or deleterious material concentrations shall be less than those which may affect public health, the natural aquatic environment, or the desirability of the water for any use.

(6) *Special Resource Water Class (SRW)—*(i) *General characteristics.* These are fresh or saline waters which comprise a special and unique resource to the Reservation. Water quality of this class will be varied and unique as determined by the Regional Administrator in cooperation with the Tribes.

(ii) *Designated uses.* The designated uses include, but are not limited to, the following:

(A) Wildlife habitat.

(B) Natural foodchain maintenance.

(iii) *Water quality criteria.*

(A) Enterococci bacteria densities shall not exceed natural conditions.

(B) Dissolved oxygen—shall not show any measurable decrease from natural conditions.

(C) Total dissolved gas shall not vary from natural conditions.

(D) Temperature—shall not show any measurable change from natural conditions.

(E) pH shall not show any measurable change from natural conditions.

(F) Settleable solids shall not show any change from natural conditions.

(G) Turbidity shall not exceed 5 NTU over natural conditions.

(H) Toxic, radioactive, or deleterious material concentrations shall not exceed those found under natural conditions.

(g) *General classifications.* General classifications applying to various surface waterbodies not specifically classified under § 131.35(h) are as follows:

(1) All surface waters that are tributaries to Class I waters are classified Class I, unless otherwise classified.

(2) Except for those specifically classified otherwise, all lakes with existing average concentrations less than 2000 mg/L TDS and their feeder streams on the Colville Indian Reservation are classified as Lake Class and Class I, respectively.

(3) All lakes on the Colville Indian Reservation with existing average concentrations of TDS equal to or exceeding 2000 mg/L and their feeder streams are classified as Lake Class and Class I respectively unless specifically classified otherwise.

(4) All reservoirs with a mean detention time of greater than 15 days are classified Lake Class.

(5) All reservoirs with a mean detention time of 15 days or less are classified the same as the river section in which they are located.

(6) All reservoirs established on pre-existing lakes are classified as Lake Class.

(7) All wetlands are assigned to the Special Resource Water Class.

(8) All other waters not specifically assigned to a classification of the reservation are classified as Class II.

(h) *Specific classifications.* Specific classifications for surface waters of the Colville Indian Reservation are as follows:

(1) Streams:

Alice Creek	Class III
Anderson Creek	Class III
Armstrong Creek	Class III
Barnaby Creek	Class II
Bear Creek	Class III
Beaver Dam Creek	Class II
Bridge Creek	Class II
Brush Creek	Class III
Buckhorn Creek	Class III
Cache Creek	Class III
Canteen Creek	Class I
Capoose Creek	Class III
Cobbs Creek	Class III
Columbia River from Chief Joseph Dam to Wells Dam.	
Columbia River from northern Reservation boundary to Grand Coulee Dam (Roosevelt Lake).	
Columbia River from Grand Coulee Dam to Chief Joseph Dam.	
Cook Creek	Class I
Cooper Creek	Class III
Cornstalk Creek	Class III
Cougar Creek	Class I
Coyote Creek	Class II
Deerhorn Creek	Class III
Dick Creek	Class III
Dry Creek	Class I
Empire Creek	Class III
Faye Creek	Class I
Forty Mile Creek	Class III
Gibson Creek	Class I
Gold Creek	Class II
Granite Creek	Class II
Grizzly Creek	Class III
Haley Creek	Class III
Hall Creek	Class II
Hall Creek, West Fork	Class I
Iron Creek	Class III
Jack Creek	Class III
Jerred Creek	Class I
Joe Moses Creek	Class III
John Tom Creek	Class III
Jones Creek	Class I
Kartar Creek	Class III
Kincaid Creek	Class III
King Creek	Class III
Klondyke Creek	Class I
Lime Creek	Class III

Little Jim Creek	Class III
Little Nespelem	Class II
Louie Creek	Class III
Lynx Creek	Class II
Manila Creek	Class III
McAllister Creek	Class III
Meadow Creek	Class III
Mill Creek	Class II
Mission Creek	Class III
Nespelem River	Class II
Nez Perce Creek	Class III
Nine Mile Creek	Class II
Nineteen Mile Creek	Class III
No Name Creek	Class II
North Nanamkin Creek	Class III
North Star Creek	Class III
Okanogan River from Reservation north boundary to Columbia River.	Class II
Olds Creek	Class I
Omak Creek	Class II
Onion Creek	Class II
Parmenter Creek	Class III
Peel Creek	Class III
Peter Dan Creek	Class III
Rock Creek	Class I
San Poil River	Class I
Sanpoil, River West Fork	Class II
Seventeen Mile Creek	Class III
Silver Creek	Class III
Sitdown Creek	Class III
Six Mile Creek	Class III
South Nanamkin Creek	Class III
Spring Creek	Class III
Stapaloop Creek	Class III
Stepstone Creek	Class III
Stranger Creek	Class II
Strawberry Creek	Class III
Swimptkin Creek	Class III
Three Forks Creek	Class I
Three Mile Creek	Class III
Thirteen Mile Creek	Class II
Thirty Mile Creek	Class II
Trail Creek	Class III
Twentyfive Mile Creek	Class III
Twentyone Mile Creek	Class III
Twentythree Mile Creek	Class III
Wannacot Creek	Class III
Wells Creek	Class I

Whitelaw Creek	Class III
Wilmont Creek	Class II
(2) Lakes:	
Apex Lake	LC
Big Goose Lake	LC
Bourgeau Lake	LC
Buffalo Lake	LC
Cody Lake	LC
Crawfish Lakes	LC
Camille Lake	LC
Elbow Lake	LC
Fish Lake	LC
Gold Lake	LC
Great Western Lake	LC
Johnson Lake	LC
LaFleur Lake	LC
Little Goose Lake	LC
Little Owhi Lake	LC
McGinnis Lake	LC
Nicholas Lake	LC
Omak Lake	SRW
Owhi Lake	SRW
Penley Lake	SRW
Rebecca Lake	LC
Round Lake	LC
Simpson Lake	LC
Soap Lake	LC
Sugar Lake	LC
Summit Lake	LC
Twin Lakes	SRW

[54 FR 28625, July 6, 1989]

§131.36 Toxics criteria for those states not complying with Clean Water Act section 303(c)(2)(B).

(a) *Scope.* This section is not a general promulgation of the section 304(a) criteria for priority toxic pollutants but is restricted to specific pollutants in specific States.

(b)(1) EPA's Section 304(a) criteria for Priority Toxic Pollutants.

A			B Freshwater		C Saltwater		D Human Health (10⁻⁶ risk for carcinogens) For consumption of:	
(#) Compound		CAS Number	Criterion Maximum Conc.[d] (µg/L) (B1)	Criterion Continuous Conc.[d] (µg/L) (B2)	Criterion Maximum Conc.[d] (µg/L) (C1)	Criterion Continuous Conc.[d] (µg/L) (C2)	Water & Organisms (µg/L) (D1)	Organisms Only (µg/L) (D2)
1	Antimony	7440360					14 a	4300 a
2	Arsenic	7440382	360 m	190 m	69 m	36 m	0.018 abc	0.14 abc
3	Beryllium	7440417					n	n
4	Cadmium	7440439	3.7 e	1.0 e	42 m	9.3 m	n	n
5a	Chromium (III)	16065831	550 e	180 e			n	n
b	Chromium (VI)	18540299	15 m	10 m	1100 m	50 m	n	n
6	Copper	7440508	17 e	11 e	2.4 m	2.4 m		
7	Lead	7439921	65 e	2.5 e	210 m	8.1 m		
8	Mercury	7439976	2.1 m	0.012 ip	1.8 m	0.025 ip	0.14	0.15
9	Nickel	7440020	1400 e	160 e	74 m	8.2 m	610 a	4600 a
10	Selenium	7782492		5 p	290 m	71 m	n	n
11	Silver	7440224	3.4 e		1.9 m			
12	Thallium	7440280					1.7 a	6.3 a
13	Zinc	7440666	110 e	100 e	90 m	81 m	700 a	220000 aj
14	Cyanide	57125	22	5.2	1	1	700,000	
15	Asbestos	1332214					7,000,000 fibers/L k	
16	2,3,7,8-TCDD (Dioxin)	1746016					0.000000013 c	0.000000014 c
17	Acrolein	107028					320	780
18	Acrylonitrile	107131					0.059 ac	0.66 ac
19	Benzene	71432					1.2 ac	71 ac
20	Bromoform	75252					4.3 ac	360 ac
21	Carbon Tetrachloride	56235					0.25 ac	4.4 ac
22	Chlorobenzene	108907					680 a	21000 aj
23	Chlorodibromomethane	124481					0.41 ac	34 ac
24	Chloroethane	75003						
25	2-Chloroethylvinyl Ether	110758						
26	Chloroform	67663					5.7 ac	470 ac
27	Dichlorobromomethane	75274					0.27 ac	22 ac
28	1,1-Dichloroethane	75343						
29	1,2-Dichloroethane	107062					0.38 ac	99 ac
30	1,1-Dichloroethylene	75354					0.057 ac	3.2 ac
31	1,2-Dichloropropane	78875						
32	1,3-Dichloropropylene	542756					10 a	1700 a
33	Ethylbenzene	100414					3100 a	29000 a
34	Methyl Bromide	74839					48 a	4000 a
35	Methyl Chloride	74873					n	n
36	Methylene Chloride	75092					4.7 ac	1600 ac
37	1,1,2,2-Tetrachloroethane	79345					0.17 ac	11 ac
38	Tetrachloroethylene	127184					0.8 c	8.85 c
39	Toluene	108883					6800 a	200000 a

No.	Compound	CAS Number							
40	1,2-Trans-Dichloroethylene	156605						n	n
41	1,1,1-Trichloroethane	71556						0.60 ac	42 ac
42	1,1,2-Trichloroethane	79005						2.7 c	81 c
43	Trichloroethylene	79016						2 c	525 c
44	Vinyl Chloride	75014							
45	2-Chlorophenol	95578						93 a	790 aj
46	2,4-Dichlorophenol	120832							
47	2,4-Dimethylphenol	105679							
48	2-Methyl-4,6-Dinitrophenol	534521						13.4	765
49	2,4-Dinitrophenol	51285						70 a	14000 a
50	2-Nitrophenol	88755							
51	4-Nitrophenol	100027							
52	3-Methyl-4-Chlorophenol	59507							
53	Pentachlorophenol	87865	20	20 f	13 f	13	7.9	0.28 ac	8.2 acj
54	Phenol	108952						21000 a	4600000 aj
55	2,4,6-Trichlorophenol	88062						2.1 ac	6.5 ac
56	Acenaphthene	83329							
57	Acenaphthylene	208968							
58	Anthracene	120127						9600 a	110000 a
59	Benzidine	92875						0.00012 ac	0.00054 ac
60	Benzo(a)Anthracene	56553						0.0028 c	0.031 c
61	Benzo(a)Pyrene	50328						0.0028 c	0.031 c
62	Benzo(b)Fluoranthene	205992						0.0028 c	0.031 c
63	Benzo(ghi)Perylene	191242							
64	Benzo(k)Fluoranthene	207089						0.0028 c	0.031 c
65	Bis(2-Chloroethoxy)Methane	111911							
66	Bis(2-Chloroethyl)Ether	111444						0.031 ac	1.4 ac
67	Bis(2-Chloroisopropyl)Ether	108601						1400 a	170000 a
68	Bis(2-Ethylhexyl)Phthalate	117817						1.8 ac	5.9 ac
69	4-Bromophenyl Phenyl Ether	101553							
70	Butylbenzyl Phthalate	85687							
71	2-Chloronaphthalene	91587							
72	4-Chlorophenyl Phenyl Ether	7005723							
73	Chrysene	218019						0.0028 c	0.031 c
74	Dibenzo(ah)Anthracene	53703						0.0028 c	0.031 c
75	1,2-Dichlorobenzene	95501						2700 a	17000 a
76	1,3-Dichlorobenzene	541731						400	2600
77	1,4-Dichlorobenzene	106467						400	2600
78	3,3'-Dichlorobenzidine	91941						0.04 ac	0.077 ac
79	Diethyl Phthalate	84662						23000 a	120000 a
80	Dimethyl Phthalate	131113						313000	2900000
81	Di-n-Butyl Phthalate	84742						2700 a	12000 a
82	2,4-Dinitrotoluene	121142						0.11 c	9.1 c
83	2,6-Dinitrotoluene	606202							
84	Di-n-Octyl Phthalate	117840							
85	1,2-Diphenylhydrazine	122667						0.040 ac	0.54 ac
86	Fluoranthene	206440						300 a	370 a
87	Fluorene	86737						1300 a	14000 a
88	Hexachlorobenzene	118741						0.00075 ac	0.00077 ac
89	Hexachlorobutadiene	87683						0.44 ac	50 ac

(#)	Compound	CAS Number	Freshwater — Criterion Maximum Conc. (µg/L) (B1)	Freshwater — Criterion Continuous Conc. (µg/L) (B2)	Saltwater — Criterion Maximum Conc. (µg/L) (C1)	Saltwater — Criterion Continuous Conc. (µg/L) (C2)	Human Health — Water & Organisms (µg/L) (D1)	Human Health — Organisms Only (µg/L) (D2)
90	Hexachlorocyclopentadiene	77474					240 a	17000 aj
91	Hexachloroethane	67721					1.9 ac	8.9 ac
92	Indeno(1,2,3-cd)Pyrene	193395					0.0028 c	0.031 c
93	Isophorone	78591					8.4 ac	600 ac
94	Naphthalene	91203						
95	Nitrobenzene	98953					17 a	1900 aj
96	N-Nitrosodimethylamine	62759					0.00069 ac	8.1 ac
97	N-Nitrosodi-n-Propylamine	621647						
98	N-Nitrosodiphenylamine	86306					5.0 ac	16 ac
99	N-Nitrosodiphenylamine	85018						
100	Phenanthrene	129000					960 a	11000 a
101	1,2,4-Trichlorobenzene	120821						
102	Aldrin	309002	3 g		1.3 g		0.00013 ac	0.00014 ac
103	alpha-BHC	319846					0.0039 ac	0.013 ac
104	beta-BHC	319857					0.014 ac	0.046 ac
105	gamma-BHC	58899	2 g	0.08 g	0.16 g		0.019 c	0.063 c
106	delta-BHC	319868						
107	Chlordane	57749	2.4 g	0.0043 g	0.09 g	0.004 g	0.00057 ac	0.00059 ac
108	4,4'-DDT	50293	1.1 g	0.001 g	0.13 g	0.001 g	0.00059 ac	0.00059 ac
109	4,4'-DDE	72559					0.00059 ac	0.00059 ac
110	4,4'-DDD	72548					0.00083 ac	0.00084 ac
111	Dieldrin	60571	2.5 g	0.0019 g	0.71 g	0.0019 g	0.00014 ac	0.00014 ac
112	alpha-Endosulfan	959988	0.22 g	0.056 g	0.034 g	0.0087 g	0.93 a	2.0 a
113	beta-Endosulfan	33213659	0.22 g	0.056 g	0.034 g	0.0087 g	0.93 a	2.0 a
114	Endosulfan Sulfate	1031078					0.93 a	2.0 a
115	Endrin	72208	0.18 g	0.0023 g	0.037 g	0.0023 g	0.76 a	0.81 aj
116	Endrin Aldehyde	7421934					0.76 a	0.81 aj
117	Heptachlor	76448	0.52 g	0.0038 g	0.053 g	0.0036 g	0.00021 ac	0.00021 ac
118	Heptachlor Epoxide	1024573	0.52 g	0.0038 g	0.053 g	0.0036 g	0.00010 ac	0.00011 ac
119	PCB-1242	53469219		0.014 g		0.03 g		
120	PCB-1254	11097691		0.014 g		0.03 g		
121	PCB-1221	11104282		0.014 g		0.03 g		
122	PCB-1232	11141165		0.014 g		0.03 g		
123	PCB-1248	12672296		0.014 g		0.03 g		
124	PCB-1260	11096825		0.014 g		0.03 g		
125a	PCB-1016	12674112		0.014 g		0.03 g		
125b	Polychlorinated biphenyls (PCBs)						0.00017 q	0.00017 q
126	Toxaphene	8001352	0.73 g	0.0002	0.21 g	0.0002	0.00073 ac	0.00075 ac
	Total Number of Criteria (h) =		24	29	23	27	85	84

a. Criteria revised to reflect current agency q_1* or RfD, as contained in the Integrated Risk Information System (IRIS). The fish tissue bioconcentration factor (BCF) from the 1980 criteria documents was retained in all cases.

b. The criteria refers to the inorganic form only.

c. Criteria in the matrix based on carcinogenicity (10^{-6} risk). For a risk level of 10^{-5}, move the decimal point in the matrix value one place to the right.

d. Criteria Maximum Concentration (CMC) = the highest concentration of a pollutant to which aquatic life can be exposed for a short period of time (1-hour average) without deleterious effects. Criteria Continuous Concentration (CCC) = the highest concentration of a pollutant to which aquatic life can be exposed for an extended period of time (4 days) without deleterious effects. μg/L = micrograms per liter.

e. Freshwater aquatic life criteria for these metals are expressed as a function of total hardness (mg/L as $CaCO_3$), the pollutant's water effect ratio (WER) as defined in §131.36(c) and multiplied by an appropriate dissolved conversion factor as defined in §131.36(b)(2). For comparative purposes, the values displayed in this matrix are shown as dissolved metal and correspond to a total hardness of 100 mg/L and a water effect ratio of 1.0.

f. Freshwater aquatic life criteria for pentachlorophenol are expressed as a function of pH, and are calculated as follows. Values displayed above in the matrix correspond to a pH of 7.8.

$$CMC = exp(1.005(pH) - 4.830)$$
$$CCC = exp(1.005(pH) - 5.290)$$

g. Aquatic life criteria for these compounds were issued in 1980 utilizing the 1980 Guidelines for criteria development. The acute values shown are final acute values (FAV) which by the 1980 Guidelines are instantaneous values as contrasted with a CMC which is a one-hour average.

h. These totals simply sum the criteria in each column. For aquatic life, there are 31 priority toxic pollutants with some type of freshwater or saltwater, acute or chronic criteria. For human health, there are 85 priority toxic pollutants with either "water + fish" or "fish only" criteria. Note that these totals count chromium as one pollutant even though EPA has developed criteria based on two valence states. In the matrix, EPA has assigned numbers 5a and 5b to the criteria for chromium to reflect the fact that the list of 126 priority toxic pollutants includes only a single listing for chromium.

i. If the CCC for total mercury exceeds 0.012 μg/l more than once in a 3-year period in the ambient water, the edible portion of aquatic species of concern must be analyzed to determine whether the concentration of methyl mercury exceeds the FDA action level (1.0 mg/kg). If the FDA action level is exceeded, the State must notify the appropriate EPA Regional Administrator, initiate a revision of its mercury criterion in its water quality standards so as to protect designated uses, and take other appropriate action such as issuance of a fish consumption advisory for the affected area.

j. No criteria for protection of human health from consumption of aquatic organisms (excluding water) was presented in the 1980 criteria document or in the 1986 Quality Criteria for Water. Nevertheless, sufficient information was presented in the 1980 document to allow a calculation of a criterion, even though the results of such a calculation were not shown in the document.

k. The criterion for asbestos is the MCL (56 FR 3526, January 30, 1991).

l. [Reserved: This letter not used as a footnote.]

m. Criteria for these metals are expressed as a function of the water effect ratio, WER, as defined in 40 CFR 131.36(c).

  CMC = column B1 or C1 value × WER
  CCC = column B2 or C2 value × WER

n. EPA is not promulgating human health criteria for this contaminant. However, permit authorities should address this contaminant in NPDES permit actions using the State's existing narrative criteria for toxics.

o. [Reserved: This letter not used as a footnote.]

p. Criterion expressed as total recoverable.

q. This criterion applies to total PCBs (*e.g.*, the sum of all congener or isomer or homolog or Aroclor analyses).

1. This chart lists all of EPA's priority toxic pollutants whether or not criteria recommendations are available. Blank spaces indicate the absence of criteria recommendations. Because of variations in chemical nomenclature systems, this listing of toxic pollutants does not duplicate the listing in Appendix A of 40 CFR Part 423. EPA has added the Chemical Abstracts Service (CAS) registry numbers, which provide a unique identification for each chemical.

2. The following chemicals have organoleptic based criteria recommendations that are not included on this chart (for reasons which are discussed in the preamble): copper, zinc, chlorobenzene, 2-chlorophenol, 2,4-dichlorophenol, acenaphthene, 2,4-dimethylphenol, 3-methyl-4-chlorophenol, hexachlorocyclopentadiene, pentachlorophenol, phenol.

3. For purposes of this rulemaking, freshwater criteria and saltwater criteria apply as specified in 40 CFR 131.36(c).

NOTE TO PARAGRAPH (b)(1): On April 14, 1995, the Environmental Protection Agency issued a stay of certain criteria in paragraph (b)(1) of this section as follows: the criteria in columns B and C for arsenic, cadmium, chromium (VI), copper, lead, nickel, silver, and zinc; the criteria in B1 and C1 for mercury; the criteria in column B for chromium (III); and the criteria in column C for selenium. The stay remains in effect until further notice.

(2) Factors for Calculating Hardness-Dependent, Freshwater Metals Criteria

$$CMC = WER \ \exp \ \{ \ m_A[\ln(\text{hardness})] + b_A \} \ \times \ \text{Acute Conversion Factor}$$

$$CCC = WER \ \exp \ \{ \ m_C[\ln(\text{hardness})] + b_C \} \ \times \ \text{Chronic Conversion Factor}$$

Final CMC and CCC values should be rounded to two significant figures.

Metal	m_A	b_A	m_C	b_C	Freshwater conversion factors	
					Acute	Chronic
Cadmium	1.128	−3.828	0.7852	−3.490	[a]0.944	[a]0.909
Chromium (III)	0.8190	3.688	0.8190	1.561	0.316	0.860
Copper	0.9422	−1.464	0.8545	−1.465	0.960	0.960
Lead	1.273	−1.460	1.273	−4.705	[a]0.791	[a]0.791
Nickel	0.8460	3.3612	0.8460	1.1645	0.998	0.997
Silver	1.72	−6.52	[b]N/A	[b]N/A	0.85	[b]N/A
Zinc	0.8473	0.8604	0.8473	0.7614	0.978	0.986

Note to table: The term "exp" represents the base e exponential function.
Footnotes to table:
[a] The freshwater conversion factors (CF) for cadmium and lead are hardness-dependent and can be calculated for any hardness [see limitations in § 131.36(c)(4)] using the following equations:
Cadmium
Acute: $CF = 1.136672 - [(\ln \text{hardness})(0.041838)]$
Chronic: $CF = 1.101672 - [(\ln \text{hardness})(0.041838)]$
Lead (Acute and Chronic): $CF = 1.46203 - [(\ln \text{hardness})(0.145712)]$
[b] No chronic criteria are available for silver.

(c) *Applicability.* (1) The criteria in paragraph (b) of this section apply to the States' designated uses cited in paragraph (d) of this section and supersede any criteria adopted by the State, except when State regulations contain criteria which are more stringent for a particular use in which case the State's criteria will continue to apply.

(2) The criteria established in this section are subject to the State's general rules of applicability in the same way and to the same extent as are the other numeric toxics criteria when applied to the same use classifications including mixing zones, and low flow values below which numeric standards can be exceeded in flowing fresh waters.

(i) For all waters with mixing zone regulations or implementation procedures, the criteria apply at the appropriate locations within or at the boundary of the mixing zones; otherwise the criteria apply throughout the waterbody including at the end of any discharge pipe, canal or other discharge point.

(ii) A State shall not use a low flow value below which numeric standards can be exceeded that is less stringent than the following for waters suitable for the establishment of low flow return frequencies (*i.e.*, streams and rivers):

AQUATIC LIFE
Acute criteria (CMC)   1 Q 10 or 1 B 3
Chronic criteria   7 Q 10 or 4 B 3
(CCC)

HUMAN HEALTH
Non-carcinogens   30 Q 5

Carcinogens   Harmonic mean flow

Where:

CMC—criteria maximum concentration—the water quality criteria to protect against acute effects in aquatic life and is the highest instream concentration of a priority toxic pollutant consisting of a one-hour average not to be exceeded more than once every three years on the average;

CCC—criteria continuous concentration—the water quality criteria to protect against chronic effects in aquatic life is the highest instream concentration of a priority toxic pollutant consisting of a 4-day average not to be exceeded more than once every three years on the average;

1 Q 10 is the lowest one day flow with an average recurrence frequency of once in 10 years determined hydrologically;

1 B 3 is biologically based and indicates an allowable exceedence of once every 3 years. It is determined by EPA's computerized method (DFLOW model);

7 Q 10 is the lowest average 7 consecutive day low flow with an average recurrence frequency of once in 10 years determined hydrologically;

4 B 3 is biologically based and indicates an allowable exceedence for 4 consecutive days once every 3 years. It is determined by EPA's computerized method (DFLOW model);

30 Q 5 is the lowest average 30 consecutive day low flow with an average recurrence frequency of once in 5 years determined hydrologically; and the harmonic mean

flow is a long term mean flow value calculated by dividing the number of daily flows analyzed by the sum of the reciprocals of those daily flows.

(iii) If a State does not have such a low flow value for numeric standards compliance, then none shall apply and the criteria included in paragraph (d) of this section herein apply at all flows.

(3) The aquatic life criteria in the matrix in paragraph (b) of this section apply as follows:

(i) For waters in which the salinity is equal to or less than 1 part per thousand 95% or more of the time, the applicable criteria are the freshwater criteria in Column B;

(ii) For waters in which the salinity is equal to or greater than 10 parts per thousand 95% or more of the time, the applicable criteria are the saltwater criteria in Column C; and

(iii) For waters in which the salinity is between 1 and 10 parts per thousand as defined in paragraphs (c)(3) (i) and (ii) of this section, the applicable criteria are the more stringent of the freshwater or saltwater criteria. However, the Regional Administrator may approve the use of the alternative freshwater or saltwater criteria if scientifically defensible information and data demonstrate that on a site-specific basis the biology of the waterbody is dominated by freshwater aquatic life and that freshwater criteria are more appropriate; or conversely, the biology of the waterbody is dominated by saltwater aquatic life and that saltwater criteria are more appropriate.

(4) *Application of metals criteria.* (i) For purposes of calculating freshwater aquatic life criteria for metals from the equations in paragraph (b)(2) of this section, the minimum hardness allowed for use in those equations shall not be less than 25 mg/l, as calcium carbonate, even if the actual ambient hardness is less than 25 mg/l as calcium carbonate. The maximum hardness value for use in those equations shall not exceed 400 mg/l as calcium carbonate, even if the actual ambient hardness is greater than 400 mg/l as calcium carbonate. The same provisions apply for calculating the metals criteria for the comparisons provided

for in paragraph (c)(3)(iii) of this section.

(ii) The hardness values used shall be consistent with the design discharge conditions established in paragraph (c)(2) of this section for flows and mixing zones.

(iii) Except where otherwise noted, the criteria for metals (compounds #2, #4-# 11, and #13, in paragraph (b) of this section) are expressed as dissolved metal. For purposes of calculating aquatic life criteria for metals from the equations in footnote m. in the criteria matrix in paragraph (b)(1) of this section and the equations in paragraphs (b)(2) of this section, the water-effect ratio is computed as a specific pollutant's acute or chronic toxicity values measured in water from the site covered by the standard, divided by the respective acute or chronic toxicity value in laboratory dilution water.

(d) *Criteria for Specific Jurisdictions—* (1) *Rhode Island, EPA Region 1.* (i) All waters assigned to the following use classifications in the Water Quality Regulations for Water Pollution Control adopted under Chapters 46–12, 42–17.1, and 42–35 of the General Laws of Rhode Island are subject to the criteria in paragraph (d)(1)(ii) of this section, without exception:

6.21 Freshwater	6.22 Saltwater:
Class A	Class SA
Class B	Class SB
Class C	Class SC

(ii) The following criteria from the matrix in paragraph (b)(1) of this section apply to the use classifications identified in paragraph (d)(1)(i) of this section:

Use classification	Applicable criteria
Class A .... Class B waters where water supply use is designated Class B waters where water supply use is not designated. Class C; Class SA; Class SB; Class SC	These classifications are assigned the criteria in Column D1—#2, 68 Each of these classifications is assigned the criteria in: Column D2—#2, 68

(iii) The human health criteria shall be applied at the 10^{-5} risk level, consistent with the State policy. To determine appropriate value for carcinogens, see footnote c in the criteria matrix in paragraph (b)(1) of this section.

(2) *Vermont, EPA Region 1.* (i) All waters assigned to the following use classifications in the Vermont Water Quality Standards adopted under the authority of the Vermont Water Pollution Control Act (10 V.S.A., Chapter 47) are subject to the criteria in paragraph (d)(2)(ii) of this section, without exception:

Class A
Class B
Class C

(ii) The following criteria from the matrix in paragraph (b)(1) of this section apply to the use classifications identified in paragraph (d)(2)(i) of this section:

Use classification	Applicable criteria
1. Classes A1, A2, B1, B2, B3	These classification are assigned the criterion in: Column B2—#105.

(iii) The human health criteria shall be applied at the State-proposed 10^{-6} risk level.

(3) *New Jersey, EPA Region 2.* (i) All waters assigned to the following use classifications in the New Jersey Administrative Code (N.J.A.C.) 7:9–4.1 et seq., Surface Water Quality Standards, are subject to the criteria in paragraph (d)(3)(ii) of this section, without exception.

N.J.A.C. 7:9–4.12(b): Class PL
N.J.A.C. 7:9–4.12(c): Class FW2
N.J.A.C. 7:9–4.12(d): Class SE1

N.J.A.C. 7:9–4.12(e): Class SE2
N.J.A.C. 7:9–4.12(f): Class SE3
N.J.A.C. 7:9–4.12(g): Class SC
N.J.A.C. 7:9–4.13(a): Delaware River Zones 1C, 1D, and 1E
N.J.A.C. 7:9–4.13(b): Delaware River Zone 2
N.J.A.C. 7:9–4.13(c): Delaware River Zone 3
N.J.A.C. 7:9–4.13(d): Delaware River Zone 4
N.J.A.C. 7:9–4.13(e): Delaware River Zone 5
N.J.A.C. 7:9–4.13(f): Delaware River Zone 6

(ii) The following criteria from the matrix in paragraph (b)(1) of this section apply to the use classifications identified in paragraph (d)(3)(i) of this section:

Use classification	Applicable criteria
1. Freshwater Pinelands, FW2	These classifications are each assigned the criteria in: i. Column B1—#2, 4, 5a, 5b, 6–11, 13. ii. Column B2—#2, 4, 5a, 5b, 6–10, 13. iii. Column D1—#125b at a 10^{-6} risk level. iv. Column D2—#125b at a 10^{-6} risk level. v. Column D2—#23, 30, 37, 42, 87, 89, 93 and 105 at a 10^{-5} risk level.
2. PL (Saline Water Pinelands), SE1, SE2, SE3, SC, Delaware Bay Zone 6.	These classifications are each assigned the criteria in: i. Column C1—#2, 4, 5b, 6–11, 13. ii. Column C2—#2, 4, 5b, 6–10, 13. iii. Column D1—#125b at a 10^{-6} risk level. iv. Column D2—#125b at a 10^{-6} risk level. v. Column D2—#23, 30, 37, 42, 87, 89, 93 and 105 at a 10^{-5} risk level.
3. Delaware River Zones 1C, 1D, 1E, 2, 3, 4, and 5	i. Column B1—none. ii. Column B2—none. iii. Column D1—none. iv. Column D2—none.
4. Delaware River Zones 3, 4, and 5	These classifications are each assigned the criteria in: i. Column C1—none. ii. Column C2—none. iii. Column D2—none.

(iii) The human health criteria shall be applied at the State-proposed 10^{-6} risk level for EPA rated Class A, B₁, and B₂ carcinogens; EPA rated Class C carcinogens shall be applied at 10^{-5} risk level. To determine appropriate value for carcinogens, see footnote c. in the matrix in paragraph (b)(1) of this section.

(4) *Puerto Rico, EPA Region 2.* (i) All waters assigned to the following use classifications in the Puerto Rico

Water Quality Standards (promulgated by Resolution Number R–83–5–2) are subject to the criteria in paragraph (d)(4)(ii) of this section, without exception.

Article 2.2.2—Class SB
Article 2.2.3—Class SC

Article 2.2.4—Class SD

(ii) The following criteria from the matrix in paragraph (b)(1) of this section apply to the use classifications identified in paragraph (d)(4)(i) of this section:

Use classification	Applicable criteria
Class SD	Column B1—# 118. Column B2—#s 8, 105, 115, 118, 119, 120, 121, 122, 123,124, 125a, 125b. Column D1—#s 12, 16, 27, 60, 61, 62, 64, 73, 74, 92, 93, 103, 104, 114, 116, 118, 119, 120, 121, 122, 123, 124, 125a, 125b.
Class SB, Class SC	Column C1—#s 5b, 112, 113, 118. Column C2—#s 5b, 8, 112, 113, 118, 119, 120, 121, 122, 123, 124, 125a, 125b. Column D2—#s 12, 16, 27, 60, 61, 62, 64, 73, 74, 87, 92, 93, 103, 104, 114, 116, 118, 119, 120, 121, 122, 123, 124, 125a, 125b.

(iii) The human health criteria shall be applied at the State-proposed 10^{-5} risk level. To determine appropriate value for carcinogens, see footnote c, in the criteria matrix in paragraph (b)(1) of this section.

(5) *District of Columbia, EPA Region 3.* (i) All waters assigned to the following use classifications in chapter 11 Title 21 DCMR, Water Quality Standards of the District of Columbia are subject to the criteria in paragraph (d)(5)(ii) of this section, without exception:

1101.2 Class C waters

(ii) The following criteria from the matrix in paragraph (b)(1) of this section apply to the use classification identified in paragraph (d)(5)(i) of this section:

Use classification	Applicable criteria
1. Class C	This classification is assigned the additional criteria in: Column B2; #10, 118, 126.

(iii) The human health criteria shall be applied at the State-adopted 10^{-6} risk level.

(6) *Florida, EPA Region 4.* (i) All waters assigned to the following use classifications in Chapter 17–301 of the Florida Administrative Code (*i.e.*, identified in Section 17–302.600) are subject to the criteria in paragraph (d)(6)(ii) of this section, without exception:

Class I
Class II
Class III

(ii) The following criteria from the matrix paragraph (b)(1) of this section apply to the use classifications identified in paragraph (d)(6)(i) of this section:

Use classification	Applicable criteria
Class I	This classification is assigned the criteria in: Column D1—#16
Class II .... Class III (marine)	This classification is assigned the criteria in: Column D2—#16
Class III (freshwater)	This classification is assigned the criteria in: Column D2—#16

(iii) The human health criteria shall be applied at the State-adopted 10^{-6} risk level.

(7)–(8) [Reserved]

(9) *Kansas, EPA Region 7.* (i) All waters assigned to the following use classification in the Kansas Department of Health and Environment regulations, K.A.R. 28–16–28b through K.A.R. 28–16–28f, are subject to the criteria in paragraph (d)(9)(ii) of this section, without exception.

Section (2)(A)—Special Aquatic Life Use Waters
Section (2)(B)—Expected Aquatic Life Use Waters
Section (2)(C)—Restricted Aquatic Life Use Waters
Section (3)—Domestic Water Supply.
Section (4)—Food Procurement Use.

(ii) The following criteria from the matrix in paragraph (b)(1) of this section apply to the use classifications identified in paragraph (d)(9)(i) of this section:

Use classification	Applicable criteria
1. Sections (2)(A), (2)(B), (2)(C), (4)	These classifications are each assigned criteria as follows: i. Column B1, #2. ii. Column D2, #12, 21, 29, 39, 46, 68, 79, 81, 86, 93, 104, 114, 118.
2. Section (3)	This classification is assigned all criteria in: Column D1, all except #1, 9, 12, 14, 15, 17, 22, 33, 36, 39, 44, 75, 77, 79, 90, 112, 113, and 115.

(iii) The human health criteria shall be applied at the State-adopted 10^{-6} risk level.

(10) *California, EPA Region 9.* (i) All waters assigned any aquatic life or human health use classifications in the Water Quality Control Plans for the various Basins of the State ("Basin Plans"), as amended, adopted by the California State Water Resources Control Board ("SWRCB"), except for ocean waters covered by the Water Quality Control Plan for Ocean Waters of California ("Ocean Plan") adopted by the SWRCB with resolution Number 90–27 on March 22, 1990, are subject to the criteria in paragraph (d)(10)(ii) of this section, without exception. These criteria amend the portions of the existing State standards contained in the Basin Plans. More particularly these criteria amend water quality criteria

contained in the Basin Plan Chapters specifying water quality objectives (the State equivalent of federal water quality criteria) for the toxic pollutants identified in paragraph (d)(10)(ii) of this section. Although the State has adopted several use designations for each of these waters, for purposes of this action, the specific standards to be applied in paragraph (d)(10)(ii) of this section are based on the presence in all waters of some aquatic life designation and the presence or absence of the MUN use designation (Municipal and domestic supply). (See Basin Plans for more detailed use definitions.)

(ii) The following criteria from the matrix in paragraph (b)(1) of this section apply to the water and use classifications defined in paragraph (d)(10)(i) of this section and identified below:

Water and use classification	Applicable criteria
Waters of the State defined as bays or estuaries except the Sacramento-San Joaquin Delta and San Francisco Bay	These waters are assigned the criteria in: Column B1—pollutants 5a and 14 Column B2—pollutants 5a and 14 Column C1—pollutant 14 Column C2—pollutant 14 Column D2—pollutants 1, 12, 17, 18, 21, 22, 29, 30, 32, 33, 37, 38, 42–44, 46, 48, 49, 54, 59, 66, 67, 68, 78–82, 85, 89, 90, 91, 93, 95, 96, 98

Water and use classification	Applicable criteria
Waters of the Sacramento—San Joaquin Delta and waters of the State defined as inland (*i.e.*, all surface waters of the State not bays or estuaries or ocean) that include a MUN use designation	These waters are assigned the criteria in: Column B1—pollutants 5a and 14 Column B2—pollutants 5a and 14 Column D1—pollutants 1, 12, 15, 17, 18, 21, 22, 29, 30, 32, 33, 37, 38, 42–48, 49, 59, 66, 67, 68, 78–82, 85, 89, 90, 91, 93, 95, 96, 98
Waters of the State defined as inland without an MUN use designation	These waters are assigned the criteria in: Column B1—pollutants 5a and 14 Column B2—pollutants 5a and 14 Column D2—pollutants 1, 12, 17, 18, 21, 22, 29, 30, 32, 33, 37, 38, 42–44, 46, 48, 49, 54, 59, 66, 67, 68, 78–82, 85, 89, 90, 91, 93, 95, 96, 98
Waters of the San Joaquin River from the mouth of the Merced River to Vernalis	In addition to the criteria assigned to these waters elsewhere in this rule, these waters are assigned the criteria in: Column B2—pollutant 10
Waters of Salt Slough, Mud Slough (north) and the San Joaquin River, Sack Dam to the mouth of the Merced River	In addition to the criteria assigned to these waters elsewhere in this rule, these waters are assigned the criteria in: Column B1—pollutant 10 Column B2—pollutant 10
Waters of San Francisco Bay upstream to and including Suisun Bay and the Sacramento-San Joaquin Delta	These waters are assigned the criteria in: Column B1—pollutants 5a, 10* and 14 Column B2—pollutants 5a, 10* and 14 Column C1—pollutant 14 Column C2—pollutant 14 Column D2—pollutants 1, 12, 17, 18, 21, 22, 29, 30, 32, 33, 37, 38, 42–44, 46, 48, 49, 54, 59, 66, 67, 68, 78–82, 85, 89, 90, 91, 93, 95, 96, 98
All inland waters of the United States or enclosed bays and estuaries that are waters of the United States that include an MUN use designation and that the State has either excluded or partially excluded from coverage under its Water Quality Control Plan for Inland Surface Waters of California, Tables 1 and 2, or its Water Quality Control Plan for Enclosed Bays and Estuaries of California, Tables 1 and 2, or has deferred applicability of those tables. (Category (a), (b), and (c) waters described on page 6 of Water Quality Control Plan for Inland Surface Waters of California or page 6 of its Water Quality Control Plan for Enclosed Bays and Estuaries of California.)	These waters are assigned the criteria for pollutants for which the State does not apply Table 1 or 2 standards. These criteria are: Column B1—all pollutants Column B2—all pollutants Column D1—all pollutants except #2

491

Water and use classification	Applicable criteria
All inland waters of the United States that do not include an MUN use designation and that the State has either excluded or partially excluded from coverage under its Water Quality Control Plan for Inland Surface Waters of California, Tables 1 and 2, or has deferred applicability of these tables. (Category (a), (b), and (c) waters described on page 6 of Water Quality Control Plan for Inland Surface Waters of California.)	These waters are assigned the criteria for pollutants for which the State does not apply Table 1 or 2 standards. These criteria are: Column B1—all pollutants Column B2—all pollutants Column D2—all pollutants except #2
All enclosed bays and estuaries that are waters of the United States that do not include an MUN designation and that the State has either excluded or partially excluded from coverage under its Water Quality Control Plan for Inland Surface Waters of California, Tables 1 and 2, or its Water Quality Control Plan for Enclosed Bays and Estuaries of California, Tables 1 and 2, or has deferred applicability of those tables. (Category (a), (b), and (c) waters described on page 6 of Water Quality Control Plan for Inland Surface Waters of California or page 6 of its Water Quality Control Plan for Enclosed Bays and Estuaries of California.)	These waters are assigned the criteria for pollutants for which the State does not apply Table 1 or 2 standards. These criteria are: Column B1—all pollutants Column B2—all pollutants Column C1—all pollutants Column C2—all pollutants Column D2—all pollutants except #2

*The fresh water selenium criteria are included for the San Francisco Bay estuary because high levels of bioaccumulation of selenium in the estuary indicate that the salt water criteria are underprotective for San Francisco Bay.

(iii) The human health criteria shall be applied at the State-adopted 10^{-6} risk level.

(11) *Nevada, EPA Region 9.* (i) All waters assigned the use classifications in Chapter 445 of the Nevada Administrative Code (NAC), Nevada Water Pollution Control Regulations, which are referred to in paragraph (d)(11)(ii) of this section, are subject to the criteria in paragraph (d)(11)(ii) of this section, without exception. These criteria amend the existing State standards contained in the Nevada Water Pollution Control Regulations. More particularly, these criteria amend or supplement the table of numeric standards in NAC 445.1339 for the toxic pollutants identified in paragraph (d)(11)(ii) of this section.

(ii) The following criteria from matrix in paragraph (b)(1) of this section apply to the waters defined in paragraph (d)(11)(i) of this section and identified below:

Water and use classification	Applicable criteria
Waters that the State has included in NAC 445.1339 where Municipal or domestic supply is a designated use	These waters are assigned the criteria in: Column B1—pollutant #118 Column B2—pollutant #118 Column D1—pollutants #15, 16, 18, 19, 20, 21, 23, 26, 27, 29, 30, 34, 37, 38, 42, 43, 55, 58–62, 64, 66, 73, 74, 78, 82, 85, 87–89, 91, 92, 96, 98, 100, 103, 104, 105, 114, 116, 117, 118
Waters that the State has included in NAC 445.1339 where Municipal or domestic supply is not a designated use	These waters are assigned the criteria in: Column B1—pollutant #118 Column B2—pollutant #118 Column D2—all pollutants except #2.

(iii) The human health criteria shall be applied at the 10^{-5} risk level, consistent with State policy. To determine appropriate value for carcinogens, see footnote c in the criteria matrix in paragraph (b)(1) of this section.

(12) *Alaska, EPA Region 10.* (i) All waters assigned to the following use classifications in the Alaska Administrative Code (AAC), Chapter 18 (*i.e.,* identified in 18 AAC 70.020) are subject to the criteria in paragraph (d)(12)(ii) of this section, without exception:

70.020.(1) (A)　Fresh Water
70.020.(1) (A)　Water Supply
　(i) Drinking, culinary, and food processing,
　(iii) Aquaculture;
70.020.(1) (B)　Water Recreation
　(i) Contact recreation,
　(ii) Secondary recreation;
70.020.(1) (C)　Growth and propagation of fish, shellfish, other aquatic life, and wildlife
70.020.(2) (A)　Marine Water
70.020.(2) (A)　Water Supply
　(i) Aquaculture,

70.020.(2) (B)　Water Recreation
　(i) contact recreation,
　(ii) secondary recreation;
70.020.(2) (C) Growth and propagation of fish, shellfish, other aquatic life, and wildlife;
70.020.(2) (D)　Harvesting for consumption of raw mollusks or other raw aquatic life.

(ii) The following criteria from the matrix in paragraph (b)(1) of this section apply to the use classifications identified in paragraph (d)(12)(i) of this section:

Use classification	Applicable criteria
(1)(A)(i)	Column D1—#s 16, 18–21, 23, 26, 27, 29, 30, 32, 37, 38, 42–44, 53, 55, 59–62, 64, 66, 68, 73, 74, 78, 82, 85, 88, 89, 91–93, 96, 98, 102–105, 107–111, 117–126.
(1)(A)(iii)	Column D2—#s 14, 16, 18–21, 22, 23, 26, 27, 29, 30, 32, 37, 38, 42–44, 46, 53, 54, 55, 59–62, 64, 66, 68, 73, 74, 78, 82, 85, 88–93, 95, 96, 98, 102–105, 107–111, 115–126.
(1)(B)(i), (1)(B)(ii), (1)(C)	Column D2—#s 14, 16, 18–21, 22, 23, 26, 27, 29, 30, 32, 37, 38, 42–44, 46, 53, 54, 55, 59–62, 64, 66, 68, 73, 74, 78, 82, 85, 88–93, 95, 96, 98, 102–105, 107–111, 115–126.
(2)(A)(i), (2)(B)(i), and (2)(B)ii, (2)(C), (2)(D)	Column D2—#s 14, 16, 18–21, 22, 23, 26, 27, 29, 30, 32, 37, 38, 42–44, 46, 53, 54, 55, 59–62, 64, 66, 68, 73, 74, 78, 82, 85, 88–93, 95, 96, 98, 102–105, 107–111, 115–126.

(iii) The human health criteria shall be applied at the State-proposed risk level of 10^{-5}. To determine appropriate value for carcinogens, see footnote c in the criteria matrix in paragraph (b)(1) of this section.

(13) [Reserved]

(14) *Washington, EPA Region 10.* (i) All waters assigned to the following use classifications in the Washington Administrative Code (WAC), Chapter 173–201 (*i.e.*, identified in WAC 173–201–045) are subject to the criteria in paragraph (d)(14)(ii) of this section, without exception:

173–201–045
　Fish and Shellfish
　Fish
　Water Supply (domestic)
　Recreation

(ii) The following criteria from the matrix in paragraph (b)(1) of this section apply to the use classifications identified in paragraph (d)(14)(i) of this section:

Use classification	Applicable criteria
Fish and Shellfish; Fish	These classifications are assigned the criteria in: Column D2—all.
Water Supply (domestic)	These classifications are assigned the criteria in: Column D1—all.
Recreation	This classification is assigned the criteria in: Column D2—Marine waters and freshwaters not protected for domestic water supply.

(iii) The human health criteria shall be applied at the State proposed risk level of 10^{-6}.

[57 FR 60910, Dec. 22, 1992]

EDITORIAL NOTE: For FEDERAL REGISTER citations affecting §131.36, see the List of CFR Sections Affected, which appears in the Finding Aids section of the printed volume and at *www.fdsys.gov.*

§131.37　California.

(a) *Additional criteria.* The following criteria are applicable to waters specified in the Water Quality Control Plan for Salinity for the San Francisco Bay/Sacramento-San Joaquin Delta Estuary, adopted by the California State Water Resources Control Board in State Board Resolution No. 91–34 on May 1, 1991:

(1) *Estuarine habitat criteria.* (i) *General rule.* (A) Salinity (measured at the surface) shall not exceed 2640 micromhos/centimeter specific conductance at 25 °C (measured as a 14-day moving average) at the Confluence of the Sacramento and San Joaquin Rivers throughout the period each year from February 1 through June 30, and shall not exceed 2640 micromhos/centimeter specific conductance at 25 °C (measured as a 14-day moving average) at the specific locations noted in Table 1 near Roe Island and Chipps Island for the number of days each month in the February 1 to June 30 period computed by reference to the following formula:

Number of days required in Month X = Total number of days in Month$x$* (1 − 1/(1+e^K)

where

$K = A + (B*$natural logarithm of the previous month's 8-River Index);

A and B are determined by reference to Table 1 for the Roe Island and Chipps Island locations;

x is the calendar month in the February 1 to June 30 period;

and e is the base of the natural (or Napierian) logarithm.

Where the number of days computed in this equation in paragraph (a)(1)(i)(A) of this section shall be rounded to the nearest whole number of days. When the previous month's 8-River Index is less than 500,000 acre-feet, the number of days required for the current month shall be zero.

TABLE 1. CONSTANTS APPLICABLE TO EACH OF THE MONTHLY EQUATIONS TO DETERMINE MONTHLY REQUIREMENTS DESCRIBED.

Month X	Chipps Island		Roe Island (if triggered)	
	A	B	A	B
Feb	−1	−1	−14.36	+2.068
Mar	−105.16	+15.943	−20.79	+2.741
Apr	−47.17	+6.441	−28.73	+3.783
May	−94.93	+13.662	−54.22	+6.571
June	−81.00	+9.961	−92.584	+10.699

[1] Coefficients for A and B are not provided at Chipps Island for February, because the 2640 micromhos/cm specific conductance criteria must be maintained at Chipps Island throughout February under all historical 8-River Index values for January.

(B) The Roe Island criteria apply at the salinity measuring station maintained by the U.S. Bureau of Reclamation at Port Chicago (km 64). The Chipps Island criteria apply at the Mallard Slough Monitoring Site, Station D–10 (RKI RSAC–075) maintained by the California Department of Water Resources. The Confluence criteria apply at the Collinsville Continuous Monitoring Station C–2 (RKI RSAC–081) maintained by the California Department of Water Resources.

(ii) *Exception.* The criteria at Roe Island shall be required for any given month only if the 14-day moving average salinity at Roe Island falls below 2640 micromhos/centimeter specific conductance on any of the last 14 days of the previous month.

(2) *Fish migration criteria*—(i) *General rule*—(A) *Sacramento River.* Measured Fish Migration criteria values for the Sacramento River shall be at least the following:

At temperatures less than below 61 °F: SRFMC = 1.35

At temperatures between 61 °F and 72 °F: SRFMC = 6.96–.092 * Fahrenheit temperature

At temperatures greater than 72 °F: SRFMC = 0.34

where SRFMC is the Sacramento River Fish Migration criteria value. Temperature shall be the water temperature at release of tagged salmon smolts into the Sacramento River at Miller Park.

(B) *San Joaquin River.* Measured Fish Migration criteria values on the San Joaquin River shall be at least the following:

For years in which the SJVIndex is >2.5: SJFMC = (−0.012) + 0.184*SJVIndex

In other years: SJFMC = 0.205 + 0.0975*SJVIndex

where SJFMC is the San Joaquin River Fish Migration criteria value, and SJVIndex is the San Joaquin Valley Index in million acre feet (MAF)

(ii) *Computing fish migration criteria values for Sacramento River.* In order to assess fish migration criteria values for the Sacramento River, tagged fall-run salmon smolts will be released into the Sacramento River at Miller Park and captured at Chipps Island, or alternatively released at Miller Park and Port Chicago and recovered from the ocean fishery, using the methodology described in this paragraph (a)(2)(ii). An alternative methodology for computing fish migration criteria values can be used so long as the revised methodology is calibrated with the methodology described in this paragraph (a)(2)(ii) so as to maintain the validity of the relative index values. Sufficient releases shall be made each year to provide a statistically reliable verification of compliance with the criteria. These criteria will be considered attained when the sum of the differences between the measured experimental value and the stated criteria value (i.e., measured value minus stated value) for each experimental release conducted over a three year period (the current year and the previous two years) shall be greater than or equal to zero. Fish for release are to be tagged at the hatchery with coded-wire tags, and fin clipped. Approximately 50,000 to 100,000 fish of smolt size (size greater than 75 mm) are released for each survival index estimate, depending on expected mortality. As a control for the ocean recovery survival index, one or two groups per season are released at Benecia or Pt. Chicago. From each upstream release of tagged fish, fish are to be caught over a period of one to two weeks at Chipps Island. Daylight sampling at Chipps Island with a 9.1 by 7.9 m, 3.2 mm cod end, midwater trawl is begun 2 to 3 days after release. When the first fish is caught, full-time trawling 7 days a week should begin. Each day's trawling consists of ten 20 minute tows generally made against the current, and distributed equally across the channel.

(A) The Chipps Island smolt survival index is calculated as:

$$SSI = R \div MT(0.007692)$$

where

R=number of recaptures of tagged fish
M=number of marked (tagged) fish released
T=proportion of time sampled vs total time tagged fish were passing the site (i.e. time between first and last tagged fish recovery)

Where the value 0.007692 is the proportion of the channel width fished by the trawl, and is calculated as trawl width/channel width.

(B) Recoveries of tagged fish from the ocean salmon fishery two to four years after release are also used to calculate a survival index for each release. Smolt survival indices from ocean recoveries are calculated as:

$$OSI = R_1/M_1 \div R_2/M_2$$

where

R_1=number of tagged adults recovered from the upstream release
M_1=number released upstream
R_2=number of tagged adults recovered from the Port Chicago release
M_2=number released at Port Chicago

(*1*) The number of tagged adults recovered from the ocean fishery is provided by the Pacific States Marine Fisheries Commission, which maintains a port sampling program.

(*2*) [Reserved]

(iii) *Computing fish migration criteria values for San Joaquin River.* In order to assess annual fish migration criteria values for the San Joaquin River, tagged salmon smolts will be released into the San Joaquin River at Mossdale and captured at Chipps Island, or alternatively released at Mossdale and Port Chicago and recovered from the ocean fishery, using the methodology described in paragraph (a)(2)(iii). An alternative methodology for computing fish migration criteria values can be used so long as the revised methodology is calibrated with the methodology described below so as to maintain the validity of the relative index values. Sufficient releases shall be made each year to provide a statistically reliable estimate of the SJFMC for the year. These criteria will be considered attained when the sum of the differences between the measured experimental value and the stated criteria value (i.e., measured value minus stated value) for each experimental release conducted over a three year period (the current year and the previous two years) shall be greater than or equal to zero.

(A) Fish for release are to be tagged at the hatchery with coded-wire tags, and fin clipped. Approximately 50,000 to 100,000 fish of smolt size (size greater than 75 mm) are released for each survival index estimate, depending on expected mortality. As a control for the ocean recovery survival index, one or two groups per season are released at Benicia or Pt. Chicago. From each upstream release of tagged fish, fish are to be caught over a period of one to two weeks at Chipps Island. Daylight sampling at Chipps Island with a 9.1 by 7.9 m, 3.2 mm cod end, midwater trawl is begun 2 to 3 days after release. When the first fish is caught, full-time trawling 7 days a week should begin. Each day's trawling consists of ten 20 minute tows generally made against the current, and distributed equally across the channel.

(B) The Chipps Island smolt survival index is calculated as:

$$SSI = R \div MT(0.007692)$$

where

R=number of recaptures of tagged fish
M=number of marked (tagged) fish released
T=proportion of time sampled vs total time tagged fish were passing the site (i.e. time between first and last tagged fish recovery)

Where the value 0.007692 is the proportion of the channel width fished by the trawl, and is calculated as trawl width/channel width.

(C) Recoveries of tagged fish from the ocean salmon fishery two to four years after release are also used to calculate a survival index for each release. Smolt survival indices from ocean recoveries are calculated as:

$$OSI = R_1/M_1 \div R_2/M_2$$

where

R_1=number of tagged adults recovered from the upstream release
M_1=number released upstream
R_2=number of tagged adults recovered from the Port Chicago release
M_2=number released at Port Chicago

(1) The number of tagged adults recovered from the ocean fishery is provided by the Pacific States Marine Fisheries Commission, which maintains a port sampling program.

(2) [Reserved]

(3) *Suisun marsh criteria.* (i) Water quality conditions sufficient to support a natural gradient in species composition and wildlife habitat characteristic of a brackish marsh throughout all elevations of the tidal marshes bordering Suisun Bay shall be maintained. Water quality conditions shall be maintained so that none of the following occurs: Loss of diversity; conversion of brackish marsh to salt marsh; for animals, decreased population abundance of those species vulnerable to increased mortality and loss of habitat from increased water salinity; or for plants, significant reduction in stature or percent cover from increased water or soil salinity or other water quality parameters.

(ii) [Reserved]

(b) *Revised criteria.* The following criteria are applicable to state waters specified in Table 1-1, at Section (C)(3) ("Striped Bass—Salinity : 3. Prisoners Point—Spawning) of the Water Quality Control Plan for Salinity for the San Francisco Bay—Sacramento/San Joaquin Delta Estuary, adopted by the California State Water Resources Control Board in State Board Resolution No. 91-34 on May 1, 1991:

Location	Sampling site Nos (I—A/RKI)	Parameter	Description	Index type	San Joaquin Valley Index	Dates	Values
San Joaquin River at Jersey Point, San Andreas Landing, Prisoners Point, Buckley Cove, Rough and Ready Island, Brandt Bridge, Mossdale, and Vernalis.	D15/RSAN018, C4/RSAN032, D29/RSAN038, P8/RSAN056, -/RSAN062, C6/RSAN073, C7/RSAN087, C10/RSAN112	Specific Conductance. @ 25 °C	14-day running average of mean daily for the period not more than value shown, in mmhos.	Not Applicable.	>2.5 MAF	April 1 to May 31.	0.44 micromhos.
San Joaquin River at Jersey Point, San Andreas Landing and Prisoners Point.	D15/RSAN018, C4/RSAN032, D29/RSAN038	Specific Conductance.	14-day running average of mean daily for the period not more than value shown, in mmhos.	Not Applicable.	≤2.5 MAF	April 1 to May 31.	0.44 micromhos.

(c) *Definitions.* Terms used in paragraphs (a) and (b) of this section, shall be defined as follows:

(1) *Water year.* A water year is the twelve calendar months beginning October 1.

(2) *8-River Index.* The flow determinations are made and are published by the California Department of Water Resources in Bulletin 120. The 8-River Index shall be computed as the sum of flows at the following stations:

(i) Sacramento River at Band Bridge, near Red Bluff;

(ii) Feather River, total inflow to Oroville Reservoir;

(iii) Yuba River at Smartville;

(iv) American River, total inflow to Folsom Reservoir;

(v) Stanislaus River, total inflow to New Melones Reservoir;

(vi) Tuolumne River, total inflow to Don Pedro Reservoir;

(vii) Merced River, total inflow to Exchequer Reservoir; and

(viii) San Joaquin River, total inflow to Millerton Lake.

(3) *San Joaquin Valley Index.* (i) The San Joaquin Valley Index is computed according to the following formula:

$$I_{SJ}=0.6X+0.2Y \text{ and } 0.2Z$$

where

I_{SJ}=San Joaquin Valley Index

X=Current year's April-July San Joaquin Valley unimpaired runoff

Y=Current year's October-March San Joaquin Valley unimpaired runoff

Z=Previous year's index in MAF, not to exceed 0.9 MAF

(ii) *Measuring San Joaquin Valley unimpaired runoff.* San Joaquin Valley unimpaired runoff for the current water year is a forecast of the sum of the following locations: Stanislaus River, total flow to New Melones Reservoir; Tuolumne River, total inflow to Don Pedro Reservoir; Merced River, total flow to Exchequer Reservoir; San Joaquin River, total inflow to Millerton Lake.

(4) *Salinity.* Salinity is the total concentration of dissolved ions in water. It shall be measured by specific conductance in accordance with the procedures set forth in 40 CFR 136.3, Table 1B, Parameter 64.

[60 FR 4707, Jan. 24, 1995]

§ 131.38 Establishment of numeric criteria for priority toxic pollutants for the State of California.

(a) *Scope.* This section promulgates criteria for priority toxic pollutants in the State of California for inland surface waters and enclosed bays and estuaries. This section also contains a compliance schedule provision.

(b)(1) Criteria for Priority Toxic Pollutants in the State of California as described in the following table:

A		B Freshwater		C Saltwater		D Human Health (10⁻⁶ risk for carcinogens) For consumption of:	
# Compound	CAS Number	Criterion Maximum Conc. d B1	Criterion Continuous Conc. d B2	Criterion Maximum Conc. d C1	Criterion Continuous Conc. d C2	Water & Organisms (μg/L) D1	Organisms Only (μg/L) D2
1. Antimony	7440360					14 a,s	4300 a,t
2. Arsenic b	7440382	340 i,m,w	150 i,m,w	69 i,m	36 i,m		
3. Beryllium	7440417					n	n
4. Cadmium b	7440439	4.3 e,i,m,w,x	2.2 e,i,m,w	42 i,m	9.3 i,m	n	n
5a. Chromium (III)	16065831	550 e,i,m,o	180 e,i,m,o			n	n
5b. Chromium (VI) b	18540299	16 i,m,w	11 i,m,w	1100 i,m	50 i,m	n	n
6. Copper b	7440508	13 e,i,m,w,x	9.0 e,i,m,w	4.8 i,m	3.1 i,m	1300	
7. Lead b	7439921	65 e,i,m	2.5 e,i,m	210 i,m	8.1 i,m	n	n
8. Mercury b	7439976	[Reserved]	[Reserved]	[Reserved]	[Reserved]	0.050 a	0.051 a
9. Nickel b	7440020	470 e,i,m,w	52 e,i,m,w	74 i,m	8.2 i,m	610 a	4600 a
10. Selenium b	7782492	[Reserved] p	5.0 q	290 i,m	71 i,m	n	n
11. Silver b	7440224	3.4 e,i,m		1.9 i,m			
12. Thallium	7440280					1.7 a,s	6.3 a,t
13. Zinc b	7440666	120 e,i,m,w,x	120 e,i,m,w	90 i,m	81 i,m		
14. Cyanide b	57125	22 o	5.2 o	1 r	1 r	700 a	220,000 a,j
15. Asbestos	1332214					7,000,000 fibers/L k,s	
16. 2,3,7,8-TCDD (Dioxin)	1746016					0.000000013 c	0.000000014 c
17. Acrolein	107028					320 s	780 t
18. Acrylonitrile	107131					0.059 a,c,s	0.66 a,c,t
19. Benzene	71432					1.2 a,c	71 a,c
20. Bromoform	75252					4.3 a,c	360 a,c
21. Carbon Tetrachloride	56235					0.25 a,c,s	4.4 a,c,t
22. Chlorobenzene	108907					680 a,s	21,000 a,j,t
23. Chlorodibromomethane	124481					0.401 a,c	34 a,c
24. Chloroethane	75003						
25. 2-Chloroethylvinyl Ether	110758						

26. Chloroform	67663					[Reserved]	[Reserved]
27. Dichlorobromomethane	75274					0.56 a,c	46 a,c
28. 1,1-Dichloroethane	75343						
29. 1,2-Dichloroethane	107062					0.38 a,c,s	99 a,c,t
30. 1,1-Dichloroethylene	75354					0.057 a,c,s	3.2 a,c,t
31. 1,2-Dichloropropane	78875					0.52 a	39 a
32. 1,3-Dichloropropylene	542756					10 a,s	1,700 a,t
33. Ethylbenzene	100414					3,100 a,s	29,000 a,t
34. Methyl Bromide	74839					48 a	4,000 a
35. Methyl Chloride	74873					n	n
36. Methylene Chloride	75092					4.7 a,c	1,600 a,c
37. 1,1,2,2-Tetrachloroethane	79345					0.17 a,c,s	11 a,c,t
38. Tetrachloroethylene	127184					0.8 c,s	8.85 c,t
39. Toluene	108883					6,800 a,s	200,000 a
40. 1,2-Trans-Dichloroethylene	156605					700 a	140,000 a
41. 1,1,1-Trichloroethane	71556					n	n
42. 1,1,2-Trichloroethane	79005					0.60 a,c,s	42 a,c,t
43. Trichloroethylene	79016					2.7 c,s	81 c,t
44. Vinyl Chloride	75014					2 c,s	525 c,t
45. 2-Chlorophenol	95578					120 a	400 a
46. 2,4-Dichlorophenol	120832					93 a,s	790 a,t
47. 2,4-Dimethylphenol	105679					540 a	2,300 a
48. 2-Methyl-4,6-Dinitrophenol	534521					13.4 s	765 t
49. 2,4-Dinitrophenol	51285					70 a,s	14,000 a,t
50. 2-Nitrophenol	88755						
51. 4-Nitrophenol	100027						
52. 3-Methyl-4-Chlorophenol	59507						
53. Pentachlorophenol	87865	19 f,w	15 f,w	13	7.9	0.28 a,c	8.2 a,c,j
54. Phenol	108952					21,000 a	4,600,000 a,j,t
55. 2,4,6-Trichlorophenol	88062					2.1 a,c	6.5 a,c
56. Acenaphthene	83329					1,200 a	2,700 a
57. Acenaphthylene	208968						
58. Anthracene	120127					9,600 a	110,000 a

59. Benzidine	92875					0.00012 a,c,s	0.00054 a,c,t
60. Benzo(a)Anthracene	56553					0.0044 a,c	0.049 a,c
61. Benzo(a)Pyrene	50328					0.0044 a,c	0.049 a,c
62. Benzo(b)Fluoranthene	205992					0.0044 a,c	0.049 a,c
63. Benzo(ghi)Perylene	191242						
64. Benzo(k)Fluoranthene	207089					0.0044 a,c	0.049 a,c
65. Bis(2-Chloroethoxy)Methane	111911						
66. Bis(2-Chloroethyl)Ether	111444					0.031 a,c,s	1.4 a,c,t
67. Bis(2-Chloroisopropyl)Ether	39638329					1,400 a	170,000 a,t
68. Bis(2-Ethylhexyl)Phthalate	117817					1.8 a,c,s	5.9 a,c,t
69. 4-Bromophenyl Phenyl Ether	101553						
70. Butylbenzyl Phthalate	85687					3,000 a	5,200 a
71. 2-Chloronaphthalene	91587					1,700 a	4,300 a
72. 4-Chlorophenyl Phenyl Ether	7005723						
73. Chrysene	218019					0.0044 a,c	0.049 a,c
74. Dibenzo(a,h)Anthracene	53703					0.0044 a,c	0.049 a,c
75. 1,2 Dichlorobenzene	95501					2,700 a	17,000 a
76. 1,3 Dichlorobenzene	541731					400	2,600
77. 1,4 Dichlorobenzene	106467					400	2,600
78. 3,3'-Dichlorobenzidine	91941					0.04 a,c,s	0.077 a,c,t
79. Diethyl Phthalate	84662					23,000 a,s	120,000 a,t
80. Dimethyl Phthalate	131113					313,000 s	2,900,000 t
81. Di-n-Butyl Phthalate	84742					2,700 a,s	12,000 a,t
82. 2,4-Dinitrotoluene	121142					0.11 c,s	9.1 c,t
83. 2,6-Dinitrotoluene	606202						
84 Di-n-Octyl Phthalate	117840						
85. 1,2-Diphenylhydrazine	122667					0.040 a,c,s	0.54 a,c,t
86. Fluoranthene	206440					300 a	370 a
87. Fluorene	86737					1,300 a	14,000 a
88. Hexachlorobenzene	118741					0.00075 a,c	0.00077 a,c
89. Hexachlorobutadiene	87683					0.44 a,c,s	50 a,c,t
90. Hexachlorocyclopentadiene	77474					240 a,s	17,000 a,j,t
91. Hexachloroethane	67721					1.9 a,c,s	8.9 a,c,t

92. Indeno(1,2,3-cd) Pyrene	193395					0.0044 a,c	0.049 a,c
93. Isophorone	78591					8.4 c,s	600 c,t
94. Naphthalene	91203						
95. Nitrobenzene	98953					17 a,s	1,900 a,j,t
96. N-Nitrosodimethylamine	62759					0.00069 a,c,s	8.1 a,c,t
97. N-Nitrosodi-n-Propylamine	621647					0.005 a	1.4 a
98. N-Nitrosodiphenylamine	86306					5.0 a,c,s	16 a,c,t
99. Phenanthrene	85018						
100. Pyrene	129000					960 a	11,000 a
101. 1,2,4-Trichlorobenzene	120821						
102. Aldrin	309002	3 g		1.3 g		0.00013 a,c	0.00014 a,c
103. alpha-BHC	319846					0.0039 a,c	0.013 a,c
104. beta-BHC	319857					0.014 a,c	0.046 a,c
105. gamma-BHC	58899	0.95 w		0.16 g		0.019 c	0.063 c
106. delta-BHC	319868						
107. Chlordane	57749	2.4 g	0.0043 g	0.09 g	0.004 g	0.00057 a,c	0.00059 a,c
108. 4,4'-DDT	50293	1.1 g	0.001 g	0.13 g	0.001 g	0.00059 a,c	0.00059 a,c
109. 4,4'-DDE	72559					0.00059 a,c	0.00059 a,c
110. 4,4'-DDD	72548					0.00083 a,c	0.00084 a,c
111. Dieldrin	60571	0.24 w	0.056 w	0.71 g	0.0019 g	0.00014 a,c	0.00014 a,c
112. alpha-Endosulfan	959988	0.22 g	0.056 g	0.034 g	0.0087 g	110 a	240 a
113. beta-Endosulfan	33213659	0.22 g	0.056 g	0.034 g	0.0087 g	110 a	240 a
114. Endosulfan Sulfate	1031078					110 a	240 a
115. Endrin	72208	0.086 w	0.036 w	0.037 g	0.0023 g	0.76 a	0.81 a,j
116. Endrin Aldehyde	7421934					0.76 a	0.81 a,j
117. Heptachlor	76448	0.52 g	0.0038 g	0.053 g	0.0036 g	0.00021 a,c	0.00021 a,c
118. Heptachlor Epoxide	1024573	0.52 g	0.0038 g	0.053 g	0.0036 g	0.00010 a,c	0.00011 a,c
119-125. Polychlorinated biphenyls (PCBs)			0.014 u		0.03 u	0.00017 c,v	0.00017 c,v
126. Toxaphene	8001352	0.73	0.0002	0.21	0.0002	0.00073 a,c	0.00075 a,c
Total Number of Criteria h		22	21	22	20	92	90

FOOTNOTES TO TABLE IN PARAGRAPH (b)(1):

a. Criteria revised to reflect the Agency q1* or RfD, as contained in the Integrated Risk Information System (IRIS) as of October 1, 1996. The fish tissue bioconcentration factor (BCF) from the 1980 documents was retained in each case.

b. Criteria apply to California waters except for those waters subject to objectives in Tables III–2A and III–2B of the San Francisco Regional Water Quality Control Board's (SFRWQCB) 1986 Basin Plan that were adopted by the SFRWQCB and the State Water Resources Control Board, approved by EPA, and which continue to apply. For copper and nickel, criteria apply to California waters except for waters south of Dumbarton Bridge in San Francisco Bay that are subject to the objectives in the SFRWQCB's Basin Plan as amended by SFRWQCB Resolution R2–2002–0061, dated May 22, 2002, and approved by the State Water Resources Control Board. EPA approved the aquatic life site-specific objectives on January 21, 2003. The copper and nickel aquatic life site-specific objectives contained in the amended Basin Plan apply instead.

c. Criteria are based on carcinogenicity of 10 (-6) risk.

d. Criteria Maximum Concentration (CMC) equals the highest concentration of a pollutant to which aquatic life can be exposed for a short period of time without deleterious effects. Criteria Continuous Concentration (CCC) equals the highest concentration of a pollutant to which aquatic life can be exposed for an extended period of time (4 days) without deleterious effects. ug/L equals micrograms per liter.

e. Freshwater aquatic life criteria for metals are expressed as a function of total hardness (mg/L) in the water body. The equations are provided in matrix at paragraph (b)(2) of this section. Values displayed above in the matrix correspond to a total hardness of 100 mg/l.

f. Freshwater aquatic life criteria for pentachlorophenol are expressed as a function of pH, and are calculated as follows:

501

Values displayed above in the matrix correspond to a pH of 7.8. CMC = exp(1.005(pH) − 4.869). CCC = exp(1.005(pH) − 5.134).

g. This criterion is based on 304(a) aquatic life criterion issued in 1980, and was issued in one of the following documents: Aldrin/Dieldrin (EPA 440/5–80–019), Chlordane (EPA 440/5–80–027), DDT (EPA 440/5–80–038), Endosulfan (EPA 440/5–80–046), Endrin (EPA 440/5–80–047), Heptachlor (440/5–80–052), Hexachlorocyclohexane (EPA 440/5–80–054), Silver (EPA 440/5–80–071). The Minimum Data Requirements and derivation procedures were different in the 1980 Guidelines than in the 1985 Guidelines. For example, a "CMC" derived using the 1980 Guidelines was derived to be used as an instantaneous maximum. If assessment is to be done using an averaging period, the values given should be divided by 2 to obtain a value that is more comparable to a CMC derived using the 1985 Guidelines.

h. These totals simply sum the criteria in each column. For aquatic life, there are 23 priority toxic pollutants with some type of freshwater or saltwater, acute or chronic criteria. For human health, there are 92 priority toxic pollutants with either "water + organism" or "organism only" criteria. Note that these totals count chromium as one pollutant even though EPA has developed criteria based on two valence states. In the matrix, EPA has assigned numbers 5a and 5b to the criteria for chromium to reflect the fact that the list of 126 priority pollutants includes only a single listing for chromium.

i. Criteria for these metals are expressed as a function of the water-effect ratio, WER, as defined in paragraph (c) of this section. CMC = column B1 or C1 value×WER; CCC = column B2 or C2 value×WER.

j. No criterion for protection of human health from consumption of aquatic organisms (excluding water) was presented in the 1980 criteria document or in the 1986 Quality Criteria for Water. Nevertheless, sufficient information was presented in the 1980 document to allow a calculation of a criterion, even though the results of such a calculation were not shown in the document.

k. The CWA 304(a) criterion for asbestos is the MCL.

l. [Reserved]

m. These freshwater and saltwater criteria for metals are expressed in terms of the dissolved fraction of the metal in the water column. Criterion values were calculated by using EPA's Clean Water Act 304(a) guidance values (described in the total recoverable fraction) and then applying the conversion factors in § 131.36(b)(1) and (2).

n. EPA is not promulgating human health criteria for these contaminants. However, permit authorities should address these contaminants in NPDES permit actions using the State's existing narrative criteria for toxics.

o. These criteria were promulgated for specific waters in California in the National Toxics Rule ("NTR"), at § 131.36. The specific waters to which the NTR criteria apply include: Waters of the State defined as bays or estuaries and waters of the State defined as inland, i.e., all surface waters of the State not ocean waters. These waters specifically include the San Francisco Bay upstream to and including Suisun Bay and the Sacramento-San Joaquin Delta. This section does not apply instead of the NTR for this criterion.

p. A criterion of 20 ug/l was promulgated for specific waters in California in the NTR and was promulgated in the total recoverable form. The specific waters to which the NTR criterion applies include: Waters of the San Francisco Bay upstream to and including Suisun Bay and the Sacramento-San Joaquin Delta; and waters of Salt Slough, Mud Slough (north) and the San Joaquin River, Sack Dam to the mouth of the Merced River. This section does not apply instead of the NTR for this criterion. The State of California adopted and EPA approved a site specific criterion for the San Joaquin River, mouth of Merced to Vernalis; therefore, this section does not apply to these waters.

q. This criterion is expressed in the total recoverable form. This criterion was promulgated for specific waters in California in the NTR and was promulgated in the total recoverable form. The specific waters to which the NTR criterion applies include: Waters of the San Francisco Bay upstream to and including Suisun Bay and the Sacramento-San Joaquin Delta; and waters of Salt Slough, Mud Slough (north) and the San Joaquin River, Sack Dam to Vernalis. This criterion does not apply instead of the NTR for these waters. This criterion applies to additional waters of the United States in the State of California pursuant to 40 CFR 131.38(c). The State of California adopted and EPA approved a site-specific criterion for the Grassland Water District, San Luis National Wildlife Refuge, and the Los Banos State Wildlife Refuge; therefore, this criterion does not apply to these waters.

r. These criteria were promulgated for specific waters in California in the NTR. The specific waters to which the NTR criteria apply include: Waters of the State defined as bays or estuaries including the San Francisco Bay upstream to and including Suisun Bay and the Sacramento-San Joaquin Delta. This section does not apply instead of the NTR for these criteria.

s. These criteria were promulgated for specific waters in California in the NTR. The specific waters to which the NTR criteria apply include: Waters of the Sacramento-San Joaquin Delta and waters of the State defined as inland (i.e., all surface waters of the State not bays or estuaries or ocean) that include a MUN use designation. This section does not apply instead of the NTR for these criteria.

t. These criteria were promulgated for specific waters in California in the NTR. The specific waters to which the NTR criteria apply include: Waters of the State defined as bays and estuaries including San Francisco Bay upstream to and including Suisun Bay and the Sacramento-San Joaquin Delta; and waters of the State defined as inland (i.e., all surface waters of the State not bays or estuaries or ocean) without a MUN use designation. This section does not apply instead of the NTR for these criteria.

u. PCBs are a class of chemicals which include aroclors 1242, 1254, 1221, 1232, 1248, 1260, and 1016, CAS numbers 53469219, 11097691, 11104282, 11141165, 12672296, 11096825, and 12674112, respectively. The aquatic life criteria apply to the sum of this set of seven aroclors.

v. This criterion applies to total PCBs, e.g., the sum of all congener or isomer or homolog or aroclor analyses.

w. This criterion has been recalculated pursuant to the 1995 Updates: Water Quality Criteria Documents for the Protection of Aquatic Life in Ambient Water, Office of Water, EPA-820-B-96-001, September 1996. See also Great Lakes Water Quality Initiative Criteria Documents for the Protection of Aquatic Life in Ambient Water, Office of Water, EPA-80-B-95-004, March 1995.

x. The State of California has adopted and EPA has approved site specific criteria for the Sacramento River (and tributaries)

above Hamilton City; therefore, these criteria do not apply to these waters.

GENERAL NOTES TO TABLE IN PARAGRAPH (b)(1)

1. The table in this paragraph (b)(1) lists all of EPA's priority toxic pollutants whether or not criteria guidance are available. Blank spaces indicate the absence of national section 304(a) criteria guidance. Because of variations in chemical nomenclature systems, this listing of toxic pollutants does not duplicate the listing in appendix A to 40 CFR Part 423–126 Priority Pollutants. EPA has added the Chemical Abstracts Service (CAS) registry numbers, which provide a unique identification for each chemical.

2. The following chemicals have organoleptic-based criteria recommendations that are not included on this chart: zinc, 3-methyl-4-chlorophenol.

3. Freshwater and saltwater aquatic life criteria apply as specified in paragraph (c)(3) of this section.

(2) Factors for Calculating Metals Criteria. Final CMC and CCC values should be rounded to two significant figures.

(i) $CMC = WER \times (Acute\ Conversion\ Factor) \times (\exp\{m_A[\ln\ (hardness)] + b_A\})$

(ii) $CCC = WER \times (Chronic\ Conversion\ Factor) \times (\exp\{m_C[\ln(hardness)] + b_C\})$

(iii) Table 1 to paragraph (b)(2) of this section:

Metal	m_A	b_A	m_C	b_C
Cadmium	1.128	−3.6867	0.7852	−2.715
Copper	0.9422	−1.700	0.8545	−1.702
Chromium (III)	0.8190	3.688	0.8190	1.561
Lead	1.273	−1.460	1.273	−4.705
Nickel	0.8460	2.255	0.8460	0.0584
Silver	1.72	−6.52		
Zinc	0.8473	0.884	0.8473	0.884

Note to Table 1: The term "exp" represents the base e exponential function.

(iv) Table 2 to paragraph (b)(2) of this section:

Metal	Conversion factor (CF) for freshwater acute criteria	CF for freshwater chronic criteria	CF for saltwater acute criteria	CF[a] for saltwater chronic criteria
Antimony	[d]	[d]	[d]	[d]
Arsenic	1.000	1.000	1.000	1.000
Beryllium	[d]	[d]	[d]	[d]
Cadmium	[b]0.944	[b]0.909	0.994	0.994
Chromium (III)	0.316	0.860	[d]	[d]
Chromium (VI)	0.982	0.962	0.993	0.993
Copper	0.960	0.960	0.83	0.83
Lead	[b]0.791	[b]0.791	0.951	0.951
Mercury				
Nickel	0.998	0.997	0.990	0.990
Selenium		[c]	0.998	0.998

Metal	Conversion factor (CF) for freshwater acute criteria	CF for freshwater chronic criteria	CF for saltwater acute criteria	CF[a] for saltwater chronic criteria
Silver	0.85	(d)	0.85	(d)
Thallium	(d)	(d)	(d)	(d)
Zinc	0.978	0.986	0.946	0.946

FOOTNOTES TO TABLE 2 OF PARAGRAPH(b)(2):

[a] Conversion Factors for chronic marine criteria are not currently available. Conversion Factors for acute marine criteria have been used for both acute and chronic marine criteria.

[b] Conversion Factors for these pollutants in freshwater are hardness dependent. CFs are based on a hardness of 100 mg/l as calcium carbonate (CaCO$_3$). Other hardness can be used; CFs should be recalculated using the equations in table 3 to paragraph (b)(2) of this section.

[c] Bioaccumulative compound and inappropriate to adjust to percent dissolved.

[d] EPA has not published an aquatic life criterion value.

NOTE TO TABLE 2 OF PARAGRAPH (b)(2): The term "Conversion Factor" represents the recommended conversion factor for converting a metal criterion expressed as the total recoverable fraction in the water column to a criterion expressed as the dissolved fraction in the water column. See "Office of Water Policy and Technical Guidance on Interpretation and Implementation of Aquatic Life Metals Criteria", October 1, 1993, by Martha G. Prothro, Acting Assistant Administrator for Water available from Water Resource Center, USEPA, Mailcode RC4100, M Street SW, Washington, DC 20460 and the note to § 131.36(b)(1).

(v) Table 3 to paragraph (b)(2) of this section:

	Acute	Chronic
Cadmium	CF=1.136672—[(ln {hardness}) (0.041838)]	CF = 1.101672—[(ln {hardness})(0.041838)]
Lead	CF=1.46203—[(ln {hardness})(0.145712)]	CF = 1.46203—[(ln {hardness})(0.145712)]

(c) *Applicability.* (1) The criteria in paragraph (b) of this section apply to the State's designated uses cited in paragraph (d) of this section and apply concurrently with any criteria adopted by the State, except when State regulations contain criteria which are more stringent for a particular parameter and use, or except as provided in footnotes p, q, and x to the table in paragraph (b)(1) of this section.

(2) The criteria established in this section are subject to the State's general rules of applicability in the same way and to the same extent as are other Federally-adopted and State-adopted numeric toxics criteria when applied to the same use classifications including mixing zones, and low flow values below which numeric standards can be exceeded in flowing fresh waters.

(i) For all waters with mixing zone regulations or implementation procedures, the criteria apply at the appropriate locations within or at the boundary of the mixing zones; otherwise the criteria apply throughout the water body including at the point of discharge into the water body.

(ii) The State shall not use a low flow value below which numeric standards can be exceeded that is less stringent than the flows in Table 4 to paragraph (c)(2) of this section for streams and rivers.

(iii) Table 4 to paragraph (c)(2) of this section:

Criteria	Design flow
Aquatic Life Acute Criteria (CMC).	1 Q 10 or 1 B 3
Aquatic Life Chronic Criteria (CCC).	7 Q 10 or 4 B 3
Human Health Criteria	Harmonic Mean Flow

NOTE TO TABLE 4 OF PARAGRAPH (c)(2): 1. CMC (Criteria Maximum Concentration) is the water quality criteria to protect against acute effects in aquatic life and is the highest instream concentration of a priority toxic pollutant consisting of a short-term average not to be exceeded more than once every three years on the average.

2. CCC (Continuous Criteria Concentration) is the water quality criteria to protect against chronic effects in aquatic life and is the highest in stream concentration of a priority toxic pollutant consisting of a 4-day average not to be exceeded more than once every three years on the average.

3. 1 Q 10 is the lowest one day flow with an average recurrence frequency of once in 10 years determined hydrologically.

4. 1 B 3 is biologically based and indicates an allowable exceedence of once every 3 years. It is determined by EPA's computerized method (DFLOW model).

5. 7 Q 10 is the lowest average 7 consecutive day low flow with an average recurrence frequency of once in 10 years determined hydrologically.

6. 4 B 3 is biologically based and indicates an allowable exceedence for 4 consecutive days once every 3 years. It is determined by EPA's computerized method (DFLOW model).

(iv) If the State does not have such a low flow value below which numeric standards do not apply, then the criteria included in paragraph (d) of this section apply at all flows.

(v) If the CMC short-term averaging period, the CCC four-day averaging period, or once in three-year frequency is inappropriate for a criterion or the site to which a criterion applies, the State may apply to EPA for approval of an alternative averaging period, frequency, and related design flow. The State must submit to EPA the bases for any alternative averaging period, frequency, and related design flow. Before approving any change, EPA will publish for public comment, a document proposing the change.

(3) The freshwater and saltwater aquatic life criteria in the matrix in paragraph (b)(1) of this section apply as follows:

(i) For waters in which the salinity is equal to or less than 1 part per thousand 95% or more of the time, the applicable criteria are the freshwater criteria in Column B;

(ii) For waters in which the salinity is equal to or greater than 10 parts per thousand 95% or more of the time, the applicable criteria are the saltwater criteria in Column C except for selenium in the San Francisco Bay estuary where the applicable criteria are the freshwater criteria in Column B (refer to footnotes p and q to the table in paragraph (b)(1) of this section); and

(iii) For waters in which the salinity is between 1 and 10 parts per thousand as defined in paragraphs (c)(3)(i) and (ii) of this section, the applicable criteria are the more stringent of the freshwater or saltwater criteria. However, the Regional Administrator may approve the use of the alternative freshwater or saltwater criteria if sci-entifically defensible information and data demonstrate that on a site-specific basis the biology of the water body is dominated by freshwater aquatic life and that freshwater criteria are more appropriate; or conversely, the biology of the water body is dominated by saltwater aquatic life and that saltwater criteria are more appropriate. Before approving any change, EPA will publish for public comment a document proposing the change.

(4) *Application of metals criteria.* (i) For purposes of calculating freshwater aquatic life criteria for metals from the equations in paragraph (b)(2) of this section, for waters with a hardness of 400 mg/l or less as calcium carbonate, the actual ambient hardness of the surface water shall be used in those equations. For waters with a hardness of over 400 mg/l as calcium carbonate, a hardness of 400 mg/l as calcium carbonate shall be used with a default Water-Effect Ratio (WER) of 1, or the actual hardness of the ambient surface water shall be used with a WER. The same provisions apply for calculating the metals criteria for the comparisons provided for in paragraph (c)(3)(iii) of this section.

(ii) The hardness values used shall be consistent with the design discharge conditions established in paragraph (c)(2) of this section for design flows and mixing zones.

(iii) The criteria for metals (compounds #1—#13 in the table in paragraph (b)(1) of this section) are expressed as dissolved except where otherwise noted. For purposes of calculating aquatic life criteria for metals from the equations in footnote i to the table in paragraph (b)(1) of this section and the equations in paragraph (b)(2) of this section, the water effect ratio is generally computed as a specific pollutant's acute or chronic toxicity value measured in water from the site covered by the standard, divided by the respective acute or chronic toxicity value in laboratory dilution water. To use a water effect ratio other than the default of 1, the WER must be determined as set forth in Interim Guidance on Determination and Use of Water Effect Ratios, U.S. EPA Office of Water,

EPA–823–B–94–001, February 1994, or alternatively, other scientifically defensible methods adopted by the State as part of its water quality standards program and approved by EPA. For calculation of criteria using site-specific values for both the hardness and the water effect ratio, the hardness used in the equations in paragraph (b)(2) of this section must be determined as required in paragraph (c)(4)(ii) of this section. Water hardness must be calculated from the measured calcium and magnesium ions present, and the ratio of calcium to magnesium should be approximately the same in standard laboratory toxicity testing water as in the site water.

(d)(1) Except as specified in paragraph (d)(3) of this section, all waters assigned any aquatic life or human health use classifications in the Water Quality Control Plans for the various Basins of the State ("Basin Plans") adopted by the California State Water Resources Control Board ("SWRCB"), except for ocean waters covered by the Water Quality Control Plan for Ocean Waters of California ("Ocean Plan") adopted by the SWRCB with resolution Number 90–27 on March 22, 1990, are subject to the criteria in paragraph (d)(2) of this section, without exception. These criteria apply to waters identified in the Basin Plans. More particularly, these criteria apply to waters identified in the Basin Plan chapters designating beneficial uses for waters within the region. Although the State has adopted several use designations for each of these waters, for purposes of this action, the specific standards to be applied in paragraph (d)(2) of this section are based on the presence in all waters of some aquatic life designation and the presence or absence of the MUN use designation (municipal and domestic supply). (See Basin Plans for more detailed use definitions.)

(2) The criteria from the table in paragraph (b)(1) of this section apply to the water and use classifications defined in paragraph (d)(1) of this section as follows:

Water and use classification	Applicable criteria
(i) All inland waters of the United States or enclosed bays and estuaries that are waters of the United States that include a MUN use designation.	(A) Columns B1 and B2—all pollutants (B) Columns C1 and C2—all pollutants (C) Column D1—all pollutants
(ii) All inland waters of the United States or enclosed bays and estuaries that are waters of the United States that do not include a MUN use designation.	(A) Columns B1 and B2—all pollutants (B) Columns C1 and C2—all pollutants (C) Column D2—all pollutants

(3) Nothing in this section is intended to apply instead of specific criteria, including specific criteria for the San Francisco Bay estuary, promulgated for California in the National Toxics Rule at § 131.36.

(4) The human health criteria shall be applied at the State-adopted 10 (−6) risk level.

(5) Nothing in this section applies to waters located in Indian Country.

(e) *Schedules of compliance.* (1) It is presumed that new and existing point source dischargers will promptly comply with any new or more restrictive water quality-based effluent limitations ("WQBELs") based on the water quality criteria set forth in this section.

(2) When a permit issued on or after May 18, 2000 to a new discharger contains a WQBEL based on water quality criteria set forth in paragraph (b) of this section, the permittee shall comply with such WQBEL upon the commencement of the discharge. A new discharger is defined as any building, structure, facility, or installation from which there is or may be a "discharge of pollutants" (as defined in 40 CFR 122.2) to the State of California's inland surface waters or enclosed bays and estuaries, the construction of which commences after May 18, 2000.

(3) Where an existing discharger reasonably believes that it will be infeasible to promptly comply with a new or more restrictive WQBEL based on the

506

water quality criteria set forth in this section, the discharger may request approval from the permit issuing authority for a schedule of compliance.

(4) A compliance schedule shall require compliance with WQBELs based on water quality criteria set forth in paragraph (b) of this section as soon as possible, taking into account the dischargers' technical ability to achieve compliance with such WQBEL.

(5) If the schedule of compliance exceeds one year from the date of permit issuance, reissuance or modification, the schedule shall set forth interim requirements and dates for their achievement. The dates of completion between each requirement may not exceed one year. If the time necessary for completion of any requirement is more than one year and is not readily divisible into stages for completion, the permit shall require, at a minimum, specified dates for annual submission of progress reports on the status of interim requirements.

(6) In no event shall the permit issuing authority approve a schedule of compliance for a point source discharge which exceeds five years from the date of permit issuance, reissuance, or modification, whichever is sooner. Where shorter schedules of compliance are prescribed or schedules of compliance are prohibited by law, those provisions shall govern.

(7) If a schedule of compliance exceeds the term of a permit, interim permit limits effective during the permit shall be included in the permit and addressed in the permit's fact sheet or statement of basis. The administrative record for the permit shall reflect final permit limits and final compliance dates. Final compliance dates for final permit limits, which do not occur during the term of the permit, must occur within five years from the date of issuance, reissuance or modification of the permit which initiates the compliance schedule. Where shorter schedules of compliance are prescribed or schedules of compliance are prohibited by law, those provisions shall govern.

(8) The provisions in this paragraph (e), Schedules of compliance, shall expire on May 18, 2005.

[65 FR 31711, May 18, 2000, as amended at 66 FR 9961, Feb. 13, 2001; 68 FR 62747, Nov. 6, 2003]

EDITORIAL NOTE: At 66 FR 9961, Feb. 13, 2001, §131.38 was amended in the table to paragraph (b)(1) under the column heading for "B Freshwater" by revising the column headings for "Criterion Maximum Concentration" and "Criterion Continuous Concentration"; under the column heading for "C Saltwater" by revising the column headings for "Criterion Maximum Concentration" and "Criterion Continuous Concentration"; and by revising entries "23." and "67.", effective Feb. 13, 2001. However, this is a photographed table and the amendments could not be incorporated into the text. For the convenience of the user, the amended text is set forth as follows:

§131.38 Establishment of Numeric Criteria for priority toxic pollutants for the State of California.

*   *   *   *   *

(b)(1) * * *

A		B Freshwater		C Saltwater		D Human health (10^{-6}) risk for carcinogens) For consumption of:	
# Compound	CAS number	Criterion maximum conc. (μg/L) ^{d} B1	Criterion continous conc. (μg/L) ^{d} B2	Criterion maximum conc. (μg/L) ^{d} C1	Criterion continous conc. (μg/L) ^{d} C2	Water & organisms (μg/L) D1	Organisms only (μg/L) D2
*		*	*	*	*	*	*
*		*	*	*	*	*	*
*		*	*	*	*	*	*
23. Chlorodibromomethane	124481	*	*	....	*	^{a,c} 0.41 *	^{a,c} 34 *
67. Bis(2-Chloroisopropyl)Ether	108601	....	....	....	*	^{a} 1,400 *	^{a,t} 170,000 *

Footnotes to table in Paragraph (b)(1):

* * * * *

^{a} Criteria revised to reflect the Agency q1* or RfD, as contained in the Integrated Risk Information System (IRIS) as of October 1, 1996. The fish tissue bioconcentration factor (BCF) from the 1980 documents was retained in each case.

^{c} Criteria are based on carcinogenicity of 10^{-6} risk.

^{d} Criteria Maximum Concentration (CMC) equals the highest concentration of a pollutant to which aquatic life can be exposed for a short period of time without deleterious effects. Criteria Continuous Concentration (CCC) equals the highest concentration of a pollutant to which aquatic life can be exposed for an extended period of time (4 days) without deleterious effects. μg/L equals micrograms per liter.

* * * * *

^{t} These criteria were promulgated for specific waters in California in the NTR. The specific waters to which the NTR criteria apply include: Waters of the State defined as bays and estuaries including San Francisco Bay upstream to and including Suisun Bay and the Sacramento-San Joaquin Delta; and waters of the State defined as inland (i.e., all surface waters of the State not bays or estuaries or ocean) without a MUN use designation. This section does not apply instead of the NTR for these criteria.

* * * * *

§131.40 Puerto Rico

(a) *Use designations for marine waters.* In addition to the Commonwealth's adopted use designations, the following waterbodies in Puerto Rico have the beneficial use designated in this paragraph (a) within the bays specified below, and within the Commonwealth's territorial seas, as defined in section 502(8) of the Clean Water Act, and 33 CFR 2.05–5, except such waters classified by the Commonwealth as SB.

Waterbody segment	From	To	Designated use
Coastal Waters	500m offshore	3 miles offshore	Primary Contact Recreation.
Guayanilla & Tallaboa Bays	Cayo Parguera	Punta Verraco	Primary Contact Recreation.
Mayaguez Bay	Punta Guanajibo	Punta Algarrobo	Primary Contact Recreation.
Ponce Port	Punta Carenero	Punta Cuchara	Primary Contact Recreation.
San Juan Port	mouth of Río Bayamón	Punta El Morro	Primary Contact Recreation.
Yabucoa Port	Punta Icacos	Punta Yeguas	Primary Contact Recreation.

(b) *Criteria that apply to Puerto Rico's marine waters.* In addition to all other Commonwealth criteria, the following criteria for bacteria apply to the waterbodies in paragraph (a) of this section:

Bacteria: The fecal coliform geometric mean of a series of representative samples (at least five samples) of the waters taken sequentially shall not exceed 200 colonies/100 ml, and not more than 20 percent of the samples shall exceed 400 colonies/100 ml. The enterococci density in terms of geometric mean of at least five representative samples taken sequentially shall not exceed 35/100 ml. No single sample should exceed the upper confidence limit of 75% using 0.7 as the log standard deviation until sufficient site data exist to establish a site-specific log standard deviation.

(c) *Water quality standard variances.* (1) The Regional Administrator, EPA Region 2, is authorized to grant variances from the water quality standards in paragraphs (a) and (b) of this section where the requirements of this paragraph (c) are met. A water quality standard variance applies only to the permittee requesting the variance and only to the pollutant or pollutants specified in the variance; the underlying water quality standard otherwise remains in effect.

(2) A water quality standard variance shall not be granted if:

(i) Standards will be attained by implementing effluent limitations required under sections 301(b) and 306 of the CWA and by the permittee implementing reasonable best management practices for nonpoint source control; or

(ii) The variance would likely jeopardize the continued existence of any threatened or endangered species listed under section 4 of the Endangered Species Act or result in the destruction or adverse modification of such species' critical habitat.

(3) A water quality standards variance may be granted if the applicant demonstrates to EPA that attaining the water quality standard is not feasible because:

(i) Naturally occurring pollutant concentrations prevent the attainment of the use;

(ii) Natural, ephemeral, intermittent or low flow conditions or water levels prevent the attainment of the use, unless these conditions may be compensated for by the discharge of sufficient volume of effluent discharges without violating Commonwealth water conservation requirements to enable uses to be met;

(iii) Human caused conditions or sources of pollution prevent the attainment of the use and cannot be remedied or would cause more environmental damage to correct than to leave in place;

(iv) Dams, diversions or other types of hydrologic modifications preclude the attainment of the use, and it is not feasible to restore the waterbody to its original condition or to operate such modification in a way which would result in the attainment of the use;

(v) Physical conditions related to the natural features of the waterbody, such as the lack of a proper substrate, cover, flow, depth, pools, riffles, and the like unrelated to water quality, preclude attainment of aquatic life protection uses; or

(vi) Controls more stringent than those required by sections 301(b) and 306 of the CWA would result in substantial and widespread economic and social impact.

(4) *Procedures.* An applicant for a water quality standards variance shall submit a request to the Regional Administrator of EPA Region 2. The application shall include all relevant information showing that the requirements for a variance have been met. The applicant must demonstrate that the designated use is unattainable for one of the reasons specified in paragraph (c)(3) of this section. If the Regional Administrator preliminarily determines that grounds exist for granting a variance, he/she shall provide public notice of the proposed variance and provide an opportunity for public comment. Any activities required as a condition of the Regional Administrator's granting of a variance shall be included as conditions of the NPDES permit for the applicant. These terms and conditions shall be incorporated into the applicant's NPDES permit through the permit reissuance process or through a modification of the permit pursuant to the applicable permit modification provisions of Puerto Rico's NPDES program.

(5) A variance may not exceed five years or the term of the NPDES permit, whichever is less. A variance may be renewed if the applicant reapplies and demonstrates that the use in question is still not attainable. Renewal of the variance may be denied if the applicant did not comply with the conditions of the original variance, or otherwise does not meet the requirements of this section.

[69 FR 3524, Jan. 26, 2004]

§ 131.41 **Bacteriological criteria for those states not complying with Clean Water Act section 303(i)(1)(A).**

(a) *Scope.* This section is a promulgation of the Clean Water Act section 304(a) criteria for bacteria for coastal recreation waters in specific States. It is not a general promulgation of the Clean Water Act section 304(a) criteria

for bacteria. This section also contains a compliance schedule provision.

(b) *Definitions.* (1) *Coastal Recreation Waters* are the Great Lakes and marine coastal waters (including coastal estuaries) that are designated under section 303(c) of the Clean Water Act for use for swimming, bathing, surfing, or similar water contact activities. Coastal recreation waters do not include inland waters or waters upstream from the mouth of a river or stream having an unimpaired natural connection with the open sea.

(2) *Designated bathing beach waters* are those coastal recreation waters that, during the recreation season, are heavily-used (based upon an evaluation of use within the State) and may have: a lifeguard, bathhouse facilities, or public parking for beach access. States may include any other waters in this category even if the waters do not meet these criteria.

(3) *Moderate use coastal recreation waters* are those coastal recreation waters that are not designated bathing beach waters but typically, during the recreation season, are used by at least half of the number of people as at typical designated bathing beach waters within the State. States may also include light use or infrequent use coastal recreation waters in this category.

(4) *Light use coastal recreation waters* are those coastal recreation waters that are not designated bathing beach waters but typically, during the recreation season, are used by less than half of the number of people as at typical designated bathing beach waters within the State, but are more than infrequently used. States may also include infrequent use coastal recreation waters in this category.

(5) *Infrequent use coastal recreation waters* are those coastal recreation waters that are rarely or occasionally used.

(6) *New pathogen discharger* for the purposes of this section means any building, structure, facility, or installation from which there is or may be a discharge of pathogens, the construction of which commenced on or after December 16, 2004. It does not include relocation of existing combined sewer overflow outfalls.

(7) *Existing pathogen discharger* for the purposes of this section means any discharger that is not a new pathogen discharger.

(c) *EPA's section 304(a) ambient water quality criteria for bacteria.* (1) Freshwaters:

A Indicator [d]	B Geometric mean	C Single sample maximum (per 100 ml)			
		C1 Designated bathing beach (75% confidence level)	C2 Moderate use costal recreation waters (82% confidence level)	C3 Light use coastal recreation waters (90% confidence level)	C4 Infrequent use coastal recreation waters (95% confidence level)
E. coli [e]	126/100 mil [a]	[b] 235	[b] 298	[b] 409	[b] 575
Enterococci [e]	33/100 ml [c]	[b] 61	[b] 78	[b] 107	[b] 151

Footnotes to table in paragraph (c)(1):

a. This value is for use with analytical methods 1103.1, 1603, or 1604 or any equivalent method that measures viable bacteria.

b. Calculated using the following: single sample maximum = geometric mean * 10 + (confidence level factor * log standard deviation), where the confidence level factor is: 75%: 0.68; 82%: 0.94; 90%: 1.28; 95%: 1.65. The log standard deviation from EPA's epidemiological studies is 0.4.

c. This value is for use with analytical methods 1106.1 or 1600 or any equivalent method that measures viable bacteria.

d. The State may determine which of these indicators applies to its freshwater coastal recreation waters. Until a State makes that determination, E. coli will be the applicable indicator.

e. These values apply to E. coli or enterococci regardless of origin unless a sanitary survey shows that sources of the indicator bacteria are non-human and an epidemiological study shows that the indicator densities are not indicative of a human health risk.

(2) Marine waters:

A Indicator	B Geometric mean	C Single sample maximum (per 100 ml)			
		C1 Designated bathing beach (75% confidence level)	C2 Moderate use coastal recreation waters (82% confidence level)	C3 Light use coastal recreation waters (90% confidence level)	C4 Infrequent use coastal recreation waters (95% confidence level)
Enterococci [c]	35/100 ml [a]	[b] 104	[b] 158	[b] 276	[b] 501

Footnotes to table in paragraph (c)(2):

a. This value is for use with analytical methods 1106.1 or 1600 or any equivalent method that measures viable bacteria.

b. Calculated using the following: single sample maximum = geometric mean * 10 + (confidence level factor * log standard deviation), where the confidence level factor is: 75%: 0.68; 82%: 0.94; 90%: 1.28; 95%: 1.65. The log standard deviation from EPA's epidemiological studies is 0.7.

c. These values apply to enterococci regardless of origin unless a sanitary survey shows that sources of the indicator bacteria are non-human and an epidemiological study shows that the indicator densities are not indicative of a human health risk.

(3) As an alternative to the single sample maximum in paragraph (c)(1) or (c)(2) of this section, States may use a site-specific log standard deviation to calculate a single sample maximum for individual coastal recreation waters, but must use at least 30 samples from a single recreation season to do so.

(d) *Applicability.* (1) The criteria in paragraph (c) of this section apply to the coastal recreation waters of the States identified in paragraph (e) of this section and apply concurrently with any ambient recreational water criteria adopted by the State, except for those coastal recreation waters where State regulations determined by EPA to meet the requirements of Clean Water Act section 303(i) apply, in which case the State's criteria for those coastal recreation waters will apply and not the criteria in paragraph (c) of this section.

(2) The criteria established in this section are subject to the State's general rules of applicability in the same way and to the same extent as are other Federally-adopted and State-adopted numeric criteria when applied to the same use classifications.

(e) *Applicability to specific jurisdictions.* (1) The criteria in paragraph (c)(1) of this section apply to fresh coastal recreation waters of the following States: Illinois, Minnesota, New York, Ohio, Pennsylvania, Wisconsin.

(2) The criteria in paragraph (c)(2) of this section apply to marine coastal recreation waters of the following States: Alaska, California (except for coastal recreation waters within the jurisdiction of Regional Board 4), Florida, Georgia, Hawaii (except for coastal recreation waters within 300 meters of the shoreline), Louisiana, Maine (except for SA waters and SB and SC waters with human sources of fecal contamination), Maryland, Massachusetts, Mississippi, New York, North Carolina, Oregon, Puerto Rico (except for waters classified by Puerto Rico as intensely used for primary contact recreation and for those waters included in § 131.40), Rhode Island, United States Virgin Islands.

(f) *Schedules of compliance.* (1) This paragraph (f) applies to any State that does not have a regulation in effect for Clean Water Act purposes that authorizes compliance schedules for National Pollutant Discharge Elimination System permit limitations needed to meet the criteria in paragraph (c) of this section. All dischargers shall promptly comply with any new or more restrictive water quality-based effluent limitations based on the water quality criteria set forth in this section.

(2) When a permit issued on or after December 16, 2004, to a new pathogen discharger as defined in paragraph (b) of this section contains water quality-based effluent limitations based on water quality criteria set forth in paragraph (c) of this section, the permittee shall comply with such water quality-based effluent limitations upon the commencement of the discharge.

(3) Where an existing pathogen discharger reasonably believes that it will be infeasible to comply immediately with a new or more restrictive water quality-based effluent limitations based on the water quality criteria set forth in paragraph (c) of this section, the discharger may request approval from the permit issuing authority for a schedule of compliance.

(4) A compliance schedule for an existing pathogen discharger shall require compliance with water quality-based effluent limitations based on water quality criteria set forth in paragraph (c) of this section as soon as possible, taking into account the dis-

charger's ability to achieve compliance with such water quality-based effluent limitations.

(5) If the schedule of compliance for an existing pathogen discharger exceeds one year from the date of permit issuance, reissuance or modification, the schedule shall set forth interim requirements and dates for their achievement. The period between dates of completion for each requirement may not exceed one year.

If the time necessary for completion of any requirement is more than one year and the requirement is not readily divisible into stages for completion, the permit shall require, at a minimum, specified dates for annual submission of progress reports on the status of interim requirements.

(6) In no event shall the permit issuing authority approve a schedule of compliance for an existing pathogen discharge which exceeds five years from the date of permit issuance, reissuance, or modification, whichever is sooner.

(7) If a schedule of compliance exceeds the term of a permit, interim permit limits effective during the permit shall be included in the permit and addressed in the permit's fact sheet or statement of basis. The administrative record for the permit shall reflect final permit limits and final compliance dates. Final compliance dates for final permit limits, which do not occur during the term of the permit, must occur within five years from the date of issuance, reissuance or modification of the permit which initiates the compliance schedule.

[69 FR 67242, Nov. 16, 2004]

§ 131.42 Antidegradation Implementation Methods for the Commonwealth of Puerto Rico.

(a) *General Policy Statement.* (1) All point sources of pollution are subject to an antidegradation review.

(2) An antidegradation review shall be initiated as part of the Section 401— "Water Quality Certification Process" of the Clean Water Act.

(3) The 401 Certification Process shall follow the procedures established by the February 2, 1989 Resolution R–89–2–2 of the Governing Board of the Puerto

Rico Environmental Quality Board (EQB).

(4) The following are not subject to an antidegradation review due to the fact that they are nondischarge systems and are managed by specific applicable Puerto Rico regulations:

(i) All nonpoint sources of pollutants.

(ii) Underground Storage Tanks.

(iii) Underground Injection Facilities.

(5) The protection of water quality shall include the maintenance, migration, protection, and propagation of desirable species, including threatened and endangered species identified in the local and federal regulations.

(b) *Definitions.* (1) All the definitions included in Article 1 of the Puerto Rico Water Quality Standards Regulation (PRWQSR), as amended, are applicable to this procedure.

(2) High Quality Waters:

(i) Are waters whose quality is better than the mandatory minimum level to support the CWA Section 101(a)(2) goals of propagation of fish, shellfish, wildlife and recreation in and on the waters. High Quality Waters are to be identified by EQB on a parameter-by-parameter basis.

(ii) [Reserved]

(3) Outstanding National Resources Waters (ONRWs):

(i) Are waters classified as SA or SE in the PRWQSR, as amended, or any other water designated by Resolution of the Governing Board of EQB. ONRWs are waters that are recreationally or ecologically important, unique or sensitive.

(ii) [Reserved]

(c) *Antidegradation Review Procedure.* (1) The antidegradation review will commence with the submission of the CWA Section 401 water quality certification request. EQB uses a parameter-by-parameter approach for the implementation of the anti-degradation policy and will review each parameter separately as it evaluates the request for certification. The 401 certification/antidegradation review shall comply with Article 4(B)(3) of the Puerto Rico Environmental Public Policy Act (Law No. 416 of September 22, 2004, as amended (12 LPRA 8001 *et seq.*)). Compliance with Article 4(B)(3) shall be conducted in accordance with the Reglamento de

la Junta de Calidad Ambiental para el Proceso de Presentación, Evaluación y Trámite de Documentos Ambientales (EQB's Environmental Documents Regulation). As part of the evaluation of the Environmental Document an alternatives analysis shall be conducted (12 LPRA 8001(a)(5), EQB's Environmental Documents Regulation, e.g., Rules 211E and 253C), and a public participation period and a public hearing shall be provided (12 LPRA 8001(a), EQB's Environmental Documents Regulation, Rule 254).

(2) In conducting an antidegradation review, EQB will sequentially apply the following steps:

(i) Determine which level of antidegradation applies

(A) Tier 1—Protection of Existing and Designated Uses.

(B) Tier 2—Protection of High Quality Waters.

(C) Tier 3—Protection of ONRWs.

(ii) [Reserved]

(3) Review existing water quality data and other information submitted by the applicant. The applicant shall provide EQB with the information regarding the discharge, as required by the PRWQSR including, but not limited to the following:

(i) A description of the nature of the pollutants to be discharged.

(ii) Treatment technologies applied to the pollutants to be discharged.

(iii) Nature of the applicant's business.

(iv) Daily maximum and average flow to be discharged.

(v) Effluent characterization.

(vi) Effluent limitations requested to be applied to the discharge according to Section 6.11 of the PRWQSR.

(vii) Location of the point of discharge.

(viii) Receiving waterbody name.

(ix) Water quality data of the receiving waterbody.

(x) Receiving waterbody minimum flow (7Q2 and 7Q10) for stream waters.

(xi) Location of water intakes within the waterbody.

(xii) In the event that the proposed discharge will result in the lowering of water quality, data and information demonstrating that the discharge is necessary to accommodate important economic or social development in the area where the receiving waters are located.

(4) Determine if additional information or assessment is necessary to make the decision.

(5) Prepare an intent to issue or deny the 401 water quality certificate and publish a notice in a newspaper of wide circulation in Puerto Rico informing the public of EQB's preliminary decision and granting a public participation period of at least thirty (30) days.

(6) Address the comments received from the interested parties and consider such comments as part of the decision making process.

(7) Make the final determination to issue or deny the requested 401 certification. Such decision is subject to the reconsideration procedure established in Law 170 of August 12, 1988, *Ley de Procedimiento Administrativo Uniforme del Estado Libre Asociado de Puerto Rico* (3 LPRA 2165).

(d) *Implementation Procedures.* (1) Activities Regulated by NPDES Permits

(i) Tier 1—Protection of Existing and Designated Uses:

(A) Tier 1 waters are:

(*1*) Those waters of Puerto Rico (except Tier 2 or Tier 3 waters) identified as impaired and that have been included on the list required by Section 303(d) of the CWA; and

(*2*) Those waters of Puerto Rico (except Tier 2 and Tier 3 waters) for which attainment of applicable water quality standards has been or is expected to be, achieved through implementation of effluent limitations more stringent than technology-based controls (Best Practicable Technology, Best Available Technology and Secondary Treatment).

(B) To implement Tier 1 antidegradation, EQB shall determine if a discharge would lower the water quality to the extent that it would no longer be sufficient to protect and maintain the existing and designated uses of that waterbody.

(C) When a waterbody has been affected by a parameter of concern causing it to be included on the 303(d) List, then EQB will not allow an increase of the concentration of the parameter of concern or pollutants affecting the parameter of concern in the waterbody. This no increase will be achieved by meeting the applicable water quality

standards at the end of the pipe. Until such time that a Total Maximum Daily Load (TMDL) is developed for the parameter of concern for the waterbody, no discharge will be allowed to cause or contribute to further degradation of the waterbody.

(D) When the assimilative capacity of a waterbody is not sufficient to ensure maintenance of the water quality standard for a parameter of concern with an additional load to the waterbody, EQB will not allow an increase of the concentration of the parameter of concern or pollutants affecting the parameter of concern in the waterbody. This no increase will be achieved by meeting the applicable water quality standards at the end of the pipe. Until such time that a TMDL is developed for the parameter of concern for the waterbody, no discharge will be allowed to cause or contribute to further degradation of the waterbody.

(ii) Tier 2—Protection of High Quality Waters:

(A) To verify that a waterbody is a high quality water for a parameter of concern which initiates a Tier 2 antidegradation review, EQB shall evaluate and determine:

(*1*) The existing water quality of the waterbody;

(*2*) The projected water quality of the waterbody pursuant to the procedures established in the applicable provisions of Articles 5 and 10 of the PRWQSR including but not limited to, Sections 5.2, 5.3, 5.4, 10.2, 10.3, 10.4, 10.5, and 10.6;

(*3*) That the existing and designated uses of the waterbody will be fully maintained and protected in the event of a lowering of water quality.

In multiple discharge situations, the effects of all discharges shall be evaluated through a waste load allocation analysis in accordance with the applicable provisions of Article 10 of the PRWQSR or the applicable provisions of Article 5 regarding mixing zones.

(B) In order to allow the lowering of water quality in high quality waters, the applicant must show and justify the necessity for such lowering of water quality through compliance with the requirements of Section 6.11 of the PRWQSR. EQB will not allow the entire assimilative capacity of a waterbody for a parameter of concern to be allocated to a discharger, if the necessity of the requested effluent limitation for the parameter of concern is not demonstrated to the full satisfaction of EQB.

(iii) Tier 3—Protection of ONRWs:

(A) EQB may designate a water as Class SA or SE (ONRWs) through a Resolution (PRWQSR Sections 2.1.1 and 2.2.1). Additionally, any interested party may nominate a specific water to be classified as an ONRW and the Governing Board of EQB will make the final determination. Classifying a water as an ONRW may result in the water being named in either Section 2.1.1 or 2.2.2 of the PRWQSR, which would require an amendment of the PRWQSR. The process for amending the PRWQSR, including public participation, is set forth in Section 8.6 of said regulation.

(B) The existing characteristics of Class SA and SE waters shall not be altered, except by natural causes, in order to preserve the existing natural phenomena.

(1) No point source discharge will be allowed in ONRWs.

(2) [Reserved]

(2) Activities Regulated by CWA Section 404 or Rivers and Harbors Action Section 10 Permits (Discharge of Dredged or Fill Material)

(i) EQB will only allow the discharge of dredged or fill material into a wetland if it can be demonstrated that such discharge will not have an unacceptable adverse impact either individually or in combination with other activities affecting the wetland of concern. The impacts to the water quality or the aquatic or other life in the wetland due to the discharge of dredged or fill material should be avoided, minimized and mitigated.

(ii) The discharge of dredged or fill material shall not be certified if there is a practicable alternative to the proposed discharge which would have less adverse impact on the recipient ecosystem, so long as the alternative does not have other more significant adverse environmental consequences. Activities which are not water dependent are presumed to have practicable alternatives, unless the applicant clearly demonstrates otherwise. No discharge

of dredged and fill material shall be certified unless appropriate and practicable steps have been taken which minimize potential adverse impacts of the discharge on the recipient ecosystem. The discharge of dredged or fill material to ONRWs, however, shall be governed by paragraph (d)(1)(iii) of this section.

[72 FR 70524, Dec. 12, 2007]

§131.43 Florida.

(a)–(d) [Reserved]

(e) *Site-specific alternative criteria.* (1) The Regional Administrator may determine that site-specific alternative criteria shall apply to specific surface waters in lieu of the criteria established for Florida waters in this section, including criteria for lakes, criteria for streams, and criteria for springs. Any such determination shall be made consistent with §131.11.

(2) To receive consideration from the Regional Administrator for a determination of site-specific alternative criteria, an entity shall submit a request that includes proposed alternative numeric criteria and supporting rationale suitable to meet the needs for a technical support document pursuant to paragraph (e)(3) of this section. The entity shall provide the State a copy of all materials submitted to EPA, at the time of submittal to EPA, to facilitate the State providing comments to EPA. Site-specific alternative criteria may be based on one or more of the following approaches.

(i) Replicate the process for developing the stream criteria in this section.

(ii) Replicate the process for developing the lake criteria in this section.

(iii) Conduct a biological, chemical, and physical assessment of waterbody conditions.

(iv) Use another scientifically defensible approach protective of the designated use.

(3) For any determination made under paragraph (e)(1) of this section, the Regional Administrator shall, prior to making such a determination, provide for public notice and comment on a proposed determination. For any such proposed determination, the Regional Administrator shall prepare and

make available to the public a technical support document addressing the specific surface waters affected and the justification for each proposed determination. This document shall be made available to the public no later than the date of public notice issuance.

(4) The Regional Administrator shall maintain and make available to the public an updated list of determinations made pursuant to paragraph (e)(1) of this section as well as the technical support documents for each determination.

(5) Nothing in this paragraph (e) shall limit the Administrator's authority to modify the criteria established for Florida waters in this section, including criteria for lakes, criteria for streams, and criteria for springs.

[75 FR 75805, Dec. 6, 2010]

EFFECTIVE DATE NOTE: At 75 FR 75805, Dec. 6, 2010, § 131.43 was revised and was effective Mar. 6, 2012. However, this action was delayed until July 6, 2012 at 77 FR 13496, Mar. 7, 2012.

§ 131.43 Florida.

(a) *Scope.* This section promulgates numeric criteria for nitrogen/phosphorus pollution for Class I and Class III waters in the State of Florida. This section also contains provisions for site-specific alternative criteria.

(b) *Definitions.*—(1) *Canal* means a trench, the bottom of which is normally covered by water with the upper edges of its two sides normally above water.

(2) *Clear, high-alkalinity lake* means a lake with long-term color less than or equal to 40 Platinum Cobalt Units (PCU) and Alkalinity greater than 20 mg/L $CaCO_3$.

(3) *Clear, low-alkalinity lake* means a lake with long-term color less than or equal to 40 PCU and alkalinity less than or equal to 20 mg/L $CaCO_3$.

(4) *Colored lake* means a lake with long-term color greater than 40 PCU.

(5) *Lake* means a slow-moving or standing body of freshwater that occupies an inland

basin that is not a stream, spring, or wetland.

(6) *Lakes and flowing waters* means inland surface waters that have been classified as Class I (Potable Water Supplies) or Class III (Recreation, Propagation and Maintenance of a Healthy, Well-Balanced Population of Fish and Wildlife) water bodies pursuant to Rule 62–302.400, F.A.C., excluding wetlands, and are predominantly fresh waters.

(7) *Nutrient watershed region* means an area of the State, corresponding to drainage basins and differing geological conditions affecting nutrient levels, as delineated in Table 2.

(8) *Predominantly fresh waters* means surface waters in which the chloride concentration at the surface is less than 1,500 milligrams per liter.

(9) *South Florida Region* means those areas south of Lake Okeechobee and the Caloosahatchee River watershed to the west of Lake Okeechobee and the St. Lucie watershed to the east of Lake Okeechobee.

(10) *Spring* means a site at which ground water flows through a natural opening in the ground onto the land surface or into a body of surface water.

(11) *State* means the State of Florida, whose transactions with the U.S. EPA in matters related to 40 CFR 131.43 are administered by the Secretary, or officials delegated such responsibility, of the Florida Department of Environmental Protection (FDEP), or successor agencies.

(12) *Stream* means a free-flowing, predominantly fresh surface water in a defined channel, and includes rivers, creeks, branches, canals, freshwater sloughs, and other similar water bodies.

(13) *Surface water* means water upon the surface of the earth, whether contained in bounds created naturally or artificially or diffused. Water from natural springs shall be classified as surface water when it exits from the spring onto the Earth's surface.

(c) *Criteria for Florida waters*—(1) *Criteria for lakes.* (i) The applicable criteria for chlorophyll *a*, total nitrogen (TN), and total phosphorus (TP) for lakes within each respective lake class are shown on Table 1.

TABLE 1

A	B	C		
Lake Color[a] and Alkalinity	Chl-a (mg/L)[b,*]	TN (mg/L)	TP (mg/L)	
Colored Lakes[c]	0.020	1.27 [1.27–2.23]	0.05 [0.05–0.16]	
Clear Lakes, High Alkalinity[d]	0.020	1.05 [1.05–1.91]	0.03 [0.03–0.09]	

TABLE 1—Continued

A	B	C	
Lake Color[a] and Alkalinity	Chl-a (mg/L)[b,*]	TN (mg/L)	TP (mg/L)
Clear Lakes, ... Low Alkalinity[e]	0.006	0.51 [0.51–0.93]	0.01 [0.01–0.03]

[a] Platinum Cobalt Units (PCU) assessed as true color free from turbidity.
[b] Chlorophyll *a* is defined as corrected chlorophyll, or the concentration of chlorophyll *a* remaining after the chlorophyll degradation product, phaeophytin *a*, has been subtracted from the uncorrected chlorophyll *a* measurement.
[c] Long-term Color > 40 Platinum Cobalt Units (PCU)
[d] Long-term Color ≤ 40 PCU and Alkalinity > 20 mg/L CaCO₃
[e] Long-term Color ≤ 40 PCU and Alkalinity ≤ 20 mg/L CaCO₃
[*] For a given waterbody, the annual geometric mean of chlorophyll *a*, TN or TP concentrations shall not exceed the applicable criterion concentration more than once in a three-year period.

(ii) Baseline criteria apply unless the State determines that modified criteria within the range indicated in Table 1 apply to a specific lake. Once established, modified criteria are the applicable criteria for all CWA purposes. The State may use this procedure one time for a specific lake in lieu of the site-specific alternative criteria procedure described in paragraph (e) of this section.

(A) The State may calculate modified criteria for TN and/or TP where the chlorophyll *a* criterion-magnitude as an annual geometric mean has not been exceeded and sufficient ambient monitoring data exist for chlorophyll *a* and TN and/or TP for at least the three immediately preceding years. Sufficient data include at least four measurements per year, with at least one measurement between May and September and one measurement between October and April each year.

(B) Modified criteria are calculated using data from years in which sufficient data are available to reflect maintenance of ambient conditions. Modified TN and/or TP criteria may not be greater than the higher value specified in the range of values in column C of Table 1 in paragraph (c)(1)(i) of this section. Modified TP and TN criteria may not exceed criteria applicable to streams to which a lake discharges.

(C) The State shall notify the public and maintain a record of these modified lake criteria, as well as a record supporting their derivation. The State shall notify EPA Region 4 and provide the supporting record within 30 days of determination of modified lake criteria.

(2) *Criteria for streams.* (i) The applicable instream protection value (IPV) criteria for total nitrogen (TN) and total phosphorus (TP) for streams within each respective nutrient watershed region are shown on Table 2.

TABLE 2

Nutrient watershed region	Instream protection value criteria	
	TN (mg/L)[*]	TP (mg/L)[*]
Panhandle West[a]	0.67	0.06
Panhandle East[b]	1.03	0.18
North Central[c]	1.87	0.30
West Central[d]	1.65	0.49
Peninsula[e]	1.54	0.12

Watersheds pertaining to each Nutrient Watershed Region (NWR) were based principally on the NOAA coastal, estuarine, and fluvial drainage areas with modifications to the NOAA drainage areas in the West Central and Peninsula Regions that account for unique watershed geologies. For more detailed information on regionalization and which WBIDs pertain to each NWR, *see* the Technical Support Document.
[a] Panhandle West region includes: Perdido Bay Watershed, Pensacola Bay Watershed, Choctawhatchee Bay Watershed, St. Andrew Bay Watershed, and Apalachicola Bay Watershed.
[b] Panhandle East region includes: Apalachee Bay Watershed, and Econfina/Steinhatchee Coastal Drainage Area.
[c] North Central region includes the Suwannee River Watershed.
[d] West Central region includes: Peace, Myakka, Hillsborough, Alafia, Manatee, Little Manatee River Watersheds, and small, direct Tampa Bay tributary watersheds south of the Hillsborough River Watershed.
[e] Peninsula region includes: Waccasassa Coastal Drainage Area, Withlacoochee Coastal Drainage Area, Crystal/Pithlachascotee Coastal Drainage Area, small, direct Tampa Bay tributary watersheds west of the Hillsborough River Watershed, Sarasota Bay Watershed, small, direct Charlotte Harbor tributary watersheds south of the Peace River Watershed, Caloosahatchee River Watershed, Estero Bay Watershed, Kissimmee River/Lake Okeechobee Drainage Area, Loxahatchee/St. Lucie Watershed, Indian River Watershed, Daytona/St. Augustine Coastal Drainage Area, St. John's River Watershed, Nassau Coastal Drainage Area, and St. Mary's River Watershed.
[*] For a given waterbody, the annual geometric mean of TN or TP concentrations shall not exceed the applicable criterion concentration more than once in a three-year period.

(ii) *Criteria for protection of downstream lakes.* (A) The applicable criteria for streams that flow into downstream lakes include both the instream criteria for total phosphorus (TP) and total nitrogen (TN) in Table 2 in paragraph (c)(2)(i) and the downstream protection value (DPV) for TP and TN derived pursuant to the provisions of this paragraph. A DPV for stream tributaries (up to the point of reaching water bodies that are not streams as defined by this rule) that flow into a downstream lake is either the allowable concentration or the allowable loading of TN and/or TP applied at the point of entry into the lake. The applicable DPV for any stream shall be determined pursuant to paragraphs (c)(2)(ii)(B), (C), or (D) of this section. Contributions from stream tributaries upstream of the point of entry location must result in attainment of the DPV at the point of entry into the lake. If the DPV is not attained at the point of entry into the lake, then the collective set of streams in the upstream watershed does not attain the DPV, which is an applicable water quality criterion for the water segments in the upstream watershed. The State or EPA may establish additional DPVs at upstream tributary locations that are consistent with attaining the DPV at the point of entry into the lake. The State or EPA also have discretion to establish DPVs to account for a larger watershed area (i.e., include waters beyond the point of reaching water bodies that are not streams as defined by this rule).

(B) In instances where available data and/or resources provide for use of a scientifically defensible and protective lake-specific application of the BATHTUB model, the State or EPA may derive the DPV for TN and/or TP from use of a lake-specific application of BATHTUB. The State and EPA are authorized to use a scientifically defensible technical model other than BATHTUB upon demonstration that use of another scientifically defensible technical model would protect the lake's designated uses and meet all applicable criteria for the lake. The State or EPA may designate the wasteload and/or load allocations from a TMDL established or approved by EPA as DPV(s) if the allocations from the TMDL will protect the lake's designated uses and meet all applicable criteria for the lake.

(C) When the State or EPA has not derived a DPV for a stream pursuant to paragraph (c)(2)(ii)(B) of this section, and where the downstream lake attains the applicable chlorophyll a criterion and the applicable TP and/or TN criteria, then the DPV for TN and/or TP is the associated ambient instream levels of TN and/or TP at the point of entry to the lake. Degradation in water quality from the DPV pursuant to this paragraph is to be considered nonattainment of the DPV, unless the DPV is adjusted pursuant to paragraph (c)(2)(ii)(B) of this section.

(D) When the State or EPA has not derived a DPV pursuant to paragraph (c)(2)(ii)(B) of this section, and where the downstream lake does not attain applicable chlorophyll a criterion or the applicable TN and/or TP criteria, or has not been assessed, then the DPV for TN and/or TP is the applicable TN and/or TP criteria for the downstream lake.

(E) The State and EPA shall maintain a record of DPVs they derive based on the methods described in paragraphs (c)(2)(ii)(B) and (C) of this section, as well as a record supporting their derivation, and make such records available to the public. The State and EPA shall notify one another and provide a supporting record within 30 days of derivation of DPVs pursuant to paragraphs (c)(2)(ii)(B) or (C) of this section.

(3) *Criteria for springs.* The applicable nitrate+nitrite criterion is 0.35 mg/L as an annual geometric mean, not to be exceeded more than once in a three-year period.

(d) *Applicability.* (1) The criteria in paragraphs (c)(1) through (3) of this section apply to lakes and flowing waters, excluding flowing waters in the South Florida Region, and apply concurrently with other applicable water quality criteria, except when:

(i) State water quality standards contain criteria that are more stringent for a particular parameter and use;

(ii) The Regional Administrator determines that site-specific alternative criteria apply pursuant to the procedures in paragraph (e) of this section; or

(iii) The State adopts and EPA approves a water quality standards variance to the Class I or Class III designated use pursuant to § 131.13 that meets the applicable provisions of State law and the applicable Federal regulations at § 131.10.

(2) The criteria established in this section are subject to the State's general rules of applicability in the same way and to the same extent as are the other Federally-adopted and State-adopted numeric criteria when applied to the same use classifications.

(e) *Site-specific alternative criteria.* (1) The Regional Administrator may determine that site-specific alternative criteria shall apply to specific surface waters in lieu of the criteria established in paragraph (c) of this section. Any such determination shall be made consistent with § 131.11.

(2) To receive consideration from the Regional Administrator for a determination of site-specific alternative criteria, an entity shall submit a request that includes proposed alternative numeric criteria and supporting rationale suitable to meet the needs for a technical support document pursuant to paragraph (e)(3) of this section. The entity shall provide the State a copy of all materials submitted to EPA, at the time of submittal to EPA, to facilitate the State providing comments to EPA. Site-specific alternative criteria may be based on one or more of the following approaches:

(i) Replicate the process for developing the stream criteria in paragraph (c)(2)(i) of this section.

(ii) Replicate the process for developing the lake criteria in paragraph (c)(1) of this section.

(iii) Conduct a biological, chemical, and physical assessment of waterbody conditions.

(iv) Use another scientifically defensible approach protective of the designated use.

(3) For any determination made under paragraph (e)(1) of this section, the Regional Administrator shall, prior to making such a determination, provide for public notice and comment on a proposed determination. For any such proposed determination, the Regional Administrator shall prepare and make available to the public a technical support document addressing the specific surface waters affected and the justification for each proposed determination. This document shall be made available to the public no later than the date of public notice issuance.

(4) The Regional Administrator shall maintain and make available to the public an updated list of determinations made pursuant to paragraph (e)(1) of this section as well as the technical support documents for each determination.

(5) Nothing in this paragraph (e) shall limit the Administrator's authority to modify the criteria in paragraph (c) of this section through rulemaking.

(f) *Effective date.* This section is effective March 6, 2012, except for §131.43(e), which is effective February 4, 2011.

PART 132—WATER QUALITY GUIDANCE FOR THE GREAT LAKES SYSTEM

AUTHORITY: 33 U.S.C. 1251 *et seq.*

SOURCE: 60 FR 15387, Mar. 23, 1995, unless otherwise noted.

§132.1 Scope, purpose, and availability of documents.

(a) This part constitutes the Water Quality Guidance for the Great Lakes System (Guidance) required by section 118(c)(2) of the Clean Water Act (33 U.S.C. 1251 *et seq.*) as amended by the Great Lakes Critical Programs Act of 1990 (Pub. L. 101–596, 104 Stat. 3000 *et seq.*). The Guidance in this part identifies minimum water quality standards, antidegradation policies, and implementation procedures for the Great Lakes System to protect human health, aquatic life, and wildlife.

(b) The U.S. Environmental Protection Agency, Great Lakes States, and Great Lakes Tribes will use the Guidance in this part to evaluate the water quality programs of the States and Tribes to assure that they are protective of water quality. State and Tribal programs do not need to be identical to the Guidance in this part, but must contain provisions that are consistent with (as protective as) the Guidance in this part. The scientific, policy and legal basis for EPA's development of each section of the final Guidance in this part is set forth in the preamble, Supplementary Information Document, Technical Support Documents, and other supporting documents in the public docket. EPA will follow the guidance set out in these documents in reviewing the State and Tribal water quality programs in the Great Lakes for consistency with this part.

(c) The Great Lakes States and Tribes must adopt provisions consistent with the Guidance in this part applicable to waters in the Great Lakes System or be subject to EPA promulgation of its terms pursuant to this part.

(d) EPA understands that the science of risk assessment is rapidly improving. Therefore, to ensure that the scientific basis for the methodologies in appendices A through D are always current and peer reviewed, EPA will review the methodologies and revise them, as appropriate, every 3 years.

(e) Certain documents referenced in the appendixes to this part with a designation of NTIS and/or ERIC are available for a fee upon request to the National Technical Information Center (NTIS), U.S. Department of Commerce, 5285 Port Royal Road, Springfield, VA 22161. Alternatively, copies may be obtained for a fee upon request to the Educational Resources Information Center/Clearinghouse for Science, Mathematics, and Environmental Education (ERIC/CSMEE), 1200 Chambers Road, Room 310, Columbus, Ohio 43212. When ordering, please include the NTIS or ERIC/CSMEE accession number.

§ 132.2 Definitions.

The following definitions apply in this part. Terms not defined in this section have the meaning given by the Clean Water Act and EPA implementing regulations.

Acute-chronic ratio (ACR) is a standard measure of the acute toxicity of a material divided by an appropriate measure of the chronic toxicity of the same material under comparable conditions.

Acute toxicity is concurrent and delayed adverse effect(s) that results from an acute exposure and occurs within any short observation period which begins when the exposure begins, may extend beyond the exposure period, and usually does not constitute a substantial portion of the life span of the organism.

Adverse effect is any deleterious effect to organisms due to exposure to a substance. This includes effects which are or may become debilitating, harmful or toxic to the normal functions of the organism, but does not include non-harmful effects such as tissue discoloration alone or the induction of enzymes involved in the metabolism of the substance.

Bioaccumulation is the net accumulation of a substance by an organism as a result of uptake from all environmental sources.

Bioaccumulation factor (BAF) is the ratio (in L/kg) of a substance's concentration in tissue of an aquatic organism to its concentration in the ambient water, in situations where both the organism and its food are exposed and the ratio does not change substantially over time.

Bioaccumulative chemical of concern (BCC) is any chemical that has the potential to cause adverse effects which, upon entering the surface waters, by itself or as its toxic transformation product, accumulates in aquatic organisms by a human health bioaccumulation factor greater than 1000, after considering metabolism and other physico-chemical properties that might enhance or inhibit bioaccumulation, in accordance with the methodology in appendix B of this part. Chemicals with half-lives of less than eight weeks in the water column, sediment, and biota are not BCCs. The minimum BAF information needed to define an organic chemical as a BCC is either a field-measured BAF or a BAF derived using the BSAF methodology. The minimum BAF information needed to define an inorganic chemical, including an organometal, as a BCC is either a field-measured BAF or a laboratory-measured BCF. BCCs include, but are not limited to, the pollutants identified as BCCs in section A of Table 6 of this part.

Bioconcentration is the net accumulation of a substance by an aquatic organism as a result of uptake directly from the ambient water through gill membranes or other external body surfaces.

Bioconcentration factor (BCF) is the ratio (in L/kg) of a substance's concentration in tissue of an aquatic organism to its concentration in the ambient water, in situations where the organism is exposed through the water only and the ratio does not change substantially over time.

Biota-sediment accumulation factor (BSAF) is the ratio (in kg of organic carbon/kg of lipid) of a substance's lipid-normalized concentration in tissue of an aquatic organism to its organic carbon-normalized concentration in surface sediment, in situations where the ratio does not change substantially over time, both the organism and its food are exposed, and the surface sediment is representative of average surface sediment in the vicinity of the organism.

Carcinogen is a substance which causes an increased incidence of benign or malignant neoplasms, or substantially decreases the time to develop neoplasms, in animals or humans. The classification of carcinogens is discussed in section II.A of appendix C to part 132.

Chronic toxicity is concurrent and delayed adverse effect(s) that occurs only as a result of a chronic exposure.

Connecting channels of the Great Lakes are the Saint Mary's River, Saint Clair River, Detroit River, Niagara River, and Saint Lawrence River to the Canadian Border.

Criterion continuous concentration (CCC) is an estimate of the highest concentration of a material in the water column to which an aquatic community can be exposed indefinitely without resulting in an unacceptable effect.

Criterion maximum concentration (CMC) is an estimate of the highest concentration of a material in the water column to which an aquatic community can be exposed briefly without resulting in an unacceptable effect.

EC50 is a statistically or graphically estimated concentration that is expected to cause one or more specified effects in 50 percent of a group of organisms under specified conditions.

Endangered or threatened species are those species that are listed as endangered or threatened under section 4 of the Endangered Species Act.

Existing Great Lakes discharger is any building, structure, facility, or installation from which there is or may be a "discharge of pollutants" (as defined in 40 CFR 122.2) to the Great Lakes System, that is not a new Great Lakes discharger.

Federal Indian reservation, Indian reservation, or *reservation* means all land within the limits of any Indian reservation under the jurisdiction of the United States Government, notwithstanding the issuance of any patent, and including rights-of-way running through the reservation.

Final acute value (FAV) is (a) a calculated estimate of the concentration of a test material such that 95 percent of the genera (with which acceptable acute toxicity tests have been conducted on the material) have higher GMAVs, or (b) the SMAV of an important and/or critical species, if the SMAV is lower than the calculated estimate.

Final chronic value (FCV) is (a) a calculated estimate of the concentration

of a test material such that 95 percent of the genera (with which acceptable chronic toxicity tests have been conducted on the material) have higher GMCVs, (b) the quotient of an FAV divided by an appropriate acute-chronic ratio, or (c) the SMCV of an important and/or critical species, if the SMCV is lower than the calculated estimate or the quotient, whichever is applicable.

Final plant value (FPV) is the lowest plant value that was obtained with an important aquatic plant species in an acceptable toxicity test for which the concentrations of the test material were measured and the adverse effect was biologically important.

Genus mean acute value (GMAV) is the geometric mean of the SMAVs for the genus.

Genus mean chronic value (GMCV) is the geometric mean of the SMCVs for the genus.

Great Lakes means Lake Ontario, Lake Erie, Lake Huron (including Lake St. Clair), Lake Michigan, and Lake Superior; and the connecting channels (Saint Mary's River, Saint Clair River, Detroit River, Niagara River, and Saint Lawrence River to the Canadian Border).

Great Lakes States and Great Lakes Tribes, or Great Lakes States and Tribes means the States of Illinois, Indiana, Michigan, Minnesota, New York, Ohio, Pennsylvania, and Wisconsin, and any Indian Tribe as defined in this part which is located in whole or in part within the drainage basin of the Great Lakes, and for which EPA has approved water quality standards under section 303 of the Clean Water Act or which EPA has authorized to administer an NPDES program under section 402 of the Clean Water Act.

Great Lakes System means all the streams, rivers, lakes and other bodies of water within the drainage basin of the Great Lakes within the United States.

Human cancer criterion (HCC) is a Human Cancer Value (HCV) for a pollutant that meets the minimum data requirements for Tier I specified in appendix C of this part.

Human cancer value (HCV) is the maximum ambient water concentration of a substance at which a lifetime of exposure from either: drinking the water, consuming fish from the water, and water-related recreation activities; or consuming fish from the water, and water-related recreation activities, will represent a plausible upper-bound risk of contracting cancer of one in 100,000 using the exposure assumptions specified in the Methodologies for the Development of Human Health Criteria and Values in appendix C of this part.

Human noncancer criterion (HNC) is a Human Noncancer Value (HNV) for a pollutant that meets the minimum data requirements for Tier I specified in appendix C of this part.

Human noncancer value (HNV) is the maximum ambient water concentration of a substance at which adverse noncancer effects are not likely to occur in the human population from lifetime exposure via either: drinking the water, consuming fish from the water, and water-related recreation activities; or consuming fish from the water, and water-related recreation activities using the Methodologies for the Development of Human Health Criteria and Values in appendix C of this part.

Indian Tribe or *Tribe* means any Indian Tribe, band, group, or community recognized by the Secretary of the Interior and exercising governmental authority over a Federal Indian reservation.

LC50 is a statistically or graphically estimated concentration that is expected to be lethal to 50 percent of a group of organisms under specified conditions.

Load allocation (LA) is the portion of a receiving water's loading capacity that is attributed either to one of its existing or future nonpoint sources or to natural background sources, as more fully defined at 40 CFR 130.2(g). Nonpoint sources include: in-place contaminants, direct wet and dry deposition, groundwater inflow, and overland runoff.

Loading capacity is the greatest amount of loading that a water can receive without violating water quality standards.

Lowest observed adverse effect level (LOAEL) is the lowest tested dose or concentration of a substance which resulted in an observed adverse effect in exposed test organisms when all higher doses or concentrations resulted in the same or more severe effects.

Method detection level is the minimum concentration of an analyte (substance) that can be measured and reported with a 99 percent confidence that the analyte concentration is greater than zero as determined by the procedure set forth in appendix B of 40 CFR part 136.

Minimum Level (ML) is the concentration at which the entire analytical system must give a recognizable signal and acceptable calibration point. The ML is the concentration in a sample that is equivalent to the concentration of the lowest calibration standard analyzed by a specific analytical procedure, assuming that all the method-specified sample weights, volumes and processing steps have been followed.

New Great Lakes discharger is any building, structure, facility, or installation from which there is or may be a "discharge of pollutants" (as defined in 40 CFR 122.2) to the Great Lakes System, the construction of which commenced after March 23, 1997.

No observed adverse effect level (NOAEL) is the highest tested dose or concentration of a substance which resulted in no observed adverse effect in exposed test organisms where higher doses or concentrations resulted in an adverse effect.

No observed effect concentration (NOEC) is the highest concentration of toxicant to which organisms are exposed in a full life-cycle or partial life-cycle (short-term) test, that causes no observable adverse effects on the test organisms (i.e., the highest concentration of toxicant in which the values for the observed responses are not statistically significantly different from the controls).

Open waters of the Great Lakes (OWGLs) means all of the waters within Lake Erie, Lake Huron (including Lake St. Clair), Lake Michigan, Lake Ontario, and Lake Superior lakeward from a line drawn across the mouth of tributaries to the Lakes, including all waters enclosed by constructed breakwaters, but not including the connecting channels.

Quantification level is a measurement of the concentration of a contaminant obtained by using a specified laboratory procedure calibrated at a specified concentration above the method detection level. It is considered the lowest concentration at which a particular contaminant can be quantitatively measured using a specified laboratory procedure for monitoring of the contaminant.

Quantitative structure activity relationship (QSAR) or structure activity relationship (SAR) is a mathematical relationship between a property (activity) of a chemical and a number of descriptors of the chemical. These descriptors are chemical or physical characteristics obtained experimentally or predicted from the structure of the chemical.

Risk associated dose (RAD) is a dose of a known or presumed carcinogenic substance in (mg/kg)/day which, over a lifetime of exposure, is estimated to be associated with a plausible upper bound incremental cancer risk equal to one in 100,000.

Species mean acute value (SMAV) is the geometric mean of the results of all acceptable flow-through acute toxicity tests (for which the concentrations of the test material were measured) with the most sensitive tested life stage of the species. For a species for which no such result is available for the most sensitive tested life stage, the SMAV is the geometric mean of the results of all acceptable acute toxicity tests with the most sensitive tested life stage.

Species mean chronic value (SMCV) is the geometric mean of the results of all acceptable life-cycle and partial life-cycle toxicity tests with the species; for a species of fish for which no such result is available, the SMCV is the geometric mean of all acceptable early life-stage tests.

Stream design flow is the stream flow that represents critical conditions, upstream from the source, for protection of aquatic life, human health, or wildlife.

Threshold effect is an effect of a substance for which there is a theoretical or empirically established dose or concentration below which the effect does not occur.

Tier I criteria are numeric values derived by use of the Tier I methodologies in appendixes A, C and D of this part, the methodology in appendix B of

this part, and the procedures in appendix F of this part, that either have been adopted as numeric criteria into a water quality standard or are used to implement narrative water quality criteria.

Tier II values are numeric values derived by use of the Tier II methodologies in appendixes A and C of this part, the methodology in appendix B of this part, and the procedures in appendix F of this part, that are used to implement narrative water quality criteria.

Total maximum daily load (TMDL) is the sum of the individual wasteload allocations for point sources and load allocations for nonpoint sources and natural background, as more fully defined at 40 CFR 130.2(i). A TMDL sets and allocates the maximum amount of a pollutant that may be introduced into a water body and still assure attainment and maintenance of water quality standards.

Tributaries of the Great Lakes System means all waters of the Great Lakes System that are not open waters of the Great Lakes, or connecting channels.

Uncertainty factor (UF) is one of several numeric factors used in operationally deriving criteria from experimental data to account for the quality or quantity of the available data.

Uptake is acquisition of a substance from the environment by an organism as a result of any active or passive process.

Wasteload allocation (WLA) is the portion of a receiving water's loading capacity that is allocated to one of its existing or future point sources of pollution, as more fully defined at 40 CFR 130.2(h). In the absence of a TMDL approved by EPA pursuant to 40 CFR 130.7 or an assessment and remediation plan developed and approved in accordance with procedure 3.A of appendix F of this part, a WLA is the allocation for an individual point source, that ensures that the level of water quality to be achieved by the point source is derived from and complies with all applicable water quality standards.

Wet weather point source means any discernible, confined and discrete conveyance from which pollutants are, or may be, discharged as the result of a wet weather event. Discharges from wet weather point sources shall include only: discharges of storm water from a municipal separate storm sewer as defined at 40 CFR 122.26(b)(8); storm water discharge associated with industrial activity as defined at 40 CFR 122.26(b)(14); discharges of storm water and sanitary wastewaters (domestic, commercial, and industrial) from a combined sewer overflow; or any other stormwater discharge for which a permit is required under section 402(p) of the Clean Water Act. A storm discharge associated with industrial activity which is mixed with process wastewater shall not be considered a wet weather point source.

§ 132.3 Adoption of criteria.

The Great Lakes States and Tribes shall adopt numeric water quality criteria for the purposes of section 303(c) of the Clean Water Act applicable to waters of the Great Lakes System in accordance with § 132.4(d) that are consistent with:

(a) The acute water quality criteria for protection of aquatic life in Table 1 of this part, or a site-specific modification thereof in accordance with procedure 1 of appendix F of this part;

(b) The chronic water quality criteria for protection of aquatic life in Table 2 of this part, or a site-specific modification thereof in accordance with procedure 1 of appendix F of this part;

(c) The water quality criteria for protection of human health in Table 3 of this part, or a site-specific modification thereof in accordance with procedure 1 of appendix F of this part; and

(d) The water quality criteria for protection of wildlife in Table 4 of this part, or a site-specific modification thereof in accordance with procedure 1 of appendix F of this part.

§ 132.4 State adoption and application of methodologies, policies and procedures.

(a) The Great Lakes States and Tribes shall adopt requirements applicable to waters of the Great Lakes System for the purposes of sections 118, 301, 303, and 402 of the Clean Water Act that are consistent with:

(1) The definitions in § 132.2;

(2) The Methodologies for Development of Aquatic Life Criteria and Values in appendix A of this part;

(3) The Methodology for Development of Bioaccumulation Factors in appendix B of this part;

(4) The Methodologies for Development of Human Health Criteria and Values in appendix C of this part;

(5) The Methodology for Development of Wildlife Criteria in appendix D of this part;

(6) The Antidegradation Policy in appendix E of this part; and

(7) The Implementation Procedures in appendix F of this part.

(b) Except as provided in paragraphs (g), (h), and (i) of this section, the Great Lakes States and Tribes shall use methodologies consistent with the methodologies designated as Tier I methodologies in appendixes A, C, and D of this part, the methodology in appendix B of this part, and the procedures in appendix F of this part when adopting or revising numeric water quality criteria for the purposes of section 303(c) of the Clean Water Act for the Great Lakes System.

(c) Except as provided in paragraphs (g), (h), and (i) of this section, the Great Lakes States and Tribes shall use methodologies and procedures consistent with the methodologies designated as Tier I methodologies in appendixes A, C, and D of this part, the Tier II methodologies in appendixes A and C of this part, the methodology in appendix B of this part, and the procedures in appendix F of this part to develop numeric criteria and values when implementing narrative water quality criteria adopted for purposes of section 303(c) of the Clean Water Act.

(d) The water quality criteria and values adopted or developed pursuant to paragraphs (a) through (c) of this section shall apply as follows:

(1) The acute water quality criteria and values for the protection of aquatic life, or site-specific modifications thereof, shall apply to all waters of the Great Lakes System.

(2) The chronic water quality criteria and values for the protection of aquatic life, or site-specific modifications thereof, shall apply to all waters of the Great Lakes System.

(3) The water quality criteria and values for protection of human health, or site-specific modifications thereof, shall apply as follows:

(i) Criteria and values derived as HCV-Drinking and HNV-Drinking shall apply to the Open Waters of the Great Lakes, all connecting channels of the Great Lakes, and all other waters of the Great Lakes System that have been designated as public water supplies by any State or Tribe in accordance with 40 CFR 131.10.

(ii) Criteria and values derived as HCV-Nondrinking and HNV-Nondrinking shall apply to all waters of the Great Lakes System other than those in paragraph (d)(3)(i) of this section.

(4) Criteria for protection of wildlife, or site-specific modifications thereof, shall apply to all waters of the Great Lakes System.

(e) The Great Lakes States and Tribes shall apply implementation procedures consistent with the procedures in appendix F of this part for all applicable purposes under the Clean Water Act, including developing total maximum daily loads for the purposes of section 303(d) and water quality-based effluent limits for the purposes of section 402, in establishing controls on the discharge of any pollutant to the Great Lakes System by any point source with the following exceptions:

(1) The Great Lakes States and Tribes are not required to apply these implementation procedures in establishing controls on the discharge of any pollutant by a wet weather point source. Any adopted implementation procedures shall conform with all applicable Federal, State and Tribal requirements.

(2) The Great Lakes States and Tribes may, but are not required to, apply procedures consistent with procedures 1, 2, 3, 4, 5, 7, 8, and 9 of appendix F of this part in establishing controls on the discharge of any pollutant set forth in Table 5 of this part. Any procedures applied in lieu of these implementation procedures shall conform with all applicable Federal, State, and Tribal requirements.

(f) The Great Lakes States and Tribes shall apply an antidegradation policy consistent with the policy in appendix E for all applicable purposes under the Clean Water Act, including 40 CFR 131.12.

(g) For pollutants listed in Table 5 of this part, the Great Lakes States and Tribes shall:

(1) Apply any methodologies and procedures acceptable under 40 CFR part 131 when developing water quality criteria or implementing narrative criteria; and

(2) Apply the implementation procedures in appendix F of this part or alternative procedures consistent with all applicable Federal, State, and Tribal laws.

(h) For any pollutant other than those in Table 5 of this part for which the State or Tribe demonstrates that a methodology or procedure in this part is not scientifically defensible, the Great Lakes States and Tribes shall:

(1) Apply an alternative methodology or procedure acceptable under 40 CFR part 131 when developing water quality criteria; or

(2) Apply an alternative implementation procedure that is consistent with all applicable Federal, State, and Tribal laws.

(i) Nothing in this part shall prohibit the Great Lakes States and Tribes from adopting numeric water quality criteria, narrative criteria, or water quality values that are more stringent than criteria or values specified in §132.3 or that would be derived from application of the methodologies set forth in appendixes A, B, C, and D of this part, or to adopt antidegradation standards and implementation procedures more stringent than those set forth in appendixes E and F of this part.

§132.5 Procedures for adoption and EPA review.

(a) Except as provided in paragraph (c) of this section, the Great Lakes States and Tribes shall adopt and submit for EPA review and approval the criteria, methodologies, policies, and procedures developed pursuant to this part no later than September 23, 1996. With respect to procedure 3.C of appendix F of this part, each Great Lakes State and Tribe shall make its submission to EPA no later than May 13, 2002.

(b) The following elements must be included in each submission to EPA for review:

(1) The criteria, methodologies, policies, and procedures developed pursuant to this part;

(2) Certification by the Attorney General or other appropriate legal authority pursuant to 40 CFR 123.62 and 40 CFR 131.6(e) as appropriate;

(3) All other information required for submission of National Pollutant Discharge Elimination System (NPDES) program modifications under 40 CFR 123.62; and

(4) General information which will aid EPA in determining whether the criteria, methodologies, policies and procedures are consistent with the requirements of the Clean Water Act and this part, as well as information on general policies which may affect their application and implementation.

(c) The Regional Administrator may extend the deadline for the submission required in paragraph (a) of this section if the Regional Administrator believes that the submission will be consistent with the requirements of this part and can be reviewed and approved pursuant to this section no later than March 23, 1997, or, for procedure 3.C of appendix F of this part, no later than November 13, 2002.

(d) If a Great Lakes State or Tribe makes no submission pursuant to this part to EPA for review, the requirements of this part shall apply to discharges to waters of the Great Lakes System located within the State or Federal Indian reservation upon EPA's publication of a final rule indicating the effective date of the part 132 requirements in the identified jurisdictions.

(e) If a Great Lakes State or Tribe submits criteria, methodologies, policies, and procedures pursuant to this part to EPA for review that contain substantial modifications of the State or Tribal NPDES program, EPA shall issue public notice and provide a minimum of 30 days for public comment on such modifications. The public notice shall conform with the requirements of 40 CFR 123.62.

(f) After review of State or Tribal submissions under this section, and following the public comment period in subparagraph (e) of this section, if any, EPA shall either:

(1) Publish notice of approval of the submission in the FEDERAL REGISTER within 90 days of such submission; or

(2) Notify the State or Tribe within 90 days of such submission that EPA has determined that all or part of the submission is inconsistent with the requirements of the Clean Water Act or this part and identify any necessary changes to obtain EPA approval. If the State or Tribe fails to adopt such changes within 90 days after the notification, EPA shall publish a notice in the FEDERAL REGISTER identifying the approved and disapproved elements of the submission and a final rule in the FEDERAL REGISTER identifying the provisions of part 132 that shall apply to discharges within the State or Federal Indian reservation.

(g) EPA's approval or disapproval of a State or Tribal submission shall be based on the requirements of this part and of the Clean Water Act. EPA's determination whether the criteria, methodologies, policies, and procedures in a State or Tribal submission are consistent with the requirements of this part will be based on whether:

(1) *For pollutants listed in Tables 1, 2, 3, and 4 of this part.* The Great Lakes State or Tribe has adopted numeric water quality criteria as protective as each of the numeric criteria in Tables 1, 2, 3, and 4 of this part, taking into account any site-specific criteria modifications in accordance with procedure 1 of appendix F of this part;

(2) *For pollutants other than those listed in Tables 1, 2, 3, 4, and 5 of this part.* The Great Lakes State or Tribe demonstrates that either:

(i) It has adopted numeric criteria in its water quality standards that were derived, or are as protective as or more protective than could be derived, using the methodologies in appendixes A, B, C, and D of this part, and the site-specific criteria modification procedures in accordance with procedure 1 of appendix F of this part; or

(ii) It has adopted a procedure by which water quality-based effluent limits and total maximum daily loads are developed using the more protective of:

(A) Numeric criteria adopted by the State into State water quality standards and approved by EPA prior to March 23, 1997; or

(B) Water quality criteria and values derived pursuant to §132.4(c); and

(3) *For methodologies, policies, and procedures.* The Great Lakes State or Tribe has adopted methodologies, policies, and procedures as protective as the corresponding methodology, policy, or procedure in §132.4. The Great Lakes State or Tribe may adopt provisions that are more protective than those contained in this part. Adoption of a more protective element in one provision may be used to offset a less protective element in the same provision as long as the adopted provision is as protective as the corresponding provision in this part; adoption of a more protective element in one provision, however, is not justification for adoption of a less protective element in another provision of this part.

(h) A submission by a Great Lakes State or Tribe will need to include any provisions that EPA determines, based on EPA's authorities under the Clean Water Act and the results of consultation under section 7 of the Endangered Species Act, are necessary to ensure that water quality is not likely to jeopardize the continued existence of any endangered or threatened species listed under section 4 of the Endangered Species Act or result in the destruction or adverse modification of such species' critical habitat.

(i) EPA's approval of the elements of a State's or Tribe's submission will constitute approval under section 118 of the Clean Water Act, approval of the submitted water quality standards pursuant to section 303 of the Clean Water Act, and approval of the submitted modifications to the State's or Tribe's NPDES program pursuant to section 402 of the Clean Water Act.

[60 FR 15387, Mar. 23, 1995, as amended at 65 FR 67650, Nov. 13, 2000]

§132.6 Application of part 132 requirements in Great Lakes States and Tribes.

(a) Effective September 5, 2000, the requirements of Paragraph C.1 of Procedure 2 in Appendix F of this Part and the requirements of paragraph F.2 of Procedure 5 in appendix F of this Part shall apply to discharges within the Great Lakes System in the State of Indiana.

(b) Effective September 5, 2000, the requirements of Procedure 3 in appendix F of this Part shall apply for purposes of developing total maximum daily loads in the Great Lakes System in the State of Illinois.

(c) Effective September 5, 2000, the requirements of Paragraphs C.1 and D of Procedure 6 in appendix F of this Part shall apply to discharges within the Great Lakes System in the States of Indiana, Michigan and Ohio.

(d) Effective November 6, 2000, §132.4(d)(2) shall apply to waters designated as "Class D" under section 701.9 of Title 6 of the New York State Codes, Rules and Regulations within the Great Lakes System in the State of New York. For purposes of this paragraph, chronic water quality criteria and values for the protection of aquatic life adopted or developed pursuant to §132.4(a) through (c) are the criteria and values adopted or developed by New York State Department of Environmental Conservation (see section 703.5 of Title 6 of the New York State Codes, Rules and Regulations) and approved by EPA under section 303(c) of the Clean Water Act.

(e) Effective November 6, 2000, the criteria for mercury contained in Table 4 of this part shall apply to waters within the Great Lakes System in the State of New York.

(f) Effective December 6, 2000, the chronic aquatic life criterion for endrin in Table 2 of this part shall apply to the waters of the Great Lakes System in the State of Wisconsin designated by Wisconsin as Warm Water Sportfish and Warm Water Forage Fish aquatic life use.

(g) Effective February 5, 2001, the chronic aquatic life criterion for selenium in Table 2 of this part shall apply to the waters of the Great Lakes System in the State of Wisconsin designated by Wisconsin as Limited Forage Fish aquatic life use.

(h) Effective December 6, 2000, the requirements of procedure 3 in appendix F of this part shall apply for purposes of developing total maximum daily loads in the Great Lakes System in the State of Wisconsin.

(i) Effective December 6, 2000, the requirements of paragraphs D and E of procedure 5 in appendix F of this part shall apply to discharges within the Great Lakes System in the State of Wisconsin.

(j) Effective December 6, 2000, the requirements of paragraph D of procedure 6 in appendix F of this part shall apply to discharges within the Great Lakes System in the State of Wisconsin.

[65 FR 47874, Aug. 4, 2000, as amended at 65 FR 59737, Oct. 6, 2000; 65 FR 66511, Nov. 6, 2000; 76 FR 57652, Sept. 16, 2011]

TABLES TO PART 132

TABLE 1—ACUTE WATER QUALITY CRITERIA FOR PROTECTION OF AQUATIC LIFE IN AMBIENT WATER

EPA recommends that metals criteria be expressed as dissolved concentrations (see appendix A, I.A.4 for more information regarding metals criteria).

(a)

Chemical	CMC (µg/L)	Conversion factor (CF)
Arsenic (III)	[a,b]339.8	1.000
Chromium (VI)	[a,b]16.02	0.982
Cyanide	[c]22	n/a
Dieldrin	[d]0.24	n/a
Endrin	[d]0.086	n/a
Lindane	[d]0.95	n/a
Mercury (II)	[a,b]1.694	0.85
Parathion	[d]0.065	n/a

[a] CMC=CMC[tr].
[b] CMC[d]=(CMC[tr]) CF. The CMC[d] shall be rounded to two significant digits.
[c] CMC should be considered free cyanide as CN.
[d] CMC=CMC[t].
Notes:
The term "n/a" means not applicable.
CMC is Criterion Maximum Concentration.
CMC[tr] is the CMC expressed as total recoverable.
CMC[d] is the CMC expressed as a dissolved concentration.
CMC[t] is the CMC expressed as a total concentration.

(b)

Chemical	m_A	b_A	Conversion factor (CF)
Cadmium [a,b]	1.128	−3.6867	0.85
Chromium (III) [a,b]	0.819	+3.7256	0.316
Copper [a,b]	0.9422	−1.700	0.960
Nickel [a,b]	0.846	+2.255	0.998
Pentachlorophenol [c]	1.005	−4.869	n/a
Zinc [a,b]	0.8473	+0.884	0.978

[a] CMC[tr]=exp $\{m_A [\ln (\text{hardness})]+b_A\}$.
[b] CMC[d]=(CMC[tr]) CF. The CMC[d] shall be rounded to two significant digits.
[c] CMC[t]=exp $m_A \{[\text{pH}]+b_A\}$. The CMC[t] shall be rounded to two significant digits.
Notes:
The term "exp" represents the base e exponential function.
The term "n/a" means not applicable.
CMC is Criterion Maximum Concentration.
CMC[tr] is the CMC expressed as total recoverable.
CMC[d] is the CMC expressed as a dissolved concentration.
CMC[t] is the CMC expressed as a total concentration.

[60 FR 15387, Mar. 23, 1995, as amended at 65 FR 35286, June 2, 2000]

TABLE 2—CHRONIC WATER QUALITY CRITERIA FOR PROTECTION OF AQUATIC LIFE IN AMBIENT WATER

EPA recommends that metals criteria be expressed as dissolved concentrations (see appendix A, I.A.4 for more information regarding metals criteria).

(a)

Chemical	CCC (µg/L)	Conversion factor (CF)
Arsenic (III)	[a,b]147.9	1.000
Chromium (VI)	[a,b]10.98	0.962
Cyanide	[c]5.2	n/a
Dieldrin	[d]0.056	n/a
Endrin	[d]0.036	n/a
Mercury (II)	[a,b]0.9081	0.85
Parathion	[d]0.013	n/a
Selenium	[a,b]5	0.922

[a] CCC=CCC[tr].
[b] CCC[d]=(CCC[tr]) CF. The CCC[d] shall be rounded to two significant digits.
[c] CCC should be considered free cyanide as CN.
[d] CCC=CCC[t].
Notes:
The term "n/a" means not applicable.
CCC is Criterion Continuous Concentration.
CCC[tr] is the CCC expressed as total recoverable.
CCC[d] is the CCC expressed as a dissolved concentration.
CCC[t] is the CCC expressed as a total concentration.

(b)

Chemical	m_c	b_c	Conversion factor (CF)
Cadmium [a,b]	0.7852	−2.715	0.850
Chromium (III) [a,b]	0.819	+0.6848	0.860
Copper [a,b]	0.8545	−1.702	0.960
Nickel [a,b]	0.846	+0.0584	0.997
Pentachlorophenol [c]	1.005	−5.134	n/a
Zinc [a,b]	0.8473	+0.884	0.986

[a] CCC[tr]=exp $\{m_c[\ln (\text{hardness})]+b_c\}$.
[b] CCC[d]=(CCC[tr]) (CF). The CCC[d] shall be rounded to two significant digits.
[c] CMC[t]=exp $\{m_A[\text{pH}]+b_A\}$. The CMC[t] shall be rounded to two significant digits.
Notes:
The term "exp" represents the base e exponential function.
The term "n/a" means not applicable.
CCC is Criterion Continuous Concentration.
CCC[tr] is the CCC expressed as total recoverable.
CCC[d] is the CCC expressed as a dissolved concentration.
CCC[t] is the CCC expressed as a total concentration.

TABLE 3—WATER QUALITY CRITERIA FOR PROTECTION OF HUMAN HEALTH

Chemical	HNV (µg/L)		HCV (µg/L)	
	Drinking	Non-drinking	Drinking	Non-drinking
Benzene	1.9E1	5.1E2	1.2E1	3.1E2
Chlordane	1.4E−3	1.4E−3	2.5E−4	2.5E−4
Chlorobenzene	4.7E2	3.2E3		
Cyanides	6.0E2	4.8E4		
DDT	2.0E−3	2.0E−3	1.5E−4	1.5E−4

TABLE 3—WATER QUALITY CRITERIA FOR PROTECTION OF HUMAN HEALTH—Continued

Chemical	HNV (µg/L)		HCV (µg/L)	
	Drink-ing	Non-drink-ing	Drink-ing	Non-drink-ing
Dieldrin	4.1E–4	4.1E–4	6.5E–6	6.5E–6
2,4-Dimethylphenol	4.5E2	8.7E3		
2,4-Dinitrophenol	5.5E1	2.8E3		
Hexachlorobenzene ...	4.6E–2	4.6E–2	4.5E–4	4.5E–4
Hexachloroethane	6.0	7.6	5.3	6.7
Lindane	4.7E–1	5.0E–1		
Mercury [1]	1.8E–3	1.8E–3		
Methylene chloride	1.6E3	9.0E4	4.7E1	2.6E3
2,3,7,8-TCDD	6.7E–8	6.7E–8	8.6E–9	8.6E–9
Toluene	5.6E3	5.1E4		
Toxaphene			6.8E–5	6.8E–5
Trichloroethylene			2.9E1	3.7E2

[1] Includes methylmercury.

[60 FR 15387, Mar. 23, 1995, as amended at 62 FR 11731, Mar. 12, 1997; 62 FR 52924, Oct. 9, 1997]

TABLE 4—WATER QUALITY CRITERIA FOR PROTECTION OF WILDLIFE

Chemical	Criteria (µg/L)
DDT and metabolites	1.1E–5
Mercury (including methylmercury)	1.3E–3
PCBs (class)	1.2E–4
2,3,7,8-TCDD	3.1E–9

[60 FR 15387, Mar. 23, 1995, as amended at 62 FR 11731, Mar. 12, 1997]

TABLE 5—POLLUTANTS SUBJECT TO FEDERAL, STATE, AND TRIBAL REQUIREMENTS

Alkalinity
Ammonia
Bacteria
Biochemical oxygen demand (BOD)
Chlorine
Color
Dissolved oxygen
Dissolved solids
pH
Phosphorus
Salinity
Temperature
Total and suspended solids
Turbidity

TABLE 6—POLLUTANTS OF INITIAL FOCUS IN THE GREAT LAKES WATER QUALITY INITIATIVE

A. Pollutants that are bioaccumulative chemicals of concern (BCCs):
    Chlordane
    4,4′-DDD; p,p′-DDD; 4,4′-TDE; p,p′-TDE
    4,4′-DDE; p,p′-DDE
    4,4′-DDT; p,p′-DDT
    Dieldrin
    Hexachlorobenzene
    Hexachlorobutadiene; hexachloro-1, 3-butadiene
    Hexachlorocyclohexanes; BHCs
    alpha-Hexachlorocyclohexane; alpha-BHC
    beta-Hexachlorocyclohexane; beta-BHC
    delta-Hexachlorocyclohexane; delta-BHC
    Lindane; gamma-hexachlorocyclohexane; gamma-BHC
    Mercury
    Mirex
    Octachlorostyrene
    PCBs; polychlorinated biphenyls
    Pentachlorobenzene
    Photomirex
    2,3,7,8-TCDD; dioxin
    1,2,3,4-Tetrachlorobenzene
    1,2,4,5-Tetrachlorobenzene Toxaphene
B. Pollutants that are not bioaccumulative chemicals of concern:
    Acenaphthene
    Acenaphthylene
    Acrolein; 2-propenal
    Acrylonitrile
    Aldrin
    Aluminum
    Anthracene
    Antimony
    Arsenic
    Asbestos
    1,2-Benzanthracene; benz[a]anthracene
    Benzene
    Benzidine
    Benzo[a]pyrene; 3,4-benzopyrene
    3,4-Benzofluoranthene; benzo[b]fluoranthene
    11,12-Benzofluoranthene; benzo[k]fluoranthene
    1,12-Benzoperylene; benzo[ghi]perylene
    Beryllium
    Bis(2-chloroethoxy) methane
    Bis(2-chloroethyl) ether
    Bis(2-chloroisopropyl) ether
    Bromoform; tribomomethane.
    4-Bromophenyl phenyl ether
    Butyl benzyl phthalate
    Cadmium
    Carbon tetrachloride; tetrachloromethane
    Chlorobenzene
    p-Chloro-m-cresol; 4-chloro-3-methylphenol
    Chlorodibromomethane
    Chlorethane
    2-Chloroethyl vinyl ether
    Chloroform; trichloromethane
    2-Chloronaphthalene
    2-Chlorophenol
    4-Chlorophenyl phenyl ether
    Chlorpyrifos
    Chromium
    Chrysene
    Copper
    Cyanide
    2,4-D; 2,4-Dichlorophenoxyacetic acid
    DEHP; di(2-ethylhexyl) phthalate
    Diazinon
    1,2:5,6-Dibenzanthracene; dibenz[a,h]anthracene
    Dibutyl phthalate; di-n-butyl phthalate

1,2-Dichlorobenzene
1,3-Dichlorobenzene
1,4-Dichlorobenzene
3,3'-Dichlorobenzidine
Dichlorobromomethane;
bromodichloromethane
1,1-Dichloroethane
1,2-Dichloroethane
1,1-Dichloroethylene; vinylidene chloride
1,2-trans-Dichloroethylene
2,4-Dichlorophenol
1,2-Dichloropropane
1,3-Dichloropropene; 1,3-dichloropropylene
Diethyl phthalate
2,4-Dimethylphenol; 2,4-xylenol
Dimethyl phthalate
4,6-Dinitro-o-cresol;   2-methyl-4,6-
dinitrophenol
2,4-Dinitrophenol
2,4-Dinitrotoluene
2,6-Dinitrotoluene
Dioctyl phthalate; di-n-octyl phthalate
1,2-Diphenylhydrazine
Endosulfan; thiodan
alpha-Endosulfan
beta-Endosulfan
Endosulfan sulfate
Endrin
Endrin aldehyde
Ethylbenzene
Fluoranthene
Fluorene; 9H-fluorene
Fluoride
Guthion
Heptachlor
Heptachlor epoxide
Hexachlorocyclopentadiene
Hexachloroethane
Indeno[1,2,3-cd]pyrene; 2,3-o-phenylene pyrene
Isophorone
Lead
Malathion
Methoxychlor
Methyl bromide; bromomethane
Methyl chloride; chloromethane
Methylene chloride; dichloromethane
Napthalene
Nickel
Nitrobenzene
2-Nitrophenol
4-Nitrophenol
N-Nitrosodimethylamine
N-Nitrosodiphenylamine
N-Nitrosodipropylamine;   N-nitrosodi-n-
propylamine
Parathion
Pentachlorophenol
Phenanthrene
Phenol
Iron
Pyrene
Selenium
Silver
1,1,2,2-Tetrachloroethane
Tetrachloroethylene
Thallium

Toluene; methylbenzene
1,2,4-Trichlorobenzene
1,1,1-Trichloroethane
1,1,2-Trichloroethane
Trichloroethylene; trichloroethene
2,4,6-Trichlorophenol
Vinyl   chloride;   chloroethylene;
chloroethene
Zinc

APPENDIX A TO PART 132—GREAT LAKES
   WATER QUALITY INITIATIVE METH-
   ODOLOGIES FOR DEVELOPMENT OF
   AQUATIC LIFE CRITERIA AND VAL-
   UES

METHODOLOGY FOR DERIVING AQUATIC LIFE
          CRITERIA: TIER I

Great Lakes States and Tribes shall adopt
provisions consistent with (as protective as)
this appendix.

I. Definitions

A. *Material of Concern.* When defining the
material of concern the following should be
considered:

1. Each separate chemical that does not
ionize substantially in most natural bodies
of water should usually be considered a separate material, except possibly for struc-
turally similar organic compounds that only
exist in large quantities as commercial mixtures of the various compounds and appar-
ently have similar biological, chemical,
physical, and toxicological properties.

2. For chemicals that ionize substantially
in most natural bodies of water (e.g., some
phenols and organic acids, some salts of phenols and organic acids, and most inorganic
salts and coordination complexes of metals
and metalloid), all forms that would be in
chemical equilibrium should usually be considered one material. Each different oxida-
tion state of a metal and each different non-
ionizable covalently bonded organometallic
compound should usually be considered a
separate material.

3. The definition of the material of concern
should include an operational analytical
component. Identification of a material simply as "sodium," for example, implies "total
sodium," but leaves room for doubt. If
"total" is meant, it must be explicitly stated. Even "total" has different operational
definitions, some of which do not necessarily
measure "all that is there" in all samples.
Thus, it is also necessary to reference or describe the analytical method that is intended. The selection of the operational analytical component should take into account
the analytical and environmental chemistry
of the material and various practical considerations, such as labor and equipment requirements, and whether the method would
require measurement in the field or would

allow measurement after samples are transported to a laboratory.

a. The primary requirements of the operational analytical component are that it be appropriate for use on samples of receiving water, that it be compatible with the available toxicity and bioaccumulation data without making extrapolations that are too hypothetical, and that it rarely result in underprotection or overprotection of aquatic organisms and their uses. Toxicity is the property of a material, or combination of materials, to adversely affect organisms.

b. Because an ideal analytical measurement will rarely be available, an appropriate compromise measurement will usually have to be used. This compromise measurement must fit with the general approach that if an ambient concentration is lower than the criterion, unacceptable effects will probably not occur, i.e., the compromise measure must not err on the side of underprotection when measurements are made on a surface water. What is an appropriate measurement in one situation might not be appropriate for another. For example, because the chemical and physical properties of an effluent are usually quite different from those of the receiving water, an analytical method that is appropriate for analyzing an effluent might not be appropriate for expressing a criterion, and vice versa. A criterion should be based on an appropriate analytical measurement, but the criterion is not rendered useless if an ideal measurement either is not available or is not feasible.

NOTE: The analytical chemistry of the material might have to be taken into account when defining the material or when judging the acceptability of some toxicity tests, but a criterion must not be based on the sensitivity of an analytical method. When aquatic organisms are more sensitive than routine analytical methods, the proper solution is to develop better analytical methods.

4. It is now the policy of EPA that the use of dissolved metal to set and measure compliance with water quality standards is the recommended approach, because dissolved metal more closely approximates the bioavailable fraction of metal in the water column that does total recoverable metal. One reason is that a primary mechanism for water column toxicity is adsorption at the gill surface which requires metals to be in the dissolved form. Reasons for the consideration of total recoverable metals criteria include risk management considerations not covered by evaluation of water column toxicity. A risk manager may consider sediments and food chain effects and may decide to take a conservative approach for metals, considering that metals are very persistent chemicals. This approach could include the use of total recoverable metal in water quality standards. A range of different risk management decisions can be justified. EPA rec-

ommends that State water quality standards be based on dissolved metal. EPA will also approve a State risk management decision to adopt standards based on total recoverable metal, if those standards are otherwise approvable under this program.

B. *Acute Toxicity.* Concurrent and delayed adverse effect(s) that results from an acute exposure and occurs within any short observation period which begins when the exposure begins, may extend beyond the exposure period, and usually does not constitute a substantial portion of the life span of the organism. (Concurrent toxicity is an adverse effect to an organism that results from, and occurs during, its exposure to one or more test materials.) Exposure constitutes contact with a chemical or physical agent. Acute exposure, however, is exposure of an organism for any short period which usually does not constitute a substantial portion of its life span.

C. *Chronic Toxicity.* Concurrent and delayed adverse effect(s) that occurs only as a result of a chronic exposure. Chronic exposure is exposure of an organism for any long period or for a substantial portion of its life span.

II. Collection of Data

A. Collect all data available on the material concerning toxicity to aquatic animals and plants.

B. All data that are used should be available in typed, dated, and signed hard copy (e.g., publication, manuscript, letter, memorandum, etc.) with enough supporting information to indicate that acceptable test procedures were used and that the results are reliable. In some cases, it might be appropriate to obtain written information from the investigator, if possible. Information that is not available for distribution shall not be used.

C. Questionable data, whether published or unpublished, must not be used. For example, data must be rejected if they are from tests that did not contain a control treatment, tests in which too many organisms in the control treatment died or showed signs of stress or disease, and tests in which distilled or deionized water was used as the dilution water without the addition of appropriate salts.

D. Data on technical grade materials may be used if appropriate, but data on formulated mixtures and emulsifiable concentrates of the material must not be used.

E. For some highly volatile, hydrolyzable, or degradable materials, it might be appropriate to use only results of flow-through tests in which the concentrations of test material in test solutions were measured using acceptable analytical methods. A flow-through test is a test with aquatic organisms in which test solutions flow into constant-volume test chambers either intermittently (e.g., every few minutes) or continuously, with the excess flowing out.

F. Data must be rejected if obtained using:

1. Brine shrimp, because they usually only occur naturally in water with salinity greater than 35 g/kg.

2. Species that do not have reproducing wild populations in North America.

3. Organisms that were previously exposed to substantial concentrations of the test material or other contaminants.

4. Saltwater species except for use in deriving acute-chronic ratios. An ACR is a standard measure of the acute toxicity of a material divided by an appropriate measure of the chronic toxicity of the same material under comparable conditions.

G. Questionable data, data on formulated mixtures and emulsifiable concentrates, and data obtained with species non-resident to North America or previously exposed organisms may be used to provide auxiliary information but must not be used in the derivation of criteria.

III. Required Data

A. Certain data should be available to help ensure that each of the major kinds of possible adverse effects receives adequate consideration. An adverse effect is a change in an organism that is harmful to the organism. Exposure means contact with a chemical or physical agent. Results of acute and chronic toxicity tests with representative species of aquatic animals are necessary so that data available for tested species can be considered a useful indication of the sensitivities of appropriate untested species. Fewer data concerning toxicity to aquatic plants are usually available because procedures for conducting tests with plants and interpreting the results of such tests are not as well developed.

B. To derive a Great Lakes Tier I criterion for aquatic organisms and their uses, the following must be available:

1. Results of acceptable acute (or chronic) tests (see section IV or VI of this appendix) with at least one species of freshwater animal in at least eight different families such that all of the following are included:

a. The family Salmonidae in the class Osteichthyes;

b. One other family (preferably a commercially or recreationally important, warmwater species) in the class Osteichthyes (e.g., bluegill, channel catfish);

c. A third family in the phylum Chordata (e.g., fish, amphibian);

d. A planktonic crustacean (e.g., a cladoceran, copepod);

e. A benthic crustacean (e.g., ostracod, isopod, amphipod, crayfish);

f. An insect (e.g., mayfly, dragonfly, damselfly, stonefly, caddisfly, mosquito, midge);

g. A family in a phylum other than Arthropoda or Chordata (e.g., Rotifera, Annelida, Mollusca);

h. A family in any order of insect or any phylum not already represented.

2. Acute-chronic ratios (see section VI of this appendix) with at least one species of aquatic animal in at least three different families provided that of the three species:

a. At least one is a fish;

b. At least one is an invertebrate; and

c. At least one species is an acutely sensitive freshwater species (the other two may be saltwater species).

3. Results of at least one acceptable test with a freshwater algae or vascular plant is desirable but not required for criterion derivation (see section VIII of this appendix). If plants are among the aquatic organisms most sensitive to the material, results of a test with a plant in another phylum (division) should also be available.

C. If all required data are available, a numerical criterion can usually be derived except in special cases. For example, derivation of a chronic criterion might not be possible if the available ACRs vary by more than a factor of ten with no apparent pattern. Also, if a criterion is to be related to a water quality characteristic (see sections V and VII of this appendix), more data will be required.

D. Confidence in a criterion usually increases as the amount of available pertinent information increases. Thus, additional data are usually desirable.

IV. Final Acute Value

A. Appropriate measures of the acute (short-term) toxicity of the material to a variety of species of aquatic animals are used to calculate the Final Acute Value (FAV). The calculated Final Acute Value is a calculated estimate of the concentration of a test material such that 95 percent of the genera (with which acceptable acute toxicity tests have been conducted on the material) have higher Genus Mean Acute Values (GMAVs). An acute test is a comparative study in which organisms, that are subjected to different treatments, are observed for a short period usually not constituting a substantial portion of their life span. However, in some cases, the Species Mean Acute Value (SMAV) of a commercially or recreationally important species of the Great Lakes System is lower than the calculated FAV, then the SMAV replaces the calculated FAV in order to provide protection for that important species.

B. Acute toxicity tests shall be conducted using acceptable procedures. For good examples of acceptable procedures see American Society for Testing and Materials (ASTM) Standard E 729, Guide for Conducting Acute Toxicity Tests with Fishes, Macroinvertebrates, and Amphibians.

C. Except for results with saltwater annelids and mysids, results of acute tests during which the test organisms were fed should not be used, unless data indicate that the food did not affect the toxicity of the test material. (NOTE: If the minimum acute-chronic ratio data requirements (as described in section III.B.2 of this appendix) are not met with freshwater data alone, saltwater data may be used.)

D. Results of acute tests conducted in unusual dilution water, e.g., dilution water in which total organic carbon or particulate matter exceeded five mg/L, should not be used, unless a relationship is developed between acute toxicity and organic carbon or particulate matter, or unless data show that organic carbon or particulate matter, etc., do not affect toxicity.

E. Acute values must be based upon endpoints which reflect the total severe adverse impact of the test material on the organisms used in the test. Therefore, only the following kinds of data on acute toxicity to aquatic animals shall be used:

1. Tests with daphnids and other cladocerans must be started with organisms less than 24 hours old and tests with midges must be started with second or third instar larvae. It is preferred that the results should be the 48-hour EC50 based on the total percentage of organisms killed and immobilized. If such an EC50 is not available for a test, the 48-hour LC50 should be used in place of the desired 48-hour EC50. An EC50 or LC50 of longer than 48 hours can be used as long as the animals were not fed and the control animals were acceptable at the end of the test. An EC50 is a statistically or graphically estimated concentration that is expected to cause one or more specified effects in 50% of a group of organisms under specified conditions. An LC50 is a statistically or graphically estimated concentration that is expected to be lethal to 50% of a group of organisms under specified conditions.

2. It is preferred that the results of a test with embryos and larvae of barnacles, bivalve molluscs (clams, mussels, oysters and scallops), sea urchins, lobsters, crabs, shrimp and abalones be the 96-hour EC50 based on the percentage of organisms with incompletely developed shells plus the percentage of organisms killed. If such an EC50 is not available from a test, of the values that are available from the test, the lowest of the following should be used in place of the desired 96-hour EC50: 48- to 96-hour EC50s based on percentage of organisms with incompletely developed shells plus percentage of organisms killed, 48- to 96-hour EC50s based upon percentage of organisms with incompletely developed shells, and 48-hour to 96-hour LC50s. (NOTE: If the minimum acute-chronic ratio data requirements (as described in section III.B.2 of this appendix) are not met

with freshwater data alone, saltwater data may be used.)

3. It is preferred that the result of tests with all other aquatic animal species and older life stages of barnacles, bivalve molluscs (clams, mussels, oysters and scallops), sea urchins, lobsters, crabs, shrimp and abalones be the 96-hour EC50 based on percentage of organisms exhibiting loss of equilibrium plus percentage of organisms immobilized plus percentage of organisms killed. If such an EC50 is not available from a test, of the values that are available from a test the lower of the following should be used in place of the desired 96-hour EC50: the 96-hour EC50 based on percentage of organisms exhibiting loss of equilibrium plus percentage of organisms immobilized and the 96-hour LC50.

4. Tests whose results take into account the number of young produced, such as most tests with protozoans, are not considered acute tests, even if the duration was 96 hours or less.

5. If the tests were conducted properly, acute values reported as "greater than" values and those which are above the solubility of the test material should be used, because rejection of such acute values would bias the Final Acute Value by eliminating acute values for resistant species.

F. If the acute toxicity of the material to aquatic animals has been shown to be related to a water quality characteristic such as hardness or particulate matter for freshwater animals, refer to section V of this appendix.

G. The agreement of the data within and between species must be considered. Acute values that appear to be questionable in comparison with other acute and chronic data for the same species and for other species in the same genus must not be used. For example, if the acute values available for a species or genus differ by more than a factor of 10, rejection of some or all of the values would be appropriate, absent countervailing circumstances.

H. If the available data indicate that one or more life stages are at least a factor of two more resistant than one or more other life stages of the same species, the data for the more resistant life stages must not be used in the calculation of the SMAV because a species cannot be considered protected from acute toxicity if all of the life stages are not protected.

I. For each species for which at least one acute value is available, the SMAV shall be calculated as the geometric mean of the results of all acceptable flow-through acute toxicity tests in which the concentrations of test material were measured with the most sensitive tested life stage of the species. For a species for which no such result is available, the SMAV shall be calculated as the geometric mean of all acceptable acute toxicity tests with the most sensitive tested life stage, i.e., results of flow-through tests in which the concentrations were not measured and results of static and renewal tests based on initial concentrations (nominal concentrations are acceptable for most test materials if measured concentrations are not available) of test material. A renewal test is a test with aquatic organisms in which either the test solution in a test chamber is removed and replaced at least once during the test or the test organisms are transferred into a new test solution of the same composition at least once during the test. A static test is a test with aquatic organisms in which the solution and organisms that are in a test chamber at the beginning of the test remain in the chamber until the end of the test, except for removal of dead test organisms.

NOTE 1: Data reported by original investigators must not be rounded off. Results of all intermediate calculations must not be rounded off to fewer than four significant digits.

NOTE 2: The geometric mean of N numbers is the Nth root of the product of the N numbers. Alternatively, the geometric mean can be calculated by adding the logarithms of the N numbers, dividing the sum by N, and taking the antilog of the quotient. The geometric mean of two numbers is the square root of the product of the two numbers, and the geometric mean of one number is that number. Either natural (base e) or common (base 10) logarithms can be used to calculate

geometric means as long as they are used consistently within each set of data, i.e., the antilog used must match the logarithms used.

NOTE 3: Geometric means, rather than arithmetic means, are used here because the distributions of sensitivities of individual organisms in toxicity tests on most materials and the distributions of sensitivities of species within a genus are more likely to be lognormal than normal. Similarly, geometric means are used for ACRs because quotients are likely to be closer to lognormal than normal distributions. In addition, division of the geometric mean of a set of numerators by the geometric mean of the set of denominators will result in the geometric mean of the set of corresponding quotients.

J. For each genus for which one or more SMAVs are available, the GMAV shall be calculated as the geometric mean of the SMAVs available for the genus.

K. Order the GMAVs from high to low.

L. Assign ranks, R, to the GMAVs from "1" for the lowest to "N" for the highest. If two or more GMAVs are identical, assign them successive ranks.

M. Calculate the cumulative probability, P, for each GMAV as R/(N+1).

N. Select the four GMAVs which have cumulative probabilities closest to 0.05 (if there are fewer than 59 GMAVs, these will always be the four lowest GMAVs).

O. Using the four selected GMAVs, and Ps, calculate

$$ S^2 = \frac{\sum\left((\ln \text{GMAV})^2\right) - \dfrac{\left(\sum(\ln \text{GMAV})\right)^2}{4}}{\sum(P) - \dfrac{\left(\sum\left(\sqrt{P}\right)\right)^2}{4}} $$

$$ L = \frac{\sum(\ln \text{GMAV}) - S\left(\sum\left(\sqrt{P}\right)\right)}{4} $$

$$ A = S\left(\sqrt{0.05}\right) + L $$

$$ \text{FAV} = e^A $$

NOTE: Natural logarithms (logarithms to base e, denoted as ln) are used herein merely

because they are easier to use on some hand calculators and computers than common (base 10) logarithms. Consistent use of either will produce the same result.

P. If for a commercially or recreationally important species of the Great Lakes System the geometric mean of the acute values from flow-through tests in which the concentrations of test material were measured is lower than the calculated Final Acute Value (FAV), then that geometric mean must be used as the FAV instead of the calculated FAV.

Q. See section VI of this appendix.

V. Final Acute Equation

A. When enough data are available to show that acute toxicity to two or more species is similarly related to a water quality characteristic, the relationship shall be taken into account as described in sections V.B through V.G of this appendix or using analysis of covariance. The two methods are equivalent and produce identical results. The manual method described below provides an understanding of this application of covariance analysis, but computerized versions of covariance analysis are much more convenient for analyzing large data sets. If two or more factors affect toxicity, multiple regression analysis shall be used.

B. For each species for which comparable acute toxicity values are available at two or more different values of the water quality characteristic, perform a least squares regression of the acute toxicity values on the corresponding values of the water quality characteristic to obtain the slope and its 95 percent confidence limits for each species.

NOTE: Because the best documented relationship is that between hardness and acute toxicity of metals in fresh water and a log-log relationship fits these data, geometric means and natural logarithms of both toxicity and water quality are used in the rest of this section. For relationships based on other water quality characteristics, such as Ph, temperature, no transformation or a different transformation might fit the data better, and appropriate changes will be necessary throughout this section.

C. Decide whether the data for each species are relevant, taking into account the range and number of the tested values of the water quality characteristic and the degree of agreement within and between species. For example, a slope based on six data points might be of limited value if it is based only on data for a very narrow range of values of the water quality characteristic. A slope based on only two data points, however, might be useful if it is consistent with other information and if the two points cover a broad enough range of the water quality characteristic. In addition, acute values that appear to be questionable in comparison with other acute and chronic data available for the same species and for other species in the same genus should not be used. For ex-

ample, if after adjustment for the water quality characteristic, the acute values available for a species or genus differ by more than a factor of 10, rejection of some or all of the values would be appropriate, absent countervailing justification. If useful slopes are not available for at least one fish and one invertebrate or if the available slopes are too dissimilar or if too few data are available to adequately define the relationship between acute toxicity and the water quality characteristic, return to section IV.G of this appendix, using the results of tests conducted under conditions and in waters similar to those commonly used for toxicity tests with the species.

D. For each species, calculate the geometric mean of the available acute values and then divide each of the acute values for the species by the geometric mean for the species. This normalizes the acute values so that the geometric mean of the normalized values for each species individually and for any combination of species is 1.0.

E. Similarly normalize the values of the water quality characteristic for each species individually using the same procedure as above.

F. Individually for each species perform a least squares regression of the normalized acute values of the water quality characteristic. The resulting slopes and 95 percent confidence limits will be identical to those obtained in section V.B. of this appendix. If, however, the data are actually plotted, the line of best fit for each individual species will go through the point 1,1 in the center of the graph.

G. Treat all of the normalized data as if they were all for the same species and perform a least squares regression of all of the normalized acute values on the corresponding normalized values of the water quality characteristic to obtain the pooled acute slope, V, and its 95 percent confidence limits. If all of the normalized data are actually plotted, the line of best fit will go through the point 1,1 in the center of the graph.

H. For each species calculate the geometric mean, W, of the acute toxicity values and the geometric mean, X, of the values of the water quality characteristic. (These were calculated in sections V.D and V.E of this appendix).

I. For each species, calculate the logarithm, Y, of the SMAV at a selected value, Z, of the water quality characteristic using the equation:

$$Y = \ln W - V(\ln X - \ln Z)$$

J. For each species calculate the SMAV at X using the equation:

$$SMAV = e^Y$$

NOTE: Alternatively, the SMAVs at Z can be obtained by skipping step H above, using the equations in steps I and J to adjust each acute value individually to Z, and then calculating the geometric mean of the adjusted values for each species individually. This alternative procedure allows an examination of the range of the adjusted acute values for each species.

K. Obtain the FAV at Z by using the procedure described in sections IV.J through IV.O of this appendix.

L. If, for a commercially or recreationally important species of the Great Lakes System the geometric mean of the acute values at Z from flow-through tests in which the concentrations of the test material were measured is lower than the FAV at Z, then the geometric mean must be used as the FAV instead of the FAV.

M. The Final Acute Equation is written as:

$$FAV = e^{(V[\ln(waterqualitycharacteristic)] = A - V[\ln Z])},$$

where:

V=pooled acute slope, and A=ln(FAV at Z).

Because V, A, and Z are known, the FAV can be calculated for any selected value of the water quality characteristic.

VI. Final Chronic Value

A. Depending on the data that are available concerning chronic toxicity to aquatic animals, the Final Chronic Value (FCV) can be calculated in the same manner as the FAV or by dividing the FAV by the Final Acute-Chronic Ratio (FACR). In some cases, it might not be possible to calculate a FCV. The FCV is (a) a calculated estimate of the concentration of a test material such that 95 percent of the genera (with which acceptable chronic toxicity tests have been conducted on the material) have higher GMCVs, or (b) the quotient of an FAV divided by an appropriate ACR, or (c) the SMCV of an important and/or critical species, if the SMCV is lower than the calculated estimate or the quotient, whichever is applicable.

NOTE: As the name implies, the ACR is a way of relating acute and chronic toxicities.

B. Chronic values shall be based on results of flow-through (except renewal is acceptable for daphnids) chronic tests in which the concentrations of test material in the test solutions were properly measured at appropriate times during the test. A chronic test is a comparative study in which organisms, that are subjected to different treatments, are observed for a long period or a substantial portion of their life span.

C. Results of chronic tests in which survival, growth, or reproduction in the control treatment was unacceptably low shall not be used. The limits of acceptability will depend on the species.

D. Results of chronic tests conducted in unusual dilution water, e.g., dilution water in which total organic carbon or particulate matter exceeded five mg/L, should not be used, unless a relationship is developed between chronic toxicity and organic carbon or particulate matter, or unless data show that organic carbon, particulate matter, etc., do not affect toxicity.

E. Chronic values must be based on endpoints and lengths of exposure appropriate to the species. Therefore, only results of the following kinds of chronic toxicity tests shall be used:

1. Life-cycle toxicity tests consisting of exposures of each of two or more groups of individuals of a species to a different concentration of the test material throughout a life cycle. To ensure that all life stages and life processes are exposed, tests with fish should begin with embryos or newly hatched young less than 48 hours old, continue through maturation and reproduction, and should end not less than 24 days (90 days for salmonids) after the hatching of the next generation. Tests with daphnids should begin with young less than 24 hours old and last for not less than 21 days, and for ceriodaphnids not less than seven days. For good examples of acceptable procedures see American Society for Testing and Materials (ASTM) Standard E 1193 Guide for conducting renewal life-cycle toxicity tests with *Daphnia magna* and ASTM Standard E 1295 Guide for conducting three-brood, renewal toxicity tests with *Ceriodaphnia dubia*. Tests with mysids should begin with young less than 24 hours old and continue until seven days past the median time of first brood release in the controls. For fish, data should be obtained and analyzed on survival and growth of adults and young, maturation of males and females, eggs spawned per female, embryo viability (salmonids only), and hatchability. For daphnids, data should be obtained and analyzed on survival and young per female. For mysids, data should be obtained and analyzed on survival, growth, and young per female.

2. Partial life-cycle toxicity tests consist of exposures of each of two more groups of individuals of a species of fish to a different concentration of the test material through most portions of a life cycle. Partial life-cycle tests are allowed with fish species that require more than a year to reach sexual maturity, so that all major life stages can be exposed to the test material in less than 15 months. A life-cycle test is a comparative study in which organisms, that are subjected to different treatments, are observed at least from a life stage in one generation to the

same life-stage in the next generation. Exposure to the test material should begin with immature juveniles at least two months prior to active gonad development, continue through maturation and reproduction, and end not less than 24 days (90 days for salmonids) after the hatching of the next generation. Data should be obtained and analyzed on survival and growth of adults and young, maturation of males and females, eggs spawned per female, embryo viability (salmonids only), and hatchability.

3. Early life-stage toxicity tests consisting of 28- to 32-day (60 days post hatch for salmonids) exposures of the early life stages of a species of fish from shortly after fertilization through embryonic, larval, and early juvenile development. Data should be obtained and analyzed on survival and growth.

NOTE: Results of an early life-stage test are used as predictions of results of life-cycle and partial life-cycle tests with the same species. Therefore, when results of a life-cycle or partial life-cycle test are available, results of an early life-stage test with the same species should not be used. Also, results of early life-stage tests in which the incidence of mortalities or abnormalities increased substantially near the end of the test shall not be used because the results of such tests are possibly not good predictions of comparable life-cycle or partial life-cycle tests.

F. A chronic value may be obtained by calculating the geometric mean of the lower and upper chronic limits from a chronic test or by analyzing chronic data using regression analysis.

1. A lower chronic limit is the highest tested concentration:

a. In an acceptable chronic test;

b. Which did not cause an unacceptable amount of adverse effect on any of the specified biological measurements; and

c. Below which no tested concentration caused an unacceptable effect.

2. An upper chronic limit is the lowest tested concentration:

a. In an acceptable chronic test;

b. Which did cause an unacceptable amount of adverse effect on one or more of the specified biological measurements; and,

c. Above which all tested concentrations also caused such an effect.

NOTE: Because various authors have used a variety of terms and definitions to interpret and report results of chronic tests, reported results should be reviewed carefully. The amount of effect that is considered unacceptable is often based on a statistical hypothesis test, but might also be defined in terms of a specified percent reduction from the controls. A small percent reduction (e.g., three percent) might be considered acceptable even if it is statistically significantly different from the control, whereas a large percent reduction (e.g., 30 percent) might be considered unacceptable even if it is not statistically significant.

G. If the chronic toxicity of the material to aquatic animals has been shown to be related to a water quality characteristic such as hardness or particulate matter for freshwater animals, refer to section VII of this appendix.

H. If chronic values are available for species in eight families as described in section III.B.1 of this appendix, a SMCV shall be calculated for each species for which at least one chronic value is available by calculating the geometric mean of the results of all acceptable life-cycle and partial life-cycle toxicity tests with the species; for a species of fish for which no such result is available, the SMCV is the geometric mean of all acceptable early life-stage tests. Appropriate GMCVs shall also be calculated. A GMCV is the geometric mean of the SMCVs for the genus. The FCV shall be obtained using the procedure described in sections IV.J through IV.O of this appendix, substituting SMCV and GMCV for SMAV and GMAV respectively. See section VI.M of this appendix.

NOTE: Section VI.I through VI.L are for use when chronic values are not available for species in eight taxonomic families as described in section III.B.1 of this appendix.

I. For each chronic value for which at least one corresponding appropriate acute value is available, calculate an ACR, using for the numerator the geometric mean of the results of all acceptable flow-through (except static is acceptable for daphnids and midges) acute tests in the same dilution water in which the concentrations are measured. For fish, the acute test(s) should be conducted with juveniles. The acute test(s) should be part of the same study as the chronic test. If acute tests were not conducted as part of the same study, but were conducted as part of a different study in the same laboratory and dilution water, then they may be used. If no such acute tests are available, results of acute tests conducted in the same dilution water in a different laboratory may be used. If no such acute tests are available, an ACR shall not be calculated.

J. For each species, calculate the SMACR as the geometric mean of all ACRs available for that species. If the minimum ACR data requirements (as described in section III.B.2 of this appendix) are not met with freshwater data alone, saltwater data may be used along with the freshwater data.

K. For some materials, the ACR seems to be the same for all species, but for other materials the ratio seems to increase or decrease as the SMAV increases. Thus the FACR can be obtained in three ways, depending on the data available:

1. If the species mean ACR seems to increase or decrease as the SMAVs increase,

the FACR shall be calculated as the geometric mean of the ACRs for species whose SMAVs are close to the FAV.

2. If no major trend is apparent and the ACRs for all species are within a factor of ten, the FACR shall be calculated as the geometric mean of all of the SMACRs.

3. If the most appropriate SMACRs are less than 2.0, and especially if they are less than 1.0, acclimation has probably occurred during the chronic test. In this situation, because continuous exposure and acclimation cannot be assured to provide adequate protection in field situations, the FACR should be assumed to be two, so that the FCV is equal to the Criterion Maximum Concentration (CMC). (See section X.B of this appendix.)

If the available SMACRs do not fit one of these cases, a FACR may not be obtained and a Tier I FCV probably cannot be calculated.

L. Calculate the FCV by dividing the FAV by the FACR.

FCV=FAV÷FACR

If there is a Final Acute Equation rather than a FAV, see also section V of this appendix.

M. If the SMCV of a commercially or recreationally important species of the Great Lakes System is lower than the calculated FCV, then that SMCV must be used as the FCV instead of the calculated FCV.

N. See section VIII of this appendix.

VII. Final Chronic Equation

A. A Final Chronic Equation can be derived in two ways. The procedure described in section VII.A of this appendix will result in the chronic slope being the same as the acute slope. The procedure described in sections VII.B through N of this appendix will usually result in the chronic slope being different from the acute slope.

1. If ACRs are available for enough species at enough values of the water quality characteristic to indicate that the ACR appears to be the same for all species and appears to be independent of the water quality characteristic, calculate the FACR as the geometric mean of the available SMACRs.

2. Calculate the FCV at the selected value Z of the water quality characteristic by dividing the FAV at Z (see section V.M of this appendix) by the FACR.

3. Use V=pooled acute slope (see section V.M of this appendix), and

L=pooled chronic slope.

4. See section VII.M of this appendix.

B. When enough data are available to show that chronic toxicity to at least one species is related to a water quality characteristic, the relationship should be taken into account as described in sections C through G below or using analysis of covariance. The two methods are equivalent and produce identical results. The manual method described below provides an understanding of this application of covariance analysis, but computerized versions of covariance analysis are much more convenient for analyzing large data sets. If two or more factors affect toxicity, multiple regression analysis shall be used.

C. For each species for which comparable chronic toxicity values are available at two or more different values of the water quality characteristic, perform a least squares regression of the chronic toxicity values on the corresponding values of the water quality characteristic to obtain the slope and its 95 percent confidence limits for each species.

NOTE: Because the best documented relationship is that between hardness and acute toxicity of metals in fresh water and a log-log relationship fits these data, geometric means and natural logarithms of both toxicity and water quality are used in the rest of this section. For relationships based on other water quality characteristics, such as Ph, temperature, no transformation or a different transformation might fit the data better, and appropriate changes will be necessary throughout this section. It is probably preferable, but not necessary, to use the same transformation that was used with the acute values in section V of this appendix.

D. Decide whether the data for each species are relevant, taking into account the range and number of the tested values of the water quality characteristic and the degree of agreement within and between species. For example, a slope based on six data points might be of limited value if it is based only on data for a very narrow range of values of the water quality characteristic. A slope based on only two data points, however, might be more useful if it is consistent with other information and if the two points cover a broad range of the water quality characteristic. In addition, chronic values that appear to be questionable in comparison with other acute and chronic data available for the same species and for other species in the same genus in most cases should not be used. For example, if after adjustment for the water quality characteristic, the chronic values available for a species or genus differ by more than a factor of 10, rejection of some or all of the values is, in most cases, absent countervailing circumstances, appropriate. If a useful chronic slope is not available for at least one species or if the available slopes are too dissimilar or if too few data are available to adequately define the relationship between chronic toxicity and the water quality characteristic, it might be appropriate to assume that the chronic slope is the same as the acute slope, which is equivalent to assuming that the ACR is independent of the water quality characteristic. Alternatively, return to section VI.H of this appendix, using the results of tests conducted under conditions and in waters similar to those commonly used for toxicity tests with the species.

E. Individually for each species, calculate the geometric mean of the available chronic values and then divide each chronic value for a species by the mean for the species. This normalizes the chronic values so that the geometric mean of the normalized values for each species individually, and for any combination of species, is 1.0.

F. Similarly, normalize the values of the water quality characteristic for each species individually.

G. Individually for each species, perform a least squares regression of the normalized chronic toxicity values on the corresponding normalized values of the water quality characteristic. The resulting slopes and the 95 percent confidence limits will be identical to those obtained in section VII.B of this appendix. Now, however, if the data are actually plotted, the line of best fit for each individual species will go through the point 1,1 in the center of the graph.

H. Treat all of the normalized data as if they were all the same species and perform a least squares regression of all of the normalized chronic values on the corresponding normalized values of the water quality characteristic to obtain the pooled chronic slope, L, and its 95 percent confidence limits.

If all normalized data are actually plotted, the line of best fit will go through the point 1,1 in the center of the graph.

I. For each species, calculate the geometric mean, M, of the toxicity values and the geometric mean, P, of the values of the water quality characteristic. (These are calculated in sections VII.E and F of this appendix.)

J. For each species, calculate the logarithm, Q, of the SMCV at a selected value, Z, of the water quality characteristic using the equation:

$$Q = \ln M - L(\ln P - \ln Z)$$

NOTE: Although it is not necessary, it is recommended that the same value of the water quality characteristic be used here as was used in section V of this appendix.

K. For each species, calculate a SMCV at Z using the equation:

$$SMCV = e^Q$$

NOTE: Alternatively, the SMCV at Z can be obtained by skipping section VII.J of this appendix, using the equations in sections VII.J and K of this appendix to adjust each chronic value individually to Z, and then calculating the geometric means of the adjusted values for each species individually. This alternative procedure allows an examination of the range of the adjusted chronic values for each species.

L. Obtain the FCV at Z by using the procedure described in sections IV.J through O of this appendix.

M. If the SMCV at Z of a commercially or recreationally important species of the Great Lakes System is lower than the calculated FCV at Z, then that SMCV shall be used as the FCV at Z instead of the calculated FCV.

N. The Final Chronic Equation is written as:

$$FCV = e^{(L\&[\ln(\text{waterqualitycharacteristic})] = \ln S - L[\ln Z])}$$

Where:

L=pooled chronic slope and S = FCV at Z.

Because L, S, and Z are known, the FCV can be calculated for any selected value of the water quality characteristic.

VIII. Final Plant Value

A. A Final Plant Value (FPV) is the lowest plant value that was obtained with an important aquatic plant species in an acceptable toxicity test for which the concentrations of the test material were measured and the adverse effect was biologically important. Appropriate measures of the toxicity of the material to aquatic plants are used to compare the relative sensitivities of aquatic plants and animals. Although procedures for conducting and interpreting the results of toxicity tests with plants are not well-developed, results of tests with plants usually indicate that criteria which adequately protect aquatic animals and their uses will, in most cases, also protect aquatic plants and their uses.

B. A plant value is the result of a 96-hour test conducted with an alga or a chronic test conducted with an aquatic vascular plant.

NOTE: A test of the toxicity of a metal to a plant shall not be used if the medium contained an excessive amount of a complexing agent, such as EDTA, that might affect the toxicity of the metal. Concentrations of EDTA above 200 μg/L should be considered excessive.

C. The FPV shall be obtained by selecting the lowest result from a test with an important aquatic plant species in which the concentrations of test material are measured and the endpoint is biologically important.

IX. Other Data

Pertinent information that could not be used in earlier sections might be available concerning adverse effects on aquatic organisms. The most important of these are data on cumulative and delayed toxicity, reduction in survival, growth, or reproduction, or any other adverse effect that has been shown to be biologically important. Delayed toxicity is an adverse effect to an organism that results from, and occurs after the end of, its exposure to one or more test materials. Especially important are data for species for which no other data are available. Data from behavioral, biochemical, physiological, microcosm, and field studies might also be available. Data might be available from tests conducted in unusual dilution water (see sections IV.D and VI.D of this appendix), from chronic tests in which the concentrations were not measured (see section VI.B of this appendix), from tests with previously exposed organisms (see section II.F.3 of this appendix), and from tests on formulated mixtures or emulsifiable concentrates (see section II.D of this appendix). Such data might affect a criterion if the data were obtained with an important species, the test concentrations were measured, and the endpoint was biologically important.

X. Criterion

A. A criterion consists of two concentrations: the CMC and the Criterion Continuous Concentration (CCC).

B. The CMC is equal to one-half the FAV. The CMC is an estimate of the highest concentration of a material in the water column to which an aquatic community can be exposed briefly without resulting in an unacceptable effect.

C. The CCC is equal to the lowest of the FCV or the FPV (if available) unless other data (see section IX of this appendix) show that a lower value should be used. The CCC is an estimate of the highest concentration of a material in the water column to which an aquatic community can be exposed indefinitely without resulting in an unacceptable effect. If toxicity is related to a water quality characteristic, the CCC is obtained from the Final Chronic Equation or FPV (if available) that results in the lowest concentrations in the usual range of the water quality characteristic, unless other data (see section IX) show that a lower value should be used.

D. Round both the CMC and the CCC to two significant digits.

E. The criterion is stated as:

The procedures described in the Tier I methodology indicate that, except possibly where a commercially or recreationally important species is very sensitive, aquatic organisms should not be affected unacceptably if the four-day average concentration of (1) does not exceed (2) μg/L more than once every three years on the average and if the one-hour average concentration does not exceed (3) μg/L more than once every three years on the average.

Where:

(1) = insert name of material
(2) = insert the CCC
(3) = insert the CMC

If the CMC averaging period of one hour or the CCC averaging period of four days is inappropriate for the pollutant, or if the once-in-three-year allowable excursion frequency is inappropriate for the pollutant or for the sites to which a criterion is applied, then the State may specify alternative averaging periods or frequencies. The choice of an alternative averaging period or frequency shall be justified by a scientifically defensible analysis demonstrating that the alternative values will protect the aquatic life uses of the water. Appropriate laboratory data and/or well-designed field biological surveys shall be submitted to EPA as justification for differing averaging periods and/or frequencies of exceedance.

XI. Final Review

A. The derivation of the criterion should be carefully reviewed by rechecking each step of the Guidance in this part. Items that should be especially checked are:

1. If unpublished data are used, are they well documented?

2. Are all required data available?

3. Is the range of acute values for any species greater than a factor of 10?

4. Is the range of SMAVs for any genus greater than a factor of 10?

5. Is there more than a factor of 10 difference between the four lowest GMAVs?

6. Are any of the lowest GMAVs questionable?

7. Is the FAV reasonable in comparison with the SMAVs and GMAVs?

8. For any commercially or recreationally important species of the Great Lakes System, is the geometric mean of the acute values from flow-through tests in which the concentrations of test material were measured lower than the FAV?

9. Are any of the chronic values used questionable?

10. Are any chronic values available for acutely sensitive species?

11. Is the range of acute-chronic ratios greater than a factor of 10?

12. Is the FCV reasonable in comparison with the available acute and chronic data?

13. Is the measured or predicted chronic value for any commercially or recreationally important species of the Great Lakes System below the FCV?

14. Are any of the other data important?

15. Do any data look like they might be outliers?

16. Are there any deviations from the Guidance in this part? Are they acceptable?

B. On the basis of all available pertinent laboratory and field information, determine if the criterion is consistent with sound scientific evidence. If it is not, another criterion, either higher or lower, shall be derived consistent with the Guidance in this part.

METHODOLOGY FOR DERIVING AQUATIC LIFE
VALUES: TIER II

XII. Secondary Acute Value

If all eight minimum data requirements for calculating an FAV using Tier I are not met, a Secondary Acute Value (SAV) for the waters of the Great Lakes System shall be calculated for a chemical as follows:

To calculate a SAV, the lowest GMAV in the database is divided by the Secondary Acute Factor (SAF) (Table A–1 of this appendix) corresponding to the number of satisfied minimum data requirements listed in the Tier I methodology (section III.B.1 of this appendix). (Requirements for definitions, data collection and data review, contained in sections I, II, and IV shall be applied to calculation of a SAV.) If all eight minimum data requirements are satisfied, a Tier I criterion calculation may be possible. In order to calculate a SAV, the database must contain, at a minimum, a genus mean acute value (GMAV) for one of the following three genera in the family Daphnidae—*Ceriodaphnia sp., Daphnia sp.*, or *Simocephalus sp.*

If appropriate, the SAV shall be made a function of a water quality characteristic in a manner similar to that described in Tier I.

XIII. Secondary Acute-Chronic Ratio

If three or more experimentally determined ACRs, meeting the data collection and review requirements of Section VI of this appendix, are available for the chemical, determine the FACR using the procedure described in Section VI. If fewer than three acceptable experimentally determined ACRs are available, use enough assumed ACRs of 18 so that the total number of ACRs equals three. Calculate the Secondary Acute-Chronic Ratio (SACR) as the geometric mean of the three ACRs. Thus, if no experimentally determined ACRs are available, the SACR is 18.

XIV. Secondary Chronic Value

Calculate the Secondary Chronic Value (SCV) using one of the following:

A. $SCV = \dfrac{FAV}{SACR}$ (use FAV from Tier I)

B. $SCV = \dfrac{SAV}{FACR}$

C. $SCV = \dfrac{SAV}{SACR}$

If appropriate, the SCV will be made a function of a water quality characteristic in a manner similar to that described in Tier I.

XV. Commercially or Recreationally Important Species

If for a commercially or recreationally important species of the Great Lakes System the geometric mean of the acute values or chronic values from flow-through tests in which the concentrations of the test materials were measured is lower than the calculated SAV or SCV, then that geometric mean must be used as the SAV or SCV instead of the calculated SAV or SCV.

XVI. Tier II Value

A. A Tier II value shall consist of two concentrations: the Secondary Maximum Concentration (SMC) and the Secondary Continuous Concentration (SCC).

B. The SMC is equal to one-half of the SAV.

C. The SCC is equal to the lowest of the SCV or the Final Plant Value, if available, unless other data (see section IX of this appendix) show that a lower value should be used.

If toxicity is related to a water quality characteristic, the SCC is obtained from the Secondary Chronic Equation or FPV, if available, that results in the lowest concentrations in the usual range of the water quality characteristic, unless other data (See section IX of this appendix) show that a lower value should be used.

D. Round both the SMC and the SCC to two significant digits.

E. The Tier II value is stated as:

The procedures described in the Tier II methodology indicate that, except possibly where a locally important species is very sensitive, aquatic organisms should not be affected unacceptably if the four-day average concentration of (1) does not exceed (2) µg/L more than once every three years on the average and if the one-hour average concentration does not exceed (3) µg/L more than once every three years on the average.

Where:

(1) = insert name of material
(2) = insert the SCC
(3) = insert the SMC

As discussed above, States and Tribes have the discretion to specify alternative averaging periods or frequencies (see section X.E. of this appendix).

XVII. Appropriate Modifications

On the basis of all available pertinent laboratory and field information, determine if the Tier II value is consistent with sound scientific evidence. If it is not, another value, either higher or lower, shall be derived consistent with the Guidance in this part.

TABLE A-1—SECONDARY ACUTE FACTORS

Number of minimum data requirements satisfied	Adjustment factor
1	21.9
2	13.0
3	8.0
4	7.0
5	6.1
6	5.2
7	4.3

APPENDIX B TO PART 132—GREAT LAKES WATER QUALITY INITIATIVE

METHODOLOGY FOR DERIVING BIOACCUMULATION FACTORS

Great Lakes States and Tribes shall adopt provisions consistent with (as protective as) this appendix.

I. Introduction

A. The purpose of this methodology is to describe procedures for deriving bioaccumulation factors (BAFs) to be used in the calculation of Great Lakes Water Quality Guidance (Guidance) human health Tier I criteria and Tier II values and wildlife Tier I criteria. A subset of the human health BAFs are also used to identify the chemicals that are considered bioaccumulative chemicals of concern (BCCs).

B. Bioaccumulation reflects uptake of a substance by aquatic organisms exposed to the substance through all routes (i.e., ambient water and food), as would occur in nature. Bioconcentration reflects uptake of a substance by aquatic organisms exposed to the substance only through the ambient water. Both BAFs and bioconcentration factors (BCFs) are proportionality constants that describe the relationship between the concentration of a substance in aquatic organisms and its concentration in the ambient water. For the Guidance in this part, BAFs, rather than BCFs, are used to calculate Tier I criteria for human health and wildlife and Tier II values for human health because they better account for the total exposure of aquatic organisms to chemicals.

C. For organic chemicals, baseline BAFs can be derived using four methods. Measured baseline BAFs are derived from field-measured BAFs; predicted baseline BAFs are derived using biota-sediment accumulation factors (BSAFs) or are derived by multiplying a laboratory-measured or predicted BCF by a food-chain multiplier (FCM). The lipid content of the aquatic organisms is used to account for partitioning of organic chemicals within organisms so that data from different tissues and species can be in-

tegrated. In addition, the baseline BAF is based on the concentration of freely dissolved organic chemicals in the ambient water to facilitate extrapolation from one water to another.

D. For inorganic chemicals, baseline BAFs can be derived using two of the four methods. Baseline BAFs are derived using either field-measured BAFs or by multiplying laboratory-measured BCFs by a FCM. For inorganic chemicals, BAFs are assumed to equal BCFs (i.e., the FCM is 1.0), unless chemical-specific biomagnification data support using a FCM other than 1.0.

E. Because both humans and wildlife consume fish from both trophic levels 3 and 4, two baseline BAFs are needed to calculate either a human health criterion or value or a wildlife criterion for a chemical. When appropriate, ingestion through consumption of invertebrates, plants, mammals, and birds in the diet of wildlife species to be protected may be taken into account.

II. Definitions

Baseline BAF. For organic chemicals, a BAF that is based on the concentration of freely dissolved chemical in the ambient water and takes into account the partitioning of the chemical within the organism; for inorganic chemicals, a BAF that is based on the wet weight of the tissue.

Baseline BCF. For organic chemicals, a BCF that is based on the concentration of freely dissolved chemical in the ambient water and takes into account the partitioning of the chemical within the organism; for inorganic chemicals, a BCF that is based on the wet weight of the tissue.

Bioaccumulation. The net accumulation of a substance by an organism as a result of uptake from all environmental sources.

Bioaccumulation factor (BAF). The ratio (in L/kg) of a substance's concentration in tissue of an aquatic organism to its concentration in the ambient water, in situations where both the organism and its food are exposed to and the ratio does not change substantially over time.

Bioconcentration. The net accumulation of a substance by an aquatic organism as a result of uptake directly from the ambient water through gill membranes or other external body surfaces.

Bioconcentration factor (BCF). The ratio (in L/kg) of a substance's concentration in tissue of an aquatic organism to its concentration in the ambient water, in situations where the organism is exposed through the water only and the ratio does not change substantially over time.

Biota-sediment accumulation factor (BSAF). The ratio (in kg of organic carbon/kg of lipid) of a substance's lipid-normalized concentration in tissue of an aquatic organism to its organic carbon-normalized concentration in surface sediment, in situations where the ratio does not change substantially over time, both the organism and its food are exposed, and the surface sediment is representative of average surface sediment in the vicinity of the organism.

Depuration. The loss of a substance from an organism as a result of any active or passive process.

Food-chain multiplier (FCM). The ratio of a BAF to an appropriate BCF.

Octanol-water partition coefficient (K_{OW}). The ration of the concentration of a substance in the n-octanol phase to its concentration in the aqueous phase in an equilibrated two-phase octanol-water system. For log K_{OW}, the log of the octanol-water partition coefficient is a base 10 logarithm.

Uptake. Acquisition of a substance from the environment by an organism as a result of any active or passive process.

III. Review and Selection of Data

A. *Data Sources.* Measured BAFs, BSAFs and BCFs are assembled from available sources including the following:

1. EPA Ambient Water Quality Criteria documents issued after January 1, 1980.

2. Published scientific literature.

3. Reports issued by EPA or other reliable sources.

4. Unpublished data.

One useful source of references is the Aquatic Toxicity Information Retrieval (AQUIRE) database.

B. *Field-Measured BAFs.* The following procedural and quality assurance requirements shall be met for field-measured BAFs:

1. The field studies used shall be limited to those conducted in the Great Lakes System with fish at or near the top of the aquatic food chain (i.e., in trophic levels 3 and/or 4).

2. The trophic level of the fish species shall be determined.

3. The site of the field study should not be so unique that the BAF cannot be extrapolated to other locations where the criteria and values will apply.

4. For organic chemicals, the percent lipid shall be either measured or reliably estimated for the tissue used in the determination of the BAF.

5. The concentration of the chemical in the water shall be measured in a way that can be related to particulate organic carbon (POC) and/or dissolved organic carbon (DOC) and should be relatively constant during the steady-state time period.

6. For organic chemicals with log K_{OW} greater than four, the concentrations of POC and DOC in the ambient water shall be either measured or reliably estimated.

7. For inorganic and organic chemicals, BAFs shall be used only if they are expressed on a wet weight basis; BAFs reported on a dry weight basis cannot be converted to wet weight unless a conversion factor is meas-

ured or reliably estimated for the tissue used in the determination of the BAF.

C. *Field-Measured BSAFs.* The following procedural and quality assurance requirements shall be met for field-measured BSAFs:

1. The field studies used shall be limited to those conducted in the Great Lakes System with fish at or near the top of the aquatic food chain (i.e., in trophic levels 3 and/or 4).

2. Samples of surface sediments (0–1 cm is ideal) shall be from locations in which there is net deposition of fine sediment and is representative of average surface sediment in the vicinity of the organism.

3. The K_{OW} s used shall be acceptable quality as described in section III.F below.

4. The site of the field study should not be so unique that the resulting BAF cannot be extrapolated to other locations where the criteria and values will apply.

5. The tropic level of the fish species shall be determined.

6. The percent lipid shall be either measured or reliably estimated for the tissue used in the determination of the BAF.

D. *Laboratory-Measured BCFs.* The following procedural and quality assurance requirements shall be met for laboratory-measured BCFs:

1. The test organism shall not be diseased, unhealthy, or adversely affected by the concentration of the chemical.

2. The total concentration of the chemical in the water shall be measured and should be relatively constant during the steady-state time period.

3. The organisms shall be exposed to the chemical using a flow-through or renewal procedure.

4. For organic chemicals, the percent lipid shall be either measured or reliably estimated for the tissue used in the determination of the BCF.

5. For organic chemicals with log K_{OW} greater than four, the concentrations of POC and DOC in the test solution shall be either measured or reliably estimated.

6. Laboratory-measured BCFs should be determined using fish species, but BCFs determined with molluscs and other invertebrates may be used with caution. For example, because invertebrates metabolize some chemicals less efficiently than vertebrates, a baseline BCF determined for such a chemical using invertebrates is expected to be higher than a comparable baseline BCF determined using fish.

7. If laboratory-measured BCFs increase or decrease as the concentration of the chemical increases in the test solutions in a bioconcentration test, the BCF measured at the lowest test concentration that is above concentrations existing in the control water shall be used (i.e., a BCF should be calculated from a control treatment). The concentrations of an inorganic chemical in a

bioconcentration test should be greater than normal background levels and greater than levels required for normal nutrition of the test species if the chemical is a micronutrient, but below levels that adversely affect the species. Bioaccummulation of an inorganic chemical might be overestimated if concentrations are at or below normal background levels due to, for example, nutritional requirements of the test organisms.

8. For inorganic and organic chemicals, BCFs shall be used only if they are expressed on a wet weight basis. BCFs reported on a dry weight basis cannot be converted to wet weight unless a conversion factor is measured or reliably estimated for the tissue used in the determination of the BAF.

9. BCFs for organic chemicals may be based on measurement or radioactivity only when the BCF is intended to include metabolites or when there is confidence that there is no interference due to metabolites.

10. The calculation of the BCF must appropriately address growth dilution.

11. Other aspects of the methodology used should be similar to those described by ASTM (1990).

E. *Predicted BCFs.* The following procedural and quality assurance requirements shall be met for predicted BCFs:

1. The K_{OW} used shall be of acceptable quality as described in section III.F below.

2. The predicted baseline BCF shall be calculated using the equation: predicted baseline BCF = K_{OW}

where:

K_{OW} = octanol-water partition coefficient.

F. *Octanol-Water Partition Coefficient (K_{OW}).* 1. The value of K_{OW} used for an organic chemical shall be determined by giving priority to the experimental and computational techniques used as follows:

Log K_{OW} < 4:

Priority	Technique
1	Slow-stir.
1	Generator-column.
1	Shake-flask.
2	Reverse-phase liquid chromatography on C18 chromatography packing with extrapolation to zero percent solvent.
3	Reverse-phase liquid chromatography on C18 chromatography packing without extrapolation to zero percent solvent.
4	Calculated by the CLOGP program.

Log K_{OW} > 4:

Priority	Technique
1	Slow Stir.
1	Generator-column.
2	Reverse-phase liquid chromatography on C18 chromatography packing with extrapolation to zero percent solvent.

Priority	Technique
3	Reverse-phase liquid chromatography on C18 chromatography packing without extrapolation to zero percent solvent.
4	Shake-flask.
5	Calculated by the CLOGP program.

2. The CLOGP program is a computer program available from Pomona College. A value of K_{OW} that seems to be different from the others should be considered an outlier and not used. The value of K_{OW} used for an organic chemical shall be the geometric mean of the available K_{OW} s with highest priority or can be calculated from the arithmetic mean of the available log K_{OW} with the highest priority. Because it is an intermediate value in the derivation of a BAF, the value used for the K_{OW} of a chemical should not be rounded to fewer than three significant digits and a value for log K_{OW} should not be rounded to fewer than three significant digits after the decimal point.

G. This methodology provides overall guidance for the derivation of BAFs, but it cannot cover all the decisions that must be made in the review and selection of acceptable data. Professional judgment is required throughout the process. A degree of uncertainty is associated with the determination of any BAF, BSAF, BCF or K_{OW}. The amount of uncertainty in a baseline BAF depends on both the quality of data available and the method used to derive the BAF.

H. Hereinafter in this methodology, the terms BAF, BSAF, BCF and K_{OW} refer to ones that are consistent with the procedural and quality assurance requirements given above.

IV. Four Methods for Deriving Baseline BAFs

Baseline BAFs shall be derived using the following four methods, which are listed from most preferred to least preferred:

A. A measured baseline BAF for an organic or inorganic chemical derived from a field study of acceptable quality.

B. A predicted baseline BAF for an organic chemical derived using field-measured BSAFs of acceptable quality.

C. A predicted baseline BAF for an organic or inorganic chemical derived from a BCF measured in a laboratory study of acceptable quality and a FCM.

D. A predicted baseline BAF for an organic chemical derived from a K_{OW} of acceptable quality and a FCM.

For comparative purposes, baseline BAFs should be derived for each chemical by as many of the four methods as available data allow.

V. Calculation of Baseline BAFs for Organic Chemicals

A. *Lipid Normalization.* 1. It is assumed that BAFs and BCFs for organic chemicals can be extrapolated on the basis of percent lipid from one tissue to another and from one aquatic species to another in most cases.

2. Because BAFs and BCFs for organic chemicals are related to the percent lipid, it does not make any difference whether the tissue sample is whole body or edible portion, but both the BAF (or BCF) and the percent lipid must be determined for the same tissue. The percent lipid of the tissue should be measured during the BAF or BCF study, but in some cases it can be reliably estimated from measurements on tissue from other organisms. If percent lipid is not reported for the test organisms in the original study, it may be obtained from the author; or, in the case of a laboratory study, lipid data for the same or a comparable laboratory population of test organisms that were used in the original study may be used.

3. The lipid-normalized concentration, C_l, of a chemical in tissue is defined using the following equation:

$$C_l = \frac{C_B}{f_l}$$

Where:

C_B=concentration of the organic chemical in the tissue of aquatic biota (either whole organism or specified tissue) (μg/g).
f_l=fraction of the tissue that is lipid.

B. *Bioavailability.* By definition, baseline BAFs and BCFs for organic chemicals, whether measured or predicted are based on the concentration of the chemical that is freely dissolved in the ambient water in order to account for bioavailability. For the purposes of this Guidance in this part, the relationship between the total concentration of the chemical in the water (i.e., that which is freely dissolved plus that which is sorbed to particulate organic carbon or to dissolved organic carbon) to the freely dissolved concentration of the chemical in the ambient water shall be calculated using the following equation:

$$C_w^{fd} = \left(f_{fd}\right)\left(C_w^t\right)$$

Where:

C_w^{fd}=freely dissolved concentration of the organic chemical in the ambient water;
C_w^t=total concentration of the organic chemical in the ambient water;
f_{fd}=fraction of the total chemical in the ambient water that is freely dissolved.

The fraction of the total chemical in the ambient water that is freely dissolved, f_{fd}, shall be calculated using the following equation:

$$f_{fd} = \frac{1}{1 + \dfrac{(DOC)(K_{ow})}{10} + (POC)(K_{ow})}$$

Where:

DOC=concentration of dissolved organic carbon, kg of dissolved organic carbon/L of water.
K_{OW}=octanol-water partition coefficient of the chemical.
POC=concentration of particulate organic carbon, kg of particulate organic carbon/L of water.

C. *Food-Chain Multiplier.* In the absence of a field-measured BAF or a predicted BAF derived from a BSAF, a FCM shall be used to calculate the baseline BAF for trophic levels 3 and 4 from a laboratory-measured or predicted BCF. For an organic chemical, the FCM used shall be derived from Table B–1 using the chemical's log K_{OW} and linear interpolation. A FCM greater than 1.0 applies to most organic chemicals with a log K_{OW} of four or more. The trophic level used shall take into account the age or size of the fish species consumed by the human, avian or mammalian predator because, for some species of fish, the young are in trophic level 3 whereas the adults are in trophic level 4.

D. *Calculation of a Baseline BAF from a Field-Measured BAF.* A baseline BAF shall be calculated from a field-measured BAF of acceptable quality using the following equation:

$$\text{Baseline BAF} = \left[\frac{\text{Measured BAF}_T^t}{f_{fd}} - 1\right]\left[\frac{1}{f_l}\right]$$

Where:

BAF_T^t=BAF based on total concentration in tissue and water.
f_l=fraction of the tissue that is lipid.

f_{fd}=fraction of the total chemical that is freely dissolved in the ambient water.

The trophic level to which the baseline BAF applies is the same as the trophic level of the organisms used in the determination of the field-measured BAF. For each trophic level, a species mean measured baseline BAF shall be calculated as the geometric mean if more than one measured baseline BAF is available for a given species. For each trophic level, the geometric mean of the species mean measured baseline BAFs shall be calculated. If a baseline BAF based on a measured BAF is available for either trophic level 3 or 4, but not both, a measured baseline BAF for the other trophic level shall be calculated using the ratio of the FCMs that are obtained by linear interpolation from Table B–1 for the chemical.

E. *Calculation of a Baseline BAF from a Field-Measured BSAF.* 1. A baseline BAF for organic chemical "i" shall be calculated from a field-measured BSAF of acceptable quality using the following equation:

$$(\text{Baseline BAF})_i = (\text{Baseline BAF})_r \cdot \frac{(\text{BSAF})_i \cdot (K_{ow})_i}{(\text{BSAF})_r \cdot (K_{ow})_r}$$

Where:

$(\text{BSAF})_i$=BSAF for chemical "i".

$(\text{BSAF})_r$=BSAF for the reference chemical "r".

$(K_{ow})_i$=octanol-water partition coefficient for chemical "i".

$(K_{ow})_r$=octanol-water partition coefficient for the reference chemical "r".

2. A BSAF shall be calculated using the following equation:

$$\text{BSAF} = \frac{C_l}{C_{SOC}}$$

Where:

C_l=the lipid-normalized concentration of the chemical in tissue.

C_{SOC}=the organic carbon-normalized concentration of the chemical in sediment.

3. The organic carbon-normalized concentration of a chemical in sediment, C_{SOC}, shall be calculated using the following equation:

$$C_{SOC} = \frac{C_S}{f_{OC}}$$

Where:

C_S=concentration of chemical in sediment (μg/g sediment).

f_{OC}=fraction of the sediment that is organic carbon.

4. Predicting BAFs from BSAFs requires data from a steady-state (or near steady-state) condition between sediment and ambient water for both a reference chemical "r" with a field-measured BAF_l fd and other chemicals "n=i" for which BSAFs are to be determined.

5. The trophic level to which the baseline BAF applies is the same as the trophic level of the organisms used in the determination of the BSAF. For each trophic level, a species mean baseline BAF shall be calculated as the geometric mean if more than one baseline BAF is predicted from BSAFs for a given species. For each trophic level, the geometric mean of the species mean baseline BAFs derived using BSAFs shall be calculated.

6. If a baseline BAF based on a measured BSAF is available for either trophic level 3 or 4, but not both, a baseline BAF for the other trophic level shall be calculated using the ratio of the FCMs that are obtained by linear interpolation from Table B–1 for the chemical.

F. *Calculation of a Baseline BAF from a Laboratory-Measured BCF.* A baseline BAF for trophic level 3 and a baseline BAF for trophic level 4 shall be calculated from a laboratory-measured BCF of acceptable quality and a FCM using the following equation:

$$\text{Baseline BAF} = (\text{FCM})\left[\frac{\text{Measured BCF}_T^t}{f_{fd}} - 1\right]\left[\frac{1}{f_l}\right]$$

Where:

BCF_T^t=BCF based on total concentration in tissue and water.

f_l=fraction of the tissue that is lipid.

f_{fd}=fraction of the total chemical in the test water that is freely dissolved.

FCM=the food-chain multiplier obtained from Table B–1 by linear interpolation for trophic level 3 or 4, as necessary.

For each trophic level, a species mean baseline BAF shall be calculated as the geometric mean if more than one baseline BAF is predicted from laboratory-measured BCFs for a given species. For each trophic level, the geometric mean of the species mean baseline BAFs based on laboratory-measured BCFs shall be calculated.

G. *Calculation of a Baseline BAF from an Octanol-Water Partition Coefficient.* A baseline BAF for trophic level 3 and a baseline BAF for trophic level 4 shall be calculated from a K_{OW} of acceptable quality and a FCM using the following equation:

Baseline BAF=(FCM) (predicted baseline BCF)=(FCM) (K_{OW})

Where:

FCM=the food-chain multiplier obtained from Table B–1 by linear interpolation for trophic level 3 or 4, as necessary.

K_{OW}=octanol-water partition coefficient.

VI. Human Health and Wildlife BAFs for Organic Chemicals

A. To calculate human health and wildlife BAFs for an organic chemical, the K_{OW} of the chemical shall be used with a POC concentration of 0.00000004 kg/L and a DOC concentration of 0.000002 kg/L to yield the fraction freely dissolved:

$$f_{fd} = \cfrac{1}{1 + \cfrac{(DOC)(K_{ow})}{10} + (POC)(K_{ow})}$$

$$= \cfrac{1}{1 + \cfrac{(0.000002 \text{ kg}/L)(K_{ow})}{10} + (0.00000004 \text{ kg}/L)(K_{ow})}$$

$$= \cfrac{1}{1 + (0.00000024 \text{ kg}/L)(K_{ow})}$$

B. The human health BAFs for an organic chemical shall be calculated using the following equations:

For trophic level 3:

$$\text{Human Health BAF}_{TL3}^{HH} = [(\text{baseline BAF})(0.0182) + 1](f_{fd})$$

For trophic level 4:

$$\text{Human Health BAF}_{TL4}^{HH} = [(\text{baseline BAF})(0.0310) + 1](f_{fd})$$

Where:

0.0182 and 0.0310 are the standardized fraction lipid values for trophic levels 3 and 4, respectively, that are used to derive human health criteria and values for the GLI.

C. The wildlife BAFs for an organic chemical shall be calculated using the following equations:

For trophic level 3:

$$\text{Wildlife BAF}_{TL3}^{WL} = [(\text{baseline BAF})(0.0646)+1](f_{fd})$$

For trophic level 4:

$$\text{Wildlife BAF}_{TL4}^{WL} = [(\text{baseline BAF})(0.1031)+1](f_{fd})$$

Where:
0.0646 and 0.1031 are the standardized fraction lipid values for trophic levels 3 and 4, respectively, that are used to derive wildlife criteria for the GLI.

VII. Human Health and Wildlife BAFs for Inorganic Chemicals

A. For inorganic chemicals, the baseline BAFs for trophic levels 3 and 4 are both assumed to equal the BCF determined for the chemical with fish, i.e., the FCM is assumed to be 1 for both trophic levels 3 and 4. However, a FCM greater than 1 might be applicable to some metals, such as mercury, if, for example, an organometallic form of the metal biomagnifies.

B. *BAFs for Human Health Criteria and Values.*

1. Measured BAFs and BCFs used to determine human health BAFs for inorganic chemicals shall be based on edible tissue (e.g., muscle) of freshwater fish unless it is demonstrated that whole-body BAFs or BCFs are similar to edible-tissue BAFs or BCFs. BCFs and BAFs based on measurements of aquatic plants and invertebrates should not be used in the derivation of human health criteria and values.

2. If one or more field-measured baseline BAFs for an inorganic chemical are available from studies conducted in the Great Lakes System with the muscle of fish:

a. For each trophic level, a species mean measured baseline BAF shall be calculated as the geometric mean if more than one measured BAF is available for a given species; and

b. For each trophic level, the geometric mean of the species mean measured baseline BAFs shall be used as the human health BAF for that chemical.

3. If an acceptable measured baseline BAF is not available for an inorganic chemical and one or more acceptable edible-portion laboratory-measured BCFs are available for the chemical, a predicted baseline BAF shall be calculated by multiplying the geometric mean of the BCFs times a FCM. The FCM will be 1.0 unless chemical-specific biomagnification data support using a multiplier other than 1.0. The predicted baseline

BAF shall be used as the human health BAF for that chemical.

C. *BAFs for Wildlife Criteria.*

1. Measured BAFs and BCFs used to determine wildlife BAFs for inorganic chemicals shall be based on whole-body freshwater fish and invertebrate data unless it is demonstrated that edible-tissue BAFs or BCFs are similar to whole-body BAFs or BCFs.

2. If one or more field-measured baseline BAFs for an inorganic chemical are available from studies conducted in the Great Lakes System with whole body of fish or invertebrates:

a. For each trophic level, a species mean measured baseline BAF shall be calculated as the geometric mean if more than one measured BAF is available for a given species.

b. For each trophic level, the geometric mean of the species mean measured baseline BAFs shall be used as the wildlife BAF for that chemical.

3. If an acceptable measured baseline BAF is not available for an inorganic chemical and one or more acceptable whole-body laboratory-measured BCFs are available for the chemical, a predicted baseline BAF shall be calculated by multiplying the geometric mean of the BCFs times a FCM. The FCM will be 1.0 unless chemical-specific biomagnification data support using a multiplier other than 1.0. The predicted baseline BAF shall be used as the wildlife BAF for that chemical.

VIII. Final Review

For both organic and inorganic chemicals, human health and wildlife BAFs for both trophic levels shall be reviewed for consistency with all available data concerning the bioaccumulation, bioconcentration, and metabolism of the chemical. For example, information concerning octanol-water partitioning, molecular size, or other physicochemical properties that might enhance or inhibit bioaccumulation should be considered for organic chemicals. BAFs derived in accordance with this methodology should be modified if changes are justified by available data.

IX. Literature Cited

ASTM. 1990. Standard Practice for Conducting Bioconcentration Tests with Fishes and Saltwater Bivalve Molluscs. Standard E 1022. American Society for Testing and Materials, Philadelphia, PA.

TABLE B–1—FOOD-CHAIN MULTIPLIERS FOR TROPHIC LEVELS 2, 3 & 4

Log K_{OW}	Trophic level 2	Trophic [1] level 3	Trophic level 4
2.0	1.000	1.005	1.000
2.5	1.000	1.010	1.002
3.0	1.000	1.028	1.007
3.1	1.000	1.034	1.007
3.2	1.000	1.042	1.009
3.3	1.000	1.053	1.012
3.4	1.000	1.067	1.014
3.5	1.000	1.083	1.019
3.6	1.000	1.103	1.023
3.7	1.000	1.128	1.033
3.8	1.000	1.161	1.042
3.9	1.000	1.202	1.054
4.0	1.000	1.253	1.072
4.1	1.000	1.315	1.096
4.2	1.000	1.380	1.130
4.3	1.000	1.491	1.178
4.4	1.000	1.614	1.242
4.5	1.000	1.766	1.334
4.6	1.000	1.950	1.459
4.7	1.000	2.175	1.633
4.8	1.000	2.452	1.871
4.9	1.000	2.780	2.193
5.0	1.000	3.181	2.612
5.1	1.000	3.643	3.162
5.2	1.000	4.188	3.873
5.3	1.000	4.803	4.742
5.4	1.000	5.502	5.821
5.5	1.000	6.266	7.079
5.6	1.000	7.096	8.551
5.7	1.000	7.962	10.209
5.8	1.000	8.841	12.050
5.9	1.000	9.716	13.964
6.0	1.000	10.556	15.996
6.1	1.000	11.337	17.783
6.2	1.000	12.064	19.907
6.3	1.000	12.691	21.677
6.4	1.000	13.228	23.281
6.5	1.000	13.662	24.604
6.6	1.000	13.980	25.645
6.7	1.000	14.223	26.363
6.8	1.000	14.355	26.669
6.9	1.000	14.388	26.669
7.0	1.000	14.305	26.242
7.1	1.000	14.142	25.468
7.2	1.000	13.852	24.322
7.3	1.000	13.474	22.856
7.4	1.000	12.987	21.038
7.5	1.000	12.517	18.967
7.6	1.000	11.708	16.749
7.7	1.000	10.914	14.388
7.8	1.000	10.069	12.050
7.9	1.000	9.162	9.840
8.0	1.000	8.222	7.798
8.1	1.000	7.278	6.012
8.2	1.000	6.361	4.519
8.3	1.000	5.489	3.311
8.4	1.000	4.683	2.371
8.5	1.000	3.949	1.663
8.6	1.000	3.296	1.146
8.7	1.000	2.732	0.778
8.8	1.000	2.246	0.521
8.9	1.000	1.837	0.345

TABLE B–1—FOOD-CHAIN MULTIPLIERS FOR TROPHIC LEVELS 2, 3 & 4—Continued

Log K_{OW}	Trophic level 2	Trophic [1] level 3	Trophic level 4
9.0	1.000	1.493	0.226

[1] The FCMs for trophic level 3 are the geometric mean of the FCMs for sculpin and alewife.

APPENDIX C TO PART 132—GREAT LAKES WATER QUALITY INITIATIVE METHODOLOGIES FOR DEVELOPMENT OF HUMAN HEALTH CRITERIA AND VALUES

Great Lakes States and Tribes shall adopt provisions consistent with (as protective as) this appendix.

I. INTRODUCTION

Great Lakes States and Tribes shall adopt provisions consistent with this appendix C to ensure protection of human health.

A. *Goal.* The goal of the human health criteria for the Great Lakes System is the protection of humans from unacceptable exposure to toxicants via consumption of contaminated fish and drinking water and from ingesting water as a result of participation in water-oriented recreational activities.

B. *Definitions.*

Acceptable daily exposure (ADE). An estimate of the maximum daily dose of a substance which is not expected to result in adverse noncancer effects to the general human population, including sensitive subgroups.

Adverse effect. Any deleterious effect to organisms due to exposure to a substance. This includes effects which are or may become debilitating, harmful or toxic to the normal functions of the organism, but does not include non-harmful effects such as tissue discoloration alone or the induction of enzymes involved in the metabolism of the substance.

Carcinogen. A substance which causes an increased incidence of benign or malignant neoplasms, or substantially decreases the time to develop neoplasms, in animals or humans. The classification of carcinogens is discussed in section II.A of appendix C to part 132.

Human cancer criterion (HCC). A Human Cancer Value (HCV) for a pollutant that meets the minimum data requirements for Tier I specified in appendix C.

Human cancer value (HCV). The maximum ambient water concentration of a substance at which a lifetime of exposure from either: drinking the water, consuming fish from the water, and water-related recreation activities; or consuming fish from the water, and water-related recreation activities, will represent a plausible upper-bound risk of contracting cancer of one in 100,000 using the exposure assumptions specified in the Methodologies for the Development of Human

Health Criteria and Values in appendix C of this part.

Human noncancer criterion (HNC). A Human Noncancer Value (HNV) for a pollutant that meets the minimum data requirements for Tier I specified in appendix C of this part.

Human noncancer value (HNV). The maximum ambient water concentration of a substance at which adverse noncancer effects are not likely to occur in the human population from lifetime exposure via either: drinking the water, consuming fish from the water, and water-related recreation activities; or consuming fish from the water, and water-related recreation activities using the Methodologies for the Development of Human Health criteria and Values in appendix C of this part.

Linearized multi-stage model. A conservative mathematical model for cancer risk assessment. This model fits linear dose-response curves to low doses. It is consistent with a no-threshold model of carcinogenesis, i.e., exposure to even a very small amount of the substance is assumed to produce a finite increased risk of cancer.

Lowest observed adverse effect level (LOAEL). The lowest tested dose or concentration of a substance which resulted in an observed adverse effect in exposed test organisms when all higher doses or concentrations resulted in the same or more severe effects.

No observed adverse effect level (NOAEL). The highest tested dose or concentration of a substance which resulted in no observed adverse effect in exposed test organisms where higher doses or concentrations resulted in an adverse effect.

Quantitative structure activity relationship (QSAR) or structure activity relationship (SAR). A mathematical relationship between a property (activity) of a chemical and a number of descriptors of the chemical. These descriptors are chemical or physical characteristics obtained experimentally or predicted from the structure of the chemical.

Relative source contribution (RSC). The factor (percentage) used in calculating an HNV or HNC to account for all sources of exposure to a contaminant. The RSC reflects the percent of total exposure which can be attributed to surface water through water intake and fish consumption.

Risk associated dose (RAD). A dose of a known or presumed carcinogenic substance in (mg/kg/day) which, over a lifetime of exposure, is estimated to be associated with a plausible upper bound incremental cancer risk equal to one in 100,000.

Slope factor. Also known as q_1^*, slope factor is the incremental rate of cancer development calculated through use of a linearized multistage model or other appropriate model. It is expressed in (mg/kg/day) of exposure to the chemical in question.

Threshold effect. An effect of a substance for which there is a theoretical or empirically established dose or concentration below which the effect does not occur.

Uncertainty factor (UF). One of several numeric factors used in operationally deriving criteria from experimental data to account for the quality or quantity of the available data.

C. *Level of Protection.* The criteria developed shall provide a level of protection likely to be without appreciable risk of carcinogenic and/or noncarcinogenic effects. Criteria are a function of the level of designated risk or no adverse effect estimation, selection of data and exposure assumptions. Ambient criteria for single carcinogens shall not be set at a level representing a lifetime upper-bound incremental risk greater than one in 100,000 of developing cancer using the hazard assessment techniques and exposure assumptions described herein. Criteria affording protection from noncarcinogenic effects shall be established at levels that, taking into account uncertainties, are considered likely to be without an appreciable risk of adverse human health effects (i.e., acute, subchronic and chronic toxicity including reproductive and developmental effects) during a lifetime of exposure, using the risk assessment techniques and exposure assumptions described herein.

D. *Two-tiered Classification.* Chemical concentration levels in surface water protective of human health shall be derived based on either a Tier I or Tier II classification. The two Tiers are primarily distinguished by the amount of toxicity data available for deriving the concentration levels and the quantity and quality of data on bioaccumulation.

II. MINIMUM DATA REQUIREMENTS

The best available toxicity data on the adverse health effects of a chemical and the best data on bioaccumulation factors shall be used when developing human health Tier I criteria or Tier II values. The best available toxicity data shall include data from well-conducted epidemiologic and/or animal studies which provide, in the case of carcinogens, an adequate weight of evidence of potential human carcinogenicity and, in the case of noncarcinogens, a dose-response relationship involving critical effects biologically relevant to humans. Such information should be obtained from the EPA Integrated Risk Information System (IRIS) database, the scientific literature, and other informational databases, studies and/or reports containing adverse health effects data of adequate quality for use in this procedure. Strong consideration shall be given to the most currently available guidance provided by IRIS in deriving criteria or values, supplemented with any recent data not incorporated into IRIS. When deviations from IRIS are anticipated or considered necessary, it is strongly recommended that such actions be communicated to the EPA Reference Dose (RfD) and/or the Cancer Risk Assessment Verification Endeavor (CRAVE) workgroup immediately. The best available bioaccumulation data shall include data from field studies and well-conducted laboratory studies.

A. *Carcinogens.* Tier I criteria and Tier II values shall be derived using the methodologies described in section III.A of this appendix when there is adequate evidence of potential human carcinogenic effects for a chemical. It is strongly recommended that the EPA classification system for chemical carcinogens, which is described in the 1986 EPA Guidelines for Carcinogenic Risk Assessment (U.S. EPA, 1986), or future modifications thereto, be used in determining whether adequate evidence of potential carcinogenic effects exists. Carcinogens are classified, depending on the weight of evidence, as either human carcinogens, probable human carcinogens, or possible human carcinogens. The human evidence is considered inadequate and therefore the chemical cannot be classified as a human carcinogen, if one of two conditions exists: (a) there are few pertinent data, or (b) the available studies, while showing evidence of association, do not exclude chance, bias, or confounding and therefore a casual interpretation is not credible. The animal evidence is considered inadequate, and therefore the chemical cannot be classified as a probable or possible human carcinogen, when, because of major qualitative or quantitative limitations, the evidence cannot be interpreted as showing either the presence or absence of a carcinogenic effect.

Chemicals are described as "human carcinogens" when there is sufficient evidence from epidemiological studies to support a causal association between exposure to the chemicals and cancer. Chemicals described as "probable human carcinogens" include chemicals for which the weight of evidence of human carcinogenicity based on epidemiological studies is limited. Limited human evidence is that which indicates that a causal interpretation is credible, but that alternative explanations, such as chance, bias, or confounding, cannot adequately be excluded. Probable human carcinogens are also agents for which there is sufficient evidence from animal studies and for which there is inadequate evidence or no data from epidemiologic studies. Sufficient animal evidence is data which indicates that there is an increased incidence of malignant tumors or combined malignant and benign tumors: (a) in multiple species or strains; (b) in multiple experiments (e.g., with different routes of administration or using different dose levels); or (c) to an unusual degree in a single experiment with regard to high incidence, unusual site or type of tumor, or early age at onset.

Additional evidence may be provided by data on dose-response effects, as well as information from short-term tests (such as mutagenicity/genotoxicity tests which help determine whether the chemical interacts directly with DNA) or on chemical structure, metabolism or mode of action.

"Possible human carcinogens" are chemicals with limited evidence of carcinogenicity in animals in the absence of human data. Limited animal evidence is defined as data which suggests a carcinogenic effect but are limited because: (a) The studies involve a single species, strain, or experiment and do not meet criteria for sufficient evidence (see preceding paragraph); or (b) the experiments are restricted by inadequate dosage levels, inadequate duration of exposure to the agent, inadequate period of follow-up, poor survival, too few animals, or inadequate reporting; or (c) the studies indicate an increase in the incidence of benign tumors only. More specifically, this group can include a wide variety of evidence, e.g., (a) a malignant tumor response in a single well-conducted experiment that does not meet conditions for sufficient evidence, (b) tumor response of marginal statistical significance in studies having inadequate design or reporting, (c) benign but not malignant tumors with an agent showing no response in a variety of short-term tests for mutagenicity, and (d) response of marginal statistical significance in a tissue known to have a high or variable background rate.

1. *Tier I:* Weight of evidence of potential human carcinogenic effects sufficient to derive a Tier I HCC shall generally include human carcinogens, probable human carcinogens and can include, on a case-by-case basis, possible human carcinogens if studies have been well-conducted albeit based on limited evidence, when compared to studies used in classifying human and probable human carcinogens. The decision to use data on a possible human carcinogen for deriving Tier I criteria shall be a case-by-case determination. In determining whether to derive a Tier I HCC, additional evidence that shall be considered includes but is not limited to available information on mode of action, such as mutagenicity/genotoxicity (determinations of whether the chemical interacts directly with DNA), structure activity, and metabolism.

2. *Tier II:* Weight of evidence of possible human carcinogenic effects sufficient to derive a Tier II human cancer value shall include those possible human carcinogens for which there are at a minimum, data sufficient for quantitative risk assessment, but for which data are inadequate for Tier I criterion development due to a tumor response of marginal statistical significance or inability to derive a strong dose-response relationship. In determining whether to derive Tier II human cancer values, additional evidence

that shall be considered includes but is not limited to available information on mode of action such as mutagenicity/genotoxicity (determinations of whether the chemical interacts directly with DNA), structure activity and metabolism. As with the use of data on possible human carcinogens in developing Tier I criteria, the decision to use data on possible human carcinogens to derive Tier II values shall be made on a case-by-case basis.

B. *Noncarcinogens.* All available toxicity data shall be evaluated considering the full range of possible health effects of a chemical, i.e., acute/subacute, chronic/subchronic and reproductive/developmental effects, in order to best describe the dose-response relationship of the chemical, and to calculate human noncancer criteria and values which will protect against the most sensitive endpoint(s) of toxicity. Although it is desirable to have an extensive database which considers a wide range of possible adverse effects, this type of data exists for a very limited number of chemicals. For many others, there is a range in quality and quantity of data available. To assure minimum reliability of criteria and values, it is necessary to establish a minimum database with which to develop Tier I criteria or Tier II values. The following represent the minimum data sets necessary for this procedure.

1. *Tier I:* The minimum data set sufficient to derive a Tier I human HNC shall include at least one well-conducted epidemiologic study or animal study. A well-conducted epidemiologic study for a Tier I HNC must quantify exposure level(s) and demonstrate positive association between exposure to a chemical and adverse effect(s) in humans. A well-conducted study in animals must demonstrate a dose response relationship involving one or more critical effect(s) biologically relevant to humans. (For example, study results from an animal whose pharmacokinetics and toxicokinetics match those of a human would be considered most biologically relevant.) Ideally, the duration of a study should span multiple generations of exposed test species or at least a major portion of the lifespan of one generation. This type of data is currently very limited. By the use of uncertainty adjustments, shorter term studies (such as 90-day subchronic studies) with evaluation of more limited effect(s) may be used to extrapolate to longer exposures or to account for a variety of adverse effects. For Tier I criteria developed pursuant to this procedure, such a limited study must be conducted for at least 90 days in rodents or 10 percent of the lifespan of other appropriate test species and demonstrate a no observable adverse effect level (NOAEL). Chronic studies of one year or longer in rodents or 50 percent of the lifespan or greater in other appropriate test species that demonstrate a lowest observable adverse effect

level (LOAEL) may be sufficient for use in Tier I criterion derivation if the effects observed at the LOAEL were relatively mild and reversible as compared to effects at higher doses. This does not preclude the use of a LOAEL from a study (of chronic duration) with only one or two doses if the effects observed appear minimal when compared to effect levels observed at higher doses in other studies.

2. *Tier II:* When the minimum data for deriving Tier I criteria are not available to meet the Tier I data requirements, a more limited database may be considered for deriving Tier II values. As with Tier I criteria, all available data shall be considered and ideally should address a range of adverse health effects with exposure over a substantial portion of the lifespan (or multiple generations) of the test species. When such data are lacking it may be necessary to rely on less extensive data in order to establish a Tier II value. With the use of appropriate uncertainty factors to account for a less extensive database, the minimum data sufficient to derive a Tier II value shall include a NOAEL from at least one well-conducted short-term repeated dose study. This study shall be of at least 28 days duration, in animals demonstrating a dose-response, and involving effects biologically relevant to humans. Data from studies of longer duration (greater than 28 days) and LOAELs from such studies (greater than 28 days) may be more appropriate in some cases for derivation of Tier II values. Use of a LOAEL should be based on consideration of the following information: severity of effect, quality of the study and duration of the study.

C. *Bioaccumulation factors (BAFs).*

1. *Tier I for Carcinogens and Noncarcinogens:* To be considered a Tier I cancer or noncancer human health criterion, along with satisfying the minimum toxicity data requirements of sections II.A.1 and II.B.1 of this appendix, a chemical must have the following minimum bioaccumulation data. For all organic chemicals either: (a) a field-measured BAF; (b) a BAF derived using the BSAF methodology; or (c) a chemical with a BAF less than 125 regardless of how the BAF was derived. For all inorganic chemicals, including organometals such as mercury, either: (a) a field-measured BAF or (b) a laboratory-measured BCF.

2. *Tier II for Carcinogens and Noncarcinogens:* A chemical is considered a Tier II cancer or noncancer human health value if it does not meet either the minimum toxicity data requirements of sections II.A.1 and II.B.1 of this appendix or the minimum bioaccumulation data requirements of section II.C.1 of this appendix.

III. PRINCIPLES FOR DEVELOPMENT OF TIER I CRITERIA OR TIER II VALUES

The fundamental components of the procedure to calculate Tier I criteria or Tier II values are the same. However, certain of the aspects of the procedure designed to account for short-duration studies or other limitations in data are more likely to be relevant in deriving Tier II values than Tier I criteria.

A. *Carcinogens.*

1. A non-threshold mechanism of carcinogenesis shall be assumed unless biological data adequately demonstrate the existence of a threshold on a chemical-specific basis.

2. All appropriate human epidemiologic data and animal cancer bioassay data shall be considered. Data specific to an environmentally appropriate route of exposure shall be used. Oral exposure should be used preferentially over dermal and inhalation since, in most cases, the exposure routes of greatest concern are fish consumption and drinking water/incidental ingestion. The risk associated dose shall be set at a level corresponding to an incremental cancer risk of one in 100,000. If acceptable human epidemiologic data are available for a chemical, it shall be used to derive the risk associated dose. If acceptable human epidemiologic data are not available, the risk associated dose shall be derived from available animal bioassay data. Data from a species that is considered most biologically relevant to humans (i.e., responds most like humans) is preferred where all other considerations regarding quality of data are equal. In the absence of data to distinguish the most relevant species, data from the most sensitive species tested, i.e., the species showing a carcinogenic effect at the lowest administered dose, shall generally be used.

3. When animal bioassay data are used and a non-threshold mechanism of carcinogenicity is assumed, the data are fitted to a linearized multistage computer model (e.g., Global '86 or equivalent model). Global '86 is the linearized multistage model, derived by Howe, Crump and Van Landingham (1986), which EPA uses to determine cancer potencies. The upper-bound 95 percent confidence limit on risk (or, the lower 95 percent confidence limit on dose) at the one in 100,000 risk level shall be used to calculate a risk associated dose (RAD). Other models, including modifications or variations of the linear multistage model which are more appropriate to the available data may be used where scientifically justified.

4. If the duration of the study is significantly less than the natural lifespan of the test animal, the slope may be adjusted on a case-by-case basis to compensate for latent tumors which were not expressed (e.g., U.S. EPA, 1980) In the absence of alternative approaches which compensate for study duration significantly less than lifetime, the permitting authority may use the process described in the 1980 National Guidelines (see 45 FR 79352).

5. A species scaling factor shall be used to account for differences between test species and humans. It shall be assumed that milligrams per surface area per day is an equivalent dose between species (U.S. EPA, 1986). All doses presented in mg/kg bodyweight will be converted to an equivalent surface area dose by raising the mg/kg dose to the 2/3 power. However, if adequate pharmacokinetic and metabolism studies are available, these data may be factored into the adjustment for species differences on a case-by-case basis.

6. Additional data selection and adjustment decisions must also be made in the process of quantifying risk. Consideration must be given to tumor selection for modeling, e.g., pooling estimates for multiple tumor types and identifying and combining benign and malignant tumors. All doses shall be adjusted to give an average daily dose over the study duration. Adjustments in the rate of tumor response must be made for early mortality in test species. The goodness-of-fit of the model to the data must also be assessed.

7. When a linear, non-threshold dose response relationship is assumed, the RAD shall be calculated using the following equation:

$$RAD = \frac{0.00001}{q_1{}^*}$$

Where:

RAD=risk associated dose in milligrams of toxicant per kilogram body weight per day (mg/kg/day).

0.00001 (1×10^{-5})=incremental risk of developing cancer equal to one in 100,000.

$q_1{}^*$=slope factor (mg/kg/day)$^{-1}$.

8. If human epidemiologic data and/or other biological data (animal) indicate that a chemical causes cancer via a threshold mechanism, the risk associated dose may, on a case-by-case basis, be calculated using a method which assumes a threshold mechanism is operative.

B. *Noncarcinogens.*

1. Noncarcinogens shall generally be assumed to have a threshold dose or concentration below which no adverse effects should be observed. Therefore, the Tier I criterion or Tier II value is the maximum water concentration of a substance at or below which a lifetime exposure from drinking the water, consuming fish caught in the water, and ingesting water as a result of participating in water-related recreation activities is likely to be without appreciable risk of deleterious effects.

For some noncarcinogens, there may not be a threshold dose below which no adverse effects should be observed. Chemicals acting as genotoxic teratogens and germline mutagens are thought to possibly produce reproductive and/or developmental effects via a genetically linked mechanism which may have no threshold. Other chemicals also may not demonstrate a threshold. Criteria for these types of chemicals will be established on a case-by-case basis using appropriate assumptions reflecting the likelihood that no threshold exists.

2. All appropriate human and animal toxicologic data shall be reviewed and evaluated. To the maximum extent possible, data most specific to the environmentally relevant route of exposure shall be used. Oral exposure data should be used preferentially over dermal and inhalation since, in most cases, the exposure routes of greatest concern are fish consumption and drinking water/incidental ingestion. When acceptable human data are not available (e.g., well-conducted epidemiologic studies), animal data from species most biologically relevant to humans shall be used. In the absence of data to distinguish the most relevant species, data from the most sensitive animal species tested, i.e., the species showing a toxic effect at the lowest administered dose (given a relevant route of exposure), should generally be used.

3. Minimum data requirements are specified in section II.B of this appendix. The experimental exposure level representing the highest level tested at which no adverse effects were demonstrated (NOAEL) from studies satisfying the provisions of section II.B of this appendix shall be used for criteria calculations. In the absence of a NOAEL, the LOAEL from studies satisfying the provisions of section II.B of this appendix may be used if it is based on relatively mild and reversible effects.

4. Uncertainty factors shall be used to account for the uncertainties in predicting acceptable dose levels for the general human population based upon experimental animal data or limited human data.

a. An uncertainty factor of 10 shall generally be used when extrapolating from valid experimental results from studies on prolonged exposure to average healthy humans. This 10-fold factor is used to protect sensitive members of the human population.

b. An uncertainty factor of 100 shall generally be used when extrapolating from valid results of long-term studies on experimental animals when results of studies of human exposure are not available or are inadequate. In comparison to a, above, this represents an additional 10-fold uncertainty factor in extrapolating data from the average animal to the average human.

c. An uncertainty factor of up to 1000 shall generally be used when extrapolating from animal studies for which the exposure duration is less than chronic, but greater than subchronic (e.g., 90 days or more in length), or when other significant deficiencies in study quality are present, and when useful long-term human data are not available. In comparison to b, above, this represents an additional UF of up to 10-fold for less than chronic, but greater than subchronic, studies.

d. An UF of up to 3000 shall generally be used when extrapolating from animal studies for which the exposure duration is less than subchronic (e.g., 28 days). In comparison to b above, this represents an additional UF of up to 30-fold for less than subchronic studies (e.g., 28-day). The level of additional uncertainty applied for less than chronic exposures depends on the duration of the study used relative to the lifetime of the experimental animal.

e. An additional UF of between one and ten may be used when deriving a criterion from a LOAEL. This UF accounts for the lack of an identifiable NOAEL. The level of additional uncertainty applied may depend upon the severity and the incidence of the observed adverse effect.

f. An additional UF of between one and ten may be applied when there are limited effects data or incomplete sub-acute or chronic toxicity data (e.g., reproductive/developmental data). The level of quality and quantity of the experimental data available as well as structure-activity relationships may be used to determine the factor selected.

g. When deriving an UF in developing a Tier I criterion or Tier II value, the total uncertainty, as calculated following the guidance of sections 4.a through f, cited above, shall not exceed 10,000 for Tier I criteria and 30,000 for Tier II values.

5. All study results shall be converted, as necessary, to the standard unit for acceptable daily exposure of milligrams of toxicant per kilogram of body weight per day (mg/kg/day). Doses shall be adjusted for continuous exposure (i.e., seven days/week, 24 hours/day, etc.).

C. *Criteria and Value Derivation.*

1. *Standard Exposure Assumptions.* The following represent the standard exposure assumptions used to calculate Tier I criteria and Tier II values for carcinogens and noncarcinogens. Higher levels of exposure may be assumed by States and Tribes pursuant to Clean Water Act (CWA) section 510, or where appropriate in deriving site-specific criteria pursuant to procedure 1 in appendix F to part 132.

BW = body weight of an average human (BW = 70kg).

WC_d = per capita water consumption (both drinking and incidental exposure) for surface waters classified as public water supplies = two liters/day.

—or—

WC_r = per capita incidental daily water ingestion for surface waters not used as human drinking water sources = 0.01 liters/day.

FC = per capita daily consumption of regionally caught freshwater fish = 0.015kg/day (0.0036 kg/day for trophic level 3 and 0.0114 kg/day for trophic level 4).

BAF = bioaccumulation factor for trophic level 3 and trophic level 4, as derived using the BAF methodology in appendix B to part 132.

2. *Carcinogens.* The Tier I human cancer criteria or Tier II values shall be calculated as follows:

$$HCV = \frac{RAD \times BW}{WC + \left[\left(FC_{TL3} \times BAF_{TL3}^{HH}\right) + \left(FC_{TL4} \times BAF_{TL4}^{HH}\right)\right]}$$

Where:

HCV=Human Cancer Value in milligrams per liter (mg/L).

RAD=Risk associated dose in milligrams toxicant per kilogram body weight per day (mg/kg/day) that is associated with a lifetime incremental cancer risk equal to one in 100,000.

BW=weight of an average human (BW=70 kg).

WC_d=per capita water consumption (both drinking and incidental exposure) for surface waters classified as public water supplies=two liters/day.

or

WC_r=per capita incidental daily water ingestion for surface waters not used as human drinking water sources=0.01 liters/day.

FC_{TL3}=mean consumption of trophic level 3 of regionally caught freshwater fish=0.0036 kg/day.

FC_{TL4}=mean consumption of trophic level 4 of regionally caught freshwater fish=0.0114 kg/day.

$BAF^{HH}{}_{TL3}$=bioaccumulation factor for trophic level 3 fish, as derived using the BAF methodology in appendix B to part 132.

$BAF^{HH}{}_{TL4}$=bioaccumulation factor for trophic level 4 fish, as derived using the BAF methodology in appendix B to part 132.

3. *Noncarcinogens.* The Tier I human noncancer criteria or Tier II values shall be calculated as follows:

$$HNV = \frac{ADE \times BW \times RSC}{WC + \left[\left(FC_{TL3} \times BAF_{TL3}^{HH}\right) + \left(FC_{TL4} \times BAF_{TL4}^{HH}\right)\right]}$$

Where:

HNV=Human noncancer value in milligrams per liter (mg/L).

ADE=Acceptable daily exposure in milligrams toxicant per kilogram body weight per day (mg/kg/day).

RSC=Relative source contribution factor of 0.8. An RSC derived from actual exposure data may be developed using the methodology outlined by the 1980 National Guidelines (see 45 FR 79354).

BW=weight of an average human (BW=70 kg).

WC_d=per capita water consumption (both drinking and incidental exposure) for surface waters classified as public water supplies=two liters/day.

or

WC_r=per capita incidental daily water ingestion for surface waters not used as human drinking water sources=0.01 liters/day.

FC_{TL3}=mean consumption of trophic level 3 fish by regional sport fishers of regionally caught freshwater fish=0.0036 kg/day.

FC_{TL4}=mean consumption of trophic level 4 fish by regional sport fishers of regionally caught freshwater fish=0.0114 kg/day.

$BAF^{HH}{}_{TL3}$=human health bioaccumulation factor for edible portion of trophic level 3 fish, as derived using the BAF methodology in appendix B to part 132.

$BAF^{HH}{}_{TL4}$=human health bioaccumulation factor for edible portion of trophic level 4 fish, as derived using the BAF methodology in appendix B to part 132.

IV. REFERENCES

A. Howe, R.B., K.S. Crump and C. Van Landingham. 1986. Computer Program to Extrapolate Quantitative Animal Toxicity Data to Low Doses. Prepared for EPA under subcontract #2–251U–2745 to Research Triangle Institute.

B. U.S. Environmental Protection Agency. 1980. Water Quality Criteria Availability, Appendix C Guidelines and Methodology Used in the Preparation of Health Effects Assessment Chapters of the Consent Decree Water Quality Criteria Documents. Available from U.S. Environmental Protection Agency, Office of Water Resource Center (WH–550A), 1200 Pennsylvania Ave., NW., Washington, DC 20460.

C. U.S. Environmental Protection Agency. 1986. Guidelines for Carcinogen Risk Assessment. Available from U.S. Environmental Protection Agency, Office of Water Resource Center (WH–550A), 1200 Pennsylvania Ave., NW., Washington, DC 20460.

APPENDIX D TO PART 132—GREAT LAKES WATER QUALITY INITIATIVE METHODOLOGY FOR THE DEVELOPMENT OF WILDLIFE CRITERIA

Great Lakes States and Tribes shall adopt provisions consistent with (as protective as) this appendix.

I. INTRODUCTION

A. A Great Lakes Water Quality Wildlife Criterion (GLWC) is the concentration of a substance which is likely to, if not exceeded, protect avian and mammalian wildlife populations inhabiting the Great Lakes basin from adverse effects resulting from the ingestion of water and aquatic prey taken from surface waters of the Great Lakes System. These criteria are based on existing toxicological studies of the substance of concern and quantitative information about the exposure of wildlife species to the substance (i.e., food and water consumption rates). Since toxicological and exposure data for individual wildlife species are limited, a GLWC is derived using a methodology similar to that used to derive noncancer human health criteria (Barnes and Dourson, 1988; NAS, 1977; NAS, 1980; U.S. EPA, 1980). Separate avian and mammalian values are developed using taxonomic class-specific toxicity data and exposure data for five representative Great Lakes basin wildlife species. The wildlife species selected are representative of avian and mammalian species resident in the Great Lakes basin which are likely to experience the highest exposures to bioaccumulative contaminants through the aquatic food web; they are the bald eagle, herring gull, belted kingfisher, mink, and river otter.

B. This appendix establishes a methodology which is required when developing Tier I wildlife criteria for bioaccumulative chemicals of concern (BCCs). The use of the equation provided in the methodology is encouraged, but not required, for the development of Tier I criteria or Tier II values for pollutants other than those identified in Table 6–A for which Tier I criteria or Tier II values are determined to be necessary for the protection of wildlife in the Great Lakes basin. A discussion of the methodology for

deriving Tier II values can be found in the Great Lakes Water Quality Initiative Technical Support Document for Wildlife Criteria (Wildlife TSD).

C. In the event that this methodology is used to develop criteria for pollutants other than BCCs, or in the event that the Tier II methodology described in the Wildlife TSD is used to derive Tier II values, the methodology for deriving bioaccumulation factors under appendix B to part 132 must be used in either derivation. For chemicals which do not biomagnify to the extent of BCCs, it may be appropriate to select different representative species which are better examples of species with the highest exposures for the given chemical. The equation presented in this methodology, however, is still encouraged. In addition, procedure 1 of appendix F of this part describes the procedures for calculating site-specific wildlife criteria.

D. The term "wildlife value" (WV) is used to denote the value for each representative species which results from using the equation presented below, the value obtained from averaging species values within a class, or any value derived from application of the site-specific procedure provided in procedure 1 of appendix F of this part. The WVs calculated for the representative species are used to calculate taxonomic class-specific WVs. The WV is the concentration of a substance which, if not exceeded, should better protect the taxon in question.

E. "Tier I wildlife criterion," or "Tier I criterion" is used to denote the number derived from data meeting the Tier I minimum database requirements, and which will be protective of the two classes of wildlife. It is synonymous with the term "GLWC," and the two are used interchangeably.

II. CALCULATION OF WILDLIFE VALUES FOR TIER I CRITERIA

Table 4 of Part 132 and Table D–1 of this appendix contain criteria calculated by EPA using the methodology provided below.

A. *Equation for Avian and Mammalian Wildlife Values.* Tier I wildlife values for the pollutants designated BCCs pursuant to part 132 are to be calculated using the equation presented below.

$$WV = \frac{\dfrac{TD}{UF_A \times UF_S \times UF_L} \times Wt}{W + \sum \left(F_{TLi} \times BAF_{TLi}^{WL} \right)}$$

Where:

WV=Wildlife Value in milligrams of substance per liter (mg/L).

TD=Test Dose (TD) in milligrams of substance per kilograms per day (mg/kg-d) for the test species. This shall be either a NOAEL or a LOAEL.

UF$_A$=Uncertainty Factor (UF) for extrapolating toxicity data across species (unitless). A species-specific UF shall be selected and applied to each representative species, consistent with the equation.

UF$_S$=UF for extrapolating from subchronic to chronic exposures (unitless).

UF$_L$=UF for LOAEL to NOAEL extrapolations (unitless).

Wt=Average weight in kilograms (kg) for the representative species.

W=Average daily volume of water consumed in liters per day (L/d) by the representative species.

F$_{TLi}$=Average daily amount of food consumed from trophic level i in kilograms per day (kg/d) by the representative species.

BAF^{WL}$_{TLi}$=Bioaccumulation factor (BAF) for wildlife food in trophic level i in liters per kilogram (L/kg), developed using the BAF methodology in appendix B to part 132, Methodology for Development of Bioaccumulation Factors. For consumption of piscivorous birds by other birds (e.g., herring gull by eagles), the BAF is derived by multiplying the trophic level 3 BAF for fish by a biomagnification factor to account for the biomagnification from fish to the consumed birds.

B. *Identification of Representative Species for Protection.* For bioaccumulative chemicals, piscivorous species are identified as the focus of concern for wildlife criteria development in the Great Lakes. An analysis of known or estimated exposure components for avian and mammalian wildlife species is presented in the Wildlife TSD. This analysis identifies three avian species (eagle, kingfisher and herring gull) and two mammalian species (mink and otter) as representative species for protection. The TD obtained from toxicity data for each taxonomic class is used to calculate WVs for each of the five representative species.

C. *Calculation of Avian and Mammalian Wildlife Values and GLWC Derivation.* The avian WV is the geometric mean of the WVs calculated for the three representative avian species. The mammalian WV is the geometric mean of the WVs calculated for the two representative mammalian species. The lower of the mammalian and avian WVs must be selected as the GLWC.

III. PARAMETERS OF THE EFFECT COMPONENT OF THE WILDLIFE CRITERIA METHODOLOGY

A. *Definitions.* The following definitions provide additional specificity and guidance in the evaluation of toxicity data and the application of this methodology.

Acceptable endpoints. For the purpose of wildlife criteria derivation, acceptable subchronic and chronic endpoints are those which affect reproductive or developmental success, organismal viability or growth, or any other endpoint which is, or is directly related to, parameters that influence population dynamics.

Chronic effect. An adverse effect that is measured by assessing an acceptable endpoint, and results from continual exposure over several generations, or at least over a significant part of the test species' projected life span or life stage.

Lowest-observed-adverse-effect-level (LOAEL). The lowest tested dose or concentration of a substance which resulted in an observed adverse effect in exposed test organisms when all higher doses or concentrations resulted in the same or more severe effects.

No-observed-adverse-effect-level (NOAEL). The highest tested dose or concentration of a substance which resulted in no observed adverse effect in exposed test organisms where higher doses or concentrations resulted in an adverse effect.

Subchronic effect. An adverse effect, measured by assessing an acceptable endpoint, resulting from continual exposure for a period of time less than that deemed necessary for a chronic test.

B. *Minimum Toxicity Database for Tier I Criteria Development.* A TD value is required for criterion calculation. To derive a Tier I criterion for wildlife, the data set shall provide enough data to generate a subchronic or chronic dose-response curve for any given substance for both mammalian and avian species. In reviewing the toxicity data available which meet the minimum data requirements for each taxonomic class, the following order of preference shall be applied to select the appropriate TD to be used for calculation of individual WVs. Data from peer-reviewed field studies of wildlife species take precedence over other types of studies, where such studies are of adequate quality. An acceptable field study must be of subchronic or chronic duration, provide a defensible, chemical-specific dose-response curve in which cause and effect are clearly established, and assess acceptable endpoints as defined in this document. When acceptable wildlife field studies are not available, or determined to be of inadequate quality, the needed toxicity information may come from peer-reviewed laboratory studies. When laboratory studies are used, preference shall be given to laboratory studies with wildlife species over traditional laboratory animals to reduce uncertainties in making interspecies extrapolations. All available laboratory data and field studies shall be reviewed to corroborate the final GLWC, to assess the reasonableness of the toxicity value used, and to assess the appropriateness of any UFs which are applied. When evaluating the studies from which a test dose is derived in general, the following requirements must be met:

1. The mammalian data must come from at least one well-conducted study of 90 days or

greater designed to observe subchronic or chronic effects as defined in this document.

2. The avian data must come from at least one well-conducted study of 70 days or greater designed to observe subchronic or chronic effects as defined in this document.

3. In reviewing the studies from which a TD is derived for use in calculating a WV, studies involving exposure routes other than oral may be considered only when an equivalent oral daily dose can be estimated and technically justified because the criteria calculations are based on an oral route of exposure.

4. In assessing the studies which meet the minimum data requirements, preference should be given to studies which assess effects on developmental or reproductive endpoints because, in general, these are more important endpoints in ensuring that a population's productivity is maintained. The Wildlife TSD provides additional discussion on the selection of an appropriate toxicity study.

C. *Selection of TD Data.* In selecting data to be used in the derivation of WVs, the evaluation of acceptable endpoints, as defined in Section III.A of this appendix, will be the primary selection criterion. All data not part of the selected subset may be used to assess the reasonableness of the toxicity value and the appropriateness of the Ufs which are applied.

1. If more than one TD value is available within a taxonomic class, based on different endpoints of toxicity, that TD, which is likely to reflect best potential impacts to wildlife populations through resultant changes in mortality or fecundity rates, shall be used for the calculation of WVs.

2. If more than one TD is available within a taxonomic class, based on the same endpoint of toxicity, the TD from the most sensitive species shall be used.

3. If more than one TD based on the same endpoint of toxicity is available for a given species, the TD for that species shall be calculated using the geometric mean of those TDs.

D. *Exposure Assumptions in the Determination of the TD.* 1. In those cases in which a TD is available in units other than milligrams of substance per kilograms per day (mg/kg/d), the following procedures shall be used to convert the TD to the appropriate units prior to calculating a WV.

2. If the TD is given in milligrams of toxicant per liter of water consumed by the test animals (mg/L), the TD shall be multiplied by the daily average volume of water consumed by the test animals in liters per day (L/d) and divided by the average weight of the test animals in kilograms (kg).

3. If the TD is given in milligrams of toxicant per kilogram of food consumed by the test animals (mg/kg), the TD shall be multiplied by the average amount of food in kilo-grams consumed daily by the test animals (kg/d) and divided by the average weight of the test animals in kilograms (kg).

E. *Drinking and Feeding Rates.* 1. When drinking and feeding rates and body weight are needed to express the TD in milligrams of substance per kilograms per day (mg/kg/d), they are obtained from the study from which the TD was derived. If not already determined, body weight, and drinking and feeding rates are to be converted to a wet weight basis.

2. If the study does not provide the needed values, the values shall be determined from appropriate scientific literature. For studies done with domestic laboratory animals, either the Registry of Toxic Effects of Chemical Substances (National Institute for Occupational Safety and Health, the latest edition, Cincinnati, OH), or Recommendations for and Documentation of Biological Values for Use in Risk Assessment (U.S. EPA, 1988) should be consulted. When these references do not contain exposure information for the species used in a given study, either the allometric equations from Calder and Braun (1983) and Nagy (1987), which are presented below, or the exposure estimation methods presented in Chapter 4 of the Wildlife Exposure Factors Handbook (U.S. EPA, 1993), should be applied to approximate the needed feeding or drinking rates. Additional discussion and recommendations are provided in the Wildlife TSD. The choice of the methods described above is at the discretion of the State or Tribe.

3. For mammalian species, the general allometric equations are:

a. $F = 0.0687 \times (Wt)^{0.82}$

Where:

F = Feeding rate of mammalian species in kilograms per day (kg/d) dry weight.
Wt = Average weight in kilograms (kg) of the test animals.

b. $W = 0.099 \times (Wt)^{0.90}$

Where:

W = Drinking rate of mammalian species in liters per day (L/d).
Wt = Average weight in kilograms (kg) of the test animals.

4. For avian species, the general allometric equations are:

a. $F = 0.0582 (Wt)^{0.65}$

Where:

F = Feeding rate of avian species in kilograms per day (kg/d) dry weight.
Wt = Average weight in kilograms (kg) of the test animals.

b. $W = 0.059 \times (Wt)^{0.67}$

Where:

W = Drinking rate of avian species in liters per day (L/d).
Wt = Average weight in kilograms (kg) of the test animals.

F. *LOAEL to NOAEL Extrapolations (UF_L).* In those cases in which a NOAEL is unavailable as the TD and a LOAEL is available, the LOAEL may be used to estimate the NOAEL. If used, the LOAEL shall be divided by an UF to estimate a NOAEL for use in deriving WVs. The value of the UF shall not be less than one and should not exceed 10, depending on the dose-response curve and any other available data, and is represented by UF_L in the equation expressed in Section II.A of this appendix. Guidance for selecting an appropriate UF_L, based on a review of available wildlife toxicity data, is available in the Wildlife TSD.

G. *Subchronic to Chronic Extrapolations (US_S).* In instances where only subchronic data are available, the TD may be derived from subchronic data. In such cases, the TD shall be divided by an UF to extrapolate from subchronic to chronic levels. The value of the UF shall not be less than one and should not exceed 10, and is represented by UF_S in the equation expressed in Section II.A of this appendix. This factor is to be used when assessing highly bioaccumulative substances where toxicokinetic considerations suggest that a bioassay of limited length underestimates chronic effects. Guidance for selecting an appropriate UF_S, based on a review of available wildlife toxicity data, is available in the Wildlife TSD.

H. *Interspecies Extrapolations (UF_A).* 1. The selection of the UF_A shall be based on the available toxicological data and on available data concerning the physicochemical, toxicokinetic, and toxicodynamic properties of the substance in question and the amount and quality of available data. This value is an UF that is intended to account for differences in toxicological sensitivity among species. Guidance for selecting an appropriate UF_A, based on a review of available wildlife toxicity data, is available in the Wildlife TSD. Additional discussion of an interspecies UF located in appendix A to the Great Lakes Water Quality Initiative Technical Support Document for Human Health Criteria may be useful in determining the appropriate value for UF_A.

2. For the derivation of Tier I criteria, a UF_A shall not be less than one and should not exceed 100, and shall be applied to each of the five representative species, based on existing data and best professional judgment. The value of UF_A may differ for each of the representative species.

3. For Tier I wildlife criteria, the UF_A shall be used only for extrapolating toxicity data across species within a taxonomic class, except as provided below. The Tier I UF_A is not intended for interclass extrapolations because of the poorly defined comparative toxicokinetic and toxicodynamic parameters between mammals and birds. However, an interclass extrapolation employing a UF_A may be used for a given chemical if it can be supported by a validated biologically-based dose-response model or by an analysis of interclass toxicological data, considering acceptable endpoints, for a chemical analog that acts under the same mode of toxic action.

IV. PARAMETERS OF THE EXPOSURE COMPONENT OF THE WILDLIFE CRITERIA METHODOLOGY

A. *Drinking and Feeding Rates of Representative Species.* The body weights (Wt), feeding rates (F_{Tli}), drinking rates (W), and trophic level dietary composition (as food ingestion rate and percent in diet) for each of the five representative species are presented in Table D–2 of this appendix. Guidance on incorporating the non-aquatic portion of the bald eagle and mink diets in the criteria calculations is available in the Wildlife TSD.

B. *BAFs.* The Methodology for Development of Bioaccumulation Factors is presented in appendix B to part 132. Trophic level 3 and 4 BAFs are used to derive Wvs because these are the trophic levels at which the representative species feed.

V. REFERENCES

A. Barnes, D.G. and M. Dourson. 1988. Reference Dose (RfD): Description and Use in Health Risk Assessments. Regul. Toxicol. Pharmacol. 8:471–486.

B. Calder III, W.A. and E.J. Braun. 1983. Scaling of Osmotic Regulation in Mammals and Birds. American Journal of Physiology. 244:601–606.

C. Nagy, K.A. 1987. Field Metabolic Rate and Food Requirement Scaling in Mammals and Birds. Ecological Monographs. 57(2):111–128.

D. National Academy of Sciences. 1977. Chemical Contaminants: Safety and Risk Assessment, *in* Drinking Water and Health, Volume 1. National Academy Press.

E. National Academy of Sciences. 1980. Problems of Risk Estimation, *in* Drinking Water and Health, Volume 3. National Academy Press.

F. National Institute for Occupational Safety and Health. Latest edition. Registry of Toxic Effects of Chemical Substances. Division of Standards Development and Technology Transfer. (Available only on microfiche or as an electronic database.)

G. U.S. EPA. 1980. Appendix C. Guidelines and Methodology Used in the Preparation of Health Effect Assessment Chapters of the Consent Decree Water Criteria Documents, pp. 79347–79357 *in* Water Quality Criteria Documents; Availability. Available from U.S. Environmental Protection Agency, Office of Water Resource Center (WH–550A), 1200 Pennsylvania Ave., NW, Washington, DC 20460.

H. U.S. EPA. 1988. Recommendations for, and documentation of, biological values for use in risk assessment. NTIS–PB88–179874.

I. U.S. EPA. 1993. Wildlife Exposure Factors Handbook, Volumes I and II. EPA/600/R–93/187a and b.

Tables to Appendix D to Part 132

TABLE D–1—TIER I GREAT LAKES WILDLIFE CRITERIA

Substance	Criterion (µg/L)
DDT & Metabolites	1.1E–5
Mercury	1.3E–3
PCBs (total)	7.4E–5
2,3,7,8-TCDD	3.1E–9

TABLE D–2—EXPOSURE PARAMETERS FOR THE FIVE REPRESENTATIVE SPECIES IDENTIFIED FOR PROTECTION

Species (units)	Adult body weight (kg)	Water ingestion rate (L/day)	Food ingestion rate of prey in each trophic level (kg/day)	Trophic level of prey (percent of diet)
Mink	0.80	0.081	TL3: 0.159; Other: 0.0177	TL3: 90; Other: 10.
Otter	7.4	0.600	TL3: 0.977; TL4: 0.244	TL3: 80; TL4: 20.
Kingfisher	0.15	0.017	TL3: 0.0672	TL3: 100.
Herring gull	1.1	0.063	TL3: 0.192; TL4: 0.0480 Other: 0.0267	Fish: 90—TL3: 80; TL4: 20. Other: 10.
Bald eagle	4.6	0.160	TL3: 0.371; TL4: 0.0929 PB: 00283; Other: 0.0121	Fish: 92—TL3: 80; TL4: 20. Birds: 8—PB: 70; non-aquatic: 30.

NOTE: TL3=trophic level three fish; TL4=trophic level four fish; PB=piscivorous birds; Other=non-aquatic birds and mammals.

APPENDIX E TO PART 132—GREAT LAKES WATER QUALITY INITIATIVE ANTIDEGRADATION POLICY

Great Lakes States and Tribes shall adopt provisions consistent with (as protective as) appendix E to part 132.

The State or Tribe shall adopt an antidegradation standard applicable to all waters of the Great Lakes System and identify the methods for implementing such a standard. Consistent with 40 CFR 131.12, an acceptable antidegradation standard and implementation procedure are required elements of a State's or Tribe's water quality standards program. Consistent with 40 CFR 131.6, a complete water quality standards submission needs to include both an antidegradation standard and antidegradation implementation procedures. At a minimum, States and Tribes shall adopt provisions in their antidegradation standard and implementation methods consistent with sections I, II, III and IV of this appendix, applicable to pollutants identified as bioaccumulative chemicals of concern (BCCs).

I. ANTIDEGRADATION STANDARD

This antidegradation standard shall be applicable to any action or activity by any source, point or nonpoint, of pollutants that is anticipated to result in an increased loading of BCCs to surface waters of the Great Lakes System and for which independent regulatory authority exists requiring compliance with water quality standards. Pursuant to this standard:

A. Existing instream water uses, as defined pursuant to 40 CFR 131, and the level of water quality necessary to protect existing uses shall be maintained and protected. Where designated uses of the waterbody are impaired, there shall be no lowering of the water quality with respect to the pollutant or pollutants which are causing the impairment;

B. Where, for any parameter, the quality of the waters exceed levels necessary to support the propagation of fish, shellfish, and wildlife and recreation in and on the waters, that water shall be considered high quality for that parameter consistent with the definition of high quality water found at section II.A of this appendix and that quality shall be maintained and protected unless the State or Tribe finds, after full satisfaction of the intergovernmental coordination and public participation provisions of the State's or Tribe's continuing planning process, that allowing lower water quality is necessary to accommodate important economic or social development in the area in which the waters are located. In allowing such degradation, the State or Tribe shall assure water quality adequate to protect existing uses fully. Further, the State or Tribe shall assure that there shall be achieved the highest statutory and regulatory requirements for all new and existing point sources and all cost-effective and reasonable best management practices for nonpoint source control. The State or Tribe shall utilize the Antidegradation Implementation Procedures adopted pursuant to the requirements of this regulation in determining if any lowering of water quality will be allowed;

C. Where high quality waters constitute an outstanding national resource, such as waters of national and State parks and wildlife refuges and waters of exceptional recreational or ecological significance, that water quality shall be maintained and protected; and

D. In those cases where the potential lowering of water quality is associated with a thermal discharge, the decision to allow such degradation shall be consistent with section 316 of the Clean Water Act (CWA).

II. ANTIDEGRADATION IMPLEMENTATION PROCEDURES

A. *Definitions.*

Control Document. Any authorization issued by a State, Tribal or Federal agency to any source of pollutants to waters under its jurisdiction that specifies conditions under which the source is allowed to operate.

High quality waters. High quality waters are water bodies in which, on a parameter by parameter basis, the quality of the waters exceeds levels necessary to support propagation of fish, shellfish, and wildlife and recreation in and on the water.

Lake Superior Basin—Outstanding International Resource Waters. Those waters designated as such by a Tribe or State consistent with the September 1991 Bi-National Program to Restore and Protect the Lake Superior Basin. The purpose of such designations shall be to ensure that any new or increased discharges of Lake Superior bioaccumulative substances of immediate concern are subject to best technology in process and treatment requirements.

Lake Superior Basin—Outstanding National Resource Waters. Those waters designated as such by a Tribe or State consistent with the September 1991 Bi-National Program to Restore and Protect the Lake Superior Basin. The purpose of such designations shall be to prohibit new or increased discharges of Lake Superior bioaccumulative substances of immediate concern from point sources in these areas.

Lake Superior bioaccumulative substances of immediate concern. A list of substances identified in the September 1991 Bi-National Program to Restore the Lake Superior Basin. They include: 2, 3, 7, 8-TCDD; octachlorostyrene; hexachlorobenzene; chlordane; DDT, DDE, and other metabolites; toxaphene; PCBs; and mercury. Other chemicals may be added to the list following States' or Tribes' assessments of environmental effects and impacts and after public review and comment.

Outstanding National Resource Waters. Those waters designated as such by a Tribe or State. The State or Tribal designation shall describe the quality of such waters to serve as the benchmark of the water quality that shall be maintained and protected. Waters that may be considered for designation as Outstanding National Resource Waters include, but are not limited to, water bodies that are recognized as:

Important because of protection through official action, such as Federal or State law, Presidential or secretarial action, international treaty, or interstate compact;

Having exceptional recreational significance;

Having exceptional ecological significance;

Having other special environmental, recreational, or ecological attributes; or waters whose designation as Outstanding National Resource Waters is reasonably necessary for the protection of other waters so designated.

Significant Lowering of Water Quality. A significant lowering of water quality occurs when there is a new or increased loading of any BCC from any regulated existing or new facility, either point source or nonpoint source for which there is a control document or reviewable action, as a result of any activity including, but not limited to:

(1) Construction of a new regulated facility or modification of an existing regulated facility such that a new or modified control document is required;

(2) Modification of an existing regulated facility operating under a current control document such that the production capacity of the facility is increased;

(3) Addition of a new source of untreated or pretreated effluent containing or expected to contain any BCC to an existing wastewater treatment works, whether public or private;

(4) A request for an increased limit in an applicable control document;

(5) Other deliberate activities that, based on the information available, could be reasonably expected to result in an increased loading of any BCC to any waters of the Great Lakes System.

b. Notwithstanding the above, changes in loadings of any BCC within the existing capacity and processes, and that are covered by the existing applicable control document, are not subject to an antidegradation review. These changes include, but are not limited to:

(1) Normal operational variability;

(2) Changes in intake water pollutants;

(3) Increasing the production hours of the facility, (e.g., adding a second shift); or

(4) Increasing the rate of production.

C. Also, excluded from an antidegradation review are new effluent limits based on improved monitoring data or new water quality criteria or values that are not a result of changes in pollutant loading.

B. For all waters, the Director shall ensure that the level of water quality necessary to protect existing uses is maintained. In order to achieve this requirement, and consistent with 40 CFR 131.10, water quality standards

561

use designations must include all existing uses. Controls shall be established as necessary on point and nonpoint sources of pollutants to ensure that the criteria applicable to the designated use are achieved in the water and that any designated use of a downstream water is protected. Where water quality does not support the designated uses of a waterbody or ambient pollutant concentrations exceed water quality criteria applicable to that waterbody, the Director shall not allow a lowering of water quality for the pollutant or pollutants preventing the attainment of such uses or exceeding such criteria.

C. For Outstanding National Resource Waters:

1. The Director shall ensure, through the application of appropriate controls on pollutant sources, that water quality is maintained and protected.

2. Exception. A short-term, temporary (i.e., weeks or months) lowering of water quality may be permitted by the Director.

D. For high quality waters, the Director shall ensure that no action resulting in a lowering of water quality occurs unless an antidegradation demonstration has been completed pursuant to section III of this appendix and the information thus provided is determined by the Director pursuant to section IV of this appendix to adequately support the lowering of water quality.

1. The Director shall establish conditions in the control document applicable to the regulated facility that prohibit the regulated facility from undertaking any deliberate action, such that there would be an increase in the rate of mass loading of any BCC, unless an antidegradation demonstration is provided to the Director and approved pursuant to section IV of this appendix prior to commencement of the action. Imposition of limits due to improved monitoring data or new water quality criteria or values, or changes in loadings of any BCC within the existing capacity and processes, and that are covered by the existing applicable control document, are not subject to an antidegradation review.

2. For BCCs known or believed to be present in a discharge, from a point or nonpoint source, a monitoring requirement shall be included in the control document. The control document shall also include a provision requiring the source to notify the Director or any increased loadings. Upon notification, the Director shall require actions as necessary to reduce or eliminate the increased loading.

3. Fact Sheets prepared pursuant to 40 CFR 124.8 and 124.56 shall reflect any conditions developed under sections II.D.1 or II.D.2 of this appendix and included in a permit.

E. *Special Provisions for Lake Superior.* The following conditions apply in addition to those specified in section II.B through II.C of this appendix for waters of Lake Superior so designated.

1. A State or Tribe may designate certain specified areas of the Lake Superior Basin as Lake Superior Basin—Outstanding National Resource Waters for the purpose of prohibiting the new or increased discharge of Lake Superior bioaccumulative substances of immediate concern from point sources in these areas.

2. States and Tribes may designate all waters of the Lake Superior Basin as Outstanding International Resource Waters for the purpose of restricting the increased discharge of Lake Superior bioaccumulative substances of immediate concern from point sources consistent with the requirements of sections III.C and IV.B of this appendix.

F. *Exemptions.* Except as the Director may determine on a case-by-case basis that the application of these procedures is required to adequately protect water quality, or as the affected waterbody is an Outstanding National Resource Water as defined in section II.A of this appendix, the procedures in this part do not apply to:

1. Short-term, temporary (i.e., weeks or months) lowering of water quality;

2. Bypasses that are not prohibited at 40 CFR 122.41(m); and

3. Response actions pursuant to the Comprehensive Environmental Response, Compensation and Liability Act (CERCLA), as amended, or similar Federal, State or Tribal authorities, undertaken to alleviate a release into the environment of hazardous substances, pollutants or contaminants which may pose an imminent and substantial danger to public health or welfare.

III. ANTIDEGRADATION DEMONSTRATION

Any entity seeking to lower water quality in a high quality water or create a new or increased discharge of Lake Superior bioaccumulative substances of immediate concern in a Lake Superior Outstanding International Resource Water must first, as required by sections II.D or II.E.2 of this appendix, submit an antidegradation demonstration for consideration by the Director. States and Tribes should tailor the level of detail and documentation in antidegradation reviews, to the specific circumstances encountered. The antidegradation demonstration shall include the following:

A. *Pollution Prevention Alternatives Analysis.* Identify any cost-effective pollution prevention alternatives and techniques that are available to the entity, that would eliminate or significantly reduce the extent to which the increased loading results in a lowering of water quality.

B. *Alternative or Enhanced Treatment Analysis.* Identify alternative or enhanced treatment techniques that are available to the entity that would eliminate the lowering of water quality and their costs relative to the cost of treatment necessary to achieve applicable effluent limitations.

C. *Lake Superior.* If the States or Tribes designate the waters of Lake Superior as Outstanding International Resource Waters pursuant to section II.E.2 of this appendix, then any entity proposing a new or increased discharge of any Lake Superior bioaccumulative substance of immediate concern to the Lake Superior Basin shall identify the best technology in process and treatment to eliminate or reduce the extent of the lowering of water quality. In this case, the requirements in section III.B of this appendix do not apply.

D. *Important Social or Economic Development Analysis.* Identify the social or economic development and the benefits to the area in which the waters are located that will be foregone if the lowering of water quality is not allowed.

E. *Special Provision for Remedial Actions.* Entities proposing remedial actions pursuant to the CERCLA, as amended, corrective actions pursuant to the Resource Conservation and Recovery Act, as amended, or similar actions pursuant to other Federal or State environmental statutes may submit information to the Director that demonstrates that the action utilizes the most cost effective pollution prevention and treatment techniques available, and minimizes the necessary lowering of water quality, in lieu of the information required by sections III.B through III.D of this appendix.

IV. ANTIDEGRADATION DECISION

A. Once the Director determines that the information provided by the entity proposing to increase loadings is administratively complete, the Director shall use that information to determine whether or not the lowering of water quality is necessary, and, if it is necessary, whether or not the lowering of water quality will support important social and economic development in the area. If the proposed lowering of water quality is either not necessary, or will not support important social and economic development, the Director shall deny the request to lower water quality. If the lowering of water quality is necessary, and will support important social and economic development, the Director may allow all or part of the proposed lowering to occur as necessary to accommodate the important social and economic development. In no event may the decision reached under this section allow water quality to be lowered below the minimum level required to fully support existing and designated uses. The decision of the Director shall be subject to the public participation requirements of 40 CFR 25.

B. If States designate the waters of Lake Superior as Outstanding International Resource Waters pursuant to section II.E.2 of this appendix, any entity requesting to lower water quality in the Lake Superior Basin as a result of the new or increased discharge of any Lake Superior bioaccumulative substance of immediate concern shall be required to install and utilize the best technology in process and treatment as identified by the Director.

APPENDIX F TO PART 132—GREAT LAKES WATER QUALITY INITIATIVE IMPLEMENTATION PROCEDURES

PROCEDURE 1: SITE-SPECIFIC MODIFICATIONS TO CRITERIA AND VALUES

Great Lakes States and Tribes shall adopt provisions consistent with (as protective as) this procedure.

A. Requirements for Site-specific Modifications to Criteria and Values. Criteria and values may be modified on a site-specific basis to reflect local environmental conditions as restricted by the following provisions. Any such modifications must be protective of designated uses and aquatic life, wildlife or human health and be submitted to EPA for approval. In addition, any site-specific modifications that result in less stringent criteria must be based on a sound scientific rationale and shall not be likely to jeopardize the continued existence of endangered or threatened species listed or proposed under section 4 of the Endangered Species Act (ESA) or result in the destruction or adverse modification of such species' critical habitat. More stringent modifications shall be developed to protect endangered or threatened species listed or proposed under section 4 of the ESA, where such modifications are necessary to ensure that water quality is not likely to jeopardize the continued existence of such species or result in the destruction or adverse modification of such species' critical habitat. More stringent modifications may also be developed to protect candidate (C1) species being considered by the U.S. Fish and Wildlife Service (FWS) for listing under section 4 of the ESA, where such modifications are necessary to protect such species.

1. *Aquatic Life.*

a. Aquatic life criteria or values may be modified on a site-specific basis to provide an additional level of protection, pursuant to authority reserved to the States and Tribes under Clean Water Act (CWA) section 510.

Guidance on developing site-specific criteria in these instances is provided in Chapter 3 of the U.S. EPA Water Quality Standards Handbook, Second Edition—Revised (1994).

b. Less stringent site-specific modifications to chronic or acute aquatic life criteria or values may be developed when:

i. The local water quality characteristics such as Ph, hardness, temperature, color, etc., alter the biological availability or toxicity of a pollutant; or

ii. The sensitivity of the aquatic organisms species that "occur at the site" differs from the species actually tested in developing the criteria. The phrase "occur at the site" includes the species, genera, families, orders, classes, and phyla that: are usually present at the site; are present at the site only seasonally due to migration; are present intermittently because they periodically return to or extend their ranges into the site; were present at the site in the past, are not currently present at the site due to degraded conditions, and are expected to return to the site when conditions improve; are present in nearby bodies of water, are not currently present at the site due to degraded conditions, and are expected to be present at the site when conditions improve. The taxa that "occur at the site" cannot be determined merely by sampling downstream and/or upstream of the site at one point in time. "Occur at the site" does not include taxa that were once present at the site but cannot exist at the site now due to permanent physical alteration of the habitat at the site resulting, for example, from dams, etc.

c. Less stringent modifications also may be developed to acute and chronic aquatic life criteria or values to reflect local physical and hydrological conditions.

Guidance on developing site-specific criteria is provided in Chapter 3 of the U.S. EPA Water Quality Standards Handbook, Second Edition—Revised (1994).

d. Any modifications to protect threatened or endangered aquatic species required by procedure 1.A of this appendix may be accomplished using either of the two following procedures:

i. If the Species Mean Acute Value (SMAV) for a listed or proposed species, or for a surrogate of such species, is lower than the calculated Final Acute Value (FAV), such lower SMAV may be used instead of the calculated FAV in developing site-specific modified criteria; or,

ii. The site-specific criteria may be calculated using the recalculation procedure for site-specific modifications described in Chapter 3 of the U.S. EPA Water Quality Standards Handbook, Second Edition—Revised (1994).

2. *Wildlife.*

a. Wildlife water quality criteria may be modified on a site-specific basis to provide an additional level of protection, pursuant to authority reserved to the States and Tribes under CWA section 510.

b. Less stringent site-specific modifications to wildlife water quality criteria may be developed when a site-specific bioaccumulation factor (BAF) is derived which is lower than the system-wide BAF derived under appendix B of this part. The modification must consider both the mobility of prey organisms and wildlife populations in defining the site for which criteria are developed. In addition, there must be a showing that:

i. Any increased uptake of the toxicant by prey species utilizing the site will not cause adverse effects in wildlife populations; and

ii. Wildlife populations utilizing the site or downstream waters will continue to be fully protected.

c. Any modification to protect endangered or threatened wildlife species required by procedure 1.A of this appendix must consider both the mobility of prey organisms and wildlife populations in defining the site for which criteria are developed, and may be accomplished by using the following recommended method.

i. The methodology presented in appendix D to part 132 is used, substituting appropriate species-specific toxicological, epidemiological, or exposure information, including changes to the BAF;

ii. An interspecies uncertainty factor of 1 should be used where epidemiological data are available for the species in question. If necessary, species-specific exposure parameters can be derived as presented in appendix D of this part;

iii. An intraspecies uncertainty factor (to account for protection of individuals within a wildlife population) should be applied in the denominator of the effect part of the wildlife equation in appendix D of this part in a manner consistent with the other uncertainty factors described in appendix D of this part; and

iv. The resulting wildlife value for the species in question should be compared to the two class-specific wildlife values which were previously calculated, and the lowest of the three shall be selected as the site-specific modification.

NOTE: Further discussion on the use of this methodology may be found in the Great Lakes Water Quality Initiative Technical Support Document for Wildlife Criteria.

3. *BAFs.*

a. BAFs may be modified on a site-specific basis to larger values, pursuant to the authority reserved to the States and Tribes under CWA section 510, where reliable data show that local bioaccumulation is greater than the system-wide value.

b. BAFs may be modified on a site-specific basis to lower values, where scientifically defensible, if:

i. The fraction of the total chemical that is freely dissolved in the ambient water is different than that used to derive the system-wide BAFs (i.e., the concentrations of particulate organic carbon and the dissolved organic carbon are different than those used to derive the system-wide BAFs);

ii. Input parameters of the Gobas model, such as the structure of the aquatic food web and the disequilibrium constant, are different at the site than those used to derive the system-wide BAFs;

iii. The percent lipid of aquatic organisms that are consumed and occur at the site is different than that used to derive the system-wide BAFs; or

iv. Site-specific field-measured BAFs or biota-sediment accumulation factor (BSAFs) are determined.

If site-specific BAFs are derived, they shall be derived using the methodology in appendix B of this part.

c. Any more stringent modifications to protect threatened or endangered species required by procedure 1.A of this appendix shall be derived using procedures set forth in the methodology in appendix B of this part.

4. *Human Health.*

a. Human health criteria or values may be modified on a site-specific basis to provide an additional level of protection, pursuant to authority reserved to the States and Tribes under CWA section 510. Human health criteria or values shall be modified on a site-specific basis to provide additional protection appropriate for highly exposed subpopulations.

b. Less stringent site-specific modifications to human health criteria or values may be developed when:

i. local fish consumption rates are lower than the rate used in deriving human health criteria or values under appendix C of this part; and/or

ii. a site-specific BAF is derived which is lower than that used in deriving human health criteria or values under appendix C of this part.

B. *Notification Requirements.* When a State proposes a site-specific modification to a criterion or value as allowed in section 4.A above, the State should notify the other Great Lakes States of such a proposal and, for less stringent criteria, supply appropriate justification.

C. *References.*

U.S. EPA. 1984. Water Quality Standards Handbook—Revised. Chapter 3 and Appendices. U.S. Environmental Protection Agency, Office of Water Resource Center (RC–4100), 1200 Pennsylvania Ave., NW., Washington, DC 20960.

PROCEDURE 2: VARIANCES FROM WATER QUALITY STANDARDS FOR POINT SOURCES

The Great Lakes States or Tribes may adopt water quality standards (WQS) variance procedures and may grant WQS variances for point sources pursuant to such procedures. Variance procedures shall be consistent with (as protective as) the provisions in this procedure.

A. *Applicability.* A State or Tribe may grant a variance to a WQS which is the basis of a water quality-based effluent limitation included in a National Pollutant Discharge Elimination System (NPDES) permit. A WQS variance applies only to the permittee requesting the variance and only to the pollutant or pollutants specified in the variance. A variance does not affect, or require

the State or Tribe to modify, the corresponding water quality standard for the waterbody as a whole.

1. This provision shall not apply to new Great Lakes dischargers or recommencing dischargers.

2. A variance to a water quality standard shall not be granted that would likely jeopardize the continued existence of any endangered or threatened species listed under Section 4 of the Endangered Species Act (ESA) or result in the destruction or adverse modification of such species' critical habitat.

3. A WQS variance shall not be granted if standards will be attained by implementing effluent limits required under sections 301(b) and 306 of the Clean Water Act (CWA) and by the permittee implementing cost-effective and reasonable best management practices for nonpoint source control.

B. *Maximum Timeframe for Variances.* A WQS variance shall not exceed five years or the term of the NPDES permit, whichever is less. A State or Tribe shall review, and modify as necessary, WQS variances as part of each water quality standards review pursuant to section 303(c) of the CWA.

C. *Conditions to Grant a Variance.* A variance may be granted if:

1. The permittee demonstrates to the State or Tribe that attaining the WQS is not feasible because:

a. Naturally occurring pollutant concentrations prevent the attainment of the WQS;

b. Natural, ephemeral, intermittent or low flow conditions or water levels prevent the attainment of the WQS, unless these conditions may be compensated for by the discharge of sufficient volume of effluent to enable WQS to be met without violating State or Tribal water conservation requirements;

c. Human-caused conditions or sources of pollution prevent the attainment of the WQS and cannot be remedied, or would cause more environmental damage to correct than to leave in place;

d. Dams, diversions or other types of hydrologic modifications preclude the attainment of the WQS, and it is not feasible to restore the waterbody to its original condition or to operate such modification in a way that would result in the attainment of the WQS;

e. Physical conditions related to the natural features of the waterbody, such as the lack of a proper substrate cover, flow, depth, pools, riffles, and the like, unrelated to chemical water quality, preclude attainment of WQS; or

f. Controls more stringent than those required by sections 301(b) and 306 of the CWA would result in substantial and widespread economic and social impact.

2. In addition to the requirements of C.1, above, the permittee shall also:

a. Show that the variance requested conforms to the requirements of the State's or Tribe's antidegradation procedures; and

b. Characterize the extent of any increased risk to human health and the environment associated with granting the variance compared with compliance with WQS absent the variance, such that the State or Tribe is able to conclude that any such increased risk is consistent with the protection of the public health, safety and welfare.

D. *Submittal of Variance Application.* The permittee shall submit an application for a variance to the regulatory authority issuing the permit. The application shall include:

1. All relevant information demonstrating that attaining the WQS is not feasible based on one or more of the conditions in section C.1 of this procedure; and,

2. All relevant information demonstrating compliance with the conditions in section C.2 of this procedure.

E. *Public Notice of Preliminary Decision.* Upon receipt of a complete application for a variance, and upon making a preliminary decision regarding the variance, the State or Tribe shall public notice the request and preliminary decision for public comment pursuant to the regulatory authority's Administrative Procedures Act and shall notify the other Great Lakes States and Tribes of the preliminary decision. This public notice requirement may be satisfied by including the supporting information for the variance and the preliminary decision in the public notice of a draft NPDES permit.

F. *Final Decision on Variance Request.* The State or Tribe shall issue a final decision on the variance request within 90 days of the expiration of the public comment period required in section E of this procedure. If all or part of the variance is approved by the State or Tribe, the decision shall include all permit conditions needed to implement those parts of the variance so approved. Such permit conditions shall, at a minimum, require:

1. Compliance with an initial effluent limitation which, at the time the variance is granted, represents the level currently achievable by the permittee, and which is no less stringent than that achieved under the previous permit;

2. That reasonable progress be made toward attaining the water quality standards for the waterbody as a whole through appropriate conditions;

3. When the duration of a variance is shorter than the duration of a permit, compliance with an effluent limitation sufficient to meet the underlying water quality standard, upon the expiration of said variance; and

4. A provision that allows the permitting authority to reopen and modify the permit based on any State or Tribal triennial water quality standards revisions to the variance.

The State shall deny a variance request if the permittee fails to make the demonstrations required under section C of this procedure.

G. *Incorporating Variance into Permit.* The State or Tribe shall establish and incorporate into the permittee's NPDES permit all conditions needed to implement the variance as determined in section F of this procedure.

H. *Renewal of Variance.* A variance may be renewed, subject to the requirements of sections A through G of this procedure. As part of any renewal application, the permittee shall again demonstrate that attaining WQS is not feasible based on the requirements of section C of this procedure. The permittee's application shall also contain information concerning its compliance with the conditions incorporated into its permit as part of the original variance pursuant to sections F and G of this procedure. Renewal of a variance may be denied if the permittee did not comply with the conditions of the original variance.

I. *EPA Approval.* All variances and supporting information shall be submitted by the State or Tribe to the appropriate EPA regional office and shall include:

1. Relevant permittee applications pursuant to section D of this procedure;

2. Public comments and records of any public hearings pursuant to section E of this procedure;

3. The final decision pursuant to section F of this procedure; and,

4. NPDES permits issued pursuant to section G of this procedure.

5. Items required by sections I.1 through I.3. of this procedure shall be submitted by the State within 30 days of the date of the final variance decision. The item required by section I.4 of this procedure shall be submitted in accordance with the State or Tribe Memorandum of Agreement with the Regional Administrator pursuant to 40 CFR 123.24.

6. EPA shall review the State or Tribe submittal for compliance with the CWA pursuant to 40 CFR 123.44, and 40 CFR 131.21.

J. *State WQS Revisions.* All variances shall be appended to the State or Tribe WQS rules.

Procedure 3: Total Maximum Daily Loads, Wasteload Allocations for Point Sources, Load Allocations for Nonpoint Sources, Wasteload Allocations in the Absence of a TMDL, and Preliminary Wasteload Allocations for Purposes of Determining the Need for Water Quality Based Effluent Limits

The Great Lakes States and Tribes shall adopt provisions consistent with (as protective as) this procedure 3 for the purpose of developing Total Maximum Daily Loads (TMDLs), Wasteload Allocations (WLAs) in the Absence of TMDLs, and Preliminary

Wasteload Allocations for Purposes of Determining the Need for Water Quality Based Effluent Limits (WQBELs), except as specifically provided.

A. Where a State or Tribe develops an assessment and remediation plan that the State or Tribe certifies meets the requirements of sections B through F of this procedure and public participation requirements applicable to TMDLs, and that has been approved by EPA as meeting those requirements under 40 CFR 130.6, the assessment and remediation plan may be used in lieu of a TMDL for purposes of appendix F to part 132. Assessment and remediation plans under this procedure may include, but are not limited to, Lakewide Management Plans, Remedial Action Plans, and State Water Quality Management Plans. Also, any part of an assessment and remediation plan that also satisfies one or more requirements under Clean Water Act (CWA) section 303(d) or implementing regulations may be incorporated by reference into a TMDL as appropriate. Assessment and remediation plans under this section should be tailored to the level of detail and magnitude for the watershed and pollutant being assessed.

B. *General Conditions of Application.* Except as provided in §132.4, the following are conditions applicable to establishing TMDLs for all pollutants and pollutant parameters in the Great Lakes System, with the exception of whole effluent toxicity, unless otherwise provided in procedure 6 of appendix F. Where specified, these conditions also apply to wasteload allocations (WLAs) calculated in the absence of TMDLs and to preliminary WLAs for purposes of determining the needs for WQBELs under procedure 5 of appendix F.

1. *TMDLs Required.* TMDLs shall, at a minimum, be established in accordance with the listing and priority setting process established in section 303(d) of the CWA and at 40 CFR 130.7. Where water quality standards cannot be attained immediately, TMDLs must reflect reasonable assurances that water quality standards will be attained in a reasonable period of time. Some TMDLs may be based on attaining water quality standards over a period of time, with specific controls on individual sources being implemented in stages. Determining the reasonable period of time in which water quality standards will be met is a case-specific determination considering a number of factors including, but not limited to: receiving water characteristics; persistence, behavior and ubiquity of pollutants of concern; type of remediation activities necessary; available regulatory and non-regulatory controls; and individual State or Tribal requirements for attainment of water quality standards.

2. *Attainment of Water Quality Standards.* A TMDL must ensure attainment of applicable water quality standards, including all numeric and narrative criteria, Tier I criteria, and Tier II values for each pollutant or pollutants for which a TMDL is established.

3. *TMDL Allocations.*

a. TMDLs shall include WLAs for point sources and load allocations (LAs) for nonpoint sources, including natural background, such that the sum of these allocations is not greater than the loading capacity of the water for the pollutant(s) addressed by the TMDL, minus the sum of a specified margin of safety (MOS) and any capacity reserved for future growth.

b. Nonpoint source LAs shall be based on:

i. Existing pollutant loadings if changes in loadings are not reasonably anticipated to occur;

ii. Increases in pollutant loadings that are reasonably anticipated to occur;

iii. Anticipated decreases in pollutant loadings if such decreased loadings are technically feasible and are reasonably anticipated to occur within a reasonable time period as a result of implementation of best management practices or other load reduction measures. In determining whether anticipated decreases in pollutant loadings are technically feasible and can reasonably be expected to occur within a reasonable period of time, technical and institutional factors shall be considered. These decisions are case-specific and should reflect the particular TMDL under consideration.

c. WLAs. The portion of the loading capacity not assigned to nonpoint sources including background, or to an MOS, or reserved for future growth is allocated to point sources. Upon reissuance, NPDES permits for these point sources must include effluent limitations consistent with WLAs in EPA-approved or EPA-established TMDLs.

d. Monitoring. For LAs established on the basis of subsection b.iii above, monitoring data shall be collected and analyzed in order to validate the TMDL's assumptions, to verify anticipated load reductions, to evaluate the effectiveness of controls being used to implement the TMDL, and to revise the WLAs and LAs as necessary to ensure that water quality standards will be achieved within the time-period established in the TMDL.

4. *WLA Values.* If separate EPA-approved or EPA-established TMDLs are prepared for different segments of the same watershed, and the separate TMDLs each include WLAs for the same pollutant for one or more of the same point sources, then WQBELs for that pollutant for the point source(s) shall be consistent with the most stringent of those WLAs in order to ensure attainment of all applicable water quality standards.

5. *Margin of Safety (MOS).* Each TMDL shall include a MOS sufficient to account for technical uncertainties in establishing the TMDL and shall describe the manner in which the MOS is determined and incorporated into the TMDL. The MOS may be

provided by leaving a portion of the loading capacity unallocated or by using conservative modeling assumptions to establish WLAs and LAs. If a portion of the loading capacity is left unallocated to provide a MOS, the amount left unallocated shall be described. If conservative modeling assumptions are relied on to provide a MOS, the specific assumptions providing the MOS shall be identified.

6. *More Stringent Requirements.* States and Tribes may exercise authority reserved to them under section 510 of the CWA to develop more stringent TMDLs (including WLAs and LAs) than are required herein, provided that all LAs in such TMDLs reflect actual nonpoint source loads or those loads that can reasonably be expected to occur within a reasonable time-period as a result of implementing nonpoint source controls.

7. *Accumulation in Sediments.* TMDLs shall reflect, where appropriate and where sufficient data are available, contributions to the water column from sediments inside and outside of any applicable mixing zones. TMDLs shall be sufficiently stringent so as to prevent accumulation of the pollutant of concern in sediments to levels injurious to designated or existing uses, human health, wildlife and aquatic life.

8. *Wet Weather Events.* Notwithstanding the exception provided for the establishment of controls on wet weather point sources in §132.4(e)(1), TMDLs shall reflect, where appropriate and where sufficient data are available, discharges resulting from wet weather events. This procedure does not provide specific procedures for considering discharges resulting from wet weather events. However, some of the provisions of procedure 3 may be deemed appropriate for considering wet weather events on a case-by-case basis.

9. *Background Concentration of Pollutants.* The representative background concentration of pollutants shall be established in accordance with this subsection to develop TMDLs, WLAs calculated in the absence of a TMDL, or preliminary WLAs for purposes of determining the need for WQBELs under procedure 5 of appendix F. Background loadings may be accounted for in a TMDL through an allocation to a single "background" category or through individual allocations to the various background sources.

a. *Definition of Background.* "Background" represents all loadings that: (1) flow from upstream waters into the specified watershed, waterbody or waterbody segment for which a TMDL, WLA in the absence of a TMDL or preliminary WLA for the purpose of determining the need for a WQBEL is being developed; (2) enter the specified watershed, waterbody or waterbody segment through atmospheric deposition or sediment release or resuspension; or (3) occur within the watershed, waterbody or waterbody segment as a result of chemical reactions.

b. *Data considerations.* When determining what available data are acceptable for use in calculating background, the State or Tribe should use best professional judgment, including consideration of the sampling location and the reliability of the data through comparison to reported analytical detection levels and quantification levels. When data in more than one of the data sets or categories described in section B.9.c.i through B.9.c.iii below exist, best professional judgment should be used to select the one data set that most accurately reflects or estimates background concentrations. Pollutant degradation and transport information may be considered when utilizing pollutant loading data.

c. *Calculation requirements.* Except as provided below, the representative background concentration for a pollutant in the specified watershed, waterbody or waterbody segment shall be established on a case-by-case basis as the geometric mean of:

i. Acceptable available water column data; or

ii. Water column concentrations estimated through use of acceptable available caged or resident fish tissue data; or

iii. Water column concentrations estimated through use of acceptable available or projected pollutant loading data.

d. *Detection considerations.*

i. Commonly accepted statistical techniques shall be used to evaluate data sets consisting of values both above and below the detection level.

ii. When all of the acceptable available data in a data set or category, such as water column, caged or resident fish tissue or pollutant loading data, are below the level of detection for a pollutant, then all the data for that pollutant in that data set shall be assumed to be zero.

10. *Effluent Flow.* If WLAs are expressed as concentrations of pollutants, the TMDL shall also indicate the point source effluent flows assumed in the analyses. Mass loading limitations established in NPDES permits must be consistent with both the WLA and assumed effluent flows used in establishing the TMDL.

11. *Reserved Allocations.* TMDLs may include reserved allocations of loading capacity to accommodate future growth and additional sources. Where such reserved allocations are not included in a TMDL, any increased loadings of the pollutant for which the TMDL was developed that are due to a new or expanded discharge shall not be allowed unless the TMDL is revised in accordance with these proceudres to include an allocation for the new or expanded discharge.

C. *Mixing Zones for Bioaccumulative Chemicals of Concern (BCCs).* The following requirements shall be applied in establishing TMDLs, WLAs in the absence of TMDLs, and preliminary WLAs for purposes of determining the need for WQBELs under procedure 5 of appendix F, for BCCs:

1. There shall be no mixing zones available for new discharges of BCCs to the Great Lakes System. WLAs established through TMDLs, WLAs in the absence of TMDLs, and preliminary WLAs for purposes of determining the need for WQBELs for new discharges of BCCs shall be set no higher than the most stringent applicable water quality criteria or values for the BCCs in question. This prohibition takes effect for a Great Lakes State or Tribe on the date EPA approves the State's or Tribe's submission of such prohibition or publishes a notice under 40 CFR 132.5(f) identifying that prohibition as applying to discharges within the State or Federal Tribal reservation.

2. For purposes of section C of procedure 3 of appendix F, new discharges are defined as: (1) a "discharge of pollutants" (as defined in 40 CFR 122.2) to the Great Lakes System from a building, structure, facility, or installation, the construction of which commences after the date the prohibition in section C.1 takes effect in that State or Tribe; (2) a new discharge from an existing Great Lakes discharger that commences after the date the prohibition in section C.1 takes effect in that State or Tribe; or (3) an expanded discharge from an existing Great Lakes discharger that commences after the date the prohibition in section C.1 takes effect in that State or Tribe, except for those expanded discharges resulting from changes in loadings of any BCC within the existing capacity and processes (e.g., normal operational variability, changes in intake water pollutants, increasing the production hours of the facility or adding additional shifts, or increasing the rate of production), and that are covered by the existing applicable control document. Not included within the definition of "new discharge" are new or expanded discharges of BCCs from a publicly owned treatment works (POTW as defined at 40 CFR 122.2) when such discharges are necessary to prevent a public health threat to the community (e.g., a situation where a community with failing septic systems is connected to a POTW to avert a potential public health threat from these failing systems). These and all other discharges of BCCs are defined as existing discharges.

3. Up until November 15, 2010, mixing zones for BCCs may be allowed for existing discharges to the Great Lakes System pursuant to the procedures specified in sections D and E of this procedure.

4. Except as provided in sections C.5 and C.6 of this procedure, permits issued on or after this provision takes effect in a Great Lakes State or Tribe shall not authorize mixing zones for existing discharges of BCCs to the Great Lakes System after November 15, 2010. After November 15, 2010, WLAs established through TMDLs, WLAs established in the absence of TMDLs, and preliminary WLAs for purposes of determining the need for WQBELs under procedure 5 of appendix F for existing discharges of BCCs to the Great Lakes System shall be equal to the most stringent applicable water quality criteria or values for the BCCs in question.

5. *Exception for Water Conservation.* Great Lakes States and Tribes may grant mixing zones for any existing discharge of BCCs to the Great Lakes System beyond the date specified in section C.4 of this procedure where it can be demonstrated, on a case-by-case basis, that failure to grant a mixing zone would preclude water conservation measures that would lead to overall load reductions in BCCs, even though higher concentrations of BCCs occur in the effluent. Such mixing zones must also be consistent with sections D and E of this procedure.

6. *Exception for Technical and Economic Considerations.* Great Lakes States and Tribes may grant mixing zones beyond the date specified in section C.4 of this procedure for any existing discharge of a BCC to the Great Lakes System upon the request of a discharger, subject to sections C.6.a through C.6.c below.

a. The State or Tribe must determine that:

i. The discharger is in compliance with and will continue to implement, for the BCC in question, all applicable requirements of Clean Water Act sections 118, 301, 302, 303, 304, 306, 307, 401, and 402, including existing National Pollutant Discharge Elimination System (NPDES) water-quality based effluent limitations; and

ii. The discharger has reduced and will continue to reduce the loading of the BCC for which a mixing zone is requested to the maximum extent possible, such that any additional controls or pollution prevention measures to reduce or ultimately eliminate the BCC discharge would result in unreasonable economic effects on the discharger or the affected community because the controls or measures are not feasible or cost-effective.

b. Any mixing zone established pursuant to this section shall:

i. Not result in any less stringent limitations than those existing prior to November 13, 2000;

ii. Be no larger than necessary to account for the technical constraints and economic effects identified pursuant to paragraph C.6.a.ii above;

iii. Meet all applicable acute and chronic aquatic life, wildlife and human health criteria and values within and at the edge of the mixing zone or be consistent with the applicable TMDL or assessment and remediation plan authorized under procedure 3.A.

iv. Be accompanied, as appropriate, by a permit condition requiring the discharger to implement an ambient monitoring plan to ensure compliance with water quality standards and consistency with any applicable TMDL or such other strategy consistent with Section A of this procedure, including the evaluation of alternative means for reducing the BCC from other sources in the watershed; and

v. Be limited to one permit term unless the permitting authority makes a new determination in accordance with this section for each successive permit application in which a mixing zone for the BCC is sought.

c. For each draft NPDES permit that would allow a mixing zone for one or more BCCs after November 15, 2010, the fact sheet or statement of basis for the draft permit that is required to be made available through public notice under 40 CFR 124.6(e) shall:

i. Specify the mixing provisions used in calculating the permit limits; and

ii. Identify each BCC for which a mixing zone is proposed.

7. Any mixing zone authorized under sections C.3, C.5 or C.6 must be consistent with sections D and E of this procedure, as applicable.

D. *Deriving TMDLs, WLAs, and LAs for Point and Nonpoint Sources: WLAs in the Absence of a TMDL; and Preliminary WLAs for Purposes of Determining the Need for WQBELs for OWGL.* This section addresses conditions for deriving TMDLs for Open Waters of the Great Lakes (OWGL), inland lakes and other waters of the Great Lakes System with no appreciable flow relative to their volumes. State and Tribal procedures to derive TMDLs under this section must be consistent with (as protective as) the general conditions in section B of this procedure, CWA section 303(d), existing regulations (40 CFR 130.7), section C of this procedure, and sections D.1. through D.4 below. State and Tribal procedures to derive WLAs calculated in the absence of a TMDL and preliminary WLAs for purposes of determining the need for WQBELs under procedure 5 of appendix F must be consistent with sections B.9, C.1, C3 through C.6, and D. 1 through D.4 of this procedure.

1. Individual point source WLAs and preliminary WLAs for purposes of determining the need for WQBELs under procedure 5 of appendix F shall assume no greater dilution than one part effluent to 10 parts receiving water for implementation of numeric and narrative chronic criteria and values (including, but not limited to human cancer criteria, human cancer values, human noncancer values, human noncancer criteria, wildlife criteria, and chronic aquatic life criteria and values) unless an alternative mixing zone is demonstrated as appropriate in a mixing zone demonstration conducted pursuant to section F of this procedure. In no case shall a mixing zone be granted that exceeds

the area where discharge-induced mixing occurs.

2. Appropriate mixing zone assumptions to be used in calculating load allocations for nonpoint sources shall be determined, consistent with applicable State or Tribal requirements, on a case-by-case basis.

3. WLAs and preliminary WLAs based on acute aquatic life criteria or values shall not exceed the Final Acute Value (FAV), unless a mixing zone demonstration is conducted and approved pursuant to section F of this procedure. If mixing zones from two or more proximate sources interact or overlap, the combined effect must be evaluated to ensure that applicable criteria and values will be met in the area where acute mixing zones overlap.

4. In no case shall a mixing zone be granted that would likely jeopardize the continued existence of any endangered or threatened species listed under section 4 of the ESA or result in the destruction or adverse modification of such species' critical habitat.

E. *Deriving TMDLs, WLAs, and LAs for Point and Nonpoint Sources; WLAs in the Absence of a TMDL; and Preliminary WLAs for the Purposes of Determining the Need for WQBELs for Great Lakes Systems Tributaries and Connecting Channels.* This section describes conditions for deriving TMDLs for tributaries and connecting channels of the Great Lakes System that exhibit appreciable flows relative to their volumes. State and Tribal procedures to derive TMDLs must be consistent with the general conditions listed in section B of this procedure, section C of this procedure, existing TMDL regulations (40 CFR 130.7) and specific conditions E.1 through E.5. State and Tribal procedures to derive WLAs calculated in the absence of a TMDL, and preliminary WLAs for purposes of determining reasonable potential under procedure 5 of this appendix for discharges to tributaries and connecting channels must be consistent with sections B.9, C.1, C.3 through C.6, and E.1 through E.5 of this procedure.

1. *Stream Design.* These design flows must be used unless data exist to demonstrate that an alternative stream design flow is appropriate for stream-specific and pollutant-specific conditions. For purposes of calculating a TMDL, WLAs in the absence of a TMDL, or preliminary WLAs for the purposes of determining reasonable potential under procedure 5 of this appendix, using a steady-state model, the stream design flows shall be:

a. The 7-day, 10-year stream design flow (7Q10), or the 4-day, 3-year biologically-based stream design flow for chronic aquatic life criteria or values;

b. The 1-day, 10-year stream design flow (1Q10), for acute aquatic life criteria or values;

c. The harmonic mean flow for human health criteria or values;

d. The 90-day, 10-year flow (90Q10) for wildlife criteria.

e. TMDLs, WLAs in the absence of TMDLs, and preliminary WLAs for the purpose of determining the need for WQBELs calculated using dynamic modelling do not need to incorporate the stream design flows specified in sections E.1.a through E.1.d of this procedure.

2. *Loading Capacity.* The loading capacity is the greatest amount of loading that a water can receive without violating water quality standards. The loading capacity is initially calculated at the farthest downstream location in the watershed drainage basin. The maximum allowable loading consistent with the attainment of each applicable numeric criterion or value for a given pollutant is determined by multiplying the applicable criterion or value by the flow at the farthest downstream location in the tributary basin at the design flow condition described above. This loading is then compared to the loadings at sites within the basin to assure that applicable numeric criteria or values for a given pollutant are not exceeded at all applicable sites. The lowest load is then selected as the loading capacity.

3. *Pollutant Degradation.* TMDLs, WLAs in the absence of a TMDL and preliminary WLAs for purposes of determining the need for WQBELs under procedure 5 of appendix F shall be based on the assumption that a pollutant does not degrade. However, the regulatory authority may take into account degradation of the pollutant if each of the following conditions are met.

a. Scientifically valid field studies or other relevant information demonstrate that degradation of the pollutant is expected to occur under the full range of environmental conditions expected to be encountered;

b. Scientifically valid field studies or other relevant information address other factors that affect the level of pollutants in the water column including, but not limited to, resuspension of sediments, chemical speciation, and biological and chemical transformation.

4. *Acute Aquatic Life Criteria and Values.* WLAs and LAs established in a TMDL, WLAs in the absence of a TMDL, and preliminary WLAs for the purpose of determining the need for WQBELs based on acute aquatic life criteria or values shall not exceed the FAV, unless a mixing zone demonstration is completed and approved pursuant to section F of this procedure. If mixing zones from two or more proximate sources interact or overlap, the combined effect must be evaluated to ensure that applicable criteria and values will be met in the area where any applicable acute mixing zones overlap. This acute WLA review shall include, but not be limited to, consideration of:

a. The expected dilution under all effluent flow and concentration conditions at stream design flow;

b. Maintenance of a zone of passage for aquatic organisms; and

c. Protection of critical aquatic habitat.

In no case shall a permitting authority grant a mixing zone that would likely jeopardize the continued existence of any endangered or threatened species listed under section 4 of the ESA or result in the destruction or adverse modification of such species' critical habitat.

5. *Chronic Mixing Zones.* WLAs and LAs established in a TMDL, WLAs in the absence of a TMDL, and preliminary WLAs for the purposes of determining the need for WQBELs for protection of aquatic life, wildlife and human health from chronic effects shall be calculated using a dilution fraction no greater than 25 percent of the stream design flow unless a mixing zone demonstration pursuant to section F of this procedure is conducted and approved. A demonstration for a larger mixing zone may be provided, if approved and implemented in accordance with section F of this procedure. In no case shall a permitting authority grant a mixing zone that would likely jeopardize the continued existence of any endangered or threatened species listed under section 4 of the ESA or result in the destruction or adverse modification of such species' critical habitat.

F. *Mixing Zone Demonstration Requirements.*

1. For purposes of establishing a mixing zone other than as specified in sections D and E above, a mixing zone demonstration must:

a. Describe the amount of dilution occurring at the boundaries of the proposed mixing zone and the size, shape, and location of the area of mixing, including the manner in which diffusion and dispersion occur;

b. For sources discharging to the open waters of the Great Lakes (OWGLs), define the location at which discharge-induced mixing ceases;

c. Document the substrate character and geomorphology within the mixing zone;

d. Show that the mixing zone does not interfere with or block passage of fish or aquatic life;

e. Show that the mixing zone will be allowed only to the extent that the level of the pollutant permitted in the waterbody would not likely jeopardize the continued existence of any endangered or threatened species listed under section 4 of the ESA or result in the destruction or adverse modification of such species' critical habitat;

f. Show that the mixing zone does not extend to drinking water intakes;

g. Show that the mixing zone would not otherwise interfere with the designated or existing uses of the receiving water or downstream waters;

h. Document background water quality concentrations;

i. Show that the mixing zone does not promote undesirable aquatic life or result in a dominance of nuisance species; and

j. Provide that by allowing additional mixing/dilution:

i. Substances will not settle to form objectionable deposits;

ii. Floating debris, oil, scum, and other matter in concentrations that form nuisances will not be produced; and

iii. Objectionable color, odor, taste or turbidity will not be produced.

2. In addition, the mixing zone demonstration shall address the following factors:

a. Whether or not adjacent mixing zones overlap;

b. Whether organisms would be attracted to the area of mixing as a result of the effluent character; and

c. Whether the habitat supports endemic or naturally occurring species.

3. The mixing zone demonstration must be submitted to EPA for approval. Following approval of a mixing zone demonstration consistent with sections F.1 and F.2, adjustment to the dilution ratio specified in section D.1 of this procedure shall be limited to the dilution available in the area where discharger-induced mixing occurs.

4. The mixing zone demonstration shall be based on the assumption that a pollutant does not degrade within the proposed mixing zone, unless:

a. Scientifically valid field studies or other relevant information demonstrate that degradation of the pollutant is expected to occur under the full range of environmental conditions expected to be encountered; and

b. Scientifically valid field studies or other relevant information address other factors that affect the level of pollutants in the water column including, but not limited to, resuspension of sediments, chemical speciation, and biological and chemical transformation.

PROCEDURE 4: ADDITIVITY

The Great Lakes States and Tribes shall adopt additivity provisions consistent with (as protective as) this procedure.

A. The Great Lakes States and Tribes shall adopt provisions to protect human health from the potential adverse additive effects from both the noncarcinogenic and carcinogenic components of chemical mixtures in effluents. For the chlorinated dibenzo-p-dioxins (CDDs) and chlorinated dibenzofurans (CDFs) listed in Table 1, potential adverse additive effects in effluents shall be accounted for in accordance with section B of this procedure.

B. *Toxicity Equivalency Factors (TEFs)/Bioaccumulation Equivalency Factors (BEFs).*

1. The TEFs in Table 1 and BEFs in Table 2 shall be used when calculating a 2,3,7,8-TCDD toxicity equivalence concentration in effluent to be used when implementing both human health noncancer and cancer criteria. The chemical concentration of each CDDs and CDFs in effluent shall be converted to a 2,3,7,8-TCDD toxicity equivalence concentration in effluent by (a) multiplying the chemical concentration of each CDDs and CDFs in the effluent by the appropriate TEF in Table 1 below, (b) multiplying each product from step (a) by the BEF for each CDDs and CDFs in Table 2 below, and (c) adding all final products from step (b). The equation for calculating the 2,3,7,8-TCDD toxicity equivalence concentration in effluent is:

$$(TEC)_{tcdd} = \sum (C)_x (TEF)_x (BEF)_x$$

where:

$(TEC)_{tcdd}$=2,3,7,8-TCDD toxicity equivalence concentration in effluent

$(C)_x$=concentration of total chemical x in effluent

$(TEF)_x$=TCDD toxicity equivalency factor for x

$(BEF)_x$=TCDD bioaccumulation equivalency factor for x

2. The 2,3,7,8-TCDD toxicity equivalence concentration in effluent shall be used when developing waste load allocations under procedure 3, preliminary waste load allocations for purposes of determining reasonable potential under procedure 5, and for purposes of establishing effluent quality limits under procedure 5.

TABLE 1—TOXICITY EQUIVALENCY FACTORS FOR CDDS AND CDFS

Congener	TEF
2,3,7,8-TCDD	1.0
1,2,3,7,8-PeCDD	0.5
1,2,3,4,7,8-HxCDD	0.1
1,2,3,6,7,8-HxCDD	0.1
1,2,3,7,8,9-HxCDD	0.1
1,2,3,4,6,7,8-HpCDD	0.01
OCDD	0.001
2,3,7,8-TCDF	0.1
1,2,3,7,8-PeCDF	0.05
2,3,4,7,8-PeCDF	0.5
1,2,3,4,7,8-HxCDF	0.1
1,2,3,6,7,8-HxCDF	0.1
2,3,4,6,7,8-HxCDF	0.1
1,2,3,7,8,9-HxCDF	0.1
1,2,3,4,6,7,8-HpCDF	0.01
1,2,3,4,7,8,9-HpCDF	0.01
OCDF	0.001

TABLE 2—BIOACCUMULATION EQUIVALENCY FACTORS FOR CDDS AND CDFS

Congener	BEF
2,3,7,8-TCDD	1.0
1,2,3,7,8-PeCDD	0.9
1,2,3,4,7,8-HxCDD	0.3
1,2,3,6,7,8-HxCDD	0.1
1,2,3,7,8,9-HxCDD	0.1

TABLE 2—BIOACCUMULATION EQUIVALENCY
FACTORS FOR CDDS AND CDFS—Continued

Congener	BEF
1,2,3,4,6,7,8-HpCDD	0.05
OCDD	0.01
2,3,7,8-TCDF	0.8
1,2,3,7,8-PeCDF	0.2
2,3,4,7,8-PeCDF	1.6
1,2,3,4,7,8-HxCDF	0.08
1,2,3,6,7,8-HxCDF	0.2
2,3,4,6,7,8-HxCDF	0.7
1,2,3,7,8,9-HxCDF	0.6
1,2,3,4,6,7,8-HpCDF	0.01
1,2,3,4,7,8,9-HpCDF	0.4
OCDF	0.02

PROCEDURE 5: REASONABLE POTENTIAL TO
EXCEED WATER QUALITY STANDARDS

Great Lakes States and Tribes shall adopt
provisions consistent with (as protective as)
this procedure. If a permitting authority determines that a pollutant is or may be discharged into the Great Lakes System at a
level which will cause, have the reasonable
potential to cause, or contribute to an excursion above any Tier I criterion or Tier II
value, the permitting authority shall incorporate a water quality-based effluent limitation (WQBEL) in an NPDES permit for the
discharge of that pollutant. When facility-
specific effluent monitoring data are available, the permitting authority shall make
this determination by developing preliminary effluent limitations (PEL) and com-
paring those effluent limitations to the projected effluent quality (PEQ) of the discharge in accordance with the following procedures. In all cases, the permitting authority shall use any valid, relevant, representative information that indicates a reasonable
potential to exceed any Tier I criterion or
Tier II value.

A. *Developing Preliminary Effluent Limitations on the Discharge of a Pollutant From a
Point Source.*

1. The permitting authority shall develop
preliminary wasteload allocations (WLAs)
for the discharge of the pollutant from the
point source to protect human health, wildlife, acute aquatic life, and chronic aquatic
life, based upon any existing Tier I criteria.
Where there is no Tier I criterion nor sufficient data to calculate a Tier I criterion, the
permitting authority shall calculate a Tier
II value for such pollutant for the protection
of human health, and aquatic life and the
preliminary WLAs shall be based upon such
values. Where there is insufficient data to
calculate a Tier II value, the permitting authority shall apply the procedure set forth in
section C of this procedure to determine
whether data must be generated to calculate
a Tier II value.

2. The following provisions in procedure 3
of appendix F shall be used as the basis for
determining preliminary WLAs in accord-

ance with section 1 of this procedure: proce-
dure 3.B.9, Background Concentrations of
Pollutants; procedure 3.C, Mixing Zones for
Bioaccumulative Chemicals of Concern
(BCCs), procedures 3.C.1, and 3.C.3 through
3.C.6; procedure 3.D, Deriving TMDLs for
Discharges to Lakes (when the receiving
water is an open water of the Great Lakes
(OWGL), an inland lake or other water of the
Great Lakes System with no appreciable
flow relative to its volume); procedure 3.E,
Deriving TMDLs, WLAs and Preliminary
WLAs, and load allocations (LAs) for Discharges to Great Lakes System Tributaries
(when the receiving water is a tributary or
connecting channel of the Great Lakes that
exhibits appreciable flow relative to its volume); and procedure 3.F, Mixing Zone Demonstration Requirements.

3. The permitting authority shall develop
PELs consistent with the preliminary WLAs
developed pursuant to sections A.1 and A.2 of
this procedure, and in accordance with existing State or Tribal procedures for converting
WLAs into WQBELs. At a minimum:

a. The PELs based upon criteria and values
for the protection of human health and wildlife shall be expressed as monthly limitations;

b. The PELs based upon criteria and values
for the protection of aquatic life from chronic effects shall be expressed as either monthly limitations or weekly limitations; and

c. The PELs based upon the criteria and
values for the protection of aquatic life from
acute effects shall be expressed as daily limitations.

B. *Determining Reasonable Potential Using
Effluent Pollutant Concentration Data.*

If representative, facility-specific effluent
monitoring data samples are available for a
pollutant discharged from a point source to
the waters of the Great Lakes System, the
permitting authority shall apply the following procedures:

1. The permitting authority shall specify
the PEQ as the 95 percent confidence level of
the 95th percentile based on a log-normal
distribution of the effluent concentration; or
the maximum observed effluent concentration, whichever is greater. In calculating the
PEQ, the permitting authority shall identify
the number of effluent samples and the coefficient of variation of the effluent data, obtain the appropriate multiplying factor from
Table 1 of procedure 6 of appendix F, and
multiply the maximum effluent concentration by that factor. The coefficient of variation of the effluent data shall be calculated
as the ratio of the standard deviation of the
effluent data divided by the arithmetic average of the effluent data, except that where
there are fewer than ten effluent concentration data points the coefficient of variation
shall be specified as 0.6. If the PEQ exceeds
any of the PELs developed in accordance

573

with section A.3 of this procedure, the permitting authority shall establish a WQBEL in a NPDES permit for such pollutant.

2. In lieu of following the procedures under section B.1 of this procedure, the permitting authority may apply procedures consistent with the following:

a. The permitting authority shall specify the PEQ as the 95th percentile of the distribution of the projected population of daily values of the facility-specific effluent monitoring data projected using a scientifically defensible statistical method that accounts for and captures the long-term daily variability of the effluent quality, accounts for limitations associated with sparse data sets and, unless otherwise shown by the effluent data set, assumes a lognormal distribution of the facility-specific effluent data. If the PEQ exceeds the PEL based on the criteria and values for the protection of aquatic life from acute effects developed in accordance with section A.3 of this procedure, the permitting authority shall establish a WQBEL in an NPDES permit for such pollutant;

b. The permitting authority shall calculate the PEQ as the 95th percentile of the distribution of the projected population of monthly averages of the facility-specific effluent monitoring data using a scientifically defensible statistical method that accounts for and captures the long-term variability of the monthly average effluent quality, accounts for limitations associated with sparse data sets and, unless otherwise shown by the effluent data set, assumes a lognormal distribution of the facility-specific effluent data. If the PEQ exceeds the PEL based on criteria and values for the protection of aquatic life from chronic effects, human health or wildlife developed in accordance with section A.3 of this procedure, the permitting authority shall establish a WQBEL in an NPDES permit for such pollutant; and

c. The permitting authority shall calculate the PEQ as the 95th percentile of the distribution of the projected population of weekly averages of the facility-specific effluent monitoring data using a scientifically defensible statistical method that accounts for and captures the long-term variability of the weekly average effluent quality, accounts for limitations associated with sparse data sets and, unless otherwise shown by the effluent data set, assumes a lognormal distribution of the facility-specific effluent data. If the PEQ exceeds the PEL based on criteria and values to protect aquatic life from chronic effects developed in accordance with section A.3 of this procedure, the permitting authority shall establish a WQBEL in an NPDES permit for such pollutant.

C. *Developing Necessary Data to Calculate Tier II Values Where Such Data Does Not Currently Exist.*

1. Except as provided in sections C.2, C.4, or D of this procedure, for each pollutant listed in Table 6 of part 132 that a permittee reports as known or believed to be present in its effluent, and for which pollutant data sufficient to calculate Tier II values for noncancer human health, acute aquatic life and chronic aquatic life do not exist, the permitting authority shall take the following actions:

a. The permitting authority shall use all available, relevant information, including Quantitative Structure Activity Relationship information and other relevant toxicity information, to estimate ambient screening values for such pollutant which will protect humans from health effects other than cancer, and aquatic life from acute and chronic effects.

b. Using the procedures specified in sections A.1 and A.2 of this procedure, the permitting authority shall develop preliminary WLAs for the discharge of the pollutant from the point source to protect human health, acute aquatic life, and chronic aquatic life, based upon the estimated ambient screening values.

c. The permitting authority shall develop PELs in accordance with section A.3 of this procedure, which are consistent with the preliminary WLAs developed in accordance with section C.1.b of this procedure.

d. The permitting authority shall compare the PEQ developed according to the procedures set forth in section B of this procedure to the PELs developed in accordance with section C.1.c of this procedure. If the PEQ exceeds any of the PELs, the permitting authority shall generate or require the permittee to generate the data necessary to derive Tier II values for noncancer human health, acute aquatic life and chronic aquatic life.

e. The data generated in accordance with section C.1.d of this procedure shall be used in calculating Tier II values as required under section A.1 of this procedure. The calculated Tier II value shall be used in calculating the preliminary WLA and PEL under section A of this procedure, for purposes of determining whether a WQBEL must be included in the permit. If the permitting authority finds that the PEQ exceeds the calculated PEL, a WQBEL for the pollutant or a permit limit on an indicator parameter consistent with 40 CFR 122.44(d)(1)(vi)(C) must be included in the permit.

2. With the exception of bioaccumulative chemicals of concern (BCCs), a permitting authority is not required to apply the procedures set forth in section C.1 of this procedure or include WQBELs to protect aquatic life for any pollutant listed in Table 6 of part 132 discharged by an existing point source into the Great Lakes System, if:

a. There is insufficient data to calculate a Tier I criterion or Tier II value for aquatic life for such pollutant;

b. The permittee has demonstrated through a biological assessment that there are no acute or chronic effects on aquatic life in the receiving water; and

c. The permittee has demonstrated in accordance with procedure 6 of this appendix that the whole effluent does not exhibit acute or chronic toxicity.

3. Nothing in sections C.1 or C.2 of this procedure shall preclude or deny the right of a permitting authority to:

a. Determine, in the absence of the data necessary to derive a Tier II value, that the discharge of the pollutant will cause, have the reasonable potential to cause, or contribute to an excursion above a narrative criterion for water quality; and

b. Incorporate a WQBEL for the pollutant into an NPDES permit.

4. If the permitting authority develops a WQBEL consistent with section C.3 of this procedure, and the permitting authority demonstrates that the WQBEL developed under section C.3 of this procedure is at least as stringent as a WQBEL that would have been based upon the Tier II value or values for that pollutant, the permitting authority shall not be obligated to generate or require the permittee to generate the data necessary to derive a Tier II value or values for that pollutant.

D. *Consideration of Intake Pollutants in Determining Reasonable Potential.*

1. *General.*

a. Any procedures adopted by a State or Tribe for considering intake pollutants in water quality-based permitting shall be consistent with this section and section E.

b. The determinations under this section and section E shall be made on a pollutant-by-pollutant, outfall-by-outfall, basis.

c. This section and section E apply only in the absence of a TMDL applicable to the discharge prepared by the State or Tribe and approved by EPA, or prepared by EPA pursuant to 40 CFR 130.7(d), or in the absence of an assessment and remediation plan submitted and approved in accordance with procedure 3.A. of appendix F. This section and section E do not alter the permitting authority's obligation under 40 CFR 122.44(d)(vii)(B) to develop effluent limitations consistent with the assumptions and requirements of any available WLA for the discharge, which is part of a TMDL prepared by the State or Tribe and approved by EPA pursuant to 40 CFR 130.7, or prepared by EPA pursuant to 40 CFR 130.7(d).

2. *Definition of Same Body of Water.*

a. This definition applies to this section and section E of this procedure.

b. An intake pollutant is considered to be from the same body of water as the discharge if the permitting authority finds that the intake pollutant would have reached the vicinity of the outfall point in the receiving water within a reasonable period had it not been removed by the permittee. This finding may be deemed established if:

i. The background concentration of the pollutant in the receiving water (excluding any amount of the pollutant in the facility's discharge) is similar to that in the intake water;

ii. There is a direct hydrological connection between the intake and discharge points; and

iii. Water quality characteristics (e.g., temperature, Ph, hardness) are similar in the intake and receiving waters.

c. The permitting authority may also consider other site-specific factors relevant to the transport and fate of the pollutant to make the finding in a particular case that a pollutant would or would not have reached the vicinity of the outfall point in the receiving water within a reasonable period had it not been removed by the permittee.

d. An intake pollutant from groundwater may be considered to be from the same body of water if the permitting authority determines that the pollutant would have reached the vicinity of the outfall point in the receiving water within a reasonable period had it not been removed by the permittee, except that such a pollutant is not from the same body of water if the groundwater contains the pollutant partially or entirely due to human activity, such as industrial, commercial, or municipal operations, disposed actions, or treatment processes.

e. An intake pollutant is the amount of a pollutant that is present in waters of the United States (including groundwater as provided in section D.2.d of this procedure) at the time it is withdrawn from such waters by the discharger or other facility (e.g., public water supply) supplying the discharger with intake water.

3. *Reasonable Potential Determination.*

a. The permitting authority may use the procedure described in this section of procedure 5 in lieu of procedures 5.A through C provided the conditions specified below are met.

b. The permitting authority may determine that there is no reasonable potential for the discharge of an identified intake pollutant or pollutant parameter to cause or contribute to an excursion above a narrative or numeric water quality criterion within an applicable water quality standard where a discharger demonstrates to the satisfaction of the permitting authority (based upon information provided in the permit application or other information deemed necessary by the permitting authority) that:

i. The facility withdraws 100 percent of the intake water containing the pollutant from the same body of water into which the discharge is made;

ii. The facility does not contribute any additional mass of the identified intake pollutant to its wastewater;

575

iii. The facility does not alter the identified intake pollutant chemically or physically in a manner that would cause adverse water quality impacts to occur that would not occur if the pollutants were left instream;

iv. The facility does not increase the identified intake pollutant concentration, as defined by the permitting authority, at the edge of the mixing zone, or at the point of discharge if a mixing zone is not allowed, as compared to the pollutant concentration in the intake water, unless the increased concentration does not cause or contribute to an excursion above an applicable water quality standard; and

v. The timing and location of the discharge would not cause adverse water quality impacts to occur that would not occur if the identified intake pollutant were left instream.

c. Upon a finding under section D.3.b of this procedure that a pollutant in the discharge does not cause, have the reasonable potential to cause, or contribute to an excursion above an applicable water quality standard, the permitting authority is not required to include a WQBEL for the identified intake pollutant in the facility's permit, provided:

i. The NPDES permit fact sheet or statement of basis includes a specific determination that there is no reasonable potential for the discharge of an identified intake pollutant to cause or contribute to an excursion above an applicable narrative or numeric water quality criterion and references appropriate supporting documentation included in the administrative record;

ii. The permit requires all influent, effluent, and ambient monitoring necessary to demonstrate that the conditions in section D.3.b of this procedure are maintained during the permit term; and

iii. The permit contains a reopener clause authorizing modification or revocation and reissuance of the permit if new information indicates changes in the conditions in section D.3.b of this procedure.

d. Absent a finding under section D.3.b of this procedure that a pollutant in the discharge does not cause, have the reasonable potential to cause, or contribute to an excursion above an applicable water quality standard, the permitting authority shall use the procedures under sections 5.A through C of this procedure to determine whether a discharge causes, has the reasonable potential to cause, or contribute to an excursion above an applicable narrative or numeric water quality criterion.

E. *Consideration of Intake Pollutants in Establishing WQBELs.*

1. *General.* This section applies only when the concentration of the pollutant of concern upstream of the discharge (as determined using the provisions in procedure 3.B.9 of appendix F) exceeds the most stringent applicable water quality criterion for that pollutant.

2. The requirements of sections D.1-D.2 of this procedure shall also apply to this section.

3. *Intake Pollutants from the Same Body of Water.*

a. In cases where a facility meets the conditions in sections D.3.b.i and D.3.b.iii through D.3.b.v of this procedure, the permitting authority may establish effluent limitations allowing the facility to discharge a mass and concentration of the pollutant that are no greater than the mass and concentration of the pollutant identified in the facility's intake water ("no net addition limitations"). The permit shall specify how compliance with mass and concentration limitations shall be assessed. No permit may authorize "no net addition limitations" which are effective after March 23, 2007. After that date, WQBELs shall be established in accordance with procedure 5.F.2 of appendix F.

b. Where proper operation and maintenance of a facility's treatment system results in removal of a pollutant, the permitting authority may establish limitations that reflect the lower mass and/or concentration of the pollutant achieved by such treatment, taking into account the feasibility of establishing such limits.

c. For pollutants contained in intake water provided by a water system, the concentration of the intake pollutant shall be determined at the point where the raw water supply is removed from the same body of water, except that it shall be the point where the water enters the water supplier's distribution system where the water treatment system removes any of the identified pollutants from the raw water supply. Mass shall be determined by multiplying the concentration of the pollutant determined in accordance with this paragraph by the volume of the facility's intake flow received from the water system.

4. *Intake Pollutants from a Different Body of Water.* Where the pollutant in a facility's discharge originates from a water of the United States that is not the same body of water as the receiving water (as determined in accordance with section D.2 of this procedure), WQBELs shall be established based upon the most stringent applicable water quality criterion for that pollutant.

5. *Multiple Sources of Intake Pollutants.* Where a facility discharges intake pollutants that originate in part from the same body of water, and in part from a different body of water, the permitting authority may apply the procedures of sections E.3 and E.4 of this procedure to derive an effluent limitation reflecting the flow-weighted average of each source of the pollutant, provided that adequate monitoring to determine compliance

can be established and is included in the permit.

F. *Other Applicable Conditions.*

1. In addition to the above procedures, effluent limitations shall be established to comply with all other applicable State, Tribal and Federal laws and regulations, including technology-based requirements and antidegradation policies.

2. Once the permitting authority has determined in accordance with this procedure that a WQBEL must be included in an NPDES permit, the permitting authority shall:

a. Rely upon the WLA established for the point source either as part of any TMDL prepared under procedure 3 of this appendix and approved by EPA pursuant to 40 CFR 130.7, or as part of an assessment and remediation plan developed and approved in accordance with procedure 3.A of this appendix, or, in the absence of such TMDL or plan, calculate WLAs for the protection of acute and chronic aquatic life, wildlife and human health consistent with the provisions referenced in section A.1 of this procedure for developing preliminary wasteload allocations, and

b. Develop effluent limitations consistent with these WLAs in accordance with existing State or Tribal procedures for converting WLAs into WQBELs.

3. When determining whether WQBELs are necessary, information from chemical-specific, whole effluent toxicity and biological assessments shall be considered independently.

4. If the geometric mean of a pollutant in fish tissue samples collected from a waterbody exceeds the tissue basis of a Tier I criterion or Tier II value, after consideration of the variability of the pollutant's bioconcentration and bioaccumulation in fish, each facility that discharges detectable levels of such pollutant to that water has the reasonable potential to cause or contribute to an excursion above a Tier I criteria or a Tier II value and the permitting authority shall establish a WQBEL for such pollutant in the NPDES permit for such facility.

PROCEDURE 6: WHOLE EFFLUENT TOXICITY REQUIREMENTS

The Great Lakes States and Tribes shall adopt provisions consistent with (as protective as) procedure 6 of appendix F of part 132.

The following definitions apply to this part:

Acute toxic unit (TU_a). $100/LC_{50}$ where the LC_{50} is expressed as a percent effluent in the test medium of an acute whole effluent toxicity (WET) test that is statistically or graphically estimated to be lethal to 50 percent of the test organisms.

Chronic toxic unit (TU_c). $100/NOEC$ or $100/IC_{25}$, where the NOEC and IC_{25} are expressed as a percent effluent in the test medium.

Inhibition concentration 25 (IC_{25}). The toxicant concentration that would cause a 25 percent reduction in a non-quantal biological measurement for the test population. For example, the IC_{25} is the concentration of toxicant that would cause a 25 percent reduction in mean young per female or in growth for the test population.

No observed effect concentration (NOEC). The highest concentration of toxicant to which organisms are exposed in a full life-cycle or partial life-cycle (short-term) test, that causes no observable adverse effects on the test organisms (i.e., the highest concentration of toxicant in which the values for the observed responses are not statistically significantly different from the controls).

A. *Whole Effluent Toxicity Requirements.* The Great Lakes States and Tribes shall adopt whole effluent toxicity provisions consistent with the following:

1. A numeric acute WET criterion of 0.3 acute toxic units (TU_a) measured pursuant to test methods in 40 CFR part 136, or a numeric interpretation of a narrative criterion establishing that 0.3 TU_a measured pursuant to test methods in 40 CFR part 136 is necessary to protect aquatic life from acute effects of WET. At the discretion of the permitting authority, the foregoing requirement shall not apply in an acute mixing zone that is sized in accordance with EPA-approved State and Tribal methods.

2. A numeric chronic WET criterion of one chronic toxicity unit (TU_c) measured pursuant to test methods in 40 CFR part 136, or a numeric interpretation of a narrative criterion establishing that one TU_c measured pursuant to test methods in 40 CFR part 136 is necessary to protect aquatic life from the chronic effects of WET. At the discretion of the permitting authority, the foregoing requirements shall not apply within a chronic mixing zone consistent with: (a) procedures 3.D.1 and 3.D.4, for discharges to the open of the Great Lakes (OWGL), inland lakes and other waters of the Great Lakes System with no appreciable flow relative to their volume, or (b) procedure 3.E.5 for discharges to tributaries and connecting channels of the Great Lakes System.

B. *WET Test Methods.* All WET tests performed to implement or ascertain compliance with this procedure shall be performed in accordance with methods established in 40 CFR part 136.

C. *Permit Conditions.*

1. Where a permitting authority determines pursuant to section D of this procedure that the WET of an effluent is or may be discharged at a level that will cause, have the reasonable potential to cause, or contribute to an excursion above any numeric WET criterion or narrative criterion within a State's or Tribe's water quality standards, the permitting authority:

a. Shall (except as provided in section C.1.e of this procedure) establish a water quality-based effluent limitation (WQBEL) or WQBELs for WET consistent with section C.1.b of this procedure;

b. Shall calculate WQBELs pursuant to section C.1.a. of this procedure to ensure attainment of the State's or Tribe's chronic WET criteria under receiving water flow conditions described in procedures 3.E.1.a (or where applicable, with procedure 3.E.1.e) for Great Lakes System tributaries and connecting channels, and with mixing zones no larger than allowed pursuant to section A.2. of this procedure. Shall calculate WQBELs to ensure attainment of the State's or Tribe's acute WET criteria under receiving water flow conditions described in procedure 3.E.1.b (or where applicable, with procedure 3.E.1.e) for Great Lakes System tributaries and connecting channels, with an allowance for mixing zones no greater than specified pursuant to section A.1 of this procedure.

c. May specify in the NPDES permit the conditions under which a permittee would be required to perform a toxicity reduction evaluation.

d. May allow with respect to any WQBEL established pursuant to section C.1.a of this procedure an appropriate schedule of compliance consistent with procedure 9 of appendix F; and

e. May decide on a case-by-case basis that a WQBEL for WET is not necessary if the State's or Tribe's water quality standards do not contain a numeric criterion for WET, and the permitting authority demonstrates in accordance with 40 CFR 122.44(d)(1)(v) that chemical-specific effluent limits are sufficient to ensure compliance with applicable criteria.

2. Where a permitting authority lacks sufficient information to determine pursuant to section D of this procedure whether the WET of an effluent is or may be discharged at levels that will cause, have the reasonable potential to cause, or contribute to an excursion above any numeric WET criterion or narrative criterion within a State's or Tribe's water quality standards, then the permitting authority should consider including in the NPDES permit appropriate conditions to require generation of additional data and to control toxicity if found, such as:

a. WET testing requirements to generate the data needed to adequately characterize the toxicity of the effluent to aquatic life;

b. Language requiring a permit reopener clause to establish WET limits if any toxicity testing data required pursuant to section C.2.a of this procedure indicate that the WET of an effluent is or may be discharged at levels that will cause, have the reasonable potential to cause, or contribute to an excursion above any numeric WET criterion or

narrative criterion within a State's or Tribe's water quality standards.

3. Where sufficient data are available for a permitting authority to determine pursuant to section D of this procedure that the WET of an effluent neither is nor may be discharged at a level that will cause, have the reasonable potential to cause, or contribute to an excursion above any numeric WET criterion or narrative criterion within a State's or Tribe's water quality standards, the permitting authority may include conditions and limitations described in section C.2 of this procedure at its discretion.

D. *Reasonable Potential Determinations.* The permitting authority shall take into account the factors described in 40 CFR 122.44(d)(1)(ii) and, where representative facility-specific WET effluent data are available, apply the following requirements in determining whether the WET of an effluent is or may be discharged at a level that will cause, have the reasonable potential to cause, or contribute to an excursion above any numeric WET criterion or narrative criterion within a State's or Tribe's water quality standards.

1. The permitting authority shall characterize the toxicity of the discharge by:

a. Either averaging or using the maximum of acute toxicity values collected within the same day for each species to represent one daily value. The maximum of all daily values for the most sensitive species tested is used for reasonable potential determinations;

b. Either averaging or using the maximum of chronic toxicity values collected within the same calendar month for each species to represent one monthly value. The maximum of such values, for the most sensitive species tested, is used for reasonable potential determinations;

c. Estimating the toxicity values for the missing endpoint using a default acute-chronic ratio (ACR) of 10, when data exist for either acute WET or chronic WET, but not for both endpoints.

2. The WET of an effluent is or may be discharged at a level that will cause, or contribute to an excursion above any numeric acute WET criterion or numeric interpretation of a narrative criterion within a State's or Tribe's water quality standards, when effluent-specific information demonstrates that:

(TU$_a$ effluent) (B) (effluent flow/ (Qad+effluent flow))>AC

Where TU$_a$ effluent is the maximum measured acute toxicity of 100 percent effluent determined pursuant to section D.1.a. of this procedure, B is the multiplying factor taken from Table F6-1 of this procedure to convert the highest measured effluent toxicity value to the estimated 95th percentile toxicity value for the discharge, effluent flow is the same effluent flow used to calculate the preliminary wasteload allocations (WLAs) for individual pollutants to meet the acute criteria and values for those pollutants, AC is the numeric acute WET criterion or numeric interpretation of a narrative criterion established pursuant to section A.1 of this procedure and expressed in TU_a, and Qad is the amount of the receiving water available for dilution calculated using: (i) the specified design flow(s) for tributaries and connecting channels in section C.1.b of this procedure, or where appropriate procedure 3.E.1.e of appendix F, and using EPA-approved State and Tribal procedures for establishing acute mixing zones in tributaries and connecting channels, or (ii) the EPA-approved State and Tribal procedures for establishing acute mixing zones in OWGLs. Where there are less than 10 individual WET tests, the multiplying factor taken from Table F6–1 of this procedure shall be based on a coefficient of variation (CV) or 0.6. Where there are 10 or more individual WET tests, the multiplying factor taken from Table F6–1 shall be based on a CV calculated as the standard deviation of the acute toxicity values found in the WET tests divided by the arithmetic mean of those toxicity values.

3. The WET of an effluent is or may be discharged at a level that will cause, have the reasonable potential to cause, or contribute to an excursion above any numeric chronic WET criterion or numeric interpretation of a narrative criterion within a State's or Tribe's water quality standards, when effluent-specific information demonstrates that:

(TU$_c$ effluent) (B) (effluent flow/Qad+effluent flow))>CC

Where TU$_c$ effluent is the maximum measured chronic toxicity value of 100 percent effluent determined in accordance with section D.1.b. of this procedure, B is the multiplying factor taken from Table F6–1 of this procedure, effluent flow is the same effluent flow used to calculate the preliminary WLAs for individual pollutants to meet the chronic criteria and values for those pollutants, CC is the numeric chronic WET criterion or numeric interpretation of a narrative criterion established pursuant to section A.2 of this procedure and expressed in TU_c, and Qad is the amount of the receiving water available for dilution calculated using: (i) the design flow(s) for tributaries and connecting channels specified in procedure 3.E.1.a of appendix F, and where appropriate procedure 3.E.1.e of appendix F, and in accordance with the provisions of procedure 3.E.5 for chronic mixing zones, or (ii) procedures 3.D.1 and 3.D.4 for discharges to the OWGLs. Where there are less than 10 individual WET tests, the multiplying factor taken from Table F6–1 of this procedure shall be based on a CV of 0.6. Where there are 10 more individual WET tests, the multiplying factor taken from Table F6–1 of this procedure shall be based on a CV calculated as the standard deviation of the WET tests divided by the arithmetic mean of the WET tests.

TABLE F6–1—REASONABLE POTENTIAL MULTIPLYING FACTORS: 95% CONFIDENCE LEVEL AND 95% PROBABILITY BASIS

Number of Samples	Coefficient of variation																			
	0.1	0.2	0.3	0.4	0.5	0.6	0.7	0.8	0.9	1.0	1.1	1.2	1.3	1.4	1.5	1.6	1.7	1.8	1.9	2.0
1	1.4	1.9	2.6	3.6	4.7	6.2	8.0	10.1	12.6	15.5	18.7	22.3	26.4	30.8	35.6	40.7	46.2	52.1	58.4	64.9
2	1.3	1.6	2.0	2.5	3.1	3.8	4.6	5.4	6.4	7.4	8.5	9.7	10.9	12.2	13.6	15.0	16.4	17.9	19.5	21.1
3	1.2	1.5	1.8	2.1	2.5	3.0	3.5	4.0	4.6	5.2	5.8	6.5	7.2	7.9	8.6	9.3	10.0	10.8	11.5	12.3
4	1.2	1.4	1.7	1.9	2.2	2.6	2.9	3.3	3.7	4.2	4.6	5.0	5.5	6.0	6.4	6.9	7.4	7.8	8.3	8.8
5	1.2	1.4	1.6	1.8	2.1	2.3	2.6	2.9	3.2	3.6	3.9	4.2	4.5	4.9	5.2	5.6	5.9	6.2	6.6	6.9
6	1.1	1.3	1.5	1.7	1.9	2.1	2.4	2.6	2.9	3.1	3.4	3.7	3.9	4.2	4.5	4.7	5.0	5.2	5.5	5.7
7	1.1	1.3	1.4	1.6	1.8	2.0	2.2	2.4	2.6	2.8	3.1	3.3	3.5	3.7	3.9	4.1	4.3	4.5	4.7	4.9
8	1.1	1.3	1.4	1.6	1.7	1.9	2.1	2.3	2.4	2.6	2.8	3.0	3.2	3.3	3.5	3.7	3.9	4.0	4.2	4.3
9	1.1	1.2	1.4	1.5	1.7	1.8	2.0	2.1	2.3	2.4	2.6	2.8	2.9	3.1	3.2	3.4	3.5	3.6	3.8	3.9
10	1.1	1.2	1.3	1.5	1.6	1.7	1.9	2.0	2.2	2.3	2.4	2.6	2.7	2.8	3.0	3.1	3.2	3.3	3.4	3.6
11	1.1	1.2	1.3	1.4	1.6	1.7	1.8	1.9	2.1	2.2	2.3	2.4	2.5	2.7	2.8	2.9	3.0	3.1	3.2	3.3
12	1.1	1.2	1.3	1.4	1.5	1.6	1.7	1.9	2.0	2.1	2.2	2.3	2.4	2.5	2.6	2.7	2.8	2.9	3.0	3.0
13	1.1	1.2	1.3	1.4	1.5	1.6	1.7	1.8	1.9	2.0	2.1	2.2	2.3	2.4	2.5	2.5	2.6	2.7	2.8	2.9
14	1.1	1.2	1.3	1.4	1.4	1.5	1.6	1.7	1.8	1.9	2.0	2.1	2.2	2.3	2.3	2.4	2.5	2.6	2.6	2.7
15	1.1	1.2	1.2	1.3	1.4	1.5	1.6	1.7	1.8	1.8	1.9	2.0	2.1	2.2	2.2	2.3	2.4	2.4	2.5	2.5
16	1.1	1.1	1.2	1.3	1.4	1.5	1.6	1.6	1.7	1.8	1.9	1.9	2.0	2.1	2.1	2.2	2.3	2.3	2.4	2.4
17	1.1	1.1	1.2	1.3	1.4	1.4	1.5	1.6	1.7	1.7	1.8	1.9	1.9	2.0	2.0	2.1	2.2	2.2	2.3	2.3
18	1.1	1.1	1.2	1.3	1.3	1.4	1.5	1.6	1.6	1.7	1.7	1.8	1.9	1.9	2.0	2.0	2.1	2.1	2.2	2.2
19	1.1	1.1	1.2	1.3	1.3	1.4	1.5	1.5	1.6	1.6	1.7	1.8	1.8	1.9	1.9	2.0	2.0	2.0	2.1	2.1
20	1.1	1.1	1.2	1.2	1.3	1.4	1.4	1.5	1.5	1.6	1.6	1.7	1.7	1.8	1.8	1.9	1.9	2.0	2.0	2.0
30	1.0	1.1	1.1	1.1	1.2	1.2	1.2	1.3	1.3	1.3	1.3	1.4	1.4	1.4	1.4	1.5	1.5	1.5	1.5	1.5
40	1.0	1.0	1.1	1.1	1.1	1.1	1.1	1.1	1.2	1.2	1.2	1.2	1.2	1.2	1.2	1.2	1.2	1.2	1.3	1.3
50	1.0	1.0	1.0	1.0	1.0	1.0	1.0	1.1	1.1	1.1	1.1	1.1	1.1	1.1	1.1	1.1	1.1	1.1	1.1	1.1
60	1.0	1.0	1.0	1.0	1.0	1.0	1.0	1.0	1.0	1.0	1.0	1.0	1.0	1.0	1.0	1.0	1.0	1.0	1.0	1.0
70	1.0	1.0	1.0	1.0	1.0	0.9	0.9	0.9	0.9	0.9	0.9	0.9	0.9	0.9	0.9	0.9	0.9	0.9	0.9	0.9
80	1.0	1.0	1.0	0.9	0.9	0.9	0.9	0.9	0.9	0.9	0.9	0.9	0.9	0.9	0.9	0.8	0.8	0.8	0.8	0.8
90	1.0	1.0	0.9	0.9	0.9	0.9	0.9	0.9	0.9	0.8	0.8	0.8	0.8	0.8	0.8	0.8	0.8	0.8	0.8	0.8
100	1.0	1.0	0.9	0.9	0.9	0.9	0.9	0.8	0.8	0.8	0.8	0.8	0.8	0.8	0.8	0.8	0.8	0.8	0.7	0.7

PROCEDURE 7: LOADING LIMITS

The Great Lakes States and Tribes shall adopt provisions consistent with (as protective as) this procedure.

Whenever a water quality-based effluent limitation (WQBEL) is developed, the WQBEL shall be expressed as both a concentration value and a corresponding mass loading rate.

A. Both mass and concentration limits shall be based on the same permit averaging periods such as daily, weekly, or monthly averages, or in other appropriate permit averaging periods.

B. The mass loading rates shall be calculated using effluent flow rates that are consistent with those used in establishing the WQBELs expressed in concentration.

PROCEDURE 8: WATER QUALITY-BASED EFFLUENT LIMITATIONS BELOW THE QUANTIFICATION LEVEL

The Great Lakes States and Tribes shall adopt provisions consistent with (as protective as) this procedure.

When a water quality-based effluent limitation (WQBEL) for a pollutant is calculated to be less than the quantification level:

A. *Permit Limits.* The permitting authority shall designate as the limit in the NPDES permit the WQBEL exactly as calculated.

B. *Analytical Method and Quantification Level.*

1. The permitting authority shall specify in the permit the most sensitive, applicable, analytical method, specified in or approved under 40 CFR part 136, or other appropriate method if one is not available under 40 CFR part 136, to be used to monitor for the presence and amount in an effluent of the pollutant for which the WQBEL is established; and shall specify in accordance with section B.2 of this procedure, the quantification level that can be achieved by use of the specified analytical method.

2. The quantification level shall be the minimum level (ML) specified in or approved under 40 CFR part 136 for the method for that pollutant. If no such ML exists, or if the method is not specified or approved under 40 CFR part 136, the quantification level shall be the lowest quantifiable level practicable. The permitting authority may specify a higher quantification level if the permittee demonstrates that a higher quantification level is appropriate because of effluent-specific matrix interference.

3. The permit shall state that, for the purpose of compliance assessment, the analytical method specified in the permit shall be used to monitor the amount of pollutant in an effluent down to the quantification level, provided that the analyst has complied with the specified quality assurance/quality control procedures in the relevant method.

4. The permitting authority shall use applicable State and Tribal procedures to average and account for monitoring data. The permitting authority may specify in the permit the value to be used to interpret sample values below the quantification level.

C. *Special Conditions.* The permit shall contain a reopener clause authorizing modification or revocation and reissuance of the permit if new information generated as a result of special conditions included in the permit indicates that presence of the pollutant in the discharge at levels above the WQBEL. Special conditions that may be included in the permit include, but are not limited to, fish tissue sampling, whole effluent toxicity (WET) tests, limits and/or monitoring requirements on internal waste streams, and monitoring for surrogate parameters. Data generated as a result of special conditions can be used to reopen the permit to establish more stringent effluent limits or conditions, if necessary.

D. *Pollutant Minimization Program.* The permitting authority shall include a condition in the permit requiring the permittee to develop and conduct a pollutant minimization program for each pollutant with a WQBEL below the quantification level. The goal of the pollutant minimization program shall be to maintain the effluent at or below the WQBEL. In addition, States and Tribes may consider cost-effectiveness when evaluating the requirements of a PMP. The pollutant minimization program shall include, but is not limited to, the following:

1. An annual review and semi-annual monitoring of potential sources of the pollutant, which may include fish tissue monitoring and other bio-uptake sampling;

2. Quarterly monitoring for the pollutant in the influent to the wastewater treatment system;

3. Submittal of a control strategy designed to proceed toward the goal of maintaining the effluent below the WQBEL;

4. Implementation of appropriate, cost-effective control measures consistent with the control strategy; and

5. An annual status report that shall be sent to the permitting authority including:

a. All minimization program monitoring results for the previous year;

b. A list of potential sources of the pollutant; and

c. A summary of all action undertaken pursuant to the control strategy.

6. Any information generated as a result of procedure 8.D can be used to support a request for subsequent permit modifications, including revisions to (e.g., more or less frequent monitoring), or removal of the requirements of procedure 8.D, consistent with 40 CFR 122.44, 122.62 and 122.63.

PROCEDURE 9: COMPLIANCE SCHEDULES

The Great Lakes States and Tribes shall adopt provisions consistent with (as protective as) procedure 9 of appendix F of part 132.

A. *Limitations for New Great Lakes Dischargers.* When a permit issued on or after March 23, 1997 to a new Great Lakes discharger (defined in Part 132.2) contains a water quality-based effluent limitation (WQBEL), the permittee shall comply with such a limitation upon the commencement of the discharge.

B. *Limitations for Existing Great Lakes Dischargers.*

1. Any existing permit that is reissued or modified on or after March 23, 1997 to contain a new or more restrictive WQBEL may allow a reasonable period of time, up to five years from the date of permit issuance or modification, for the permittee to comply with that limit, provided that the Tier I criterion or whole effluent toxicity (WET) criterion was adopted (or, in the case of a narrative criterion, Tier II value, or Tier I criterion derived pursuant to the methodology in appendix A of part 132, was newly derived) after July 1, 1977.

2. When the compliance schedule established under paragraph 1 goes beyond the term of the permit, an interim permit limit effective upon the expiration date shall be included in the permit and addressed in the permit's fact sheet or statement of basis. The administrative record for the permit shall reflect the final limit and its compliance date.

3. If a permit establishes a schedule of compliance under paragraph 1 which exceeds one year from the date of permit issuance or modification, the schedule shall set forth interim requirements and dates for their achievement. The time between such interim dates may not exceed one year. If the time necessary for completion of any interim requirement is more than one year and is not readily divisible into stages for completion, the permit shall require, at a minimum, specified dates for annual submission of progress reports on the status of any interim requirements.

C. *Delayed Effectiveness of Tier II Limitations for Existing Great Lakes Discharges.*

1. Whenever a limit (calculated in accordance with Procedure 3) based upon a Tier II value is included in a reissued or modified permit for an existing Great Lakes discharger, the permit may provide a reasonable period of time, up to two years, in which to provide additional studies necessary to develop a Tier I criterion or to modify the Tier II value. In such cases, the permit shall require compliance with the Tier II limitation within a reasonable period of time, no later than five years after permit issuance or modification, and contain a reopener clause.

2. The reopener clause shall authorize permit modifications if specified studies have been completed by the permittee or provided by a third-party during the time allowed to conduct the specified studies, and the permittee or a third-party demonstrates, through such studies, that a revised limit is appropriate. Such a revised limit shall be incorporated through a permit modification and a reasonable time period, up to five years, shall be allowed for compliance. If incorporated prior to the compliance date of the original Tier II limitation, any such revised limit shall not be considered less-stringent for purposes of the anti-backsliding provisions of section 402(o) of the Clean Water Act.

3. If the specified studies have been completed and do not demonstrate that a revised limit is appropriate, the permitting authority may provide a reasonable additional period of time, not to exceed five years with which to achieve compliance with the original effluent limitation.

4. Where a permit is modified to include new or more stringent limitations, on a date within five years of the permit expiration date, such compliance schedules may extend beyond the term of a permit consistent with section B.2 of this procedure.

5. If future studies (other than those conducted under paragraphs 1, 2, or 3 above) result in a Tier II value being changed to a less stringent Tier II value or Tier I criterion, after the effective date of a Tier II-based limit, the existing Tier II-based limit may be revised to be less stringent if:

(a) It complies with sections 402(o) (2) and (3) of the CWA; or,

(b) In non-attainment waters, where the existing Tier II limit was based on procedure 3, the cumulative effect of revised effluent limitation based on procedure 3 of this appendix will assure compliance with water quality standards; or,

(c) In attained waters, the revised effluent limitation complies with the State or Tribes' antidegradation policy and procedures.

[60 FR 15387, Mar. 23, 1995, as amended at 63 FR 20110, Apr. 23, 1998; 65 FR 67650, Nov. 13, 2000]

PART 133—SECONDARY TREATMENT REGULATION

AUTHORITY: Secs. 301(b)(1)(B), 304(d)(1), 304(d)(4), 308, and 501 of the Federal Water

Pollution Control Act as amended by the Federal Water Pollution Control Act Amendments of 1972, the Clean Water Act of 1977, and the Municipal Wastewater Treatment Construction Grant Amendments of 1981; 33 U.S.C. 1311(b)(1)(B), 1314(d) (1) and (4), 1318, and 1361; 86 Stat. 816, Pub. L. 92–500; 91 Stat. 1567, Pub. L. 95–217; 95 Stat. 1623, Pub. L. 97–117.

Source: 49 FR 37006, Sept. 20, 1984, unless otherwise noted.

§ 133.100 Purpose.

This part provides information on the level of effluent quality attainable through the application of secondary or equivalent treatment.

§ 133.101 Definitions.

Terms used in this part are defined as follows:

(a) *7-day average.* The arithmetic mean of pollutant parameter values for samples collected in a period of 7 consecutive days.

(b) *30-day average.* The arithmetic mean of pollutant parameter values for samples collected in a period of 30 consecutive days.

(c) *Act.* The Clean Water Act (33 U.S.C. 1251 *et seq.*, as amended).

(d) *BOD.* The five day measure of the pollutant parameter biochemical oxygen demand (BOD).

(e) $CBOD_5$. The five day measure of the pollutant parameter carbonaceous biochemical oxygen demand ($CBOD_5$).

(f) *Effluent concentrations consistently achievable through proper operation and maintenance.* (1) For a given pollutant parameter, the 95th percentile value for the 30-day average effluent quality achieved by a treatment works in a period of at least two years, excluding values attributable to upsets, bypasses, operational errors, or other unusual conditions, and (2) a 7-day average value equal to 1.5 times the value derived under paragraph (f)(1) of this section.

(g) *Facilities eligible for treatment equivalent to secondary treatment.* Treatment works shall be eligible for consideration for effluent limitations described for treatment equivalent to secondary treatment (§ 133.105), if:

(1) The BOD_5 and SS effluent concentrations consistently achievable through proper operation and maintenance (§ 133.101(f)) of the treatment

works exceed the minimum level of the effluent quality set forth in §§ 133.102(a) and 133.102(b),

(2) A trickling filter or waste stabilization pond is used as the principal process, and

(3) The treatment works provide significant biological treatment of municipal wastewater.

(h) *mg/l.* Milligrams per liter.

(i) *NPDES.* National Pollutant Discharge Elimination System.

(j) *Percent removal.* A percentage expression of the removal efficiency across a treatment plant for a given pollutant parameter, as determined from the 30-day average values of the raw wastewater influent pollutant concentrations to the facility and the 30-day average values of the effluent pollutant concentrations for a given time period.

(k) *Significant biological treatment.* The use of an aerobic or anaerobic biological treatment process in a treatment works to consistently achieve a 30-day average of a least 65 percent removal of BOD_5.

(l) *SS.* The pollutant parameter total suspended solids.

(m) *Significantly more stringent limitation* means BOD_5 and SS limitations necessary to meet the percent removal requirements of at least 5 mg/l more stringent than the otherwise applicable concentration-based limitations (e.g., less than 25 mg/l in the case of the secondary treatment limits for BOD_5 and SS), or the percent removal limitations in §§ 133.102 and 133.105, if such limits would, by themselves, force significant construction or other significant capital expenditure.

(n) *State Director* means the chief administrative officer of any State or interstate agency operating an "approved program," or the delegated representative of the State Director.

[49 FR 37006, Sept. 20, 1984; 49 FR 40405, Oct. 16, 1984, as amended at 50 FR 23387, June 3, 1985]

§ 133.102 Secondary treatment.

The following paragraphs describe the minimum level of effluent quality attainable by secondary treatment in terms of the parameters—BOD_5, SS and pH. All requirements for each parameter shall be achieved except as provided for in §§133.103 and 133.105.

(a) *BOD₅*.

(1) The 30-day average shall not exceed 30 mg/l.

(2) The 7-day average shall not exceed 45 mg/l.

(3) The 30-day average percent removal shall not be less than 85 percent.

(4) At the option of the NPDES permitting authority, in lieu of the parameter BOD₅ and the levels of the effluent quality specified in paragraphs (a)(1), (a)(2) and (a)(3), the parameter CBOD₅ may be substituted with the following levels of the CBOD₅ effluent quality provided:

(i) The 30-day average shall not exceed 25 mg/l.

(ii) The 7-day average shall not exceed 40 mg/l.

(iii) The 30-day average percent removal shall not be less than 85 percent.

(b) *SS*. (1) The 30-day average shall not exceed 30 mg/l.

(2) The 7-day average shall not exceed 45 mg/l.

(3) The 30-day average percent removal shall not be less than 85 percent.

(c) *pH*. The effluent values for pH shall be maintained within the limits of 6.0 to 9.0 unless the publicly owned treatment works demonstrates that: (1) Inorganic chemicals are not added to the waste stream as part of the treatment process; and (2) contributions from industrial sources do not cause the pH of the effluent to be less than 6.0 or greater than 9.0.

[49 FR 37006, Sept. 20, 1984; 49 FR 40405, Oct. 16, 1984]

§133.103 Special considerations.

(a) *Combined sewers*. Treatment works subject to this part may not be capable of meeting the percentage removal requirements established under §§133.102(a)(3) and 133.102(b)(3), or §§133.105(a)(3) and 133.105(b)(3) during wet weather where the treatment works receive flows from combined sewers (*i.e.*, sewers which are designed to transport both storm water and sanitary sewage). For such treatment works, the decision must be made on a case-by-case basis as to whether any attainable percentage removal level can be defined, and if so, what the level should be.

(b) *Industrial wastes*. For certain industrial categories, the discharge to navigable waters of BOD₅ and SS permitted under sections 301(b)(1)(A)(i), (b)(2)(E) or 306 of the Act may be less stringent than the values given in §§133.102(a)(1), 133.102(a)(4)(i), 133.102(b)(1), 133.105(a)(1), 133.105(b)(1) and 133.105(e)(1)(i). In cases when wastes would be introduced from such an industrial category into a publicly owned treatment works, the values for BOD₅ and SS in §§133.102(a)(1), 133.102(a)(4)(i), 133.102(b)(1), 133.105(a)(1), 133.105(b)(1), and 133.105(e)(1)(i) may be adjusted upwards provided that: (1) The permitted discharge of such pollutants, attributable to the industrial category, would not be greater than that which would be permitted under sections 301(b)(1)(A)(i), 301(b)(2)(E) or 306 of the Act if such industrial category were to discharge directly into the navigable waters, and (2) the flow or loading of such pollutants introduced by the industrial category exceeds 10 percent of the design flow or loading of the publicly owned treatment works. When such an adjustment is made, the values for BOD₅ or SS in §§133.102(a)(2), 133.102(a)(4)(ii), §133.102(b)(2), 133.105(a)(2), 133.105(b)(2), and 133.105(e)(1)(ii) should be adjusted proportionately.

(c) *Waste stabilization ponds*. The Regional Administrator, or, if appropriate, State Director subject to EPA approval, is authorized to adjust the minimum levels of effluent quality set forth in §133.105 (b)(1), (b)(2), and (b)(3) for treatment works subject to this part, to conform to the SS concentrations achievable with waste stabilization ponds, provided that: (1) Waste stablization ponds are the principal process used for secondary treatment; and (2) operation and maintenance data indicate that the SS values specified in §133.105 (b)(1), (b)(2), and (b)(3) cannot be achieved. The term "SS concentrations achievable with waste stabilization ponds" means a SS value, determined by the Regional Administrator, or, if appropriate, State Director subject to EPA approval, which is equal to the effluent concentration achieved 90 percent of the time within a State or appropriate contiguous geographical area by waste stabilization ponds that are achieving the levels of effluent quality for BOD_5 specified in § 133.105(a)(1). [cf. 43 FR 55279].

(d) *Less concentrated influent wastewater for separate sewers.* The Regional Administrator or, if appropriate, State Director is authorized to substitute either a lower percent removal requirement or a mass loading limit for the percent removal requirements set forth in §§ 133.102(a)(3), 133.102(a)(4)(iii), 133.102(b)(3), 102.105(a)(3), 133.105(b)(3) and 133.105(e)(1)(iii) provided that the permittee satisfactorily demonstrates that: (1) The treatment works is consistently meeting, or will consistently meet, its permit effluent concentration limits but its percent removal requirements cannot be met due to less concentrated influent wastewater, (2) to meet the percent removal requirements, the treatment works would have to achieve significantly more stringent limitations than would otherwise be required by the concentration-based standards, and (3) the less concentrated influent wastewater is not the result of excessive I/I. The determination of whether the less concentrated wastewater is the result of excessive I/I will use the definition of excessive I/I in 40 CFR 35.2005(b)(16) plus the additional criterion that inflow is nonexcessive if the total flow to the POTW (i.e., wastewater plus inflow plus infiltration) is less than 275 gallons per capita per day.

(e) Less concentrated influent wastewater for combined sewers during dry weather. The Regional Administrator or, if appropriate, the State Director is authorized to substitute either a lower percent removal requirement or a mass loading limit for the percent removal requirements set forth in §§ 133.102(a)(3), 133.102(a)(4)(iii), 133.102(b)(3), 133.105(a)(3), 133.105(b)(3) and 133.105(e)(1)(iii) provided that the permittee satisfactorily demonstrates that: (1) The treatment works is consistently meeting, or will consistently meet, its permit effluent concentration limits, but the percent removal requirements cannot be met due to less concentrated influent wastewater; (2) to meet the percent removal require-

ments, the treatment works would have to achieve significantly more stringent effluent concentrations than would otherwise be required by the concentration-based standards; and (3) the less concentrated influent wastewater does not result from either excessive infiltration or clear water industrial discharges during dry weather periods. The determination of whether the less concentrated wastewater results from excessive infiltration is discussed in 40 CFR 35.2005(b)(28), plus the additional criterion that either 40 gallons per capita per day (gpcd) or 1500 gallons per inch diameter per mile of sewer (gpdim) may be used as the threshold value for that portion of the dry weather base flow attributed to infiltration. If the less concentrated influent wastewater is the result of clear water industrial discharges, then the treatment works must control such discharges pursuant to 40 CFR part 403.

[49 FR 37006, Sept. 20, 1984, as amended at 50 FR 23387, June 3, 1985; 50 FR 36880, Sept. 10, 1985; 54 FR 4228, Jan. 27, 1989]

§ 133.104 Sampling and test procedures.

(a) Sampling and test procedures for pollutants listed in this part shall be in accordance with guidelines promulgated by the Administrator in 40 CFR part 136.

(b) Chemical oxygen demand (COD) or total organic carbon (TOC) may be substituted for BOD_5 when a long-term BOD:COD or BOD:TOC correlation has been demonstrated.

§ 133.105 Treatment equivalent to secondary treatment.

This section describes the minimum level of effluent quality attainable by facilities eligible for treatment equivalent to secondary treatment (§ 133.101(g)) in terms of the parameters—BOD_5, SS and pH. All requirements for the specified parameters in paragraphs (a), (b) and (c) of this section shall be achieved except as provided for in § 133.103, or paragraphs (d), (e) or (f) of this section.

(a) *BOD_5.* (1) The 30-day average shall not exceed 45 mg/l.

(2) The 7-day average shall not exceed 65 mg/l.

(3) The 30-day average percent removal shall not be less than 65 percent.

(b) *SS.* Except where SS values have been adjusted in accordance with § 133.103(c):

(1) The 30-day average shall not exceed 45 mg/l.

(2) The 7-day average shall not exceed 65 mg/l.

(3) The 30-day average percent removal shall not be less than 65 percent.

(c) *pH.* The requirements of § 133.102(c) shall be met.

(d) *Alternative State requirements.* Except as limited by paragraph (f) of this section, and after notice and opportunity for public comment, the Regional Administrator, or, if appropriate, State Director subject to EPA approval, is authorized to adjust the minimum levels of effluent quality set forth in paragraphs (a)(1), (a)(2), (b)(1) and (b)(2) of this section for trickling filter facilities and in paragraphs (a)(1) and (a)(2) of this section for waste stabilization pond facilities, to conform to the BOD_5 and SS effluent concentrations consistently achievable through proper operation and maintenance (§ 133.101(f)) by the median (50th percentile) facility in a representative sample of facilities within a State or appropriate contiguous geographical area that meet the definition of facilities eligible for treatment equivalent to secondary treatment (§ 133.101(g)).

(The information collection requirements contained in this rule have been approved by OMB and assigned control number 2040–0051)

(e) *$CBOD_5$* limitations:

(1) Where data are available to establish $CBOD_5$ limitations for a treatment works subject to this section, the NPDES permitting authority may substitute the parameter $CBOD_5$ for the parameter BOD_5 In §§ 133.105(a)(1), 133.105(a)(2) and 133.105(a)(3), on a case-by-case basis provided that the levels of $CBOD_5$ effluent quality are not less stringent than the following:

(i) The 30-day average shall not exceed 40 mg/l.

(ii) The 7-days average shall not exceed 60 mg/l.

(iii) The 30-day average percent removal shall not be less than 65 percent.

(2) Where data are available, the parameter $CBOD_5$ may be used for effluent quality limitations established

under paragraph (d) of this section. Where concurrent BOD effluent data are available, they must be submitted with the CBOD data as a part of the approval process outlined in paragraph (d) of this section.

(f) *Permit adjustments.* Any permit adjustment made pursuant to this part may not be any less stringent than the limitations required pursuant to § 133.105(a)–(e). Furthermore, permitting authorities shall require more stringent limitations when adjusting permits if: (1) For existing facilities the permitting authority determines that the 30-day average and 7-day average BOD_5 and SS effluent values that could be achievable through proper operation and maintenance of the treatment works, based on an analysis of the past performance of the treatment works, would enable the treatment works to achieve more stringent limitations, or

(2) For new facilities, the permitting authority determines that the 30-day average and 7-day average BOD_5 and SS effluent values that could be achievable through proper operation and maintenance of the treatment works, considering the design capability of the treatment process and geographical and climatic conditions, would enable the treatment works to achieve more stringent limitations.

[49 FR 37006, Sept. 20, 1984; 49 FR 40405, Oct. 16, 1984]

PART 135—PRIOR NOTICE OF CITIZEN SUITS

Subpart A—Prior Notice Under the Clean Water Act

Subpart B—Prior Notice Under the Safe Drinking Water Act

AUTHORITY: Subpart A, issued under Sec. 505, Clean Water Act, as amended 1987; Sec. 504, Pub. L. 100–4; 101 Stat. 7 (33 U.S.C. 1365).

Subpart B, issued under Sec. 1449, Safe Drinking Water Act (42 U.S.C. 300j-8).

SOURCE: 38 FR 15040, June 7, 1973, unless otherwise noted.

Subpart A—Prior Notice Under the Clean Water Act

§ 135.1 Purpose.

(a) Section 505(a)(1) of the Clean Water Act (hereinafter the Act) authorizes any person or persons having an interest which is or may be adversely affected to commence a civil action on his own behalf to enforce the Act or to enforce certain requirements promulgated pursuant to the Act. In addition, section 505(c)(3) of the Act provides that, for purposes of protecting the interests of the United States, whenever a citizen enforcement action is brought under section 505(a)(1) of the Act in a court of the United States, the Plaintiff shall serve a copy of the complaint on the Attorney General and the Administrator. Section 505(c)(3) also provides that no consent judgment shall be entered in any citizen action in which the United States is not a party prior to 45 days following the receipt of a copy of the proposed consent judgment by the Attorney General and the Administrator.

(b) The purpose of this subpart is to prescribe procedures governing the giving of notice required by section 505(b) of the Act as a prerequisite to the commencing of such actions, and governing the service of complaints and proposed consent judgments as required by section 505(c)(3) of the Act.

[56 FR 11515, Mar. 19, 1991]

§ 135.2 Service of notice.

(a) Notice of intent to file suit pursuant to section 505(a)(1) of the Act shall be served upon an alleged violator of an effluent standard or limitation under the Act, or an order issued by the Administrator or a State with respect to such a standard or limitation, in the following manner:

(1) If the alleged violator is an individual or corporation, service of notice shall be accomplished by certified mail addressed to, or by personal service upon, the owner or managing agent of the building, plant, installation, vessel, facility, or activity alleged to be in violation. A copy of the notice shall be mailed to the Administrator of the Environmental Protection Agency, the Regional Administrator of the Environmental Protection Agency for the region in which such violation is alleged to have occurred, and the chief administrative officer of the water pollution control agency for the State in which the violation is alleged to have occurred. If the alleged violator is a corporation, a copy of such notice also shall be mailed to the registered agent, if any, of such corporation in the State in which such violation is alleged to have occurred.

(2) If the alleged violator is a State or local agency, service of notice shall be accomplished by certified mail addressed to, or by personal service upon, the head of such agency. A copy of such notice shall be mailed to the chief administrative officer of the water pollution control agency for the State in which the violation is alleged to have occurred, the Administrator of the Environmental Protection Agency, and the Regional Administrator of the Environmental Protection Agency for the region in which such violation is alleged to have occurred.

(3) If the alleged violator is a Federal agency, service of notice shall be accomplished by certified mail addressed to, or by personal service upon, the head of such agency. A copy of such notice shall be mailed to the Administrator of the Environmental Protection Agency, the Regional Administrator of the Environmental Protection Agency for the region in which such violation is alleged to have occurred, the Attorney General of the United States, and the Chief administrative officer of the water pollution control agency for the State in which the violation is alleged to have occurred.

(b) Service of notice of intent to file suit pursuant to section 505(a)(2) of the Act shall be accomplished by certified mail addressed to, or by personal service upon, the Administrator, Environmental Protection Agency, Washington, DC 20460. A copy of such notice shall be mailed to the Attorney General of the United States.

(c) Notice given in accordance with the provisions of this subpart shall be deemed to have been served on the postmark date if mailed, or on the date of receipt if served personally.

§135.3 Contents of notice.

(a) *Violation of standard, limitation or order.* Notice regarding an alleged violation of an effluent standard or limitation or of an order with respect thereto, shall include sufficient information to permit the recipient to identify the specific standard, limitation, or order alleged to have been violated, the activity alleged to constitute a violation, the person or persons responsible for the alleged violation, the location of the alleged violation, the date or dates of such violation, and the full name, address, and telephone number of the person giving notice.

(b) *Failure to act.* Notice regarding an alleged failure of the Administrator to perform any act or duty under the Act which is not discretionary with the Administrator shall identify the provision of the Act which requires such act or creates such duty, shall describe with reasonable specificity the action taken or not taken by the Administrator which is alleged to constitute a failure to perform such act or duty, and shall state the full name, address and telephone number of the person giving the notice.

(c) *Identification of counsel.* The notice shall state the name, address, and telephone number of the legal counsel, if any, representing the person giving the notice.

§135.4 Service of complaint.

(a) A citizen plaintiff shall mail a copy of a complaint filed against an alleged violator under section 505(a)(1) of the Act to the Administrator of the Environmental Protection Agency, the Regional Administrator of the EPA Region in which the violations are alleged to have occurred, and the Attorney General of the United States.

(b) The copy so served shall be of a filed, date-stamped complaint, or shall be a conformed copy of the filed complaint which indicates the assigned civil action number, accompanied by a signed statement by the plaintiff or his attorney as to when the complaint was filed.

(c) A citizen plaintiff shall mail a copy of the complaint on the same date on which the plaintiff files the complaint with the court, or as expeditiously thereafter as practicable.

(d) If the alleged violator is a Federal agency, a citizen plaintiff must serve the complaint on the United States in accordance with relevant Federal law and court rules affecting service on defendants, in addition to complying with the service requirements of this subpart.

[56 FR 11515, Mar. 19, 1991]

§135.5 Service of proposed consent judgment.

(a) The citizen plaintiff in a citizen enforcement suit filed against an alleged violator under section 505(a)(1) of the Act shall serve a copy of a proposed consent judgment, signed by all parties to the lawsuit, upon the Administrator, Environmental Protection Agency, Washington, DC 20460, and the Attorney General, Department of Justice, Citizen Suit Coordinator, Room 2615, Washington, DC 20530. The plaintiff shall serve the Administrator and the Attorney General by personal service or by certified mail (return receipt requested.) The plaintiff shall also mail a copy of a proposed consent judgment at the same time to the Regional Administrator of the EPA Region in which the violations were alleged to have occurred.

(b) When the parties in an action in which the United States is not a party file or lodge a proposed consent judgment with the court, the plaintiff shall notify the court of the statutory requirement that the consent judgment shall not be entered prior to 45 days following receipt by both the Administrator and the Attorney General of a copy of the consent judgment.

(1) If the plaintiff knows the dates upon which the Administrator and the Attorney General received copies of the proposed consent judgment, the plaintiff shall so notify the court.

(2) If the plaintiff does not know the date upon which the Administrator and Attorney General received copies of the proposed consent judgment, the plaintiff shall so notify the court, but upon receiving such information regarding the dates of service of the proposed

consent judgment upon the Administrator and Attorney General, the plaintiff shall so notify the court of the dates of service.

[56 FR 11515, Mar. 19, 1991]

Subpart B—Prior Notice Under the Safe Drinking Water Act

SOURCE: 54 FR 20771, May 12, 1989, unless otherwise noted.

§ 135.10   Purpose.

Section 1449 of the Safe Drinking Water Act (the Act) authorizes any person to commence a civil action to enforce the Act against an alleged violator of any requirements prescribed by or under the Act, or against the Administrator for failure to perform any duty which is not discretionary under the Act. No citizen suit may be commenced prior to sixty days after giving notice of the alleged violation to the Administrator, any alleged violator, and to the State. The purpose of this subpart is to prescribe procedures for giving the notice required by section 1449(b).

§ 135.11   Service of notice.

(a) Notice of intent to file suit pursuant to section 1449(a)(1) of the Act shall be served in the following manner upon an alleged violator of any requirement prescribed by or under the Act:

(1) If the alleged violator is an individual or corporation, service of notice shall be accomplished by certified mail, return receipt requested, addressed to, or by personal service upon, such individual or corporation. If a public water system or underground injection well is alleged to be in violation, service shall be upon the owner or operator. A copy of the notice shall be sent by certified mail, return receipt requested, to the Administrator of the Environmental Protection Agency, the Regional Administrator of the Environmental Protection Agency for the region in which such violation is alleged to have occurred, the chief administrative officer of the responsible state agency (if any), and the Attorney General for the State in which the violation is alleged to have occurred. If the alleged violator is a corporation, a copy of the notice shall also be sent by certified mail, return receipt requested, to the registered agent (if any) of the corporation in the State in which the violation is alleged to have occurred.

(2) If the alleged violator is a State or local agency, service of notice shall be accomplished by certified mail, return receipt requested, addressed to, or by personal service upon, the head of such agency. A copy of the notice shall be sent by certified mail, return receipt requested, to the Administrator of the Environmental Protection Agency, the Regional Administrator of the Environmental Protection Agency for the region in which the violation is alleged to have occurred, the chief administrative officer of the responsible state agency (if any), and the Attorney General for the State in which the violation is alleged to have occurred.

(3) If the alleged violator is a Federal agency, service of notice shall be accomplished by certified mail, return receipt requested, addressed to, or by personal service upon, the head of the Federal agency. A copy of the notice shall be sent by certified mail, return receipt requested, to the Administrator of the Environmental Protection Agency, the Regional Administrator of the Environmental Protection Agency for the region in which the violation is alleged to have occurred, the Attorney General of the United States, the chief administrative officer of the responsible state agency (if any), and the Attorney General for the State in which the violation is alleged to have occurred.

(b) Service of notice of intent to file suit pursuant to section 1449(a)(2) of the Act shall be accomplished by certified mail, return receipt requested, addressed to, or by personal service upon, the Administrator of the Environmental Protection Agency, Washington, DC 20460. A copy of the notice shall be sent by certified mail to the Attorney General of the United States.

(c) Notice given in accordance with the provisions of this subpart shall be deemed to have been given on the date of receipt of service, if served personally. If service was accomplished by mail, the date of receipt will be considered to be the date noted on the return receipt card.

§135.12 Contents of notice.

(a) *Violation of standard or requirement.* Notice regarding an alleged violation of any requirement prescribed by or under the Act shall include sufficient information to permit the recipient to identify the specific requirement alleged to have been violated, the activity alleged to constitute a violation, the person or persons responsible for the alleged violation, the location of the alleged violation, the date or dates of the alleged violation, and the full name, address, and telephone number of the person giving notice.

(b) *Failure to act.* Notice regarding an alleged failure of the Administrator to perform any act or duty under the Act which is not discretionary with the Administrator shall identify the provision of the Act which requires the act or creates the duty, and shall describe with reasonable specificity the action taken or not taken by the Administrator which is alleged to constitute a failure to perform such act or duty, and shall state the full name, address, and telephone number of the person giving notice.

(c) *Identification of counsel.* All notices shall include the name, address, and telephone number of the legal counsel, if any, representing the person giving notice.

§135.13 Timing of notice.

No action may be commenced under section 1449(a)(1) or (a)(2) until the plaintiff has given each of the appropriate parties sixty days notice of intent to file such an action. Actions concerning injection wells disposing of hazardous waste which allege jurisdiction solely under section 7002(c) of the Resource Conservation and Recovery Act may proceed immediately after notice to the appropriate parties.

FINDING AIDS

A list of CFR titles, subtitles, chapters, subchapters and parts and an alphabetical list of agencies publishing in the CFR are included in the CFR Index and Finding Aids volume to the Code of Federal Regulations which is published separately and revised annually.

Table of CFR Titles and Chapters
(Revised as of July 1, 2012)

Title 1—General Provisions

Title 2—Grants and Agreements

Title 2—Grants and Agreements—Continued

Title 3—The President

Title 4—Accounts

Title 5—Administrative Personnel

Title 5—Administrative Personnel—Continued

Title 6—Domestic Security

Title 7—Agriculture

Title 7—Agriculture—Continued

Title 8—Aliens and Nationality

Title 9—Animals and Animal Products

Title 13—Business Credit and Assistance

Title 14—Aeronautics and Space

Title 15—Commerce and Foreign Trade

Title 16—Commercial Practices

Title 17—Commodity and Securities Exchanges

Title 18—Conservation of Power and Water Resources

Title 19—Customs Duties

Title 20—Employees' Benefits

Title 21—Food and Drugs

Title 22—Foreign Relations

Title 23—Highways

Title 24—Housing and Urban Development

601

Title 25—Indians

Title 25—Indians—Continued

Title 26—Internal Revenue

Title 27—Alcohol, Tobacco Products and Firearms

Title 28—Judicial Administration

Title 29—Labor

Title 29—Labor—Continued

604

Title 35 [Reserved]

Title 36—Parks, Forests, and Public Property

Title 37—Patents, Trademarks, and Copyrights

Title 38—Pensions, Bonuses, and Veterans' Relief

Title 39—Postal Service

Title 40—Protection of Environment

Title 41—Public Contracts and Property Management

Title 42—Public Health

Title 42—Public Health—Continued

Title 43—Public Lands: Interior

Title 44—Emergency Management and Assistance

Title 45—Public Welfare

609

Title 49—Transportation

610

611

Alphabetical List of Agencies Appearing in the CFR

(Revised as of July 1, 2012)

Agency	CFR Title, Subtitle or Chapter
Appalachian Regional Commission	5, IX
Architectural and Transportation Barriers Compliance Board	36, XI
Arctic Research Commission	45, XXIII
Armed Forces Retirement Home	5, XI
Army Department	32, V
Engineers, Corps of	33, II; 36, III
Federal Acquisition Regulation	48, 51
Bilingual Education and Minority Languages Affairs, Office of	34, V
Blind or Severely Disabled, Committee for Purchase from People Who Are	41, 51
Broadcasting Board of Governors	22, V
Federal Acquisition Regulation	48, 19
Bureau of Ocean Energy Management, Regulation, and Enforcement	30, II
Census Bureau	15, I
Centers for Medicare & Medicaid Services	42, IV
Central Intelligence Agency	32, XIX
Chemical Safety and Hazardous Investigation Board	40, VI
Chief Financial Officer, Office of	7, XXX
Child Support Enforcement, Office of	45, III
Children and Families, Administration for	45, II, III, IV, X
Civil Rights, Commission on	5, LXVIII; 45, VII
Civil Rights, Office for	34, I
Court Services and Offender Supervision Agency for the District of Columbia	5, LXX
Coast Guard	33, I; 46, I; 49, IV
Coast Guard (Great Lakes Pilotage)	46, III
Commerce Department	2, XIII; 44, IV; 50, VI
Census Bureau	15, I
Economic Affairs, Under Secretary	37, V
Economic Analysis, Bureau of	15, VIII
Economic Development Administration	13, III
Emergency Management and Assistance	44, IV
Federal Acquisition Regulation	48, 13
Foreign-Trade Zones Board	15, IV
Industry and Security, Bureau of	15, VII
International Trade Administration	15, III; 19, III
National Institute of Standards and Technology	15, II
National Marine Fisheries Service	50, II, IV
National Oceanic and Atmospheric Administration	15, IX; 50, II, III, IV, VI
National Telecommunications and Information Administration	15, XXIII; 47, III, IV
National Weather Service	15, IX
Patent and Trademark Office, United States	37, I
Productivity, Technology and Innovation, Assistant Secretary for	37, IV
Secretary of Commerce, Office of	15, Subtitle A
Technology, Under Secretary for	37, V
Technology Administration	15, XI
Technology Policy, Assistant Secretary for	37, IV
Commercial Space Transportation	14, III
Commodity Credit Corporation	7, XIV
Commodity Futures Trading Commission	5, XLI; 17, I
Community Planning and Development, Office of Assistant Secretary for	24, V, VI
Community Services, Office of	45, X
Comptroller of the Currency	12, I
Construction Industry Collective Bargaining Commission	29, IX
Consumer Financial Protection Bureau	12, X
Consumer Product Safety Commission	5, LXXI; 16, II
Copyright Office	37, II
Copyright Royalty Board	37, III
Corporation for National and Community Service	2, XXII; 45, XII, XXV
Cost Accounting Standards Board	48, 99
Council on Environmental Quality	40, V
Court Services and Offender Supervision Agency for the District of Columbia	5, LXX; 28, VIII

Agency	CFR Title, Subtitle or Chapter
Customs and Border Protection	19, I
Defense Contract Audit Agency	32, I
Defense Department	2, XI; 5, XXVI; 32, Subtitle A; 40, VII
Advanced Research Projects Agency	32, I
Air Force Department	32, VII
Army Department	32, V; 33, II; 36, III, 48, 51
Defense Acquisition Regulations System	48, 2
Defense Intelligence Agency	32, I
Defense Logistics Agency	32, I, XII; 48, 54
Engineers, Corps of	33, II; 36, III
National Imagery and Mapping Agency	32, I
Navy Department	32, VI; 48, 52
Secretary of Defense, Office of	2, XI; 32, I
Defense Contract Audit Agency	32, I
Defense Intelligence Agency	32, I
Defense Logistics Agency	32, XII; 48, 54
Defense Nuclear Facilities Safety Board	10, XVII
Delaware River Basin Commission	18, III
District of Columbia, Court Services and Offender Supervision Agency for the	5, LXX; 28, VIII
Drug Enforcement Administration	21, II
East-West Foreign Trade Board	15, XIII
Economic Affairs, Under Secretary	37, V
Economic Analysis, Bureau of	15, VIII
Economic Development Administration	13, III
Economic Research Service	7, XXXVII
Education, Department of	2, XXXIV; 5, LIII
Bilingual Education and Minority Languages Affairs, Office of	34, V
Civil Rights, Office for	34, I
Educational Research and Improvement, Office of	34, VII
Elementary and Secondary Education, Office of	34, II
Federal Acquisition Regulation	48, 34
Postsecondary Education, Office of	34, VI
Secretary of Education, Office of	34, Subtitle A
Special Education and Rehabilitative Services, Office of	34, III
Vocational and Adult Education, Office of	34, IV
Educational Research and Improvement, Office of	34, VII
Election Assistance Commission	2, LVIII; 11, II
Elementary and Secondary Education, Office of	34, II
Emergency Oil and Gas Guaranteed Loan Board	13, V
Emergency Steel Guarantee Loan Board	13, IV
Employee Benefits Security Administration	29, XXV
Employees' Compensation Appeals Board	20, IV
Employees Loyalty Board	5, V
Employment and Training Administration	20, V
Employment Standards Administration	20, VI
Endangered Species Committee	50, IV
Energy, Department of	2, IX; 5, XXIII; 10, II, III, X
Federal Acquisition Regulation	48, 9
Federal Energy Regulatory Commission	5, XXIV; 18, I
Property Management Regulations	41, 109
Energy, Office of	7, XXIX
Engineers, Corps of	33, II; 36, III
Engraving and Printing, Bureau of	31, VI
Environmental Protection Agency	2, XV; 5, LIV; 40, I, IV, VII
Federal Acquisition Regulation	48, 15
Property Management Regulations	41, 115
Environmental Quality, Office of	7, XXXI
Equal Employment Opportunity Commission	5, LXII; 29, XIV
Equal Opportunity, Office of Assistant Secretary for	24, I
Executive Office of the President	3, I
Administration, Office of	5, XV

617

618

Agency	CFR Title, Subtitle or Chapter
Employee Benefits Security Administration	29, XXV
Employees' Compensation Appeals Board	20, IV
Employment and Training Administration	20, V
Employment Standards Administration	20, VI
Federal Acquisition Regulation	48, 29
Federal Contract Compliance Programs, Office of	41, 60
Federal Procurement Regulations System	41, 50
Labor-Management Standards, Office of	29, II, IV
Mine Safety and Health Administration	30, I
Occupational Safety and Health Administration	29, XVII
Office of Workers' Compensation Programs	20, VII
Public Contracts	41, 50
Secretary of Labor, Office of	29, Subtitle A
Veterans' Employment and Training Service, Office of the Assistant Secretary for	41, 61; 20, IX
Wage and Hour Division	29, V
Workers' Compensation Programs, Office of	20, I
Labor-Management Standards, Office of	29, II, IV
Land Management, Bureau of	43, II
Legal Services Corporation	45, XVI
Library of Congress	36, VII
Copyright Office	37, II
Copyright Royalty Board	37, III
Local Television Loan Guarantee Board	7, XX
Management and Budget, Office of	5, III, LXXVII; 14, VI; 48, 99
Marine Mammal Commission	50, V
Maritime Administration	46, II
Merit Systems Protection Board	5, II, LXIV
Micronesian Status Negotiations, Office for	32, XXVII
Millennium Challenge Corporation	22, XIII
Mine Safety and Health Administration	30, I
Minority Business Development Agency	15, XIV
Miscellaneous Agencies	1, IV
Monetary Offices	31, I
Morris K. Udall Scholarship and Excellence in National Environmental Policy Foundation	36, XVI
Museum and Library Services, Institute of	2, XXXI
National Aeronautics and Space Administration	2, XVIII; 5, LIX; 14, V
Federal Acquisition Regulation	48, 18
National Agricultural Library	7, XLI
National Agricultural Statistics Service	7, XXXVI
National and Community Service, Corporation for	2, XXII; 45, XII, XXV
National Archives and Records Administration	2, XXVI; 5, LXVI; 36, XII
Information Security Oversight Office	32, XX
National Capital Planning Commission	1, IV
National Commission for Employment Policy	1, IV
National Commission on Libraries and Information Science	45, XVII
National Council on Disability	34, XII
National Counterintelligence Center	32, XVIII
National Credit Union Administration	12, VII
National Crime Prevention and Privacy Compact Council	28, IX
National Drug Control Policy, Office of	21, III
National Endowment for the Arts	2, XXXII
National Endowment for the Humanities	2, XXXIII
National Foundation on the Arts and the Humanities	45, XI
National Highway Traffic Safety Administration	23, II, III; 47, VI; 49, V
National Imagery and Mapping Agency	32, I
National Indian Gaming Commission	25, III
National Institute for Literacy	34, XI
National Institute of Food and Agriculture	7, XXXIV
National Institute of Standards and Technology	15, II
National Intelligence, Office of Director of	32, XVII
National Labor Relations Board	5, LXI; 29, I
National Marine Fisheries Service	50, II, IV
National Mediation Board	29, X

List of CFR Sections Affected

All changes in this volume of the Code of Federal Regulations that were made by documents published in the FEDERAL REGISTER since January 1, 2001, are enumerated in the following list. Entries indicate the nature of the changes effected. Page numbers refer to FEDERAL REGISTER pages. The user should consult the entries for chapters and parts as well as sections for revisions.

Title 40 was established at 36 FR 12213, June 29, 1971. For the period before January 1, 2001, see the "List of CFR Sections Affected, 1964–1972, 1973–1985, and 1986–2000," published in 10 separate volumes.

List of CFR Sections Affected

40 CFR—Continued

74 FR
Page

Chapter I—Continued
(a)(2), (b)(2) and (d)(1)(vi) removed; (a)(1), (b)(3) and (d)(1)(vii) redesignated as (a), new (b)(2) and new (d)(1)(vi); new (a), (b)(1), new (b)(2), new (d)(1)(vi) and (g)(2) revised 58809
112.5 Regulation at 73 FR 74301 eff. date delayed...............................5900
Regulation at 73 FR 74301 eff. date delayed to 1–14–10...............14736
(b) and (c) removed; (d) and (e) redesignated as new (b) and (c); new (b) revised 58810
112.6 Regulation at 73 FR 74302 eff. date delayed...........................5900
Regulation at 73 FR 74302 eff. date delayed to 1–14–10...............14736
(a)(1)(vii), (b)(1)(vii), (3)(iii) and (4)(ii) revised.............................58810
112.7 Regulation at 73 FR 74303 eff. date delayed..............................5900
Regulation at 73 FR 74303 eff. date delayed to 1–14–10...............14736
112.7 (a)(3) introductory text and (h) introductory text revised...58810
112.8 Regulation at 73 FR 74304 eff. date delayed..............................5900
Regulation at 73 FR 74304 eff. date delayed to 1–14–10...............14736
112.9 Regulation at 73 FR 74304 eff. date delayed..............................5900
Regulation at 73 FR 74304 eff. date delayed to 1–14–10...............14736
(c)(6) revised 58810
112.12 Regulation at 73 FR 74305 eff. date delayed 5900
Regulation at 73 FR 74305 eff. date delayed to 1–14–10...............14736
Regulation at 73 FR 74306 eff. date delayed 5900
Regulation at 73 FR 74306 eff. date delayed to 1–14–10...............14736

40 CFR—Continued

74 FR
Page

Chapter I—Continued
112 Appendix G revised..................58811

2010

40 CFR

75 FR
Page

Chapter I
112 Policy statement 79961
112.3 (a), (b) and (c) revised 63102
112.22 Added; interim (temporary)....................................37719
124.10 (c) introductory text revised; (c)(1)(xi) added 77286
131.32 Removed............................29901
131.43 Added 75805
Revised; eff. 3–6–12.......................75805

2011

40 CFR

76 FR
Page

Chapter I
112.1 (d)(2)(ii)(F) and (12) added.......................................21660
112.3 (c) removed..........................21660
(a)(3) added 64248
(a)(3) revised 72124
112 Appendix E corrected; CFR correction 18894
116.4 Tables A and B amended........55584
124 Policy statement 56982
132.6 (f) and (g) revised 57652

2012

(Regulations published from January 1, 2012, through July 1, 2012)

40 CFR

77 FR
Page

Chapter I
131.43 Regulation at 75 FR 75805 eff. date delayed to 7-6-12...........13496
627